Witold Abramowicz (Ed.)

T0189995

# Business Information Systems

10th International Conference, BIS 2007
Poznań, Poland, April 25-27, 2007
Proceedings

Springer

Volume Editor

Witold Abramowicz
Poznań University of Economics
Department of Information Systems
Al. Niepodleglosci 10
60-967 Poznań, Poland
E-mail: Witold@Abramowicz.pl

Library of Congress Control Number: 2007924603

CR Subject Classification (1998): H.3.5, H.4, J.1, H.5.3, K.4, K.6

LNCS Sublibrary: SL 3 – Information Systems and Application, incl. Internet/Web and HCI

ISSN 0302-9743
ISBN-10 3-540-72034-0 Springer Berlin Heidelberg New York
ISBN-13 978-3-540-72034-8 Springer Berlin Heidelberg New York

This work is subject to copyright. All rights are reserved, whether the whole or part of the material is concerned, specifically the rights of translation, reprinting, re-use of illustrations, recitation, broadcasting, reproduction on microfilms or in any other way, and storage in data banks. Duplication of this publication or parts thereof is permitted only under the provisions of the German Copyright Law of September 9, 1965, in its current version, and permission for use must always be obtained from Springer. Violations are liable to prosecution under the German Copyright Law.

Springer is a part of Springer Science+Business Media

springer.com

© Springer-Verlag Berlin Heidelberg 2007

Typesetting: Camera-ready by author, data conversion by Scientific Publishing Services, Chennai, India
Printed on acid-free paper SPIN: 12050581 06/3180 5 4 3 2 1 0

# Preface

BIS 2007 was the tenth in a series of international conferences on business information systems. The conference took place in Poznan, where it started in 1997. From the very beginning it has been recognized by professionals as a forum for the exchange and dissemination of topical research in the development, implementation, application and improvement of computer systems for business processes.

The theme of this jubilee conference was "Semantic Integration of Data and Processes across Enterprises and Societies." The material collected in this volume covers research trends as well as current achievements and cutting-edge developments in the area of modern business information systems. A set of 51 papers were selected for the presentation during the main event and grouped around conference topics: Business Process Management, Web Services, Ontologies, Information Retrieval, System Design, Agents and Mobile Applications, Decision Support, Social Issues, Specific MIS Issues.

The Program Committee consisted of almost 100 members that carefully evaluated all the submitted papers. This year their work was significantly harder as the number of submissions doubled compared to BIS 2006. As in previous years, they were supported by an online transparent review system. Once again we observed an increase in the quality of the reviews, thus sustaining the fairness of the final acceptance decisions. It not only raised the quality of the conference, but also positively affected the work of authors.

The regular program was complemented by outstanding keynote speakers. We are proud that BIS 2007 hosted John Domingue (Open University, UK), Alois Ferscha (Johannes Kepler University of Linz, Austria), Lutz Heuser (SAP AG, Germany), Wei-Ying Ma (Microsoft Research Asia, China), Wasim Sadiq (SAP Research, Australia), and A Min Tjoa (Vienna University of Technology, Austria).

Back-to-back with BIS the International Research Forum Eastern Europe was organized by SAP in co-operation with Poznan University of Economics on the topic: Service-Oriented Enterprise Computing – The Way to New Business Transformation.

BIS 2007 was kindly supported by the ACM Special Interest Group on Management Information Systems, GI, EMISA, and the Polish Society for Information Systems.

April 2007 Witold Abramowicz

# Organization

## Program Committee

General chair: Witold Abramowicz, Poznan University of Economics, Poland

| | |
|---|---|
| Ajith Abraham | Chung-Ang University, Korea |
| Nabil R. Adam | Rutgers University, USA |
| Richard Bonner | Mälardalen University, Sweden |
| Kalina Bontcheva | University of Sheffield, UK |
| Laszlo Böszörmenyi | University of Klagenfurt, Austria |
| Hans-Dieter Burkhard | Humboldt University, Germany |
| Peter Buxmann | Technical University of Darmstadt, Germany |
| Wojciech Cellary | Poznań University of Economics, Poland |
| Peter Chamoni | University of Duisburg-Essen, Germany |
| Peter Dadam | University of Ulm, Germany |
| Stefan Decker | National University of Ireland, Ireland |
| Tommaso Di Noia | Technical University of Bari, Italy |
| Ying Ding | University of Innsbruck, Austria |
| Klaus Dittrich | University of Zurich, Switzerland |
| John Domingue | Open University, UK |
| Dieter Fensel | University of Innsbruck, Austria |
| Bogdan Franczyk | University of Leipzig, Germany |
| Ulrich Frank | University of Duisburg-Essen, Germany |
| Johann-Christoph Freytag | Humboldt University, Germany |
| Jan Goliński | Warsaw School of Economics, Poland |
| Manfred Grauer | University of Siegen, Germany |
| Norbert Gronau | University of Potsdam, Germany |
| Volker Gruhn | University of Leipzig, Germany |
| Oliver Günther | Humboldt University, Germany |
| Hele-Mai Haav | Tallinn University of Technology, Estonia |
| Ulrich Hasenkamp | University of Marburg, Germany |
| Manfred Hauswirth | National University of Ireland, Ireland |
| Armin Heinzl | University of Mannheim, Germany |
| Martin Hepp | University of Innsbruck, Austria |
| Matthias Jarke | RWTH Aachen, Germany |
| Paweł J. Kalczyński | University of Toledo, USA |
| Dimitris Karagiannis | University of Vienna, Austria |
| Uzay Kaymak | Erasmus University Rotterdam, The Netherlands |
| Pradeep Khosla | Carnegie Mellon University, USA |
| Jerzy Kisielnicki | Warsaw University, Poland |
| Gary Klein | University Of Colorado At Colorado Springs, USA |
| Gerhard F. Knolmayer | University of Bern, Switzerland |

| | |
|---|---|
| Jacek Koronacki | Polish Academy of Sciences, Poland |
| Ryszard Kowalczyk | Swinburne University of Technology, Australia |
| Helmut Krcmar | TU München, Germany |
| Sanjay Kumar | XLRI School of Management, India |
| Dennis Kundisch | Albert Ludwig University of Freiburg, Germany |
| Dominik Kuropka | University of Potsdam, Germany |
| Henry Linger | Monash University, Australia |
| Peter C. Lockemann | University of Karlsruhe, Germany |
| Peter Loos | University of Saarbrücken, Germany |
| Leszek Maciaszek | Macquarie University, Australia |
| Heinrich C. Mayr | University of Klagenfurt, Austria |
| Hao Min | Fudan University, China |
| Marie-Francine Moens | Katholieke Universiteit Leuven, Belgium |
| Günter Müller | University of Freiburg, Germany |
| Ludwig Nastansky | University of Paderborn, Germany |
| Adam Nowicki | The Wroclaw University of Economics, Poland |
| Markus Nüttgens | University of Hamburg, Germany |
| Andreas Oberweis | University of Karlsruhe, Germany |
| Mitsunori Ogihara | University of Rochester, USA |
| Józef Oleński | University of Warsaw, Poland |
| Maria E. Orlowska | The University of Queensland, Australia |
| Marcin Paprzycki | Polish Academy of Sciences, Poland |
| Eric Paquet | National Research Council, Canada |
| Witold Pedrycz | University of Alberta, Canada |
| Euripides Petrakis | Technical University of Crete, Greece |
| Arnold Picot | University of Munich, Germany |
| Jakub Piskorski | EU, JRC, Web and Language Technology, Italy |
| Jaroslav Pokorný | Charles University, Czech Republic |
| Maria Raffai | Szechenyi Istvan University, Hungary |
| Václav Repa | VSE Praha, Czech Republic |
| Narcyz Roztocki | State University of New York at New Paltz, USA |
| Shazia Sadiq | The University of Queensland, Australia |
| Alexander Schill | Dresden University of Technology, Germany |
| Elmar J. Sinz | University of Bamberg, Germany |
| Janice C. Sipior | Villanova University, USA |
| Henk G. Sol | University of Groningen, The Netherlands |
| Martin Spann | University of Passau, Germany |
| Bogdan Stefanowicz | Warsaw School of Economics, Poland |
| Kilian Stoffel | University of Neuchâtel, Switzerland |
| York Sure | University of Karlsruhe, Germany |
| Witold Suryn | École de technologie supérieure, Canada |
| Vojtech Svatek | VSE Praha, Czech Republic |
| Joseph Urban | Arizona State University, USA |
| Susan Urban | Arizona State University, USA |
| Thaddeus W. Usowicz | San Francisco State University, USA |

| | |
|---|---|
| Olegas Vasilecas | Vilnius Gediminas Technical University, Lithuania |
| Zygmunt Vetulani | Adam Mickiewicz University, Poland |
| Herna Viktor | University of Ottawa, Canada |
| Christof Weinhardt | University of Karlsruhe, Germany |
| Mathias Weske | Hasso Plattner Institute, Potsdam, Germany |
| Krzysztof Węcel | Poznań University of Economics, Poland |
| Maria Wimmer | University of Koblenz, Germany |
| Viacheslav E. Wolfengagen | Institute JurInfoR-MSU, Russia |
| Stanislaw Wrycza | University of Gdansk, Poland |
| Slawomir Zadrozny | Polish Academy of Sciences, Poland |
| Mohammed J. Zaki | Rensselaer Polytechnic Institute, USA |
| Arkady Zaslavsky | Monash University, Australia |

## Additional Reviewers

Acker, Hilmar
Blom, Sören
Boonstra, Albert
Bortoluzzi, Mariana Kessler
Brosch, Christian
Brzostowski, Jakub
Burmester, Lars
Carter, Belinda
Chaoji, Vineet
Chikova, Pavlina
Christopeit, Dietrich
Colasuonno, Francesco
Colucci, Simona
Fabian, Benjamin
Fettke, Peter
Fischmann, Matthias
Glavic, Boris
González, Rafael
Grosan, Crina
Hasan, Mohammad
Henke, Jan
Hesse, Bernd
Höfferer, Peter
Iqbal, Kashif
Ivantysynova, Lenka
Jossen, Claudio
Kaczmarek, Tomasz
Kozlova, Elizaveta
Kruk, Sebastian Ryszard
Lenz, Mario
Li, Xue
Lux, Mathias
Minor, Mirjam
Mueller, Ingo
Nemetz, Martin
Panigrahi, Prabin
Ploch, Holger
Puhlmann, Frank
Ragone, Azzurra
Rommelspacher, Jonas
Ronaghi, Faribors
Rykowski, Jarogniew
Salem, Saeed
Schuschel, Hilmar
Seel, Christian
Stollberg, Michael
Strykowski, Sergio
Subasu, Ionut Emanuel
Tanev, Hristo
Tinelli, Eufemia
Wang, Xia
Wassing, Steffen
Wittges, Holger
Wojciechowski, Grzegorz
Wolf, Petra
Wolff-Marting, Vincent
Zaremba, Maciej
Ziegler, Patrick
Ziekow, Holger
Zwicker, Jörg

## Supporters

ACM Special Interest Group on Management Information Systems, USA
GI - Gesellschaft für Informatik, Austria, Germany, Switzerland
NTIE - Polish Society for Information Systems

## Organizer

Poznań University of Economics
Department of Information Systems
Al. Niepodległości 10
60-967 Poznan
POLAND
tel. +48(61)854-3381
fax +48(61)854-3633
http://www.kie.ae.poznan.pl

## Organizing Committee

Konstanty Haniewicz
Tomasz Kaczmarek
Wojciech Rutkowski
Krzysztof Węcel

# Table of Contents

## Keynote



## Business Process Management



## Web Services

## Ontologies

## Information Retrieval



## System Design

## Agents And Mobile Applications



## Decision Support



## Social Issues

## Specific MIS Issues

# Security Issues for the Use of Semantic Web in E-Commerce

Andreas Ekelhart[1], Stefan Fenz[1], A Min Tjoa[2], and Edgar R. Weippl[2]

[1] Secure Business Austria, A-1040 Vienna, Austria
{aekelhart,sfenz}@securityresearch.at
http://www.securityresearch.at

[2] Institute for Software Technology and Interactive Systems - Vienna University of Technology, A-1040 Vienna, Austria
{atjoa,eweippl}@ifs.tuwien.ac.at
http://www.ifs.tuwien.ac.at

**Abstract.** As the ontologies are the pivotal element of the Semantic Web in E-Commerce, it is necessary to protect the ontology's integrity and availability. In addition, both suppliers and buyers will use an ontology to store confidential knowledge pertaining to their preferences or possible substitutions for certain products. Thus, parts of an ontology will need to be kept confidential. We propose to use well established standards of XML access control. E-commerce processes require the confidentiality of customer information, the integrity of product offers and the availability of the vendors' servers. Our main contribution-the introduction of a Security Ontology-helps to structure and simulate IT security risks of e-commerce players that depend on their IT infrastructure.

## 1 Introduction

We emphasize on the large potential of applying the semantic web technology to electronic commerce. Autonomous or semi autonomous agents can use the semantic information to search for and compare products or suppliers and negotiate with other agents [GTM99] [TBP02] [Sch03]. Generalizing previous work we propose the following short definition for semantic e-commerce:

> Semantic e-commerce is the processing of buying and selling via the semantic web.

Even though concepts of solutions already exist for years, they were not successful on the market. Thus till today information asymmetries still exist [Gup02] and one of the resulting shortcomings is the fact that the better informed buyer increasingly gets a better value for his money. Unfortunately searching is still a costly task and due to current data structures often an inefficient, economic activity. Research projects such as [ebS06] attempt to address these issues. The aim of this innovative project is to offer suppliers the option to publish their products and services in a machine-readable language based on open-source,

W. Abramowicz (Ed.): BIS 2007, LNCS 4439, pp. 1–13, 2007.
© Springer-Verlag Berlin Heidelberg 2007

domain specific structures i.e. an ontology. Such semantically enriched descriptions enable intelligent software agents to query and read product information autonomously and prepare it for human customers in an appropriate way.

## 2 Introducing Semantic E-Commerce

### 2.1 Architecture

Customers and suppliers are confronted with a very diversified market environment. Figure 1 shows the typical situation of a customer/supplier who intends to buy/sell a certain article over the world wide web. Compared to the conventional real life market environment, tools such as comparison shopping portals (e.g. www.geizhals.at) and search engines ease the search for the favored product and give suppliers the possibility to offer their products on a central marketplace. Despite these tools the customer is usually still overwhelmed with a big amount of offers and different product descriptions. Even though comparative-shopping-portals offer the possibility to search within specific product groups the customer still has to compare the different product descriptions to figure out which article matches his requirements most.

Figure 2 shows a possible scenario of a centralized semantic e-commerce environment. The product ontology provides as a central element the knowledge about defined product groups and their specific attributes (e.g. for mobile phones: display size, memory and organizer capabilities). The supplier agent uses the ontology data to dynamically build a user interface for the human supplier who is then able to feed the supplier agent with relevant product and price information. The last step requires the supplier agent to register itself at a central directory with its virtual location and offered product groups. On the customer

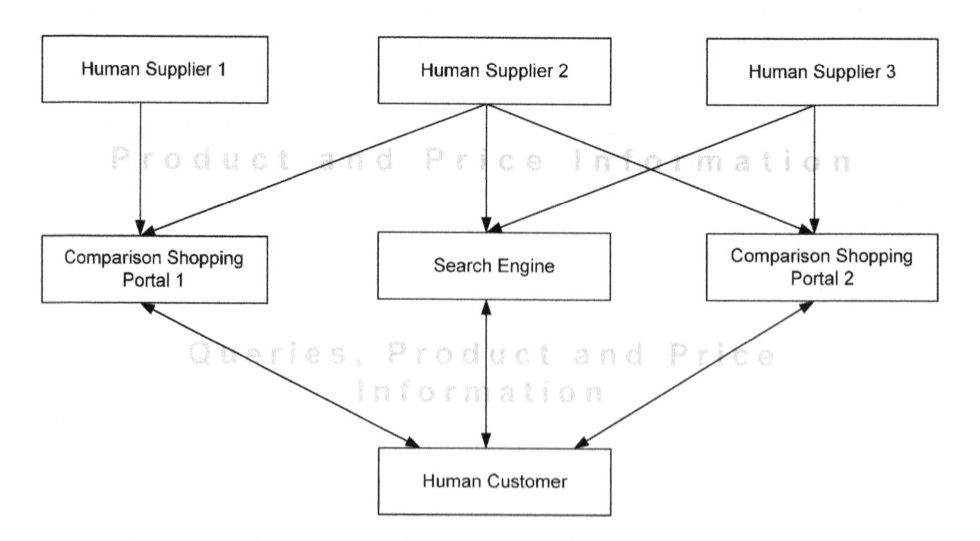


**Fig. 1.** e-Commerce - state of the art

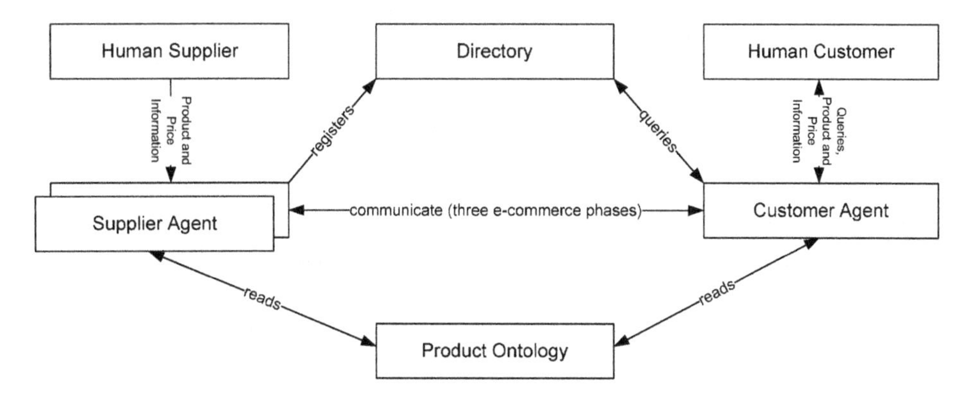


**Fig. 2.** e-Commerce - the centralized semantic approach

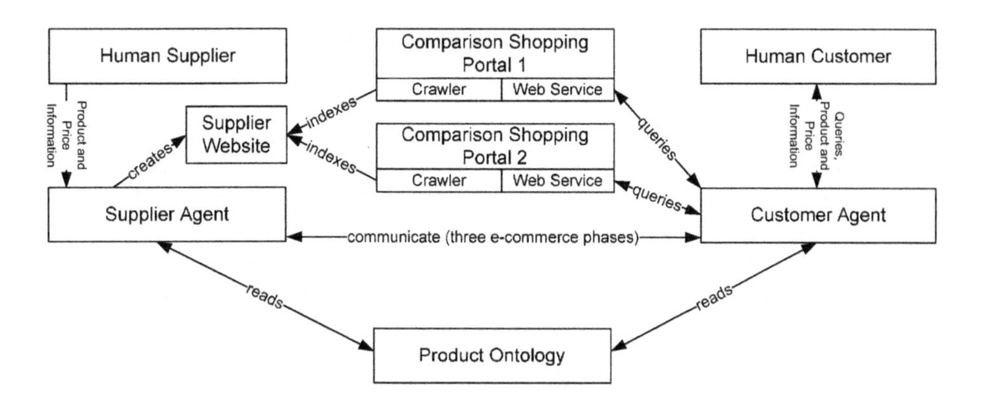


**Fig. 3.** e-Commerce - the decentralized semantic approach

side the process is almost identical. Depending on the desired product group the customer agent reads out the proper product ontology and creates a user interface which is capable to find out customer's requirements regarding a specific product. A mixture of questions and checklists could be used to find out what the customer really requires. After the requirement specification the customer agent queries a central directory to find supplier agents which offer the right product group. With a list of all available supplier agents the customer agent is able to start the communication (the three e-commerce phases) with each supplier agent.

One shortcoming of the centralized directory approach, is the fact that there has to be a central authority which maintains the directory service. With the utilization of a central ontology and semantic (in the sense of product and price descriptions) websites a more decentralized architecture which uses web crawlers to identify possible semantic e-commerce websites will be possible (compare Figure 3). Of course these websites have to use the classification of the central product ontology to ensure compatibility with the consumer agents. In realistic terms it will not be possible that every consumer runs its own crawler that processes large parts of the world wide web. Thus some kind of services (e.g. extensions to established comparison shopping portals) which run their own crawlers have to be established and the consumer agent looks for possible supplier agents at these sites to start the three-phase e-commerce communication.

### 2.2 E-Commerce Phases

E-commerce transactions, which take place between businesses and customers, consist of three phases: search, negotiate and fulfillment [Pet00] [SKLQ01]. In the following, each of these phases will be discussed in detail, describing the current situation and security relevant issues:

**Search.** Usually an e-commerce transactions starts with a user or business searching for potential trading partners. For this task two general approaches exist: (1) searching for a company with specific characteristics or (2) looking for goods with particular features and subsequently for companies which offer them. Initially all product characteristics are often not specified or not yet known and therefore this phase should result in a list of potential trading partners, each offering products of interest.

We distinguish general-purpose search engines (e.g. Google) and domain-specific portals (e.g. MEDLINEplus) on the Web as proposed by [BCJ+03]. In both cases, facing purely syntactic information, only keyword-based search can be conducted, which is known to be inefficient [Sch03]. The obvious need for semantic search approaches has been realized [KB04], and nowadays search portals, taking advantage of proprietary, lightweight semantic definitions, up to companies, offering sound product descriptions based on shared domain specifications in OWL [OWL04], exist. In this paper we concentrate on this last-mentioned newly approach, matchmaking by ontological product descriptions by reason that it is flexible and offers the most accurate search results. Pertaining to the semantic e-Commerce approaches, depicted in Figure 2 and Figure 3, autonomous agents carry out the search instead of the human customer itself. Initially the search parameters are provided to the agent which subsequently queries for supplier agents. Concerning the CIA triad (confidentiality, integrity and availability), ontological product descriptions and offers sometimes have to be confidential (encrypted parts for example), the integrity has to be maintained to counter fraud and availability is necessary for successful matchmaking. Security solutions regarding ontological descriptions, mostly available in XML (RDF or OWL), will be discussed in Section 3.

**Negotiate.** Once potential business partners have been identified in the search phase, the second phase of transaction, namely electronic negotiation, starts. This is performed through an interchange of negotiation proposals describing constraints on an acceptable deal and results in an agreement (which is transformed into a legally binding contract), specifying the terms that both parties consider acceptable. These terms could include the product or service description, the price, delivery date, etc. [TBP02]

Negotiation relies on a shared terminology to guarantee efficient interactions and to avoid misunderstandings and conflicts. Ontologies can provide definitions of concepts and relations, describing the domain of interest as well as negotiation specific concepts. [SBQ+02] state that ontology-based negotiation approaches enable efficient, complex and unambiguous exchanges that result in business contracts.

Confidentiality and integrity are of main concern during the negotiation phase pertaining to security. Especially the exchange of private information (including credit card numbers) demands a high level of security and trust and furthermore, non-repudiation must be enforced.

**Fulfillment.** After a contract is agreed upon, the promises set in the negotiation phase and specified in the contract are carried out. Usually automatic workflows are executed to initiate payments or delivery processes which are (automatically) monitored to control and sometimes enforce the correct fulfillment of the contracts. Automic reasoning on contract obligation fulfillment or non-fulfillment demands formal contract definitions as well as formal transaction information to show the relevant context in which it occurs.

The fulfillment processes and corresponding resources and monitoring installations in place pose as potential targets for attacks, especially pertaining to fraud.

Agent based e-commerce aims to support the whole transaction process by autonomous means. By using sound semantic descriptions it is possible for agents, given a set of initial parameters, to find products and services automatically. Also the negotiation phase can be carried out by agents if the terms are defined and negotiating agents understand each other (using the same vocabulary, which can be achieved by common ontologies). "Intelligent", autonomous agents can unburden users in their daily, time-consuming and complex tasks and even reach better results but legal questions and security issues, including trust between agents, are a crucial point and will be discussed in Section 4.

Another aspect of (semantic) e-commerce security is the business crucial IT-environment, comprising (web-)servers hosting company information and agent services, databases with product and private user-information, ontological file storages for products and domain specific knowledge, etc. Only in a well protected and maintained IT-environment reliable and secure e-commerce can be conducted, which is often overseen, especially by small- and medium-sized enterprises. [Hau00] summarized the problems of SMEs regarding the IT-Security aspect: (1) Smaller IT budget, relative to total budget as well as in absolute figures (2) Less IT knowledge, information technology is often looked after by employees from other departments (3) IT is not considered as important as within larger enterprises although more and more core processes are processed by IT elements (4) IT environments are not homogeneous. To overcome these problems we introduced a security ontology approach for holistic IT-infrastructure security [EFKW06] and Section 5 refers to the technical details of the security ontology approach.

## 3 Security Within Ontologies

Ontologies are at the focus of our approach. We thus need to protect their confidentiality and integrity.

### 3.1 Access Control

While the proposed product ontology remains public to ensure a shared vocabulary among the market participants, each supplier derives its own ontology, filled with concrete values such as price and delivery information, which has to be secured against unauthorized reading or writing attempts.

Due to the fact that each OWL- or RDF-based ontology uses XML as surface syntax [OWL04], access control models for XML documents can be also applied to OWL- or RDF-based ontologies.

Research in the field of XML access control models is already mature and several approaches for securing ontologies already exist: [FCG04] propose the concept of security views which provide for each user group an XML view consisting of that information that the users are authorized to access. The approach requires a XML query-execution engine that implements the DTD-based access control model. [DdVPS02] present a language for the specification of access restrictions on XML-based files and the corresponding system architecture for access control which should enforce its usage. The proposed XML Access Control Processor (ACP) takes as input a valid XML document requested by the user and the XML Access Sheet listing the associated access authorizations at the instance level. The ACP generates a valid XML document, including only the information the user is allowed to access [DdVPS02]. [BF02] extend the approach by fine-grained XML document encryption and proper key distribution mechanisms to ensure confidentiality within shared XML documents. The Semantic Access Control Language (SACL) proposed by [QA03] is suitable to express concept-level access authorizations within OWL-based ontologies.

Such mechanisms are suitable for enhanced and implemented ontology access control approaches and especially in the semantic e-commerce field with its various actors and different relationships we have to enforce a strong access control technology.

### 3.2 Integrity

Since the very central product-ontology and the derived supplier ontologies with their price information play an important role in a possible semantic e-commerce scenario there have to be proper mechanisms which ensure the integrity of the ontology structure and its content. Especially the derived supplier ontologies act as a storage for price and delivery information which are used by the agents at the negotiation phase. Therefore the integrity of these data elements is crucial for the long-term establishment of semantic e-commerce systems and due to the XML-based syntax of OWL- and RDF-ontologies we are able to use established standards such as XML Digital Signature [xml02] and XML Key Management Systems (XKMS) [xkm01] to ensure data integrity.

## 4 Trust Issues

Trust is one of the main issues pertaining to e-Commerce, based on the following reasons: 1) a potential buyer has no physical access to the product of interest, 2) seller or buyer may not abide by the agreement reached at the electronic marketplace [Zac99]. Agent based systems add another layer of indirection between a buyer and a seller, resulting in a more complex framework and new trust issues.

[Gam00] defines trust as *a particular level of the subjective probability with which an agent assesses that another agent or group of agents will perform a particular action, both before he can monitor such action (or independently of his capacity ever to be able to monitor it) and in a context in which it affects his own action.*

We distinguish between two fundamental trust models which are (1) either built on an agent's direct experience of an interaction partner (interaction trust) (2) or reports provided by third parties about their experiences with a partner (witness reputation) [HJS06]. Nowadays, taking eBay [EBA07] as an example, traders receive a feedback (such as +1, 0 or -1) for their reliability in each auction. Furthermore textual comments can be submitted to describe the customer's experience pertaining to the seller. Besides trust based on previous transactions (if they exist), customer feedback (feedback scores and comments) is a crucial element of trust in a seller. According to companies, independend third party evaluation and certification is another possibility to convince customers of their trustworthiness. Concerning to the centralized semantic e-Commerce approach in Figure 2, we identified the following trust issues and possible methods of resolution:

In the first place the human interacion partner has to trust his agent, viz the software system - the underlying lines of code created by the system developer. The agent has to fulfill the promised functionality and should not have any vulnerabilities. Certified providers as well as certified agent systems help to establish the trust needed.

Each agent has to "know" its communication partner before reputation can be considered, thus authentication mechanisms have to be implemented. As a principle an agent has to provide his idendity, usually in form of a public key certificate, issued by a certification authory (CA).

If agents have the ability to purchase products (on the behalf of the agent's principal), the risks can be minimized by only granting a limited payment capability [CPV03]. Furthermore, if digital signatures are required, the use of the private key should be limited to the agent. [RS99] for example propose proxy certificates: in this approach only a new, lifetime limited key pair is handed to the agent. This makes it difficult for malicious hosts to discover the private key before the certificate expires. Additionally, arbitrary transations can be constrained. To avoid contract repudiation—especially users denying that an agent acted on their behalf—the user instruction parameters should be collected and digitally signed.

The Directory service, shown in Figure 2, should only register and subsequently mediate trustworthy agents. Besides looking for available certificates, cumstomer agents have the possibility to rate their experiences with supplier agents. SPORAS [Zac99] is a possible model for an agent based, centralized rating system.

## 5 The Security Ontology

Beside the very deep going aspects of securing ontologies and communication between various agents, we also have to consider the IT-Security regarding the company's physical environment. Servers hosting company information and agent services, databases with private user-information or files containing ontological product information have to be secured to ensure a reliable and secure e-commerce service. Especially small- and medium-sized enterprises often oversee the need for a holistic IT-Security approach and thus we developed a Security Ontology [EFKW06] to provide a proper knowledge base about threats and the corresponding countermeasures. In [EFKW07] we extended the threat simulation approach with risk analysis methods to improve quantitative risk analysis methods. The current section summarizes the research results and proposes the implementation of the Security Ontology to enhance the overall IT-Security level.

The most important parts of the Security Ontology are represented by the sub-ontologies *Threat*, *ThreatPrevention* and *Infrastructure*:

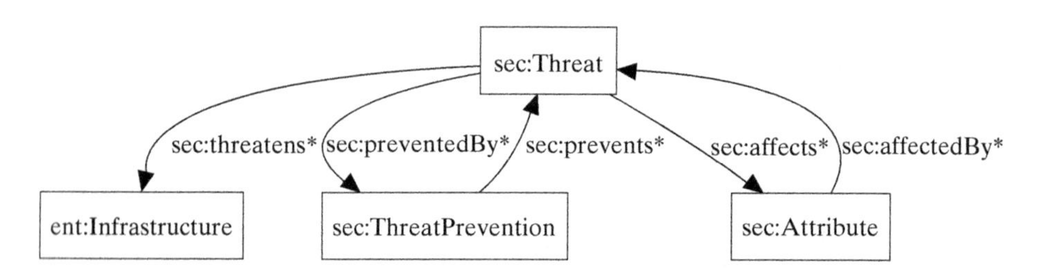


**Fig. 4.** Sub-ontology: Threat

Figure 4 shows the threat ontology with its corresponding relations: (1) To model the threats which endanger certain infrastructure elements we introduced the *sec:threatens* relation (every threat threatens $n$ infrastructure elements) (2) Of course we want to mitigate the threats and so we created the *sec:preventedBy* and *sec:prevents* relation respectively (3) To enable companies to optimize their IT-Security approach to certain IT-Security attributes such as confidentiality or availability we assigned affected attributes to each threat by the *sec:affects* and its inverse relation.

Figure 5 shows the security ontology's infrastructure area. The building, with its corresponding floors and rooms, can be described using the infrastructure framework. To map the entire building plan exactly on the security ontology, each room is described by its position within the building. The ontology *knows* in which building and on which floor a certain room is located. The

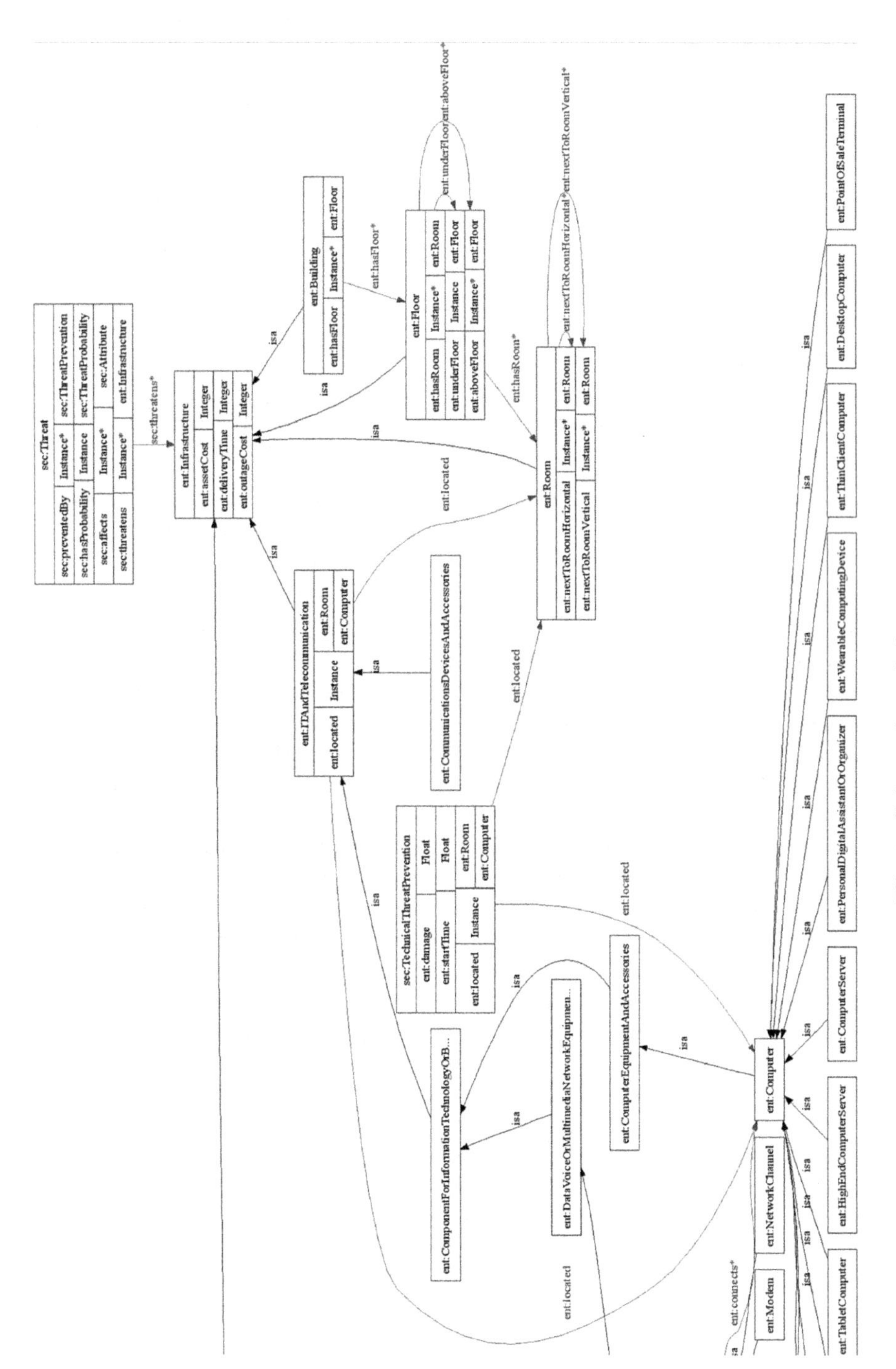


**Fig. 5.** Sub-ontology: Infrastructure

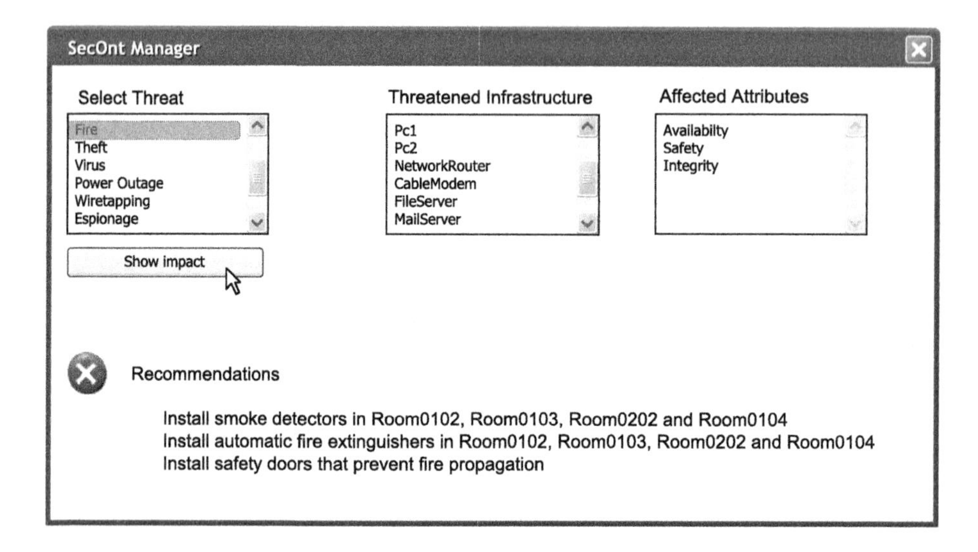


**Fig. 6.** SecOnt Manager Prototype

attributes *ent:nextToRoomHorizontal* and *ent:nextToRoomVertical* describe the exact location of each room. Each instance of *ent:ITAndTelecommunication* and *sec:TechnicalThreatPrevention* is located in a particular room. A room can, of course, also contain more concepts. The current ontology uses a flexible and easily extendable structure: additional concepts can be included without effort. The concept *ent:TechnicalThreatPrevention* is subdivided into *ent:CounterMeasure* and *ent:Detector*, which are used to model detectors (fire, smoke, noise, etc.) and their corresponding countermeasures (fire extinguisher, alarm system, etc.).

Figure 6 shows the prototype with its four main user interface elements: (1) Selection of a threat: The user is able to choose a certain threat and the SecOnt Manager shows the impact of that threat (2) Threatened infrastructure: The ontology provides an extendable framework for various infrastructure elements to enable the user to create instances of concrete and real infrastructure elements which enables the ontology to show which infrastructure elements are threatened by a certain threat scenario (3) Affected attributes: Works like the threatened infrastructure where the ontology *knows* which threats are affecting certain security attributes (4) Recommendations: Are the most important part for the user, because it gives concrete recommendations to prevent a certain threat. Figure 6 shows an example application for the fire threat and we see that the ontology has to store the whole infrastructure, including the building with its floors and rooms, to make location-based recommendations possible.

So why are we using an ontology instead of a database solution which has various advantages over a file-based ontology? The main advantage of an ontology is the possibility of inferring new knowledge by utilizing a reasoning engine which considers existing facts and rules.

$$
\begin{aligned}
sec: affectsOS(?x, ?z) \ \wedge \ ent: hasOS(?y, ?z) \wedge \ sec: AntiVirusProgram(?c) \\
\wedge \ \neg sec: installedOn(?c, ?y) \ \wedge \ sec: prevents(?c, ?x) \\
\rightarrow \ sec: threatens(?x, ?y)
\end{aligned}
\tag{1}
$$

Equation 1 illustrates a possible axiom which formalizes the *sec:threatens* relation between a computer virus and a computer device. First *sec:affectsOS* determines which operating systems are endangered by a certain virus and in the second step *ent:hasOS* looks up for all computers and their corresponding operating systems. Variable *?c* stores all available anti virus programs and looks with ¬*sec:installedOn* for computers that have not installed such a program. With *sec:prevents* it is possible to determine which anti virus protection is useful to a certain virus and so the ontology, equipped with a proper reasoning engine, is able to identify those computers that are directly threatened by a certain virus.

## 6 Conclusion

In this paper we covered the three phases of e-business (search, negotiation, and fulfillment) and investigated how semantic information and ontologies can support and improve these processes. Moreover, we explored the case for protecting the ontology which is the central element of this approach. Mechanisms of XML access control are used to protect the confidentiality, integrity and availability of ontologies. Finally, we presented how the introduced Security Ontology can be used to secure all assets required by IT-centered companies to ensure CIA (confidentiality, integrity and availability) of information processed in their business processes.

## Acknowledgements

This work was performed at the Research Center Secure Business Austria funded by the Federal Ministry of Economics and Labor of the Republic of Austria (BMWA) and the federal province of Vienna.

## References

[BCJ+03] Suresh K. Bhavnani, Bichakjian K. Christopher, Timothy M. Johnson, Roderick J. Little, Frederick A. Peck, Jennifer L. Schwartz, and Victor J. Strecher. Strategy hubs: next-generation domain portals with search procedures. In *CHI '03: Proceedings of the SIGCHI conference on Human factors in computing systems*, pages 393–400, New York, NY, USA, 2003. ACM Press.

[BF02] Elisa Bertino and Elena Ferrari. Secure and selective dissemination of xml documents. *ACM Trans. Inf. Syst. Secur.*, 5(3):290–331, 2002.

[CPV03] Joris Claessens, Bart Preneel, and Joos Vandewalle. (how) can mobile agents do secure electronic transactions on untrusted hosts? a survey of the security issues and the current solutions. *ACM Trans. Inter. Tech.*, 3(1):28–48, 2003.

[DdVPS02] Ernesto Damiani, Sabrina De Capitani di Vimercati, Stefano Paraboschi, and Pierangela Samarati. A fine-grained access control system for xml documents. *ACM Trans. Inf. Syst. Secur.*, 5(2):169–202, 2002.

[EBA07] ebay. http://www.ebay.com/, 2007.

[ebS06] ebsemantics. www.ebsemantics.org, 2006.

[EFKW06] Andreas Ekelhart, Stefan Fenz, Markus Klemen, and Edgar R. Weippl. Security ontology: Simulating threats to corporate assets. In Aditya Bagchi and Vijayalakshmi Atluri, editors, *Information Systems Security*, volume 4332 of *Lecture Notes in Computer Science*, pages 249–259. Springer, Dec 2006.

[EFKW07] Andreas Ekelhart, Stefan Fenz, Markus Klemen, and Edgar R. Weippl. Security ontologies: Improving quantitative risk analysis. In *Proceedings of the 40th Hawaii International Conference on System Sciences (HICSS 2007)*, Jan 2007.

[FCG04] Wenfei Fan, Chee-Yong Chan, and Minos Garofalakis. Secure xml querying with security views. In *SIGMOD '04: Proceedings of the 2004 ACM SIGMOD international conference on Management of data*, pages 587–598, New York, NY, USA, 2004. ACM Press.

[Gam00] Diego Gambetta. Can we trust trust? In Diego Gambetta, editor, *Trust: Making and Breaking Cooperative Relatioins*, chapter 13, pages 213–237. Published Online, 2000.

[GTM99] Robert J. Glushko, Jay M. Tenenbaum, and Bart Meltzer. An xml framework for agent-based e-commerce. *Commun. ACM*, 42(3):106–ff., 1999.

[Gup02] *Reduction of price dispersion through Semantic E-commerce*, volume 55 of *CEUR Workshop Proceedings*. CEUR-WS.org, 2002.

[Hau00] Hans Eduard Hauser. Smes in germany, facts and figures 2000. Institut für Mittelstandsforschung, Bonn, 2000.

[HJS06] Trung Dong Huynh, Nicholas R. Jennings, and Nigel R. Shadbolt. Certified reputation: how an agent can trust a stranger. In *AAMAS '06: Proceedings of the fifth international joint conference on Autonomous agents and multiagent systems*, pages 1217–1224, New York, NY, USA, 2006. ACM Press.

[KB04] Mark Klein and Abraham Bernstein. Toward high-precision service retrieval. *IEEE Internet Computing*, 8(1):30–36, 2004.

[OWL04] Owl web ontology language. http://www.w3.org/TR/owl-features/, 2004.

[Pet00] Ralf Peters. Elektronische märkte und automatisierte verhandlungen. *Wirtschaftsinformatik*, 42(5):413–421, 2000.

[QA03] Li Qin and Vijayalakshmi Atluri. Concept-level access control for the semantic web. In *XMLSEC '03: Proceedings of the 2003 ACM workshop on XML security*, pages 94–103, New York, NY, USA, 2003. ACM Press.

[RS99] Artur Romao and Miguel Mira Da Silva. Proxy certificates: A mechanism for delegating digital signature power to mobile agents. In *IAT99 Workshop on Agents in Electronic Commerce*, 1999.

[SBQ+02] Mareike Schoop, Andreas Becks, Christoph Quix, Thomas Burwick, Christoph Engels, and Matthias Jarke. Enhancing decision and negotiation support in enterprise networks through semantic web technologies. In *XML Technologien für das Semantic Web - XSW 2002, Proceedings zum Workshop*, pages 161–167. GI, 2002.

[Sch03] Mareike Schoop. Semantic web technology for electronic commerce. In *Proceedings of the The Tenth Research Symposium on Emerging Electronic Markets*, 2003.

[SKLQ01] Mareike Schoop, Joerg Koeller, Thomas List, and Christoph Quix. A three-phase model of electronic marketplaces for software components in chemical engineering. In *I3E '01: Proceedings of the IFIP Conference on Towards The E-Society*, pages 507–522, Deventer, The Netherlands, The Netherlands, 2001. Kluwer, B.V.

[TBP02] D. Trastour, C. Bartolini, and C. Priest. Semantic web support for the business-to-business e-commerce lifecycle, 2002.

[xkm01] Xml key management specification (xkms). http://www.w3.org/TR/xkms/, 2001.

[xml02] Xml-signature syntax and processing. http://www.w3.org/TR/xmldsig-core/, 2002.

[Zac99] Giorgos Zacharia. Trust management through reputation mechanisms. In *Third International Conference on Autonomous Agents (Agents '99)*, New York, NY, USA, May 1999. ACM Press.

# Automatic Merging of Work Items in Business Process Management Systems

Dat C. Ma, Joe Y.-C. Lin, and Maria E. Orlowska

School of Information Technology and Electrical Engineering
The University of Queensland
Brisbane, Australia
{datma,jlin,maria}@itee.uq.edu.au

**Abstract.** Current functional requirements for Process Management Systems are often demanding accommodation of rapid changes of business processes as well as simple adoption to new business requirements. This paper identifies and distinguishes between two new ways to handle system's assisted multiple work items execution; work item merging and work item grouping. We show that traditional workflow management systems technology has limitations for automatic item merging while Harmonized Messaging Technology (HMT) (Sadiq, Orlowska, Sadiq et al., 2004b) is better positioned for such item merging behaviour. For completeness of presentation, we provide a brief introduction to the basic concepts in HMT and demonstrate how automatic merging can be specified in this environment. We conclude by investigating the impact on process related data when adopting such merging extension into process enforcement technology.

**Keywords:** Merging, Grouping, Activity instances, Work items, Flexibility, Workflow, Business Process Management.

## 1 Introduction

Business process management systems (BPMS) today face more challenges than in the past as they must not only be used to automate and manage complex business processes but also handle various new functionalities stemming from business requirements. Among the technologies for BPMS, workflow technology has been considered being the most common technology in supporting the automating of business processes. Basically in workflow, the structure of a process is modelled as a directed graph, which connects *activities* together using *coordinators* (such as *AND-Split*, *AND-Join*, *XOR-Split*, *XOR-Join*, *Begin*, *End*) (WfMC, 1999). During execution, a *process instance* within the process represents a particular case of the process; an *activity instance* (or *task*) represents an activity within a (single) enactment of a process (i.e. within a process instance); *workflow performers* (or *participants*) perform the work represented by a workflow activity instance by selecting a *work item*; work item and activity instance are two perspectives of activity execution. A detailed description of those concepts can be referred to in (WfMC, 1999).

The popularity of workflow technology can be seen as the result of the simplicity in process models and the clear separation between aspects of a process model such as

W. Abramowicz (Ed.): BIS 2007, LNCS 4439, pp. 14–28, 2007.
© Springer-Verlag Berlin Heidelberg 2007

control, data, and resource. However, the technology is often criticised for its lack of "flexibility". Under the "flexibility" umbrella, many research efforts, such as (Kappel, Rausch-Schott, & Retschitzegger, 1996), (Sadiq, 2000), (Bae, Bae, Kang et al., 2004), (Müller, Greiner, & Rahm, 2004), (Adams, Hofstede, Edmond et al., 2005), aim to address the exception handling issue. In those literature, different types of exceptions in workflows are considered, which include *system* related exceptions (i.e. hardware or software failures) and *logical* exceptions (i.e. events that effect the normal execution of workflows). In addition to exception handling, works investigating control flow, data flow, and resource aspects of workflow, such as (Kiepuszewski, 2002), (Aalst, Hofstede, Kiepuszewski et al., 2003), (Russell, Hofstede, Edmond et al., 2004a), (Russell, Hofstede, Edmond et al., 2004b) are also presented, which show the limitations of current workflow specifications to support various new functionalities.

This paper is motivated by the requirement to extend BPMS with a new functionality to handle multiple work items across multiple process instances of the same activity, the way some business policy would be enforced during process execution. Current workflow management systems (or WFMS) maintain process instances such that generally there is at most one work item per activity of a process instance, if no loop is involved. However, some business rules (or policies) require accumulating multiple work items across different process instances.

We differentiate two ways of accumulating work items, *grouping* and *merging*. In the case of work item grouping, a "wrapper" is produced to wrap several work items together to be processed at the same time and the unity of work items will remain after grouping. When merging work items, a new work item is created in place of several work items and the new work item has its content as the collated contents of all the merged work items. For example, in purchasing processes, shipments to the same address are grouped together to reduce the delivery cost; while several orders from the same supplier can be merged into one single order to get better discount. The main difference between these two operations is that grouped shipments can still be processed individually afterwards, however orders being merged cannot. Such functionalities of grouping and merging work items can be very useful for business, therefore need to be considered as an extension to existing BPMS.

In (Ma, Lin, & Orlowska, 2006), we investigate the extension of the grouping function in WFMS. This paper primarily focuses on merging work items. In particular, we demonstrate that current workflow specification find it hard to meet the requirement of the merging function without major extension to the data specification and the engine implementation. Therefore, an alternative rule-based workflow called Harmonized Messaging Technology (HMT) is proposed. In section 3, we introduce basic related concepts in HMT and show how to specify merging using HMT. In section 4, we discuss the requirement for deploying such merging extension. The conclusion and future work are presented in section 5.

## 2 Grouping and Merging Work Items

As mentioned above, grouping and merging are two different concepts. The main difference between these functions can be illustrated as in Figure 1, which represents a typical result of executing two consecutive activities within a purchasing workflow.

In Figure 1, PR and PO represent multiple work items that need to be executed by the performers (or participants) of activities: Create Purchase Request and Create Purchase Order. Each work item has an associated content. For example, the work item PR2 requests to order two products C and D with their respective values (i.e. 30 and 20) as its content.

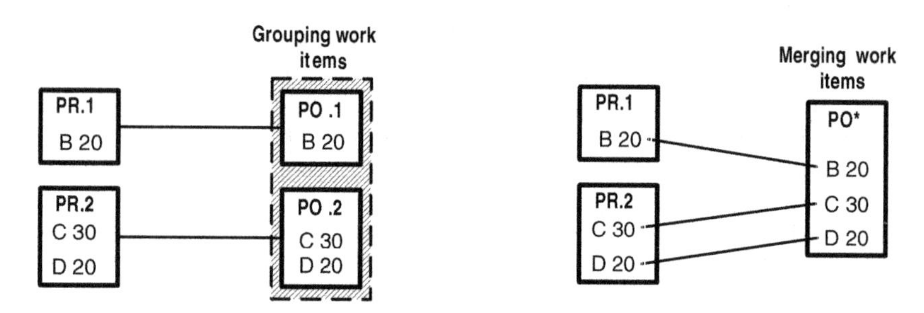


**Fig. 1.** Grouping and Merging of work items

As illustrated in Figure 1, when grouping, a group of work items is created (i.e. a group formed by PO.1and PO.2). While in the case of merging work items, a new work item is created as a result of merging (i.e. PO* is the corresponding purchase order for both PR.1 and PR.2).

If we denote $A_i$ as an activity and $WI_i$ the set of all work items of activity $A_i$, then we define the *Group* and *Merge* functions for work items as follows.

**Group:** $WI_i \rightarrow 2^{WI_i}$, where $2^{WI_i}$ is the power set of $WI_i$. When the group function is applied on $WI_i$, it will necessarily produce a partition of $WI_i$. After partitioning, each part of the set $WI_i$ is called a *group* iff it has more than one element (i.e. one work item).

**Merge:** $2^{WI_i} \rightarrow WI_i$.When the merge function is applied on a subset of $WI_i$, it will replace the subset by creating a new work item $wi_t \in WI_i$ (different from the case of grouping work items, where no new work item is created). The newly created work item $wi_t$ has its content as the collated contents of all the work items it merged.

In practice, the group function offers a mechanism allowing activity performers to group items as a whole in a preferred way for the purpose of being *activated*, *suspended*, *terminated*, or *completed* (WfMC, 1999) at the same time. The item group can also be used to preserve works to be accessed and executed by dedicated performers. A detailed investigation on the impact, benefits, and potential implementation of extending the grouping function, supporting preferred work practices, in WFMS is presented in (Ma, Lin, & Orlowska, 2006).

On the other hand, merging of work items does not aim to provide preferred work practices, but the function allows BPMS to handle specific business rules or policies. The merging feature allows specific work items to be accumulated and collated into a new work item in order to facilitate certain business requirements, such as requirements to implement discount policy, payment procedure, or delivery method. We observed that current WFMS has limitation in supporting the merging feature according to the following two main reasons.

Firstly, when work items are merged based on their contents, it is required that the workflow engine must have the access to their contents before merging. However, in current WFMS the content of a work item is hidden to the engine.

Secondly, creating a new merged work item and populating its content cannot be specified by current workflow specifications. Thus it requires either extension to workflow specifications or WFMS's applications (Sadiq, Orlowska, Sadiq et al., 2005).

In the next section, we will introduce Harmonized Messaging Technology (or HMT) as an alternative of workflow technology in business process enforcement and show that HMT can support the automatic merging function.

## 3 Harmonized Messaging Technology

Unlike workflow technology that considers a business process as an ordering collection of activities, Harmonized Messaging Technology (or HMT) visions process as a communication phenomenon. Where process *participants* communicate with each other by exchanging *messages* of predefined *message types* (Sadiq, Orlowska, Sadiq et al., 2004b), (Sadiq, Orlowska, Sadiq et al., 2004a), (Sadiq, Orlowska, & Sadiq, 2005). A Harmonized Messaging Management System (or HMMS) (Ma, Carter, Sadiq et al., 2006) facilitates the communication interactions between participants using *harmonizing rules* (or *rules*).

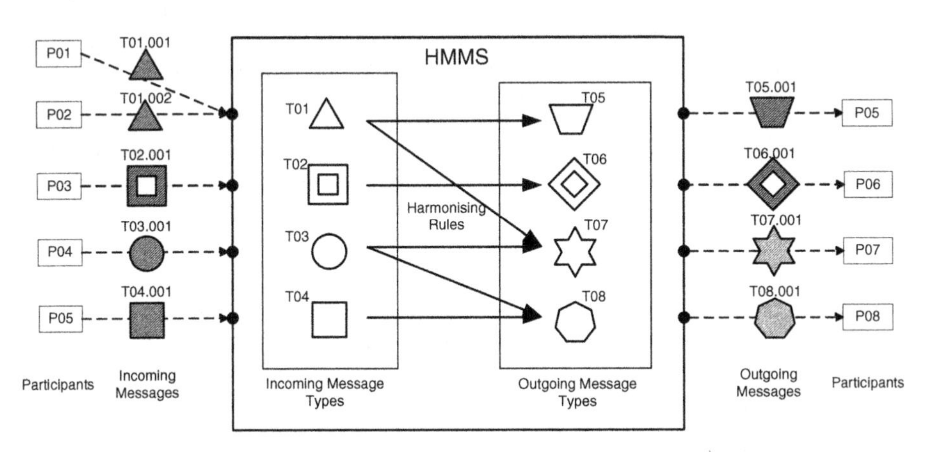


**Fig. 2.** An HMMS execution

Figure 2 envisages an example of an HMMS execution. Rules are used to specify the relationships between message types (e.g. T01, T02, or T08). Based on those rules, when participants (e.g. P01, P02, or P05) send incoming messages (e.g. T01.001, T01.002, or T04.001) to the HMMS, the HMMS then generates outgoing messages (e.g. T05.001, T06.001, or T08.001) to dispatch to other participants (e.g.

P05, P06, or P08). In the following section, we will give a description of HMT basic concepts.

### 3.1 HMT Basic Concepts

The most fundamental concept in HMT, required for process modelling, is Collaboration Space (or CS), which is a 3-tuple of $<\mathcal{P}, \Re, \Im>$. We denote $\mathcal{P}$ as a set of Participants, $\Re$ as a set of Rules, and $\Im$ as a set of Message Types (Ma, Orlowska, & Sadiq, 2005).

**Participants.** A participant represents a component process, an organization, or a program participating in the process. Participants can send and/or receive messages to/from the system (HMMS).

**Message Types.** All the communication messages must conform to predefined message types. Each message type defines the *fields* that it contains (e.g. message receiver, or message ID), the *fields domains* (i.e. the set of values that the fields can accept), the *fields constraints* (i.e. other restrictions on fields such as mandatory/optional, subtype or subset constraints). Message types are classified into two classes, namely *Incoming Message Type* (or *IMT*) and *Outgoing Message Type* (or *OMT*), based on the receipt and dispatch functions of the HMMS.

We use the notations $T_n$, $T_n.F_k$ to denote an IMT (or OMT) and its field; where n, k are the IDs/names of the message type and the field respectively.

**Rules.** HMT rules are empowered by the concept of *events*, which specify the reaction of HMMS when it detects the arrival/dispatch of specific messages and/or certain time occurrences. In general, harmonizing rules are of the form *Condition* $\rightarrow$ *Action*.

- *Conditions.* The condition of a rule is a first order formula, which is built on *events* using Boolean operators ( $\neg$, $\wedge$, $\oplus$ (or XOR)) as well as constants, field values of the message pertaining to an *event* using $\forall$, $\exists$, $\neg$, $\wedge$, $\oplus$; where HMT event refers to either the *Arrival* (or *Dispatch*) of a specific *message*, or a *Clock* event (i.e. the occurrence of a specific time point).
- *Actions.* When the condition part of a rule is evaluated to be TRUE, the HMMS will react by performing a communication action. HMT considers three types of actions: *Associate* (i.e. associates specific event occurrences with an OMT), *Assign* (or *Populate*) (i.e. assigns a value to a message field), and *Send* (i.e. sends the populated message to a participant). In addition, there are three rule types. Each deals with one action type, namely *rule type 1*, *rule type 2*, and *rule type 3*.

A detailed description of HMT rule specification and its formal foundation is presented in (Ma, Orlowska, & Sadiq, 2005). In (Ma, Orlowska, & Sadiq, 2005), the safety and expressive power of HMT specification have also been investigated.

### 3.2 Modelling Process Using HMT

In a simplified sale order processing workflow (Figure 3), the HMMS passes messages between performers (i.e. HMT participants) of activities: Create Purchase Request,

Create Purchase Order, Organize Shipment, Send Invoice, and Receipt Payment, in order to manage the execution order between these activities.

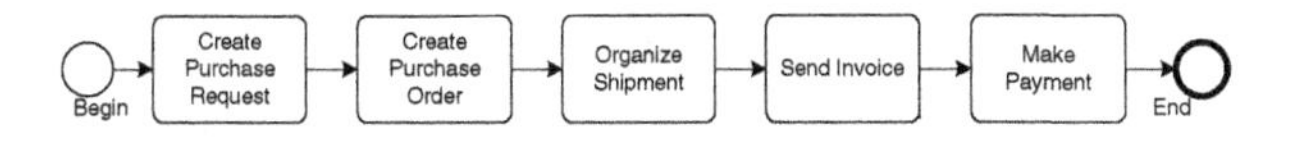


**Fig. 3.** Order processing workflow

Typically, an outgoing message that is sent by the HMMS to a participant represents a request for a work (i.e. a work item) to be done by the participant; and after completing the work item the participant will return an incoming message. Thus a work item can be modelled as a pair of an outgoing and an incoming message, and a workflow activity $t_i$ is modelled in HMT as a pair of an OMT and an IMT, which are denoted as $T_{Request-i}$ and $T_{Confirm-i}$ respectively.

Message Type: $T_{PR}$ *(F0)*

PURCHASE REQUEST

Reference No.: PR-00021 *(F1)*

Customer: ITEE – UQ *(F2)*
Customer ID: CUST-0000037 *(F3)*
Account No.: ACC-5600075 *(F4)*
Email: itee@uq.edu.au *(F5)*

Time, date sent: 08:10:20 - 06/09/04 *(F6)*

Vendor: Dell Australia Company *(F7)*
Email: sales@dell.com.au *(F9)*

Vendor ID: V001 *(F8)*
Need by: 10/09/04 *(F10)*
Shipto: GP Building, St Lucia *(F11)*

*(Tab1)*

| Item *(Tab1.F1)* | Description *(Tab1.F2)* | Quantity *(Tab1.F3)* | Unit Price - AUD *(Tab1.F4)* | Sub total - AUD *(Tab1.F5)* |
|---|---|---|---|---|
| 1 | PC DELL - Dimension 8400 | 10 | 1750 | 17500 |
| 2 | Server DELL-Precision 470DT | 3 | 3200 | 9600 |
| | | | Total - AUD *(Tab1.F6)* | 27100 |

Note: *(F13)*
Signature: ALEX *(F14)*

Message Type: $T_{Request-CreatePO}$ *(F0:=* $T_{Request-CreatePO}$*)*

REQUEST – CREATE PURCHASE ORDER

Message ID: R-PO-00021 *(F1:=SYSTEM)*

Participant: Purchasing department *(F2:= "Purchasing department")*

Reference task: Create purchase order *(F3:= "Create purchase order")*

Reference messages: PR-00021 *(F4:=* $T_{PR}$*.F1)*

Time, date issued: 09:10:25-08/09/04 *(F5:= SYSTEM)*

**Fig. 4.** A Purchase Request and its corresponding Request to Create Purchase Order

For example, in Figure 3, the Create Purchase Order activity is modelled as $< T_{Re\,quest-CreatePO}, T_{Confirm-CreatePO} >$ and the transition between two activities Create Purchase Request and Create Purchase Order is facilitated by harmonizing rules that generate a $T_{Re\,quest-CreatePO}$ message in response to the occurrence of a Purchase Request (i.e. a $T_{PR}$ message). Figure 4 illustrates an example of a message of type $T_{PR}$ and its corresponding message of type $T_{Re\,quest-CreatePO}$.

Tree types of rules to generate $T_{Re\,quest-CreatePO}$ messages in Figure 4 are:

**Rule type 1. IF *cond* THEN ASSOCIATE** $T_i$, where *cond* is the rule's condition and $T_i$ is an OMT.

For example, IF $T_{PR}$ THEN ASSOCIATE $T_{Re\,quest-CreatePO}$ (i.e. associate OMT *request to create a purchase order* when a *purchase request* arrives).

**Rules type 2.** $T_i .F$ := **IF *cond₁* THEN *value-expression₁***
**IF *cond₂* THEN *value-expression₂***
**...**
**IF *condₘ* THEN *value-expressionₘ***

Where $T_i .F$ is a field, *condition$_k$* is a condition, and *value-expression$_k$* refers to constants, system generated values, or values derived from previously received messages.

For example, in the message "Request to Create a Purchase Order" (Figure 4), the field "Time, date issued" with ID F5 has a system generated value; the field "Reference messages" with ID F4 takes the value of a field (i.e. $T_{PR}$.F1) in a previously received purchase request.

**Rule type 3. IF *cond* THEN SEND** $T_i$, where $T_i$ is an OMT.

For example, IF $T_{Re\,quest-CreatePO}$ THEN SEND $T_{Re\,quest-CreatePO}$ (i.e. a *request to create a purchase order* is sent when all its mandatory fields have valid values).

**Table 1.** Modelling process structure using HMT

| Activities, transitions, and constructs | Message types and interaction templates |
|---|---|
| 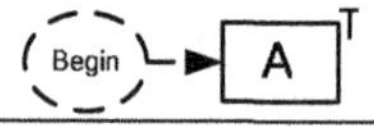 Begin → A ($T_i$) | $T_i$ |
| 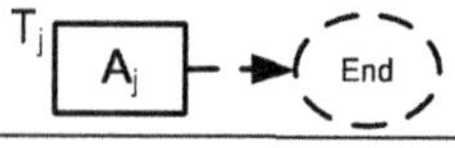 $T_j$ $A_j$ → End | $T_j$ |
| Activity $A_k$<br>$A_l$ → $A_k$ → $A_s$ ($T_{Request-k}$, $T_{Confirm-k}$) 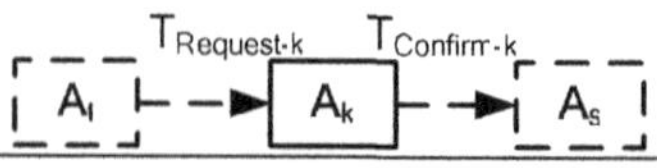 | $<T_{Request-k}, T_{Confirm-k}>$ |

**Table 1.** *(continued)*

| | |
|---|---|
| Null activity $A_k$ 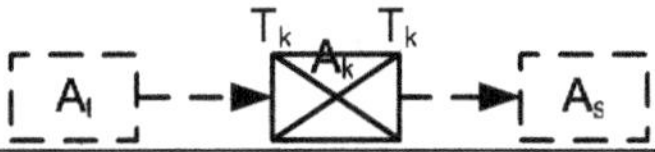 | $<T_k , T_k>$ |
| Transition 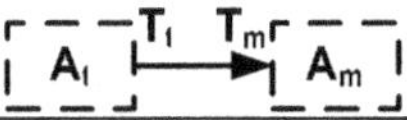 | $IT_m$ such that<br>$r_{i1}$: IF $T_t$ THEN ASSOCIATE $T_m$ |
| XOR-Split 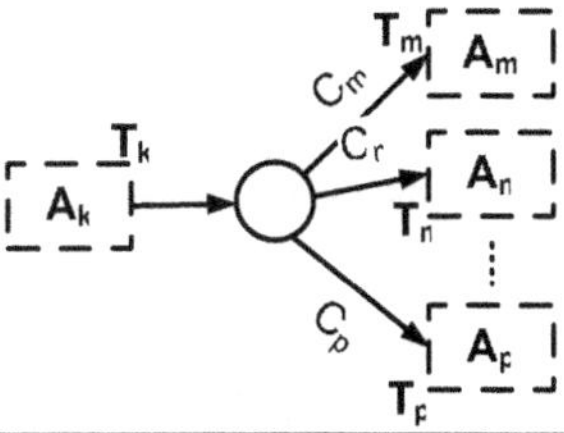 | $\forall i \in \{m, n, \ldots, p\}$, $IT_i$ such that<br>$r_{i1}$: IF ($T_k \wedge C_i$) THEN ASSOCIATE $T_i$<br>where $C_i$ is the XOR-Split condition |
| AND-Split 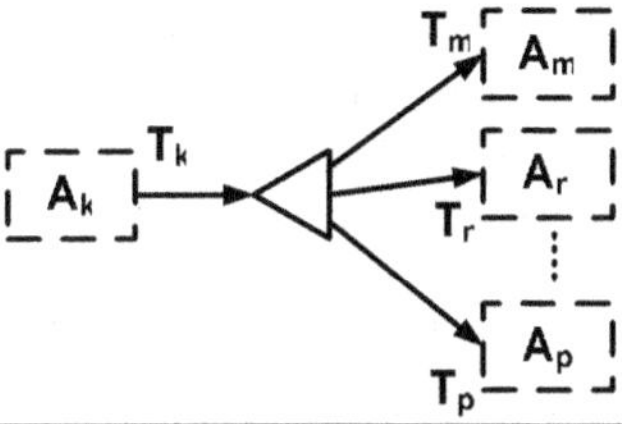 | $\forall i \in \{m, n, \ldots, p\}$, $IT_i$ such that<br>$r_{i1}$: IF $T_k$ THEN ASSOCIATE $T_i$ |
| XOR-Join 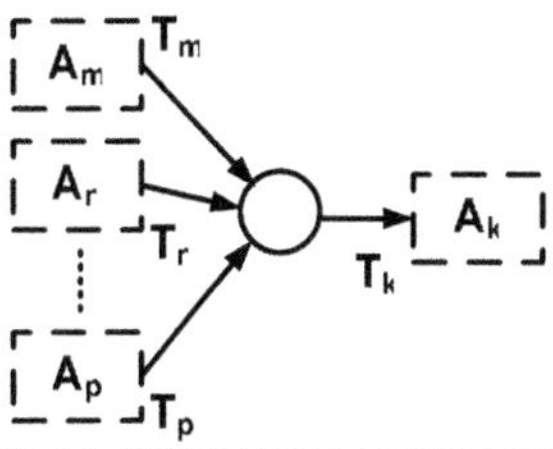 | $IT_k$ such that<br>$r_{k1}$: IF ($T_m \oplus T_n \oplus \ldots \oplus T_p$) THEN ASSOCIATE $T_k$ |
| AND-Join 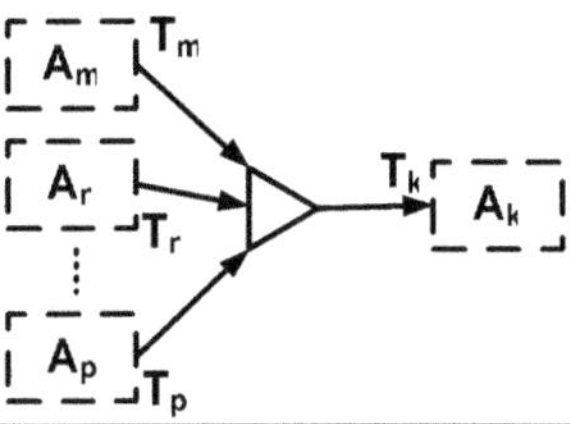 | $IT_k$ such that<br>$r_{k1}$: IF ($T_m \wedge T_n \wedge \ldots \wedge T_p$) THEN ASSOCIATE $T_k$ |

A formal specification of HMT rules is presented in (Ma, Orlowska, & Sadiq, 2005). In addition, to reduce the complexity of rule specification, we abstract rules that manipulate the associating, populating, and sending messages of an OMT as an **interaction template**. For an OMT $\mathbf{T_i}$, let $\mathbf{r_{i1}}$ be the rule type 1 to associate $T_i$, $\mathbf{R_{i2}}$ be the set of all rules type 2 to assign field values, $\mathbf{r_{i3}}$ be the rule to send $T_i$ messages. Then the interaction template $\mathbf{IT_i}$ is defined as a 3-tuple of $\mathbf{<r_{i1}, R_{i2}, r_{i3}>}$. Interaction templates are used to model workflow structures as illustrated in Table 1.

In addition, HMT rules (especially rules type 2) offer a mechanism to specify how to derive data from previously received messages to put into the contents of newly created outgoing messages. Thus, they communicate data between process activities and certain levels of control on the process's data flow can be specified.

### 3.3 Merging Work Items in HMT

As mentioned in section 2, in order to merge work items, BPMS need to create a new work item and populate its content. We observe that one could specify merging in HMT based on the following two reasons:

- The first reason is that in HMT a work item can be presented by a pair of incoming message notifying the completion of a work item, and an outgoing message requesting the creation of another work item. Then applying the conjunction on several incoming message events in a rule type 1, associating an outgoing message type, is necessary to correlate the completion of multiple work items to generate a unique (merged) request for a new work item. Based on this characteristic, HMT rules type 1 can be used to create a new work item.

For example, as in Figure 4, each *purchase request* will have a corresponding *purchase order*. However, we assume that it is required to merge every two *purchase orders* from the same customer into one *purchase order*. In this case, the HMMS can perform the merging task by associating two *purchase requests* with the OMT *request to create a purchase order*, following a rule of type 1:

$$\text{IF } (T_{PR}{}^{i} \wedge T_{PR}{}^{j}) \wedge (T_{PR}{}^{i}.Customer = T_{PR}{}^{i}.Customer) \quad (1)$$
$$\text{THEN ASSOCIATE } T_{Request-CreatePO}$$

- The second reason is that the task of manipulating the content of a work item can be specified by rules. This is because the content of a work item is represented by corresponding message contents; and rules type 2 can specify how to populate message contents.

For example, rules of types 2 in the message type $T_{Request-CreatePO}$ are deployed to collate the contents of the two components (i.e. $T_{PR}{}^{i}$ and $T_{PR}{}^{j}$ in (1)) as illustrated in Figure 5.

In Figure 5, the field "Reference messages" with ID F4 has the value {PR-00021, PR-00101}, which is used to refer to the contents of *purchase requests* with IDs PR-00021 and PR-00101. Therefore, the newly created *request to create a purchase order* corresponds to two *purchase requests*.

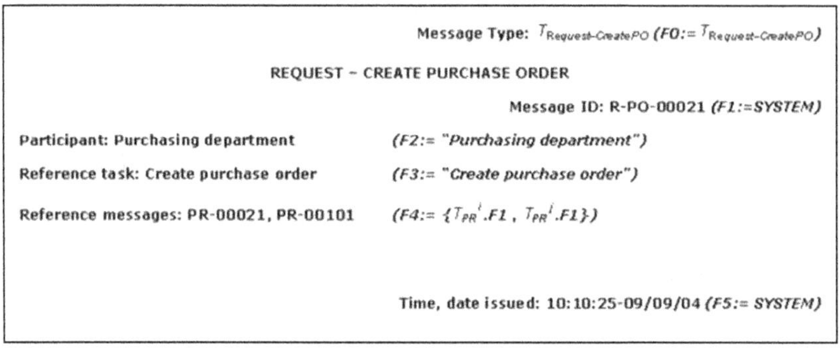

Message Type: $T_{Request\text{-}CreatePO}$ *(F0:= $T_{Request\text{-}CreatePO}$)*

REQUEST – CREATE PURCHASE ORDER

Message ID: R-PO-00021 *(F1:=SYSTEM)*

Participant: Purchasing department *(F2:= "Purchasing department")*

Reference task: Create purchase order *(F3:= "Create purchase order")*

Reference messages: PR-00021, PR-00101 *(F4:= {$T_{PR}^{i}$.F1 , $T_{PR}^{j}$.F1})*

Time, date issued: 10:10:25-09/09/04 *(F5:= SYSTEM)*

**Fig. 5.** A "Merged" Request to Create Purchase Order

In summary, the multi-aspect nature of HMT rule specifications has the advantage of specifying not only the control flow but also the data flow aspect of process models. Specifically, HMT rules can be used to specify how to create and populate contents of work items in the case of merging. In the next section, we will consider the issues in deployment of such merging feature in BPMS.

# 4 Deployment of Merging

Automatic merging of work items, although, can be implemented in BPMS (for example, in HMMS and in WFMS with certain extension). However, specific consideration should be taken when implementing the new feature. In this section, we first investigate the abnormalities in process execution when merging work items, and then consider the management of item IDs to avoid such abnormalities.

## 4.1 Process Execution Abnormalities

Extending BPMS with the merging feature may introduce some abnormalities in workflow execution as illustrated in Figure 6 and Figure 7.

Figure 6 illustrates a typical result of executing a part of a workflow, which consists of four activities (i.e. A1, A2, A3, and A4) with their corresponding work items (A1.1, A1.2, A2.1, A2.2, A3.1, A4.1, and A4.2) and an XOR-Split coordinator (with decision condition "Total > 50") in two situations, i.e. when there is no merging and when there is merging at activity A2.

In figure 6, in the case that there is no merging, after executions of activity A2, activity A4 is selected to execute as the result of evaluating the "Total" value of the contents of A2.1 and A2.2 (i.e. 20 and 50 respectively). However, a different result occurs when merging is performed at A2. In this case, instead of A4, activity A3 is selected to execute as a result of evaluating the "Total" value (i.e. 20+30+20=70) of the merged work item A2.1*.

Figure 7 illustrates a result of executing a part of a workflow, consisting of five activities with an AND-Join coordinator in two situations, i.e. when there is merging and no merging at activity A2.

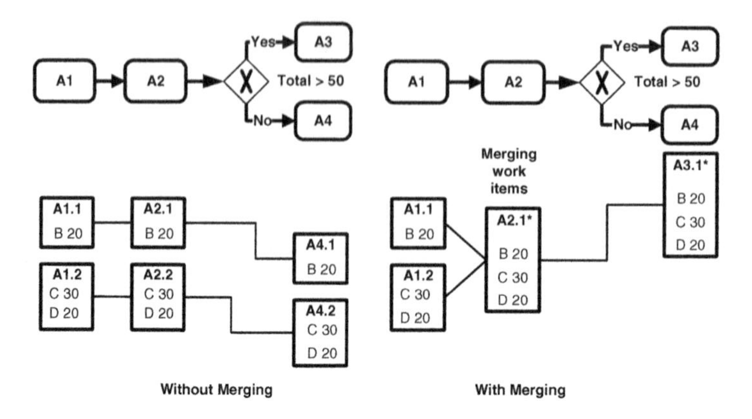


**Fig. 6.** Merging with XOR-Split

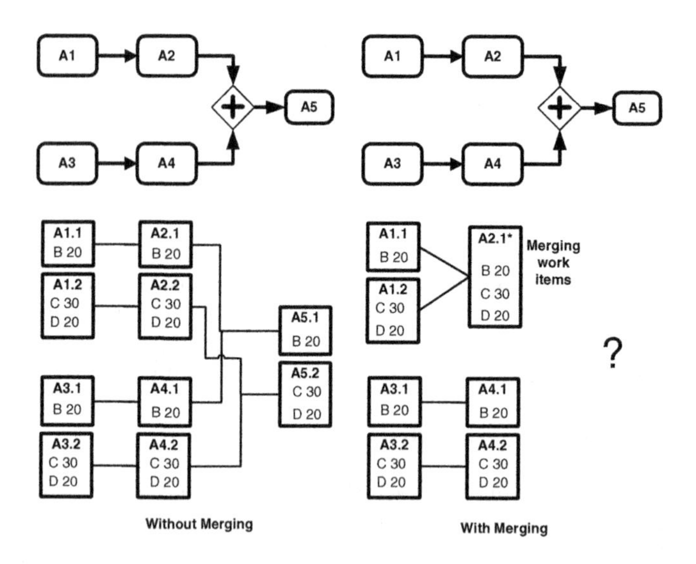


**Fig. 7.** Merging with AND-Join

In figure 7, work item A5.1 is created as a result of completion both A2.1 and A4.1. Similarly, the completion of both A2.2 and A4.2 triggers the execution of A5.2.

However, in the case of merging work items (the right hand side of Figure 7), after the completion of A2.1* and A4.1, it is not desired to execute A5, since there is a non-correspondence between contents of A2.1* and A4.1 (similarly for A2.1* and A4.2). For clarity, if A2, A4, and A5 refer to the activities of Receive Purchase Order, Receive Payment, and Send Goods respectively, then A5 should not be executed when there is a mismatch between the contents of a purchase order and the payment.

Such abnormalities in process execution emerge as a consequence of the violation of the coupled relationship between a work item and a process instance. Originally, each work item (represented as messages in HMT) belongs to at most one process instance. However, in the case of merging, this relationship does not hold. When merging, a new work item (or a new message) is created in replacement of specific work items (or specific messages) that originally belong to multiple process instances. Thus the newly created work item (or message) now belongs to multiple process instances.

In order to avoid such abnormalities in process execution, specific considerations must be given. In particularly, the problem when incorporating the merging feature resides in the ability to manage the contents of messages during process execution, such that the above mismatch can be resolved. In the next section, we propose to deal with such abnormalities by managing of message IDs.

### 4.2 ID Management

In practice, there is a requirement to manage the component item/message signatures when merging/unmerging happens during process execution. Therefore, each component work item or message is uniquely identified by an ID. When merging happens, a new merged work item or merged message (i.e. a merge) is created and associated with a new ID. Thus merging at some stage in a process introduces a relationship between IDs of components and the ID of the merge. A merging solution can only be practical and beneficial when the relationships of the IDs can be stored and managed efficiently and effectively; we consider issues of ID management in the following aspects: *capturing*, *matching*, *converting*, and *tracking*.

**Capturing.** In the simplest form, the ID relationship between merged items/messages and their components can be presented as a 2-attribute relation (i.e. merged item ID and component ID). However, in real practice the ID relationship is more complicated as extra dimensions are introduced. For example, incorporating the process instance IDs of components, time dimension as well as allowing nested merging (i.e. new merged items are built from previously merged items).

**Matching.** In order to avoid abnormality in process execution (as in Figure 7), it is required that process enforcement engines need to match IDs (thus contents) between work items/messages. For example, before passing an AND-Join, process engines must check the correspondence between merged and non-merged items/messages, before allowing them to be proceeded. Consideration should take into account the constraint on matching, for instance, the system needs to wait for all the corresponding items/messages of a merge to come (called *total matching*) or allowing part of the merge to go further (called *partial matching*). Figure 8, illustrates an example of total matching and partial matching.

In Figure 8, in the case of total matching, the process enforcement engine waits until all items A2.1*, A4.1, and A4.2 finish and then generates item A5.1*. While in the case of partial matching, if items A2.1* and A4.1 finish first, then the engine will create the item A5.1* without waiting for the occurrence of item A4.2. The flexibility of the merging feature can be specified by its matching function and the key issue resides in the way we define the matching function.

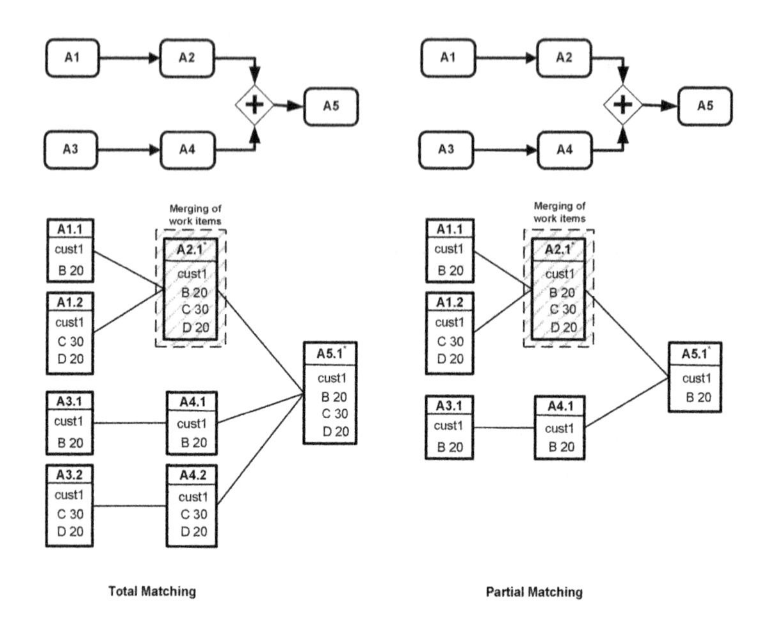


**Fig. 8.** Matching constraint

**Converting.** It is more practical to consider that work items (or messages) are merged based on different merging conditions along a process. In addition, in order to prevent abnormality in process execution, as in Figure 6, reverting the affect of the merging function (i.e. un-merging function) is required. ID of a previously merged work item (or merged message) is converted into IDs of its components and the components can be merged again later. For example, purchase orders can be merged when they belong to the same supplier, but order deliveries could be merged based on the locations of the warehouse dispatching the goods and the delivery address. In the case of allowing a nested merge, previously merged work items (or messages) can be merged again to introduce a higher nested level. Then controlling the nested level when unmerging remains an issue to process engine builders.

**Tracking.** Tracking process execution is an essential function of BPMS and makes the merging functionality attractive to many applications. Queries such as "trace the purchase order of shipment ID S1001" or "find out the status of all the purchase requests sent by customer A during last month" can be found quite often in practice. Tracking process executions remains a challenge in today's business processes. These are characterised to be long running, dynamic, and high volume in nature. Merging extension puts even more load on the tracking function.

In particular, since the relationship between a merge item/message and its components is always a parent-child relationship in nature; performing tracking queries typically requires recursive computation on IDs. In current WFMS, the concept of process instance plays a critical role in avoiding such expensive computation by anchoring all the work items within a case. However, in the case of merging, process instance cannot be used to link merge work items (or merged messages). Thus it requires a careful consideration in the design of the tracking function and data structure storing IDs.

## 5 Conclusion

Workflow technology has long been considered the most commonly used Business Process Management (BPM) technology. The clear separation between aspects of a process model such as control, data, and resource in specifications can be considered as one of the most critical advantages of the technology. However, the strength of workflow technology in some cases may become its weakness due to different process requirements.

In this paper, we considered an extension of BPM systems with the merging of work items facility to reflect useful business requirements. We identified the distinction between merging and grouping of work items, and showed that workflow technology cannot model items merging due to the separation between aspects of processes in workflow specifications.

We introduced Harmonized Messaging Technology (HMT) as an alternative of workflow technology in supporting business process automation, which designs and facilitates process coordination through message communication using rules. The multi-aspect nature of HMT rule specifications facilitate specific business requirements such as merging work items.

The paper concludes with an investigation into the consequence on the data aspect management of such merging extension onto process automation technology in general, regardless of the specific technology used. The data management issues in merging lead to the managing of work item/message IDs. We identified the major challenges for ID management, which fall into ID capturing, matching, converting, and tracking. The consideration of proposing data structures to store IDs, the managing of those structures, and the visual methods for HMT rules will form the basis of our future research.

## References


1. Aalst, W. M. P. v. d., Hofstede, A. H. M. t., Kiepuszewski, B., & Barros, A. P. (2003). Workflow patterns. *Distributed and Parallel Databases*, 14(3), 5-51.
2. Adams, M., Hofstede, A. H. M. t., Edmond, D., & Aalst, W. M. P. v. d. (2005). *Facilitating Flexibility and Dynamic Exception Handling in Workflows through Worklets*. Paper presented at the Proceedings of the 17th Conference on Advanced Information Systems Engineering (CAiSE05), Porto, Portugal.
3. Bae, J., Bae, H., Kang, S.-H., & Kim, Y. (2004). Automatic Control of Workflow Processes Using ECA Rules. *IEEE Transactions on Knowledge and Data Engineering*, 16(8), 1010-1023.
4. Kappel, G., Rausch-Schott, S., & Retschitzegger, W. (1996). *Coordination in Workflow Management Systems - A Rule-Based Approach*. Paper presented at the Coordination Technology for Collaborative Applications - Organizations, Processes, and Agents [ASIAN 1996 Workshop], Singapore.

5. Kiepuszewski, B. (2002). *Expressiveness and Suitability of Languages for Control Flow Modelling in Workflows.* PhD thesis, Queensland University of Technology, Brisbane, Australia.
6. Ma, D. C., Carter, B., Sadiq, S. W., & Orlowska, M. E. (2006). *Enterprise Integration Architecture based on Harmonized Messaging.* In Enterprise Systems Architecture in Practice: Idea Group Inc.
7. Ma, D. C., Lin, J. Y.-C., & Orlowska, M. E. (2006). *On Consideration of Grouping and Merging of Workflow Activities Instances* (Technical Report No. 466). Brisbane: The University of Queensland.
8. Ma, D. C., Orlowska, M. E., & Sadiq, S. W. (2005). Formal Considerations of Rule-based Messaged for Business Process Integration. *Cybernetics and Systems: An International Journal*, 37, 171-196.
9. Müller, R., Greiner, U., & Rahm, E. (2004). AGENTWORK: A Workflow-System Supporting Rule-Based Workflow Adaptation. *Data and Knowledge Engineering*, 51(2), 223-256.
10. Russell, N., Hofstede, A. H. M. t., Edmond, D., & Aalst, W. M. P. v. d. (2004a). *Workflow data patterns.* Brisbane: Queensland University of Technology.
11. Russell, N., Hofstede, A. H. M. t., Edmond, D., & Aalst, W. M. P. v. d. (2004b). *Workflow Resource Patterns.* Eindhoven, Netherlands: Eindhoven University of Technology.
12. Sadiq, S. (2000). *On Capturing Exceptions in Workflow Process Models.* Paper presented at the Proceedings of the 4th International Conference on Business Information Systems (BIS00). Poznan, Poland.
13. Sadiq, S., Orlowska, M., & Sadiq, W. (2005). *Role of Messaging in Collaborative Business Processes.* Paper presented at the IRMA International Conference (Business Process Management Track), San Diego, California, USA.
14. Sadiq, S., Orlowska, M., Sadiq, W., & Schulz, K. (2004a). Collaborative Business Process Management through Harmonized Messaging. *Journal of Information and Organizational Sciences*, 28(1 & 2).
15. Sadiq, S., Orlowska, M., Sadiq, W., & Schulz, K. (2004b). *Facilitating Business Process Management with Harmonized Messaging.* Paper presented at the 6th International Conference on Enterprise Information Systems (ICEIS 2004), Universidade Portucalense, Porto, Portugal.
16. Sadiq, S., Orlowska, M., Sadiq, W., & Schulz, K. (2005). *When workflows will not deliver – The case of contradicting work practice.* Paper presented at the the 8th International Conference on Business Information Systems (BIS 2005), Poznan, Poland.
17. WfMC. (1999). *Workflow Management Coalition Terminology & Glossary.* Retrieved June, 2006, from http://www.wfmc.org/standards/docs/TC-1011_term_glossary_v3.pdf

# Complex Events in Business Processes

Alistair Barros[1], Gero Decker[2], and Alexander Grosskopf[1]

[1] SAP Research Centre Brisbane, Australia
{alistair.barros,alexander.grosskopf}@sap.com
[2] Hasso-Plattner-Institute, Potsdam, Germany
gero.decker@hpi.uni-potsdam.de

**Abstract.** Flow-oriented process modeling languages have a long tradition in the area of Business Process Management and are widely used for capturing activities with their behavioral and data dependencies. Individual events were introduced for triggering process instantiation and activities. However, real-world business cases drive the need for also covering complex event patterns as they are known in the field of Complex Event Processing. Therefore, this paper puts forward a catalog of requirements for handling complex events in process models, which can be used as reference framework for assessing process definition languages and systems. An assessment of BPEL and BPMN is provided.

## 1 Introduction

In order to flexibly adapt changing business requirements, companies are in need of IT systems that allow rapid reconfiguration. Business Process Management (BPM) puts process models into the center of attention capturing the activities that have to be carried out as well as their behavioral and data dependencies. The functionality of the available IT systems is invoked by process execution engines, which turns process models into central configuration artifacts for the enterprise systems.

Over the past years different process definition languages have been proposed. They tackle different levels of detail ranging from high-level models for business analysts to executable process models. Prominent examples are e.g. the Business Process Modeling Notation (BPMN [2]), UML 2.0 Activity Diagrams ([1]) and the Business Process Execution Language (BPEL [3]). All of these languages incorporate the notion of events for triggering process instantiation or steps within a process instance. Events in the form of message exchanges or timeouts are very common in executable languages. In the case of higher-level modeling languages there is the possibility to also consider coarser grained business events, such as "goods have arrived".

Events are a way to loosely interconnect different process instances: Events produced in one process instance are consumed by one or several other process instances. Furthermore, composite events, i.e. the combination of different interrelated events, must be handled in process models, too. As activities can normally be decomposed into flows of sub-activities, we also need the possibility to handle

W. Abramowicz (Ed.): BIS 2007, LNCS 4439, pp. 29–40, 2007.
© Springer-Verlag Berlin Heidelberg 2007

different levels of events in process models. As an example the high-level event "goods have arrived" might be decomposed into a number of "line item has been stored in warehouse" events.

Process definition languages have been benchmarked for their suitability regarding common scenarios in processes. However, a benchmark regarding the eventing capabilities of process definition languages is still missing. Therefore, this paper puts forward requirements derived from real-world business cases.

The remainder of this paper is structured as follows: Section 2 highlights related work before section 3 recapitulates on how events are consumed in business processes. In section 4 a catalog of patterns for composite events in business processes is presented and assessments for BPEL and BPMN are given. Finally, section 5 concludes and gives an outlook to future work.

## 2 Related Work

The field of Complex Event Processing (CEP) comes with a set of languages and architectures for describing and efficiently executing complex event rules. A good reference for CEP is the book by Luckham [6], where he also introduces Rapide, an event pattern language. A framework for detecting complex event patterns can be found e.g. in [7]. Considerable work on event pattern languages can also be found in the field of active databases. In [5] the event algebra Snoop is introduced and compared with other event languages.

It is argued that process definition languages could be superseded by event pattern languages. However, flow-oriented languages, i.e. languages where the flow relation between activities is at the center of attention, have a long tradition in the field of Business Process Management and their suitability has been studied extensively [11,8]. The fact that the most commonly used formalism in the field of business processes, namely Petri nets ([10]), is also flow-oriented underlines the importance of the flow-oriented paradigm.

Seeing message exchanges between process instances as the main event type in service-oriented architectures, the Correlation Patterns introduced in [4] describe the relationship between these communication events and the structure of process instances. The relationship between individual patterns in this paper to respective Correlation Patterns will be given in section 4.

## 3 Event Consumption in Business Processes

As already mentioned in the introduction, state-of-the-art process definition languages include the notion of events. E.g. in BPEL `invoke`, `receive` and `onMessage` activities specify production and consumption of message events. An initial `receive` and `onMessage` activity with the attribute `createInstance` set to `yes` defines a WSDL port type / operation combination that is relevant for the instantiation of BPEL processes: As soon as a message of that particular combination arrives, a process instance is created.

We can find three typical steps for the consumption of events in process instances (depicted in Figure 1): i) A *subscription* to events (e.g. incoming messages) is initiated. ii) An *event occurs* (e.g. a message arrives). iii) The *event is matched* by a subscription. This determines that the event will be consumed by a particular process instance. Either it is consumed by an already existing process instance or a new instance is created as result of that event.

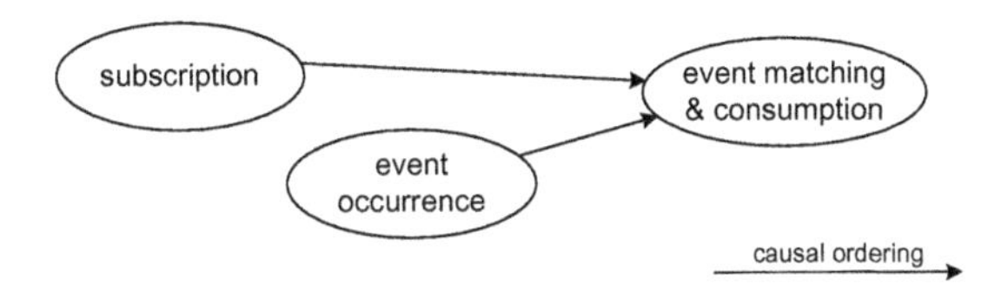


**Fig. 1.** Event handling in process instances: three steps

As a forth step we could also consider unsubscription. If an event is not awaited any longer the subscription is taken back. A typical scenario could be that at a given moment in the process a message of one out of a set of different types could be consumed. In this case, there is a subscription for every type and as soon as one message arrives no message of the other types is waited for.

We know that the consumption of an event cannot happen before the occurrence of that event nor before a subscription. These causal ordering constraints are also depicted in Figure 1. However, we leave open at this stage, whether the subscription has to precede the event occurrence. The architectural implication for allowing subscriptions after the actual event occurrence, is that the system has to store the event for later consumption. Such an architecture might not be desired since it is hard to tell how long an event should be stored.

The enlisted steps can also be found in BPEL. Subscriptions for those messages that lead to process instantiation are initiated at deployment-time of a process definition. Subscriptions for those messages that are consumed by a running process instance are normally initiated as soon as the respective `receive` or `onMessage` activity is reached. Normally, the next two steps happen at once in the case of BPEL: as soon as a message has been matched it is routed to a process instance. However, the specification leaves it open to the process engine implementers if also messages can be consumed by a `receive` or `onMessage` activity that have arrived before that activity was reached.

BPEL only considers individual events: It is checked on a per-message basis if a message matches a registered subscription (based on the port type, operation and correlation sets) and only one message is consumed in a `receive` or `onMessage` activity. This is different to what is supported in event rules in the field of Complex Event Processing. Event rules specify patterns of events that have to be matched. E.g. it is required that five corresponding messages of type customer complaint are present within a given timeframe for a given event rule to fire. Event rules enable hierarchical event architectures: Several low level events are matched in event rules producing higher-level events. In business scenarios at least two different levels of events should be present. At a low level we find

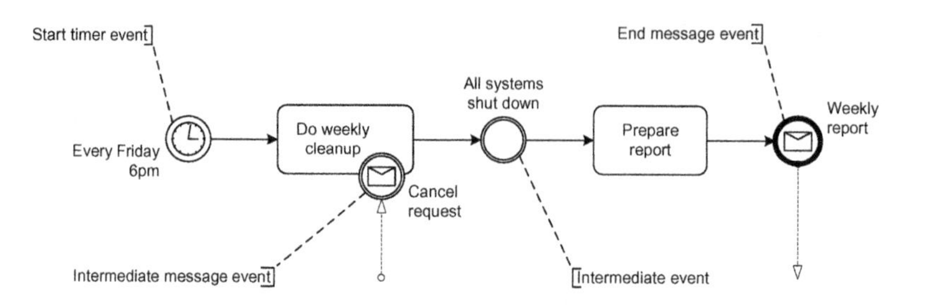


**Fig. 2.** Event consumption and production in BPMN

individual events, e.g. the arrival of a container detected by an RFID station, whereas on a higher level business events such as "sufficient containers available for shipment" are considered within process models. Unfortunately, such event hierarchies are not present in BPEL, only one level of events is considered. A remedy could be using event aggregation components as a separate architectural component in addition to a process execution engine. However, a seamless way of modeling processes and event aggregation is necessary in order to provide a consistent view to the process experts.

BPMN does not have the capability to express different levels of events, either. However, it allows more event types than BPEL: In addition to messages, timeouts and exceptions that are also present in BPEL, BPMN also comes with rule events and it even allows to extend the language with custom defined event types. Since BPMN is basically a graphical notation without defined execution semantics, it is unclear how and when subscription for events is handled in BPMN. Anyway the BPMN assumes that all messages are persistent. Thus they are kept until a process instance is ready to consume them. As it is the case for BPEL, BPMN distinguishes start events and intermediate events as two kinds of event consumption: start events lead to process instantiation and intermediate events are consumed by a running process instance. BPMN also allows to specify message consumption and production through send and receive activities.

The BPMN specification does not explicitly state how often an event can be consumed by process instances. But as the specification intends to map BPMN to BPEL, we assume that every event is only consumed once like in BPEL.

## 4 Patterns for Composite Events

This section introduces a set of patterns as reference framework for assessing process definition languages regarding their support for composite events in business processes. We have seen that both event consumption and production are present in process languages. The patterns enlisted in this section only focus on the consumption side. Each of the patterns comes with a short description, examples and an assessment of BPEL and BPMN.

### 4.1 Co-occurrence Patterns

This set of patterns describes scenarios where several events have to be considered in order to decide whether a pattern matches or not.

**1. Event Conjunction.** Two or more events have to have occurred in order to be matched. The order of occurrence is irrelevant. This pattern is similar to the Atomic Consumption pattern from [4].

*Examples.* (a) As part of the management of a shipment delivery to transient storage nodes, when the number of carriers collectively yielding the minimal outbound shipment has arrived, an event is raised to alert the relevant carriers to confirm delivery pick-up.

(b) If goods have been delivered which were previously canceled, a delivery exception event is raised.

*Assessment of BPEL.* Only one event is consumed at a time in BPEL processes. A workaround for some scenarios can be that e.g. several `receive` activities are placed within a `flow` constructs (i.e. in parallel). For process instantiation scenarios a `pick` representing the alternative occurrence sequences could be used. In this solution, we run into the problem that the instance is created as soon as the first message arrives. Event Conjunction demands atomicity: Only in the presence of all demanded events, an action in the process should be taken. Therefore, there is no direct support for this pattern in BPEL.

*Assessment of BPMN.* The situation for BPMN is similar to that of BPEL. Atomic consumption of several events is not possible in BPMN processes. Therefore, no support for this pattern, either. Similar workarounds like in BPEL are possible, though.

**2. Event Cardinality.** A specified number of events of the same type that are all subject to the same constraints have to have occurred in order to be matched. Event Cardinality is a special case of Event Conjunction. There are two flavors of Event Cardinality: (i) a fixed number is specified (ii) a range of numbers is specified. In the latter case a set of events can be matched as soon as the minimum number of events are present. However, if more events are available at the moment of matching, all available events are matched (as a maximum number the upper limit of the range). The fixed number and the range of numbers might be known at design-time or only at runtime.

*Examples.* (a) GSM stations send status report events. Some events indicate errors due to minor technical malfunctions. If more than a threshold number of errors are reported, an event is raised for trouble-shooting.

(b) Requests for purchase of small items are not processed immediately but are batched, and subsequently trigger ordering when a certain number of purchase requests is reached.

*Assessment of BPEL.* There is no support for this pattern in BPEL since we find the same problem like in the case of Event Conjunction: only one event is consumed at a time. Similar workarounds would be needed to implement event cardinality (e.g. using while constructs) but the constraint of atomic matching can not be fulfilled.

*Assessment of BPMN.* In analogy to the case of BPEL, there is no support for this pattern in BPMN.

**3. Event Disjunction.** There are alternatives of events that have to have occurred in order to be matched. The Workflow Pattern "Deferred Choice" ([9]) is a special case of this pattern, where alternative individual events are waited for.

*Examples.* (a) The shipment planning process is started either automatically at a scheduled time before or at an earlier time determined by the shipment scheduler (e.g. in case of additional stock variances).

(b) The ordering process of an online video-on-demand marketplace uses one of several payment instruments obtained from a customer's profile. When the one of these clears credit check, the transaction phase of ordering can proceed.

*Assessment of BPEL.* The `pick` construct in BPEL allows to define alternative event types. The first matching event is consumed and the process resumes. This semantics implements the Deferred Choice pattern. We conclude that there is direct support for Event Disjunction in BPEL.

*Assessment of BPMN.* The event-based gateway has similar semantics like the `pick` construct in BPEL. Moreover, BPMN has a multiple event type. It can be triggered by alternative events and used for process instantiation scenarios. Hence, there is also direct support in BPMN.

**4. Inhibiting Event.** An event can only be matched in the absence of another specified event. This inhibiting event is not consumed.

*Examples.* (a) A fraud alert is raised if an invoice paid event was detected without a corresponding invoice approved event.

(b) A passenger's seat allocation on a flight is flagged if the passenger cannot be located through the search/alert passenger process and the departure gate of the flight is closed. Commencement of seat cancellation triggers retrieval of the passenger's baggage, although the passenger may still be allowed to board the flight if the baggage has not yet been retrieved. Baggage retrieval signifies completion of seat cancellation and the passenger is not allowed to board the flight from that point.

*Assessment of BPEL.* Matching of messages to subscriptions is done on a per-message basis in BPEL. I.e. no other messages are considered when deciding whether it matches or not. This leads to the situation that there is no direct support for the Inhibiting Event pattern in BPEL. However, in some cases the notion of cancellation can emulate an inhibiting event: in a certain scope of the process, incoming messages are dealt with by event handlers throwing an exception which in turns causes the scope to be canceled. This does not work for the case of process instantiation. The inhibiting event should cause an instance not to be created.

*Assessment of BPMN.* Like in the case of BPEL, there is no direct support for this pattern in BPMN. However, we can also think of workarounds using intermediate events that are attached to activities. A running activity would then be canceled as soon as the specified event occurs.

### 4.2 Time Relation Patterns

These patterns describe common time-related constraints for event patterns. The moment an event occurs might be relevant for deciding whether a pattern matches a certain group of events.

**5. Event – Event Time Relation.** Two events can only be matched if their occurrence happens within or outside a given timeframe. A special case is where the event of one type always has to have occurred before the other. This pattern only appears as additional constraint for Event Conjunction.

*Examples.* (a) For a supermarket chain, suppliers of certain categories of stock are responsible for replenishing stock to predefined thresholds, monitored by suppliers. If an event is raised that line-item sub-category falls within a certain threshold and has order notification from the supplier for replenishment within a certain time since the threshold was reached, the replenishment process terminates without exception.

(b) Customers in an online video-on-demand marketplace are served by media content brokers. A broker who cannot fulfill a request may forward it to other brokers. Forwarded requests need to be fulfilled within a certain time of the request issued by the customer, otherwise the request is no longer current.

(c) Buy and sell events arising from the stock market of a customer portfolios are automatically correlated within a certain time of their occurrence by an investment management process, otherwise they are ignored.

*Assessment of BPEL.* Since this pattern always requires the presence of an Event Conjunction there is no support for this pattern in BPEL. Possible workarounds could include the usage of timeouts, i.e. `onAlarm` events: As soon as the first message is consumed, the timer is started. If the timeout occurs before the arrival and consumption of the second event, cancellation takes place. Such a workaround only works if only those messages are consumed that arrived after the corresponding `receive` activity was reached. In the other cases it is not known how much earlier the first message arrived before reaching the `receive` activity.

*Assessment of BPMN.* In analogy to BPEL there is no support for this pattern in BPMN. A similar workaround could be used: an intermediate timer event is attached to an activity containing the second intermediate event.

**6. Event – Subscription Time Relation.** An event can only be matched if it occurs within a given timeframe relative to the moment of subscription, e.g. an event must occur within 5 minutes after the moment of subscription. Alternatively it is specified that it occurs outside a given timeframe relative to the moment of subscription, e.g. an event must have occurred at least 10 days before the moment of subscription. This pattern is only relevant for cases where events are to be consumed by already running process instances.

*Examples.* (a) A company assessment process determines business properties. When it reaches the point of stock prize evaluation it consumes all stock prize updates of the last 2 months for that company to calculate average growth and variance.

(b) A process supporting tax return applications accepts input for the application up until the point where a final version of the tax return is prepared. Thus, any information relating to tax returns such as exemptions or investments can be asynchronously consumed during the process.

*Assessment of BPEL.* It is not specified when the subscription to a message is initiated in BPEL. We only know that the latest moment of subscription is when the `receive` or `onMessage` activity is reached. If we use a `pick` construct in combination with an `onMessage` and `onAlarm`, we can define a duration how long the engine should maximally wait for the message starting at the moment of reaching the `onAlarm` activity. This covers the case where the message should arrive within a given time after the subscription. All other cases are not directly supported in BPEL. We conclude that there is partial support for this pattern.

*Assessment of BPMN.* The event-based gateway in BPMN directly corresponds to the `pick` construct in BPEL. Therefore, we also find partial support for this pattern in BPMN.

**7. Event – Consumption Time Relation.** An event can only be matched if it occurs at least a certain time before the moment of consumption. Alternatively it is specified that it occurs at most a certain time before consumption. This pattern is especially important for process instantiation scenarios.

*Examples.* (a) Ad-hoc stock requests which occur between regular replenishment cycles are processed together for allocation to existing shipments of the next cycle.

(b) Within a certain time of stock pick-up for the next shipment time, the relevant carriers are expected to report pick-up and commencement of delivery. Outstanding carriers are contacted for determination of whether alternative transportation should be triggered.

*Assessment of BPEL.* There is no support for this pattern in BPEL.

*Assessment of BPMN.* There is no support for this pattern in BPMN.

**8. Event – Absolute Time Relation.** An event can only be matched if it occurs before or after an absolute point in time.

*Examples.* (a) Incoming calls into a service hotline during business hours are handled by the local support center. Calls outside business hours are directed to global support centers.

(b) A scheduled upgrade of a system is set for a specific time. Warning are sent out to users that log in within that specific time frame.

(c) During fixed times on weekdays, processes are triggered for major roads with interchangeable lanes (for left/right side) to have their lane allocations set.

*Assessment of BPEL.* The `onAlarm` construct in BPEL has both a duration semantics (the timeout occurs after a certain time period has passed) and a deadline semantics (the timeout occurs as soon as a certain deadline is reached). The latter can be used to easily express one flavor of the pattern: A `pick` construct in combination with an `onMessage` and an `onAlarm` can implement cases where the message has to arrive before a given deadline. Since this implementation does not cover the situation where a message has to arrive after a given point in time, we opt for partial support for this pattern.

*Assessment of BPMN.* In analogy to BPEL we can use a combination of an event-based gateway with an intermediate event and an intermediate timer event. The other flavor of the pattern is not directly supported, either. Therefore, we also have partial support for this pattern in BPMN.

### 4.3 Data Dependency Patterns

For the following set of patterns we assume that events carry additional data. For instance, incoming messages carry message content and a "goods have arrived" event carries information about the supplier and the corresponding order. The patterns describe how data dependencies constrain the matching of events.

**9. Event – Event Data Dependency.** Two events can only be matched if their data is in a specified relation. This pattern only appears as additional constraint for Event Conjunction. It is similar to Key-based and Property-based Correlation in [4].

*Examples.* (a) An incoming order confirmation or order rejection event is matched with the outgoing order request event with the same OrderID.

(b) An investigation for provision of land tenure (e.g. land planned for a school site) involves complex searches for related tenure applications and future land actions planned, based on geographic locality (e.g. railway line planned or environmental regulations in the locality) and "neighborhood" parcels of land within a geographic locality (shopping center planned in the same block). Keys for data correlation accordingly vary.

*Assessment of BPEL.* Since this pattern always requires the presence of an Event Conjunction there is no support for this pattern in BPEL. In general, correlation sets can be defined for constraining the messages matched for a running process instance. However, since only a combination of two or more events should be matched that fulfill the data constraint, it would be invalid to accept a first message independently of the data constraint. Therefore, receiving any message of a desired port type / operation and then using this message for initializing the correlation set which in turn is used for matching the second one, cannot be a valid workaround.

*Assessment of BPMN.* Unlike BPEL, BPMN does not provide any support for correlation. Therefore, data dependencies between two events cannot be expressed in BPMN.

**10. Event – Process Instance Data Dependency.** An event can only be matched if its data is in a specified relation to the control data of the subscribing process instance.

*Examples.* (a) A Customer Payment Details event is consumed by the process that reference the same Customer within its process context.

(b) A reply to an asynchronous request is routed to the process instance that holds correlation data within its process context matching the event data.

(c) An Order Cancellation affects the process instance that holds the correct Order identifier in its context.

*Assessment of BPEL.* Correlation sets are a means to restrict subscriptions to messages with specific content. E.g. a customer ID can be included in a message and only those messages are accepted that belong to a particular customer. We conclude that there is direct support for this pattern in BPEL.

*Assessment of BPMN.* BPMN simply lacks the notion of correlation and therefore does not support this pattern.

**11. Event – Environment Data Dependency.** An event can only be matched if its data is in a specified relation to data that can be accessed by different process instances. This pattern is especially important for process instantiation.

*Examples.* (a) Email requests from premium customers start premium handling processes, other start normal handling processes.

(b) A shipper processes shipment request events only if they come from known business partners.

(c) Only invoices referencing a valid order start an approval process.

*Assessment of BPEL.* Whether a message triggers process instantiation is only decided based on the port type / operation combination. BPEL does not allow to further constrain such consumption using correlation information. Therefore, there is no direct support for the pattern in BPEL. A workaround could be that messages first trigger process instantiation and are then checked for their content. If the content does not fulfill the constraint the process instance is terminated.

*Assessment of BPMN.* It is not possible to constrain the consumption of events based on data attached to them. Therefore, there is no support for this pattern, either.

### 4.4 Consumption Patterns

The Consumption Patterns describe how often an event can be consumed.

**12. Consume Once.** An event can only be consumed at most once by one out of all process instances.

*Examples.* (a) An order request event is to be processed exactly once.

(b) Activation for a newly provided credit card event is required only once.

(c) An event notifying the breakdown of a carrier should be consumed once by a specific breakdown-service agent.

*Assessment of BPEL.* Every incoming message is routed to at most one process instance and is consumed by at most one `receive` or `onMessage` activity. Hence, there is direct support for this pattern.

*Assessment of BPMN.* In the previous section we mentioned the assumption that every event is consumed by at most one BPMN process instance. This results in direct support for this pattern.

**13. Consume Multiple Times.** An event is consumed several times (possibly within the same process instance). This pattern is similar to the Multiple Consumption pattern from [4].

*Examples.* (a) Events signifying share buy/sell recommendations are consumed many times by an investment monitoring process for different customer portfolios.

(b) A traffic monitoring system provides traffic updates for its subscriber carriers involved in delivering goods.

(c) Changes to a shipments are broadcasted to its stakeholders including different carriers, storage nodes, final consignments and stock-to-shelf dispatchers.

*Assessment of BPEL.* In the case of BPEL, every message is consumed at most once. Therefore, BPEL does not support this pattern.

*Assessment of BPMN.* Due to our assumption that every event is consumed by at most one BPMN process instance, we conclude that there is no support for this pattern in BPMN.

## 5 Conclusion and Outlook

This paper has introduced a set of patterns describing common eventing scenarios in business processes. In analogy to other sets of patterns in the field of Business Process Management, they can be used to evaluate process definition languages and systems. Table 1 summarizes the assessment of BPEL and BPMN that we carried out in section 4. Direct support for a pattern is denoted as "+", partial support as "+/–" and no support as "–". It turns out that BPEL and BPMN support similar patterns, while a wide range of patterns are not supported by both languages. This underlines the initial assumption that only very basic eventing scenarios can be captured. We argued that modeling process logic and describing complex event patterns should not occur independently of each other since both aspects are essential for process experts to capture the overall process context. As a result, we see the need to closely integrate event pattern descriptions into executable process definition languages such as BPEL and higher-level modeling languages such as BPMN.

**Table 1.** Composite Event Pattern support in BPEL and BPMN

| *Composite Event Patterns* | BPEL | BPMN |
|---|---|---|
| 1. Event Conjunction | – | – |
| 2. Event Cardinality | – | – |
| 3. Event Disjunction | + | + |
| 4. Inhibiting Event | – | – |
| 5. Event – Event Time Relation | – | – |
| 6. Event – Subscription Time Relation | +/– | +/– |
| 7. Event – Consumption Time Relation | – | – |
| 8. Event – Absolute Time Relation | +/– | +/– |
| 9. Event – Event Data Dependency | – | – |
| 10. Event – Process Instance Data Dependency | + | – |
| 11. Event – Environment Data Dependency | – | – |
| 12. Consume Once | + | + |
| 13. Consume Multiple Times | – | – |

Future will especially focus on integrating more sophisticated eventing mechanisms into BPMN. As part of that, graphical representations for event patterns will be proposed.

## References


1. UML 2.0 Superstructure Specification. Technical report, Object Management Group (OMG), August 2005.
2. Business Process Modeling Notation (BPMN) Specification, Final Adopted Specification. Technical report, Object Management Group (OMG), February 2006. http://www.bpmn.org/.
3. T. Andrews, F. Curbera, H. Dholakia, Y. Goland, J. Klein, F. Leymann, K. Liu, D. Roller, D. Smith, S. Thatte, I. Trickovic, and S. Weerawarana. Business Process Execution Language for Web Services, version 1.1. Technical report, OASIS, May 2003. http://www-106.ibm.com/developerworks/webservices/library/ws-bpel.
4. A. Barros, G. Decker, M. Dumas, and F. Weber. Correlation Patterns in Service-Oriented Architectures. In *Proceedings of the 9th International Conference on Fundamental Approaches to Software Engineering (FASE)*, Braga, Portugal, March 2007.
5. S. Chakravarthy and D. Mishra. Snoop: An expressive event specification language for active databases. *Data Knowledge Engineering*, 14(1):1–26, 1994.
6. D. Luckham. *The Power of Events: An Introduction to Complex Event Processing in Distributed Enterprise Systems.* Addison-Wesley, 2001.
7. P. R. Pietzuch, B. Shand, and J. Bacon. A Framework for Event Composition in Distributed Systems. In *Proceedings of the 4th International Conference on Middleware (MW'03)*, Rio de Janeiro, Brazil, 2003.
8. N. Russell, W. M. van der Aalst, A. ter Hofstede, and P. Wohed. On the Suitability of UML 2.0 Activity Diagrams for Business Process Modelling. In *Proceedings 3rd Asia-Pacific Conference on Conceptual Modelling (APCCM 2006)*, volume 53 of *CRPIT*, pages 95–104, Hobart, Australia, 2006.
9. W. M. P. van der Aalst, A. H. M. ter Hofstede, B. Kiepuszewski, and A. P. Barros. Workflow Patterns. *Distributed and Parallel Databases*, 14(1):5–51, 2003.
10. W. v. d. van der Aalst and K. v. van Hee. *Workflow Management: Models, Methods, and Systems (Cooperative Information Systems).* The MIT Press, January 2002.
11. P. Wohed, W. M. van der Aalst, M. Dumas, A. ter Hofstede, and N. Russell. On the Suitability of BPMN for Business Process Modelling. In *Proceedings 4th International Conference on Business Process Management (BPM 2006)*, LNCS, Vienna, Austria, 2006. Springer Verlag.

# Collaborative E-Business Process Modelling: A Holistic Analysis Framework Focused on Small and Medium-Sized Enterprises

Volker Hoyer[1,2] and Oliver Christ[1]

[1] SAP Research CEC St. Gallen, Switzerland
[2] Institute of Industrial Management, University of Hamburg, Germany
{volker.hoyer,oliver.christ}@sap.com

**Abstract.** In this work, we propose a holistic analysis framework for collaborative e-Business process modelling approaches that takes into account the specific challenges small and medium-sized enterprises (SME) are facing with regard to modelling inter-organizational processes. Based on concepts of the management approach Balanced Scorecard (BSC) four different perspectives are derived from empirical studies, conceptual research results and completed with modelling experiences of an EU-funded project. By considering concepts of Management Theories, Business Process Management (BPM) and Service-Oriented Architecture (SOA) paradigms, requirements for collaborative modelling approaches are presented according to the four perspectives financial, working process, innovation/ learning and user. Finally, a strategy map describes the complex interactions of the identified requirements between the perspectives.

**Keywords:** Collaborative Business Process, Analysis Framework, e-Business, SME, Business Process Management, Service-Oriented Architecture.

## 1 Introduction

Concepts of borderless enterprises [1] have been discussed for years and highlight the increased relevance of collaboration among enterprises and their business environment. According to the management approach of Business Process Reengineering (BPR) [2], enterprises introduce the internal process concept to overcome the functional-oriented organizational structure. By extending the process-oriented way of doing business across enterprise borders [3], seamless processes and real-time businesses [4] are the new challenges in adaptive business networks. This next wave of process-oriented enterprises is enabled by a wide penetration of e-Business and rapid technology innovations. According to popular business opinion, Information Technology (IT), especially hardware and software, will be transformed into a commodity meaning a common infrastructure like telephone or power grids [5]. Combined with focusing more and more on core competences of corporations [6], new concepts like Software as a Service (SaaS) on a technical level and Business Process Outsourcing (BPO) on a business level will become

W. Abramowicz (Ed.): BIS 2007, LNCS 4439, pp. 41–53, 2007.
© Springer-Verlag Berlin Heidelberg 2007

more important for enterprises, in particular for small and medium-sized enterprises (SMEs) characterized by missing IT literacy and limited financial resources to access traditional enterprise software applications. At this central interface, modelling of collaborative e-Business processes will take on a mediating role and binds internal and external processes as well as local and global knowledge and IT-systems. As a result, performing collaborative e-Business processes will be a new competitive differentiator with strategic importance for SMEs. Due to the fact that there does not yet exist a requirement analysis for such a modelling approach in literature a designed holistic analysis framework in this work closes this research gap following the research approach "Innovation through Recombination". As shown in Figure 1 we combine the intersection of Balanced Scorecard (BSC), Management Theories, Business Process Management (BPM), and Service-Oriented Architecture (SOA) to design the analysis framework. Thereby empirical studies, research results and modelling experience of the EU-funded project GENESIS [7] prove the identified requirements.

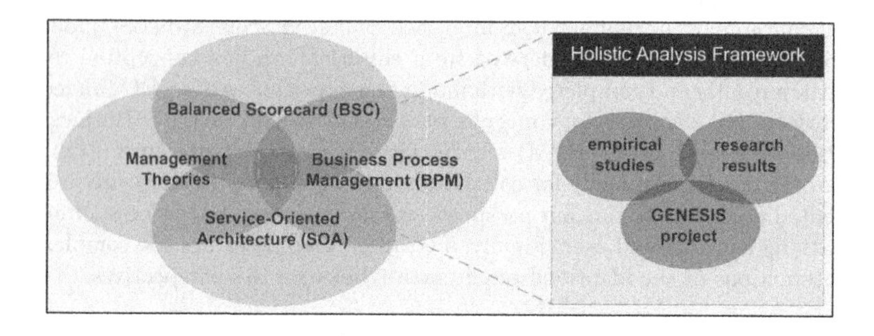


**Fig. 1.** Research Approach "Innovation through Recombination"

The remainder of the paper is structured as follows. First, we outline specific characteristics of SMEs that fundamentally distinguish SMEs from large enterprises. After a brief introduction of electronic collaboration among enterprises in Chapter 3, we present in Chapter 4 a holistic analysis framework for collaborative e-Business process modelling approaches based on concepts of the Balanced Scorecard. Identified requirements are presented and a strategy map describes their complex interactions. Chapter 5 illustrates some relevant research activities that deal with approaches to modell collaborative e-Business processes, while Chapter 6 gives a short summary and an overview of future work.

## 2 Characteristics of Small and Medium-Sized Enterprises (SMEs)

In the European Union - but in most areas of the world as well - SMEs are the predominant form of enterprises. Defined as companies that employ fewer than 250 persons and have an annual turnover not exceeding 50 million Euro and/ or an annual balance sheet total not exceeding 43 million Euro [8], SMEs are not necessarily miniature versions of larger enterprises [9]. Related to their environment, structure, strategy and decision process, and psycho sociological context such as the dominant role of owner-manager, the following characteristics distinguish SMEs from large enterprises:

- **Specialization and Individuality.** First of all, SMEs act on business markets that are not covered by large enterprises. Characterized by a high specialization and individuality many SMEs pursue a segmentation or niche strategy that leads to a certain strength in competition [10].
- **Proximity to markets.** Compared with large enterprises, SMEs are strongly focused on their end-user allowing a high proximity to markets. Instead of focusing on exchangeable products or services for anonymous markets like large enterprises SMEs provide services oriented at the customer's needs.
- **Flexibility.** Quickness to react and reorient themselves on business changes is a major characteristic of SMEs [11]. This flexibility in decision making and implementing organizational changes is archived by preferring simplicity and flexibility regarding their processes and organizational structures [12].
- **Limited resources.** As mentioned before SMEs are limited like all companies by tight resources, especially missing IT literacy and financial resources [11]. Missing know-how can be compensated with basis knowledge of many areas due to the fact that employees at SMEs are generally "all-rounders" and are good at multi-tasking.
- **Technical heterogeneity.** Smaller firms often lack coherent Information and Communication Technology (ICT) strategy or the related skills. For instance, IT landscapes consist of heterogeneous systems, reaching from Enterprise Resource Planning Systems (ERP) to spreadsheet-based island applications for conducting their every-day business transaction [12].
- **Globalization.** The growing internationalization of markets since one decade affects the strategy of SMEs due to deregulation and liberalisation of former market barriers. The chance of new potential business partners involves an enormous adaptation pressure for SMEs which have less experiences than large enterprises in global electronic business [13].

# 3 Collaborative E-Business Processes

## 3.1 E-Business and Collaboration

During the last years, e-Business has become widely accepted in many different industry segments. By adopting systems that allow for business transaction to be conducted electronically rather than paper-based, enterprises can significantly reduce the effort for data-processing, increase business data accuracy and may even discover new business models or partners [14]. However, electronic collaboration is characterized by a low penetration among the above mentioned SMEs [15] since the introduction of Electronic Data Interchange (EDI) in the late 1960s. Due to huge technical complexity and missing global accepted e-Business standards [16], electronic collaboration among SMEs is currently limited to portal technology on presentation level. Thereby users are confronted with diverse user interfaces and working processes provided by the business partners (1:n relation). New standardization approaches like ebXML or United Nations Centre for Trade Facilitation and Electronic Business (UN/CEFACT) Core Component Technical Specification (CCTS) [17] try to integrate business processes and semantic aspects on application level [18] enabling real networkability and n:m relation. As a result standardized business processes lead to next generation e-Business frameworks [19] build on SOA and BPM concepts.

In the frame of the EU-funded GENESIS project [7], a consortium of several partners from across Europe proposes such a holistic framework for performing seamless Business-to-Business (B2B) and Business-to-Government (B2G) processes focused on the business environment of SMEs in Eastern Europe. Typical business processes considered in GENESIS are transactional processes like order, invoice, VAT declaration or bank transfer. The main goal is the research, development and pilot application of the needed methodologies, infrastructure and software components creating a living e-Business platform. Project partners include universities, software vendors, governmental institutions and end-users (SMEs) which have played a major role during the user modelling phase of inter-organizational business processes. These first modelling experiences are assimilated in the holistic analysis framework presented in Chapter 4.

### 3.2 Business Process Modelling Layers

Inter-organizational business processes are performed by multiple independent parties. Since organization borders usually represent boundaries for system interactions and information flows, a number of particularities arise in comparison to company-internal (private) business processes. To achieve seamless business processes across enterprise borders the heterogeneity of different terminologies and modelling notations used within the organizations have to be overcome. However, autonomy of the different business partners has to be taken into account meaning that an organization should be able to flexibly participate in business relations. Important contributions to handle these challenges with regard to inter-organizational business processes come from workflow management, e.g. the Public-To-Private Approach [20] and the Process-View-Model [21]. These approaches distinguish between internal process (private process) and cross-organizational interaction (collaborative process), as depicted in Figure 2.

On a **private process level**, organizations model their internal business processes according to a modelling approach or notation that is most suitable for internal demands independently of the modelling methodologies used by the business partners. As shown in Figure 2, for instance, SME A uses the Unified Modelling Language (UML) for modelling internal processes whereas SME B models with Event-Driven-Process Chains (EPC). A comparison addressing the heterogeneity of business process models can be found in [22]. As a response, abstraction concepts hide details of the internal business process from external business partners on the **public process level**. According to the SOA paradigm

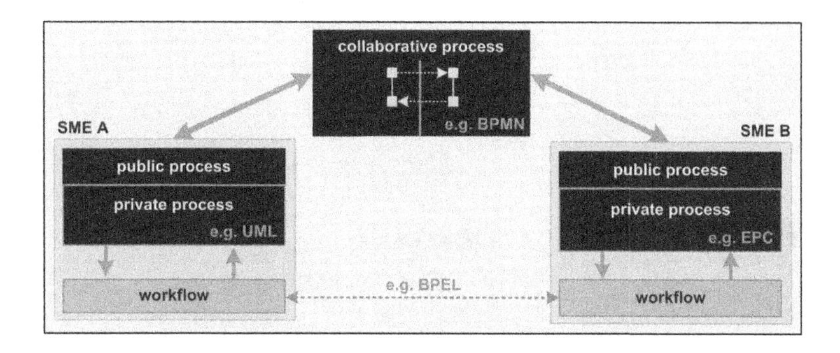


**Fig. 2.** Business Process Modelling Layers

[23] public processes are comparable to the Web Service Definition Language (WSDL) and can be interpreted as a mediator and interface. By hiding the internal process implementation like the specific modelling language and protecting also critical internal information [24] public processes provided by an organization connect private processes to a **collaborative business process**, the third modelling layer. This level defines the interactions of two or more business entities taking place between the defined public processes. One possible language for modelling collaborative processes could be the Business Process Modeling Notation (BPMN) that consolidates ideas from divergent notations into a single standard notation. Examples of notations or methodologies that were reviewed are: UML Activity Diagram, UML EDOC Business Processes, ebXML Business Process Specification Scheme (BPSS), Activity-Decision Flow (ADF) Diagram, RosettaNet, and EPC [25].

## 4 A Holistic Analysis Framework

### 4.1 Perspectives

In the following Chapter, we present a holistic analysis framework for modelling collaborative e-Business processes focused on SMEs. Based on the Balanced Scorecard (BSC) [26] four perspective and their indicators are derived from empirical studies, conceptual research results and complemented with modelling experience taken in the EU-project GENESIS. In contrast to the original perspectives proposed by Kaplan and Norton (financial, customer, internal working process and innovation/ learning) the specific importance of collaboration among enterprises leads to four adaptive perspectives that are depicted in Figure 3. This kind of thinking in different perspectives reduces an unbalanced point of view by identifying challenges SMEs are confronted with modelling collaborative e-Business processes. The following paragraphs present these perspectives and their identified indicators in detail.

**User.** The first perspective focuses on the users who model the collaborative e-Business processes, no matter whether a business or IT specialist is involved.

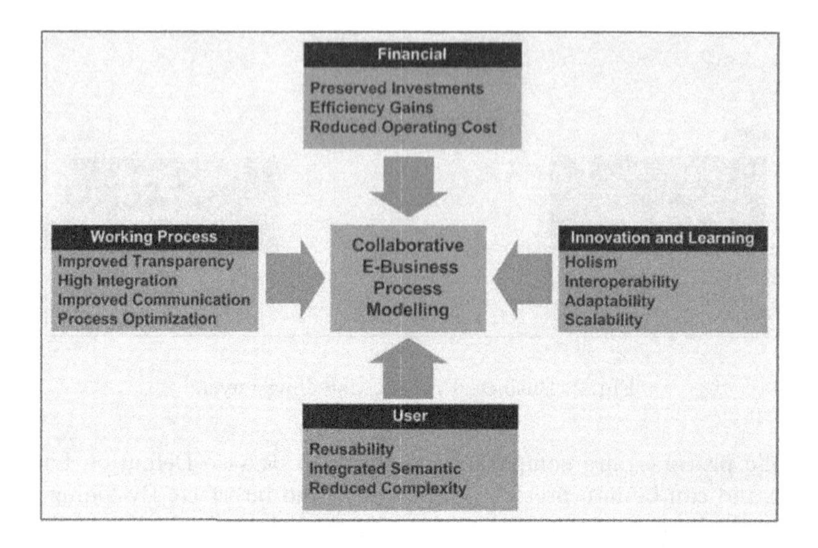


**Fig. 3.** Perspectives of the Holistic Analysis Framework

User modelling experiences taken in the GENESIS project have emphasized three major requirements.

First, **reusability** of knowledge compensates the missing modelling know-how of employees at SMEs who are generally characterized as all-rounder or multi-task worker. In contrast to large enterprises employing specialists for modelling business processes SMEs do not have these resources. Therefore by means of templates and best practices the user is relieved from routine modelling activities. Additionally by supporting a new concept transforming the software paradigm Design Pattern [27] to collaborative processes it renders assistance to model process transactions among business partners. Besides this so-called business transaction design pattern [28] a central repository for process and data building blocks enables a high reusability by adapting, adding or creating collaborative business processes.

Search criteria combined with a specified context, for instance the country, business partner role or industry, lead to the second requirement of the user perspective, the **integrated semantic**. A common understanding with regard to description of processes and data across enterprise and even department borders is a prerequisite to solve this business dilemma [29]. A concept handling the accustomed terminology of the different user groups (e.g. a business expert speaks another language than an IT specialist) is required to reduce misunderstandings and potential sources of error. Especially the growing international orientation of SMEs increases this phenomenon due to regional, cultural or language differences as modelling experiences in GENESIS have shown. As a result of this additional information overload the modelling approach has to provide only the relevant information according the users' specific context. Unnecessary information has to be hidden.

Combined with the above presented requirement reusability, an integrated semantic leads to a **reduced complexity**, the third and last requirement of the user perspective [30]. Different levels of abstraction and granularity as well as different views on business processes provide modelling environments corresponding to the respective user group. In this sense business people see a company as a set of processes that generate and consume different kind of flows and are carried out by resources. From another perspective, IT people interpret a company as a set of information systems or services [31].

**Innovation and Learning.** Focused on innovation and learning the second perspective of our holistic analysis framework regards collaborative e-Business processes from a strategic point of view. Due to the fact that business processes are subject to permanent change, the specific SME characteristic of high flexibility to react and adapt to business changes is a major challenge in this perspective. In this context innovation is not interpreted as product innovation but as process innovation. First of all, the widespread thinking in island approaches at SMEs has to be overcome by a **holistic** modelling approach. Derived from the management concept of Continuous Improvement Process (CIP) that is a never ending effort to adapt and improve processes, modelling of collaborative e-Business processes is a cycle using small steps improvement, rather than implementing one huge radical improvement as proposed by the BPR approach [2]. Figure 4 depicts a designed cycle with the identified phases of collaborative e-Business processes. In the first phase, SMEs discover potential business partners as well as their provided business processes. Followed by a modelling and design phase where local processes are mapped to collaborative processes, the third phase implements private processes to executable workflows. The fourth and last phase closes the continuous improvement cycle by executing and evaluating collaborative e-Business processes. In addition, an integrated data and process modelling approach has to take into account the traditional strong focus on data interchange in context of e-Business and EDI [32].

This will lead to the second requirement regarding the innovation and learning perspective: **interoperability**, meaning the ability of ICT systems and of the business processes they support to exchange data and to enable the sharing of information and knowledge [33]. Divided into organizational (laws, processes),

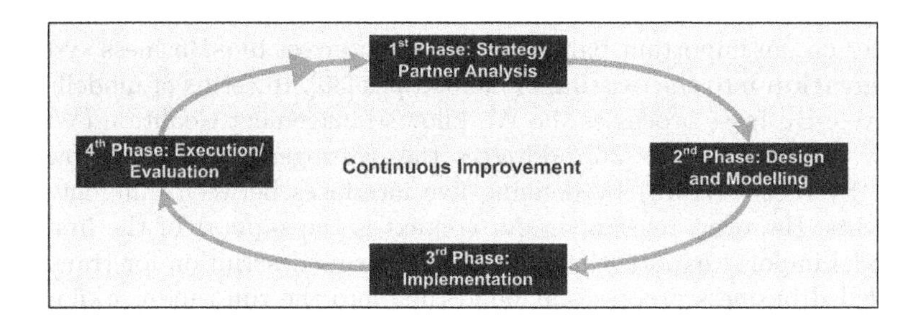


**Fig. 4.** Continuous Improvement Process Cycle

semantic (data structures, services) and technical (protocols, technical interfaces) interoperability [33], globally accepted standards are a prerequisite to integrate business partners into the enterprise value chain without a huge effort [30]. Especially the increased globalization implies a strong focus on extending the integrated semantic approach of the user perspective by linking local and global knowledge as well as combining the technical SOA paradigm with BPM concepts.

Furthermore interoperability fosters the agility to react and reorient on business changes and generates a competitive advantage for the enterprise [11]. As a consequence a model-driven approach has to bridge the gap between IT and business to achieve full **adaptability** in the dynamic business environments. According to the ISO Open-EDI reference model [34] modelled business processes on the so-called Business Operational View (BOV) are mapped automatically to executable workflows on the Functional Service View (FSV) with the workflow patterns of van der Aalst [35] in mind. Caused by rapid technology innovations technical changes are no longer hurdles for SMEs characterized by their missing know-how in technology issues due to the separation of specification and implementation.

A flexible modelling architecture has to escort the organizational evolution and growth with a huge level of **scalability**. Open interfaces within a modelling approach support a modular architecture for collaborative e-Business processes [24] and allow outsourcing of non-core competences according to the Resource-Based View (RBV) [6]. Especially new concepts like Business Process Outsourcing (BPO) on business level or Software as a Service (SaaS) on technical level have to be taken into account.

**Working Process.** In contrast to the previous perspective focused on strategy challenges our third perspective regards on operational aspects of modelling collaborative e-Business processes. According the three modelling layers presented in Chapter 3.2 a process step must be mapped **transparently** between the private and public process layer regarding the heterogeneous modelling approaches and notations used on private process level [30]. The challenge is an automatic and bidirectional linking without losing information, but providing information hiding simultaneously [24]. As already mentioned before, critical information is not published to all potential business partners. Besides a black or white box approach as provided by BPMN [25] a detailed differentiated grey box approach is required.

Further on, an important barrier for SMEs to invest in e-Business systems is the **integration** into the existing IT landscapes [30]. In terms of modelling collaborative e-Business processes the Workflow Management Coalition (WfMC)'s workflow reference model [36] addresses the heterogeneity of Workflow Management Systems (WfMS) by defining five interfaces between different WfMS components. However, relevant in our context is the support of the first interface 'model import/ export' that specifies a process description for transferring the modelled business processes at build time into the run time workflow environment. Thereby workflow languages like Business Process Execution Language (BPEL) focus in general only on the orchestration of activities exposed as Web

Services. In addition, process definition typically incorporates people as an additional possible type of participant. A BPEL extension called BPEL4People tries to close the missing integration of human interactions in workflows [37].

The third identified requirement in the working process perspective is **improved communication** across department and enterprise borders [30]. Allowing by the holistic modelling approach with the integration of semantic aspects, communication activities are no longer located on a technical level. Instead of negotiating the technical details business processes establish the communication basis for conducting e-Business processes over the Internet. As a consequence, process models serve as discussion basis and abstract from technically oriented standards.

To measure the related **process optimization** business processes are not only limited to describing the information flow and to assigning organizations units. Also key performance indicators (KPI) have to be included allowing an analysis of business processes. Especially, the increased relevance of BPO and SaaS require an analysis support to validate agreed limit values in the frame of Service Level Agreements (SLA). Besides this post analysis, capabilities for simulation can reduce the learning time for SMEs.

**Financial.** The fourth and last dimension of our analysis framework is the financial perspective. Characterized by limited financial resources [11] SMEs indicate the **preserved investments** as an important aspect in the context of e-Business [30]. In consideration of reuse of already modelled private business processes, existing modelling know-how at SMEs has to be protected meaning a new modelling notation on private process level would implicate additional training session for the employees. For this reason using of well-known modelling languages on private process level is to prefer.

In addition, **efficiency gains** can be achieved by elimination of manual activities during the four modelling phases, such as semi-automatic mapping between public and private processes. One positive side effect of this automation is a gain of time for strategic issues and thus an increase of value-added productivity by error-free processing and reduced sources of error. For that reason, SMEs indicate efficiency gains as the main driver for future IT investments [30].

Last but not least, a better operating efficiency influences the **operating costs** [11]. By reducing transaction costs SMEs have new liquid resources at one's disposal to strengthen their innovation capability and their growth archiving by minimization of payroll and system costs.

### 4.2 Strategy Map

After presenting the identified requirements within the four perspectives of the anaylsis framework a designed strategy map describes the complex interactions by connecting the criteria in explicit transitive cause-and-effect relationships to each other [26]. By focusing on the important interactions, alignment can be created around the requirements which make a successful implementation more easy

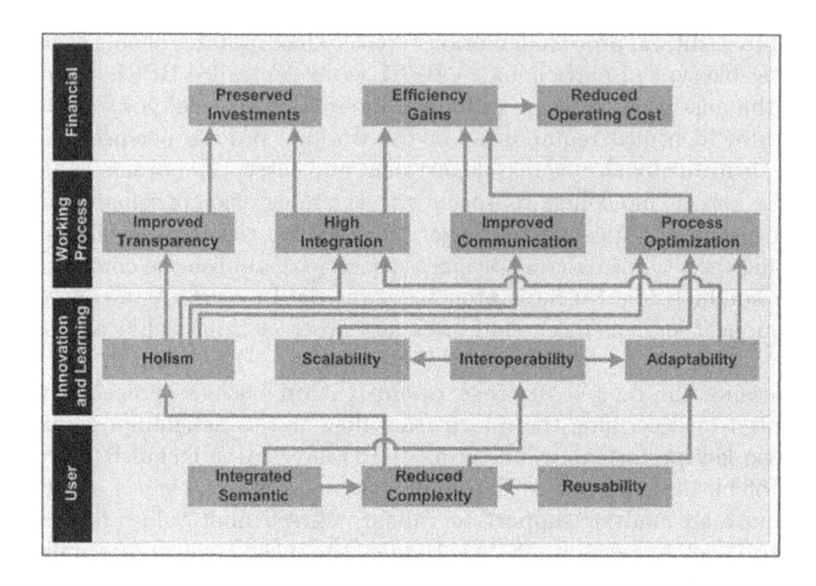


**Fig. 5.** Strategy Map

and represents the heart of our analysis framework. Figure 5 depicts a clear diagram to deeply analyse collaborative e-Business process modelling approaches.

Starting from the user perspective all criteria can be derived. A reduced complexity supported by a huge degree of knowledge reusability as well as an integrated semantic approach to interpret knowledge across departmental and organizational borders influences directly the adaptability of enterprises. Fast reaction and reorientation on business environment changes are only possible if the complexity is hidden from the user and all modelling phases are covered by a holistic modelling approach. As Figure 5 illustrates by the number of outgoing arrows, holism plays a central role with respect to collaborative e-business processes. In fact, all criteria of the working process perspective are caused by the four identified modelling phases. First, transparent bidirectional mapping between private and public processes combined with information hiding is required to preserve investments by using well-known process languages on private process level. Second, the integration of existing IT landscapes as well as mapping human interactions to workflow languages like BPEL brides the gap between IT and business supported by a model-driven approach related to the adaptability requirement. Separation of specification and implementation (SOA/ Open-EDI reference model) abstracts from technical interoperability issues and leads to a "Business drives IT" concept located on process level. Third, an improved communication concerning internal and external partners is reachable by a common understanding of different issues beyond all modelling phases. Last of all, process optimization measured by KPI and generated by continuous improvements leads to efficiency gains due to a faster reaction and error-free processing by eliminating manual activities. Hence employees at SMEs can work more on strategic issues associated with a value-added productivity and reduced operating costs.

## 5 Related Work

Modelling collaborative e-Business processes are discussed from several perspectives in literature. Existing approaches are limited to just one process modelling layer. For instance, a widespread method on private process level is the Architecture for Integrated Information Systems (ARIS) that separates business processes into five views: organization, data, control, function, and service view [38]. To describe the dynamics of the business processes and to link the different views, ARIS uses a modelling language known as Even-Driven-Process Chains (EPC) in the center of the ARIS House of Business Engineering (HOBE). This semi-formal modelling languages combines process and analysis elements and is designed for business aspects like the whole ARIS concept. New research activities divide the ARIS HOBE into a vertical axis of global knowledge (available for all network participants) and a horizontal axis of local knowledge. Also a new modelling language named Process Module Chain is introduced to describe inter-organizational processes [39]. Further modelling approaches on private process level are the Zachmann framework [40] or Business Engineering [41].

A methodology dealing with collaborative processes is the UN/CEFACT Modelling Methodology (UMM) [28]. According to the Open-EDI reference model [34] it specifies collaborative business processes involving information exchange in a technology-neutral, implementation-independent manner. Based on UML and Rational Unified Process (RUP) UMM is a methodology similar to a software process and supports components to capture business process knowledge. It combines an integrated process (UML) and data (CCTS) modelling approach [17] as well as modeling business collaboration in context [42].

## 6 Conclusion and Further Work

In the frame of this work, we have presented a framework that identifies the challenges SME are facing with regard to modelling collaborative e-Business processes, the next wave of conducting business in a process-oriented inter-organizational way. Following the management concept Balanced Scorecard the requirements are considered from different viewpoints and generate a holistic understanding. Based on the identified criteria, further work will deal with the design of a modelling architecture for collaborative e-Business processes that extends current IT-driven SOA paradigms to a real business-driven inter-organizational SOA/BPM. Further on, the above mentioned existing modelling methodologies ARIS limited on private process level and UMM on collaborative process level should be considered complemented with the proposed approaches of this paper.

**Acknowledgments.** This paper has been created closely to research activities during the EU-funded project GENESIS (Contract Number FP6-027867) [7].

## References

1. Picot, A., Reichwald, R., Wigand, R.T.: Information Organization and Management: Expanding Markets and Corporate Boundaries. John Wiley and Sons, New York (1999)
2. Hammer, M., Champy, J.: Reengineering the corporation. Brealey, London (1993)
3. Champy, J.: X-Engineering the Corporation: Reinventing your Business in the Digital Age. Warner Books, New York (2002)
4. Alt, R., Oesterle, H.: Real-Time Business. Springer, Berlin et al. (2004)
5. Carr, N.G.: IT doesn't matter. Harvard Business Review **81**(5) (2003) 41–49
6. Hamel, G., Prahalad, C.K.: The core competence of the corporation. Harvard Business Review (1990) 79–91
7. GENESIS: EU project GENESIS (FP6-027867). http://www.genesis-ist.eu (2006)
8. European Commission: The new SME definition - user guide and model declaration. http://ec.europa.eu/enterprise (2003)
9. Chen, J.C., Williams, B.C.: The impact of electronic data interchange (EDI) on SMEs: Summary of eigth british case studies. Journal of Small Business Management **36** (1998) 68–72
10. Porter, M.E.: Competitive Strategy. The Free Press, New York (1980)
11. The Economist Intelligence Unit: Thinking big. midsize companies and the challenges of growth. http://www.eiu.com (2006)
12. European Commission: The european e-business report 2005 edition - a portrait of e-business in 10 sectors of the EU economoy. http://ec.europa.eu/enterprise (2005)
13. Ibielski, D.: SMEs on the move of globalization. http://www.unece.org/indust/sme/ibielski.htm (2002)
14. United Nations Economic Commission for Europe: A roadmap towards paperless trade. http://www.unece.org/cefact (2006)
15. Beck, R., Weitzel, T.: Some economics of vertical standards: Intergrating smes in EDI supply chain. Electronic Markets **15**(4) (2005) 313–322
16. Zhao, K., Xia, M., Shaw, M.J.: Vertical e-business standards and standards developing organizations. Electronic Markets **15**(4) (2005) 289–300
17. UN/CEFACT: Core component technical specification v2.01. http://www.untmg.org (2003)
18. Janner, T., Schmidt, A., Schroth, C., Stuhec, G.: From EDI to UN/CEFACT: An evolutionary path towards a next generation e-business framework. In: Proceedings of the 5th International Conference on e-Business (NCEB2006). (2006)
19. Hoyer, V., Janner, T., Mayer, P., Raus, M., Schroth, C.: Small and medium enterprise's benefits of next generation e-business platforms. The Business Review, Cambridge (BRC) **6**(1) (2006) 285–291
20. van der Aalst, W.M.P., Weske, M.: The P2P approach to interorganizational workflows. Lecture Notes in Computer Science **2068** (2001) 140–156
21. Shen, M., Liu, D.R.: Coordinating interorganizational workflows based on process-views. In: Proceedings of the 12th International Conference on Database and Expert Systems Applications (DEXA'01). (2001) 274–283
22. Mendling, J., Neumann, G., Nuettgens, M.: A comparison of XML interchange format for business process modelling. In: Proceedings of EMISA 2004 - Information Systems in E-Business and E-Government. (2004)
23. Alonso, G., Casati, F., Kuno, H., Machiraju, V.: Web Services Concepts, Architectures and Applications. Springer, Berlin et al. (2004)
24. Parnas, D.L.: On the criteria to be used in decomposing systems into modules. Communications of the ACM **15**(12) (1972) 1053–1058

25. Object Management Group: Business process modeling notation specification, OMG final adopted specification. http://www.bpmn.org (2006)
26. Kaplan, R.S., Norton, D.P.: The balanced scorecard - measures that drive performance. Harvard Business Review **70**(1) (1992) 71–79
27. Gamma, E., Helm, R., Johnson, R., Vlissides, J.: Design Patterns. 2nd edn. Addison-Wesley, Reading et al. (1994)
28. UN/CEFACT: UN/CEFACT Modeling methodology (UMM). http://www.unece.org/cefact/umm (2006)
29. Stuhec, G.: How to solve the business standards dilemma - the context driven business exchange. SAP Developer Network (SDN), http://sdn.sap.com (2006)
30. Fricke, M., Goetz, K., Renner, T., Polz, A.: Studie e-business barometer 2006/2007. http://www.wegweiser.de (2006)
31. Anaya, V., Ortiz, A.: How enterprises can support integration. In: Proceedings of the First International Workshop on Interoperability of Heterogeneous Information Systems. (2005)
32. Glushko, R.J., McGrath, T.: Document engineering - analyzing and designing the semantics of business service networks. In: Proceedings of the IEEE EEE05 international workshop on Business Service Networks. (2005)
33. European Commission: European interoperability framework for pan-european eGovernment services. http://europa.eu.int/idabc (2004)
34. ISO/IEC: Information technology - Open-EDI reference model, ISO/IEC standard 14662:2004(e). http://www.iso.org (2004)
35. van der Aalst, W.M.P., ter Hofstede, A.H.M., Kiepuszewski, B., Barros, A.P.: Workflow pattern. Distributed and Parallel Databases **14**(3) (2003) 5–51
36. Hollingsworth, D.: The workflow reference model. http://www.wfmc.org/standards/docs/tc003v11.pdf (1995)
37. Kloppmann, M., Koenig, D., Leymann, F., Pfau, G., Rickayzen, A., von Riegen, C., Schmidt, P., Trickovic, I.: WS-BPEL extension for people - BPEL4People. ftp://www6.software.ibm.com/software/developer/library/ws-bpel4people.pdf (2005)
38. Scheer, A.W.: ARIS, Business Process Frameworks. Springer, Berlin et al. (1999)
39. Keller, M., Scherer, R.J., Menzel, K., Theling, T., Vanderhaeghen, D., Loos, P.: Support of collaborative business process networks in AEC. Journal of Information Technology in Construction **11** (2006) 449–465
40. Zachmann, J.A.: A framework for information system architecture. IBM Systems Journal **26**(3) (1987) 276–292
41. Oesterle, H., Back, A., Winter, R.: Business Engineering. Springer, Berlin et al. (2004)
42. Hofreiter, B., Huemer, C.: Modeling business collaboration in context. Lecture Notes in Computer Science **2889** (2003) 829–844

# A Conceptual Framework for Composition in Business Process Management

Ingo Weber, Ivan Markovic, and Christian Drumm

SAP Research, Karlsruhe, Germany
{firstname.lastname}@sap.com

**Abstract.** In this work, we present a conceptual framework for deriving executable business process models from high-level, graphical business process models based on the paradigm of Service-Oriented Architectures and Semantic Web technology. We hereby envision a direct, but implicit link from a business analyst's view on a process model to its execution driven by an IT system. This linkage enables the derivation of an execution-level model for newly created business process models as well as adaptation of the execution model after re-engineering processes, possibly under certain re-design goals (such as quality, cost, execution time, flexibility, or others).

The framework includes a component architecture and an algorithm that describes how to combine executable artifacts, such as (Semantic) Web services, in order to find an implementation that matches a given business process model. An extensible set of criteria can be used for validating the composition.[1]

## 1 Introduction

One of the promises of the Service-Oriented Architecture (SOA) paradigm is increased flexibility by coupling components loosely. In the area of enterprise applications, Web services can accordingly be used to encapsulate business functionality. The loose coupling of business functions aims to increase the flexibility in executable business processes: the process flow can then be separated to a large degree from the implementation of business functions. This increased flexibility in process modeling and enactment is highly desired: in today's business world the business models[2] are changed at an ever increasing frequency in order to react to changing market situations. Allowing for a swift adaptation of operational business processes is a key requirement for modern enterprise application systems.

Among other challenges, the question of how to leverage process modeling at the execution level is a key question for the uptake of SOA in enterprise software solutions [15]. This paper addresses the question how changes can be propagated from the process modeling level to the execution level. That is, if a new process model is created or an existing model is changed, then the respective implementation

[1] This work has in part been funded through the European Union's 6th Framework Programme, within Information Society Technologies (IST) priority under the SUPER project (FP6-026850, http://www.ip-super.org).

[2] Under business model we refer to the concept of how an enterprise makes money.

W. Abramowicz (Ed.): BIS 2007, LNCS 4439, pp. 54–66, 2007.
© Springer-Verlag Berlin Heidelberg 2007

has to be created or modified to reflect the changes in the process model[3] at the execution level (i.e. the information technology (IT) infrastructure).

A solution to this problem would provide an enterprise with the opportunity to react faster to changes in their respective environments, e.g., changes in regulations or business models, and would allow for leveraging small windows of opportunities. Such a solution would basically serve for re-arranging available capabilities in an enterprise in order to meet a changed business goal, while missing capabilities or services can be identified. An identified lack of capabilities could potentially be compensated by extending the application infrastructure or outsourcing concerned parts of a process.

We envision a solution that is based on the novel combination of a number of known techniques, namely Web service composition, discovery, mediation, and orchestration, structural process translation, model checking and process validation, as well as machine-accessible semantics. Ontologies[4] aim at making the semantics, i.e. the meaning of terms in a domain of discourse, machine-accessible, which enables reasoners to infer logical conclusions from the presented facts. This functionality can for example be applied to the discovery of Web services. Also, by modeling the relevant domain as an ontology that captures the potential states of the world, state-based composition approaches [3][13] can be employed. The output of the composition of Web services is then expressed as an orchestration of Web service calls, e.g., in Business Process Execution Language (BPEL) [1]. Ultimately, model checking and process validation add to the approach by checking the output under various criteria. The advantage over performing those functions on a high-level process model lies in the increased degree of formalism of the executable process. This paper presents a coarse-grained algorithm for applying these techniques to the given problem and describes the particularities of each point where they are used. The modular approach of the composition component enables different usage scenarios of the components and flexible inclusion of additional ways to automatically validate and provide feedback for the derived executable process model.

The presented work can be used as follows: A business expert models a business process on a high level[5] of abstraction, which should be made executable on the available technical infrastructure. For this purpose, the composition component is called, which attempts to combine available executable artifacts and returns an executable process model or a failure note. The suggested executable process can subsequently be validated or directly get deployed for enactment. Benefits of this approach are increased reuse of artifacts, easier accessibility of the process space in an organization, increased flexibility through simpler change management, the potential for a more fine-grained evaluation of the validity, correctness, and compliance of a process model under various viewpoints, lower costs in maintaining the enterprise application infrastructure, and more.

---

[3] Note that the opposite direction of the problem is also of high interest: How to adapt a process model to changes on the execution level.

[4] We here refer to ontologies in notations such as OWL [25] and WSMO [20].

[5] Commonly used graphical notations for business processes on this level are for instance the Business Process Modeling Notation (BPMN), the Event-driven Process Chains (EPC), and UML Activity Diagrams.

This paper describes a conceptual framework for composition in the described context. It builds upon the requirements analysis formulated in [18] and contains an outline of the most important components to instantiate the framework as well as an algorithm that describes the interplay between those components. The most relevant functions span from the discovery of artifacts, their actual composition, the compatibility of the data exchange between the artifacts (handled through mediators), to the validation of the composed process. As the paper addresses our current work, the granularity of the description of each individual building block is rather high-level. The goal is to formulate on an abstract level how the different techniques can be brought together in order to realize the larger vision. However, the current status of the work is purely conceptual.

The remainder of the paper is organized as follows: The following section describes the addressed problem in more detail. The conceptual framework in Section 3 explains the presented solution in terms of a component overview, a composition algorithm, and a further investigation of the usage of discovery, mediation, and validation techniques. Section 4 examines related work and Section 5 concludes.

## 2 Problem Description

When modeling a business process today, the modeler usually creates the process manually in a graphical tool. The outcome is a process model that reflects the business expert's view on real-world activities or a to-be process. Subsequently, this process model is implemented by IT experts - or, rather, its control and data flow are mapped to implemented artifacts in information systems. The relationship between the business-level model and its IT-level implementation is oftentimes weak. Consequently, the process implementation can deviate substantially from the model, and changes on either one of the levels cannot be easily propagated to the respective other level.

The approach of automated composition which we pursue attempts to bridge the gap between the modeled processes and their implementation by finding program parts which can be used for the implementation of a process model and defining their usage in an executable process model. For this vision to become reality, several needs must be met: Application program fragments with business functionality must be available for remote invocation as encapsulated and executable IT artifacts that can be enacted through an IT infrastructure. Examples for such executable IT artifacts include: Web services with simple and complex interfaces, partial processes and sub-processes.

In order to allow the automation of tasks such as composition, these artifacts are annotated with formal, machine-accessible semantics. In particular, we assume that the executable artifacts are annotated by semantic descriptions of their functionality and non-functional properties, e.g., by linking the meaning of the used terms to the content of ontologies which model the domain of interest. In addition, we assume that process tasks are annotated with goals as formalizations of desired task functionalities and, optionally, quality requirements. This allows a composition component to find suitable artifacts in a repository and evaluate their applicability and compatibility in the context of other artifacts.

While other recent work addressed similar issues, there is a notable difference to this paper: we are looking at composition from the viewpoint of an enterprise, not an end user. While business processes in business-to-consumer (B2C) scenarios can be a point of contact with end users, service composition on the consumer's end is not the focus of our work. Our work is rather placed in the context of enterprise application software and enterprise application integration. In the scope of this work, the final results must be correct and compliant with both internal policies and external regulations such as Basel II or Sarbanes-Oxley (SOX). Thus, the focus is on design-time composition in a known domain, which simplifies the problem in two ways: Firstly, design-time composition can always be checked and approved manually before deploying and enacting it. Thus, unanticipated side effects[6] can be avoided because in the manual control step the decisions from the automated composition can be overruled. And secondly, in our known domain - the enterprise - we can assume to own the artifacts and can thus enforce a uniform way and formalism of their description.

On the above basis, the problem addressed by this paper can be restated in the following way: Given a high-level process model and a set of previously modeled artifacts in a repository, all semantically annotated, a composition approach should come up with the required set of artifacts from the repository, orchestrating the artifacts in a way that reflects the business process model's structure and business semantics.

In preliminary work [18] we examined the requirements on a solution of this problem in more depth, providing a list of 14 such requirements. In this paper, we describe a conceptual framework as a general solution strategy, which does not yet address all of the listed requirements completely. The framework rather serves as an extensible base structure on which solutions to the individual requirements from [18] can be combined. The following section describes the framework.

# 3 Conceptual Framework for BPM Composition

The framework described in the following addresses the problem laid out in the preceding section. It describes an architecture in terms of the required components (Fig. 1), an algorithm explaining the interplay between the components, and the most important functions.

## 3.1 Component Overview and Explanation

**The Process Modeling Environment** is used for designing process models and serves as the Graphical User Interface (GUI) to the rest of the architecture. In this modeling environment, the user must be provided with a convenient way to attach semantic annotations to his process models, which are part of the necessary input for the composition approach. At any point during the modeling phase or after finishing the design of the process, the user may request the composition of an executable

[6] Cf. [7] for such a side effect: a composite process that satisfies the pre-condition of having a credit card prior to the execution of a service by directly applying for a credit card.

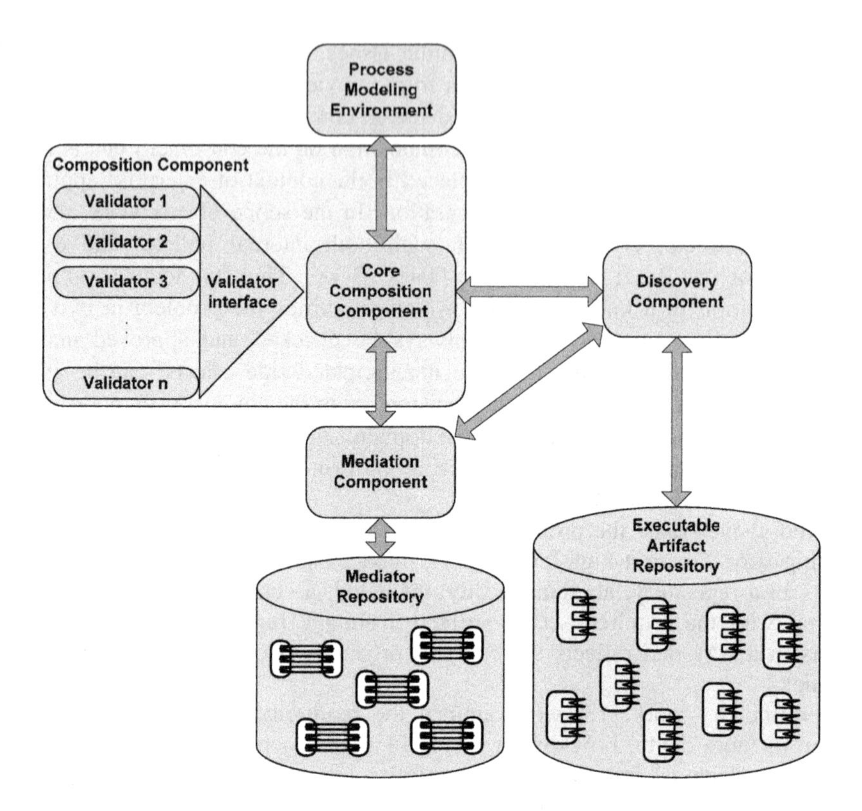


**Fig. 1.** Component architecture and interactions

process for his process model or parts thereof, which triggers an interaction between the Modeling Environment and the Composition Component.

**The Executable Artifact Repository** stores descriptions of the available artifacts. For each artifact this repository contains the description of its functionality, its non-functional properties and in particular how it can be accessed.

**The Discovery Component** serves as a clever interface to the Executable Artifact Repository, in that it answers simple and complex requests for artifacts by matching requests to semantic descriptions of artifacts in the repository.

**The Mediation Component** provides a simple querying interface for the **Mediator Repository**. It enables the other components to retrieve the available mediators for a given mediation problem. The Mediator Repository contains descriptions of the available mediators.

**The Composition Component** interacts with the Process Modeling Environment and the Discovery Component in order to provide the composition of executable artifacts. It holds the Core Composition Component, which implements the composition algorithm described below and handles the interactions with the other components, along with a Validator Interface and the required Validator Plug-Ins. Validation is desired in order to provide an achievable degree of automatic assertion. The suggested structure allows for flexibly plugging in Validators with respect to the current context of the application.

### 3.2 Composition Algorithm

After describing the high-level functionality of the components involved in our framework we will in this section focus on the details of the composition component. The composition component performs the following steps in order to provide the modeling environment with the desired functionality of composing an executable process for a given business process model. The flowchart in Figure 2 corresponds to this algorithm:

1. Identification of existing target-process parts (e.g., in re-engineering, parts of a process may stay unchanged, which implies that the respective executable process does not necessarily have to be changed in those parts, either.)
2. Translating the process structure, which may be necessary, e.g., if there is a change in the underlying process description paradigm[7].
3. For each task / step in the source process: derivation of executable artifacts
   - 3.1. Discovery of single artifacts that implements a task. Pre-existing research results for service discovery such as [8][9][11][16][17] can be adapted for this purpose.
   - 3.2. If no single artifact can achieve the required functionality:
     - 3.2.1. Composition of artifacts for implementing a single task. Here, known approaches to Web service composition, such as [2][3][10][12][13] can be adapted.
     - 3.2.2. Consistency checking of the composed artifacts on the level of this single task implementation (data flow, detection of inconsistencies such as goal invalidation, adherence to policies and regulations...). Note that this step can potentially be integrated to a degree with the previous step.
4. Combination of the task implementation to form the complete executable process. If an inconsistency is detected, another solution for affected task implementations is searched by calling step 3 again.
   - 4.1. Global control flow vs. local (artifact-level) constraint checking. Potentially there are inconsistencies between the global control flow as defined in the process model and constraints over the artifacts implementing the individual tasks. They may be resolved by adding additional ordering constraints over certain activities. If the problem is not solvable in another way, the task implementations are changed.
   - 4.2. Validation through the validator plug-ins, e.g., sequentially based on a prioritization.

---

[7] Many high-level process modeling notations are graph-based, while execution-level process descriptions as BPEL are often block-based. In this step, the graph-based structure, i.e., the control flow "skeleton" of the process, would be translated to an according block-based process structure. [22] and [21] deal with such translations.

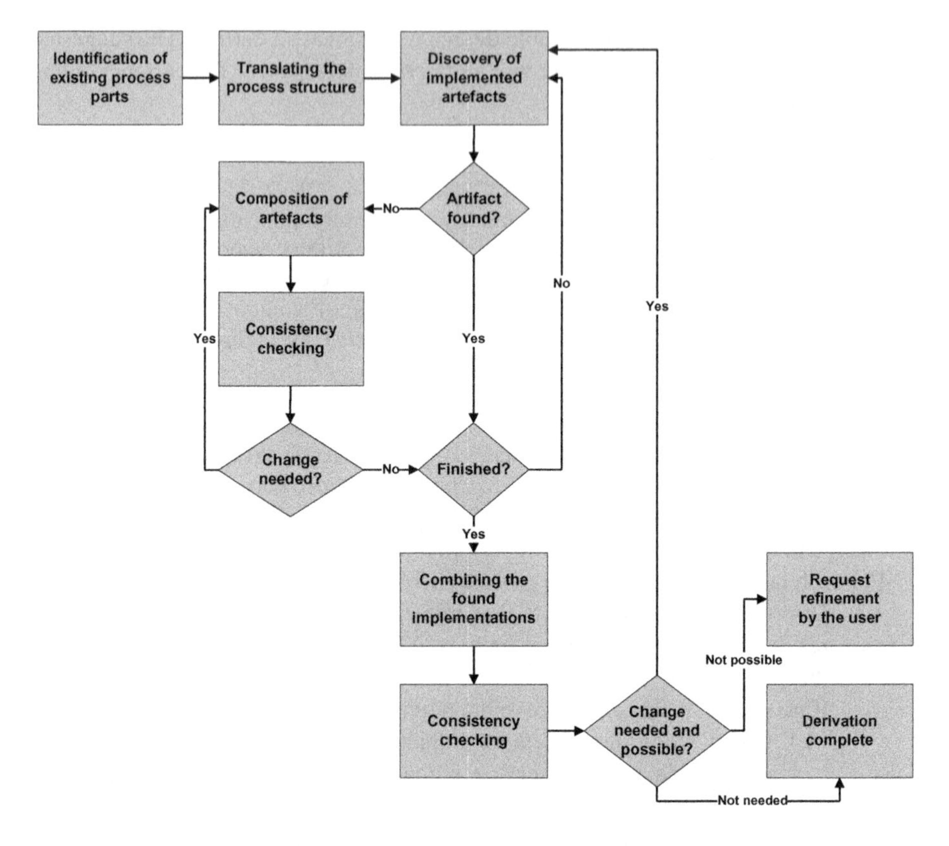


**Fig. 2.** Flowchart depicting the Composition Algorithm

In order to explain the steps of the algorithm in a more tangible way, Figures 3 (a) – 3 (d) depict how the resulting composed process evolves with the steps of the algorithm. Fig. 3 (a) shows a BPMN process model with tasks, connectors, and events. Fig. 3 (b) presents the same process after discovering single services that can implement a full task in the process (step 3.1 in the algorithm). Fig 3 (c) shows additionally groups of services that jointly implement tasks (as composed together by step 3.2.1 in the algorithm). Finally, Fig 3 (d) depicts the process in terms of only the services after the combination (step 4 in the algorithm).

### 3.3 Underlying Composition Approaches

In step 3.2.1 of the Algorithm, we plan to build on existing approaches to Web service composition, such as AI Planning[8]. The applicability of this technology for service composition has been examined, e.g. in [3][10][13], and is rather well understood. Certain approaches can be adapted for the usage in the algorithm above and extended towards directly handling a subset of the requirements from [18].

---

[8] An overview over Artificial Intelligence (AI) planning can be found in [14], recent publications in the field at [3].

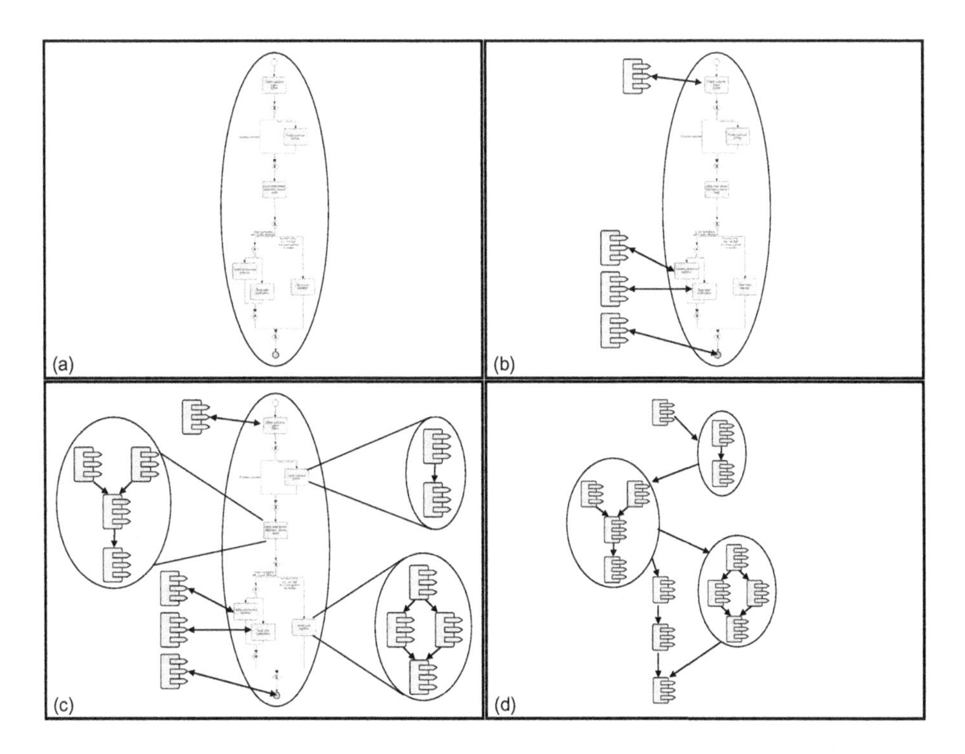


**Fig. 3.** Results from the steps of the Composition Algorithm

Our ongoing work in the SUPER project[1] includes specifying and implementing forward search AI planning guided by heuristics. We research using domain ontologies as background theories in the planning. The approach builds on the Fast-Forward (FF) planning algorithm [24].

### 3.4 Discovery

Two steps of the composition algorithm make use of the discovery functionality: in step 3.1 an attempt is made to discover an individual artifact that can implement a task, while in step 3.2.1 a set of artifacts is composed together and must be discovered beforehand. For this task we plan to adapt and extend the Semantic Web service discovery techniques proposed in [11][16][17] to find single artifacts that implement process tasks.

The annotations of executable artifacts and process tasks serve as an input to the discovery component that decides whether an executable artifact fits to the goal of a process task. Discovery of executable artifacts from the repository is realized by utilizing matchmaking techniques for comparing their semantic descriptions against the semantic descriptions of process tasks. The matching artifact descriptions are then ranked according to their relevance w.r.t. the goal.

Functionality-based discovery of a single artifact for process task implementation consists of two phases. In the first phase, we utilize semantic descriptions of artifacts to filter only invocable artifacts. Here, by invocable we mean artifacts whose preconditions for execution in the current state are satisfied. In the second phase, the resulting set of executable artifacts is queried against the desired goal of a process task. The final set of discovered artifacts for each process task is then ordered by using the widely used degrees of match introduced in [9] and [8]. If a single artifact with required functionality can not be found, the discovery component will provide a set of invocable artifacts as an input for artifact composition.

In the business domain, besides discovering the artifacts meeting the functional requirements, it is important to discover artifacts which best meet the non-functional (quality) requirements of a process task [17]. We envision that the before-mentioned functionality-based discovery will take place at design time. As a result of this process, the semantic description of each task will contain a set of links that associate a goal with the discovered artifacts. During process execution, ranking of the artifacts using the current values of selected non-functional requirements (e.g., price, execution-time, availability, etc.) can be performed.

In case that a single artifact fully matches the desired goal, the top ranked artifact is selected for process task implementation. Otherwise the discovery component returns a set of artifacts for composition that both provide the desired functionality and comply with the non-functional requirements.

Note that the described discovery process can be performed only if the goals and artifacts are annotated using the same ontology. If this is not the case, the discovery component first has to find a mediator for translating between different ontologies, as described in the following section.

### 3.5 Mediators

Mediators play an important role in two areas of our framework. Firstly they are required in order to compose different independently developed Web service into one executable process. In this context it is necessary to mediate between the different message formats used by these services. Secondly the discovery component needs mediators to cope with artifacts annotated using different ontologies as described in the previous section. In the context of this composition framework we do not want to focus on how necessary mediators are developed. Therefore we assume that the mediators have already been defined beforehand and furthermore have been deployed to a mediator repository [19]. In addition to that we also assume that a mediator is available through a standard Web service interface.

In our framework two types of mediators are necessary: i) mediators capable of translating between syntactic messages formats of Web service and ontology instances and ii) mediators capable of mediating between ontologies. Depending on the type of the mediator invoking the associated Web service interface requires different input data. Invoking a mediator of the first type will for example require an instance of an XML message as an input whereas invoking a mediator of the second type will require instances of concepts according to some ontology. Note that our notion of mediators is broader then the one used by WSMO[20]. WSMO mediators are only concerned with the mediation on the ontology level whereas our notion of mediators also takes the mediation between syntactic message formats and ontologies into account.

Based on these assumptions we are now able to differentiate between two possible usage scenarios of our framework. In the first scenario we assume that all artifacts are annotated using the same ontology whereas we assume in the second one that the artifacts are annotated using different, not integrated ontologies. Each of these scenarios results in certain requirements on the integration of mediation and is detailed in the following subsections.

**Annotation using a single ontology:** If all artifacts are annotated using a single ontology no mediators need to be executed during the composition of the process. In addition the composition step doesn't need to insert any ontology mediation steps into the process. However, in order to execute the composed process, mediators of the first type might need to be inserted before and after calls to the Web services implementing tasks or parts of them. The necessary calls to the mediators therefore need to be inserted into the process before it is deployed onto the runtime.

**Annotation using different ontologies:** If all artifacts are annotated using different ontologies the situation becomes much more complex. In order to discover artifacts capable of implementing a given task the discovery components need to execute mediators of the second type as the tasks and each of the artifacts might be annotated using different ontologies. Therefore the discovery needs to interact with the mediator repository in order to locate the required mediators – given they exist. However, this approach might result in the need for the execution of a very large number of mediators. Therefore a pre-selection might be necessary in order to identify promising candidates before executing the mediator. How such a pre-selection could be performed efficiently is currently an open research question and will not be further discussed in this paper.

After the discovery component has found suitable artifacts for a given task it will return this artifact as well as the mediator for mediating between the ontology in which the task is described and the ontology in which the artifact is described to the composition component. The composition component will then use this mediator to create the composed process by inserting it before and after the discovered artifact. Note that this approach results in the usage of the ontology in which the tasks are described as a central hub format. If this is not desired it would also be possible to query the mediator repository for suitable mediators during the composition step again.

### 3.6 A Sample of Validators

The following two examples show possible validators for the composition component, as depicted in Figure 1. Note that these validators correspond to common requirements from [18].

- Validation w.r.t. policies and regulations: Organizations typically have internal policies and are subject to laws or other regulations. If these policies and regulations are expressible in a machine-accessible way, e.g., in Business Rule Engines, they can potentially be used to evaluate in how far the composed executable process is compliant with them.

- Taking into account transactional requirements: A certain set of artifacts might have an interrelationship with respect to a joint outcome, e.g., either all artifacts achieve a positive outcome or the effects of all artifacts' executions have to be undone [10]. Such a setting would require i) an understanding for the joint outcome, i.e., which outcomes denote success and which ones represent failures; ii) using services that provide cancellation actions; and iii) case-based cancellation policies, depending on which actions actually have to be undone in which cases. Detecting such situations, evaluating and correcting the composition could be achieved by a validator. Also, besides atomic behavior (yes-or-no outcomes only), more complex transactional properties can be presented and handled.

Together, the components and the algorithm form an extensible framework for the problem described in Section 2. The modeling environment serves as the user entry point to the system. The composition component executes the composition algorithm, which explains how and when discovery is used for finding suitable available artifacts; the mediation component bridges heterogeneities; and the validators are executed for evaluating the results of the composition.

## 4 Related Work

Recently service composition has been an active area of research [2][3][10][12][13]. However, this paper addresses a different problem than most of the current work around Web service composition, as argued in Section 2.

More related issues address mixed-initiative approaches to composition, such as [13] or [23], where composition of services is performed in an interleaved fashion with human modeling steps. Still, the two mentioned works operate on end-to-end composition scenarios, i.e., the automated composition fills in the missing services between the anticipated start and end states. In comparison to our approach, [13] and [23] would perform steps 3.1 and 3.2.1 of the algorithm in Section 3.2 automatically and leave the rest of the work to the process engineer.

Business-Driven Development (BDD) [5][6] is also related, but assumes many manual steps and is more directed to traditional software engineering, in contrast to the service-assumption made here. Probably it is not feasible to automate all the tasks in BDD, but the use of explicit, formal semantics could provide many desired features. The framework in this paper might serve the advancement of automation in BDD.

So far, we have not encountered other approaches that address composition of services over a complete business process model, in particular not if the granularity of the items that have to be composed is not necessarily that of a Web service.

## 5 Conclusion and Outlook

In this paper we have presented a conceptual framework for composition in business process management. We have introduced the necessary components and described their interactions. Furthermore we have presented a composition algorithm that details how and when these components are used. The described components and the validator plug-ins in combination with the composition algorithm should allow the implementation of business process models by composing an executable process out of executable artifacts.

The main contribution of this paper is the composition component including the validation technique and the algorithm. Additional contributions are the novel combination of existing technology and the introduction of business requirements into composition present additional contributions.

In future work, we plan to implement the framework and develop more details of how the underlying composition technology can directly be extended towards further business requirements. Also, the seamless integration of the presented approach with a modeling environment is going to be addressed.

## References

[1] Alexandre Alves et al. Web Services Business Process Execution Language Version 2.0, OASIS Public Review Draft, 23rd August, 2006, http://docs.oasis-open.org/wsbpel/2.0/

[2] Daniela Berardi, Diego Calvanese, Giuseppe De Giacomo, Richard Hull, and Massimo Mecella. Automatic Composition of Transition-based Semantic Web Services With Messaging. In VLDB '05: Proceedings of the 31st international conference on Very Large Data Bases, pages 613–624. VLDB Endowment, 2005.

[3] International Conference on Automated Planning and Scheduling Systems, ICAPS 2003 – 2006, proceedings at http://www.aaai.org/Library/ICAPS/icaps-library.php

[4] Sheila McIlraith and Tran Cao Son. Adapting Golog for Composition of Semantic Web Services. In Proc. of the 8th Int. Conf. on Principles and Knowledge Representation and Reasoning (KR-02), Toulouse, France, 2002.

[5] Jana Koehler, Rainer Hauser, Jochen Küster, Ksenia Ryndina, Jussi Vanhatalo, and Michael Wahler. The Role of Visual Modeling and Model Transformations in Business-driven Development. In Fifth International Workshop on Graph Transformation and Visual Modeling Techniques, April 2006.

[6] Jochen Küster, Jana Koehler, and Ksenia Ryndina. Improving Business Process Models with Reference Models in Business-Driven Development. In Proc. 2nd Workshop on Business Processes Design (BPD'06), LNCS, Springer-Verlag, 2006.

[7] Ulrich Küster, Mirco Stern, and Birgitta König-Ries. A Classification of Issues and Approaches in Automatic Service Composition. In Proc. of the First International Workshop on Engineering Service Compositions (WESC'05), pages 25–33, December 2005.

[8] L. Li and I. Horrocks. A Software Framework For Matchmaking Based on Semantic Web Technology. In Proceedings of the 12th World Wide Web Conference, 2003.

[9] M. Paolucci, T. Kawamura, T. Payne, and K. Sycara. Semantic Matching of Web Service Capabilities. In Proceedings of the 1st International Semantic Web Conference (ISWC), 2002.

[10] Marco Pistore, Paolo Traverso, and Piergiorgio Bertoli. Automated Composition of Web Services by Planning in Asynchronous Domains. In Proceedings of the International Conference on Automated Planning and Scheduling, 2005.

[11] C. Preist. A Conceptual Architecture for Semantic Web Services. In Proceedings of the 3rd International Semantic Web Conference (ISWC), 2004.

[12] Jinghai Rao. Semantic Web Service Composition via Logic-based Program Synthesis. PhD thesis, Norwegian University of Science and Technology, 2004.

[13] Jinghai Rao, Dimitar Dimitrov, Paul Hofmann, and Norman Sadeh. A Mixed Initiative Approach to Semantic Web Service Discovery and Composition: SAP's Guided Procedures Framework. In Proceedings of the 2006 IEEE International Conference on Web Services (ICWS 2006), Chicago, USA, September 18 - 22, 2006.
[14] Stuart J. Russell and Peter Norvig. Artificial Intelligence: A Modern Approach. Pearson Education, 2003.
[15] Jim Sinur and Janelle B. Hill. Align BPM and SOA Initiatives Now to Increase Chances of Becoming a Leader by 2010. Gartner Predicts 2007. 10 November 2006.
[16] I. Toma, K. Iqbal, M. Moran, D. Roman, T. Strang, and D. Fensel: An Evaluation of Discovery approaches in Grid and Web services Environments. NODe/GSEM 2005: 233-247
[17] Vu, L.-H., Hauswirth, M., Porto, F., and Aberer, K. A Search Engine for QoS-enabled Discovery of Semantic Web Services. Special Issue of the International Journal on Business Process Integration and Management (IJBPIM), 2006, to be published.
[18] Ingo Weber, Requirements for the Implementation of Business Process Models through Composition of Semantic Web Services. Proceedings of the 3rd International Conference on Interoperability for Enterprise Software and Applications (I-ESA) March 2007, Funchal, Portugal
[19] L. Cabral, C. Drumm, J. Domingue, C. Pedrinaci, A. Goyal: D5.6 Mediator Library. Deliverable of the DIP project, July 2006
[20] Digital Enterprise Research Institute (DERI): Web Service Modelling Ontology, http://www.wsmo.org. 2004.
[21] C. Ouyang, W.M.P. van der Aalst, M. Dumas, and A.H.M. ter Hofstede. Translating BPMN to BPEL. BPM Center Report BPM-06-02, BPMcenter.org, 2006.
[22] Jan Mendling, Kristian Bisgaard Lassen, Uwe Zdun: Transformation Strategies between Block-Oriented and Graph-Oriented Process Modelling Languages. In: F. Lehner, H. Nösekabel, P. Kleinschmidt, eds.: Multikonferenz Wirtschaftsinformatik 2006, Band 2, XML4BPM Track, GITO-Verlag Berlin, 2006, ISBN 3-936771-62-6, pages 297-312.
[23] J. Schaffner, H. Meyer, C. Tosun: A Semi-automated Orchestration Tool for Service-based Business Processes. In: Proceedings of the 2nd International Workshop on Engineering Service-Oriented Applications: Design and Composition, Chicago, USA (to appear)
[24] J. Hoffmann, FF: The Fast-Forward Planning System. In: AI Magazine, Volume 22, Number 3, 2001, Pages 57 - 62.
[25] World Wide Web Consortium (W3C), Web Ontology Language (OWL), W3C Recommendation 10 February 2004, http://www.w3.org/2004/OWL/

# Process Dependencies and Process Interference Rules for Analyzing the Impact of Failure in a Service Composition Environment

Yang Xiao and Susan D. Urban

Department of Computer Science and Engineering
Arizona State University
PO Box 878809 Tempe, AZ, 85287-8809 USA
{yang.xiao,s.urban}@asu.edu

**Abstract.** This paper presents a process dependency model for dynamically analyzing data dependencies among concurrently executing processes in an autonomous, distributed service composition environment. Data dependencies are derived from incremental data changes captured at each service execution site. Deltas are then used within a rule-based recovery model to specify how failure recovery of one process can potentially affect another process execution based on application semantics. This research supports relaxed isolation and application-dependent semantic correctness for concurrent process execution, with a unique approach to resolving the impact of process failure recovery on other processes, using data dependencies derived from distributed, autonomous services.

**Keywords:** Data Dependencies, Process Interference, Service Composition.

## 1 Introduction

Web Services and Grid Services have become widely used for B2B integration. However, the loosely-coupled, autonomous nature of services poses new challenges for the correctness of concurrent execution of global processes that are composed of distributed services. Most processes that are composed of Web Services must execute using a relaxed notion of isolation since individual service invocations can unilaterally commit before a process completes [15]. Relaxed isolation leads to dirty reads and writes and also calls for a more user-defined approach to the correctness of concurrent execution. Previous work with advanced transaction models [4] and transactional workflows [19] has used compensation to semantically undo a process. However, existing research has not fully addressed the *process interference* [20] that is introduced when the recovery of one process affects other concurrently-executing processes due to data changes introduced by the compensation of a failed process. A robust service composition environment should not only recover a failed process, but should also make sure process interference is properly handled based on application semantics.

This research has defined a process dependency model to dynamically analyze the data dependencies among concurrently executing processes that are composed of

W. Abramowicz (Ed.): BIS 2007, LNCS 4439, pp. 67–81, 2007.
© Springer-Verlag Berlin Heidelberg 2007

distributed services, providing a rule-based approach to resolve process interference based on application-dependent semantic correctness. The research has been conducted in the context of the DeltaGrid project, where services known as *Delta-Enabled Grid Services (DEGS)* [2] are extended with the capability of recording incremental data changes, known as *deltas*. Deltas from distributed service execution sites are merged to create a global, time-ordered schedule of delta values within a *Process History Capture System (PHCS)* [20, 21]. The merged schedule of deltas then provides a means for tracking data dependencies among concurrently executing processes to determine how the failure of one process can potentially affect other processes. Once the potential process interference is identified, deltas can also be queried using *process interference rules (PIR)* to apply user-defined semantic correctness conditions that determine whether an affected process should keep running or invoke its own recovery procedures.

Based on the service composition model presented in [20, 22], the focus of this paper is on the specification of the process dependency model, the definition of process interference rules, and the illustration of the way in which process interference rules can be used to query deltas to impose user-defined correctness conditions as part of the recovery process [20]. This research contributes towards ensuring a semantically robust, concurrent process execution environment for distributed, autonomous service composition, by dynamically analyzing data dependencies among concurrently executing processes and by providing a rule-based approach to resolve process interference based on application semantics.

The rest of this paper is organized as follows. After outlining related work in Section 2, the paper provides an overview of the DeltaGrid abstract execution model in Section 3. Section 4 presents the process dependency model, while Section 5 elaborates on the use of process interference rules. The paper concludes in Section 6 with a summary of our implementation of the global execution history and process interference rules as well as a discussion of future research.

## 2 Related Work

Research on exception handling in a service composition environment has primarily focused on implementing ACID transaction semantics. Open nested transactions over Web Services are supported in [12], contingency is applied to forward recover a composite service in [16], and WS-Transaction [3] defines processes as either Atomic Transactions with ACID properties or Business Activities with compensation capabilities. An agent-based transaction model (ABT) [7] integrates agent technologies in coordinating Web Services to form a transaction. To avoid the cost of compensation, tentative holding is used in [9] to achieve a tentative commit state for transactions over Web Services. Acceptable Termination States (ATS) [1] are used to ensure user-defined failure atomicity of composite services. Active rules have been used to handle service exceptions independent of application logic, such as service availability, selection, and enactment [14, 23], or search for substitute services when an application exception occurs [10]. But the question of how the recovery of a composite process could possibly affect other concurrently executing processes has not been addressed.

Advanced transaction models support relaxed isolation for long running transactions composed of subtransactions. Sagas [6] can be backward recovered by compensating each task in reverse order. The flexible transaction model [23] executes an alternative path when the original path fails. The backward recovery of a failed transaction causes cascaded rollback or compensation of other transactions that are read or write dependent on the failed transaction. Instead of cascaded rollback, only transactions dependent on tainted data produced by a flawed transaction are removed in [11]. However this work is conducted in a centralized database system which relies on an extended database engine to capture read dependency information.

Research in the transactional workflow area has adopted compensation as a backward recovery technique [5, 8, 18] and explored the handling of data dependencies among workflows [8, 18]. The ConTract model [18] uses pre-conditions integrated into a workflow script to determine whether a compensation of a step affects another step execution. However, read dependency is not considered. WAMO [5] defines a flexible recovery framework for a workflow without considering process interference. In CREW [8], a static specification on the equivalence of data items across workflows is required to track data dependencies.

The research presented in this paper supports relaxed isolation and application-dependent semantic correctness for concurrent process execution, with a rule-based approach to resolving the impact of process failure and recovery on other concurrently executing processes [20]. Instead of statically specifying data dependencies, this research dynamically analyzes write dependencies and potential read dependencies among concurrently executing processes by capturing and merging data changes from distributed service execution, providing an intelligent approach to discovering dependencies among processes in an autonomous service composition environment and using data dependencies to support failure recovery.

# 3 Overview of the DeltaGrid Abstract Execution Model

Our research has developed the DeltaGrid system as a semantically robust execution environment for distributed processes executing over Delta-Enabled Grid Services (DEGS). A DEGS is a Grid Service that has been enhanced with an interface for accessing the deltas that are associated with service execution [2]. The work in [2] has demonstrated the manner in which incremental data changes can be captured from data-centric services, using features such as triggers as well as facilities for monitoring and externalizing database log files, such as Oracle Streams. Deltas collected from DEGSs are forwarded to a Process History Capture System (PHCS) [21] to form a global execution history where data dependencies among concurrently executing processes can be dynamically analyzed. The merged deltas from distributed sites also form the basis for analyzing process interference, determining the effect that the failure recovery of one process can have on other concurrently executing processes using application semantics.

In support of our research, we have defined an abstract execution model for studying the manner in which data changes from DEGS can be used to analyze the dependencies that exist among concurrent processes. Fig. 1 shows the DeltaGrid abstract execution model, which is composed of three components: the *service composition and recovery*

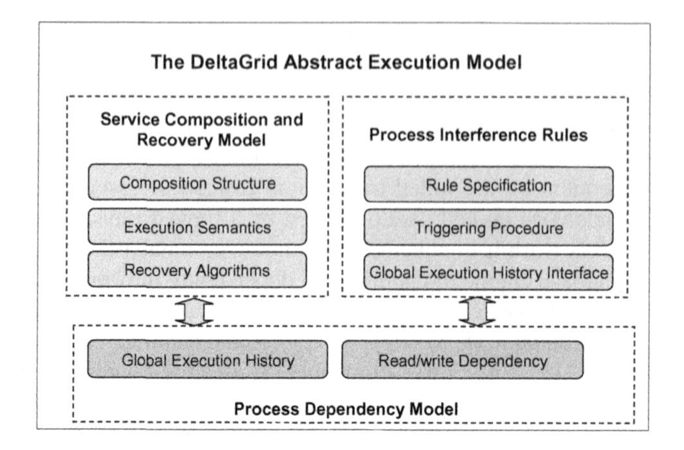


**Fig. 1.** The DeltaGrid abstract execution model

*model*, the *process dependency model*, and *process interference rules*. The service composition model defines a hierarchical service composition structure as well as the semantics of execution entities for the handling of operation execution failure occurring at any compositional level. The recovery model resolves a service execution failure within a process. Recovery techniques, such as compensation for logical backward recovery, contingency for forward recovery, and a process known as delta-enabled rollback for physical rollback of a completed service, are applied at different composition levels to maximize the recovery of a failed process. The details of the composition model and recovery model can be found in [20, 22].

The process dependency model further defines the relationships that exist among concurrently executing processes in terms of write dependencies and potential read dependencies. The process dependency model also defines how data dependencies can be dynamically analyzed from a global process execution history. Process interference rules are active rules that query the data changes from a failed process and its read and write dependent processes, using application semantics to determine if the recovery of a failed process has an affect on active processes that are read and/or write dependent on the failed process. The rest of this paper presents the process dependency model and illustrates the use process interference rules in the recovery process.

# 4 The Process Dependency Model

As described in the previous section, deltas from distributed service executions are forwarded to a Process History Capture System. These deltas are merged to form a time-ordered sequence of data changes that are used to analyze data dependencies [21]. This section presents the process dependency model. Section 4.1 defines the global execution history, integrating process execution context and deltas from distributed sites. Section 4.2 defines write dependency and how to derive write dependencies from deltas. Section 4.3 defines potential read dependency and how to analyze potential read dependency from the process execution context. Section 4.4 then provides a case study to illustrate the use of the process dependency model to discover write dependencies among concurrently executing processes.

### 4.1 Global Execution History

The global execution history of concurrently executing, distributed processes is composed of 1) the execution context of each process $p_i$, and 2) the merged deltas from each DEGS operation $op_{ij}$ invoked from $p_i$. In the definitions that follow, a pair of square brackets [] indicates a partially ordered list of elements ordered by the timestamp associated with each delta. Each $op_{ij}$ is also associated with one DEGS through a degsID.

**Definition 1 (Delta):** A delta represents an incremental value change on an attribute of an object generated by execution of a DEGS operation. A delta is a six-element tuple, denoted as $\Delta(oID, a, V_{old}, V_{new}, ts_n, op_{ij})$, representing an object's attribute value change produced by an operation at a specific time. A delta contains an object identifier (oID) indicating the changed object, an attribute name (a) indicating the changed attribute, the old value of the attribute ($V_{old}$) before the execution of the operation, the new value of the attribute ($V_{new}$) created by the operation, a timestamp ($ts_n$) indicating the time of the new attribute value that is generated, and the identifier of the operation ($op_{ij}$) that has created this delta.

As an autonomous entity, a DEGS produces a local execution history.

**Definition 2 (DEGS Local Execution History):** A DEGS local execution history lh(degsID) is a three-element tuple, denoted as $lh(degsID) = <ts_s, ts_e, \delta(degsID)>$, where:

-$ts_s$ and $ts_e$ are the start time and end time of a DEGS execution history, respectively.

-$\delta(degsID)$ is a time-ordered sequence of deltas that are generated by operations that are executed at a specific DEGS during the time frame formed by $ts_s$ and $ts_e$, denoted as $\delta(degsID) = [\Delta(oID_A, a, V_{old}, V_{new}, ts_x, op_{ij}) \mid op_{ij}.degsID = degsID \text{ and } ts_s \leq ts_x \leq ts_e]$, such that the order of the list is based on the delta creation time $ts_x$.

A DEGS operation execution creates an execution context for an operation.

**Definition 3 (Operation Execution Context):** An operation execution context is a five-element tuple, denoted as $ec(op_{ij}) = <ts_s, ts_e, I, O, S>$. The operation execution context $ec(op_{ij})$ contains an execution start time ($ts_s$), end time ($ts_e$), input (I), output (O), and an execution state (S) as defined in the operation execution semantics in the service composition model [20, 22].

As the enclosing entity of an operation, a process also has its own execution context $ec(p_i) = <ts_s, ts_e, I, O, S>$ that corresponds to the definition of the operation execution context, but defines the context at the process level.

Each DEGS local execution history together with the process execution context of each process forms the global execution context.

**Definition 4 (Global Execution Context):** A global execution context is a time-ordered sequence of runtime context information for operations and processes that occur within a certain time frame formed by $ts_s$ and $ts_e$ ($ts_s < ts_e$). The global execution context is denoted as gec = [ec(entity) | (entity=$op_{ij}$ or entity=$p_i$) and ($ts_s \leq$ ec(entity).$ts_s <$ ec(entity).$ts_e \leq ts_e$)], where entity represents either an operation $op_{ij}$ or a process $p_i$.

Integrating local execution histories of DEGSs and the global execution context, the global execution history provides the basis to analyze write dependencies among concurrently executing processes by the provision of a time-ordered object access schedule.

**Definition 5 (Global Execution History):** A global execution history is an integration of individual DEGS execution histories within the time frame formed by start time $ts_s$ and end time $ts_e$, denoted as gh = <$ts_s$, $ts_e$, δg, gec>, where δg is a time-ordered sequence of deltas that are generated by distributed operation execution, denoted as δg = [Δ($oID_A$, a, $V_{old}$, $V_{new}$, $ts_x$, $op_{ij}$) | $ts_s \leq ts_x \leq ts_e$], and gec is the global execution context within the given time frame.

**Definition 6 (System Invocation Event Sequence):** A system invocation event sequence is an ordered list of events that are associated with the invocation of an execution entity based on the event occurrence timestamp. A system invocation event is denoted as $e_{entity}$, such that e indicates an invocation event and entity indicates an execution entity as the event source. System events include the invocation of a process ($e_{pi}$), an operation ($e_{opij}$), or a failure recovery event such as compensation of an operation ($e_{copij}$). A system invocation event sequence is denoted as $E_{seq}$ = [$e_{entity}$ | entity = $op_{ij}$ or entity = $p_i$].

Fig. 2 illustrates a process execution scenario with two concurrently executing processes $p_1$ and $p_2$ shown at the top of the Figure. The process $p_1$ is composed of operations $op_{11}$, $op_{12}$, $op_{13}$ and $op_{14}$. The process $p_2$ contains $op_{21}$ and $op_{22}$. These operations execute on two different sites $DEGS_1$ and $DEGS_2$. The operations $op_{11}$, $op_{21}$, $op_{13}$ and $op_{14}$ are provided by $DEGS_1$, while $op_{12}$ and $op_{22}$ are provided by $DEGS_2$. The horizontal coordinate indicates the time frame starting from $ts_s$ to $ts_e$.

The bottom part of the diagram shows the deltas generated by each operation, ordered by the timestamp of each delta. $DEGS_1$ contains deltas for objects X and Y, and $DEGS_2$ contains deltas for object Z. Each item represents a delta for a specific attribute of an object. For example, $x_2$ indicates a delta for object X where the subscript identifies the sequence of the delta with respect to the creation time. The process execution scenario illustrates:

- $DEGS_1$'s local execution history lh($DEGS_1$) contains all of the deltas that are created by operations that execute on $DEGS_1$, denoted as lh($DEGS_1$) = <$ts_s$, $ts_e$, δ($DEGS_1$)>, where δ($DEGS_1$) = [Δ(X, $attr_1$, $x_0$, $x_1$, $ts_1$, $op_{11}$), Δ(X, $attr_2$, $x_1$, $x_2$, $ts_2$, $op_{21}$), Δ(Y, $attr_1$, $y_0$, $y_1$, $ts_2$, $op_{21}$), Δ(X, $attr_1$, $x_2$, $x_3$, $ts_4$, $op_{13}$), Δ(X, $attr_1$, $x_3$, $x_4$, $ts_6$, $op_{14}$)].
- $DEGS_2$'s local execution history lh($DEGS_2$) = <$ts_s$, $ts_e$, δ($DEGS_2$)> , where δ($DEGS_2$) = [Δ(Z, $attr_1$, $z_0$, $z_1$, $ts_3$, $op_{12}$), Δ(Z, $attr_2$, $z_1$, $z_2$, $ts_5$, $op_{22}$)].
- The global execution history gh = <$ts_s$, $ts_e$, δg, gec>, where δg integrates δ($DEGS_1$) and δ($DEGS_2$), ordering each delta in time order. δg = [Δ(X, $attr_1$, $x_0$, $x_1$, $ts_1$, $op_{11}$), Δ(X, $attr_2$, $x_1$, $x_2$, $ts_2$, $op_{21}$), Δ(Y, $attr_1$, $y_0$, $y_1$, $ts_2$, $op_{21}$), Δ(Z, $attr_1$, $z_0$, $z_1$, $ts_3$, $op_{12}$), Δ(X, $attr_1$, $x_2$, $x_3$, $ts_4$, $op_{13}$), Δ(Z, $attr_2$, $z_1$, $z_2$, $ts_5$, $op_{22}$), Δ(X, $attr_1$, $x_3$, $x_4$, $ts_6$, $op_{14}$)].
- The system invocation event sequence indicates the invocation of each operation, ordered by operation start time. $E_{seq}$ = [$e_{op11}$, $e_{op21}$, $e_{op12}$, $e_{op13}$, $e_{op22}$, $e_{op14}$].

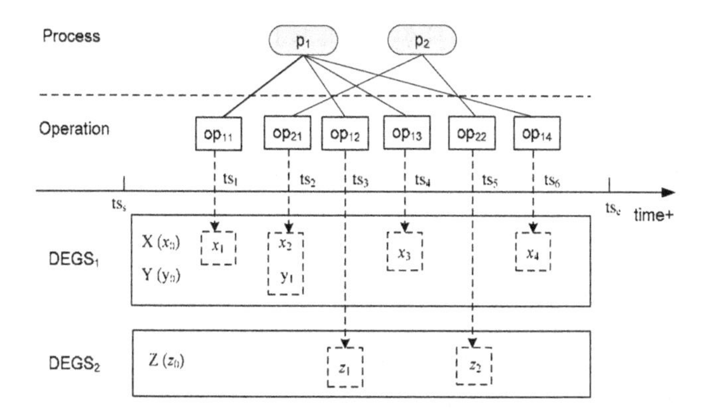


**Fig. 2.** A process execution scenario

### 4.2 Write Dependency

The merged deltas $\delta g$ of the global execution history can be used to identify the active processes that are dependent on a failed process.

**Definition 7 (Process-Level Write Dependency):** A process-level write dependency exists if a process $p_i$ writes an object $x$ that has been written by another process $p_j$ before $p_j$ completes ($i \neq j$). In this case, $p_i$ is write dependent on $p_j$ with respect to $x$, denoted as $p_i \rightarrow_w p_j$.

**Definition 8 (Operation-Level Write Dependency):** An operation-level write dependency exists if an operation $op_{ik}$ of process $p_i$ writes an object that has been written by another operation $op_{jl}$ of process $p_j$, denoted as $op_{ik} \rightarrow_w op_{jl}$. Operation-level write dependency can exist between two operations within the same process ($i = j$).

The operations that are write dependent on a specific operation $op_{jl}$ form $op_{jl}$'s write dependent set.

**Definition 9 (Operation-Level Write Dependent Set for an Operation):** An operation $op_{jl}$'s operational-level write dependent set is the set of all operations that are write dependent on $op_{jl}$, denoted as $wd_{op}(op_{jl}) = \{op_{ik} \mid op_{ik} \rightarrow_w op_{jl}\}$.

If $op_{ik}$ is write dependent on $op_{jl}$ ($op_{ik} \rightarrow_w op_{jl}$ ($i \neq j$)), the enclosing process of $op_{ik}$ is also write dependent on $op_{jl}$($p_i \rightarrow_w op_{jl}$).

Operation-level write dependencies can be derived from the global execution history gh. Assume two operations $op_{ik}$ and $op_{jl}$ have modified the same object $oID_A$. In gh, we observe $\delta g = [\ldots, \Delta(oID_A, a, V_{old}, V_{new}, ts_x, op_{jl}), \ldots, \Delta(oID_A, b, V_{old}, V_{new}, ts_y, op_{ik}), \ldots]$, where $ts_x < ts_y$. Then $\delta g$ indicates that at the operation level, $op_{ik} \in wd_{op}(op_{jl})$.

### 4.3 Potential Read Dependency

Since a DEGS does not capture read information, the global execution context can be used to reveal *potential* read dependency among operations. An operation $op_{ik}$ is potentially read dependent on another operation $op_{jl}$ under the following conditions:

1) $op_{ik}$ and $op_{jl}$ execute on the same DEGS, denoted as $ec(op_{ik}).degsID = ec(op_{jl}).degsID$.
2) the execution duration of $op_{ik}$ and $op_{jl}$ overlaps, or $op_{ik}$ is invoked after the termination of $op_{jl}$.

Fig. 3 shows various execution duration overlap scenarios of $op_{ik}$ and $op_{jl}$. In (a) and (b), $op_{ik}$ can be read dependent on $op_{jl}$. If $op_{ik}$ terminates before $op_{jl}$ is invoked, as shown in (c), it is not possible for $op_{ik}$ to be read dependent on $op_{jl}$.

The read dependency conditions in (a) and (b) can be expressed using operation execution start time and end time. For (a), $op_{ik}$ starts before $op_{jl}$ completes ($ec(op_{ik}).ts_s < ec(op_{jl}).ts_e$), and ends after $op_{jl}$ starts ($ec(op_{ik}).ts_e > ec(op_{jl}).ts_s$). For (b), $op_{ik}$ must start after $op_{jl}$ completes ($ec(op_{ik}).ts_s > ec(op_{jl}).ts_e$). Thus $op_{ik}$ is potentially read dependent on $op_{jl}$ if: 1) $ec(op_{ik}).ts_s < ec(op_{jl}).ts_s$ and $ec(op_{ik}).ts_e > ec(op_{jl}).ts_s$, or 2) $ec(op_{ik}).ts_s \geq ec(op_{jl}).ts_e$.

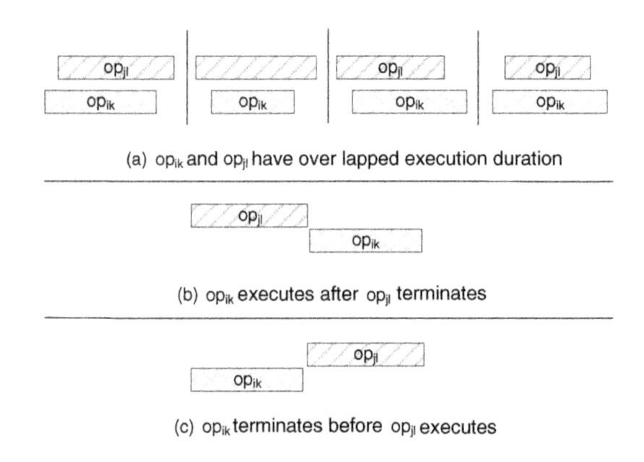


**Fig. 3.** Execution scenarios of $op_{ik}$ and $op_{jl}$

Potential read dependency can be defined at the process or operation levels.

**Definition 10 (Process-Level Read Dependency):** A process-level read dependency exists if a process $p_i$ reads an object $x$ that has been written by another process $p_j$ before $p_j$ completes ($i \neq j$). In this case, $p_i$ is read dependent on $p_j$ with respect to $x$, denoted as $p_i \rightarrow_r p_j$.

**Definition 11 (Operation-Level Read Dependency):** An operation-level read dependency exists if an operation $op_{ik}$ of process $p_i$ reads an object that has been written by another operation $op_{jl}$ of process $p_j$, denoted as $op_{ik} \rightarrow_r op_{jl}$ ($i \neq j$).

The operations that are potentially read dependent on an operation $op_{jl}$ form a set referred to as $op_{jl}$'s *read dependent set.*

**Definition 12 (Operation-Level Read Dependent Set for an Operation):** An operation $op_{jl}$'s operational-level read dependent set is a set of all operations that are potentially read dependent on $op_{jl}$, denoted as $rd_{op}(op_{jl}) = \{op_{ik} \mid op_{ik} \rightarrow_r op_{jl}\}$.

If $op_{ik}$ is read dependent on $op_{jl}$ ($op_{ik} \rightarrow_r op_{jl}$ ($i \neq j$)), the enclosing process of $op_{ik}$ is also read dependent on $op_{jl}$ ($p_i \rightarrow_r op_{jl}$).

Assume an operation $op_{ij}$ is compensated and the evaluation of process interference requires the identification of processes that are potentially read dependent on $op_{ij}$.

Suppose in the global execution context, we observe the system invocation event sequence $E_{seq}$ = [..., $e_{opij}$, $e_{opkm}$, $e_{opxy}$, $e_{copij}$]. $E_{seq}$ shows that $op_{km}$ and $op_{xy}$ are invoked after $op_{ij}$ starts ($ec(op_{km}).ts_s \geq ec(op_{ij}).ts_s$). If $op_{km}$ and $op_{ij}$ execute on the same DEGS ($ec(op_{km}).degsID = ec(op_{ij}).degsID$), then $op_{km} \rightarrow_r op_{ij}$. Thus the enclosing process $p_k$ of $op_{km}$ is potentially read dependent on $op_{ij}$.

### 4.4 Case Study

An online shopping case study has been used to illustrate the tracking of write dependencies among concurrent processes. The case study will be used again in Section 5 to illustrate the use of process interference rules.

The online shopping application contains typical business processes that describe the activities conducted by shoppers, the store, and vendors. For example, the process placeClientOrder is responsible for invoking services that place client orders and decrease the inventory quantity. The process replenishInventory invokes services that increase the inventory quantity when vendor orders are received.

Fig. 4 shows write dependency between two instances of the process placeClientOrder ($p_{c1}$ and $p_{c2}$) and an instance of the process replenishInventory ($p_r$). The top part of the diagram shows the executed operations in time sequence order. The bottom part shows deltas generated by an operation's execution. To keep the case simple, we use three objects (I from $DEGS_1$, CA and CB from $DEGS_2$) to demonstrate the existence of write dependency. All three process instances are related with the same inventory item identified by object I. Process instances $p_{c1}$ and $p_{c2}$ have created two different orders, identified as CA and CB.

The processes $p_{c1}$, $p_{c2}$, and $p_r$ start at different times. Process $p_{c1}$'s current operation is packOrder and $p_{c2}$'s current operation is decInventory. Process $p_r$ is in the process of backward recovery since the items that have entered the inventory by operation incInventory are recalled by the vendor due to quality problems. Process $p_r$'s recovery procedure contains compensating operations packBackOrder, decInventory and incInventory.

The global execution history for this scenario shows that write dependency exists among $p_{c1}$, $p_{c2}$, and $p_r$. The process $p_{c2}$ is write dependent on $p_{c1}$ and $p_r$. The process $p_r$ is write dependent on $p_{c1}$ and $p_{c2}$. The process $p_{c1}$ is not write dependent on $p_{c2}$ or $p_r$. The operation $p_{c2}$.decInventory decreases the quantity of the inventory item I that has been increased by $p_r$.decInventory. Then $p_r$.decInventory's compensation $p_r$.cop:decInventory modified the value of I by decreasing the value. This modification could potentially affect the execution of $p_{c2}$ if the number of required inventory items for the client order is no longer available. If the recovery procedure of $p_r$ does affect $p_{c2}$'s execution (i.e., process interference exists between $p_r$ and $p_{c2}$), $p_{c2}$ needs to be recovered. Otherwise $p_{c2}$ can keep running. Whether a cascading recovery is necessary is determined by process interference rules defined in Section 5.

Space does not permit the presentation of a full example of identifying read dependencies. It is important to remember that read dependencies are not derived from the schedule of delta values but are instead derived from the execution context, which identifies the overlapping execution timeframe of concurrent processes on the same service. As such, we refer to read dependencies as *potential* read dependencies since information about the specific data items that have been read is not available.

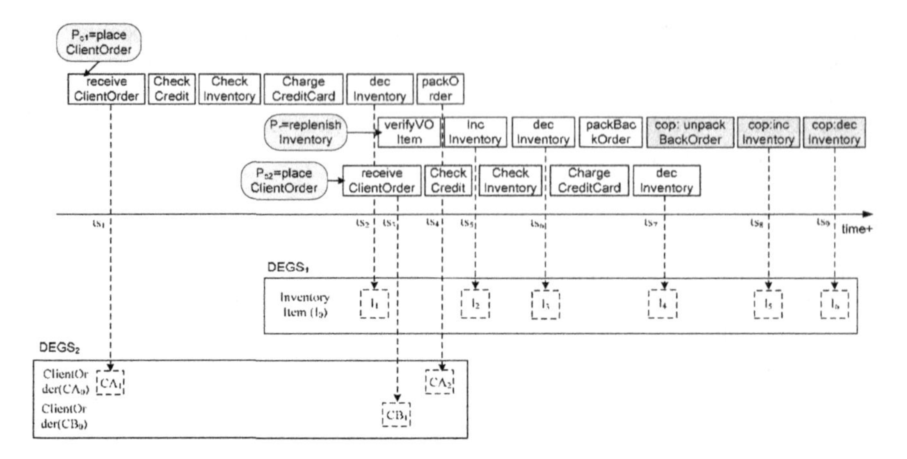


**Fig. 4.** Write dependency among multiple processes

Process interference rules as described in the next section can be used to determine if the failed process has accessed and modified any of the data items that are critical to the execution of the potentially read dependent process.

# 5 Process Interference Rules

The previous section illustrated the analysis of write dependencies and potential read dependencies from the global execution history. In a typical environment that supports relaxed isolation, the failure of one process causes cascaded recovery of other dependent processes. In the DeltaGrid environment, we use process interference rules (PIRs) to determine if the cascaded recovery of a dependent process is necessary. Process interference rules query the delta values of the global execution history, using user-defined application semantics to determine if an affected process should keep running or invoke its own recovery procedures. Section 5.1 defines the structure of a PIR and the way in which it queries the global execution history object model. Section 5.2 then presents a PIR example.

## 5.1 PIR Definition

A PIR is written from the perspective of an executing process ($p_e$) that is interrupted by the recovery of an unknown failed process ($p_f$). The interruption occurs because $p_e$ is identified as being write dependent or potentially read dependent on $p_f$. PIRs are expressed using an extended version of the Integration Rule Language (IRL) [13, 17], which was originally defined to provide a rule-based approach to component integration.

A process interference rule has four elements: event, define, condition, and action, as shown in Fig. 5. The triggering event of a PIR is a write dependent or a read dependent event. A write dependent event is triggered after the backward recovery of a failed process (failedProcess) if the failedProcess has at least one write dependent process. The format of a write dependent event is: <writeDependentProcessName>WriteDependency (failedProcess, wdProcess). The write dependent event contains the name of the write dependent process instance (writeDependentProcessName), and two parameters: the identifier of the failed process (failedProcess) and the identifier of the write dependent process instance (wdProcess). In Fig. 4, after the compensation of the failed process replenishInventory ($p_r$), a write dependency event placeClientOrderWriteDependency ($p_r$, $p_{c2}$) will be raised.

**create rule** *ruleName*
**event** *failureRecoveryEvent*
**define** [*viewName* as <OQL expression>]
**condition** [**when** condition]
**action** recovery commands

**Fig. 5.** Process interference rule structure

Similarly, a potential read dependent event is triggered if the failed process has at least one potentially read dependent process. The format of a read dependent event is <readDependentProcessName>ReadDependency (failedProcess, rdProcess), such that readDependentProcessName is the name of the read dependent process, and rdProcess is the identifier of the read dependent process instance.

Define declares variables to support condition evaluation. A condition is a Boolean expression to determine the existence of process interference based on application semantics. Define and condition use the object model presented below as an interface to access the global execution history.

Action is a list of recovery commands. The command could invoke backward recovery of a process (deepCompensate), re-execution of a process, backward recovery of an operation, or re-execution of an operation.

Fig. 6 presents the global execution history object model as the interface to access the global execution history in the declare and condition clauses of a PIR. The object classes in the model include Process, Operation, and Delta. The methods of each class as well as the attributes and relationships among these classes provide the basis for retrieving the deltas associated with a specific process and/or operation, and also for identifying read and write dependent processes and operations. For example, the method getDeltas(className) is used to retrieve all the deltas of a given class name that are created by an operation or a process. Similarly getDeltas(className, attrName) further limits the returned deltas to be of a given attribute name identified by attrName.

Since the condition evaluation of a process interference rule contains a query over deltas generated by the normal execution and recovery activities of a process, a process has several types of operations to support querying over deltas. For example, getDeltasBeforeRecovery returns deltas that are created by the process before any recovery activity is performed, and getDeltasByRecovery returns deltas that are created by the recovery activities of a process. Class name (className) and attribute name (attrName) can be used as parameters for these methods to limit the returned deltas to those associated

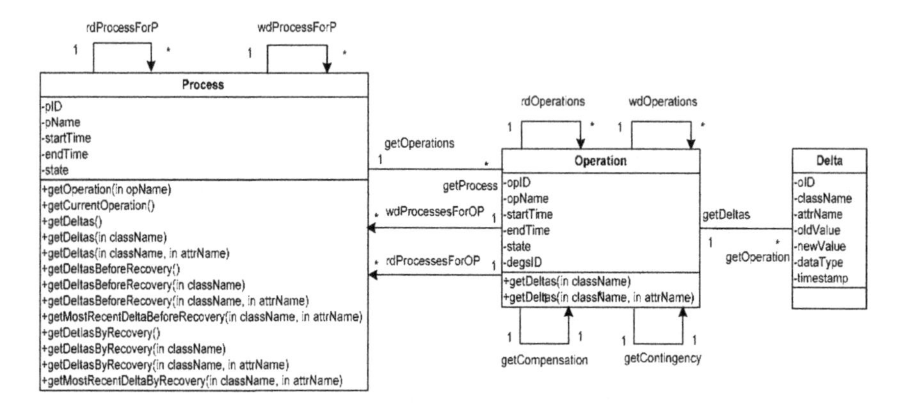


**Fig. 6.** Global Execution History Object Model

with a given class name and attribute name. getMostRecentDeltaBeforeRecovery(className, attrName) returns the most recent delta created by a process before the execution of any recovery activity, associated with a given class name and attribute name. Similarly, getMostRecentDeltaByRecovery(className, attrName) returns the most recent delta created by the recovery activities of a process.

## 5.2 Example

In Fig. 4, recall that the placeClientOrder process ($p_{c2}$) is write dependent on the process replenishInventory ($p_r$). As a result, $p_{c2}$ may process an order for a client based on an increase in inventory caused by $p_r$. If $p_r$ fails, however, after increasing the inventory for an item x that is also accessed in $p_{c2}$, then $p_{c2}$ may also be affected if there will no longer be enough items in inventory for item x. As a result, $p_{c2}$ may need to be backward recovered. Fig. 7 presents a PIR expressing the above application constraint.

The event is placeClientOrderWriteDependency(failedProcess, wdProcess). In this particular scenario, compensation of the process replenishInventory($p_r$) raises the event placeClientOrderWriteDependency ($p_r$, $p_{c2}$) which matches the event defined in the rule.

The define clause creates bindings for inventory items (decreasedItems) that have been decreased by the recovery of failedProcess. The select statement finds deltas related to the quantity of an inventory item (getDeltasByRecovery("InventoryItem", "quantity")) created by the recovery activities of failedProcess, finding items with decreased quantity by accumulating the changes on quantity for each item.

The condition has a when statement checking to determine if any decreased item (decItem) appears in the current client order. The evaluation is conducted by checking if any decItem is the same as the old of deltas associated with class InventoryItem and attribute quantity created by wdProcess.

The action is to backward recover wdProcess from the current operation. The action will be performed if the when statement in the condition evaluates to true.

Similar rules can be developed to check user-defined conditions in the case of potential read dependencies. In this case, a PIR can be used to 1) determine if the failed process intersects with the potentially read dependent process on critical data

```
Create rule   inventoryDecrease

Event         placeClientOrderWriteDependency(failedProcess, wdProcess)

Define        decreasedItems as
              select  item: fd.old
              from    fd in failedProcess.getDeltasByRecovery("InventoryItem", "quantity")
              group by item: fd.old
              having  sum(fd.newValue – fd.oldValue) < 0

Condition     when    exists decItem in decreasedItems:
                      decItem in
                      (select  d
                       from    d in wdProcess.getDeltas("InventoryItem", "quantity"))

Action        deepCompensate(wdProcess);
```

**Fig. 7.** PIR decreaseInventory for Process placeClientOrder

items by querying deltas on the process execution history, and 2) express application-specific conditions on the status of the critical data items after recovery of the failed process. If application constraints are violated, the read dependent process can invoke recovery procedures. Otherwise the read dependent process can resume execution. Further examples of PIRs for read and write dependencies can be found in [20].

# 6 Conclusion and Future Directions

This paper has presented a process dependency model together with the notion of process interference rules to resolve how the failure recovery of one process can potentially affect other concurrently executing processes in a distributed service composition environment. The resolution process is dependent on collecting information about data changes from delta-enabled grid services that are capable of forwarding these changes to a process history capture system [2]. We have implemented the PHCS [20, 21], which constructs a global execution history by merging deltas and process execution context. We have also fully implemented the PHCS object model to determine read and write dependencies and to invoke PIRs to determine if concurrent processes are affected by recovery of a failed process [20]. The research in [20] provides further details about simulation of the complete environment, with a performance evaluation of the implemented components. Our future work is focused on developing a distributed, peer-to-peer approach to the management and communication of deltas in resolving read and write dependencies, instead of forwarding deltas to a central process execution history component. We are also investigating the methodological issues associated with the use of process interference rules, as well as more dynamic approaches to the use of events and rules in the composition and recovery of distributed processes.

## References


1. Bhiri, S., O. Perrin, and C. Godart, "Ensuring required failure atomicity of composite Web Services," Proc. of the 14th int. conference on World Wide Web, 2005.
2. Blake, L., Design and implementation of Delta-Enabled Grid Services, MS Thesis, Deptment of Computer Science and Engineering, Arizona State Univ., (2005).
3. Web Services Transaction (WS-Transaction), http://www.ibm.com/developerworks/library/ws-transpec/, 2002.
4. de By, R., W. Klas, and J. Veijalainen, Transaction management support for cooperative applications. 1998: Kluwer Academic Publishers.
5. Eder, J. and W. Liebhart, "The Workflow Activity Model WAMO," Proc. of the 3rd Int. Conference on Cooperative Information Systems (CoopIs), 1995.
6. Garcia-Molina, H. and K. Salem, "Sagas," Proc. of the ACM SIGMOD Int. Conference on Management of Data, 1987.
7. Jin, T. and S. Goschnick, "Utilizing web services in an Agent Based Transaction Model (ABT)," Proc. of the the 1st Int. Workshop on Web Services and Agent-based Engineering (WSABE' 2003) held in conjunction with the 2nd Int. Joint Conference on Autonomous Agents and Multi-Agent Systems, 2003. Melbourne, Australia.
8. Kamath, M. and K. Ramamritham, "Failure handling and coordinated execution of concurrent workflows," Proc. of the IEEE Int. Conference on Data Engineering, 1998.
9. Limthanmaphon, B. and Y. Zhang, "Web Service composition transaction management," Proc. of the 15th Australasian database conference, 2004. Dunedin, New Zealand: Australian Computer Society, Inc.
10. Lin, F., Chang, H., "B2B E-commerce and enterprise Integration: the development and evaluation of exception handling mechanisms for order fulfillment process based on BPEL4WS," Proc. of the 7th IEEE Int. Conference on Electronic commerce, 2005.
11. Lomet, D., Z. Vagena, and R. Barga, "Recovery from "bad" user transactions," Proc. of the ACM SIGMOD Int. Conference on Management of Data, 2006.
12. Mikalsen, T., S. Tai, and I. Rouvellou, "Transactional attitudes: reliable composition of autonomous Web Services," Proc. of the Workshop on Dependable Middleware-based Systems (WDMS), part of the Int. Conference on Dependable Systems and Networks (DSN), 2002.
13. Peri, R.S., Compilation of the Integration Rule Language, MS Report, Department of Computer Science and Engineering, Arizona State University, (2002).
14. Shi, Y., L. Zhang, and B. Shi, "Exception handling of workflow for Web Services," Proc. of the 4th Int. Conference on Computer and Information Technology, 2004.
15. Singh, M.P. and M.N. Huhns, Service-Oriented computing. 2005: Wiley.
16. Tartanoglu, F., et al., "Dependability in the Web Services architecture," Proc. of the Architecting Dependable Systems, LNCS 2677, 2003.
17. Urban, S.D., et al., "The IRules Project: using active rules for the integration of distributed software components," Proc. of the 9th IFIP 2.6 Working Conference on Database Semantics: Semantic Issues in E-Commerce Systems, 2001.
18. Wachter, H. and A. Reuter, "The ConTract model," in Database transaction models for advanced applications, A. Elmagarmid, Editor. 1992.
19. Worah, D. and A. Sheth, "Transactions in transactional workflows," in Advanced transaction models and architectures, S. Jajodia and L. Kershberg, Editors. 1997, Springer.
20. Xiao, Y., Using deltas to support semantic correctness of concurrent process execution, Ph.D Dissertation, Department of Computer Science and Engineering, Arizona State Univ., (2006).

21. Xiao, Y., S.D. Urban, and S.W. Dietrich, "A process history capture system for analysis of data dependencies in concurrent process execution," Proc. of the 2nd Int. Workshop on Data Engineering Issues in E-Commerce and Services, 2006.
22. Xiao, Y., S.D. Urban, and N. Liao, "The DeltaGrid abstract execution model: service composition and process interference handling," Proc. of the 25th Int. Conference on Conceptual Modeling, 2006.
23. Zeng, L., et al., "Policy-driven exception-management for composite Web Services," Proc. of the 7th IEEE Int. Conference on E-Commerce Technology, 2005.

# A Survey of Comparative Business Process Modeling Approaches

Ruopeng Lu and Shazia Sadiq

School of Information Technology and Electrical Engineering, The University of Queensland, Brisbane, QLD, 4072, Australia
{ruopeng,shazia}@itee.uq.edu.au

**Abstract.** There has been a huge influx of business process modeling langu ages as business process management (BPM) and process-aware information systems continue to expand into various business domains. The origins of process modeling languages are quite diverse, although two dominant approaches can be observed; one based on graphical models, and the other based on rule specifications. However, at this time, there is no report in literature that specifically targets a comparative analysis of these two approaches, on aspects such as the relative areas of application, power of expression, and limitations. In this paper we have attempted to address this question. We will present both a survey of the two approaches as well as a critical and comparative analysis.

**Keywords:** business process management, workflows, business process modeling and analysis, rule-based workflows, graph-based workflows.

## 1 Introduction

Business process management (BPM) solutions have been prevalent in both industry products and academic prototypes since the late 1990s. It has been long established that automation of specific functions of enterprises will not provide the productivity gains for businesses unless support is provided for overall business process control and monitoring. Business process modeling is the first and most important step in BPM lifecycle [5], which intends to separate process logic from application logic, such that the underlying business process can be automated [32]. Typically, process logic is implemented and managed through a business process management system (BPMS) and application logic through underlying application components.

Business process modeling is a complicated process and it is obvious that different modeling approaches have their strengths and weaknesses in different aspects due to the variety of their underlying formalisms. There are many well-known problems regarding process modeling methodologies, such as the classic tradeoff between expressibility of the modeling language and complexity of model checking. Some languages offer richer syntax sufficient to express most relevant business activities and their relationships in the process model, while some provide more generic

W. Abramowicz (Ed.): BIS 2007, LNCS 4439, pp. 82–94, 2007.
© Springer-Verlag Berlin Heidelberg 2007

modeling constructs which facilitate efficient verification of the process model at design time. These have been prevalent in research prototypes (e.g., FlowMake [30], $ADEPT_{flex}$ [29], YAWL [3]), in commercial products (e.g., Tibco Staffware Process Suite [34], Oracle BPEL Process Manager [28], ILOG BPM [15]), as well as in industrial standard modeling languages (BPEL4WS [26], BPMN [27]).

Among the huge options of modeling languages, there have been methodical investigations in literature that attempt to address a variety of issues. These investigations involve a number of comparison techniques. First, in [33], an **empirical study** on process modeling success in industry is presented, where success factors of process modeling are generalized from multiple case studies of industry applications and the measure for effective process modeling is derived. Second, **ontological comparison** techniques utilize the semantic richness of an appropriate ontology as the benchmark for comparing process modeling languages. In [12], the interoperability of a business process specification (in particular, ebXML) is studied through a mapping from constructs in Bunge-Wand-Weber (BWW) ontology model to constructs in ebXML. Lastly, [20] presents a framework for selecting appropriate process modeling tools based on the **heuristics** collected from process modeling and business domain experts. The heuristics is used to provide quantifiable measure for indicating preferences on modeling tools selection.

We have conducted a study on the comparative business process modeling languages [23] based on a different comparison criteria. The scope of the comparison is on the most critical dimension of business process models, namely **control flow** perspective [30], from a selection of modeling approaches based on different theoretical foundations. The two most dominant foundations can be found in models bases on graphs and rules. The goal of comparison is to investigate, through the language representatives, the strengths and limitations of different theoretical foundations when being applied in business process modeling.

The focus of this paper is to summarize the comparison results and critical remarks reported in [23], and to facilitate future investigations and developments on business process modeling. In what follows, a survey of business process modeling approaches is first presented in section 2 to provide insights into current process modeling practices, based on which the comparison methodology is discussed in section 3. The comparison results, along with critical remarks are presented in section 4. Process modeling techniques in current commercial BPMS products are also briefly discussed in section 5 to present industry developments and trends. We conclude this paper and discuss possible future work in section 6.

## 2 A Survey on Business Process Modeling Approaches

The objective of process modeling is to provide high-level specification independent from the implementation of such specification. In this paper, we use the following definition for process modeling languages: A **process modeling language** provides appropriate syntax and semantics to precisely specify business process requirements, in order to support automated process verification, validation, simulation and process automation. The syntax of the language provides grammar to specify objects and their dependencies of the business process, often represented as a language-specific **process model**, while the semantics defines consistent interpretation for the process model to reflect the underlying process logic.

It is essential that a process model is properly defined, analyzed, and refined before being deployed in the execution environment. In a narrower scope, business process modeling can be referred to as workflow modeling, as workflow management systems (WFMS) provide equivalent functionalities to BPMS in business process modeling, analysis and enactment.

It has been found that there are two most predominant formalisms on which process modeling languages are developed, namely **graph-based** formalism and **rule-based** formalism. A graph-based modeling language has its root in graph theory or its variants, while a rule-based modeling language is based on formal logic.

## 2.1 Graph-Based Process Modeling Approaches

In a graph-based modeling language, process definition is specified in graphical process models, where activities are represented as nodes, and control flow and data dependencies between activities as arcs. The graphical process models provide explicit specification for process requirements.

Most graph-based languages have their root in Petri Net theory, which was applied in workflow modeling for the first time in 1977 by Zisman [40]. Many process modeling languages have been proposed based on different variants of Petri Nets to provide extra expressibility and functionality since then, including High Level Petri Nets [11], Low Level Petri Nets [36], and Colored Petri Nets [24]. More details can be found in a survey on Petri Net applications in workflow modeling by Jenssens et al [16].

The strengths of Petri Net based modeling approaches include formal semantics despite the graphical nature, and abundance of analysis techniques [1]. Formal methods [2] have been provided for specifying, analyzing and verifying the properties of static workflow structures (e.g. state transitions, deadlocks).

On the other hand, there are many graph-based modeling languages that carry similar advantages of Petri Net based languages, which also have simple and easy-to-understand syntax and semantics.

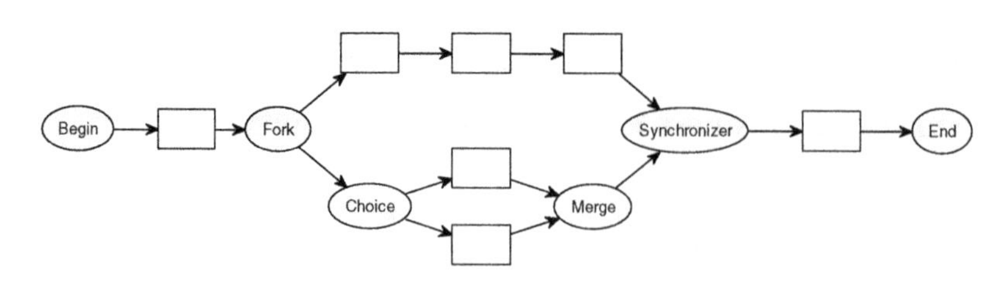


**Fig. 1.** A graph-based process model in FlowMake [30] syntax

For example, the syntax of FlowMake language [30] (cf. Fig. 1) contains 3 types of objects, task, coordinator and transition. A **task** represents a unit of work to be done (denoted by a rectangle). A **coordinator** is used to define how tasks are scheduled (denoted by an oval), which is further divided into begin, end, choices, merge, fork and synchronizer. A **transition** links any two nodes (task or coordinator) and is represented by a directed arc.

Table 1 provides a list of representative graph-based modeling approaches.

**Table 1.** Graph-based modeling approaches

| Authors | Approach | Brief Description |
|---|---|---|
| Sadiq & Orlowska [30] | FlowMake | FlowMake is a design and analysis methodology for workflow modeling, which includes a set of constraints to verify the syntactic correctness of the graphical workflow specifications by a graph-reduction algorithm. |
| Reichert & Dadam [29] | $\text{ADEPT}_{\text{flex}}$ | $\text{ADEPT}_{\text{flex}}$ is a graph-based modeling methodology which supports ad hoc changes to process schema. A complete and minimal set of change rules is given to preserve the correctness and consistency property, which provides a comprehensive solution for applying complex and dynamic structural changes to a workflow instance during its execution. |
| Casati et al [8] | Conceptual Modeling | The conceptual modeling approach divides a business process into workflow tasks (WT) and workflow (WF). A workflow execution architecture is proposed, which supports syntax-directed translation from workflow definition to executable active-rules, and provides operational semantics and an implementation scheme for many components of WFMS. |
| van der Aalst et al [3] | YAWL | YAWL is a Petri Net based workflow language, which supports specification of the control flow and the data perspective of business processes. The language has formal semantics that encompasses workflow patterns [26] to guarantee language expressibility. |
| Casati et al [9] | WIDE | WIDE is designed to support next-generation workflow management functionality in a distributed environment. The architecture is based on a commercial database management system as the implementation platform, with extended transaction management and active rule support. |
| Liu & Pu [22] | ActivityFlow | ActivityFlow provides a uniform workflow specification interface and helps increase the flexibility of workflow changes by supporting reasoning about correctness and security of complex workflow activities independently from the underlying implementation mechanism. |

## 2.2 Rule-Based Process Modeling Approaches

Rule-based formalisms have a wide area of applications in BPM domain, such as workflow coordination [18, 19] and exception handling [6]. Our consideration is limited to rule-based approaches where logical rules are used to represent structural, data and/or resource dependencies between task executions in business processes.

In a typical rule-based approach, process logic is abstracted into a set of rules, each of which is associated with one or more business activity, specifying properties of the activity such as the pre and post conditions of execution. The processing entity (process enactment mechanism) is a rule inference engine. At runtime, the engine examines data and control conditions and determines the best order for executing relevant business activities according to pre-defined rules. The typical enactment mechanism for general rule-based workflows is the rule inference engine, which is capable for evaluating current process events and triggering further actions (i.e., analogous to a workflow engine in common understanding [37]).

In [13], rule-based systems have been proposed as the first practical methodology to capture and refine human expertise, and to automate reasoning for problem solving. In 1990s, active database systems with the basic mechanism of Event-Condition-Action (**E-C-A**) rules have been applied to WFMS for coordinating task execution [14, 19]. The E-C-A paradigm has since been the foundation for many but not most rule-based process modeling languages. A basic E-C-A rule has the following syntax:

**ON** event **IF** condition **DO** action

An **event** specifies the triggering operation when a rule has to be evaluated, which indicates the transformation from one execution state to the other. An event can also be a simple change of task/process execution state (e.g., task "submit maintenance request" completes execution), or upon complex business process events (e.g. 2 days after product delivery, or on every 50 funding cases approved by the same project manager). The **condition** is the pre-condition to be checked before triggering any subsequent action, typically be the availability or of certain value of some process relevant data (e.g., requested funding $\geq$ 50,000). An **action** can be the execution of certain tasks, or triggering evaluation of other E-C-A rules. The result of executing an action can as well raise an event. A simple E-C-A rule can be extended with an additional **Else Action** to indicate the subsequent activities if the condition is not satisfied.

At the same time, software agent technology has been applied to model and enact workflows to provide flexibility and expressibility in business process automation. An agent is a piece of autonomous software that can perform certain actions to fulfill the design goal. An agency is a collection of autonomous artificial agents that communicate and work collaboratively to realize the process goal [17]. In the agent-based approach, the processing entity is an agency (i.e., a collection of software agents), and logical expressions are used to regulate the behaviors of autonomous agents. Rules are used to regulate actions of agents, or serve as the shared knowledge base among collaborating agents.

Table 2 lists some representative rule-based modeling approaches.

**Table 2.** Rule-based modeling approaches

| Authors | Approach | Brief Description |
|---|---|---|
| Knolmayer et al [21] | E-C-A Based Business Rules | The approach provides an E-C-A rule-based process model to serve as an integration layer between multiple process modeling languages, as well as functionality to support refinement of business rules. |
| Zeng et al [38] | $PLM_{flow}$ | $PLM_{flow}$ provides a set of business inference rules which is designed to dynamically generate and execute work-flows. The process definition is specified in business rule templates, which include backward-chain rules and forward-chain rules. The process instance schema is determined by the rule engine using backward-chain and forward-chain inference at runtime. |
| Kappel et al. [18] | Object-Rule-Role approach | The proposed framework supports reusability and adaptability using E-C-A rules to allocate tasks and resources in workflows. |
| Jennings et al [17] | ADEPT | The ADEPT system is an infrastructure for designing and implementing multi-agent systems for workflows. Process logic is expressed in the service definition language (SDL), which specifies services that give the agents sufficient freedom to take alternative execution paths at run-time to complete the process goal. |
| Müller et al [25] | AgentWork | AgentWork is a WFMS prototype based on agent technology, where agents are used for monitoring exceptional events. Reactive and predictive adaptations to workflow exceptions are defined through temporal E-C-A rules and automated by agents. A rule model is proposed for managing temporal E-C-A rules for workflows. |
| Zeng et al [39] | AgFlow | AgFlow contains a workflow specification model and the agent-based workflow architecture. The process definition is specified by defining the set of tasks and the workflow process tuple, where the control flow aspects can be reflected in the task specific E-C-A rules. |

## 3 Comparison Methodology

The following methodology has been developed to conduct the comparative analysis presented in this paper [23]. First, a selection of process modeling approaches have been identified in literature, and classified according to two most prominent formalisms (graph-based and rule-based) in process modeling, through which a variety of control flow functionalities are displayed (cf. Section 2).

Second, a minimal set of comparison criteria is identified for charactering the functionality of **control flow** capabilities, considering the design time (process modeling) and runtime (execution) requirements. The most important criteria are briefly discussed as follows:

- **Expressibility:** The expressive power of a process modeling language that is governed by its ability to express specific process requirements reflecting the purpose of process modeling and execution. A process model is required to be complete, which should contain structure, data, execution, temporal, and transactional information of the business process [30, 32].
- **Flexibility:** The ability of the business process to execute on the basis of a loosely, or partially specified model, where the full specification is made at runtime [31].
- **Adaptability:** Which is the ability of the workflow processes to react to exceptional circumstances, which may or may not be foreseen, and generally would affect one or a few process instances [31].
- **Dynamism:** The ability of the workflow process to change when the business process evolves. This evolution may be slight as for process improvements, or drastic as for process innovation or process reengineering [31].
- **Complexity:** The measures of the difficulty to model, analyze, and deploy a process model [7], as well as the support for the dynamic and changing business process.

Third, a functional comparison, followed by an empirical comparison is carried out. The **functional** comparison uses workflow control patterns [4] as the benchmark for examining the expressibility and complexity of the modeling languages. Workflow control patterns address the requirements for the modeling languages instead of the overall methodology to model business processes, which provide a mean to examine the expressibility of a particular process modeling language. FlowMake [30], $PLM_{flow}$ [38] and ADEPT [17] have been chosen as language representatives, where mappings from workflow patterns (including basic control flow, advanced branching and synchronization patterns [4]) to the language-specific model constructs of each language are carried out. While in the **empirical** comparison, the selected languages are used to model a real-life business process which involves complex control flow structures including multiple choices, parallel executions and indefinite looping.

Lastly, based on the results of comparisons, each language representative is analyzed according to the identified comparison criteria. The analysis results are then cross-referenced when critical remarks are given for the graph-based and rule-based modeling approach.

## 4 Comparative Analysis

In this section, key results for the comparative analysis are presented according to the study in [23].

When considering **design time** characteristics, graph-based languages have formal foundation in graph theory, which provides rich mathematical properties for the syntax and semantics and theoretical support. The process definition is robust and structurally sound. Besides, graph theory is well-known and has been well-studied, and most importantly, it is visual and hence intuitive and useful for all kinds of workflow designers (with or without technical background). On the other hand, while having their root in formal logic, rule-based languages are competitive to the graph-based rival in terms of mathematical soundness, model robustness and myriad of model checking techniques. However, rule-based modeling languages are inevitably more **complex**, reflected by the effort to specify, reason about and manage a large number of rules for complex business processes, which require reasonable proficiency in rule-based formalism.

In terms of **expressibility**, the richness of the graph-based language syntax allows explicit specifications for complex workflow constructs. The mapping from workflow control patterns [4] to FlowMake constructs is straightforward. The goal for graph-based process modeling is to provide a structurally and semantically correct process definition that is suitable for business process automation [32]. To ensure structural correctness all possible execution paths must be defined at design time and verified. Rule-based languages are able to represent all considered workflow control patterns, in that simple rule expressions connected by AND and OR operators are capable to express same constructs as those specified by basic and advanced graphical operators (e.g., choice, merge, fork, synchronizer, discriminator [4, 32]). Furthermore, rule-based languages are slightly more expressive than graph-based languages. An obvious example is the ability to specify the temporal requirement in addition, e.g., the relative deadline for a task execution.

When considering **runtime** characteristics such as ad hoc modification to workflow schemas and exception handling, the rule-based approach takes the advantage. The rigidity in graph-based models incurs problems of lack of **flexibility**, **dynamism** and **adaptability**, which compromise the ability of the graph-based processes to react to dynamic changes in business process and exceptional circumstances. Although there have been proposals [29, 31] to cope with such, e.g., to define a set of operation rules such that runtime modifications to current process model do not introduce any conflicts, the incorporation of change operations contributes to the overall modeling **complexity**. In the rule-based approach, the completeness requirement for process models is relaxed, which provides the ability to deploy partially-specified process definitions (in rules). This is supported by the enactment mechanism, the rule engine which performs logical inferences at runtime, i.e., to determine what to execute by evaluating relevant rules on certain process event. In addition, process logic of underlying business processes is externalized from the execution environment. As a result, runtime modifications to process definition can be realized by amending the existing set of rules (i.e., modify, insert and delete existing rules to reflect changes in process logic or to implement process improvement) without impacting the executing process instances.

Table 3 provides an overview of the above discussion.

**Table 3.** Summary of comparative approaches

| Criteria | Graph-Based Language | Rule-Based Language |
|---|---|---|
| Expressibility | – Able to express structure, data, and execution requirements.<br>– Most examined workflow patterns [23] can be expressed by the graph-based representative. | – Able to express structure, data, execution, as well as temporal requirements.<br>– Rule expressions can represent more workflow patterns than graph-based languages. |
| Flexibility | – Processes can only be executed on complete process models, in which all possible execution scenarios are explicitly specified. The conditions for each scenario must also be articulated in the process model a priori. | – More flexible as incomplete specification for task dependency is supported, e.g., in ADEPT [17], it is possible to specify *task T2 and T3 must execute after T1 has finished, either in parallel or serially* in a single rule expression without explicit specification on the conditions for parallel or serial execution. |
| Adaptability | – Exceptions rise if some executional behavior occurred that has not been defined in the process model. Exception handlings need to be defined through an additional set of policies (rules). | – Anticipated exceptions can be handled by specifying additional rules besides that for expressing regular process logic. |
| Dynamism | – Support for ad hoc changes to process model after deployment is limited.<br>– Requires defining a complete and minimal set of change operations in order to preserve the structural dependencies of the process model and running process instances. | – Rule expressions can be revised at runtime to realize ad hoc changes to process logic.<br>– Impact on running process instances is minimal as process logic is isolated from the executional environment.<br>– Process refinement can be rapidly implemented. |
| Complexity | – Modeling languages have more abstract syntax and simpler semantics, thus less complexity in model representation and verification.<br>– Runtime support for dynamic workflows is more complex as extra modeling constructs and verification effort is required. | – Model languages have a logical syntax and required some expertise when modeling.<br>– Process models have no visual appeal, verification process is more complex (however can be automated by logic reasoning engine).<br>– Change to process logic is realized by rule modification and hence less complex. |

## 5 Process Modeling Support in Commercial Products

Industry researches show that many process modelers are business process owners rather than technical personnel [10]. Graph-based languages have the visual appeal of being intuitive and explicit, even for those who have little or no technical background. However, rule-based modeling languages such as in ADEPT and $PLM_{flow}$ require good understanding of propositional logic and the syntax of logical expressions, thus are less attractive from the usability point of view. For this reason, most commercial BPMS, as well as current industrial standard including BPMN [27] by Business Process Management Initiative (BPMI) endorse the graph-based formalism as the process definition language. Examples of many commercial BPMS that use graphical process definition include SAP NetWeaver, Tibco Staffware Process Suite [34], and Utimus [35]).

On the other hand, the rule-based approach is often referred to as business rule management systems (BRMS). The common objective of BRMS is to integrate complex process logic into process model as rules, but externalize from BPMS in order to support dynamic changes. BRMS provides well-supported functions such that rule evaluation can be performed efficiently [15]. Example BRMS include Tibco iProcess Decisions [34], and ILOG JRules [15].

Many vendors advocate that rule-driven BPMS presents best solutions under current business requirements as a compromise of graphical appeal and the power of rules. A rule-driven BPMS is a superset of BRMS and BPMS, which provides rich development options for many aspects of BPMS development, of both procedural (graph) and declarative (rule) approach. The technique to realize this is to maintain a set of business rules in the external BRMS which can automate complex decision making in BPMS (i.e., when a complex decision has to be made to choose execution path). An example can be ILOG BPM [15], which is a BPMS enhanced by the functionality of the ILOG Rule Engine. Business processes are modeled in the BPM component, while the BRM component is responsible for formulate, compile, manage, and update business rules (in the E-C-A paradigm), invoke rule evaluation, and communicate with BPM component upon request to automate decision making.

## 6 Summary and Future Work

We proposed the use of control flow capabilities as the starting point for evaluating business process models based on two dominant approaches, namely graph-based and rule-based. This paper has presented the result of a critical and comprehensive analysis of these two prominent modeling approaches for business processes, with the focus on control flow capabilities. The presented survey gives an overview of process modeling languages developed using these two formalisms. The analysis of two approaches reviews their strengths and weakness in terms of expressibility, flexibility, adaptability, dynamism and complexity considerations. The intended future work includes a more detail empirical comparison on a selection of typical business scenarios, and with the focus on capabilities, other than control flow, of the process modeling approaches.

## References

1. van der Aalst. W.M.P.: Three Good reasons for Using a Petri-net-based Workflow Management System. In proceedings of the International Working Conference on Information and Process Integration in Enterprises (IPIC'96), Cambridge, Massachusetts (1996)
2. van der Aalst, W.M.P.: Verification of Workflow Nets. In proceedings of Application and Theory of Petri Nets. Lecture Notes in Computer Science, Vol. 1248 (1997) 407 – 426
3. van der Aalst, W.M.P., ter Hofstede, A.H.M.: YAWL - Yet Another Workflow Language. Information Systems, Vol. 30(4) (2005) 245-275
4. van der Aalst, W.M.P., ter Hofstede, A.H.M. Kiepuszewski, B., Barros, A.P.: Workflow Patterns. Distributed and Parallel Databases, Vol. 14(3) (2003) 5-51
5. van der Aalst, W.M.P., t. Hofstede, A.H.M., Weske, M.: Business Process Management: A Survey. In proceedings of Conference on Business Process Management (BPM 2003), Eindhoven, The Netherlands (2003)
6. Bae, J., Bae, H., Kang, S., Kim, Y.: Automatic Control of Workflow Processes Using ECA Rules. In IEEE Transactions on Knowledge and Data Engineering, Vol. 16(8) (2004)
7. Cardoso, J.: How to Measure the Control-Flow Complexity of Web Processes and Workflows. In: Fischer, L. (ed.) The Workflow Handbook, WfMC (2005) 199-212
8. Casati, F., Ceri, S., Pernici, B., Pozzi, G.: Conceptual Modeling of Workflows. In Proceedings of 14th International Conference of Object-Oriented and Entity-Relationship Modeling (OOER'95), Gold Coast, Australia (1995)
9. Casati, F., Grefen, P., Pernici, B., Pozzi, G., Sánchez, G.: A Specification Language for the WIDE Workflow Model. Technical report, University of Twente (1996)
10. Delphi Group: BPM 2005 Market Milestone Report. (2005) URL: http://www.delphigroup.com/research/whitepapers.htm.
11. Ellis, C.A., Nutt, G.J.: Modeling and Enactment of Workflow Systems. in Proceedings of 14th International Conference of Application and Theory of Petri Nets, Chicago, USA (1993)
12. Green, P., Rosemann, M, Indulska, M.: Ontological Evaluation of Enterprise Systems Interoperability Using ebXML. In IEEE Transactions on Knowledge and Data Engineering, Vol. 17(5) (2005)
13. Hayes-Roth, F.: Rule-Based Systems. Communications of the ACM, Vol. 28(9) (1985) 921–932
14. Herbst, H., Knolmayer, G., Myrach, T., Schlesinger, M.: The Specification Of Business Rules: A Comparison of Selected Methodologies. In Verrijn-Stuart A.A., Olle, T.W. (eds.) Methods and Associated Tools for the Information System Life Cycle, North-Holland, IFIP–18 (1994)
15. ILOG: ILOG Components for Business Process Management Solutions. White Paper, ILOG (2006) URL: www.ilog.com
16. Jenssens, G.K., Verelst, J., Weyn, B.: Techniques for Modeling Workflows and Their Support of Reuse. In Business Process Managements - Models, Techniques and Empirical Studies, Lecture Notes in Computer Science, Vol. 1806 (2000) 1-15
17. Jennings, N.R., Faratin, P., Norman, T.J., O'Brien, P., Odgers, B., Alty, J.L.: Implementing a Business Process Management System using ADEPT: a Real-World Case Study, International Journal of Applied Artificial Intelligence, Vol. 14 (2000) 421-463
18. Kappel, G., Rausch-Schott, S., Retschitzegger, W.: A Framework for Workflow Management Systems Based on Objects, Rules and Roles. In ACM Computing Surveys, Vol. 32 (2000)

19. Kappel, G., Rausch-Schott, S., Retschitzegger, W.: Coordination in Workflow Management Systems – a Rule-Based Approach. In Conen, W., Neumann, G. (eds.) Coordination Technology for Collaborative Applications - Organizations, Processes, and Agents, Springer LNCS 1364 (1998) 99-120
20. Kaschek, R., Pavlov, R., Shekhovtsov, V.A., Zlatkin, S.: Towards Selecting Among Business Process Modeling Methodologies. In Proceedings of 9th International Conference on Business Information Systems (BIS2006), Klagenfurt, Austria (2006)
21. Knolmayer, G., Endl, R., Pfahrer, M.: Modeling Processes and Workflows by Business Rules. In van der Aalst W.M.P (Eds,) Business Process Management, LCNS 1806, Springer-Verlag Berlin Heidelberg (2000) 16-29
22. Liu, L., Pu, C.: ActivityFlow: Towards Incremental Specification and Flexible Coordination of Workflow Activities. In proceedings of 16th International Conference on Conceptual Modeling / the Entity Relationship Approach (ER 97), Los Angeles, USA (1997) 169-182
23. Lu, R.: Comparison of Workflow Modeling Approaches. Honors Thesis, School of Information Technology and Electrical Engineering, The University of Queensland, Brisbane (2004)
24. Merz, M., Moldt, D., Muller K., Lamersdorf, W.: Workflow Modeling and Execution with Colored Petri Nets in COSM. In proceedings of the Workshop on Applications of Petri Nets to Protocols within the 16th International Conference on Application and Theory of Petri Nets (1995) 1-12
25. Müller, R., Greiner, U., Rahm, Erhard.: AgentWork: a Workflow System Supporting Rule-Based Workflow Adaptation. In Data & Knowledge Engineering, Vol. 51(2) (2004) 223-256
26. OASIS: Business Process Execution Language for Web Services Version 1.1 (BPEL4WS 1.1) Specification. (2006)
27. Object Management Group: Business Process Modeling Notation (BPMN) Specification 1.0 (2006)
28. Oracle: Building Flexible Enterprise Processes Using Oracle Business Rules and BPEL Process Manager, An Oracle White Paper, Oracle (2005) URL: www.oracle.com
29. Reichert, M., Dadam, P.: $\text{ADEPT}_{\text{flex}}$ - Supporting Dynamic Changes of Workflows without Losing Control. Journal of Intelligent Information Systems, Special Issue on Workflow Management, Vol. 10 (1998) 93-129
30. Sadiq W., Orlowska, M.: On Capturing Process Requirements of Workflow Based Business Information System. In proceedings of 3rd International Conference on Business Information Systems (BIS '99), Poznan, Poland (1999)
31. Sadiq, S., Sadiq, W., Orlowska, M.: A Framework for Constraint Specification and Validation in Flexible Workflows. Information Systems, Vol. 30(5) (2005)
32. Sadiq, W., Orlowska, M.: On Correctness Issues in Conceptual Modeling of Workflows. In proceedings of European Conference on Information Systems (ECIS '97), Cork, Ireland (1997)
33. Sedera, W., Rosemann, M, Doebeli, G.: A Process Modeling Success Model: Insights from a Case Study. In Ciborra CU, Mercurio R, de Marco M, Martinez M, Carignani A (eds.), Proceedings of the 18th European Conference on Information Systems, Naples, Italy (2003)
34. Tibco: Enhancing BPM with a Business Rule Engine, Tibco White Paper, Tibco, (2006) URL: http://www.tibco.com
35. Ultimus: Adaptive Discovery: Accelerating the Deployment and Adaptation of Automated Business Processes. White Paper, Ultimus Inc. (2004) URL: www.ultimus.com
36. Wikarski, D.: An Introduction to Modular Process Nets. Technical Report TR-96-019 International Computer Science Institute (ICSI), Berkeley, CA, USA (1996)

37. Workflow Management Coalition: Workflow Process Definition Interface - XML Process Definition Language, (2006) URL: http://www.wfmc.org/standards/docs.htm
38. Zeng, L., Flaxer, D., Chang, H., Jeng, J.: $PLM_{flow}$: Dynamic Business Process Composition and Execution by Rule Inference. In proceedings of 3rd VLDB Workshop on Technologies for E-Services (TES'02), Hong Kong, China (2002)
39. Zeng, L., Ngu, A., Benatallah, B., O'Dell, M.: An Agent-Based Approach for Supporting Cross-Enterprise Workflows. In proceedings of the 12th Australasian Database Conference (ADC2001) (2001)
40. Zisman, M.D.: Representation, Specification and Automation of Office Procedures. PhD Thesis, Wharton School of Business, University of Pennsylvania (1977)

# Web Service Discovery Based on Past User Experience

Natallia Kokash, Aliaksandr Birukou, and Vincenzo D'Andrea

DIT - University of Trento, Via Sommarive, 14, 38050 Trento, Italy
{kokash,birukou,dandrea}@dit.unitn.it

**Abstract.** Web service technology provides a way for simplifying interoperability among different organizations. A piece of functionality available as a web service can be involved in a new business process. Given the steadily growing number of available web services, it is hard for developers to find services appropriate for their needs. The main research efforts in this area are oriented on developing a mechanism for semantic web service description and matching. In this paper, we present an alternative approach for supporting users in web service discovery. Our system implements the implicit culture approach for recommending web services to developers based on the history of decisions made by other developers with similar needs. We explain the main ideas underlying our approach and report on experimental results.

**Keywords:** Web Service Discovery, Recommendation Systems, Implicit Culture.

## 1 Introduction

The state-of-the-art in business integration is defined by implementation of the service-oriented vision using web service technology. Web services are loosely coupled, distributed entities that can be described, published, discovered and invoked via the web infrastructure. Three main standards in this area include Web Service Description Language (WSDL) for presenting service interfaces, Universal Description, Discovery and Integration (UDDI) registries for publishing, and Simple Object Access Protocol (SOAP) for message transporting.

With ever increasing number of available web services it is problematic to find a service with required functionality and appropriate quality characteristics. Most of the proposals in the area of web service discovery rely on logically precise semantic descriptions of web services by providers [1][2][3]. Such approaches are efficient only if providers publish exhaustive service specifications. Tools for automatic or semi-automatic semantic annotation can significantly reduce required amount of work, but, in principle, the consumer must trust the provider to deliver the service fully compliant with the description. Additionally, web services or providers can be evaluated by a trusted party, i.e., by a specialized unbiased agency that tests web services, verifies their descriptions (whether there is a discrepancy between specified and implemented features), publishes Quality

W. Abramowicz (Ed.): BIS 2007, LNCS 4439, pp. 95–107, 2007.
© Springer-Verlag Berlin Heidelberg 2007

of Service (QoS) data, etc. This solution is relatively expensive and inefficient due to its rather static nature. Automated central monitors are complex, and either provide limited monitoring facilities or require involvement of domain-specific logic for verifying web service behavior [4][5].

On the other hand, there are service clients who already have experience in using web services and therefore can help in selecting services with adequate quality. This principle is widely used by (collaborative) recommendation and reputation systems [6][7]. Often web services are oriented not on public use but aim at enabling easy information exchange between a set of partner organizations. Since web services belong to different domains, only a specific set of web services is interesting for a particular consumer. A group of clients with common interests form a virtual *community* where they can exchange the experience, i.e., the knowledge gained after having interaction with a web service. Being a member of such a community can help to reduce the information overload and enhance web service discovery and selection facilities.

In this paper, we present a system for discovery of web services. The system is based on the implicit culture framework [8] and helps developers make a decision about which services to use by getting suggestions from the community. The implicit culture framework has been implemented in the form of a domain-independent meta-recommendation service, the *IC-Service*, that uses web service technology and can be tuned via configuration interface [9]. In our approach, no communication between members of the community is needed, and no explicit ratings of web services are required.

The paper is organized as follows. In Section 2, the basic idea of implicit culture is presented and the configuration of the *IC-Service* for our application is explained. Section 3 provides implementation details, while Section 4 presents experimental results. Related work is analyzed in Section 5, and Section 6 draws conclusions and outlines future work.

## 2 Implicit Culture

This section presents an overview of the general idea of the implicit culture framework and the System for Implicit Culture Support (SICS). The SICS provides the basis for the *IC-Service* we have used to provide recommendation facilities.

The behavior of a person in an unknown environment is far from optimal. There exist many situations where it is difficult to take the right decision due to the lack of knowledge. This might not be the case for experienced people who have previously encountered similar problems and have acquired the necessary knowledge. The knowledge about acting effectively in the environment is often implicit (i.e., highly personalized) and specific to the community. Therefore, this knowledge could be referred to as a *community culture*. The idea behind the implicit culture framework is that it is possible to elicit the community culture by observing the interactions of people with the environment and to encourage the newcomer(s) to behave in a similar way. Implicit culture assumes that agents perform actions on objects, and the actions are taken in the context of situations, so agents perform situated actions. The "culture" contains information about actions and their relation to situations, namely which actions are usually taken by the observed group and in which situations. This information is then used to produce recommendations for other agents. When newcomers start to behave similarly to the community culture, it means that a *knowledge transfer* occurred. The goal of the SICS is to perform such transfer of knowledge.

The basic architecture for the SICS [8] consists of the following three components: the *observer* module, which records the actions performed by the client during the use of the system; the *inductive* module, which analyzes the stored observations and implements data mining techniques to discover behavior patterns; the *composer* module, which exploits the information collected by the observer module and analyzed by the inductive module in order to produce recommendations.

In terms of our problem domain, the observer saves the following information: the user request (a textual description and characteristics of a required web service), the context in which the request occurred, the services proposed as a solution, the service chosen and invoked by the user, and, finally, the result of the invocation (successful web service invocation, exception raised, etc.). Then, the request-solution pairs that indicate which web services are selected for which requests could be determined by analyzing the interaction history between users and the system. This step is performed by the inductive module and it is now omitted. Finally, the composer module matches the user request with web services by calculating the similarity between the request given by the user and the requests that users provided previously, and by selecting web services chosen for the most similar past requests.

The described schema is implemented within the *IC-Service* [9] that provides recommendation facilities based on implicit culture. The configuration of the *IC-Service* for our application is shown in Figure 1. Along with the *SICS core*, which forms the main part of the service, the *IC-Service* includes the other two important components: the *remote client* and the *remote module*. The recommendation system is available as a web service that can be accessed via the remote client. The remote client presents a wrapper that hides protocols used for information exchange with the SICS. The remote module defines protocols for information exchange with the remote client from the direction of the SICS and converts the objects of the SICS core in the format compatible with these protocols. We refer the reader interested in the details of the modules to the description of the *IC-Service* [9].

In the current version of our system we do not use the inductive module to infer new behavior patterns, but predefine them manually. The *IC-Service* allows for the adjustment of a recommendation strategy through configuring theory rules. A theory rule is defined as follows:

$$\text{if } \textit{consequent}(\text{predicates}) \text{ then } \textit{antecedent}(\text{predicates}),$$

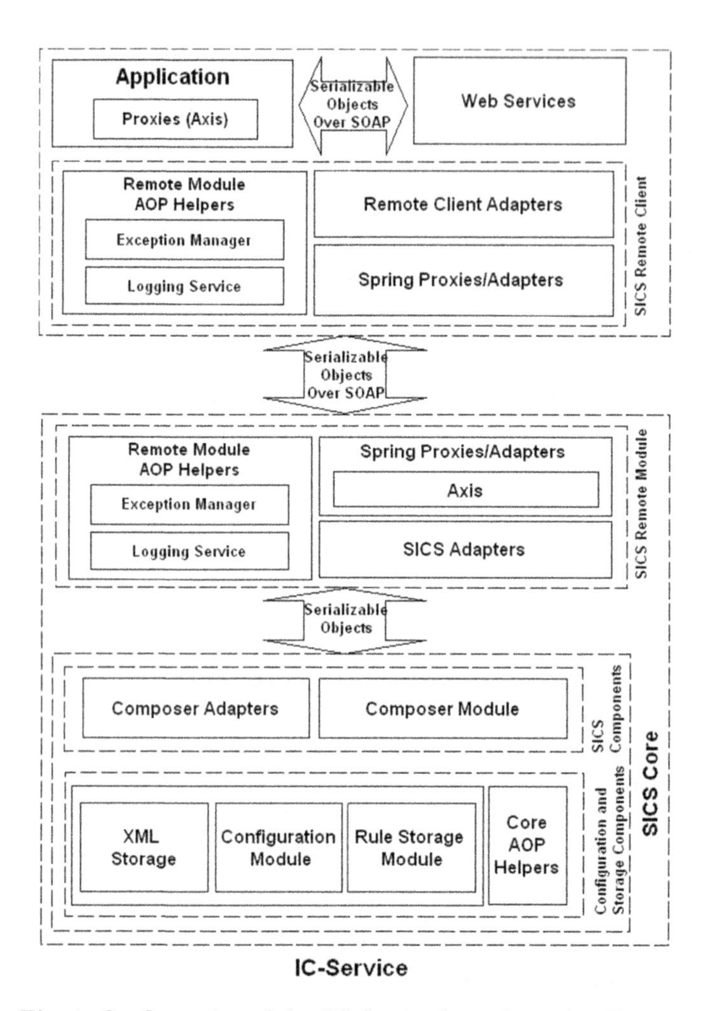


**Fig. 1.** Configuration of the *IC-Service* for web service discovery

where predicates describe either conditions on observations (action-predicates) or conditions on time (temporal-predicates). Each predicate may include several action-rules, which specify patterns on actions, agents, objects, scenes[1], and their attributes. Observations from the SICS storage are analyzed by the composer module according to these patterns. For matching of the discovered observations a similarity algorithm must be defined. The *IC-Service* provides a simple matching method that compares pairs of observations using predefined similarity values for their elements (actions, objects, etc.). These values can be configured for each particular type or for each particular instance of the element. A similarity threshold for matching also can be tuned. In addition, a plug-in mechanism enables the possibility of involving other similarity algorithms.

[1] A *scene* is the architectural abstraction of a situation.

## 3 A System for Web Service Discovery

This section gives a description of the process of web service discovery within the system.

The motivation for adopting the implicit culture approach for web service discovery stems from the difficulty of developers in finding and selecting web services suitable for their applications [10]. The system is intended for the use by a virtual community, giving suggestions about web services suitable for this community. In our domain, developers and their applications perform actions on web services. Types of actions analyzed by the SICS are presented in Table 1 and will be explained later in more detail. Actions, agents and objects also may have multiple attributes, i.e., features helpful for their analysis. For example, information about a web service (id, name, provider, etc.) is stored as an attribute of an object *operation*. The description of the complete set of the stored attributes is not needed for understanding the current paper and therefore is omitted here.

**Table 1.** Actions observed by the system

| Action | Agents | Objects |
|---|---|---|
| invoke | application | operation, input |
| get_response | application | operation, output |
| raise_exception | application | operation, input |
| provide_feedback | application, developer | operation, rate |
| submit_request | application, developer | request |

In order to use the system, each user must install a remote client. The goal of the remote client is to communicate with the SICS, in particular, forward user requests and store observations about user actions, applications and behavior of web services. To enable observations of interaction with a web service we have extended the `JavaStubsWriter` class of the open-source Apache Axis framework[2]. This class generates stubs for web service invocation. The modification that we implemented allows the stubs to report the information about the communication between a user application and a web service to the *IC-Service*, using the remote client. Thus, to join a virtual community that shares experience in retrieving of web services, the user must (1) install the remote client, (2) generate stubs for service invocations using the modified version of the Axis tool. No further intervention, user-to-user or user-to-system communications are required, except for submitting requests. If the user does not need to search for new web services, the system can be used for service monitoring on the client side. Run-time web service monitoring is essential for real-world service-oriented systems where control of service quality is needed [11][12].

The SICS remote client provides an interface for the user to access the system by submitting requests. Request may include textual description of the goal, name of the desired operation, description of its input/output parameters, description of a desired web service and its features (provider, etc.). By configuring

[2] http://xml.apache.org/axis/

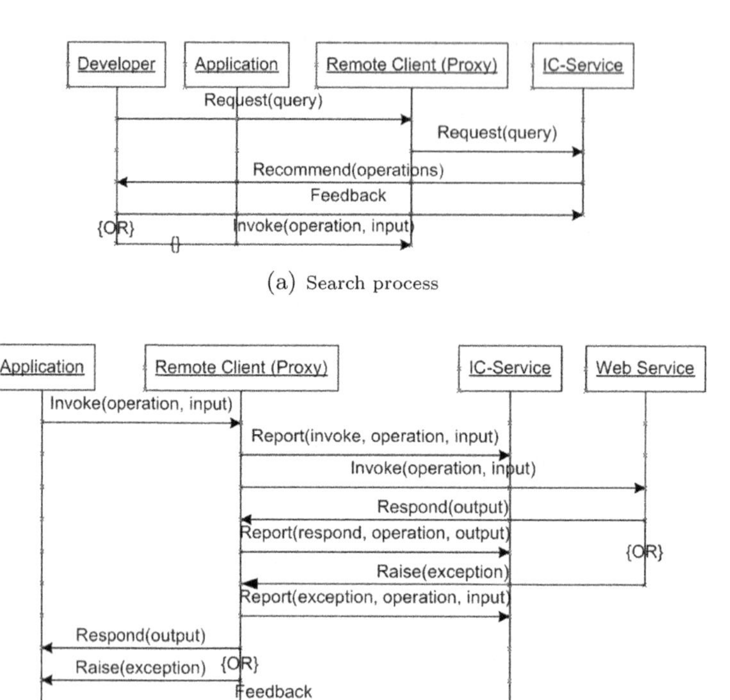


(a) Search process

(b) Monitoring process

**Fig. 2.** Sequence diagrams

the similarity algorithm it is possible to define whether these requirements are considered as strict (only services that meet them are recommended) or as preferred (services that better fit the request than others are recommended).

The search scenario is given in Figure 2(a). A user submits a request via the remote client, from where the request is forwarded to the *IC-Service*, and a list of recommended services is returned. The feedback is collected via the optional *provide_feedback* action, which expresses the level of user satisfaction with the result, or through the *invoke* action, which marks a service as suitable for the request. If the user decides to use one of the services, the further information is acquired. The *get_response* action marks a service as available and the *raise_exception* action signals that the service is not available or faulty. The monitoring process is shown in Figure 2(b). In short, when the application invokes some operation provided by a web service, the remote client reports to the *IC-Service* on the *invoke* action. Similarly, when the web service sends a response message or raises an exception, the remote client reports to the *IC-Service* on the *get_response* or *raise_exception* action, respectively. Having received a response message, the user application can generate a feedback based on an extra-knowledge about the expected result (e.g., the feedback is positive if the meaningful output has been obtained, etc.)

The *IC-Service* processes the query from the system within two steps. In the first step, the action contained in the query, i.e. the *submit_request* action, is matched with the theory to determine the next action that must follow, i.e. the *invoke* action. In the second step, the SICS finds situations where the invoke action has been previously performed, determining web service operations used for similar requests in the past. In this step, the similarity between the current user request and the previously submitted requests is calculated. As a result, the *IC-Service* returns a set of services that have been used for similar requests in the past.

Let us illustrate how the search process takes place in our example. The user submits the request represented by the following query:

Goal : Get weather forecast for Rome (this is in Italy);
Operation : Get weather;
Input : City name, country name;
Output : Weather forecast (temperature, humidity, etc.).

The *IC-Service* matches the request action with the theory, which contains rules of the following form:

if *submit_request*(request) then *invoke*(operation-X(service-Y), request).

This means that the *invoke* action must follow the *submit_request* action and both actions are related to the same query. The SICS matches the request action with the request part of the theory, and searches for situations where the invoke action has been performed. It finds the following situations:

| ID | Action | Goal | Operation |
|---|---|---|---|
| 1 | invoke | get weather report for all major cities around the world | getWeather (service = GlobalWeather) |
| 2 | invoke | get conversion rate from one currency to another currency | conversionRate (service = CurrencyConvertor) |
| 3 | invoke | return the weather for a given US postal code | getWeatherByZip (service = DOTSFastWeather) |

As a result, the SICS recommends that the user invokes either getWeather operation of the GlobalWeather web service or getWeatherByZip operation of the DOTSFastWeather web service. Having analyzed the proposed results, the user invokes the former operation. After observing the *invoke* action, this service will be marked as suitable for the above query. Further, it may be considered relevant for requests asking for information about Italy.

Note, that instead of the *invoke* action, the *get_response* action can be put in the theory. In this case, only web services invoked successfully at least once will be considered. The same mechanism can be used for reputation-based web service filtering: users can explicitly rate services using *provide_feedback* action.

The *IC-Service* enables saving various information and defining inferring rules and similarity measures on them. In the context of the presented system, the *IC-Service* is used to collect reports of service invocations by clients, to keep previous user requests, and to define similarities between users based on the information about the services they use. The implemented schema can be extended to store other important information about web services, such as cases of Service Level Agreement (SLA) [13] violation or measurements of QoS parameters.

This information further can be dynamically involved in the refined web service discovery and selection through defining new theory rules. The meta-data about web services, augmented with the help of our system, can be further used for hybrid web service matching algorithms [14]. In the simplest scenario, the remote client can submit user requests to the UDDI registry through `UDDI4J`[3] API in order to get information about recently appeared web services.

# 4 Experimental Evaluation

The goal of the experiment is to evaluate the performance of the system. We have defined user profiles in order to simulate the behavior of real users. A user profile contains a set of queries and a set of web service operations relevant to these queries. The set of queries is exploited to simulate the request-generation behavior by choosing and submitting a query randomly, while the set of web service operations is used to simulate the result-selection behavior by selecting one of the operations. Each query consists of a brief natural language description of the desired operation. The intuition behind the user profile is as follows: the user submits a request for a service operation. After getting suggestions, (s)he will invoke one of the operations (s)he considers relevant. This invocation is monitored by the remote client of the *IC-Service*. The choice of the user that submits a request to the system in a given moment is random.

**Table 2.** Experimental collection

| Category | Web service | Operation |
|---|---|---|
| Currency | {http://www.webserviceX.NET/}CurrencyConvertor | conversionRate |
| | {http://www.xmethods.net/sd/}CurrencyExchangeService | getRate |
| | {http://www.myasptools.com/}currencyWS | getRate |
| | {http://www.xignite.com/services/}XigniteCurrencies | getLatestCrossRate |
| DNA | {http://www.themindelectric.com/wsdl/Blast/}Blast | searchSimple |
| | {http://www.themindelectric.com/wsdl/Fasta/}Fasta | searchSimple |
| | {http://www.themindelectric.com/wsdl/TxSearch/}TxSearch | searchSimple |
| | {http://www.themindelectric.com/wsdl/SRS/}SRS | searchSimple |
| SMS | {http://www.webserviceX.NET}SendSMSWorld | sendSMS |
| | {http://www.sms.mio.it/webservices/sendmessages.asmx} | sendSMS |
| | {http://ws.AcrossCommunications.com/}SMS | SendEx |
| | {http://SMSServer.dotnetISP.com}ServiceSMS | sendSmsText |
| Weather | {http://www.webserviceX.NET}GlobalWeather | getWeather |
| | {http://ejse.com/WeatherService/}Service | getWeatherInfo |
| | {http://www.myasptools.com/}WeatherFetcher | getWeather |
| | {http://www.serviceobjects.com/}DOTSFastWeather | getWeatherByZip |
| ZIP | {http://www.jasongaylord.com/webservices/zipcodes}ZipCodes | zipCodesFromCityState |
| | {http://ripedev.com/xsd/ZipCodeResults.xsd}ZipCode | cityToZipCode |
| | {http://webservices.eraserver.net/}ZipCodeResolver | shortZipCode |
| | {http://www.webserviceX.NET}USZip | getInfoByCity |

The quality of recommendations is measured using the precision, recall and F-measure [15]:

$$\text{Precision} = \frac{\text{Relevant} \cap \text{Retrieved}}{\text{Retrieved}} \quad \text{Recall} = \frac{\text{Relevant} \cap \text{Retrieved}}{\text{Relevant}} \quad \text{F} = \frac{2 * \text{Precision} * \text{Recall}}{\text{Precision} + \text{Recall}}.$$

[3] http://uddi4j.sourceforge.net/doc.html

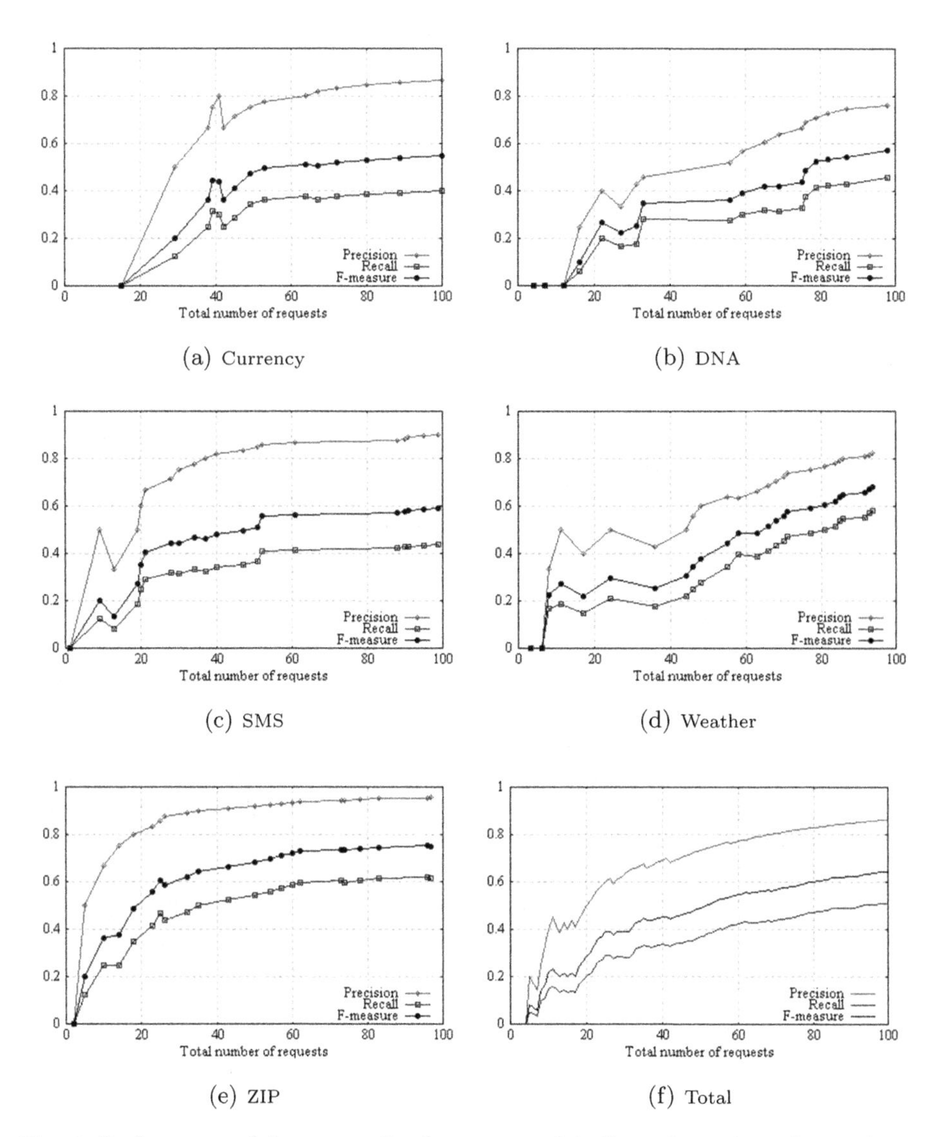


**Fig. 3.** Performance of the system For five groups of similar web services, the precision, recall and F-measure are calculated for the first 100 requests submitted to the system by four clients

The precision measures the fraction of relevant items among those recommended. The recall measures the fraction of the relevant items included in the recommendations. The F-measure is a tradeoff between these two metrics.

Since internally suggestions are filtered by the composer module of the SICS within the *IC-Service*, the precision in our case depends on the similarity measure adopted in the composer module. The recall in our settings demonstrates how the system learns from past experience.

In the experiment we used the Vector Space Model with Term Frequency - Inverse Document Frequency (TF-IDF) metric in the composer module to calculate the similarity between queries. More precisely, a query in this model is represented as a sequence of terms, $q = (t_1, t_2, ..., t_{|q|})$, where $|q|$ is the length of the query and $t_j \in T, j = \overline{1, |q|}$. $T$ is a vocabulary of terms, containing all terms from the collection of queries $Q = \{q_1, ..., q_n\}$ submitted to the system, where $n$ is a total number of queries. For each term $t_j$ let $n_{ij}$ denote the number of occurrences of $t_j$ in $q_i$, and $n_j$ the number of queries that contain $t_j$ at least once. For calculating the TF-IDF weight of the term $t_j$ in the query $q_i$ and defining the similarity between queries $q_i$ and $q_k$ the following formulas are used:

$$w_{ij} = \frac{n_{ij}}{|q_i|} * \log(\frac{n}{n_j}), \quad cos(w_i, w_k) = \frac{{w_i}^T w_k}{\sqrt{{w_i}^T w_i}\sqrt{{w_k}^T w_k}}.$$

Here $w_i = (w_{i1}, ..., w_{im}), w_k = (w_{k1}, ..., w_{km})$ denote vectors of TF-IDF weights corresponding to the queries $q_i$ and $q_k$, and $m$ is the length of the vocabulary.

In the experiment we used a collection of 20 web services from XMethods.com, divided into five topic categories (see Table 2). For each category we chose four semantically equivalent operations and formed 20 queries based on their short natural language descriptions from WSDL files. The number of the users in the experiment is equal to four and the number of requests submitted to the system is equal to 100.

The results of the simulations are given in Figure 3. The precision, recall, and F-measure of the recommendations of the SICS for each of the five groups are given and the average performance of the system for all requests is drawn. According to these results, the precision, recall and F-measure of the system tend to increase with the number of user requests. This is justified by the fact that the number of observations about past selections in the system also increases and, as a result, the SICS has more information for the analysis. We can see that just after 20 searches the precision reaches and maintains a quite high level.

## 5 Related Work

The idea of applying collaborative filtering to web service selection appeared in the literature several times, see papers by Kerrigan [16] and Sherchan [17] for example. Most of the approaches consider ratings of service providers based on subjective opinions of web service users. Manikrao and Prabhakar [18] describe a web service selection framework which combines a recommendation system with semantic matching of service requirements. The approach is based on user feedback and collaborative filtering techniques and is oriented towards helping a user to select a web service from a set of similar services. When the user invokes a web service, the system asks the user to rate the service. However, the previous research has shown that users are very unlikely to provide explicit ratings [19].

Alternatively, user profiles can be obtained by implicitly observing user interactions with the system. Maximilien et al. [20] propose an agent-based framework where agents act as proxies to collect information and to build the reputation of semantic web services. Agents are used to manage available service resources: an agent acting on behalf of the owner looks for services and evaluates possible choices. Three-level ontology is proposed to model quality issues of services. Another multi-agent framework for QoS-based web service selection is proposed by Wang et al. [21]. The authors present a distributed reputation assessment algorithm for QoS support.

The problem of unfair ratings is typical for such kind of systems. However, there exist also approaches that can successfully eliminate ratings from malicious agents [22]. The underlying idea in this approach is to associate ratings with some level of quality and ignore the ratings with associated quality below a certain threshold and ratings from the clients that have a credibility below a certain threshold. For example, credibility of a newcomer may grow up with number of the reports compliant with the reports from other clients. Sherchan et al. [17] analyze user rating behavior to infer the rationale for ratings in a web services environment.

Casati et al. [23] present a system for dynamic web service selection based on data mining techniques. The authors analyze past executions of the composite web services and build a set of context-sensitive selection models to be applied at each stage in the composite service execution.

The idea of using monitored data and/or past user experience for collaborative QoS-driven web service selection is examined in several research works. In our system, we use such an approach to match web services with user requests. The fact that the recommended web services are exploited by other users guarantees a certain level of their quality. The idea behind this is that web services with low quality do not have many clients. Our system allows for the reuse of experience by new service consumers through considering past behavior of users in similar situations. In this way, non-existent, often unavailable, incomplete or faulty services (whose descriptions, however, may be published in the registries) are filtered.

## 6 Conclusions and Future Work

We have presented a recommendation system that facilitates the discovery of web services satisfying user needs. The system is based on the implicit culture framework that uses the history of user-system interactions and client-service communication logs to provide recommendations on web services. It can be used to enhance the retrieval API of service registries.

Future work includes the implementation and evaluation of more complex recommendation scenarios such as collaborative service testing through mining dependencies between exceptions of a specified type and sort of input data, web service monitoring and QoS-based selection. We are planning to run experiments with a semantic method for matching observations implemented in the system. In perspective, the ability of inductive module to infer behavior patterns for web service discovery will be evaluated.

## Acknowledgements

We are grateful to Prof. Enrico Blanzieri for his interest in our work and many fruitful discussions we had together.

## References

1. Akkiraju, R., Farrell, J., Miller, J., Nagarajan, M., Schmidt, M., Sheth, A., Verma, K.: Web service semantics - WSDL-S, available at http://lsdis.cs.uga.edu/library/download/WSDL-S-V1.pdf (2005)
2. Keller, U., Lara, R., Polleres, A.: WSMO web service discovery. WSMO working draft, available at http://www.wsmo.org/2004/d5/d5.1/. (2004)
3. Martin, D., Burstein, M., Hobbs, J., Lassila, O., McDermott, D., University, Y., et al.: OWL-S: Semantic markup for web services. W3C member submission, available at http://www.w3.org/Submission/OWL-S/ (2004)
4. Piccinelli, G., Stefanelli, C., Trastour, D.: Trusted mediation for e-service provision in electronic marketplaces. In: Proceedings of the International Workshop on Electronic Commerce. Volume 2232 of LNCS., Springer (2001) 39 – 50
5. Mahbub, K., Spanoudakis, G.: A framework for requirements monitoring of service based systems. In: Proceedings of the International Conference on Service-Oriented Computing (ICSOC), ACM Press (2004) 84 – 93
6. Herlocker, J.L., Konstan, J.A., Riedl, J.: Explaining collaborative filtering recommendations. In: Proceedings of ACM Conference on Computer Supported Cooperative Work, ACM Press (2000) 241–250
7. Maximilien, E.M., Singh, M.P.: Conceptual model of web service reputation. SIGMOD Record **31**(4) (2002) 36–41
8. Blanzieri, E., Giorgini, P., Massa, P., Recla, S.: Implicit culture for multi-agent interaction support. In: Proceedings of the International Conference on Cooperative Information Systems (CooplS). Volume 2172 of LNCS., Springer (2001) 27–39
9. Birukou, A., Blanzieri, E., D'Andrea, V., Giorgini, P., Kokash, N., Modena, A.: *IC-Service*: A service-oriented approach to the development of recommendation systems. In: Proceedings of ACM Symposium on Applied Computing. Special Track on Web Technologies, ACM Press (2007)
10. Garofalakis, J., Panagis, Y., Sakkopoulos, E., Tsakalidis, A.: Web service discovery mechanisms: Looking for a needle in a haystack? In: International Workshop on Web Engineering. (2004)
11. Tian, M., Gramm, A., Ritter, H., Schiller, J.: Efficient selection and monitoring of QoS-aware web services with the WS-QoS framework. In: Proceedings of the IEEE/WIC/ACM International Conference on Web Intelligence, IEEE Computer Society (2004) 152 – 158
12. Bostrom, G., Giambiagi, P., Olsson, T.: Quality of service evaluation in virtual organizations using SLAs. In: International Workshop on Interoperability Solutions to Trust, Security, Policies and QoS for Enhanced Enterprise Systems (IS-TSPQ). (2006)
13. Dan, A., Davis, D., Kearney, R., et al.: Web services on demand: WSLA-driven automated management. IBM Systems Journal **43**(1) (2004) 136–158
14. Kokash, N., van den Heuvel, W.J., D'Andrea, V.: Leveraging web services discovery with customizable hybrid matching. In: Service-Oriented Computing - ICSOC 2006. Volume 4294 of LNCS., Springer (2006) 522–528

15. Baldi, P., Frasconi, P., Smyth, P.: Modeling the Internet and the Web: Probabilistic Methods and Algorithms. Wiley (2003)
16. Kerrigan, M.: Web service selection mechanisms in the web service execution environment (WSMX). In: Proceedings of the ACM Symposium on Applied Computing (SAC), ACM Press (2006) 1664–1668
17. Sherchan, W., Loke, S.W., Krishnaswamy, S.: A fuzzy model for reasoning about reputation in web services. In: Proceedings of ACM Symposium on Applied Computing, ACM Press (2006) 1886 – 1892
18. Manikrao, U.S., Prabhakar, T.: Dynamic selection of web services with recommendation system. In: Proceedings of the International Conference on Next Generation Web Services Practices (NWESP), IEEE Computer Society (2005) 117
19. Claypool, M., Le, P., Wased, M., Brown, D.: Implicit interest indicators. In: International Conference on Intelligent User Interfaces, ACM Press (2001) 33–40
20. Maximilien, E.M., Singh, M.P.: A framework and ontology for dynamic web services selection. IEEE Internet Computing **8**(5) (2004) 84–93
21. Wang, H., Yang, D., Zhao, Y., Gao, Y.: Multiagent system for reputation–based web services selection. In: International Conference on Quality Software (QSIC), IEEE Computer Society (2006) 429–434
22. Xu, P., Gao, J., Guo, H.: Rating reputation: A necessary consideration in reputation mechanism. In: Proceedings of the Fourth International Conference on Machine Learning and Cybernetics, IEEE Computer Society (2005)
23. Casati, F., Castellanos, M., Dayal, U., Shan, M.C.: Probabilistic, context-sensitive, and goal-oriented service selection. In: Proceedings of the Internationa Conference on Service-Oriented Computing (ICSOC), ACM Press (2004) 316–321

# The Service Modeling Process Based on Use Case Refactoring

Yukyong Kim and Kyung-Goo Doh*

Dept. of Computer Science and Engineering, Hanyang University, Ansan, Korea
{yukyong,doh}@hanyang.ac.kr

**Abstract.** Service-Oriented Architecture (SOA) is an architecture for a system or application that is built using services that implement business functionality with proper granularity. If the granularity of a service is finer, the flexibility and reusability of the service is lower. Therefore, it is critically important to identify what pieces of functionality will become services and to define the interfaces of those services. In this paper, we define a process to identify services by use case refactoring. Task trees are defined to restructure use cases, and five refactoring rules are introduced along with a running example. Because this modeling process can choose the correct levels of abstraction and granularity, it can be helpful in identifying coarse-grained services.

**Keywords:** Service modeling, service-oriented architecture, use case models, refactoring.

## 1 Introduction

Recently, the most difficult challenge in on-demand computing is keeping up with rapid changes in technology. The use of an Service-Oriented Architecture (SOA) is now in widespread use as an architectural approach for various types of systems.

SOA is an architecture for a system or application that is built using a set of services. Though SOA's are defined in many ways, most definitions of an SOA agree that an SOA defines application functionality as a set of shared, reusable services [1].

A service is an implementation of a well-defined piece of business functionality with a published interface that is discoverable and usable by service consumers. The services within the application are loosely coupled and reusable across multiple applications. The level of granularity for services tends to be coarser than those of objects or components [2]. A service typically exposes a single, discrete business process. The service definition should thus encapsulate the business process well enough to be coarse-grained. The granularity of service is also important, because having many fine-grained services can result in high message traffic between service users and providers. Services are recommended to be coarse-grained and, like other large pieces of software, may need to be architected themselves [3].

* Corresponding author.

W. Abramowicz (Ed.): BIS 2007, LNCS 4439, pp. 108–120, 2007.
© Springer-Verlag Berlin Heidelberg 2007

In the SOA development lifecycle, most projects undergo a modeling phase in which the critical service components are defined from the requirements created during domain analysis. Once created, the services may be deployed. Deployment may also involve multiple phases as services are brought into a test environment prior to live deployment [4]. A particular problem may occur in the modeling phase when deriving services from the classes or components, because existing approaches rely on Object-Oriented (OO) or Component-Based Development (CBD). However, the service has a different abstract level from the object or component and has a business workflow. It needs an appropriate approach for identifying and defining the services.

In our research, the services are derived from business use cases throughout the refactoring process. Refactoring has been successfully used in source code reorganization. The method was extended to various models in 2001 called cascaded refactoring [5]. Refactoring has as its goal the selection of the correct levels of abstraction and granularity, identification of commonality, elimination of redundancy, and reuse of functionality [6] [7]. In this paper, we define a process for identifying the services from business use case models: Task trees are generated, and then use case refactoring is invoked. Task trees are used to discover common behaviors in business use case models created during domain analysis. Business use cases describe business processes. This process is documented as a sequence of actions that provide observable value to a business actor [8]. Deriving the services from business use cases is hence adequate for identifying coarse-grained services.

The benefit of the proposed identification method is that it helps to define coarse-grained service candidates rapidly, because business process models may be available in an enterprise, while business use case models have to be developed by performing interviews. Business process models are often available in the form of working instructions or administrative handbooks. It can thus be more efficient to define services by extracting coarse-grained services from restructured use case models than by using the object model.

The paper is organized as follows: We start with a brief introduction of SOA, including business use case models and existing research for service modeling. Section 3 defines the process for identifying the services. We present the steps of the process and then introduce five refactoring rules. Section 4 describes a running example based on a simplified Automated Teller Machine (ATM). Finally, section 5 presents conclusions and a description of future work.

## 2 Related Work

This section introduces the concepts of SOA and business use case models. We discuss existing approaches and their problems.

### 2.1 SOA

SOA is a design methodology whose goal is maximizing the reuse of application-neutral services to increase IT adaptability and efficiency. While these concepts have existed for decades, the adoption of SOA is accelerating due to the emergence of standards-based integration technologies like Web services and XML. A service in SOA is an exposed piece of functionality with three properties:

(1) The interface contract to the service is platform-independent.
(2) The service can be dynamically located and invoked.
(3) The service is self-contained, which means the service maintains its own state.

SOA allows the consumer of a service to ask a third-party registry for the service that matches its criteria. If the registry has such a service, it gives the consumer a contract and an endpoint address for the service. The service provider is the network-addressable entity that accepts and executes requests from consumers. The service provider publishes its contract in the registry for access by service consumers. SOA solutions are composed of reusable services with well-defined, published, and standard-compliant interfaces. SOA provides a mechanism for integration of existing legacy applications regardless of their platform or language.

In SOA terms, a business process consists of a series of operations which are executed in an ordered sequence according to a set of business rules. From a modeling standpoint, the resulting challenge is how well-designed operation, service, and process abstractions can be characterized and constructed systematically [9].

High-level business process functionality is externalized for large-grained services. The business use case model provides a software-independent description of the business process. The business process describes the tasks that have to be carried out and their order. A task is the smallest unit of work that makes sense to a user. The detail associated with a business use case is documented in a business use case specification. This will include texts as well as one or more UML use case diagrams and activity diagrams. In a business use case specification, name, brief description, performance goals, workflow, special requirements, and relationships are normally included. Figure 1 shows a sample of a business use case model [8].

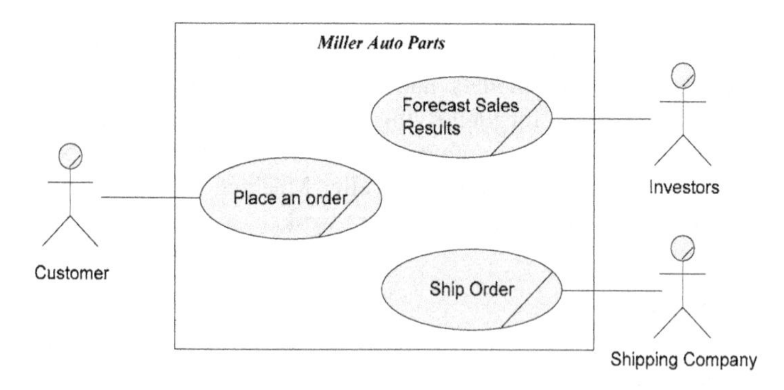


**Fig. 1.** A business use case model for order fulfillment

## 2.2 Current Approaches

The main issue of modeling practices with respect to SO is granularity. Existing modeling disciplines such as OO or CBD depend on the class. OO analysis is a powerful, proven approach. However, class resides at too low of a level of abstraction for business service modeling. Strong associations such as inheritance create a tight coupling between the involved parties. In contrast, the SO paradigm attempts to promote flexibility and agility through loose coupling. These considerations make OO difficult to align with SOA straightaway.

Currently, most approaches rely on OO and CBD to identify and develop services. In 2005, Thomas Erl introduced service-oriented analysis [1]. He defined a service modeling process to produce service and operation candidates. This process provides steps for the modeling of an SOA consisting of application, business, and orchestration service layers. The Gartner Research Group defined Service-Oriented Development Architecture (SODA) [10]. The concept of SODA is the intersection of SOA, distributed development, and CBD. IBM introduced Service-Oriented Modeling and Architecture (SOMA) [11]. SOMA illustrates the activities for service modeling and the importance of activities from the service consumer and provider perspectives. It is an enhanced, interdisciplinary service modeling approach that extends existing development processes and notations such as OO analysis and design, Enterprise Architecture (EA) frameworks, and Business Process Modeling (BPM). SOMA provides a template that can be used for architectural decisions in each layer of the SOA. Knual Mittal suggested the Service Oriented Unified Process (SOUP), which takes the best elements from RUP and XP [12]. SOUP has six phases, and the process is divided into two parts: SOA deployment and SOA management. The deployment part is based on RUP and the management part on XP.

These approaches have key limitations. One is that they lack details about how to identify and define services from the business domain. The other is that they are based on object models or component models. To remedy this limitation, these approaches usually employ classes and components grouping techniques to identify services. Services in general are coarser-grained than objects and components because a service implemented as objects has one or more coarse-grained objects that act as distributed facades. A service typically exposes a single discrete business process. Moreover, a business use case is a description of the business process. Thus, the best way to define services is using use case models. We introduce a process to identify services from use case models using the refactoring technique.

# 3 Service Modeling Via Refactoring

This section demonstrates how services can be defined from use case models. We shall start by presenting the service modeling process.

## 3.1 The Service Modeling Process

The business system can be decomposed into a set of services and the collaboration among them. A service is reused by multiple business systems and a service is implemented and executed by a set of computing systems.

Therefore, it is best to perform identification services starting with use cases of the business rather than objects or components. This paper advocates a phased approach to service development as illustrated in Figure 2.

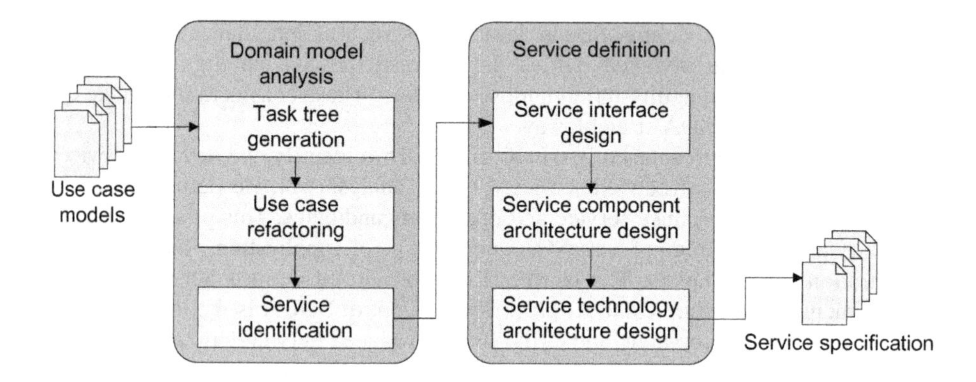


**Fig. 2.** Overview of the service modeling process

Step 1. Task tree generation
A use case can be described in terms of tasks. A task is the smallest unit of work that makes sense to a user. Each task represents the parts of the dialogues in the scenario that perform the use case. Refactoring use case models refers to the re-distribution of behaviors in the form of tasks from one use case to another. A task tree represents a workflow of the use case in terms of tasks. In this step, task trees are created from use case models and functional requirements.

Step 2. Use case refactoring
Using the task trees, use case models are restructured according to the refactoring rules described in the next section. After repeated refactoring, refined use case models are produced.

Step 3. Service identification
First, service candidates can be derived from the use case model created in the previous step. This step appreciates the dependencies between service candidates. The architecture of the service components is determined by analysis of the dependencies.

Step 4. Service interface design
The interface of the service is defined in order to refine identified services and to make clear to users the role and function of each service (more explicitly, the interface describes the operations and signatures of the service). Analyzing the interaction between services likewise makes clear the precedence of services. As needed, existing services can be modified or eliminated and new services created. The repeated application of this refinement produces a list of final services.

Step 5. Service component architecture design
This step defines the service architecture, which in turn describes the dependencies among services based on business requirements. It can be derived from the architecture generated in Step 4 through refinement using design patterns or implementation patterns.

Step 6. Service technology architecture design
In this step, service implementation technologies are selected and mapped. They can be Web service standards such as SOAP, WSDL, WS-Security, etc. based on XML.

Subsequently, specifications for identified services, and the architecture that represents the structure of services and their corresponding technologies are generated from use case models.

### 3.2 Refactoring Rules

We introduce five refactoring rules for extraction of service candidates from task trees. The functionality of a system can be defined as the set of tasks in the use cases of the system. We clarify the following definitions for refactoring of use case models.

***Definition 1.*** A refactoring of a use case model $U$ preserves the set of tasks of $U$. Assume $t_i$ be a task of $U$ and $T$ the set of tasks of $U$. Then for $\forall t_i \in T$, $t_i$ exists in the post-refactoring model $U'$.

***Definition 2.*** To apply a use case model refactoring $R$ to a use case model $U$, the precondition of $R$ must be satisfied by $U$.

A refactoring rule will be defined in a 4-tuple (Parameters, Preconditions, Postconditions, Process) where parameters are entities involved in the refactoring, preconditions are the context that must be satisfied to execute the refactoring, postconditions are the context that must be satisfied after applying the refactoring, and process is workflows of the refactoring.

**Decomposition Refactoring**
When the use case is too complicated, it can break up, that is, the task tree of the use case can break into smaller task trees. From these smaller trees, new use cases are created.

**Table 1.** The decomposition refactoring rule

| | |
|---|---|
| Parameters | $u$ : a use case to be broken up |
| | $t$ : a task tree of $u$ |
| | $t_i$ : a subtree such that $t_i \subset t$ |
| Preconditions | The use case $u$ has its task tree $t$ |
| | Subtree $t_i$ of $t$ is functionally independent of $u$ |
| Postconditions | A new use case $u'$ is generated with task tree $t_i'$ |
| Process | The subtree $t_i$ of $t$ is replaced by a pseudo task tree $t_i'$ |
| | A new use case $u'$ is generated with task tree $t_i'$ |
| | A precedence relationship between $u$ and $u'$ is created |

**Equivalence Refactoring**
If two use cases have equivalent task trees, we conclude that the two use cases have the same behavior and can be treated as one.

**Table 2.** The equivalence refactoring rule

| | |
|---|---|
| Parameters | $u_1$, $u_2$ : use cases that have equivalent task trees |
| Preconditions | The task trees of two use cases are behaviorally the same |
| Postconditions | A use case has no relationship with other use cases or actors |
| Process | The use case $u_1$ is replaced by $u_2$ |
| | All relationships with $u_1$ are taken over by $u_2$ |

**Generalization Refactoring**

If two or more use cases share common primitive tasks, there is a task tree of which the task trees of the use cases are special subsets.

**Table 3.** The generalization refactoring

| | |
|---|---|
| Parameters | $u_1$, $u_2$ : chosen use cases |
| | $t_1$, $t_2$ : task trees of chosen use cases, respectively |
| Preconditions | The use case $u_1$ has its task tree $t_1$ and $u_2$ has its task tree $t_2$ |
| | $t_1$ and $t_2$ share common primitive tasks $t = \{e_1, e_2, \ldots, e_n\}$ |
| Postconditions | A new use case $u$ is generated with task tree $t$ |
| Process | A new use case $u$ is generated with task tree $t$ |
| | Associations from $u_1$ and $u_2$ to the use case $u$ are generated |
| | Tasks $\{e_1, e_2, \ldots, e_n\}$ are removed from use cases $u_1$ and $u_2$ |
| | The common relationship between $\{u_1, u_2\}$ and other use cases or actors are taken over by $u$ from $\{u_1, u_2\}$ |

**Merge Refactoring**

If a use case is too specific compared with another, it is better to merge the cases. As a result, the relationship is removed. The task tree of the use case is inserted into the merged use case.

**Table 4.** The merge refactoring rule

| | |
|---|---|
| Parameters | $u$ : the base use case |
| | $u'$ : the associated use case |
| | $r$ : the precedence relationship between $u$ and $u'$ |
| Preconditions | The use case $u$ has its task model $t$ |
| | $u$ is associated with $u'$ |
| | The use case $u'$ has its task model $t'$ |
| Postconditions | Two use cases are substituted by one merged use case |
| Process | $u'$ is merged into $u$ |
| | The task model $t'$ is inserted in $t$ |
| | The relationship $r$ is removed |

**Delete Refactoring**
When a use case is defined but does not participate in any relationships with other use cases or actors, the case is a redundant, and can be deleted from the system.

**Table 5.** The delete refactoring rule

| | |
|---|---|
| Parameters | $u$ : a use case that has no relationship with other use cases or actors |
| Preconditions | The use case $u$ has no relationship with any other use cases or actors |
| Postconditions | The use case $u$ is deleted from the system |
| Process | It deletes the use case $u$ from the system |

Let us choose **Generalization refactoring** as an example to illustrate the behavior preservation of use case model refactorings. For a task set $t$, the change relates to use cases including the tasks. Prior to the refactoring, task set $t$ is included in $u_1$ and $u_2$. After the refactoring, $t$ is not deleted from the use cases, but moved from $u_1$ and $u_2$ to a new use case $u$. The set of tasks of the use case model $U$ are not changed before and after the refactoring. Thus, the behavior of the use case model is preserved.

## 4 A Running Example: ATM

To demonstrate the service modeling process, we select the simplified ATM as a running example (Fig. 3).

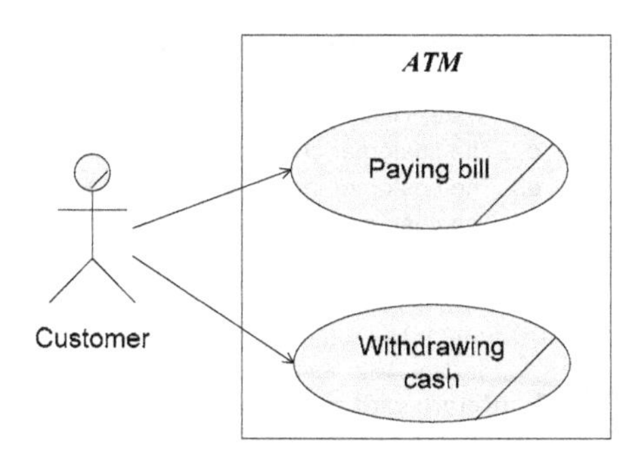


**Fig. 3.** Business use case model for ATM

Step 1. Task tree generation
First, using the description of the use case, the task tree is created. For example, the use case "Paying bill" has the following description:

T1. ID verification
  e1 The customer inserts a card
  e2 The system validates the card
  e3 The customer enters PIN
  e4 The system validates the PIN
T2. Pay bill
  e1 The system lists registered bill
  e2 The customer chooses the bill to pay
  e3 The customer inputs amount to pay
  e4 The customer chooses from which account to pay
  e5 The system checks if balance is greater than payment
    e5.1 Yes, payment succeeds
    e5.2 No, return error messages
T3. Return card and print receipt
  e1 The system returns card
  e2 The system outputs ending messages

The task tree $t_1$ is as follows:

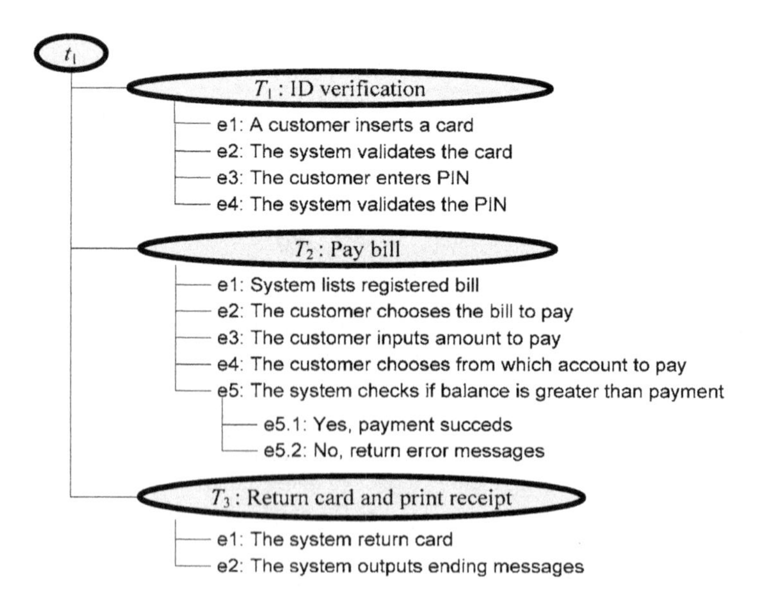


The use case "Withdrawing cash" has the following description:

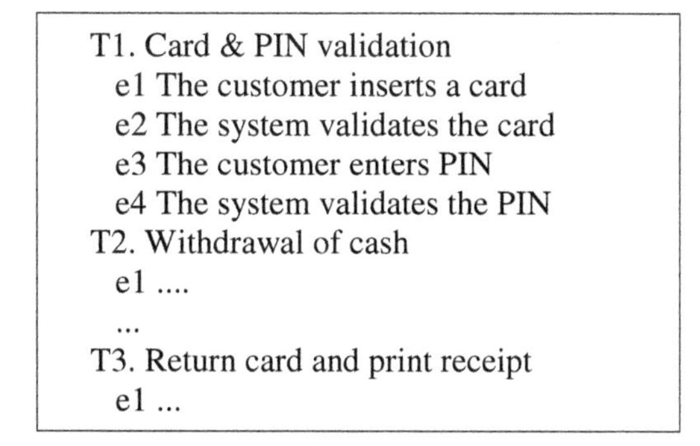
T1. Card & PIN validation
e1 The customer inserts a card
e2 The system validates the card
e3 The customer enters PIN
e4 The system validates the PIN
T2. Withdrawal of cash
e1 ....
...
T3. Return card and print receipt
e1 ...

Its task tree can be generated as above.

Step 2. Use case refactoring
The refactoring process based on refactoring rules takes place as follows:

Let's define the use case "Paying bill" as $u_1$ and "Withdrawing cash" as $u_2$. From the task tree $t_1$ of $u_1$ and $t_2$ of $u_2$, we notice that the task T1 "ID verification" and "Card & PIN validation" describe relatively independent functions. Thus the decomposition rule is applied to break up the use case into smaller use cases. After decomposition refactoring, the use case model is shown in Figure 4.

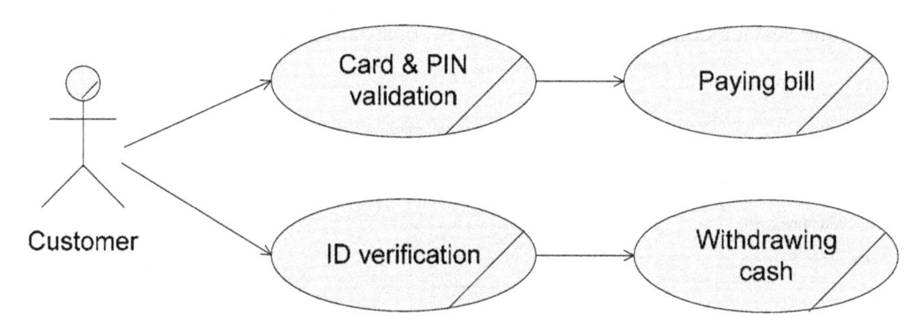


**Fig. 4.** Decomposition refactoring

Since "ID verification" and "Card & PIN validation" are equivalent in behavior, the equivalence refactoring is applied. After equivalence refactoring, the use case model is shown in Figure 5.

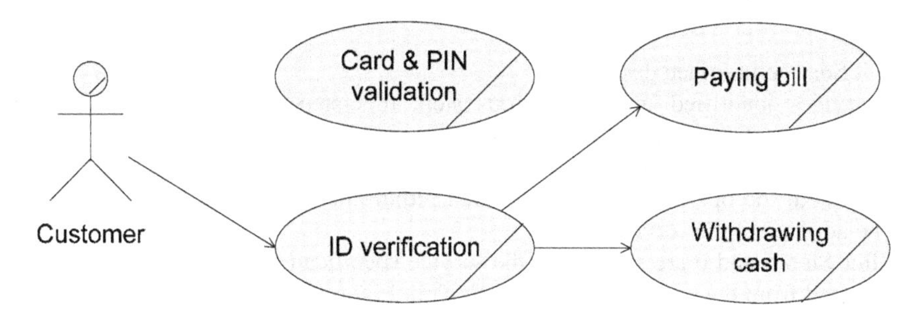


**Fig. 5.** Equivalence refactoring

After applying the equivalence refactoring, the use case "Card & PIN validation" has no relationship to any other use cases or actors. Therefore, it can be deleted from the system. And we notice that $u_1$ and $u_2$ have common tasks. Thus the generalization refactoring is applied to reduce redundancy. After applying these refactorings, a final use case model for ATM is derived (the case is illustrated in Figure 6).

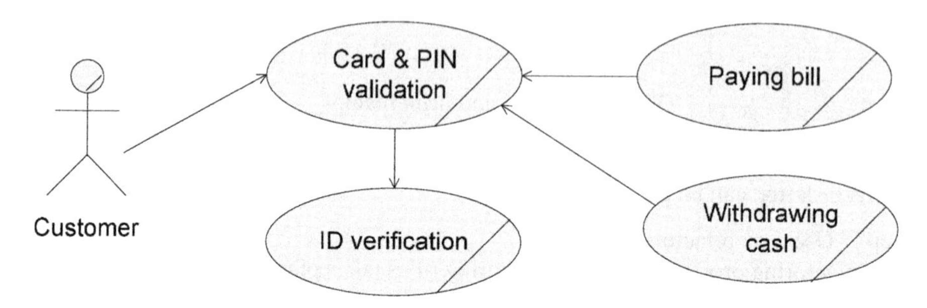


**Fig. 6.** Final use case model for ATM

Step 3. Service identification
From the final use case model, we can define four services: Process transaction, Paying bill, ID verification and Withdrawing cash. By analyzing the dependencies among services, the service component architecture is constructed as shown in Figure 7.

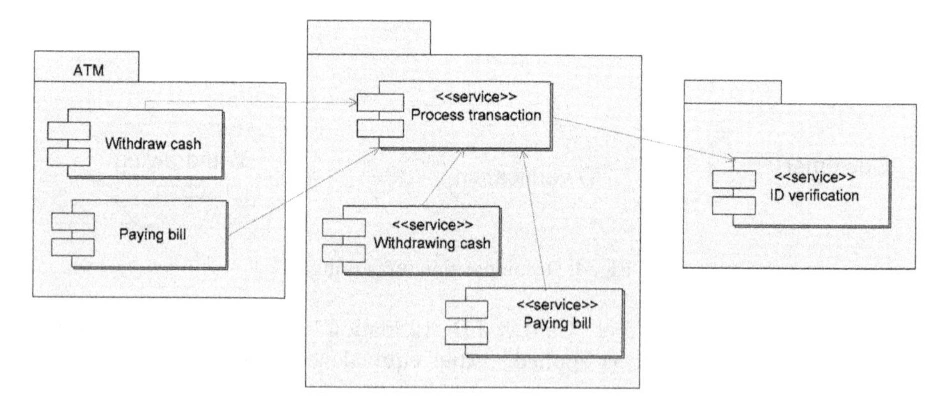


**Fig. 7.** Service components architecture

Step 4. Service interface design
For services identified in the previous step, this step defines the interface and determines the description. It includes the service name, a simple explanation of the service, implementation strategy about whether the service is implemented, wrapped, or requested, the operations, non-functional requirements, and the technology which will be adopted in the service.

After Step 5 and 6 are processed, the service specification, as shown in Table 6, is constructed from the service "ID verification" in Figure 7.

**Table 6.** An example of the service specification

<table>
<tr><td>Name</td><td colspan="3">ID verification</td></tr>
<tr><td>Explanation</td><td colspan="3">The service for verifying the user</td></tr>
<tr><td>Implementation strategy</td><td colspan="3">Request the published service</td></tr>
<tr><td rowspan="2">Operations</td><td rowspan="2">getUserID</td><td>Parameters</td><td>String userID</td></tr>
<tr><td>Return values</td><td>Boolean</td></tr>
<tr><td>Non-functional requirements</td><td colspan="3">Service security,<br>Service description and messaging</td></tr>
<tr><td>Related Standards</td><td colspan="3">SOAP, WSDL, WS-Security</td></tr>
</table>

## 5 Conclusion

Our research demonstrated a phased approach to defining services from use case models. We introduced the steps for defining services as well as rules for refactoring. Refactoring use case models helps to control service granularity and maintain a similar level of abstraction among services. In this paper, we focus on refactoring with the decomposition of a use case and the reorganization of relationships among use cases. We showed in a running example that the application of the modeling process results in services and service architectures via refactoring.

Future research should focus on the refactoring rules themselves. More refactoring rules and formalization of the refactoring rules are expected. It will be necessary in particular to make the rules more accurate and clear. Furthermore, we will focus on the tool support for the automated refactoring process. Since performing these refactorings manually can be time-consuming and error-prone, we will develop a tool for use case refactoring. Currently, we are working on several larger case studies to ensure that our approach can help define services quickly and easily.

**Acknowledgments.** This research has been supported by the Ministry of Education and Human Resources Development, South Korea, under the grant of the second stage of BK21 project.

## References

1. Thomas Erl, Service-Oriented Architecture: Concepts, Technology, and Design, Prentice-Hall, New York (2005)
2. James McGovern, Sameer Tyagi, Michael Stevens, Sunil Mathew, Java Web Services Architecture, Morgan Kaufmann Publishers, USA (2003)
3. Liam O'Brien, Len Bass, Paulo Merson, Quality Attributes and Service-Oriented Architectures, Technical Note CMU/SEI-2005-TN-014, Carnegie Mellon University (2005)
4. Granville Miller, Automating Business Processes with WSFL and BPEL4WS, WebSphere Developer's Journal, Vol. 2(2), Websphere (2003)

5. Greg Butler, Lugang Xu, Cascaded Refactoring for Framework Evolution, In SSR'01, International Symposium on Software Reusability, ACM press (2001)
6. Kexing Rui, Greg Butler, Refactoring Use Case Models: The Metamodel, In ACSC'03, 16th Australasian Computer Science Conference, Australasian Computer Society (2003)
7. S. Ren, K Rui G. Butler, Refactoring the Scenario Specification: a Message Sequence Chart Approach, Lecture Notes in Computer Science, Vol. 2817, Springer-Verlag, Berlin Heidelberg New York (2003)
8. Jim Heumann, Introduction to business modeling using the Unified Modeling Language, IBM DeveloperWorks Article, IBM (2003)
9. Olaf Z., Pal Krogdahl, Clive Gee, Elements of Service-Oriented Analysis and Design, IBM DeveloperWorks Article, IBM (2004)
10. Samir Nigam, Service Oriented Development of Applications(SODA) in Sybase Workspace, Sybase Inc. whitepaper, Sybase Inc (2005)
11. Ali Arsanjani, Service-Oriented Modeling and Architecture (SOMA), IBM DeveloperWorks Article, IBM (2005)
12. K. Mittal, Service Oriented Unified Process (SOUP), IBM Journal, Vol. 6, IBM (2006)
13. Bart O, Jian Yang, and Mike P., Model Driven Service Composition, Lecture Notes in Computer Science, Vol. 2910, Springer-Verlag, Berlin Heidelberg New York (2003)
14. Martin Fowler, Refactoring: Improving the Design of Existing Code, Addison-Wesley, New York (1999)

# Scenarios and Techniques for Choreography Design

Gero Decker[1,⋆] and Michael von Riegen[2,⋆⋆]

[1] Hasso-Plattner-Institute, University of Potsdam, Germany
gero.decker@hpi.uni-potsdam.de
[2] University of Hamburg, Germany
riegen@informatik.uni-hamburg.de

**Abstract.** Choreography description languages have been put forward for capturing sets of interactions and their control and data dependencies, seen from a global perspective. Choreographies serve as starting point for generating interface processes for the different participants which in turn are used for implementing new services or adapting existing ones. However, such top-down approaches are not sufficient for scenarios where given implementations cannot be changed and are to be used as a starting point for choreography design. This paper identifies and classifies three categories of choreography design: *choreography identification*, *choreography context expansion* and *collaboration unification*. Each category is motivated through an example from the eGovernment domain. Existing techniques needed for the individual design categories are discussed and missing techniques are highlighted.

## 1 Introduction

Services are more and more used to support long-running business processes. This trend runs alongside with a shift from merely considering simple interaction behavior of services, like request-response interaction patterns manifested in standards like SOAP and WSDL, towards conversational services that engage in long-running conversations with other services.

In order to cope with the complexity of these conversations, a new viewpoint on interacting services was introduced. It describes interactions from a global point of view, i.e. from the perspective of an ideal observer who is able to see all interactions and their flow and data dependencies. The resulting global interaction models are called service choreographies. Standards such as the Web Service Choreography Description Language (WS-CDL [8]) were put forward for describing choreographies of web services. The main motivation for introducing such an abstraction layer was to enable a model-driven approach for service design and implementation. These top-down approaches, like e.g. presented in [7], propose choreographies as a starting point for generating interface processes for

⋆ Research conducted while with SAP Research Center Brisbane.
⋆⋆ Research is sponsored under the EU IST-2004-026650 project "R4eGov".

W. Abramowicz (Ed.): BIS 2007, LNCS 4439, pp. 121–132, 2007.
© Springer-Verlag Berlin Heidelberg 2007

each service which are then the skeletons for implementing new services or for adapting existing services.

As the idea of global interaction models matures, more and more application domains adapt choreography languages and methodologies. However, it turns out that many application scenarios cannot be covered using top-down approaches. Existing services and processes are the starting point for identifying already existing choreographies or for creating to-be choreographies. In the course of this paper we will present several use cases from the EU funded project "Research for eGovernment" (R4eGov[1]) to display the need of different choreography design approaches. The project centers around interoperability and security in inter-organizational and even cross-border processes. Its main challenges are coping with heterogeneous systems and preserving local system and process ownership ([1]). Within this project, heterogeneity issues are tackled by resorting to service-oriented architecture concepts and the usage of web services. Thus, service choreography methods and techniques can be used for managing the peer-to-peer like collaborative processes.

The next section sets the scene for the paper by introducing different viewpoints for modeling collaborative processes. Section 3 presents *choreography identification*, section 4 presents *choreography context expansion* and section 5 *collaboration unification*. All three choreography modeling categories are illustrated using a use case and the corresponding techniques are discussed. Finally, a conclusion is drawn and future work is sketched. Related work will be mentioned in the course of each section.

## 2 Classification Framework

Figure 1 presents different perspectives on inter-organizational collaboration. *Process implementations* are the intra-organizational process definitions that are executed within one participant. In service-oriented architectures these definitions are called service orchestrations. While they also include internal actions that are not to be shown to other participants, the *provided interface processes* only describe the publicly visible behavior of a participant. I.e. only those actions are included in an interface process that directly relate to message exchanges with the outside world. Like it was the case for the process implementations, interface processes describe collaboration from the perspective of one single participant (endpoint-centric view). A set of interconnected interface processes describes all interaction taking place in a collaboration. As means to describe such collaboration from a truly global point of view (i.e. within one process definition), choreographies were introduced.

Furthermore, there are two different kinds of choreographies that we want to distinguish: An *observable choreography* describes the actually observable interaction behavior of a set of collaborating partners. One could imagine an ideal observer tracing all interactions belonging to a collaboration. In contrast to this we find *normative choreographies* prescribing behavior for the participants in

[1] See http://www.r4egov.info/

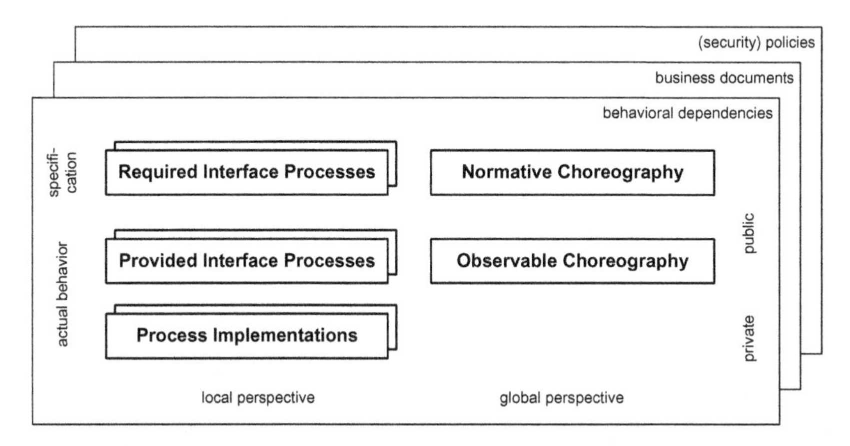


**Fig. 1.** Perspectives for describing collaborative processes

the collaboration. These choreographies are minimal sets of constraints in the collaborations and serve as contractual basis. Conformance between the actual interaction behavior of participants (the observable choreography) and the corresponding normative choreography can be checked. Most normative choreographies equal to a set of *required interface processes* constraining the interaction behavior of an individual participant. Complying to a required interface process is a prerequisite for a participant taking part in the collaboration. The required interface process defines what behavior other participants can expect from the implementing participant. If a participant violates a constraint in his process implementation, the other participant will be faced with unexpected behavior. However, participants still have different possibilities to implement a required interface process. This explains why there can be different provided interface processes that are all conforming to the same required interface process.

In addition to these different viewpoints we find three dimensions within each viewpoint.*Behavioral dependencies* between interactions cover the control flow in choreographies. The *business document* dimension takes care of the content and structure of the messages being exchanged and the data flow dependencies between different interactions. Content of business documents also influences branching decisions in the control flow dimension. Finally, *(security) policies* describe non-functional configurations in the collaboration.

Dijkman and Dumas have introduced some of the perspectives in [7]. However, they only focused on the control flow perspective and did not make a distinction between observable and normative choreographies.

Within the next sections we will display three approaches to do choreography design using this framework for describing the design procedure.

## 3 Choreography Identification

In the case of *choreography identification*, different participants have working process implementations and already use them to collaborate with each other.

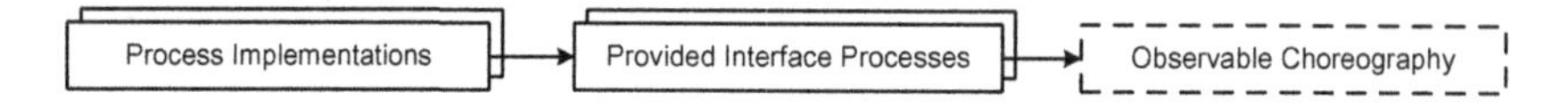


**Fig. 2.** Choreography identification

However, every participant only knows about the interactions he is directly involved in (i.e. his own provided interface process). Therefore, the goal is to identify the observable choreography the participants already engage in so that everybody has a global view of the collaboration (see figure 2). The main motivation behind choreography identification is an optimization of the overall collaboration. The overall costs for the collaboration have to be reduced and/or the performance of the collaboration has to be ameliorated. Only having the global picture at hand, partners see what interactions and dependencies exist globally and which of these might be removed or organized differently. Finally, the changes in the choreography are pushed down again into the process implementations. The next section motivates this procedure with a use case.

### 3.1 Use Case: Eurojust/Europol Collaboration

Eurojust (European Judicial Cooperation Unit) and Europol (European Police Office) have been set up to help the EU member states to cooperate in the fight against cross-border organized crime. An objective is to establish a secure connection between Eurojust and Europol to insure collaboration and effective information exchange between both parties [12]. Both organizations already manage their information with computer systems but these are completely independent. On the one hand, Europol has three information systems: The Europol Information System that supports all intelligence activities within the Europol framework, the Europol Analysis System that is only available to the analysts employed within each analysis work file, and the Information Exchange System that enables bilateral exchange of data between Member States without necessarily involving Europol. On the other hand, Eurojust has only one system which can be used within collaboration: The Case Management System that enables to deal with case-related activities.

It should be noted that there are already different kinds of collaboration between Eurojust and Europol existing. Examples are the request for mutual legal assistance for witness protection during criminal proceedings, the execution of European arrest warrants (EAW) or an EAW with a rogatory letter. From now on, these existing channels should be supported by electronic ways of communication which means that paper-based communication should be converted to a secure electronic conversation. After the identification of existing interaction, the processes need to be built around the existing infrastructure. Furthermore, it is also the case that a participant might need to know all dependencies between others to be able to react to all possible scenarios or exceptions. To do this, he needs to identify the overall choreography. A choreography can be one part of the contractual grounding of electronic conversation between the parties.

### 3.2 Techniques

In the case of *choreography identification* the observable choreography has to be extracted from the existing collaborating processes. This can be done either

1. based on the process implementations and provided interface processes or
2. based on the runtime behavior of the processes.

Option 1 involves the extraction of provided interface processes if only the process implementations are given. Once the provided interface processes are known they can be interconnected for retrieving the overall collaboration process. E.g. in [11] different workflow modules are interconnected to a workflow net which can then be used for reasoning on the overall process. Other formalisms where proposed for capturing choreographies and interface processes. In [4] a formalism for choreographies and another for interface processes are presented together with a bisimulation-like conformance relation. However, it is not mentioned for this formalism how to retrieve the choreography from a set of given interface processes. The same holds true for other interaction modeling languages such as WS-CDL ([8]) and *Let's Dance* ([17]).

Option 2 can be done by looking at the traces produced in the actual collaboration. If a sufficient number of conversations have been traced, process mining techniques (cf. [15]) can be used to retrieve the observable choreography model.

## 4 Choreography Context Expansion

Normative choreographies are normally limited to a certain business context. Assumptions are present among the participants which are (only) valid for the specific context. In order to broaden the reach of the choreography, i.e. make the choreography applicable to a broader context, these assumptions have to be removed from the choreography and therefore also from the required interface processes. This might result in a situation where individual participants cannot conform to the choreography for all covered contexts any longer. To broaden a choreography, we have to consider the provided interface processes and the normative choreography in order to get an expanded normative choreography like depicted in figure 3. The next section motivates context expansion with a use case.

### 4.1 Use Case: Electronic Procurement Schema for European Public Administrations

This scenario describes the challenges in cross-border exchanges in public procurement [3]. Normally, each country has its own public procurement system where local companies can bid. If a company wants to bid in another EU member state, it has to substantiate its legal existence and the buyer must be able to trust the received digital data. Due to the national differences in legal obligations and basis, this problem is getting really complex. There are at least

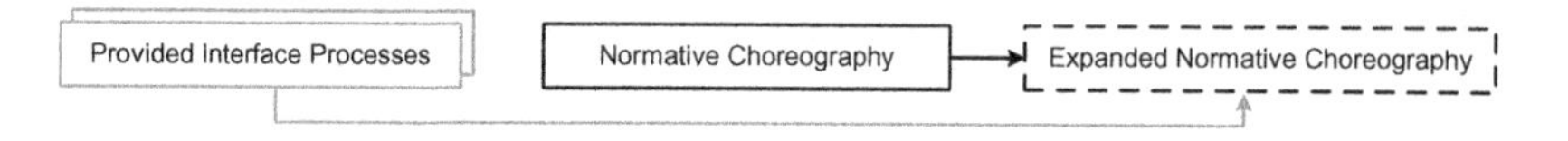


**Fig. 3.** Choreography context expansion

two main issues that have to be solved within this context to have companies participating in cross-border public procurement: The electronic certification of companies willing to bid in another country and the diffusion of legal information; whenever a company makes changes in its status, it has to report it to the trade register.

The European Commission released a legislative package consisting of two directives, 2004/18/CE and 2004/17/CE. These directives introduce a framework for open e-procurement standards and necessary conditions. The main problem is that an adoption of these directives will differ from country to country because of different technical standards, different legal requirements, different tax obligations, different laws on labor, different electronic certification processes, etc.

A collaboration model based on choreography can help to model the exchange of information between different countries without having to modify their national procedures. Thus, existing collaborations have to be opened for a wider audience. This is a scenario where we have to identify the minimal requirements that a new participant has to abide in order to join the cross-border collaboration. The process to accept cross-border participants has to be built around the existing collaboration. In contrast to this, a classic top-down redesign might change the local processes and interfaces and this would tamper the existing collaboration.

The example in figures 4(a) to 4(e) illustrates a part of the procurement use case using the choreography modeling language *Let's Dance* (cf. [17]). The French administrations have a provided interface process where a set of potential bidders are notified about a call for tender. Then, they expect a notification from the trade register that the bidder's proof of evidence is ok as well as the actual bid from the bidder (see figure 4(a)). The French bidders receive the call for tender, then they provide proof of evidence as well as the reference of the administration to the trade register. After being notified about a correct evidence, the bidder places the bid (see figure 4(b)). Finally, the trade register has a generic interface process for handling evidence cases. After a request containing the proof of evidence comes in, the trade register checks if the company is already registered or if further certification is required. If this is the case, interaction regarding the certification takes place with the requesting company. Finally, an "evidence ok" notification is sent back to the company as well as to another institution if given as "cc" (see figure 4(c)). In the case where only French participants are involved, the additional certification is not required since all French companies are registered. The normative choreography for a French-only context is illustrated in figure 4(d). When broadening the context to Europe, the normative

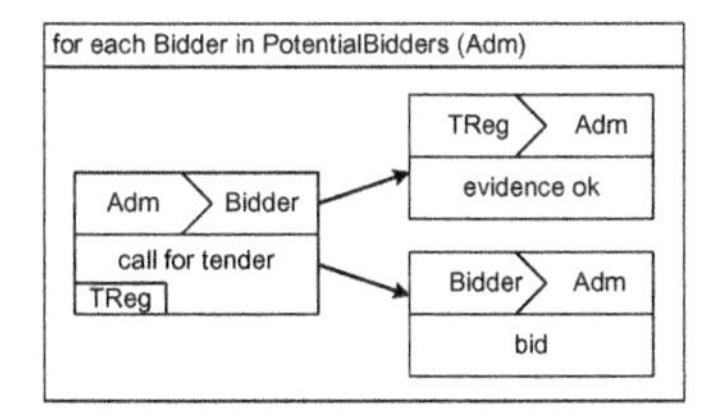


(a) Provided interface of the French administrations

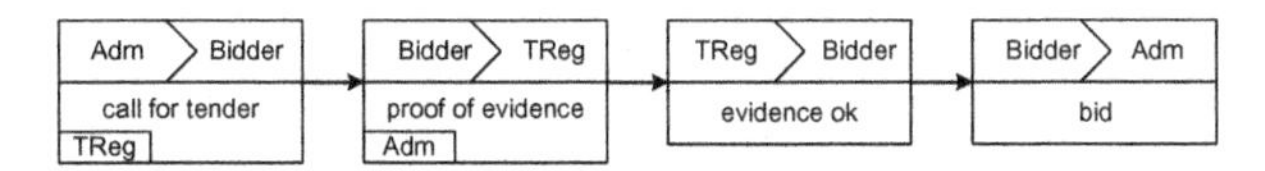


(b) Provided interface of the French bidders

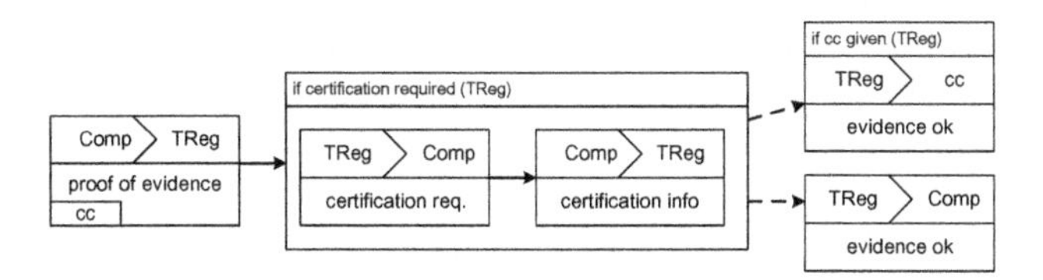


(c) Provided interface of the trade register

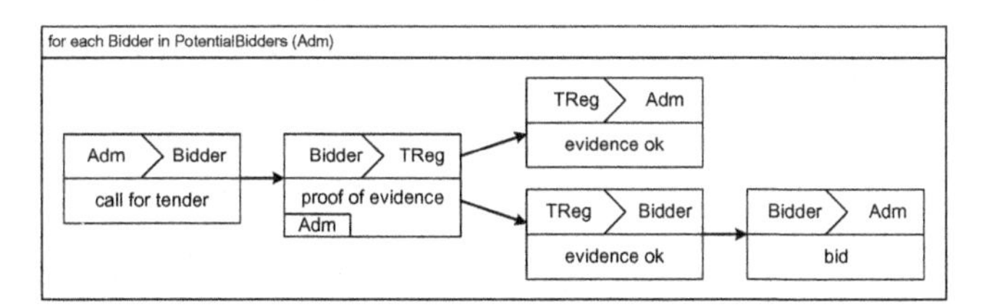


(d) Normative choreography for the context "Bidding in France"

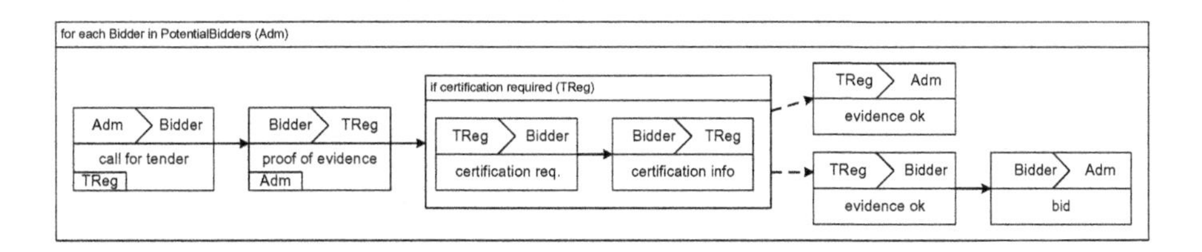


(e) Normative choreography for the context "Bidding in the European Union"

**Fig. 4.** eProcurement use case

choreography includes the potential additional certification interactions. In this context, the French bidders cannot participate any longer unless they extend their provided interface process (see figure 4(e)).

### 4.2 Techniques

In the example we find that the French bidders can only operate in the context "bidding in France". They either do not know about the decision the trade register has to make or they make an assumption about how the trade register decides. The trade register on the other side is not limited to this context and can deal with the broader context "bidding in Europe" by considering the case that a company is not yet registered.

The situation that different partners can already deal with different contexts is typical for this choreography modeling category. The strategy is now to identify what parts of interface processes apply to which context and to then include corresponding interactions into the expanded normative choreography. We therefore need process model synthesis techniques. Techniques can be found in the space of object-oriented computing, where different scenario descriptions have to be merged to a single state machine describing the complete behavior of an object (cf. [9]). In the field of message sequence charts we find different scenarios of interacting system components that are to be merged into one global interaction model (cf. [14]).

The notion of business context is a central concept in ebXML's Core Component Technical Specification ([5]). It defines how to describe business document specifications and introduces "core components" as opposed to "business information entities". Business information entities are based on core components but are limited to a specific business context. There has not been any work on business contexts in choreography models. However, results from the field of process family engineering (e.g. [13]) could be used as a starting point.

Furthermore, conformance techniques are essential for checking if an existing provided interface process that conformed to the original normative choreography with the limited context still conforms to the expanded normative choreography. Many existing conformance checking techniques like protocol and projection inheritance in [2] and conformance in [4] are based on bi-simulation. [6] highlights that conformance relations are tightly linked to compatibility relations and shows that bi-simulation is too restrictive for common compatibility notions. It provides a means to check whether a conformance relation is optimal but it does not define a concrete relation fulfilling the criteria.

## 5 Collaboration Unification

In the case of *collaboration unification*, different observable choreographies exist for the same domain. A typical reason for the evolution of such "islands of collaboration" is that there are disjoint groups of collaborators each of which has its own history for the interaction protocol. Now the goal is to enable the collaboration between participants from different islands through a unified normative choreography for all participants (See figure 5). A motivating use case will be described within the next section.

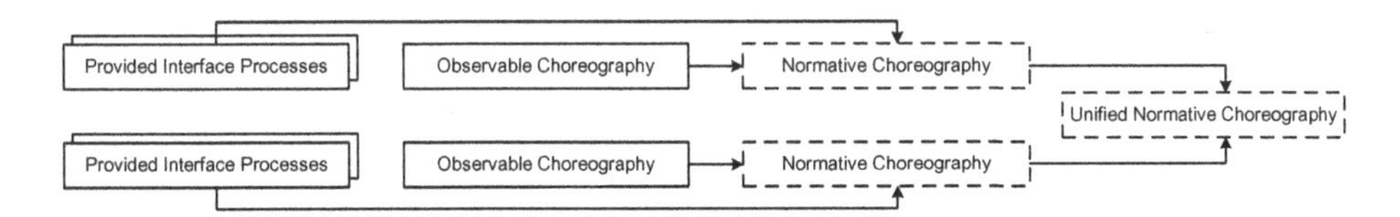


**Fig. 5.** Collaboration unification

### 5.1 Use Case: eErasmus eHigher Education (eEH)

eErasmus is an international exchange program of higher education institutes among EU countries [10]. Students joining this program can take courses at foreign universities and might have the selected courses acknowledged by the home university. For a student, it is difficult to get the acknowledgment of foreign courses and examinations by the home university. The other way round, it is also difficult for a student to get an approval for courses without having any documents from the foreign host university. eErasmus should help a student by proposing the right courses or the right documents to get approved by the host university and should help the universities during the exchange process of grades and examination results.

This simple example depicts two high-level use cases. On the one hand, we have a *preparation process* where a student has to prepare the residence at a host university. He has to get approved by the host university in order to have courses. On the other hand, we have an *acknowledgment process* where the host university passes exams and grade results back to the home university. These have to be acknowledged by the home university if the student wants to continue studying at his home university.

The main problem within eErasmus is the setup of collaboration between different universities which are using different administration systems and different grading schemata for students. For this, some universities already have a working collaboration, others do not. This scenario displays the need of collaboration between the home and host university and is an example for collaborations that already exist on some islands within the same domain. These collaboration processes have to be merged or unified and can be adopted by other universities that do not have a collaboration running.

### 5.2 Techniques

As depicted in figure 5 collaboration unification consists of two steps:

1. The minimal interaction constraints in each island of collaboration have to be identified, leading to a normative choreography for each island.
2. The different normative choreographies have to be merged.

The first step mentioned in point 1 is to do identify normative choreographies. As already mentioned in section 2, normative choreographies allow more interaction scenarios than what observable choreographies describe. If a participant could participate in more interaction scenarios than what is captured in the observable choreography, this might be a hint where to extend an observable choreography into the direction of a normative choreography. Figure 6 illustrates an example from the preparation process within the eErasmus case study. A home university X and host university Y form an island of collaboration. The provided interface process of X consists of sending student details, sending the learning agreement (a document containing the list of classes the student wants to enroll in at the host university) and receiving an acknowledgment. In contrast to this, the provided interface process of Y includes the possibility to receive updates of the learning agreement during a time period of 2 weeks until the acknowledgment is sent. The observable choreography of the collaboration between X and Y is equal to the provided interface process of X. No update will ever be sent/received. In contrast to this, Y could also interact with other home universities that send agreement updates within two weeks. Therefore, the provided interface process of Y could be used as the normative choreography while X would still conform to it (see section 4.2 for a short discussion on conformance). Having broadened the possible interaction scenarios by extending the current observable choreography to the normative choreography, more home universities can join in.

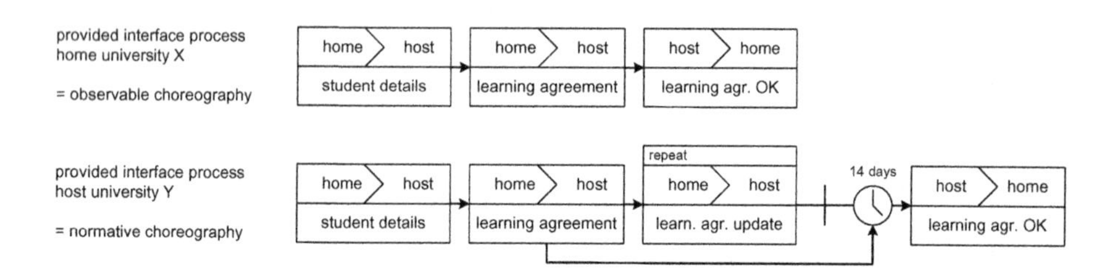


**Fig. 6.** Extracting a normative choreography

We see that identifying a normative choreography from given process implementations is the direct opposite of implementing conforming processes from a given normative choreography. During implementation certain decisions can be made (within the possibilities of conformance). E.g. assuming the normative choreography existed in the example, we could interpret that X has chosen not to send updates although it had the chance to do so. As a result, we have to remove these implementation decisions when identifying a normative choreography. Once we know what kind of decisions can be made during implementation while still preserving conformance, we can introduce techniques for identifying possible results of decisions and reverting the decision in given process implementations.

The second step mentioned in point 2 is to do a merge. Merging different normative choreographies requires for techniques that identify common structures in different process models and help to handle those parts of the process that are different. In the area of version control software these two functions are called *diff* and *merge*. The notion of process inheritance was introduced to reason on whether a changed process definition inherits properties from the original definition. Inheritance-preserving transformation rules are proposed in [16].

## 6 Conclusion

This paper has motivated the need for techniques supporting different categories of choreography design that were derived from eGovernment scenarios. As part of that the distinction between *observable choreographies* and *normative choreographies* was made. We have discussed existing techniques. However, there are still techniques missing for each of the three categories, namely for choreography identification, choreography context expansion and choreography unification. In the case of *choreography identification*, techniques for generating choreographies out of interconnected interface processes are missing for several languages. In the case of *choreography context expansion* extensions for choreography languages are missing that introduce the notion of explicit variability points, where different variants of the choreography can be defined for different business contexts. Process model synthesis techniques have to be conceived for integrating potential interactions into choreographies. The topic of conformance has to be revisited since bi-simulation-based techniques are too restrictive. Finally, in the case of *collaboration unification*, we have seen that techniques for extracting minimal constraints in a choreography and for merging conflicting process model structures into a unified normative choreography are missing. Introducing new techniques for filling the identified gaps and validating them using the presented use cases will be subject to future work.

## References

1. R4eGov - Towards e-Administration in the large: Project Description, March 2006.
2. T. Basten and W. M. P. van der Aalst. Inheritance of behavior. *JLAP*, 47(2): 47–145, 2001.
3. M.-C. Berneron, E. Deserevill, T. V. Cangh, O. Aydogmus, and A. Boujraf. SIXTH FRAMEWORK PROGRAMME, Information Society Technologies, R4eGov, Deliverable WP3-D5, Case Study: Electronic procurement schema for European public administrations, July 2006.
4. N. Busi, R. Gorrieri, C. Guidi, R. Lucchi, and G. Zavattaro. Choreography and Orchestration: A Synergic Approach for System Design. In *B. Benatallah, F. Casati, and P. Traverso (Eds.): ICSOC 2005, LNCS 3826*, pages 228–240. Springer Verlag, 2005.
5. M. Crawford et al. ebXML Core Components Technical Specification 2.01. Technical report, UN/CEFACT, November 2003.
6. G. Decker and M. Weske. Behavioral Consistency for B2B Process Integration. In *Proceedings of the 19th International Conference on Advanced Information Systems Engineering (CAiSE)*, LNCS, Trondheim, Norway, June 2007. Springer Verlag.
7. R. Dijkman and M. Dumas. Service-oriented Design: A Multi-viewpoint Approach. *International Journal of Cooperative Information Systems*, 13(4):337–368, 2004.

8. N. Kavantzas, D. Burdett, G. Ritzinger, and Y. Lafon. Web Services Choreography Description Language Version 1.0, W3C Candidate Recommendation. Technical report, November 2005. `http://www.w3.org/TR/ws-cdl-10`.
9. K. Koskimies and E. Makinen. Automatic synthesis of state machines from trace diagrams. *Software - Practice and Experience*, 24(7):643–658, 1994.
10. J. Lodge and R. Vermer. SIXTH FRAMEWORK PROGRAMME, Information Society Technologies, R4eGov, Deliverable WP3-D4, Case Study e Erasmus eHigher Education (eEH), July 2006.
11. A. Martens. Analyzing Web Service based Business Processes. In M. Cerioli, editor, *Proceedings of Intl. Conference on Fundamental Approaches to Software Engineering (FASE'05), Part of the 2005 European Joint Conferences on Theory and Practice of Software (ETAPS'05)*, volume 3442 of *Lecture Notes in Computer Science*, Edinburgh, Scotland, April 2005. Springer-Verlag.
12. P.-E. Schmitz, T. V. Cangh, and A. Boujraf. SIXTH FRAMEWORK PROGRAMME, Information Society Technologies, R4eGov, Deliverable WP3-D2, Case Study Eurojust / Europol collaboration, July 2006.
13. A. Schnieders and F. Puhlmann. Variability Mechanisms in E-Business Process Families. In *W. Abramowicz, H. Mayr (Eds.): 9th International Conference on Business Information Systems (BIS 2006)*, volume P-85 of *LNI*, pages 583–601, Bonn, Germany, 2006.
14. S. Uchitel, J. Kramer, and J. Magee. Synthesis of behavioral models from scenarios. *IEEE Trans. Softw. Eng.*, 29(2):99–115, 2003.
15. W. van der Aalst and A. Weijters. *Process-Aware Information Systems: Bridging People and Software through Process Technology*, chapter Process Mining, pages 235–255. Wiley & Sons, 2005.
16. W. M. P. van der Aalst and T. Basten. Inheritance of workflows: an approach to tackling problems related to change. *Theor. Comput. Sci.*, 270(1-2):125–203, uary.
17. J. M. Zaha, A. Barros, M. Dumas, and A. ter Hofstede. A Language for Service Behavior Modeling. In *Proceedings 14th International Conference on Cooperative Information Systems (CoopIS 2006)*, Montpellier, France, Nov 2006. Springer Verlag.

# Web Service-Based Specification and Implementation of Functional Components in Federated ERP-Systems

Nico Brehm and Jorge Marx Gómez

Department of Computer Science, Carl von Ossietzky University Oldenburg, Ammerländer Heerstrasse 114-118, 26129 Oldenburg, Germany
{nico.brehm,jorge.marx.gomez}@uni-oldenburg.de

**Abstract.** ERP systems consist of many software components which provide specific functionality. However, these ERP systems are designed as an all-in-one solution, often implementing functionality not needed. The paper presents the reference architecture of a federated ERP system which allows the distribution of ERP system components on the basis of Web Services and shows how these Web Services can be developed and provided by different software vendors. The architecture draws upon a hierarchical standardization model of data and service types. The model advances the reusability of data types and reduces the necessity of data transformation functions in business process descriptions. Furthermore the concrete architecture of a prototype implementation on the bases of open source software components is described.

**Keywords:** Federated ERP systems, FERP, Web Service, SOA, Workflow.

## 1 Introduction

One of the main reasons which increased the demand of ERP system technology in the last two decades results from its data-centric view. This paradigm forms the basis for the internal architecture of ERP systems as well as the structure of the business functionality. ERP systems facilitate the view of enterprises as a whole. The application of ERP systems mainly aims at the improvement of the collaboration between the different departments of an enterprise. An *ERP system* is a standard software system which provides functionality to integrate and automate the business practices associated with the operations or production aspects of a company. The integration is based on a common data model for all system components and extents to more than one enterprise sectors [1, 2, 3, 5].

Modern ERP systems consist of many software components which are related to each other. Currently these components are administered on a central application server. In connection to the ERP system complexity several problems appear:

- The price- performance ratio is dependent to the potential benefit an ERP system is able to generate.
- Not all installed components are needed.
- High-end computer hardware is required.
- Customizing is expensive.

W. Abramowicz (Ed.): BIS 2007, LNCS 4439, pp. 133–146, 2007.
© Springer-Verlag Berlin Heidelberg 2007

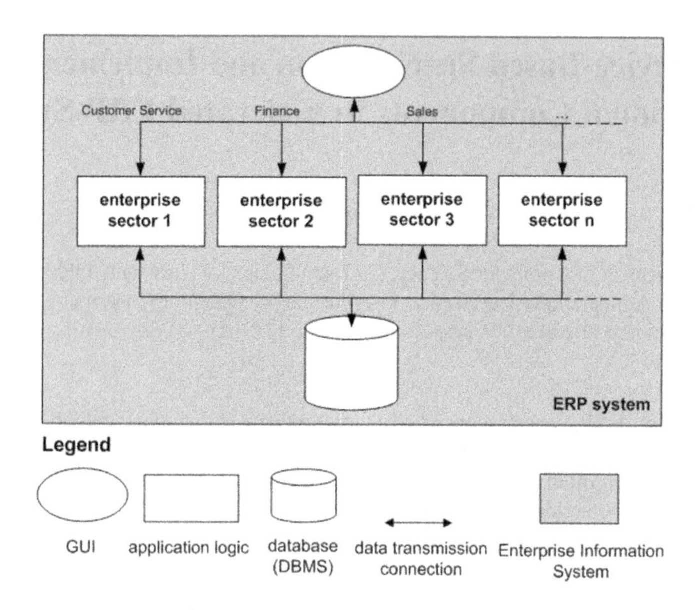


**Fig. 1.** Architecture of a conventional ERP system

Due to the expensive installation [11] and maintenance proceedings of ERP systems normally only large enterprises can afford complex ERP systems which cover all enterprise sectors. In practice small- and medium sized enterprises (SME) deploy several business application systems in parallel. Often many of these systems approach the definition of an ERP system. However, the full potential of each system is not exploited. Furthermore, the parallel operation of business application systems causes problems which jointly arise from insufficient system integration [1, 4].

A new solution to face these problems is the application of a distributed ERP system which makes its components available over a network of services providers. This component ensemble (federated system) still appears as single ERP system to the end-user, however it consists of different independent elements which exist on different computers. Based on this construction it is possible for an enterprise to access on-demand functionality (business components) as services[1] of other network members over a P2P network. This approach solves the mentioned problems as follows:

- The total application costs conform to the effective use of software functions.
- Due to the separation of local and remote functions, no local resources are wasted for unnecessary components.
- Single components are executable on small computers.
- Due to decreasing complexity of the local system also installation and maintenance costs subside.

[1] In this term, a service is a software component that encapsulates one or more functions, has a well defined interface that includes a set of messages that the service receives and sends and a set of named operations [6].

A *federated ERP system* (FERP system) is an ERP system which allows a variable assignment of business application functions to software providers. The overall functionality is provided by an ensemble of standardized subsystems that all together appear as a single ERP system to the user. Different business components can be developed by different vendors.

In this paper we present an FERP system based on Web Services. The main idea follows the multi-layer paradigm of modern information systems which aims at the separation of the application logic from the presentation layer and the database layer. In our approach the application logic of ERP systems is encapsulated in a multiplicity of Web Services which can be provided either locally or remotely. The vision of this approach is to allow the application of business logic components in a distributed manner. In order to facilitate a vendor-independent development and provision of those components the approach considers the standardization of Web Services as well as GUI descriptions and database interactions. The standardization process is supposed to be advanced by a consortium of ERP vendors, utilizing enterprises and scientific institutions (*FERP standardization consortium*).

## 2 FERP Reference Architecture

Figure 2 gives a survey of the reference architecture of a Web Service-based FERP system. The architecture consists of several subsystems which are interconnected. Because one of the main objectives of an FERP system is to integrate business components of different vendors, all components have to comply with standards. In this approach these standards are described as XML schema documents. In order to separate the three different layers of a typical layered architecture of conventional ERP systems each layer is assigned to its own standard.

The subsystems of the proposed architecture are the following:

**FERP Workflow System (FWfS)**
The FWfS coordinates all business processes which have to be described in an appropriate XML-based workflow language. A workflow in this context is a plan of sequentially or in parallel chained functions as working steps in the meaning of activities which lead to the creation or utilization of business benefits. Workflows implicitly contain the business logic of the overall system. Figure 3 shows the component structure of the FWfS. The function types a workflow in FERP systems can consist of are the following:

- model based user interface functions, e.g. show, edit, select, control
- database access functions, e.g. read, update
- application tasks which are connected to Web Service calls

**FERP User System (FUS)**
The FUS is the subsystem which implements functions for the visualization of graphical elements and coordinates interactions with end users. This subsystem is able to generate user screens at runtime. Screen descriptions which have to comply with the *FERP UI standard* are transformed to an end device-readable format, e.g. HTML in case of web browsers. Figure 4 shows the component structure of the FUS.

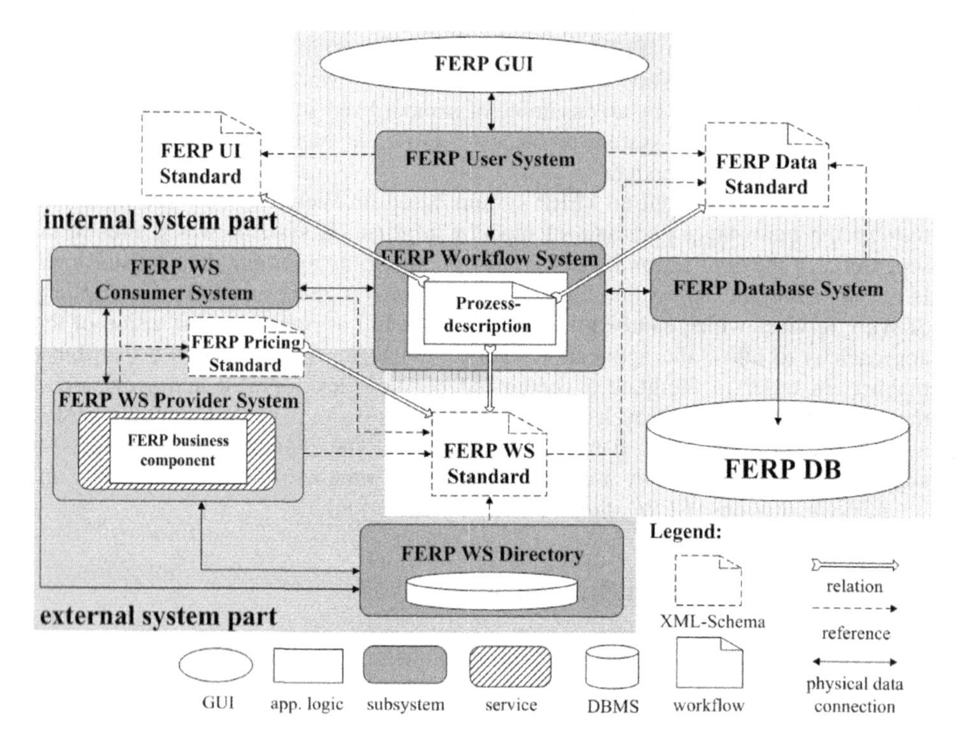


**Fig. 2.** Reference architecture of an FERP system with relation to necessary standards

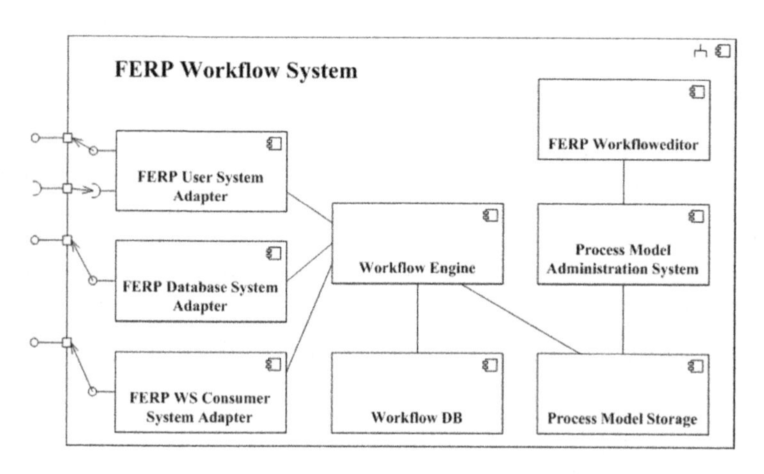


**Fig. 3.** Component structure of the FWfS

**FERP Database System (FDS)**

The FDS is the subsystem which implements functions for the communication with the FERP database. This subsystem is able to interpret XML structures which comply with the *FERP data standard*. The interface differentiates between two kinds of requests. Database update requests contain object oriented representations of business

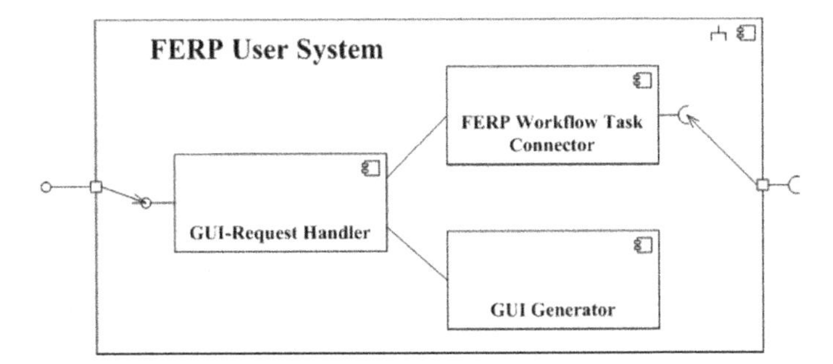


**Fig. 4.** Component structure of the FUS

entities as XML trees. Database read requests contain XML expressions specifying portions of data to be extracted. In both cases the request parameters have to be transformed into different types of request statements that vary depending on the type of database management system (DBMS) which is used. Assumed that a relational DBMS (RDBMS) is used the underlying data model also has to comply with the FERP data standard which means that the corresponding table structure has to reflect the XML-Schema specifications respectively. The java.net project *hyperjaxb2*[2] provides a solution to generate SQL statements on the basis of XML schema definitions. Another solution is the application of native XML databases or XML-enabled RDBMS. Figure 5 shows the component structure of the FDS.

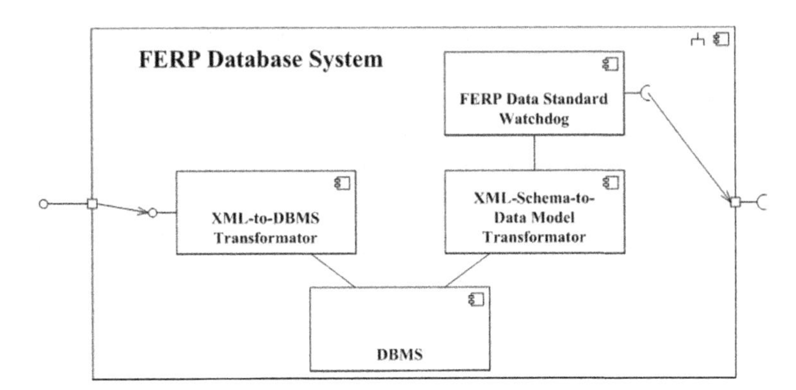


**Fig. 5.** Component structure of the FDS

**FERP Web Service Consumer System (FWCS)**

The business logic of FERP systems is encapsulated in so called FERP business components which are wrapped by a Web Service. The FWCS is the subsystem which provides functions for the invocation of Web Services. All possible types of FERP Web Services are specified by the *FERP WS standard*. This standard contains XML schema definitions which describe Web Service operations as well as input and output

---

[2] Hyperjaxb2 – relational persistence for JAXB objects: https://hyperjaxb2.dev.java.net/ (last visit October 2006).

z

messages. A Web Service references these types in its WSDL description. Furthermore this subsystem is able to search for Web Services which are defined by a unique identifier. By this it is possible that different Web Service providers implement the same business component type as Web Service. Beside the implementation of Web Service invocation and search functions this subsystem is responsible for the interpretation and consideration of non-functional parameters. Examples for those parameters are: security policies, payment polices or Quality of Service (QoS) requirements on the part of Web Service consumers. The architecture references the FERP pricing standard which is supposed to enable Web Service providers to specify Web Service invocation costs. Figure 6 shows the component structure of the FWCS.

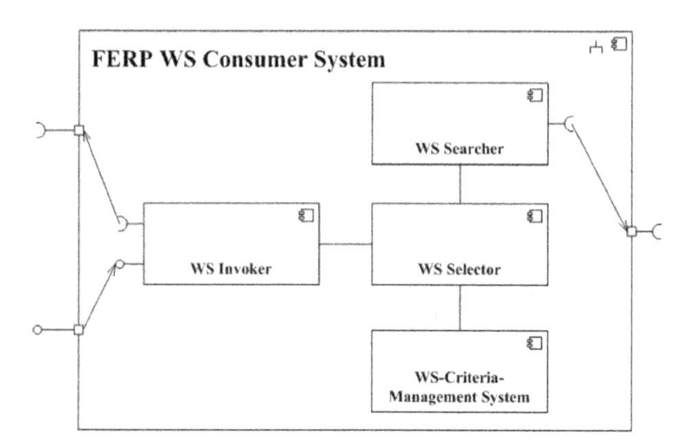


**Fig. 6.** Component structure of the FWCS

**FERP Web Service Provider System (FWPS)**

The FWPS is the subsystem which implements functions for the provision of Web Services which comply with the FERP WS Standard. The subsystem includes a Web Server which is responsible for the interpretation of incoming and outgoing HTTP

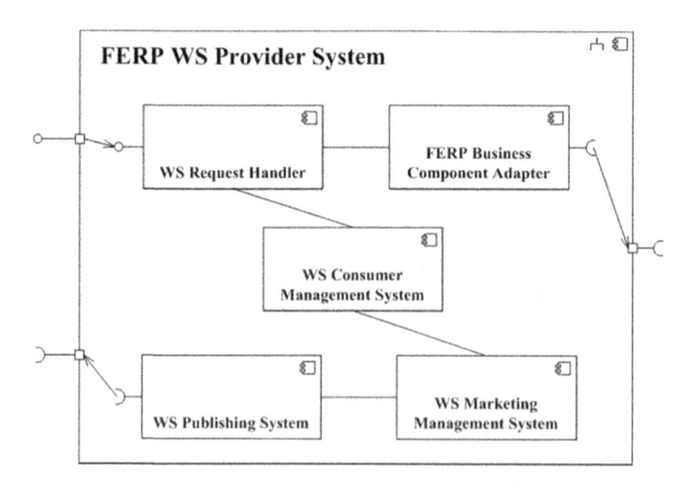


**Fig. 7.** Component structure of the FWPS

requests which in turn encapsulate SOAP requests. The subsystem provides business components of the FERP system as Web Services. A connection to the *FERP Web Service Directory* allows the publication of Web Services. Furthermore this subsystem is responsible for the negotiation of common communication policies such as e.g. security protocols or usage fees with the requesting client. Figure 7 shows the component structure of the FWPS.

**FERP Web Service Directory (FWD)**

The FWD provides an interface for the publication and the searching of FERP Web Services. The structure of this directory leans on the FERP WS standard. In this standard Web Services are assigned to categories mirroring the predetermined functional organization of enterprises. Figure 8 shows the component structure of the FWD.

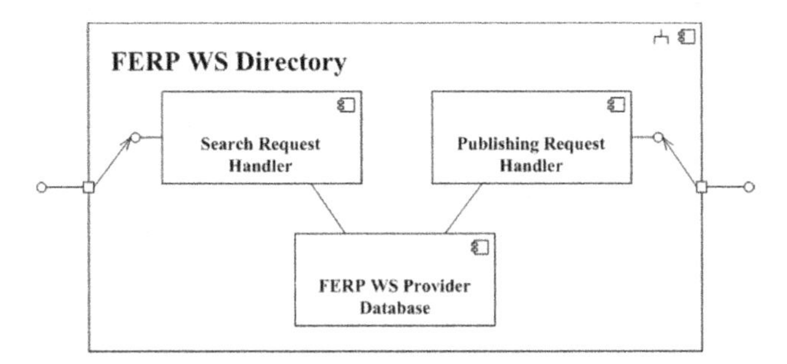


**Fig. 8.** Component structure of the FWD

# 3 Hierarchical XML Schema Structure

The proposed architecture is dependent to the specification of different standards. The next paragraph focuses the standardization of FERP data types and Web Service operations in the context of FERP systems. Because of the complexity of enterprise data models and the difficulty to standardize a completed data model we propose a hierarchical standardization model which allows different abstraction levels. This model uses XML namespaces for the representation of hierarchical levels and XML schema documents for the definition of data types and their relationships. The reason for the usage of XML schema documents is their compatibility with Web Service Description Language (WSDL) which is the common standard for the description of Web Services and is already well supported by tools. The interoperability between FERP Web Services and the FDS is achieved by a transformation of XML schema-based data model descriptions to SQL-based data model descriptions. Web Service Descriptions in WSDL reference the FERP data standard by including the appropriate XML schema documents of the standard. In order to standardize the input and output messages of FERP Web Services we propose the usage of XML schema documents as well.

Figure 9 shows the hierarchical XML schema structure of an FERP system and shows the influence on the systems activities. The left hand side represents different enterprise sectors which are assigned to XML namespaces. This hierarchy can be

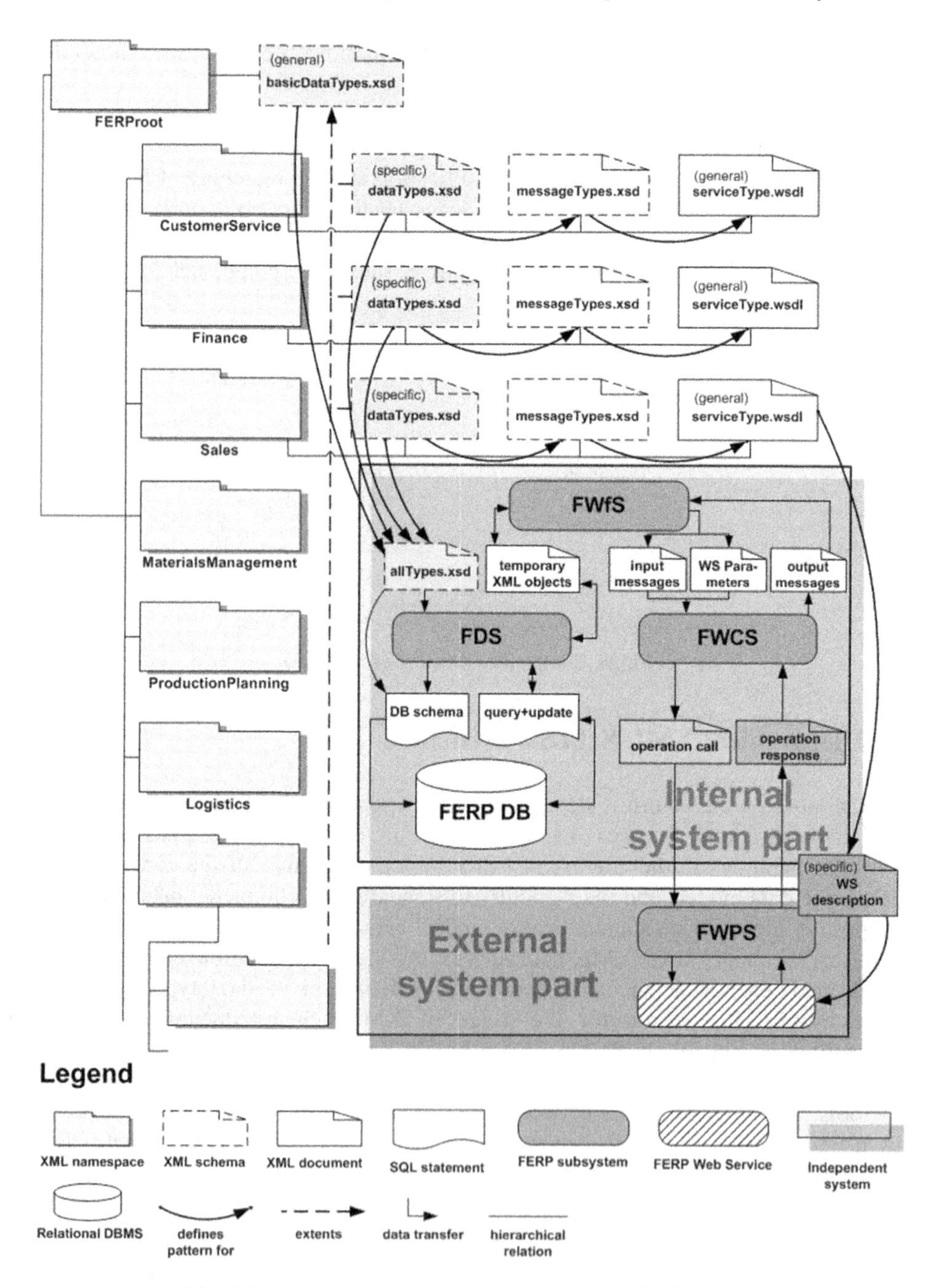


**Fig. 9.** Hierarchical XML schema structure of an FERP system

compared to the internal structure of the application logic of conventional ERP systems which is often mirrored to the navigation structure of their GUI.

The upper half of figure 9 shows the relationships between XML schema documents and concrete Web Service descriptions. Standardized Web Service input and output messages (defined in *messageTypes.xsd*) build the basis for the standardization of Web Service types (described in *serviceTypes.wsdl*). The lower half of figure 9 shows the interactions between the different subsystems of the FERP system. The system internally creates a new XML schema document (*allTypes.xsd*) which includes a copy of all standardized data types that are used in process definitions. The system has a connection to the server of the FERP standardization consortium and will be notified in the case that the standard changed. Those changes are only allowed in terms of extensions. Thereby old versions will be supported during the whole lifetime of the standard. The hierarchical structure provides a useful foundation for this requirement because it is already field-proved in the context of object oriented programming paradigms like polymorphism, generalization and specialization. The local XML schema representation will be transformed to a relational representation of the data model as SQL statement list. In addition to the schema transformation the FDS is able to transform SQL result sets to XML documents that comply with the FERP data standard in the case of *DATABASE_LOAD requests*. On the other hand XML documents will be transformed to SQL INSERT or UPDATE statements in the case of *DATABASE_STORE requests*. Both LOAD and STORE functions are provided by the FDS and can be used by the FWfS.

Web Service calls are initiated by the FWfS as well (see figure 9). Therefore the FWfS sends a standardized XML representation of the appropriate input message to the FWCS. A second XML document contains configuration parameters which specify the concrete Web Service provider to be chosen by the FWCS. Those parameters include either a URL for a static Web Service call or requirements for a dynamic call like e.g. a maximum price. An alternative way for the specification of requirements for dynamic calls is a centralized mapping between Web Service types and requirements. Once the FWCS chose an appropriate Web Service provider it will repack this message to a *SOAP operation request* which includes the standardized name of the Web Service operation to be invoked. This request will be sent to the FWPS. After having finished the processing of the business logic the FWPS will return a SOAP operation response which includes a standardized response message. Figure 9 shows how this response message is going to be sent back to the FWfS that primarily initiated the Web Service call.

# 4 Prototype

The following paragraph briefly describes a first implementation of the proposed reference architecture which is based on open source software components. Figure 10 shows the architecture of our prototype. For the implementation of the FWfS we chose the workflow engine of the *YAWL* project[3]. The FUS was implemented on the basis of *Apache Struts*[4]. Our FDS is mainly based on the API of the Hyperjaxb2

[3] http://yawlfoundation.org/

[4] http://struts.apache.org/

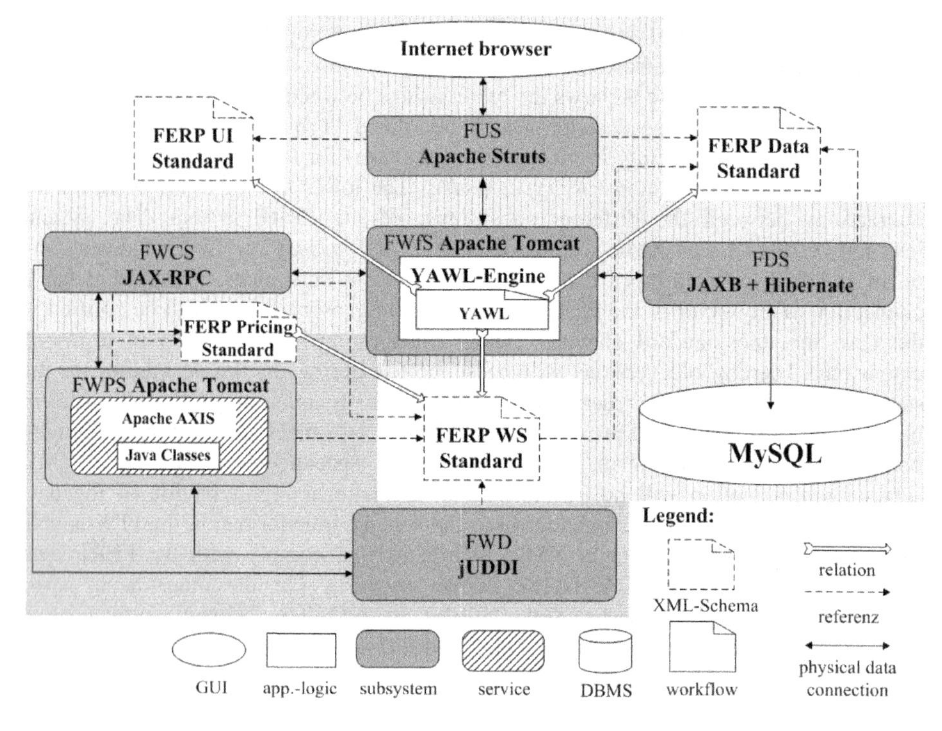


**Fig. 10.** Architecture of the prototype

project which in turn uses *JAXB*[5] and *Hibernate*[6]. *jUDDI* [7] served as basis for the implementation of the FWD. The FWCS uses *JAX-RPC*[8] (Java API for XML-based RPC) which is provided by the SUN Developer Network (SDN). Our FWPS uses *Apache AXIS*[9] as basis for the provision of Web Services.

Figure 11 shows an example process model in YAWL. Tasks in our process definitions can be assigned to one of the three function types:

- Database communication (in figure 11 indicated as DB-task)
- End-user communication (in figure 11 indicated as GUI-task)
- Web Service communication (in figure 11 indicated as WS-task)

All other symbols comply with the graphical notation of YAWL. The example process model demonstrates a workflow for the creation of a purchase order[10]. The example includes only one Web Service call which is responsible for the calculation of the total sum of a purchase order which consists of one or more order items. Order

[5] https://jaxb.dev.java.net/
[6] http://www.hibernate.org/
[7] http://ws.apache.org/juddi/
[8] http://java.sun.com/webservices/jaxrpc/
[9] http://ws.apache.org/axis/
[10] In order to improve understandability the process was simplified. Changes of entered data and order items are not supported.

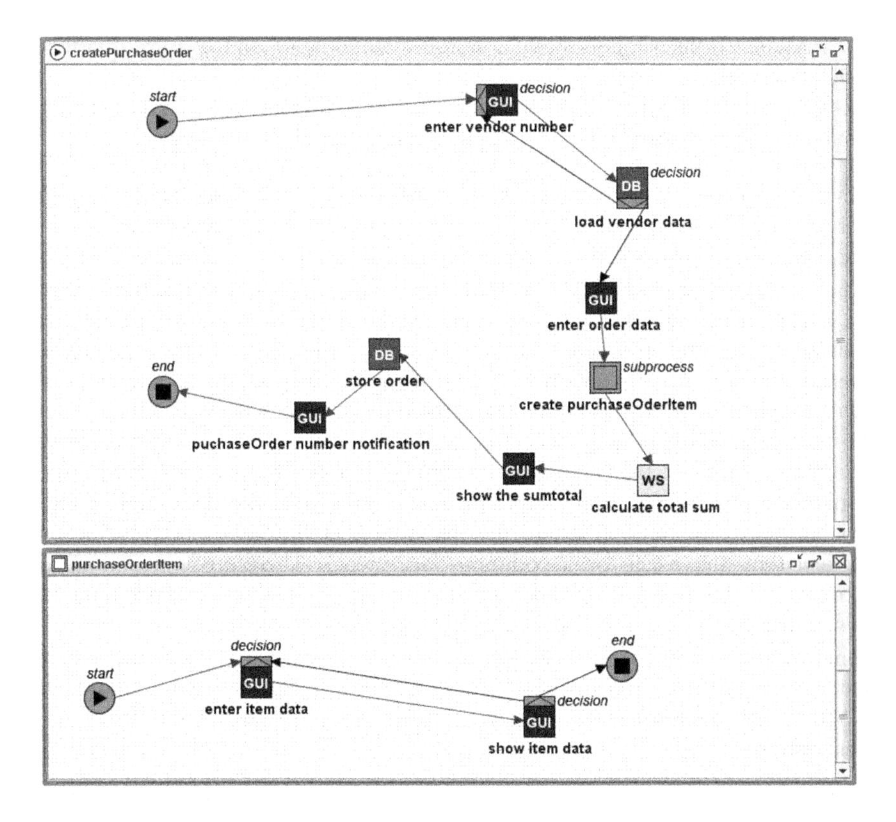


**Fig. 11.** Process model in YAWL as simplified example for the creation of a purchase order

items include a price and an amount. The Web Service receives the whole order as XML document without total sum. Having finished the calculation of the total sum the Web Service returns the completed order as XML document. The next workflow task visualizes this XML document. After the user agreed the XML document is transmitted to the FERP database system which transforms it to an *SQL-INSERT* statement in the next workflow task.

# 5 Related Works

According to the 3-tier architecture of business application systems today's EAI (Enterprise Application Integration) platforms support the integration over all three tiers. Enterprise portals mainly provide a basis of the consolidation of existing software systems on the user interface level which means that portals feature a user-centric orientation [8]. The Web Services paradigm implements a Service-oriented architecture which presupposes a middleware for the management of services. Search and publication requests are processed by this middleware. Business Process Management (BPM) platforms support the orchestration of such services. Thus it is possible to centralize Web Service accesses in business process definitions. In some cases EAI platforms support both a portal functionality and a BPM platform in combination with each other.

Disadvantages of domain-independent BPM platforms as foundation for the implementation of Web Service-based ERP systems are the following:

- XML object representations have to be transformed to different data models (XML schemas) when independent Web Services are used in one process because in most cases no common standard[11] is referenced in independent Web Service descriptions.
- XML object representations of Web Service return values have to be transformed to SQL statements[12] in the case that return values have to be stored in an external database. This transformation has to be part of the business process definition. Because of this problem Web Services often are assigned to their own database which is directly accessed by the business logic of a Web Service. The problem of this solution is that Web Services cannot be exchanged if they are provided externally[13] because the connected database will be not available anymore. Another problem is that such a solution would not comply with the definition of an ERP system (see paragraph 1) where the integration of different enterprise sectors is achieved by the usage of a common data model. Therefore conventional ERP systems use a central DBMS whereby also the management of database transactions is simplified.
- Input values for Web Services which have been extracted from the enterprise database have to be transformed from a database result set representation to an XML object representation which complies with the respective Web Service description.

The presented approach of a Web Service-based FERP system offers the following advantages that all together address actual challenges of business process modelling approaches[14]:

- Output values of Web Services can be directly used as input values for other Web Services because all Web Service definitions reference the same standard.
- Output values of Web Services can be transmitted to the FDS directly because both implement the same data model.
- XML object representations which have been extracted from the FDS can be used straightforward as input values of Web Services because both, the data model of the FDS and the parameter description of each Web Service comply with the same standard.

Another neighbouring working area is represented by Federated Database Systems [9] which are a type of Meta-DBMS. Those systems integrate multiple autonomous DBMS to a single system. This integration is achieved by schema mapping

[11] No common standard means that the BPM platform is domain-independent and transformations have to be included in the process definition.
[12] In case of using an RDBMS.
[13] Externally means that Web Services can be provided outside the enterprise's intranet by independent software vendors.
[14] Currently business processes include both abstract business logic and technical constructs in an unstructured manner which complicates traceability [10].

techniques. One solution for the data integration in Web Service-based ERP systems could be the utilization of a Federated Database System in order to consolidate independent DBMS which are directly assigned to Web Services. The Federated Database System would represent a central entry point to a decentralized DBMS structure which in turn would comply with the definition of an ERP system. This solution has the following disadvantages in comparison to the presented approach:

- Enterprises are dependent to Web Service providers who also provide one part of the database federation. In the case that different providers offer the same Web Service type a migration from one provider to another implicitly necessitates data migration.
- A global schema[15] indeed can define a normalized data model but redundancies in the overall network of independent DBMS are possible anyway. Because Web Services would directly access their local DBMS duplicate entries in the DBMS federation could lead to complications when a process or another Web Service accesses the central Federated Database System.

In comparison this approach, an FERP-system has the following advantages:

- A migration from one Web Service provider to another does not influence the data view because all data is stored in a central database which can only be accessed by a local[16] process. Web Services have no direct database connection.
- Because the hierarchical FERP standard considers the combination of Web Services' duties and affected data each level in the hierarchy is assigned to unique operation and data types. Furthermore the inheritance support allows a reutilization of general data types. Thereby redundancies in the database can be avoided because on the one hand existing data type definitions can be reused for new Web Service definitions and on the other hand standardized Web Services which use existing data types will create redundancy-free[17] data.

# 6 Conclusions and Outlook

Comparing distributed ERP systems and ERP systems running on only one computer, the distributed systems offer a lot of advantages. Particularly small- and medium sized Enterprises (SMB) benefit from using shared resources. However, the design of distributed system architectures is subject to a number of problems. The paper addresses the problem of redundant data in business application systems of independent vendors presents a basis for the standardization of ERP system components that are provided as Web Services. A standardized data model builds the basis for message and service standardization. The hierarchical structure of the presented standard advances the reuse of existing data types. Furthermore we presented a reference architecture of FERP

[15] In the Local as View (LaV) mapping direction the local schemas of independent DBMS are defined in terms of the global schema. In the Global as View (GaV) mapping direction the global schema is defined in terms of the underlying schemas.

[16] Local in this context means that the FWfS is directly connected to the central database.

[17] New Objects update old objects of the same type and the same identity because a central DBMS is used.

systems which reduces the necessity of data transformation functions in business process descriptions. The standardization of the syntactic level is only the first step. Behaviour, synchronization and quality of Web Services must flow into the definition of an overall ERP system standard. The future work must pick up these problems to realize the vision of a loosely coupled ERP system which allows the dynamic outsourcing of applications [5, 7] and the combination of software components of different providers.

## References

1. Abels, S., Brehm, N., Hahn, A., Marx Gómez, J.: Change management issues in federated ERP-systems – An approach for identifying requirements and possible solutions. International Journal of Information Systems and Change Management (IJISCM) 1 (2007) 318-335
2. Brehm, N., Marx Gómez, J.: Secure Web Service-based resource sharing in ERP networks. International Journal on Information Privacy and Security (JIPS) 1 (2005) 29-48
3. Brehm, N., Marx Gómez, J.: Standardization approach for Federated ERP systems based on Web Services. 1st International Workshop on Engineering Service Compositions, Amsterdam (2005)
4. Brehm, N., Marx Gómez, J.: Federated ERP-Systems on the basis of Web Services and P2P networks. International Journal of Information Technology and Management (IJITM) (2007)
5. Brehm, N., Marx Gómez, J., Rautenstrauch, C.: An ERP solution based on web services and peer-to-peer networks for small and medium enterprises. International Journal of Information Systems and Change Management (IJISCM) 1 (2005) 99-111
6. Cuomo, G.: IBM SOA "on the Edge". ACM SIGMOD international conference on Management of data. ACM Press, Baltimore, Maryland (2005) 840-843
7. Dan, A., Davis, D., Kearney, R., Keller, A., King, R., Kuebler, D., Ludwig, H., Polan, M., Speitzer, M., Youssef, A.: Web services on demand: WSLA-driven automated management. IBM SYSTEMS JOURNAL 43 (2004) 136-158
8. Sheth, A.P., Larson, J.A.: Federated database systems for managing distributed, heterogeneous, and autonomous databases. ACM Computing Surveys (CSUR) 22 (1990) 183 - 236
9. Smith, M.A.: Portals: toward an application framework for interoperability. Communications of the ACM 47 (2004) 93 - 97
10. Verner, L.: BPM: The Promise and the Challenge. QUEUE (ACM Press) 2 (2004) 82-91
11. Vogt, C.: Intractable ERP: a comprehensive analysis of failed enterprise-resource-planning projects. Software Engineering Notes 27 (2002) 62-68

# SEEMP: Meaningful Service-Based Collaboration Among Labour Market Actors

Emanuele Della Valle[1], Dario Cerizza[1], Irene Celino[1], Jacky Estublier[2], German Vega[2], Michael Kerrigan[3], Jaime Ramírez[4], Boris Villazon[4], Pascal Guarrera[5], Gang Zhao[5], and Gabriella Monteleone[6]

[1] CEFRIEL – Politecnico of Milano, Via Fucini 2, 20133 Milano, Italy
[2] Equipe Adele, LSR, Université Joseph Fourier, F-38041 Grenoble Cedex 9, France
[3] DERI, University of Innsbruck, Technikerstraße 21a, 6020 Innsbruck, Austria
[4] Universidad Politecnica de Madrid, 28660 Boadilla del Monte, Madrid, Spain
[5] Le Forem, Boulevard Tirou 104, 6000 Charleroi, Belgium
[6] TXT e-solutions, via Frigia 27, 21126 Milano, Italy

**Abstract.** SEEMP is an European Project that promotes increased partnership between labour market actors and the development of closer relations between private and public employment services, making optimal use of the various actors' specific characteristics, thus providing job-seekers and employers with better services". The need for such a flexible collaboration gives rise to the issue of interoperability in both data exchange and share of services. SEEMP proposes a solution that relies on the concepts of services and semantics in order to provide a *meaningful service-based collaboration among labour market actors.*

## 1 Introduction

European Member States have introduced major reforms to make the labour market more flexible, transparent and efficient (compliance with the European Employment Strategy guidelines). Such major reforms include decentralization, liberalization of the mediation market (competition between public and private), and quality monitoring of Employment Service (ES) staff and services. As an effect ESs understood the need for making available on-line a one-stop shop for the employment. This results in an increased used of ICT and in a boost in differentiating and personalizing the services they offer (e.g., Borsa Lavoro Lombardia[1], Le FOREM[2], etc.).

Current employment market is characterized by high heterogeneity of models and actors; in particular we can distinguish between Public Employment Services (PES) and Private Employment Services (PRES)[3]. The ICT systems in ES can serve different purposes: facilitating job-matching and job mobility for job seekers

[1] http://www.borsalavorolombardia.net/
[2] http://www.leforem.be/
[3] In the rest of the paper we use ES when we refer both to public and private actors, whereas we us PES and PRES referring respectively to public and private actors.

W. Abramowicz (Ed.): BIS 2007, LNCS 4439, pp. 147–162, 2007.
© Springer-Verlag Berlin Heidelberg 2007

and employers; improving the functioning of labour markets; coordination and exchange of information; allowing a more efficient management of ES internal services; monitoring of local market trends; personalized services, etc..

The need of reconciling local, regional and national policies is increasing and it concerns the combination of services and data provided by different actors.

SEEMP project[4] (IST-4-027347-STP) aims at designing and implementing in a prototype an Interoperability infrastructure for PESs and PRESs. More specifically, SEEMP is developing an EIF-compliant Architecture [1] to allow collaboration among the employment services that exist in Europe. The resulting European Employment Marketplace will overcome the national barriers complying, at the same time, with the local policies of each Member States. Thanks to SEEMP, which promotes increased partnership between labour market actors and the development of closer relations between private and public employment services, job-seekers and employers will have better services that operate at European scale. For instance, the matching process between job offers and CVs across all Europe will become possible increasing, eventually, labour hiring and workforce mobility.

The paper is structured as follows: Section 1 presents the problems to be solved and introduce a running example that will be discussed throughout the rest of the paper; Section 3 explains the interoperability issue that arises in SEEMP project; Section 4 presents the SEEMP approach to support meaningful service-based collaboration among ESs; Sections 5 and 6 outline SEEMP solution architecture and components rooting them to the related work in Web Services and Semantic Web community; Section 7 briefly discusses the approach of SEEMP comparing it with already implemented approaches; and finally Section 8 presents future work.

## 2 Problems to Be Solved

In order to fulfil SEEMP ambitious goal several problems must be solved at organizational and technical level.

**At an organizational level,** the business model of SEEMP has to be catchy for all ESs. The main reason for a ES to buying in is creating added value for its local users (both job seekers and employers) by offering interconnections with other ESs. Today it is normal for all users to insert CV and Job Offers in many ESs and collect, laboriously, the results personally. When SEEMP will be in place each ES will be able to collaborate with other ESs. From the perspective of the end user, the add-value is the outreach to other niches of the job market without 'being stretched out". End users could insert the CV and Job Offers once and collect pan-European results. From ESs perspective it will results in increase both the number of users and their faithfulness to each ES, thus an increase in transaction volume.

[4] http://www.seemp.org/

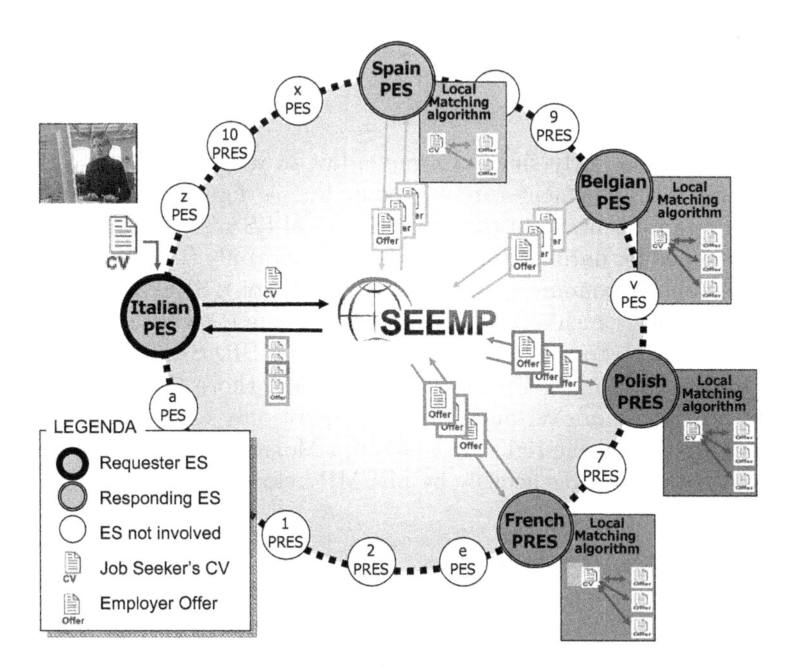


**Fig. 1.** The running example of distributed matching of CVs and job offers

**At technical level,** the need for such flexible collaboration between ESs, gives rise to the issue of interoperability in both data exchange and share of services. The technical approach of SEEMP relies on the concepts of Web Services and semantics. Web Services, exploited in a Software Engineering manner, enable an easier maintenance of the integration. Semantics, encoded in the systems by the means of ontologies and mediators, allows for reconciliation of the hundreds local professional profiles and taxonomies.

SEEMP solution will expose, following the well established Software Engineering approach of Mélusine [2], a single consistent set of abstract services each ES can invoke. Such abstract services will provide a multilateral interoperability solution that delegates the execution of the services to the local ESs (in accordance with the subsidiarity principle) and aggregates the results before sending the response back to the invoker. Moreover, following the innovative Web Service Modeling Ontology [3] approach, we will capture the semantics shared among ESs in a single consistent model. Such model includes a reference ontology in which and from which the local semantics is mapped as well as semantic description of the local Web Services for their automatic use. A set tools will be provided to each ES for modeling its local ontology and for aligning the local ontology with the reference one. As a technical result *SEEMP will enable a meaningful service-based collaboration among ESs.*

**A e-Employment Running Example.** For the discussion of this paper we will consider a running example derived by the user requirements of the SEEMP project:

*Job seekers (companies) put their CVs (Job Offers) on a local PES and ask to match them with the Job Offers (CVs) other users put in different PESs and PRESs through SEEMP.*

It may look like a fairly simple example, but to reach its potential EU-wide audience, this e-Employment running example (see figure 1) needs to fulfil a wider set of requirements than the respective local ES service. A local matching service is designed for national/regional requirements only (i.e., central database, single professional taxonomy, single user language, etc.). SEEMP has to be able to send the request, which an end-user submits to the local PES (the Italian ES on left in the figure), to the all the other PESs and PRESs in the marketplace. In order to avoid asking "all" ESs, it has to select those that most likely will be able to provide an answer and send the request only to them (the two PESs and the two PRESs on the right in the figure). Moreover, the answers should be merged and ranked homogeneously by SEEMP before they are sent back.

## 3 Interoperability Issues

The running example presented in section 1 highlights the need for a system that covers the whole EU and subsumes hundreds of real **heterogeneous systems** existing in EU countries and regions. It implies by-passing:

- *language* heterogeneity, e.g., an Italian Java Analyst Programmer may be looking for job offers written in all the different European languages;
- *CVs and Job Offers structur*al heterogeneity, i.e., the use of standards like HR-XML[5] is not wide spread and a multitude of local formats exists
- *CVs and Job Offers content* description heterogeneity, i.e., European level occupation classifications like ISCO-88[6] exist, but they do not reflect legitimate differences and perspectives of political economic, cultural and legal environments; and
- system heterogeneity in terms of *service interface and behavior*, i.e., no standard exists for e-employment services thus each ES implemented them differently.

All those are typical interoperability issues that SEEMP helps in solving. The need for interoperability at European Level among e-Government services has been perceived since 1999 [4] with the adoption of a series of actions and measures for pan-European electronic interchange of data between administrations, businesses and citizens (IDABC) [5].

The main results of IDABC is the European Interoperability Framework (EIF) [1]. EIF follows the principle of subsidiarity[7] in addressing the interoperability problem at all levels: organizational, semantic and technical. One crucial

[5] http://www.hr-xml.org/

[6] http://www.warwick.ac.uk/ier/isco/isco88.html

[7] The principle of subsidiarity recommends not to interfere with the internal workings of administrations and EU Institutions.

aspect, deriving from the principle of subsidiarity, is to keep responsibility decentralized; in other words each partner should be able to keep its own business process almost unchanged[8] and to provide externally point of exchange for its processes. EIF names these points "business interoperability interfaces" (BII).

EIF does not prescribe any solution, but it rather recommends the principles to be considered for any e-Government service to be set up at a pan-European level: accessibility, multilingualism, security, privacy, use of open standards and of open source software (whenever feasible) and, last but not least, use of multilateral solutions.

SEEMP proposes itself as an implementation of EIF in the domain of e-employment.

## 4 The SEEMP Approach

**SEEMPS Relies on the Concept of Service.** Following the EIF, each ES locally must expose its BII as Web Services. All these Web Services differ but they are fairly similar. SEEMP, as marketplace, models a single consistent set of Web Service out of those exposed by the ESs. Therefore the services exposed by SEEMP become the actual standard for the distributed independent service providers.

For instance, for the running example provided in section 1 each ES should expose two Web Services: match an external CV against the job offers stored locally and match an external job offer against the CVs stored locally. Therefore a service that subsume all the local heterogeneous ones can be modeled in SEEMP for each of the two families of similar Web Service. These two abstract service are those that SEEMP, as a marketplace, offers to the ESs.

SEEMP uses Mélusine [2] as tool for modeling those abstract services and orchestrating the process of delegating the execution to the distributed independent service providers.

**Moreover, SEEMPS Relies on the Concept Semantics (Both Ontologies and Mediators).** As for the service, each local ES has its own *local ontology* for describing at a semantic level the Web Services it exposes, and the structure/content of the messages it exchanges. All these ontologies differ but they are fairly similar, because a common knowledge about employment exists as well as the needs for exchange (i.e., you don't exchange on things with no equivalence calculus). So, SEEMP, as marketplace, models a single consistent ontology out of those exposed by the ESs. Therefore the reference ontology of SEEMP becomes the actual standard for the ESs that should provide the *mediators* for translating from the local ontologies to the reference one and vice versa.

[8] Quoting from IDABC: "it is unrealistic to believe that administrations from different Member States will be able to harmonize their business processes because of pan-European requirements".

For instance, for the running example each ES should model in a local ontology the structure/content of its CV/job offers and the way they are exchanged via Web Services. A reference ontology that subsumes all the local heterogeneous ones can be modeled in SEEMP and it becomes the source of shared understanding. Each ES has to provide its mediator for local-reference semantic mapping.

SEEMP adopts WSMO [3] a way to semantically describe Web Service, ontologies and mediators, WSML [6] as concrete syntax for encoding those descriptions and Methodology [7] as methodology for developing and maintaining those semantic descriptions.

**Minimal Shared Commitment.** In implementing SEEMP approach particular attention is paid in keeping a "win-win" situation among all ESs. The commitment (both at services and semantics level) of each ES should be minimal. ESs care about being able to share while maintaining all the necessary and unnecessary disagreements. It may appear counter-intuitive, but the most suitable set of services and ontology is the one that enables ESs to "agree while disagreeing". Both the reference set of services and the reference ontology must cover the various aspects of the employment market with an *acceptable level of details that leaves leeways of disagreement.*

Considering the running example of section 1, the minimal shared commitment the SEEMP consortium agreed upon consists in sharing a subset of the CV named *candidacy* and a subset of job offer named *vacancy*. Candidacy (vacancy) enables matching without reveling how to contact the job seeker (employer) that is, instead, in the CV (job offer) stored in the ESs and that can be contacted by invoking a service of the ES responsible for the CV (job offer). In this way SEEMP approach also takes into consideration privacy issues of collaborative network, which is a technical issue as well as a business constraint.

## 5 The SEEMP Solution Architecture

SEEMP solution is composed of *a reference part* (all the dark components in figure 2), which reflects the "minimal shared commitment" both in terms of services and semantics, and by *the connectors* toward the various local actors (the components in shading colors in figure 2).

### 5.1 Structural Overview

**The reference part of SEEMP solution** is made up of the central abstract machine, named EMPAM (Employment Market Place Abstract Machine) and a set of SEEMP services.

*The EMPAM* is an abstract machine, in that it does not perform directly any operation, but rather offers abstract services that are made concrete by delegation: when the abstract service is invoked, the EMPAM delegates its execution to the appropriate ES by invoking the correspondent concrete services. It acts as

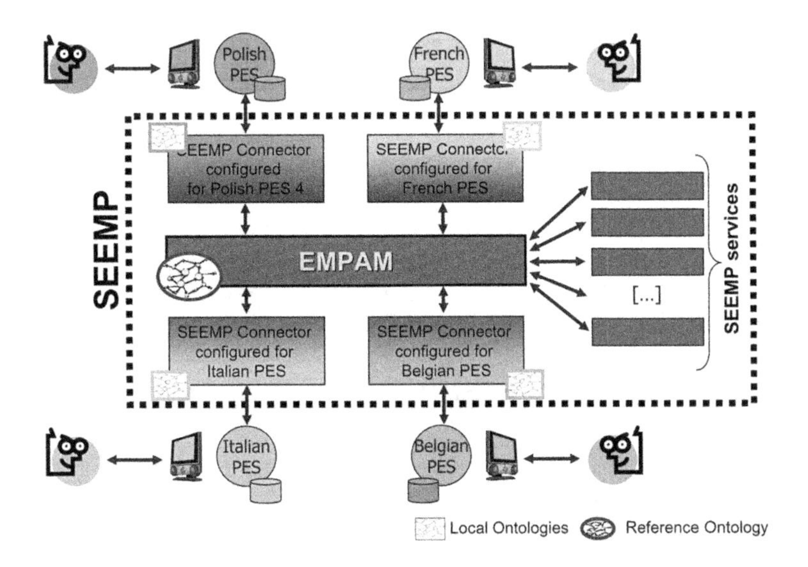


**Fig. 2.** An overview of the SEEMP solution

a multilateral solution (as request by EIF), in which all the services connected to the EMPAM are made available to all other ESs, i.e. they ensure a pan-European level of services without interfering with the Business processes of each ES.

*The SEEMP services* are meant to support EMPAM execution. The running example requires two SEEMP service: discovery and ranking. The *discovery* service is offered by Glue [8]. The EMPAM invokes Glue Discovery Engine before delegating the execution to the concrete services exposed by the ESs. Glue analyzes the CV sent by the invoking ES and it selects among all ESs those that most likely would be able to return relevant job offers. The ranking service is invoked by the EMPAM after all the concrete services have answered and it merges the results providing an homogeneous ranking of the returned job offers. It also delete possible duplicated job offers, which different ESs may have returned.

**The SEEMP connectors** enables all collaboration that occurs between the EMPAM and a given ES. A SEEMP connector will exist for each of the ESs that are connected to the EMPAM and have two main responsibilities:

- *Lifting and Lowering*: when communicating with the ES any outgoing (or incoming) data which is exchanged by the means of Web Services must be lifted form XML to WSML in terms of the local ontologies of the ES (or lowered back to the XML level from WSML).
- *Resolving Heterogeneity*: each ES has its own local ontology that represents its view on the employment domain. The SEEMP connector is responsible for resolving these heterogeneity issues by converting all the ontologized content (the content lifted from the XML received from the ES) into content in terms of the reference ontology shared by all partners and vice versa.

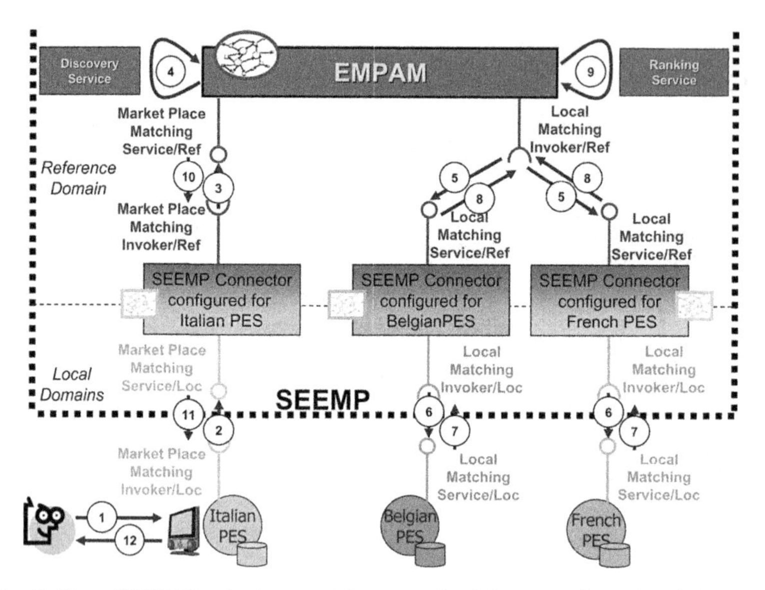


**Fig. 3.** How SEEMP solution enables meaningful service-based collaboration

## 5.2 Functional Overview

By combining the EMPAM and the connectors SEEMP solution enables a *meaningful service-based collaboration* among ESs. Figure 3 illustrates how such meaning collaboration takes place in running the example of section 1:

1. the user inserts a CV into the Italian PES and requests relevant job offers,
2. the Italian ES invokes the marketplace matching service passing the CV encoded in the Italian ontology,
3. the SEEMP connector translates the CV from the Italian ontology to the reference one,
4. the discovery service analyzes the CV and selects among all ESs those that most likely would be able to return relevant job offers,
5. the EMPAM invokes in parallel the local matching service of selected ESs,
6. the various connectors translate the CV from the reference ontology to the local ontology (i.e., the Belgian and the French ESs) and invoke the local service,
7. the Belgian PES and the French PRES compute the matching locally and returns a set of job offers,
8. the connector translates the job offers from each local ontology to the reference one,
9. the ranking service merges, at a semantic level, the responses and ranks the job offers homogeneously,
10. the job offers are sent back in the reference ontology to the Italian connector that translate them in the Italian ontology,
11. the connector responds to the Italian ES, and
12. finally the ES displays the job offers to the user.

# 6 The SEEMP Solution Components

## 6.1 Reference and Local Ontology for E-Employment

The Reference Ontology is a core component of the system. It acts as a common "language" in the form of a set of controlled vocabularies to describe the details the employment sector. The Reference Ontology has to be rich enough to support the semantic needs of all the ES (Local Ontologies) involved currently and in the future. The Reference Ontology also has to be a scalable, adaptable and maintainable ontology. For all those reason SEEMP follow some of the identified tasks of the ontology development methodology METHONTOLOGY [7].

In the case of the reference ontology, the building process consisted of:

1. Specifying, using competency questions, the necessities that the ontology has to satisfy in SEEMP.
2. Selecting the standards that cover most of the identified necessities.
3. Semantic enrichment of the chosen standard.
4. Evaluating the Ontology content
5. Integrating the resultant ontology in the SEEMP platform.

In order to build the Reference Ontology, the standards identified are:

- Currency: ISO 4217.
- Driving License: the 12 levels recognized by the European legislation.
- Economic Activity: NACE Rev. 1.1
- Occupation: ISCO-88 (COM), ONET and ED taxonomy of occupations.
- Education: FOET and ISCED97.
- Geography: ISO 3166 country codes.
- Labour Regulatory: LE FOREM classifications Contract Types and Work RuleTypes.
- Language: ISO 6392 and the Common European Framework of Reference.
- Skill: European Dynamics Skill classification.
- Time Ontology: is based on DAML ontology, and expressed in OWL.

Based on the proposed SEEMP architecture, the possible options for building the local ontologies in SEEMP ranges between to extreme options: building local ontologies taking as a seed the reference ontology and building local ontologies as a reverse engineering process from ES schema sources.

**In building local ontologies taken as a seed the reference ontology,** the concepts in the local ontology are extension in depth of the concepts already present in the reference ontology. By extension we mean including application dependent concepts that appear in each ES schema source.

The exchange of job offers and CV (once ontologized), required by the running example of section 1, is easy because all local ontologies extends the same reference vocabulary. On the contrary mappings complexity are on local ontology and schema mappings (cf. section 6.3 for more details).

**Building local ontologies as a reverse engineering process from ES schema sources,** is the easiest way for ontologizing ESs. Each concept in the local ontology is the semantic expression of a relevant concept in the ES. In this way ESs becomes ontology-based applications that are more efficient because mappings between local ontologies and schema sources should not be complex, but complex mappings appear between the local and reference ontology. Therefore data exchange will require more time, in comparison to the previous option, due to the execution of two complex mappings.

**The SEEMP way** is keeping close to the first option in the beginning, when few ESs are in the marketplace and the union of ES semantics is the most straight forward solution. Then while more and more ESs would be added to the marketplace, the solution would move toward the second option. The equilibrium between the two extreme solutions is related to the need for a "minimal shared commitment" explained in section 4.

### 6.2 An Employment Market Place Abstract Machine

The EMPAM machine is implemented as a Mélusine application, which means it is structured following the Mélusine approach in three layers (cf. Figure 4).

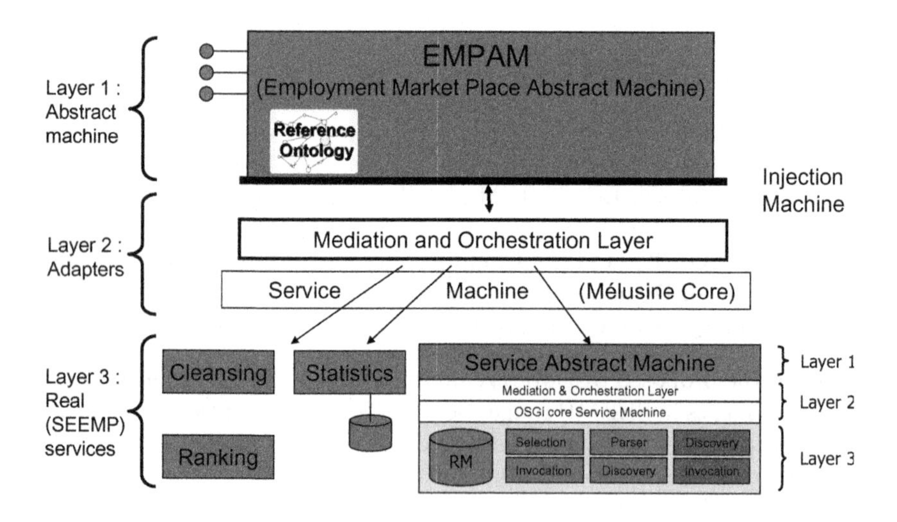


**Fig. 4.** The levels that made up the EMPAM as a Mélusine application

**Layer 1:** The abstract domain. The higher EMPAM machine layer is a java abstrct program where abstract classes represent the concepts present in our SEEMP public employment service. EMPAM acts as a ES covering completely EU, i.e. it acts as if all the CV and vacancies were present in its repositories.

However the EMPAM is abstract since, in fact, it does not have any information locally, but delegates to real ESs the duty to process part of the job. The EMPAM program defines functions like repository access and matching that are, indeed, not implemented at all, or only sketching what they are supposed to do.

**Layer 2:** The adapters. The second layer duty is to do in such a way that empty or dummy methods, found in the abstract layer, really perform what they are supposed to perform. To that end this layer is itself structured in three layers:

- *The injection machine*, whose duty is to capture those methods that need to be completed or implemented, and to transfer the call to the mediation and orchestration layer.
- *The mediation and orchestration layer* which is in charge of transforming a single abstract method call into a potentially complex orchestration of real lower level services that together will perform the required function.
- *The Service machine*, whose duty is to transparently find and load the required SEEMP service and to call them. In SEEMP, this service machine is the core Mélusine service machine (an open source implementation of the OSGi[9]).

**Layer 3:** The SEEMP services, which are OSGi services, and are called accordingly by the Mélusine. This solution ensures optimal performance to the EMPAM, while allowing large facilities to future extensions (new SEEMP service) and even dynamic changes (dynamic loading/unloading of services). Two classes of services have been identified:

- those dedicated to calling a Web Service exposed by a ES through the Service Abstract Machine (SAM). Most of the issues raised by EMPAM are related to discovering, selecting, parsing, and finally invoking a remote service; more exactly a ES wrapped as a web service connector. This part is delegated to a specific service, SAM, which is itself a Mélusine application and therefore contains itself an abstract layer in which are defined the fundamental concepts and functions of a service machine. This layer is captured and delegated to an orchestration layer that calls local services which, in the scope of SEEMPS, are WSMX components [9], wrapped as OSGi services.
- the other service; currently these services, in SEEMP, include the cleansing, ranking and statistic functions, and will include, in the future, the implementation of the specific functions and repositories of the EMPAS machine i.e. those functions and information not available in the ESs. Functions and information available in ESs are available calling the SAM service.

### 6.3 SEEMP Connectors

The SEEMP connectors behave as the mechanism through which all collaboration occurs between the EMPAM and a given ES. A SEEMP connector will

[9] `http://www.osgi.org/`

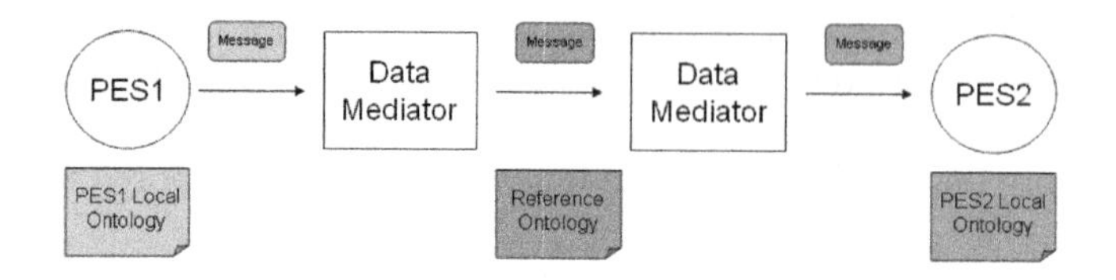


**Fig. 5.** Converting between two local ontologies via the reference ontology

exist for each of the ESs that are connected to the EMPAM and have two main responsibilities within the SEEMP architecture: Lifting/Lowering and Resolving Heterogeneity.

**Lifting and Lowering:** The ESs involved in the SEEMP marketplace only deal in terms of structured XML content and do not deal in terms of ontologies. Within the SEEMP marketplace it is important that all content is ontologized so that it can be reasoned about, thus the SEEMP connector must lift all messages received from a given ES to the ontology level. This is done by converting the XML content received to WSML in terms of the local ontologies of the ES. When communicating with the ES any outgoing data must be lowered back to the XML level so that the ES can understand the content.

Since WSMO elements can be serialized in a RDF format, this task could be done by converting XML content to RDF first, and then converting RDF to WSML. In SEEMP this task is achieved by the means of an extension to R2O language [10], which enables to describe mappings between XML schemas and ontologies, and to its related processor ODEMapster [11].

**Resolving Heterogeneity:** Each ES talks in its own language, essentially having its own local ontology that represents its view on the employment domain. The SEEMP connector is responsible for resolving these heterogeneity issues by converting all the ontologized content (the content lifted from the XML received from the ES) into content in terms of the reference ontology shared by all partners. By doing this all the ESs in the marketplace talk in the same language, and thus heterogeneity issues are resolved. Similar to the lowering back to XML, when communicating with a given ES the SEEMP connector is also responsible for converting back from the reference ontology to the local ontology of the given ES.

As described in section 6.1 the reference ontology represents the bridge, or common vocabulary, that the ESs will communicate through. Rather than managing mappings between every possible ontology pair, which essentially becomes unmanageable once a number of ESs have joined the marketplace, each ES need only maintain mappings to and from the reference ontology. These mappings represent a set of instructions (or rules) on how to convert an instance from the local ontology to an instance of the reference ontology (and vice versa). This process can be seen in figure 5, when PES1 wishes to communicate with PES2 it is necessary to convert the message from PES1's local ontology to the reference ontology and then to convert the message to PES2's local ontology.

Technologically this is achieved using the WSMX Data Mediation [12]. This work is made up of two components, the first being the design time component, within which the ES will describe the mappings between their local ontology and the reference ontology, and the second being the run time component, which is responsible for executing the mappings at run time to transform the messages between ontologies.

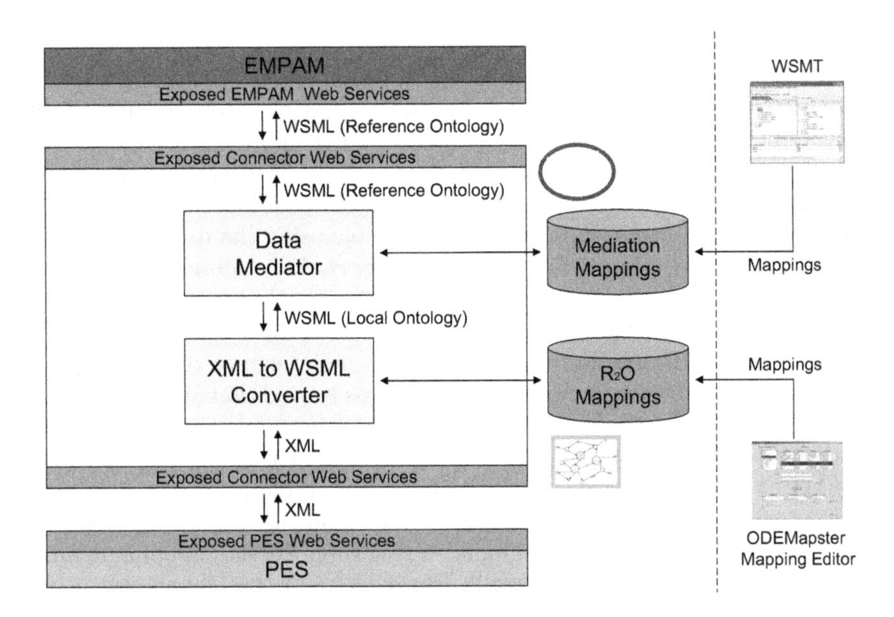


**Fig. 6.** The SEEMP Connector Architecture

**A reusable SEEMP connector** is built by bringing together the functionality described above. The architecture of the SEEMP Connector outlined in the figure 6 shows the ES communicating with the connector using XML via the exposed web services. This XML is then lifted to the local ontology using the R20 mappings stored in the repository and furthermore converted to the reference ontology using the data mediation mappings. Ultimately the EMPAM is invoked using messages in the reference ontology via its exposed web services. Communication also occurs in the opposite direction.

Each of the ESs joining the marketplace will require its own SEEMP connector, however the only difference between any two connectors is the code for executing the ESs exposed Web Services as each ES will expose services in a different way. The need for individual SEEMP connectors could be removed through the use of WSMO Choreography [3] to describe the interfaces of the ES services and the integration of the WSMX choreography engine [13] and invoker into the SEEMP Connector, however this was not considered for the first prototype of the SEEMP solution.

## 7 Comparing SEEMP with Other Approaches

In order to draw a comparison between SEEMP and other approaches we selected two case studies: private employment networks (e.g. Adecco) and hierarchical network (e.g. Borsa Lavoro Lombardia, EURES). Moreover we consider the differences both from the point of view of CEO (the decision makers) and CTO (the IT experts).

Compared to other approaches SEEMP solution offers CEO a way to enforce subsidiarity principle, therefore valuing each ES contribution in the marketplace. In private networks the subsidiarity principle is not applicable, while in hierarchical networks most of the nodes are passive actors. Moreover the marketplace creates added value by increasing the number of interconnections, hence resulting in more faithful users (more JO/CV accessible using the user language) and in more transactions. Many job offers that today could be found only at the cost of inserting the CV multiple times and merging manually the results of different ESs, becomes available through the interface of each ES.

For CTO SEEMP solution enable an easier maintenance of the integration with other ESs and minor integration costs. It was proved that Web Services used in a Service Oriented Architecture easies integration and maintenance. Moreover semantics make mapping different terminology easier because tools (such as WSMT [14]) can analyzed local and reference ontology (e.g., by comparing sub-structures and by searching for synonymies) and can guide the IT Administrator in drawing the mappings. Thank to this support, the mapping definition process requires less time or, eventually, it provides more precise mappings in the same amount of time. That support comes out with a minor integration costs.

In order to achieve this benefit CEO has to develop "partnership", i.e., the ability to collaborate with other peers, ES or staffing industries. The partnerships are different in the two case studies. In private network everything is agreed in advance. In hierarchical network partnership are necessary, but no peer to peer decision taking is possible. Decisions are institutionally imposed top-down. Moreover SEEMP supports CTO with comprehensive set of tools and methodologies for service and semantic interoperability.

Concerning services CTO has to expose ES APIs as Local Web Services and has to provide support for invoking EMPAM services. However, they don't have to understand interfaces and behavior of other ES (as in hierarchical solutions) because the connector presents the market place as if the ES was invoking its own services.

Concerning semantics, CTO has to model data structure and content and has to defining mappings with the reference ontology, but, as discussed above, this is easier and more precise than it is nowadays without ontologies and mediators.

What has to be built, and SEEMP alone won't be able to do, is a comprehensive reference ontology and abstract service machine that encompasses several employment domains. Developing and maintenance this reference part of SEEMP is not a ICT problem; it is a matter of reaching agreement at organizational level. As already discussed in section 4 the goal of SEEMP is reaching a "minimal shared commitment" in which ESs *agree* on high-level aspects, allowing for collaboration among them, *while disagreeing* on minor details that differentiate one ES from the others.

## 8 Conclusions and Future Work

This paper presented the SEEMP approach in supporting *meaningful service-based collaboration* among public and private employment services. The following results have been shown:

- **services and semantics** are the key concepts for abstracting from the hundreds of heterogeneous systems already in place that are evolving separately. They provide a straight forward way to implement the subsidiarity principle of EIF.
- **the combination of an abstract service machine with a reference ontology** is a technically sound approach to multi-laterality for marketplace implementation. Each actor in the marketplace has to care only about integrating with the marketplace. The marketplace will offer services to support the interaction with the other actors.
- **a mix of Software Engineering and Semantic approach** is required to achieve flexibility. The two approaches nicely complement each other. By means of "conventional" software engineering design SEEMP build an abstract machine that can run on "conventional" technology and at the same time embeds semantics both in the form of ontology/mediator and in the form of semantic-aware components (i.e. ODE mapster, WSMX data mediation, Glue).

Currently SEEMP consortium is running a pilot that shows the integration of EURES and Borsa Lavoro Lombardia. This integration has allowed for so far testing the functional aspects of SEEMP approach. The next step is integrating Le FOREM ES as a validation case. We expect, the reference ontology and the abstract machine to be so well designed that Le FOREM introduction would have no impact on them.

Future work includes extending the number of abstract services included in the EMPAM and the respective concepts in the reference and local ontologies. For instance, one essential service of SEEMP should be the possibility to have regularly (monthly, weekly, daily, ...) a set of key indicators regarding labour market in all participant regions (job seekers, job offers, training, etc.), in a common and comparable language, both in terms of methods (definitions, calculation of indicators, etc.) and in terms of technical requirements.

## References

1. European Communities: European interoperability framework for pan-european egovernment services. Technical report, Office for Official Publications of the European Communities (2004)
2. Estublier, J., Vega, G.: Reuse and variability in large software applications. In: ESEC/SIGSOFT FSE. (2005) 316–325
3. Fensel, D., Lausen, H., Polleres, A., de Bruijn, J., Stollberg, M., Roman, D., Domingue, J.: Enabling Semantic Web Services – The Web Service Modeling Ontology. Springer (2006)
4. 1720/1999/EC: (Decision of the European Parliament and of the Council of 12 July 1999)
5. 2004/387/EC: (Decision of the European Parliament and of the Council on Interoperable Delivery of pan-European Services to Public Administrations, 2004)
6. de Bruijn, J., Lausen, H., Polleres, A., Fensel, D.: The web service modeling language: An overview. In: Proceedings of the 3rd European Semantic Web Conference (ESWC2006), Budva, Montenegro, Springer-Verlag (2006)
7. Gómez-Pérez, A., Fernández-López, M., Corcho, O.: Ontological Engineering. Springer Verlag (2003)
8. Della Valle, E., Cerizza, D.: The mediators centric approach to automatic web service discovery of glue. In: MEDIATE2005. Volume 168 of CEUR Workshop Proceedings., CEUR-WS.org (2005) 35–50
9. Haller, A., Cimpian, E., Mocan, A., Oren, E., Bussler, C.: WSMX - A Semantic Service-Oriented Architecture. In: ICWS. (2005) 321–328
10. Barrasa, J., Corcho, O., Gómez-Pérez, A.: R2O, an extensible and semantically based database-toontology mapping language. In: Second International Workshop on Semantic Web and Databases. (2004)
11. Rodriguez, J.B., Gómez-Pérez, A.: Upgrading relational legacy data to the semantic web. In: WWW '06: Proceedings of the 15th international conference on World Wide Web, New York, NY, USA, ACM Press (2006) 1069–1070
12. Mocan, A., Cimpian, E., Kerrigan, M.: Formal model for ontology mapping creation. In: International Semantic Web Conference. (2006) 459–472
13. Cimpian, E., Mocan, A.: WSMX Process Mediation Based on Choreographies. In: Business Process Management Workshops. (2005) 130–143
14. Mocan, A., Cimpian, E.: Mappings creation using a view based approach. In: MEDIATE2005. Volume 168 of CEUR Workshop Proceedings., CEUR-WS.org (2005) 97–112

# Towards a Digital Content Services Design Based on Triple Space

David de Francisco[1], Noelia Pérez[1], Doug Foxvog[2], Andreas Harth[2], Daniel Martin[3], Daniel Wutke[3], Martin Murth[4], and Elena Paslaru Bontas Simperl[5]

[1] Telefónica Investigación y Desarrollo
{ddf268,npc}@tid.es
[2] DERI, National University of Ireland, Galway
{doug.foxvog,andreas.harth}@deri.org
[3] University of Stuttgart
{daniel.martin,daniel.wutke}@iaas.uni-stuttgart.de
[4] Vienna University of Technology
mm@complang.tuwien.ac.at
[5] Free University of Berlin
simperl@inf.fu-berlin.de

**Abstract.** Digital Asset Management is an emerging business for telecommunication companies, especially when applied to the entertainment market. Current implementations try to overcome the integration needs from each actor participating in the business processes by using Enterprise Application Integration. Triple Space is a space-based communication infrastructure which provides semantic mediation between actors involved in a dialogue.

This paper presents a Digital Asset Management use case in which Triple Space will be applied to fulfill the inherent needs of this business domain through the use of this new semantic communication paradigm.

**Keywords:** Triple Space, Digital Asset Management, Enterprise Application Integration, Space-based Computing.

## 1 Introduction

Digital content constitutes an important commercialization resource[1]. The Internet, and more specifically the Web, have enabled content providers to distribute content on a world-wide scale with very little cost, since content is distributed digitally. With the appearance of new wireless and mobile technologies, additional ways of consuming content have become available. For digital content,

[1] See figures in http://www.naa.org/technews/tn981112/editorial.html

W. Abramowicz (Ed.): BIS 2007, LNCS 4439, pp. 163–179, 2007.
© Springer-Verlag Berlin Heidelberg 2007

the full workflow can be carried out online, from offering content, browsing catalogs, delivery of goods, to final payment. This enables a vendor to get huge cost-savings by using the Internet (or mobile Internet) as a distribution channel for digital content.

For telecommunication companies such as Telefónica[2], digital asset management and digital content distribution offers a large business market. Telecommunication providers can offer services based on digital content, primarily delivering content to end users by providing network communication infrastructure plus value-added services in the digital asset management domain. Telecommunication providers can leverage their existing bandwidth services, both in fixed and mobile networks, and offer distribution channels from providers of digital content to end users.

New services must be designed and created in very short amounts of time to satisfy possible market needs as soon as possible. This paper examines a specific scenario in the Digital Asset Management arena. The intent is to supplement a sports news site, which is offered to customers of DSL lines free of charge, with multimedia content (audio and video) as a premium, paid-for service. To be able to quickly put together such service offerings, companies require methods to partially or fully automate the service creation process.

This paper makes the following contributions:

- It describes a scenario in the area of Digital Asset Management which deals with creating new service offerings based on integrating offerings of multiple partners into one coherent end-user service offering.
- It presents a software architecture based on Triple Spaces (Triple Space Communication) which enables the companies involved to implement a system which is able to fulfill the use case.

The remainder of this paper is organized as follows: Section 2.1 presents the Digital Asset Management (DAM) business domain. Section 2.2 describes the use case in detail. Section 2.3 shows possible architectures rooted in traditional Enterprise Application Information (EAI) techniques. Section 3.1 presents the Triple Space paradigm, which will be used in the use case to provide benefits pointed out in Sect. 3.2. Section 3.3 presents a high level architecture to realize the use case. Section 4 summarises the paper and presents conclusions.

## 2 Digital Asset Management (DAM)

Digital Asset Management (DAM), also known as Multimedia Asset Management (MAM), has the aim of organizing, protecting and distributing digital multimedia content in an efficient way, with the aim of improving business industries based on multimedia [1].

[2] http://www.telefonica.com/home_eng.shtml

## 2.1 DAM Business Domain

Within the DAM business domain many actors interact collaboratively to engage in e-commerce with multimedia content. This paper follows the eTOM e-business reference model [10] – the most widely accepted standard for business practices in the telecommunications industry – to identify the actors relevant to the DAM business domain.

A *service provider* provides content services to final clients by storing a catalogue of proffered media content which it offers through a service portal. All the billing processes, customer care, user management, and user security issues are performed by or subcontracted out by the service provider.

This service provider needs media content to provide these content services. A *content provider* is a supplier which owns media content and wants to trade with it.

In order to address the unauthorized copying issue, a service provider needs to use a Digital Rights Management (DRM) system [6]. A *DRM provider* provides security functions such as privacy, integrity, or authentication which are especially suited for multimedia content [20]. The DRM provider plays the role of a complementary provider in the eTOM business reference model.

The service provider needs to distribute the selected content in a secure way to the final clients that purchased it. The *content distributor* is an intermediary that not only provides the client access to the media content, possibly using different communication networks; it might also sell value-added services to the content provider, providing storage and bandwidth functionalities.

Mobility, security, quality of service, rights management and interoperability are key features offered by a service provider supplier. For that reason it is necessary for all the business processes related to DAM environments to be easily adaptable and configurable so that new business possibilities can become available in a fast and effective way. The DAM challenge is to provide agile and seamless interaction between these actors, apart from bridging different proprietary multimedia environments – the so-called content ecosystems in DAM domains – integrating DRM technologies [15].

Due to the inherent heterogeneity of actors, content, and services involved in a content service design, multiple transactions are usually held between service provider and different suppliers. These transactions should be held in parallel to save time and be maintained in a transparent way, so that service providers need not know which vendor is able to provide the content and services they need. To manage this heterogeneity all the actors involved should agree on a common communication language.

Communication between these actors should be reliable and ensure confidentiality since business data from different enterprises is being dealt with. Messages in such communications would derive from the service definition and should be easily accessed by actors involved in the content service due to the frequent modifications made to content services.

These desired features of a content services design system are illustrated in concrete scenarios below.

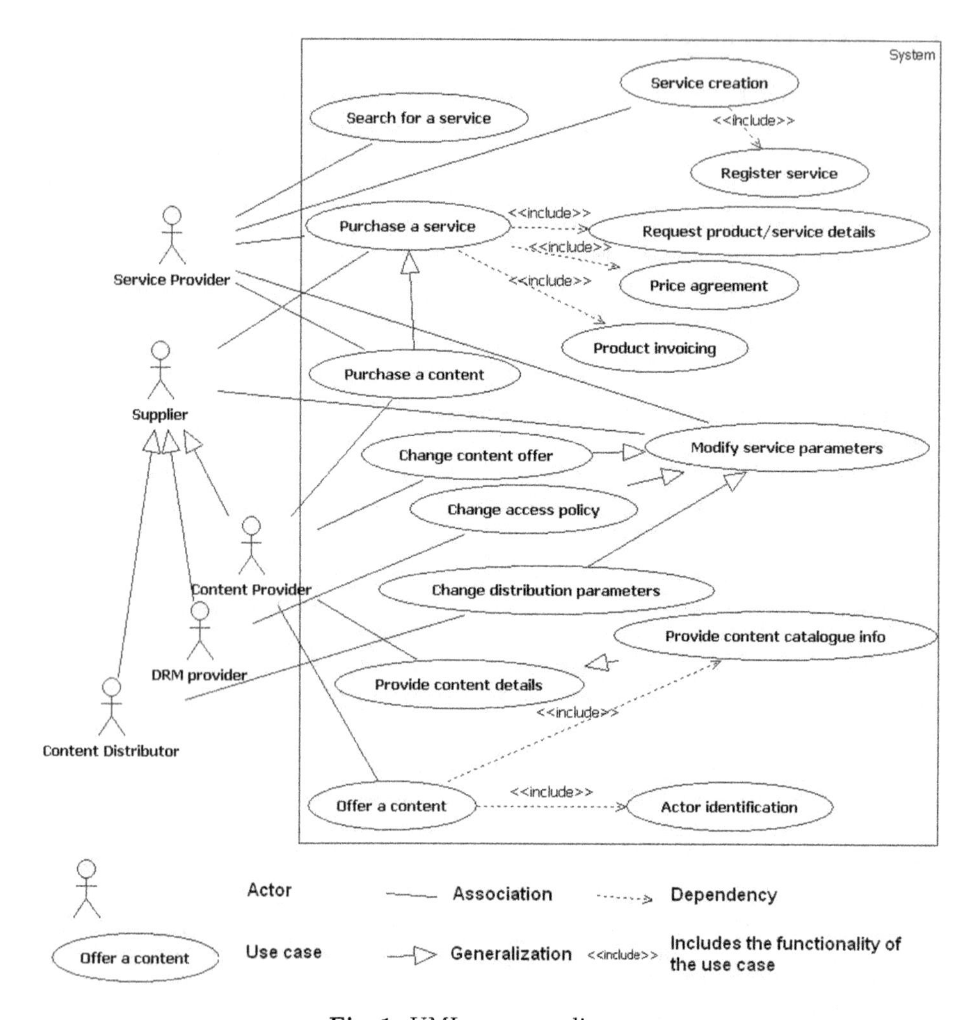


**Fig. 1.** UML use case diagram

## 2.2 Definition of the Use Cases

In order to develop an improved communication model for content services design, a concrete working flow for a complete DAM business scenario has been defined as a use case.

A content service is modeled as a structure of the components that comprise and uniquely define it: the content provided by this service, access policies defined for this content, and distribution mechanisms employed to provide this content to final users. As depicted in Fig. 1, content offering and contracting will be considered in the final use case. However, only service contracting will be considered initially. Service offering is considered redundant since business interactions are basically the same than content offering.

In this use case, Terra Sports is a Telefónica-dependant website which offers sports news for free to users who have contracted Telefónica DSL services. These news feeds present a brief summary from sports newspapers and give the users a general overview of the main news about sports events over the previous week. Given that Terra users are very interested in sports-related material, Terra has decided to provide its users with a premium sport news service "Terra Total Sports". This premium service will consist of selected multimedia content from different content providers which complements the free textual content.

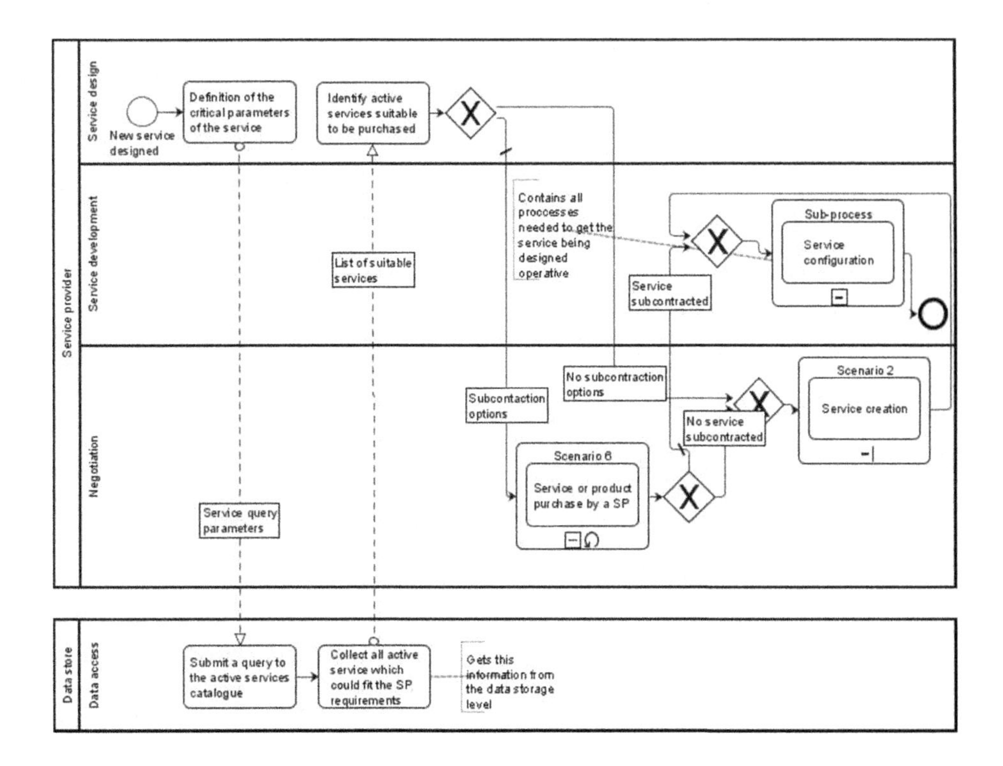


**Fig. 2.** Service Creation Flow

In the design phase of this service, Terra will define all requirements for offering this service and check if any existing services are already registered and ready in its system to which the task may be subcontracted (as shown in Fig. 2). The use case assumes no similar service is available and Terra has to negotiate the purchase of the required contents and services.

In order to get the desired content, Terra will query digital content catalogues stored already available in a data store for a service that already exists and fulfills all requirements. The most natural way to identify this service is to describe its content request which is done by using associative addressing ("template" mechanism). This request defines desired features of the content, like the kind of content you are interested in, the format or the quality. When a suitable content is found, a negotiation with the registered owner for each desired set of media content will be started until terms - including a price - are agreed. After that, Terra will send the content provider an order request, which may be amended by Terra in case its needs change. Once an order has been processed and delivered, the content provider will send an invoice to Terra to finish the transaction.

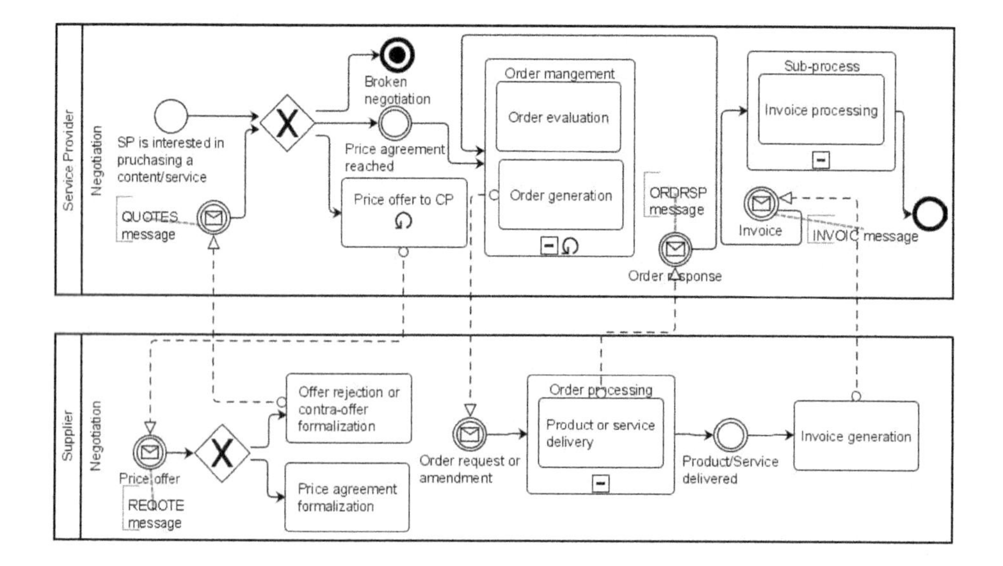


**Fig. 3.** Content and service negotiation flow

The whole process presented in Fig. 3 will be repeated for other services needed by Terra including distribution, and DRM services hiring. This negotiation proccess ends when Terra has all the contents, access policies and distribution channels that needs to run the service, the ones Terra lacked when the service design started.

After the combined services supplied are assembled, the content service would be registered in a service register inside the data store as shown in Fig. 2 making it available to be offered to final clients.

In order to integrate both data and communication flow between these actors current DAM solutions use EAI technology.

## 2.3 State of the Art in EAI Technology

The primary goal of EAI technology can be described as the "unrestricted sharing of data and business processes among any connected applications and data sources in the enterprise" [18]. In this context the term "business process" refers to a sequence of activities to be carried out to reach a well defined goal, which can be either a material product or a piece of information [17]. Business processes – even inside an enterprise – are usualy not realized using a single monolithic application, but involve multiple independent applications from different vendors, based on different platforms which need to be glued together. Such applications are naturally heterogeneous, autonomous, distributed, and immutable; they have their own data and process models, are designed to run independently, operate on local data stores and have limited adaptability to an overall IT infrastructure because they don't provide a formal description for their interfaces[4]. Their integration towards a feasible support for the execution of business processes within and among enterprises comprises three different aspects: *communication infrastructure*, *common message formats and protocols*, and *agreed-upon data semantics*. Each of these three dimensions are elaborated below.

**Communication Infrastructure.** Reliable messaging over message-oriented middleware (MOM) is the state-of-the-art communication infrastructure technology for EAI [12]. There are two major forms of message-based application integration: the point-to-point and the broker/bus-based integration.

*Point-to-Point Integration* describes the simplest possible form of integration. Each component directly communicates with each other component, requiring $O(n^2)$ message transformations for $n$ participating applications. The messaging middleware directly passes the messages to the target application, without necessitating intermediary components as in the hub and spoke or bus-based styles (see below). Point-to-point integration architectures thus have severe drawbacks with respect to scalability and maintenance.

*Broker/Bus-based Integration* was explicitly conceived to cope with these problems (see Figure 4). This integration style extends MOM with message routing and transformation functionality [12]. The key benefit is the introduction of a target application-independent, neutral message format (e.g., EDIFACT [3]) so as to reduce the number of required message transformations from $O(n^2)$ to $O(n)$.

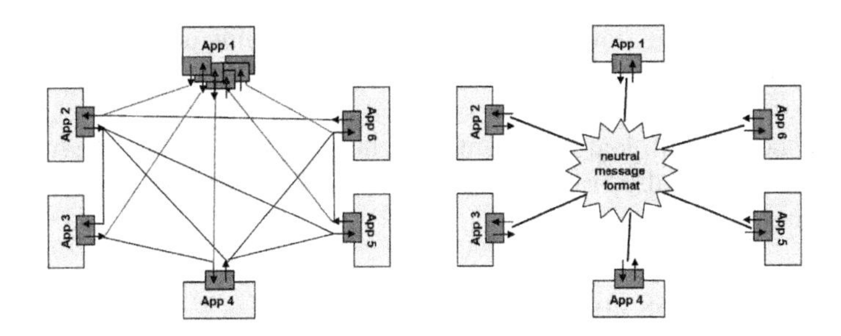


**Fig. 4.** Common Forms of Message-based Application Integration

**Common Message Format.** Traditional EAI solutions stipulate a common data format to avoid syntax transformation between heterogeneous data formats and syntaxes. There is a whole range of possible data formats and standards to chose from for this purpose. The most widely used Electronic Data Interchange (EDI) systems are EDIFACT (a UN recommendation predominant in Europe) and its U.S. counterpart X.12. Both systems were originally based on ASCII data formats, but now have XML serializations as well. Numerous newer systems based on native XML have been created with RosettaNet and ebXML being two of the most popular. The challenge with XML-based formats is that merging of messages requires *a priori* schema knowledge and custom-built XSL transformations. Knowledge of the applied schemas is particularly problematic when operating in open environments such as the Web, which are characterized by heterogeneous data formats and vocabularies.

We have chosen to use the EDIFACT system for messaging since it is the most commonly used EDI system in Europe and has a lot of experience with the message types which we need.

**Agreed-upon Data Semantics.** While EDIFACT – as a neutral, application-independent schema for business data – provides a standardized container for exchanging information between communication partners, it does not describe the semantics of the information encapsulated in the messages in a machine-understandable way. Due to the lack of this description, mediation between different representations of the message payload is necessary. For example, the IMD segment in the EDIFACT request for the REQOTE (request for quote) message[3] allows the description of "products or services that cannot be fully identified by a product code or article number" partially or totally using plain text. Use case partners that do not know each other typically would not share a common understanding of plain text terms used in this segment.

Ontologies, controlled vocabularies that formally describe a business domain in terms of domain-specific concepts and the relations between them, provide a widely-accepted instrument to cope with this problem [8]. References to commonly agreed ontological concepts can to be transmitted alongside the message payload, enriching the semantics of the original textual descriptions. If multiple ontologies for the same domain are available, their joint usage can be achieved via user-defined mediators, in form of formal ontology mapping languages or custom code [13,21].

With the help of ontologies and ontology mapping methods, an EAI system is capable of automatically transforming between different representations of the same meaning.

## 3 Realization of the Use Cases

This section introduces Triple Spaces, a coordination middleware for the Semantic Web. We motivate the major benefits of this novel technology against related

[3] http://www.unece.org/trade/untdid/d06b/trmd/reqote_c.htm

solutions in the field of EAI and demonstrate its relevance to the realization of the use cases defined in Sect. 2.2.

### 3.1 Triple Spaces – A Semantic Coordination Infrastructure

The primary objective of *Triple Spaces* is providing a novel, highly scalable, semantically-enhanced type of communication middleware for the next generation Web applications, which integrates principles of three core technologies: coordination systems, the Semantic Web, and Web services [9]. In the following we briefly introduce these technologies and their role in the context of a Triple Space infrastructure (cf. Fig. 5).

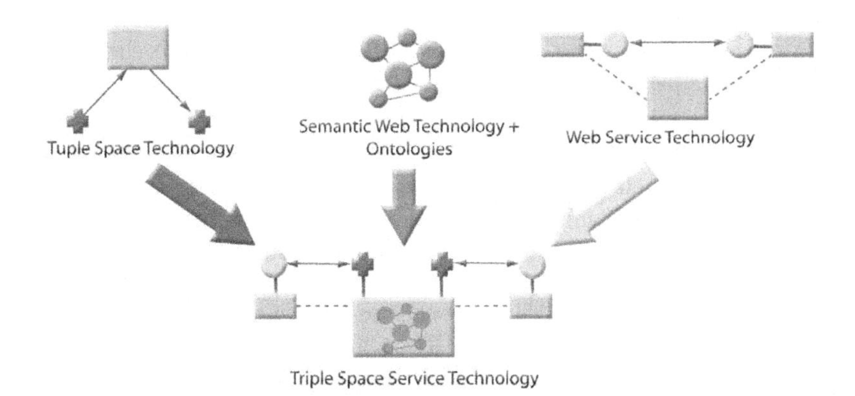


**Fig. 5.** Triple Space

**Coordination Systems.** The coordination language Linda [11] has its origins in parallel computing and was developed as a means to inject the capability of concurrent programming into sequential programming languages. It consists of coordination operations (*the coordination primitives*) and a shared data space (*the Tuple Space*) which contains data (*the tuples*). The Tuple Space is a shared data space which acts as an associative memory for a group of agents. A tuple is an ordered list of typed fields. The coordination primitives are a small yet elegant set of operations that permit agents to emit a tuple into the Tuple Space (operation *out*) or associatively retrieve tuples from the Tuple Space either removing those tuples from the space (operation *in*) or not (operation *rd*). Retrieval is governed by matching rules. Tuples are matched against a template, which is a tuple which contains both literals and typed variables. The basic matching rule requires that the template and the tuple are of the same length, that the field types are the same and that the value of literal fields are identical. Given the tuple (“N70241”,EUR,22.14) - three fields containing a string, a pre-defined type (here, currency codes) and a float - it will match the template (“N70241”,?currency,?amount) and bind to the variables currency and amount the values EUR and 22.14 respectively. The retrieval operations are blocking, i.e. they return results only when a matching tuple is found.

As identified by [5], core features of the Linda model of coordination have been mentioned as attractive for programming open distributed applications such as those encountered in the EAI field:

- The model uncouples interacting processes both in space and in time. In other words, the producer of a tuple and the consumer of that tuple do not need to know one another's location nor need they exist concurrently.
- The model permits associative addressing, i.e., the data is accessed in terms of the kind of data that is requested, rather than which specific data is referenced.
- The model supports asynchrony and concurrency as an intrinsic part of the tuplespace abstraction.
- The model separates the coordination implementation from characteristics of the host implementation environment.

**Web Services.** The aforementioned features are highly relevant for the field of Web services as well. As pointed out in [14], current Web services technology depends largely on synchronous communication links between information producers and consumers. As a matter of fact, instead of following the "persistently publish and read" paradigm of the Web, traditional Web services establish a tightly coupled communication cycle, most frequently using a synchronous HTTP transaction to transmit data. URIs, which are meant as unique and persistent identifiers for resources on the Web, are used only for the identification of the participants, whereas the information is not identifiable, but hidden in the exchanged messages. These flaws motivate the choice of a space-based communication paradigm.

**Semantic Web.** The Semantic Web [19] extends the Web with machine-processable semantic data. The representation of knowledge in a machine-readable manner through ontologies and open standards (RDF(S),OWL) on the Web is a powerful basis for the integration of heterogeneous data. While ontologies are envisioned as means for a shared knowledge understanding, it is unrealistic to expect that, in the decentralized distributed environment of the Web, different users will use *the same vocabularies* for annotating their data. The usage of various ontologies on the Web requires the definition of semantic matchings describing explicit relationships between terms in different ontologies. However this data integration is solved only at the data level. At the network level, metadata (semantic annotations of data), ontologies, and matches between ontologies are distributed across the network generally without explicit connections to one another. For integration to take place, current approaches assume that an author *explicitly defines* his own matching procedure or *is able to discover* existing data sources and the corresponding matches/mappings. Tuple spaces are an applicable model for the Semantic Web because they realize global places where information can be *published* and *persistently stored* just as in the Web architecture.

**Combining the Three Technologies.** In order to apply the Linda-based coordination paradigm to the requirements of the Semantic Web and Semantic Web services we need to re-design the underlying data model and the associated coordination primitives. Tuples should contain data which is expressed using formal knowledge representation languages typical of the Semantic Web (RDF(S), OWL, WSML, etc.) and referenced using URIs and namespaces. The coordination model needs to be revised in order to provide methods for using the virtual shared space according to common Web interaction patterns.

These topics are addressed in the context of the EU STReP project TripCom[4]. The primary objective of the project is the realization of a highly-scalable, semantics-aware communication infrastructure according to principles of tuple space computing. The emerging system will be tailored to solve the challenging integration problems encountered in areas such as eHealth, Semantic Web Services or EAI. The architecture of this paradigm will be the subject of a publication in the near future.

To summarize, employing Triple Spaces for integrating enterprise applications promises to provide the following benefits [14]:

- **Homogeneous coordination model:** Triple Spaces provide a flexible coordination model which can be used to describe and realize robust, recoverable business processes, ad-hoc workflows, and collaborative work.
- **Schema autonomy:** By adopting Semantic Web technologies to describe the meaning of data, a number of heterogeneity issues typically arising in EAI environments can be solved, e.g. data schema mediation, business process mediation, and goal-based Web service discovery.
- **Referential decoupling:** Communication parties in EAI environments do not have to know each other explicitly.
- **Temporal decoupling:** Information providers can publish data at any time, even if some or all interested consumers are disconnected from the system.
- **Spatial decoupling:** Information always resides on a virtual space and neither information providers nor information consumers need to know its physical location. Once published, data becomes independent from the EAI system which originated it.

### 3.2 Benefits of the Space-Based Approach in the DAM Scenario

The technology used to realize the DAM scenario presented in Sect. 2.2 needs to satisfy a series of core requirements:

- It needs to be able to cope with an arbitrarily high number of previously unknown and loosely connected parties originating from different administrative domains.
- It needs to integrate heterogeneous message and data formats, requiring mediation between communication partners. The integration should be performed according to a broker/bus-based approach so as to reduce the number of message and data transformations (see Sect. 2.3).

[4] `http://www.tripcom.org`

- The technology should support the non-destructive consumption of data. This means that the application data should be persistently stored so that it is accessible to other communication partners as well.
- It should support a reliable publish-subscribe style communication, with partners using message templates to describe the content of the messages they want to receive.
- It should include feasible instruments for the management of the access rights of the communicating business parties to the published information. This holds for the authentication mechanisms for both information providers and consumers.
- The integration infrastructure should ensure the security of the communication among the participants and with external applications.

State-of-the-art EAI technology – mostly *Web services* on top of *message-oriented middleware* solutions as outlined in Sect. 2.3 – shows a number of shortcomings in providing a platform for integrating the parties involved in the use case scenario and fulfilling the aforementioned requirements. While current MOM products offer reliable communication based on the paradigm of publication and subscription, they lack scalability in terms of the number of interacting partners [2] and do not provide straightforward support for non-destructive consumption of messages or feasible means for content-based data access. Furthermore, mediation between communication parties is limited to syntactic message and data transformation methods like XSLT, which primarily rely on simple text comparisons, rather than comparing the underlying concepts. For instance a field named "price" may or may not include VAT in the schemas of the data to be integrated, but a potential mismatch would not be identified at the syntactic level. These drawbacks are explicitly addressed by a Triple Space-based integration platform.

The *data mediation* functionality of the Triple Spaces allows different suppliers and service providers to have a common understanding of both the business terms of the transactions being negotiated, and the products and services mentioned in the transactions. This is achieved using semantic technologies, in particular formal ontologies and automatic mediation services. We differentiate between two ontology levels. The first ontology level includes an ontology for electronic business transactions, based on EDIFACT standards. The ontology defines the format of the messages exchanged between the actors involved in the DAM use cases in a machine-understandable manner. A second level is concerned with an ontology that models the business domain, in this case DAM-specific contents and services. Using the semantic mediation capabilities of the Triple Space, a common data schema for both suppliers and service providers can be automatically generated, thus allowing communicating parties to preserve their local autonomy with respect to the schemes used to store and manage the exchanged information. The usage of semantic technologies in this context further enables actors to perform reasoning and semantic validation tasks within the business transaction process, which leads to more accurate results in discovering content and services than pure syntactic matching mechanisms.

A second core feature of Triple Space technology is its *reliable transport mechanism* for Web services [16], which is an essential requirement for the electronic business transactions in our scenario. This communication is dynamic, allowing ad-hoc relationships between business parties, which can join or exit a negotiation depending on the satisfaction of business rules, verified by the Triple Space using semantic mediation. This behavior is highly desired within a DAM area, in which the inherent dynamism of the business domain results in frequent availability and functionality changes of content and services.

From a technical perspective, the implementation of the use cases using Triple Spaces follows the *publish-subscribe* paradigm. Message consumers express their interest in messages by describing their content, rather than listening on certain topics. This is an important difference between this and related state-of-the-art approaches. In existing Enterprise Integration Architectures [12] based on MOM and/or Web services [22] messages are *pushed* to certain destinations – identified by e.g. queue or topic names. This means that an endpoint reference has to be explicitly specified as destination of a message and exchanged before the actual message exchange is carried on. By contrast message publication in Triple Spaces can be done without knowing the receiver of a message because data is simply published to a space not addressing a communication partner directly. This is necessary when dealing with DAM business processes becausee service providers may not know the suppliers of media content andd services in advance. In the space-based approach, message receivers *pull* messages from the space by describing their content. This interaction paradigm greatly simplifies ad-hoc communication between previously unconnected parties because communicating partners do not need to share any *a priori* information about each other.

In the DAM context suppliers prefer to offer their content or services to potentially interested service providers. This requires a mechanism that enables *multiple, non-destructive consumption* of messages. In the Triple Space-based approach, this is achieved by persistently storing messages that represent a supplier's content and service descriptions in the space, thus enabling all interested parties to non-destructively retrieve them (read operation) in a way similar to broadcast communication. In turn, service providers need to be able to retrieve information about already existing content offers in order to provide the services they are designing. Order, time, and publisher of this information are irrelevant. Current MOM solutions typically deliver messages in FIFO or priority order. In terms of the DAM scenario this would require each service provider to locally store all content offers and keep them up to date. By contrast, space-based technology offers persistent storage of these messages, classifying the information of each message according to the its content's semantics, not depending on its publication time or sender. Using such storage policies, the middleware is able to implement random access and advanced query mechanisms to retrieve relevant information for each service provider.

**Table 1.** Benefits of applying Triple Space Computing

| REQUIREMENTS | MOM | TRIPLE SPACE |
|---|---|---|
| Arbitrary number of communicating parties | - | + |
| Semantically-aware data and message mediation | - | + |
| Non-destructive consumption of data | - | + |
| Reliable publish-subscribe | + | + |
| Associative (content-based) addressing | - | + |
| Access rights management | product-specific | + |
| Security of communication | product-specific | + |

To summarize, it can be said that existing data and communication integration technologies can not fulfill the requirements of the business domain and the associated use cases to a satisfactory extent. By comparison, Triple Space computing offers interesting advantages that are worth further investigation (cf. Table 1). The next section elaborates on a possible high-level architecture of a space-based approach to digital asset management.

### 3.3 High-Level Architecture

The integration model will be a bus-based architecture which will be implemented using Triple Space architecture which will be based on Web services standards.

In order to benefit from Triple Space semantic mediation, these Web services from each actor will be semantically enriched by using two ontology levels, which stand for the two EAI levels performed by the Triple Space in this use case. As shown in Fig. 6, a domain ontology is needed to define the semantics of the business domain (DAM) to assist performing data integration and a negotiation ontology is needed to integrate messages exchanged by actors when negotiating transactions. This second ontology will be based on the EDIFACT standard.

All semantically enriched supplier Web services will access the Triple Space for offering their products and services (only media content for our simplified prototype) by publishing their service descriptions (offers) into a service catalogue managed by the Triple Space. New offers from different content providers with their own content catalogue format can be integrated to provide a common perspective of available content to service providers.

Semantic Web services using WSMO [7] from both service providers and suppliers of any type will be able to negotiate purchases of content and services by using Triple Space as a space-based message mediator. These negotiations, performed by EDIFACT message exchange, will be temporally stored inside the tuple space of the Triple Space. Once all negotiations are successfully ended and the service provider has all the components needed to exploit the service being designed, the Triple Space will store the service settings inside the service register. This register will allow other service providers to search for existing services

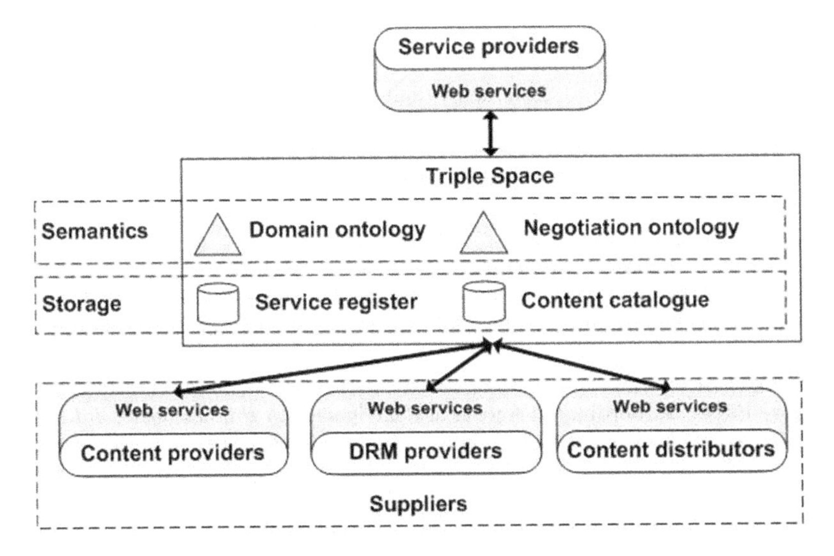


**Fig. 6.** High level architecture

similar to those they are designing and subcontract them without having to own additional infrastructure; thus allowing reuse of digital content services.

## 4 Conclusion

Digital asset management is an emerging business for telecommunication companies, especially when applied to the entertainment market. This paper analyzes a typical use case in this field from a business and technological perspective. Triple Space computing is introduced as a novel middleware technology and an explanation is provided for how such an approach can cope with the integration needs of the communicating partners at data and application levels.

TripCom activities in the EAI area aim at demonstrating the capacity of Triple Space computing to address the DAM scenario. Space-based middleware is a feasible alternative to traditional solutions to Enterprise Application Integration, since it allows agents to publish and retrieve information in an uncoupled manner in terms of space and time. By extending tuple spaces with Semantic Web technology, applications are able to store and exchange machine-understandable information in a decentralized and distributed manner, while taking advantage of powerful coordination mechanisms. This combination provides a new level of integration of the highly heterogeneous data and message formats exchanged in a DAM scenario.

## Acknowledgements

This work is supported by EU funding under the TripCom project (FP6 - 027324).

## References


1. E. Altman, S. Goyal, and S. Sahu. A digital media asset ecosystem for the global film industry. *Journal of Digital Asset Management*, 2:6–16, January 2006.
2. M. Astley, J. Auerbach, S. Bhola, G. Buttner, M. Kaplan, K. Miller, R. Saccone Jr, R. Strom, D.C. Sturman, M.J. Ward, et al. Achieving Scalability and Throughput in a Publish/Subscribe System. *IBM Research Report RC23103 (W0402-026)*, February 2004.
3. J. Berge. *The EDIFACT standards*. NCC Blackwell, 1991.
4. C. Bussler. The role of semantic web technology in eai. *Bulletin of the IEEE Computer Society Technical Committee on Data Engineering*, 26:62–68, December 2003.
5. P. Ciancarini, A. Knoche, D. Rossi, and R. Tolksdorf. Redesigning the Web: From Passive Pages to Coordinated Agents in PageSpaces. In *Proceedings of 3rd IEEE International Symposium on Autonomous Decentralized Systems ISADS*, pages 337–384, 1997.
6. Coral Consortium. Coral consortium whitepaper. `http://www.coral-interop.org/main/news/Coral.whitepaper.pdf`, February 2006.
7. Jos de Bruijn, Dieter Fensel, Uwe Keller, and Rubén Lara. Using the web service modeling ontology to enable semantic e-business. *Commun. ACM*, 48(12):43–47, 2005.
8. D. Fensel. *Ontologies: A Silver Bullet for Knowledge Management and Electronic Commerce*. Springer, 2001.
9. D. Fensel. Triple space computing: Semantic web services based on persistent publication of information. In *Proceedings of IFIP International Conference on Intelligence in Communication Systems*, 2004.
10. Tele Management Forum. Enhanced Telecom Operations Map (eTOM): the business process framework. Release 5.0, 2005.
11. D. Gelernter. Generative Communication in Linda. *ACM Transactions on Programming Languages and Systems*, 7(1):80–112, 1985.
12. G. Hohpe and B. Woolf. *Enterprise Integration Patterns: Designing, Building, and Deploying Messaging Solutions*. Addison-Wesley Professional, 2003.
13. Y. Kalfoglou and M. Schorlemmer. Ontology mapping: the state of the art. *Knowledge Engineering Review*, 18(1):1–31, 2003.
14. Reto Krummenacher, Martin Hepp, Axel Polleres, Christoph Bussler, and Dieter Fensel. WWW or What is Wrong is with Web Services. In Welf Löwe and Jean-Philippe Martin-Flatin, editors, *Proc. 3rd European Conf. on Web Services*, pages 235–243. IEEE Computer Society, November 2005.
15. G. Larose. DAM and interoperable DRM: Maintaining agility in the world of evolving content ecosystems. *Journal of Digital Asset Management*, 2:17–25, January 2006.
16. F. Leymann. Space-based Computing and Semantics: A Web Service Purist's Point-Of-View. Technical Report Fakultätsbericht Nr. 2006/05, Universität Stuttgart, Fakultät Informatik, Elektrotechnik und Informationstechnik, March 2006.
17. F. Leymann and D. Roller. *Production Workflow*. Prentice Hall PTR, 2002.
18. D. Linthicum. *Enterprise Application Integration*. Addison-Wesley Professional, 1999.
19. T.Berners-Lee, J. Hendler, and O. Lassila. The semanticweb. *Scientific American*, 5:34–43, 2001.

20. A. Uhl and A. Pommer. *Image And Video Encryption - From Digital Rights Management To Secured Personal Communication.* Springer, 2005.
21. H. Wache, T. Vögele, T. Visser, U. Stuckenschmidt, H. Schuster, G. Neumann, and S. Hübner. Ontology-based integration of information - a survey of existing approaches. In *Proceedings of the IJCAI-01 Workshop: Ontologies and Information Sharing*, pages 108–117, 2001.
22. S. Weerawarana, F. Curbera, F. Leymann, T. Storey, and D. F. Ferguson. *Web Services Platform Architecture: SOAP, WSDL, WS-Policy, WS-Addressing, WS-BPEL, WS-Reliable Messaging, and More.* Prentice Hall PTR, 2005.

# Evaluating Quality of Web Services: A Risk-Driven Approach

Natallia Kokash and Vincenzo D'Andrea

DIT - University of Trento, Via Sommarive, 14, 38050 Trento, Italy
{kokash,dandrea}@dit.unitn.it

**Abstract.** Composing existing web services to obtain new functionalities is important for e-business applications. Deficiencies of aggregated web services can be compensated involving a redundant number of them for critical tasks. Key steps lie in Quality of Service (QoS) evaluation and selection of web services with appropriate quality characteristics in order to avoid frequent and severe faults of a composite web service. This paper, first, surveys the existing approaches for QoS-driven web service selection. Then, it proposes a novel approach for evaluating quality of redundant service compositions through analysis of risk related to the use of external web services. Finally, we describe an improved selection algorithm that takes into account success rate, response time and execution cost of involved web services.

**Keywords:** Web services, Quality of Service, Selection, Composition, Risk Analysis.

## 1 Introduction

Web services are software applications with public interfaces described in XML. According to the established standards, web service interfaces are defined in Web Service Description Language (WSDL). Published in Universal Description, Discovery and Integration (UDDI) directory web services can be discovered and invoked by other software systems. These systems interact with web services using XML-based messages conveyed by Simple Object Access Protocol (SOAP). Web services are considered a promising technology for Business-to-Business (B2B) integration. A set of services from different providers can be composed together to provide new functionalities. One of the most notable efforts in the area of web service composition is the Business Process Execution Language for Web Services (BPEL4WS). BPEL4WS is a language for describing service-based business processes and specifying interaction protocols for involved services.

Web service composition is a complex process involving analysis of process requirements, semantics and behavior of existing services, service testing, adaptation, contracting and management. Despite all the efforts problems both on technical and semantical levels may appear. Modifications of the involved services and their unexpected faults may affect a client application. Erroneous services can be replaced with analogues to allow for correct behavior of client applications in such situations. Since much work is required to safely introduce a new

W. Abramowicz (Ed.): BIS 2007, LNCS 4439, pp. 180–194, 2007.
© Springer-Verlag Berlin Heidelberg 2007

component in a system, alternative services must be known in advance. In *redundant* service compositions a set of services are not normally used but cater for *fault-tolerance*, i.e. the ability of a system to behave in a well-defined manner once faults occur.

**Definition 1.** *A web service composition $c$ is said to be* redundant *iff for all executions $E$ of $c$ in which no faults occur, the set $S$ of all services of $c$ contains services that are not invoked in $E$.*

Due to unsteadiness of business environments service-based systems require run-time monitoring. Statistics about user experience with web services may be used to select well-behaved services. Quality of Service (QoS) is a set of parameters such as service execution cost, performance, reliability, robustness and the like. In this paper, we will refer to quality of composite web services as to Quality of Composition (QoC).

Analysis of QoS of web services is of paramount importance. Multiple proposals aiming at QoS evaluation and selection of better services have appeared because of multi-dimensionality and volatility of QoS parameters. They will be surveyed in the next section. In this paper, we present a novel web service selection algorithm. In contrast to existing work, it does not rely on a simple additive weighting technique for involving QoS parameters such as success rate, response time and execution cost into an objective function. A generalized strategy for QoC evaluation inspired from risk analysis is proposed. We apply our strategy to evaluate quality of redundant compositions, assuming failures of atomic services and regarding composition structure.

The paper is structured as follows. Section 2 discusses related work. In section 3, a notation for modelling redundant service compositions is explained. Section 4 discusses risk management and its application to analysis of QoC. Sections 5 studies failures of composed web services and evaluates impact of these failures on the service composition. In Section 6, an example is given that helps better understand how QoC is calculated. Our service selection algorithm is presented in Section 7. Section 8 presents experimental results. Conclusions and future work are sketched in the last section.

## 2 Related Work

Description of quality characteristics of web services can be found in [1]. Among them are *throughput* (the number of requests served in a given time period), *capacity* (a limit of concurrent requests for guaranteed performance), *response time* (the time taken by a service to process its sequence of activities), *execution cost* (the amount of money for a single service execution), *availability* (the probability that a service is available), *reliability* (stability of a service functionality, i.e., ability of a service to perform its functions under stated conditions), and so on.

There are several proposals aiming at measuring and specifying QoS for web services. Tosic et al. [2] developed a Web Services Offering Language (WSOL) that allows a service provider to specify five QoS-related constructs: constraints (functional constraints, QoS and access rights), statements, constraint groups, constraint group templates and service offerings. Maximilen and Singh [3] propose an agent-based framework and ontology for QoS measurement. In this approach service providers publish their services to registries and agencies, and service consumers use their agents in order to discover the desired service. The metrics concept is absent in this ontology. A QoS ontology proposed in [4] fills this gap. It consists of three layers: the QoS Profile Layer, used for matchmaking; the QoS Property Definition Layer, used to present the property's domain and range constraints; and the Metrics Layer that provides measurement details. Multiple QoS profiles can be attached to one service profile in this approach.

An interesting task is how to choose web services to be used by a new (composite) web service in order to have a guarantee that required quality level of the composition is reached. Cardoso et al. [5] introduced several models for QoS measurement in workflows. In particular, the authors evaluate expected response time, execution cost and reliability of a workflow applying sequential, parallel, conditional, loop and fault-tolerant system reduction rules. For example, the expected execution cost of a composition including two parallel services $s_1$ and $s_2$ is $q_{cost}(s_1) + q_{cost}(s_2)$ while its response time equals $max(q_{time}(s_1), q_{time}(s_2))$. Lakhal et al. [6] extends this work by reviewing the estimation of reliability and response time of fault-tolerant compositions.

Service compositions that embed low-quality services inherit their drawbacks as well. This poses a challenge for software developers that build new systems on the basis of available components. One can compensate deficiency of such systems if many web services with compatible functionality exist. Elaborating this idea, a good number of QoS-driven service selection algorithms have appeared. One of the first works in this direction is done by Zeng et al. [7]. They consider web service selection as a global optimization problem. Linear programming is applied to find the solution that represents the service composition optimizing the target function. The target function is defined as a linear combination of five parameters: availability, successful execution rate, response time, execution cost and reputation. In [8] the service selection is considered as a mixed integer linear program where both local and global constraints are specified. The model by Yu and Lin [9] comes to the complex multi-choice multi-dimension 0-1 knapsack problem. In this approach, the practice of offering different quality levels by services is taken into consideration. Gao et al. [10] apply integer programming to dynamic web service selection in the presence of inter service dependencies and conflicts. Wang et al. [11] consider the measurement of non-numerical qualities. Accuracy, security and exception handling are taken into account. As in the previous work, QoS-driven web service selection is based on assessment of a linear combination of scaled QoS parameters. Yang et al. [12] turn QoS factors to a form following the ascent property. Along with the five QoS attributes used in [7] a service matching degree is analyzed. Matching degree defines a compliance between composed services and, in principle, is a functional parameter. The Multiple Criteria Decision Making (MCDM) technique is used to give an overall evaluation for a composite web service.

The above solutions depend strongly on user weights assigned to each parameter. There is no clear mechanism allowing a user to set up these weights in order to obtain the desired result. Several approaches try to avoid a user involvement in the selection procedure. For example, in [13] service selection is formulated as Multiple Attribute Decision Making (MADM) problem. Four modes for determining relative weights for QoS attributes are proposed: subjective, single, objective and subjective-objective. Claro et al. [14] follows the quality model proposed in [7] with several improvements. The first extension concerns the concept of reputation that is ranked based on the user's knowledge of the service domain. Secondly, a multi-objective problem is considered as opposed to Zeng's aggregation functions. It is resolved using multi-objective genetic algorithm called NSGA-II, without giving any weight to any quality criterion. Canfora et al. [15] extend works by Cardoso [5] and Zeng [7] in a similar way. A genetic algorithm for QoS-aware workflow re-planning is proposed. An interesting approach is taken by Martin-Diaz et al. [16] who propose a constraint programming solution for procurement of web services with temporal-aware demands and offers. Bonatti and Festa [17] formalize three kinds of service selection problems to optimize the quality of the overall mapping between multiple requests and multiple offers with respect to the preferences associated to services and invocations. In particular, they prove that the problem of cost minimization is NP-hard by reduction from the Uncapacitated Facility Location Problem. Exact and approximated algorithms to solve the formulated problems are proposed. Several works experiment with expressing user QoS preferences in a fuzzy way. Mou et al. [18] set up a QoS requirement model with support of fuzzy metrics for expressing user requirements on target service QoS. In [19] the service selection problem is formalized as a Fuzzy Constraint Satisfaction Problem. Deep-first branch-and-bound method is applied to search for an appropriate web service composition.

Assuming that the failure of any individual web service causes the failure of the composite service, the overall reliability of composite service is the product of the reliability of constituent web services. Therefore, one unreliable web service can decrease the overall reliability to a very low level. The upper bound of overall reliability is often determined by the weakest constituent web services. Jaeger and Ladner [20] consider identification of such weak points. For each weak point they identify alternative candidates that meet the functional requirements. Three possible replacement patterns are analyzed: Additional Alternative Candidate, AND-1/n and XOR-XOR. Dealing with the analogous problem, Stein et al. [21] apply an algorithm for provisioning workflows that achieve higher success probability in uncertain environments through varying the number of providers provisioned for each task.

Diversity of quality metrics, their value ranges and measurements makes it difficult to provide a single QoS-driven selection algorithm for web services. The existing approaches cover a wide range of methods for multi-objective optimization, but fail to provide a valid formalization of the problem that would sufficiently reflect the real world conditions. A consistent analysis of pros and cons of the listed algorithms is out of the scope of this work. Among the negative characteristics of state-of-the-art algorithms that will be addressed are:

- *Choice of an objective function.* Dependency between different QoS factors is not considered by the existing solutions. Suppose that two weather forecasting services are available: the first one provides reliable forecasts but, as a consequence, it is expensive and rather slow, whereas the second one is cheap and fast but generates forecasts at random. Algorithms based on a simple additive technique are likely to select the second web service despite the fact that the service response time and execution cost are not important if the service is unreliable. Another grave drawback of the methods comparing a weighted sum of relative scores for each quality factor is that bigger compositions are likely to have higher total score. On the other hand, constraint satisfaction algorithms consider QoS separately and do not reason about overall quality of a service.
- *Absence of redundancy.* Despite the efforts aimed at insuring web service reliability (e.g., contracts), service composition failures are almost inevitable. Nevertheless, they can be gently treated without leading to the composition breakdown. As failure-tolerance can be reached through composition redundancy, an important characteristic of a service is the number and quality of services compatible with it in a particular environment. This implies that services must be selected with respect to their context within the composite service and its structure. So, the problem must not be reduced to the selection of a simple execution path where only one web service is assigned to each task.

## 3 Modelling Redundant Web Service Compositions

Composite web services can be defined using a set of workflow patterns [22]:

- *Sequence.* Several services are executed in a sequence.
- *Loop.* The execution of a service is repeated several times.
- *AND split followed by AND join.* Several services are invoked in parallel and all services must be executed successfully.
- *AND split followed by m-out-of-n join.* Several services ($n$) are invoked simultaneously, but only $m \leq n$ of them must be executed successfully.
- *XOR split followed by XOR join.* Only one service is invoked from a set of available services. The synchronizing operation considers only the invoked service.
- *OR split followed by OR join.* Several services ($n$) from all available ($k$) are invoked and all of the invoked services are required to be finished successfully for synchronization.
- *OR split followed by m-out-of-n join.* Several services ($n$) from all available ($k$) are invoked and $m \leq n$ services must be executed successfully.

The above workflow patterns form a set of functional and structural requirements which cover most service-based flow languages (e.g., BPEL4WS). We suppose that sequential composition prescribes an order for the execution of services.

**Table 1.** A notation for representing composite web services

| Graphical | Syntactical | Description |
|---|---|---|
| $s_i$ | $s_i$ | A web service. |
| | | The start and the end states. |
| $\rightarrow$ | $(s_1; s_2; ...; s_k)$ | Sequential operator. |
| $\|^m_n$ | $(s_1 \| s_2 \| ... \| s_k)^m_n$ | Parallel operator. Indices $m$ and $n$ are used to represent AND split followed by m-out-of-n join (bottom index $n = k$ and upper index $m = n$ can be omitted). |
| $+^m_n$ | $(s_1 + s_2 + ... + s_k)^m_n$ | Choice operator. Indices $m$ and $n$ are used to represent OR split followed by m-out-of-n join (bottom index $n = k$ and upper index $m = 1$ can be omitted). |

The situation when the execution of a set of services can be performed in an arbitrary order is also possible in practice. It can be modelled as XOR split followed by XOR join of all alternative sequences with a prescribed order. Loop can be seen as a special case of sequential composition. For specifying composite web services with redundancy we will use a notation drawn in Table 1.

Several services are composed in an application that can be available as a new web service (see Fig. 1). The provider of this service is in a difficult situation as (s)he must guarantee a certain level of QoS to end users, and in the same time, quality of the provided service depends on agreements established among the partners and quality of the involved services. For example, one of the possible problems is a limited capacity of the atomic services. A composite service will be forced to pay penalties to its clients because it cannot satisfy all requests in a required time. To avoid such bottlenecks, the maximum capacity of the composition must be controlled. A set of run-time changes in the composition model should be taken into account:

- *Service capacity correction.* It reflects changes in the monitored service performance related to the increase/decrease of service load by external clients, problems in a communication network or middleware, etc.
- *Service deletion.* Some web service is not available for the invocation.
- *Service addition.* A new service is introduced into a composition.
- *State deletion.* All services that can be invoked from this state are deleted.
- *Sub-composition deletion.* Any service can be deleted if there is no path between the start state and the end state that includes this service. Iteratively repeating state and service deletion we can delete a sub-composition.
- *Sub-composition addition.* The reverse operation to the sub-composition deletion arises if a new sub-composition is involved in the model.

Distributions of the service capacity and the expected number of concurrent requests must be compared to guarantee a stable execution of the composite web service. Generally, we may speak about risk that quality of a composite web service will be affected because of problems with involved services. If this risk is significant we must try to mitigate it, for example, negotiating quality of service with partners or adopting other services.

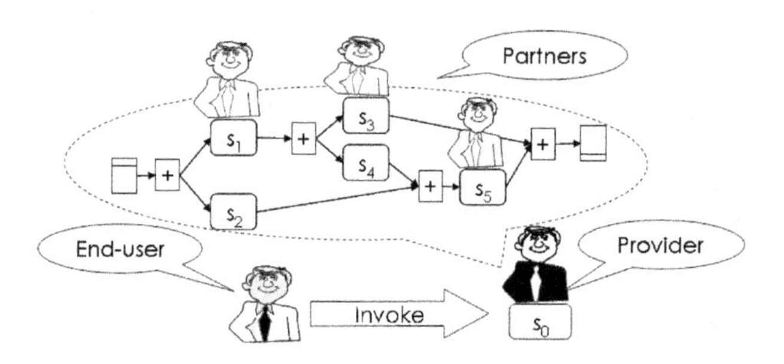


**Fig. 1.** An example of a service-based workflow

In the next section we discuss risk management and its application to QoS-driven web service selection.

## 4 Risk Management

The purpose of risk management is to reduce or neutralize potential risks and offer opportunities for some positive improvements. Risk management operates with notions of *threats* (danger sources), *probabilities* of threats (likelihoods that given threats will be triggered) and their *impacts* on the organization (monetary loss, breach of reputation, etc.). Risk $r$ can be expressed mathematically as a probability $p$ of a threat $e$ multiplied by a respective magnitude $q$ of its impact: $r(e) = p(e)q(e)$. Three main steps of risk management include *identification,* which is needed for surfacing risks before they become problems, *analysis,* when identified risk data is converted into decision-making information, and *control,* which consists of monitoring the status of risk and actions taken to reduce them. Appropriate risk metrics must be developed to enable the evaluation of the risk status and mitigation plans.

Risk management covers all steps of software development and business process modelling lifecycle. Several risk management frameworks [23][24] have been created to keep software projects within established time and budget constraints. Other related work concerns risk analysis for business processes [25][26]. A *business process* is a structured set of activities which are designed to produce a specified output for a particular customer or market. Business processes are subject to errors in each of their components: to enable the successful completion of a business process it is important to manage risks associated with each process activity and with the overall process. Defining a business process, a company can rely on *outsourcing*, i.e., a formal agreement with a third party to perform a service for an organization. In this case, the risk management process requires evaluation of proposals, identification of best providers and sources, evaluation of supplier financial viability, country risk assessment, etc. along with costs of failure or non-delivery of outsourced functional modules [27].

Further we will consider risks specific for execution of service-based business processes to that extent as they may affect composition design. Such risks can be divided into three basic categories:

- *Inter-organizational risks* are caused by providers of web services used in a service composition. Into this category we can put risks related to such events as disposal of a service by the provider, changes in interface and behavioral logics of a service, contract violation, obtrusion of a new contract with worse conditions, intentional disclosure of private user information, etc.
- *Technical risks* are related to technical aspects of distributed systems such as network or service failures.
- *Management risks* may be caused by the use of automatic management systems. For example, requests from some unprivileged clients may be ignored or delayed because of the limited capacity of a web service, etc.

Risk analysis uses two basic types of techniques, namely *quantitative* and *qualitative*. Qualitative analysis involves the extensive use of mathematics and reports based on probabilities of threats and their estimated costs. Qualitative analysis is a verbal report based on system knowledge, experience, and judgment. Statistical information about service behavior is essential for risk analysis. The assessment has to be ongoing and evaluations of the probability of events happening revised as the system is used and evolves. Without the benefit of a quantitative assessment risk analysis is subjective and has to be based largely on common sense and experience. For example, services provided by a large well-known company can be considered less risky than services of a small unknown company.

In ideal risk management a prioritization process is followed: the higher risks are handled first, and the lower risks are handled later. Each external web service can be seen as a black box with a certain QoS. Several actions are possible to manage risks caused by use of external web services:

- Communicate with the service provider in order to establish an agreement that can help to mitigate the risk.
- Mitigate the impact of the risk by identifying a triggering event and developing a contingency plan.
- Try to avoid risks by changing the design of the application. In particular, functionality of unreliable services can be (1) implemented from scratch, (2) taken from open source projects, (3) provided by software components that are deployed locally.
- Accept the risk and take no further actions, thus, accepting the consequences.
- Study the risk further to acquire more information and better determine the characteristics of the risk to enable decision making. For example, conditions when failures of external services are more likely can be discovered.

As standard protocols simplify involvement of new web services, we can try to reduce risks through web service selection. However, if too many services are included in the composition, its cost increases. A composition that maximizes the overall profit must be selected. As risks define expected loss in some period of time, the problem can be formalized as selection of a composition $c_0$, such that

$$Q_{profit}(c_0) = Q_{income}(c_0) - R(c_0) = \max_{c \in C}(Q_{income}(c) - R(c)),$$

where $Q_{income}(c)$ is an income expected by the provider, and

$$R(c) = \sum_{e_j \in E(c)} r(e_j) = \sum_{e_j \in E(c)} p(e_j) Q_{loss}(e_j)$$

is an estimated risk of the composition $c$ internally used by the composite service. Here, $C$ denotes a set of all available compositions, $E(c)$ is a set of independent risk-related events identified for a composition $c$, $p(e_j)$ is a probability that an event $e_j$ will occur and $Q_{loss}(e_j)$ is an estimated loss function of this event. In a ProRisk Management Framework [28] other strategies for risk assessment are provided, given that some events are not totally independent of others. For example, assuming that there exist one dominating threat, the following fuzzy model for risk assessment is valid[1]:

$$R(c) = \max_{e_j \in E(c)} r(e_j) = \max_{e_j \in E(c)} p(e_j) q(e_j).$$

## 5 Failure Risk

*Failure risk* is a characteristic considering probability that some fault will occur and the resulting impact of this fault on the composite service. For an atomic service $s_i$ it equals

$$r(s_i) = p(\overline{s}_i) q(s_i),$$

where $p(\overline{s}_i)$ is a failure probability of service $s_i$, and $q(s_i)$ is a *loss function*. Service $s_1$ is better than service $s_2$ if it has a smaller failure risk $r(s_1) < r(s_2)$.

Let $c = (s_1; s_2; ...; s_k)$ be a sequential composition. If a service $s_i$ fails the results of services $\{s_1, s_2, ..., s_{i-1}\}$ will be lost as well whereas their response time and execution cost increase the total expenses to satisfy a user request. These expenses are included in a loss function of a service $s_i$ failure. Hence, a failure risk of a service $s_i$ in a sequential composition equals:

$$r(s_1; ...; s_i) = p(s_1; ...; s_{i-1}; \overline{s}_i) q(s_1; ...; s_i) = \prod_{j=1}^{i-1} p(s_j) p(\overline{s}_i) \sum_{j=1}^{i} q(s_j).$$

Let us consider an example. Suppose that a user needs to translate a text from Belarusian to Turkish provided that five translation web services are available: *b-e* translates from Belarusian to English, *b-g* from Belarusian to German, *g-t* from German to Turkish, *e-t* from English to Turkish, and *g-e* from German to

[1] The original formula includes also weighting coefficients to scale the impact of the threat into an appropriate utility measure.

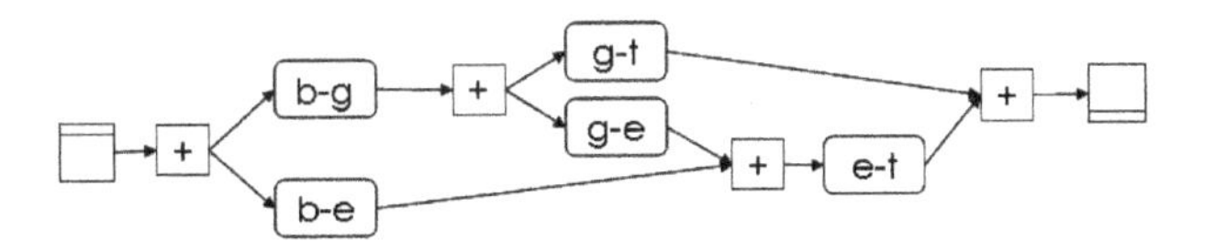


**Fig. 2.** A redundant service composition with three possible execution paths

English (see Fig. 2). There are three configurations that can fulfill the user goal, i.e., the text can be initially translated (1) from Belarussian to English and then from English to Turkish, (2) from Belarussian to German and then from German to Turkish, (3) from Belarussian to German, then from German to English and finally from English to Turkish. Suppose we have chosen the first composition. If the service *e-t* fails, the task will not be completed and the translation done by the service *b-e* will be lost. Instead, if the second composition is chosen, in case of a *g-t* failure, the task still can be completed successfully by switching to the third composition without rollback.

In our example, $r(e\text{-}t) = p(b\text{-}e)p(\overline{e\text{-}t})(q(b\text{-}e)+q(e\text{-}t))$, where $p(e\text{-}t)$ is a failure probability of the service *e-t*. Loss function $q(.)$ can refer either to execution cost of the services *b-e* and *e-t* or to their expected response time. In the latter case, the estimation of time loss will be obtained. It may be important for tasks with deadlines, i.e., with limits on the latest time for a task to be accomplished.

In complex service oriented systems calculation of a loss function may involve analysis of transactional aspects of the process. The loss function of the AND split followed by AND join pattern with service sequences in each branch $c = (s_1; ...; s_n)|(t_1; ...; t_m)$ depends on the service coordination mechanism. We may distinguish *centralized* and *decentralized* compositions. In the first case, parallel branches can be interrupted immediately after the fault detection, therefore loss functions of a service $s_i$ failure are:

$$q_{time}(c|\overline{s}_i) = \sum_{j=1}^{i} q_{time}(s_j), \quad q_{cost}(c|\overline{s}_i) = \sum_{j=1}^{i} q_{cost}(s_j) + \sum_{j=1}^{k} q_{cost}(t_j),$$

where $k\,(1 \leq k \leq m)$ is the number of services executed before the $s_i$ failure has been detected. In the second case, additional expenses can be involved since additional time is required to forward an error message to a place where it can be correctly processed.

## 6 Failure Risk of Redundant Service Compositions

In this section, we provide an example of failure risk management for redundant service compositions.

Redundant compositions include one or more XOR/OR split followed by XOR/OR/m-out-of-n join patterns that define alternative ways to accomplish some tasks. In our scenario, end users invoke a composite web service, which

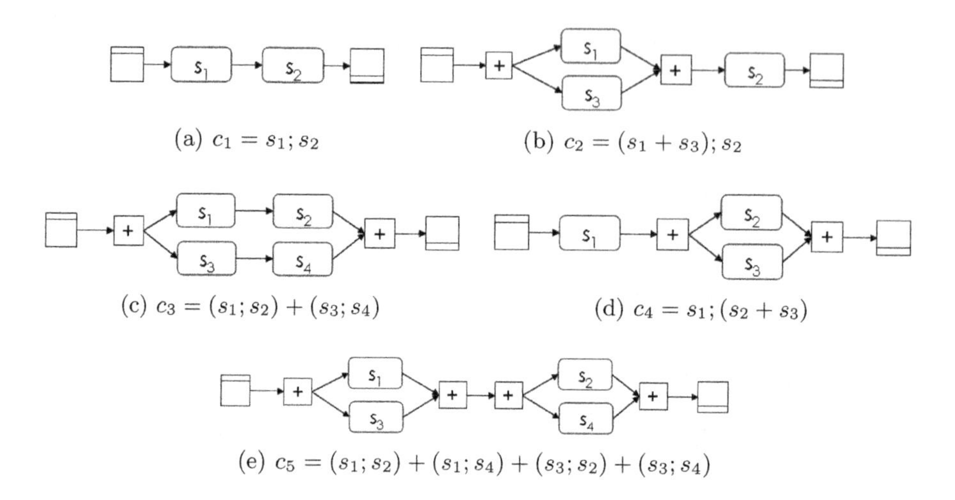

(a) $c_1 = s_1; s_2$

(b) $c_2 = (s_1 + s_3); s_2$

(c) $c_3 = (s_1; s_2) + (s_3; s_4)$

(d) $c_4 = s_1; (s_2 + s_3)$

(e) $c_5 = (s_1; s_2) + (s_1; s_4) + (s_3; s_2) + (s_3; s_4)$

**Fig. 3.** Composite web services with different structure

invokes a set of other services to fulfill user requests. If the user task is not satisfied, the provider of the composite service pays a compensation. Similarly, if an atomic service fails, the provider of the composite service receives a compensation from the provider of the failed service. A Service Level Agreement (SLA) with the end user can be established in such a way that a service composition will satisfy the constraints on response time and execution cost provided that no faults occur. Unexpected failures of component services lead to resource loss and may cause a violation of negotiated parameters. If there are stand-by resources (the maximum budget for a task is not reached, there is time before a task execution deadline), a user task can be completed by other web services. Therefore, it is reasonable to create a contingency plan in order to improve fault resistance of a composite web service. For redundant compositions a *contingency plan* is a set of triples $\langle a_k, t_j, c_i \rangle$, expressing the fact that a sub-composition $c_i$ is started from a state $t_j$ after an event $a_k$. Here, $t_j$ is one of the XOR/OR split states, and $a_k$ refers to the actions discussed in Section 4.

Let $q_{cost}(s_i)$ be an execution cost and $q_{pnlt}(s_i)$ be a penalty of an atomic service $s_i$. We will denote a difference between the service execution cost and the penalty paid if the service $s_i$ fails by $q_{df}(s_i) = q_{cost}(s_i) - q_{pnlt}(s_i)$. Let also $q_{pnlt}(c)$ be a penalty paid by the provider of the composite web service in case of its failure. Figure 3 shows five web service compositions with different internal structure. Their failure risk is shown in Table 2. It is assumed that service failures are independent, i.e., $p(\overline{s}_i|\overline{s}_j) = p(\overline{s}_i)$, $1 \leq i, j \leq n, i \neq j$. Failure risk defines an expected amount of money the provider will lose exploiting external services with certain QoS, provided different levels of redundancy: in the first case, only one service is assigned to each task; in the last case, two services are assigned to each task. An alternative combination is used only if the previous one fails to complete the process.

**Table 2.** Failure risk for compositions in Fig. 3. The risk values are given for the following parameters: $p(s_i) = p(\overline{s}_i) = 0.5$, $q_{cost}(s_i) = q_{pnlt}(s_i) = 1$, $q_{pnlt}(c) = 2$.

| $c_i$ | **Failure risk calculation formula** | $R(c_i)$ |
|---|---|---|
| $c_1$ | $p(\overline{s}_1)\big(q_{df}(s_1) + q_{pnlt}(c)\big) + p(s_1)p(\overline{s}_2)\big(q_{cost}(s_1) + q_{df}(s_2) + q_{pnlt}(c)\big)$. | 1.75 |
| $c_2$ | $p(\overline{s}_1)\big(q_{df}(s_1)+p(\overline{s}_2)(q_{df}(s_2)+q_{pnlt}(c))+p(s_2)p(\overline{s}_3)(q_{cost}(s_2)+q_{df}(s_3)+ q_{pnlt}(c)\big) + p(s_1)p(\overline{s}_3)\big(q_{cost}(s_1) + q_{df}(s_3) + q_{pnlt}(c)\big)$. | 1.625 |
| $c_3$ | $p(\overline{s}_1)\big(q_{df}(s_1)+p(\overline{s}_3)(q_{df}(s_3)+q_{pnlt}(c))+p(s_3)p(\overline{s}_4)(q_{cost}(s_3)+q_{df}(s_4)+ q_{pnlt}(c))\big) + p(s_1)p(\overline{s}_2)\big(q_{cost}(s_1) + q_{df}(s_2) + p(\overline{s}_3)(q_{df}(s_3) + q_{pnlt}(c)) + p(s_3)p(\overline{s}_4)(q_{cost}(s_3) + q_{df}(s_4) + q_{pnlt}(c))\big)$. | 1.5625 |
| $c_4$ | $p(\overline{s}_1)\big(q_{df}(s_1)+q_{pnlt}(c)\big)+p(s_1)p(\overline{s}_2)\big(q_{df}(s_2)+p(\overline{s}_3)(q_{cost}(s_1)+q_{df}(s_3)+ q_{pnlt}(c))\big)$. | 1.375 |
| $c_5$ | $p(\overline{s}_1)\big(q_{df}(s_1)+p(\overline{s}_3)(q_{df}(s_3)+q_{pnlt}(c))+p(s_3)p(\overline{s}_4)(q_{cost}(s_3)+q_{df}(s_4)+ q_{pnlt}(c))\big) + p(s_1)p(\overline{s}_2)\big(q_{df}(s_2) + p(\overline{s}_4)(q_{cost}(s_1) + q_{df}(s_4) + q_{pnlt}(c))\big)$. | 1.25 |

Failure risk is a compound measure considering probability of constituent service failures, their response time and/or execution cost along with the structure of a composition graph. Intuitively, compositions with many OR branches are more reliable. However, which configuration will be selected depends on the balance between the above parameters. For example, if only two web services can accomplish some task and one of them failed, it might be better for the composite service to stop the execution instead of trying a second service if it is too expensive.

# 7 Web Service Selection Algorithm

In this section we present a modification of the method by Zeng et al. [7] for web service selection based on the above ideas.

Let us consider a composite service that invokes a set of external web services. *Success rate* of a service $s$ can be defined statistically as $p(s) = N_{suc}(s)/N_{total}(s)$, where $N_{suc}$ is the number of successful service responses and $N_{total}$ is the total number of observed invocations. A service invocation is considered *successful* if the user goal is satisfied or we can proceed along with the execution, i.e., (1) the service is available, (2) the successful response message is received within an established timeout, (3) no errors are detected automatically in the output, (4) service effects are satisfied, (5) preconditions of a subsequent service are met. If necessary, we can distinguish the above situations and consider several metrics for service successful invocation. Success rate defines the *probability of success* $p(s)$ for future service invocations. Along with the probability of success we can consider the *probability of failure* $p(\overline{s}) = 1 - p(s)$.

The main idea of the service selection algorithm is to search for a simple path $(s_1; ...; s_k)$ between the start and the end states in the composition graph that maximizes the following target function:

$$f(c) = p(c)(q^{max} - q(c)) = p(s_1; ...; s_k)(q^{max} - q(s_1; ...; s_k)) = \\ = \prod_{i=1}^{k} p(s_i)(q^{max} - \sum_{i=1}^{k} q(s_i)),$$

where $q^{max}$ defines the resource limit, taken from an SLA or chosen big enough to guarantee the positive value of $f(c)$. This formula can be inferred from the more general one (see Section 4) assuming that the provider pays for external web services only if all of them are executed successfully, and no penalty is paid to a user if the composite service fails. Thus, to maximize the profit, the percentage of the successful invocations must be maximized and cost of each invocation must be minimized. Here $q(.)$ can refer to response time, execution cost, or a function including both of them. For instance, we can use a linear combination of execution cost and response time: $q = w_1 q_{time} + w_2 q_{cost} \mid w_1 + w_2 = 1,\ 0 \leq w_1, w_2 \leq 1$. Service availability is not considered explicitly as in [7], however, according to our definition, this aspect is characterized by the service success rate.

## 8 Experimental Evaluation

In order to analyze our approach empirically, we compared the proposed service selection algorithm with the linear programming approach by Zeng et al. [7]. We developed a simulation of a web service composition engine and generated a large number of random service compositions. For the data presented in this paper, we used 100 compositions of 10 atomic web services. Such a relatively small number of services included in one composition is chosen to follow the realistic scenarios. For each atomic web service its execution cost, response time and success rate are defined randomly with uniform distribution, from 0 to $maxCost = 1000\$$ for execution cost, from 0 to $timeout = 1000ms$ for response time, and from 0.5 to 1 for success rate (values greater than 0.5 are generated to avoid services with very low success rate). We compared the performance of our method with the performance of the linear programming approach by recording the expected response time, execution cost and success rate of the compositions chosen by these two methods. We also simulated invocation of the chosen compositions and compared their real success rates. We assigned weights $w_i = \frac{1}{3}$ for each of the three parameters for the linear programming approach, and $w_i = 0.5$ for response time and execution cost in our approach. The solutions proposed by our modified algorithm had better response time and execution cost than the solutions found by the linear programming approach, in 96% and 89% of tests, correspondingly. In the same time, expected success rates of these solutions were always better. More details about the presented results are available in our technical report [29].

## 9 Conclusion

We have proposed a risk-driven methodology for QoS evaluation. This approach may help to simplify web service selection by considering cost equivalent of various QoS factors. We have demonstrated how risk analysis can be used to measure impact of atomic service failures on a service composition. Our experiments prove that the difference between expected income and expenses better characterizes the problem of QoS-driven web service selection from provider's perspective than a linear combination of scores for various QoS factors.

An obvious drawback of our metric is that different redundant compositions require risk recalculation, which makes the approach computationally less efficient than methods relying on the QoS evaluation of well defined patterns [20]. In our previous work [30] a polynomial-time greedy heuristic selecting a less risky sub-composition in each XOR split state was proposed.

Service oriented systems are open to various risks. Different techniques might be needed for their identification, analysis and control. In our future work we are going to systematize and elaborate the above ideas.

## References


1. Ran, S.: A model for web services discovery with QoS. ACM SIGecom Exchanges **4**(1) (2003) 1–10
2. Tosic, V., Pagurek, B., Patel, K.: WSOL - a language for the formal specification of classes of service for web services. In: Proceedings of the ICWS, CSREA Press (2003) 375–381
3. Maximilien, M., Singh, M.: A framework and ontology for dynamic web services selection. IEEE Internet Computing **8**(5) (2004) 84–93
4. Zhou, C., Chia, L., Lee, B.: DAML-QoS ontology for web services. In: Proceedings of the ICWS, IEEE Computer Society (2004) 472 – 479
5. Cardoso, J., Sheth, A., Miller, J., Arnold, J., Kochut, K.: Quality of service for workflows and web service processes. Journal of Web Semantics **1**(3) (2004) 281–308
6. Lakhal, N.B., Kobayashi, T., Yokota, H.: A failure-aware model for estimating the efficiency of web service compositions. In: Proceedings of the IEEE Pacific Rim International Symposium on Dependable Computing, IEEE Computer Society (2005) 114–121
7. Zeng, L., Benatallah, B., Ngu, A.H., Dumas, M., Kalagnanam, J., Chang, H.: QoS-aware middleware for web services composition. IEEE Transactions on Software Engineering **30**(5) (2004) 311–327
8. Ardagna, D., Pernici, B.: Global and local QoS constraints guarantee in web service selection. In: Proceedings of the ICWS, IEEE Computer Society (2005) 805–806
9. Yu, T., Lin, K.: Service selection algorithms for composing complex services with multiple QoS constraints. In: Proceedings of the ICSOC. Volume 3826 of LNCS., Springer (2005) 130 – 143
10. Gao, A., Yang, D., Tang, S., Zhang, M.: QoS-driven web service composition with inter service conflicts. In: Frontiers of WWW Research and Development - APWeb: Asia-Pacific Web Conference, Springer Berlin Heidelberg (2006) 121 – 132
11. Wang, X., Vitvar, T., Kerrigan, M., Toma, I.: A QoS-aware selection model for semantic web services. In: Proceedings of the ICSOC. Volume 4294 of LNCS., Springer (2006) 390–401

12. Yang, L., Dai, Y., Zhang, B., Gao, Y.: Dynamic selection of composite web services based on a genetic algorithm optimized new structured neural network. In: Proceedings of the International Conference on Cyberworlds, IEEE Computer Society (2005) 515 – 522
13. Hu, J., Guo, C., Wang, H., Zou, P.: Quality driven web services selection. In: Proceedings of the ICEBE, IEEE Computer Society (2005)
14. Claro, D., Albers, P., Hao, J.K.: Selecting web services for optimal composition. In: Proceedings of the International Workshop on Semantic and Dynamic Web Processes. (2005) 32–45
15. Canfora, G., Penta, M.D., Esposito, R., Villani, M.L.: QoS-aware replanning of composite web services. In: Proceedings of the ICWS, IEEE CS Press (2005)
16. Martin-Diaz, O., Ruize-Cortes, A., Duran, A., Muller, C.: An approach to temporal-aware procurement of web services. In: Proceedings of the ICSOC, Springer (2005) 170–184
17. Bonatti, P., Festa, P.: On optimal service selection. In: Proceedings of the International WWW Conference, ACM Press (2005) 530–538
18. Mou, Y., Cao, J., Zhang, S., Zhang, J.: Interactive web service choice-making based on extended QoS model. In: Proceedings of the CIT, IEEE Computer Society (2005) 1130–1134
19. Lin, M., Xie, J., Guo, H., Wang, H.: Solving QoS-driven web service dynamic composition as fuzzy constraint satisfaction. In: Proceedings of the Int. Conference on e-Technology, e-Commerce and e-Service, IEEE Computer Society (2005) 9–14
20. Jaeger, M., Ladner, H.: Improving the QoS of WS compositions based on redundant services. In: Proceedings of the NWeSP, IEEE Computer Society (2005)
21. Stein, S., Gennings, N.R., Payne, T.R.: Flexible provisioning of service workflows. In: Proceedings of the ECAI, IOS Press (2006) 295–299
22. van der Aalst, W.M., ter Hofstede, A.H., Kiepuszewski, B., Barros, A.: Workflow patterns. Distributed and Parallel Databases **14**(3) (2003) 5–51
23. Verdon, D., McGraw, G.: Risk analysis in software design. IEEE Security and Privacy (2004) 33–37
24. Freimut, B., Hartkopf, S., et al.: An industrial case study of implementing software risk management. In: Proceedings of the ESEC/FSE, ACM Press (2001) 277–287
25. Muehlen, M., Ho, D.T.Y.: Integrating risks in business process models. In: Proceedings of Australasian Conference on Information Systems (ACIS). (2005)
26. Neiger, D., Churilov, L., Muehlen, M., Rosemann, M.: Integrating risks in business process models with value focused process engineering. In: Proceedings of the European Conference on Information Systems (ECIS). (2006)
27. O'Keeffe, P., Vanlandingham, S.: Managing the risks of outsourcing: a survey of current practicies and their effectiveness. White paper, Protivity Independent Risk Consulting, http://www.protiviti.com/downloads/PRO/pro-us/product_sheets/business_risk/Protiviti_ORM_WhitePaper.pdf (2004)
28. Roy, G.: A risk management framework for software engineering practice. In: Proceedings of the ASWEC, IEEE Computer Society (2004) 60 – 67
29. Kokash, N., D'Andrea, V.: Evaluating quality of web services: A risk-driven approach. Technical Report DIT-06-099, DIT-University of Trento, Italy (2006)
30. Kokash, N.: A service selection model to improve composition reliability. In: International Workshop on AI for Service Composition, University of Trento (2006) 9–14

# Proposal of an Architecture for a Biometrics Grid*

Anlong Ming and Huadong Ma

Beijing Key Laboratory of Intelligent Telecommunications Software and Multimedia, School of Computer Sci. and Tech., Beijing Univ. of Posts and Telecommunications, Beijing 100876, China
anthonyming@gmail.com, mhd@bupt.edu.cn
http://bklab.cs.bupt.cn/

**Abstract.** Biometric technology is critical to the rapidly growing suite of civilian applications. In this paper, we offer a broader scope of biometrics by a novel concept, the biometrics grid. The architecture of biometrics grid aims to overcome/resolve three main problems of existing biometric technology using the grid computing: 1) the need for cooperation in different fields, 2) duplicated works and 3) the consideration of application demands in different levels. The proposed architecture of biometrics grid provides a Web portal based virtual environment as well as features like biometric applications meeting QoS goals. Experimental results are given in order to show the feasibility to deploy the idea of grid computing in biometric applications.

**Keywords:** Grid technology, Biometrics, Face recognition, Grid application.

## 1 Introduction

In this age of digital impersonation, biometric techniques are being used increasingly as a hedge against identity theft. The premise is that a biometrica measurable physical characteristic or behavioral traitis a more reliable indicator of identity than legacy systems such as passwords and PINs. Biometric systems have been defined by the U.S. National Institute of Standards and Technology (NIST) [1, 2] as systems exploiting "automated methods of recognizing a person based on physiological or behavioral characteristics" (biometric identifiers, also called features). Physiological biometrics is based on data derived from direct measurement of a body part (i.e., fingerprints, face, retina, iris), while behavioral biometrics is based on measurements and data derived from a human action [2] (i.e., gait and signature).

* The work is supported by the National Natural Science Foundation of China (90612013), the National High Technology Research and Development Program of China under Grant No.2006AA01Z304, the Specialized Research Fund for the Doctoral Program of Higher Education (20050013010) and the NCET Program of MOE, China.

W. Abramowicz (Ed.): BIS 2007, LNCS 4439, pp. 195–208, 2007.
© Springer-Verlag Berlin Heidelberg 2007

Biometric systems are being used to verify identities and restrict access to buildings, computer networks, and other secure sites [3]. Recent global terrorism is pushing the need for secure, fast, and non-intrusive identification of people as a primary goal for homeland security. As commonly accepted, biometrics seems to be the first candidate to efficiently satisfy these requirements. For example, from October 2004, the United States control the accesses to/from country borders by biometric passports [4, 5]. Biometrics systems are traditionally used for three different applications [6, 7]: *physical access control* for the protection against unauthorized person to access to places or rooms, *logical access control* for the protection of networks and computers, and *time and attendance control.*

Biometric technology, an integral part of law enforcement, is critical to the rapidly growing suite of civilian applications, such as citizen ID cards, e-passports, and driver license. These systems not only need advanced biometric technology interfaces but also the ability to deal with security and privacy issues. The integration of biometrics with access control mechanisms and information security is another area of growing interest. The challenge to the research community is to develop integrated solutions that address the entire problems from sensors and data acquisition, to biometric data analysis and systems design.

However, biometric technology suffers problems in its way of research and applications:

- *Cooperation in diverse fields.* From a technical viewpoint, biometrics spans various technologies, such as fingerprint and face recognition, mathematics and statistics, performance evaluation, integration and system design, integrity, and last but not least, privacy and security. Therefore, there is a need for scientists and practitioners from the diverse fields of computing, sensor technologies, law enforcement and social sciences to exchange ideas research challenges and results.
- *Duplicated works.* Currently, most biometric technology research in offered production are either actually intra-organizational or operated by application domains, such as FaceVACS-SDK produced by Cognitec. It is wasteful with duplicated efforts in building test databases as well as difficulty in providing uniform performance standards. For example, face recognition researchers spend great efforts in building face databases (i.e., FERET, PIE, BANCA, CAS-PEAL, AR) while these databases are not easily accessed by others.
- *Large scale databases.* The population in a database can significantly affect performance [8]. In a system with a large scale database, the ordinary recognition processes perform poorly: with the increase of the database scale, the identification rates of most algorithms may decline rapidly; meanwhile, querying in a large scale database may be quite time-consuming. So how to deal with a large scale database has been a difficult problem faced by researchers on biometric technology in recent years. Su et al presented a face recognition system framework constructed on the client-server architecture [9]. A distributed and parallel architecture was introduced to this system. The clients and servers are connected by $1000MB$ networking switch. Although this system has gained good performance: querying one face image

in 2,560,000 faces costs only 1.094s and the identification rate is above 85% in most cases, it is limited in accessing and extending due to its C/S framework.
- *Application demands in different levels.* One important aspect of biometrics that has not been adequately addressed by the research community thus far is that of large scale applications. There is a dichotomy between established and newer biometric approaches. Some biometric technologies, such as fingerprint identification and speaker verification, are relatively mature and have drawn considerable attention over 40 years, from both a research and a development viewpoint. Other, less mature biometric technologies, such as recognition based on face, hand, palmprint and iris, have brought much recent innovation and excitement and are starting to be successfully deployed in commercial systems.

Grid systems [10, 11] are the gathering of distributed and heterogeneous resources (CPU, disk, network, etc.). A grid is a high-performance hardware and software infrastructure providing scalable, dependable and secure access to the distributed resources. Unlike distributed computing and cluster computing, the individual resources in grid computing maintain administrative autonomy and are allowed system heterogeneity; this aspect of grid computing guarantees scalability and vigor. Therefore, the grids resources must adhere to agreed-upon standards to remain open and scalable. They are promising infrastructures for executing large scale applications and to provide computational power to everyone. To promote both biometric technology and grid computing, this paper aims to present a biometrics grid (BMG).

- We present a description of the architecture of BMG mainly based on Globus Toolkit components. BMG provides a Web portal based virtual environment, biometric applications with different QoS demands. The Globus Toolkit, currently at version 4, is an open source toolkit for building computing grids provided by the Globus Alliance.
- A Web portal based virtual environment. Considering *cooperation in diverse fields*, BMG creates a virtual collaborative environment linking distributed centers, users, models, and data. The virtual collaborative environment provides advantages to urge cooperations in diverse fields.
- Biometric applications with different QoS demands. Considering *application demands in different levels* and *Large scale databases*, biometric applications with different QoS demands or large scale databases can be solved by grid computing. Users can easily enjoy these applications because BMG is based on the Globus Toolkit and the WS-Resource Framework (WSRF). WSRF is a set of six Web Services specifications that define what is termed the *WS-Resource approach* to model and manage state in a Web Services context, WSRF enables the discovery of, introspection on, and interaction with stateful resources in standard and interoperable ways.

The remainder of the paper is organized as follows: the architecture is described in Section 2. The BMG-specific features are presented in Section 3. The discussion is given in Section 4. We give experimental results in Section 5 and conclude our work in Section 6.

## 2 Concepts

The goal of the BMG system is to simplify both the resource management task and support mature biometric applications with different QoS demands, by making it easier for resource managers to make resource available to others, and the resource access task, by making biometric data as easy to access as Web pages via a Web browser.

### 2.1 Architecture

We present an architecture of BMG in Fig. 1 (a). The major components involved in the BMG system are described as follows:

- Databases and servers: They are the basic resources on which BMG is constructed, the Web Services container is used to host the BMG portal services.
- Globus/Grid infrastructure: This provides remote, authenticated access to the shared BMG resources. All these components are specified by WSRF.
- High-level and the BMG-specific services: These services span the BMG resources and provide capabilities such as services (submitted by users) deployment, site to site data movement, distributed and reliable metadata access, and data aggregation and filtering.
- Client applications: All biometric applications are wrapped to Web Services specified by WSRF and deployed into the Web Services container. The Web portal control and render the user interfaceCinteraction.

Moreover, such efforts are wasteful, with duplicated work in building test databases as well as difficulty in providing uniform performance standards. So we attempt to give an uniform dataset whose data structure can be defined as follows:

```
struct Person{
    /*personal information*/
    name,
    address,
    nationality,
    ......
    /*physiological characteristics (NULL when absent)*/
    fingerprints,
    face,
    retina,
    iris,
    ....
    /*behavioral characteristics (NULL when absent)*/
    gait,
    signature
    ......
}
```

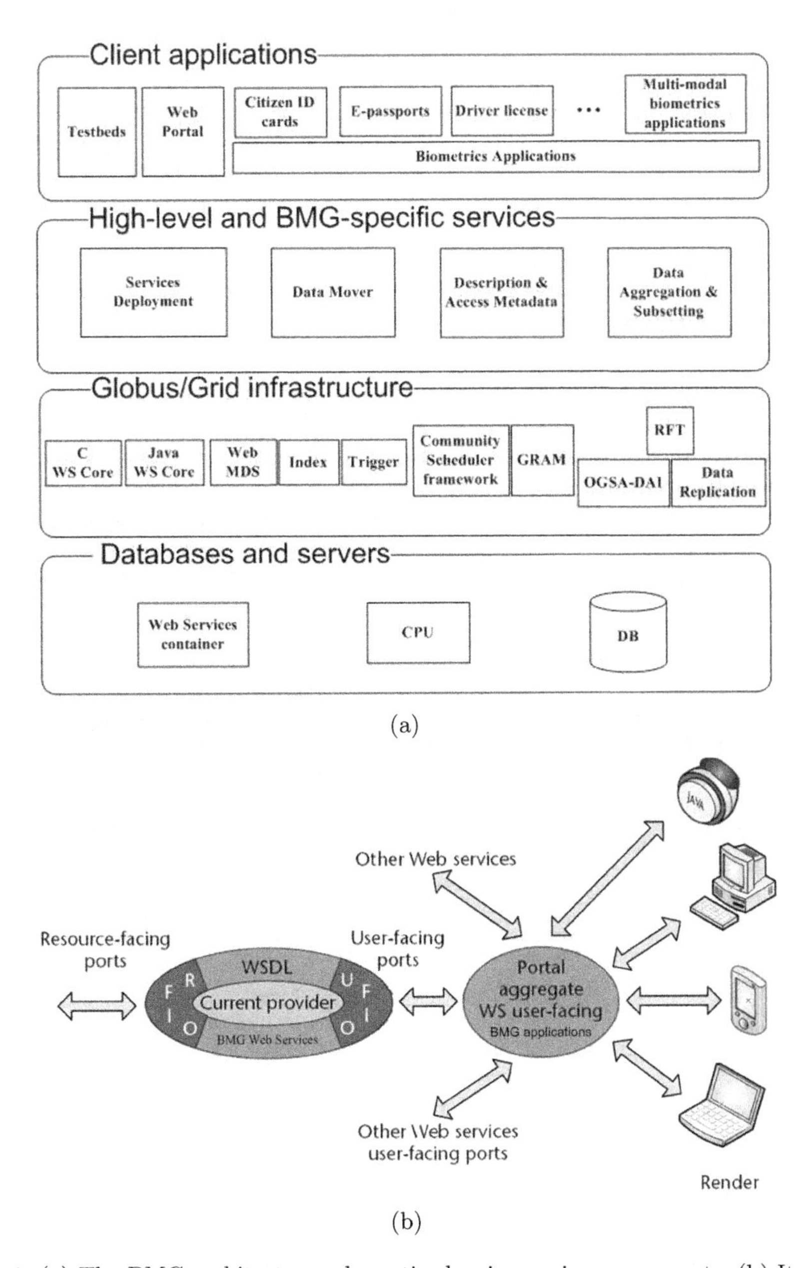


**Fig. 1.** (a) The BMG architecture schematic showing major components. (b) Its illustration shows a portal providing an aggregation service for the BMG applications of a content providing Web Service.

According to the mentioned definition, all kinds of licensed biometric characteristics can be aggregated for further applications without replicated work. However, a basic requirement is for tools that allow data managers to make licensed "person" data available to the BMG community. These tools include the means to create searchable databases of persons, provide catalogs of the data that locate a given piece of data on an archival system or online storage, and make catalogs and data accessible via the Web. Prior to the advent of the grid, these capabilities did not exist, so potential users of the model data had to contact the data managers personally and begin the laborious process of retrieving the data they wanted.

## 3 The BMG-Specific Features

As a typical nugget programming challenge, we must take into account the needed latency and bandwidth of application and network constraints (firewalls) to decide the most appropriate communication mechanism between nuggets. This runtime specification of a service-service interaction's implementation has no agreed-upon approach. Given the bandwidth and capacity constraints of computation nodes, BMG provides biometric services with different QoS demands meeting with the QoS goals.

### 3.1 Providing Services with Different QoS Demands

The service deployment component take charge of deploying the biometric Web Services on BMG.

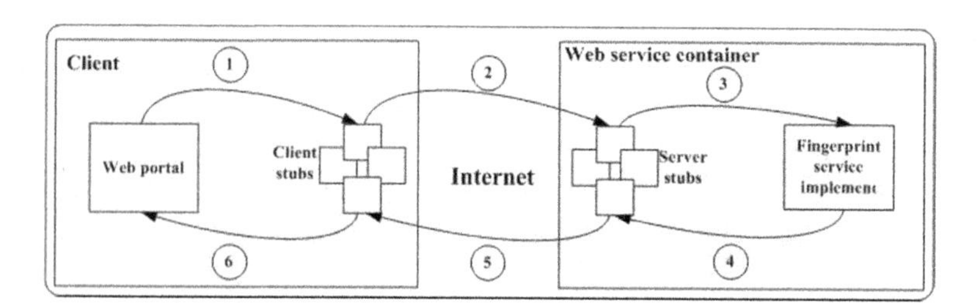


**Fig. 2.** An example of a fingerprint service invocation

Some biometric technologies, such as fingerprint identification and speaker verification, are relatively mature and can be firstly deployed on BMG for commercial applications. Globus provides tools for us to wrap applications to Web Services. For example, a fingerprint application can be easily wrapped into a Web Service (see Fig. 2):

1. Whenever the client fingerprint application needs to invoke the fingerprint Web Service, it will really call the client stub. The client stub will turn this 'local invocation' into a proper SOAP request. This is often called the marshaling or serializing process.

2. The SOAP request is sent over a network using the HTTP protocol. The server receives the SOAP requests and hands it to the server stub. The server stub will convert the SOAP request into something the service implementation can understand.
3. Once the SOAP request has been deserialized, the server stub invokes the fingerprint service implementation, which then carries out the work it has been asked to do.
4. The result of the requested operation is handed to the server stub, which will turn it into a SOAP response.
5. The SOAP response is sent over a network using the HTTP protocol. The client stub receives the SOAP response and turns it into something the client application can understand.
6. The fingerprint application receives the result of the Web Service invocation and uses it.

We can use grid computing to distribute computing applications on the Internet [13] with meeting the QoS goals, given the bandwidth and capacity constraints of computation nodes.

Consider a set of $M$ applications $A = \{a_1, \ldots, a_M\}$ and a set of $K$ computation nodes $C = \{c_1, \ldots, c_K\}$. The bandwidth of the network connection for each computation node $c_j$ in $C$ is $B_j$ bits per second. We can deploy more than one application to any computation node, but a given application runs entirely on a single computation node. Not all possible allocations are feasible. If the network demands of the applications allocated to a given computation node exceed that bandwidth of computation node.

QoS goals are associated with each application. Such goals for application $a_i$ include its maximum average execution time $r_i^{max}$ and its minimum throughput $x_i^{min}$. The workload intensity associated with the application $a_i$ is equal to $\lambda_i$ requests per second. The resource demand of application $a_i$ is characterized by the tuple $(p_i, d_i, n_i)$, where $p_i$ is the processing time of application at one of the computation nodes in $C$, taken as a reference computation node; $d_i$ is the application's I/O demand measured in seconds of disk service time; and $n_i$ is the applications network demand measured in bps. Let us consider that all computation nodes have the same capacity.

Not all possible allocations are feasible. If the network demands of the applications allocated to a given computation node exceed that bandwidth of computation node:

$$B_j \geq \sum_{a_i \in \Omega(c_j)} f(a_i), \quad j = 1, ..., K \tag{1}$$

where $\Omega(c_j)$ is the set of applications allocated to $c_j$, $f(a_i) = n_i$.

An allocation isn't feasible if at least one of the QoS goals for the application is violated. Let $r_i$ be the average response time of application $a_i$. Then, for any feasible allocation, the following relationship must hold:

$$r_i \leq r_i^{max}, \quad i = 1, ..., M \tag{2}$$

The average response time $r_i$ of application $a_i$ depends on its resource demands, its workload intensity, and the resource demands $p_i$ and $d_i$ and workload intensities of all applications allocated to the same computation node as $a_i$. BMG can use a multi-class open queuing network model [14] to compute the average response time for each application in a given allocation. BMG need to create a queuing network model for each computation node, with the applications allocated to that computation node for a given allocation constituting the various classes of model.

### 3.2 A Web Portal Based Virtual Environment

A key architectural of Web portal is shown in Fig. 1 (b). Generally, portals let you take multiple Web pages, automatically produce controls to link between them, and let subsets of them be displayed on a single Web page. All biometric applications are wrapped to Web Services (each with user-facing ports) are aggregated for the user into a single client environment.

We assume that all material presented to users originates from a Web service, called a content provider in this case. This content could come from a simulation, data repository, or stream from an instrument. Each Web Service has resource- or service-facing ports that communicate with other services [12]. However, we are more concerned with the user-facing ports, which produce content for users and accept input from client devices. These ports use an extension of the Web Services Definition Language (WSDL), which is being standardized by the Organization for the Advancement of Structured Information Standards (OASIS). Most user interfaces need information from more than one content provider. For example, a computing portal could feature separate panels for job submittal, job status, visualization, and other services. We could integrate this in a custom, application-specific Web Service, but providing a generic aggregation service is more attractive. This lets users and administrators choose which content providers to display and what portion of the display real estate the content will occupy.

### 3.3 Large Scale Biometric Database Supported

Grid computing represents the latest and most exciting technology to evolve from the familiar realm of parallel, peer-to-peer, and client-server models that can address the problem of fault-tolerant storage for backup and recovery of biometric data. A grid is a high-performance hardware and software infrastructure providing scalable, dependable and secure access to the distributed resources. Unlike distributed computing and cluster computing, the individual resources in grid computing maintain administrative autonomy and are allowed system heterogeneity; this aspect of grid computing guarantees scalability and vigor. Therefore, the grids resources must adhere to agreed-upon standards to remain open and scalable. They are promising infrastructures for executing large scale applications and to provide computational power to everyone.

So, BMG can conquer disadvantages of C/S framework because in the heterogeneous grid environments, we can hide the heterogeneity of computational resources and networks by providing Globus Toolkit Services and can implement the distributed parallel computing of a large scale problem by taking full advantage of Internet resources. According to the applied demand, there are two implementations for grid-based distributed parallel computing, which are Loosely Coupled Parallel Services and grid MPI Parallel Program. Loosely coupled parallel services combine the service-oriented architecture with parallel batch model to implement the parallel computing; that is to say, the parallel computing can be achieved by parallel services. Grid MPI parallel program is offered for specialized applications. Its design sustains and integrates closely with parallel processing from the bottom, so it can be applied in different applications.

## 4 Discussions

BMG offers advantages over the classical biometric system. However, Some challenges stand in the way of the BMG building. We will discuss three of them here.

**Standardization.** Standards are key to the industry's growth and a technology's acceptance [15]. Biometrics is a relatively young industry with immense potential and is thus a key target for standardization (e.g., according to an uniform definition, all kinds of licensed biometric characteristics can be aggregated for further applications without replicated work). Its roots lie in law enforcement and other government applications, in which custom integration was the norm, so it is not surprising that the standards to adapt biometrics to off-the-shelf applications are still evolving. Tracing the lineage of the current effort toward a biometric application programming interface standard provides insight into its architecture, components, functions, and data.

Several technical issues await resolution [16] (e.g., how will the consortium implement scoring and thresholding, will they include normalization, how will they handle model adaptation), either because they werent adequately addressed in any of the three earlier APIs or because these APIs addressed each issue in different ways.

**Liberties.** BMG poses threats to civil liberties. In biometric systems, the degree of similarity between templates required for a positive match depends on a decision threshold, a user-defined system parameter. The user can specify high security, in which case innocent subjects might be caught when the system casts a broader net. Alternatively, the user might specify low security, in which case terrorists could escape. Setting this parameter thus directly affects the false positive rate, which in turn directly affects subjects' privacy. Another important civil liberty issue involves the potential for biometric systems to locate and physically track airline passengers. People being scanned and possibly tracked may not be aware of the system and thus cannot control it.

**Who are the BMG Makers.** Obviously, the BMG designers and developers must be aware of lots of issues such as development of grid computing, their work's civil liberty implications, cost, and the tremendous effect their decisions and actions can have on society as a whole. So, the BMG makers are undetermined.

# 5 Experiments

In our experiments, we carry out a face recognition process on grid with the same interfaces using four different face recognition methods. The face recognition approaches we used are listed as: the line based face recognition algorithm [17], the improved line based face recognition algorithm [18], PCA and PCA+LDA [19].

## 5.1 The Experimental Environment

We build a grid platform for the implementation of a biometric paradigm. The schematic architecture of our grid platform is shown as Fig. 3 (a), the flowchart of a face recognition job is denoted in Fig. 3 (b), the Web portal we used for face recognition is shown in Fig. 4 (a). Table 1 shows our experimental environment.

**Table 1.** Experimental Environment

| Grid computing nodes | P4 2.93GHz |
|---|---|
| OS | Linux Fedora Core 4 |
| Toolkit | Globus Tookit 3.2 |
| DBMS | MySQL 5.0 |
| Web server platform | Apache Tomcat 5.0 |
| Others | HTML, JSP, XML, Java Bean |

We use a face database established by ourselves to evaluate the performance of our algorithm. Pictures of 35 persons are taken by a standard camera (6 pictures per person) under different environments of illumination intensity (weak, medium and strong). We select 3 views of each person for training, and the other 3 views (in weak, medium, and strong illumination intensity respectively) is used for test.

## 5.2 A Face Recognition Process on Grid

We define 3 simple interfaces, which are executable files of c language in Linux platform, to run a face recognition job.

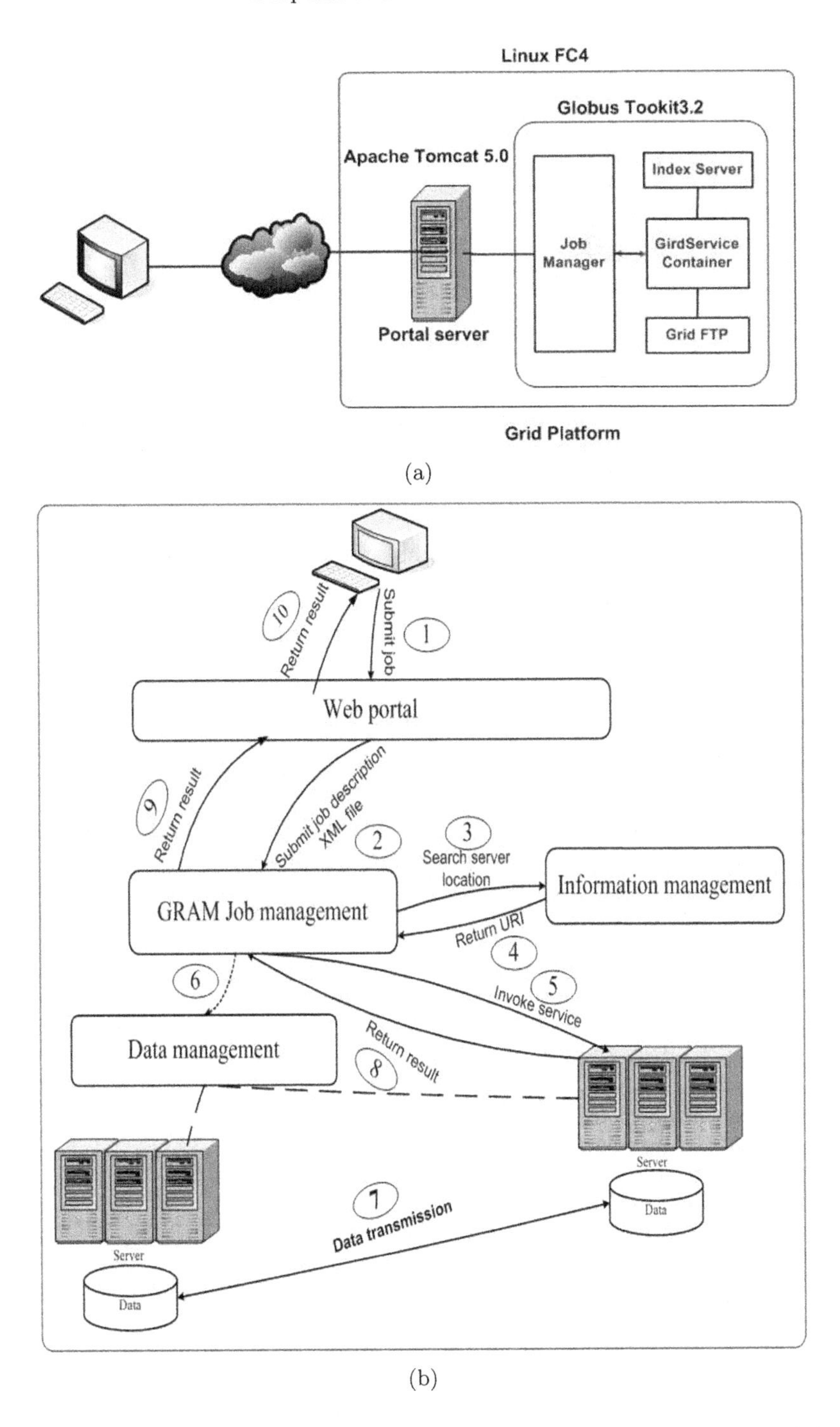


**Fig. 3.** (a) The schematic architecture of our experimental grid platform. (b) The flowchart of a face recognition job (URI represents Uniform Resource Identifier).

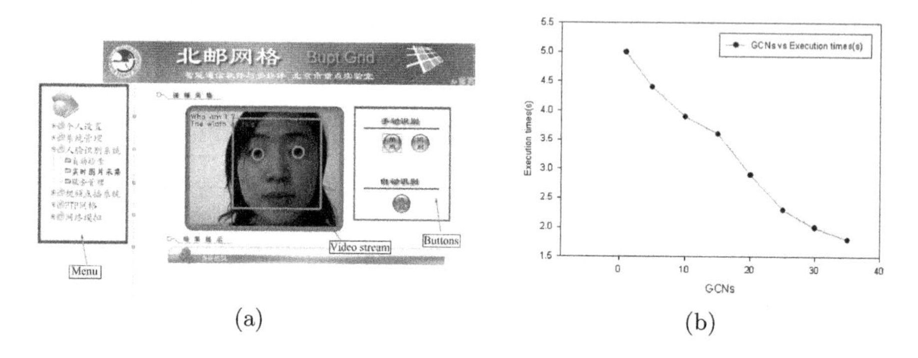


**Fig. 4.** (a) The Web portal; (b) The relation between the execution times and the numbers of grid computation nodes (GCNs) of the improved line based face recognition algorithm

- *Interface 1* Training.exe, a executable file for face images training, can be invoked as follows:

$$Training.exe \quad face\_image\_1 \quad face\_image\_2...$$

- *Interface 2* FeatureExt.exe, a executable file for extracting feature vectors using training results, can be invoked as follows:

$$FeatureExt.exe \quad face\_image$$

- *Interface 3* FeatureMat.exe, a executable file for matching feature vectors between two face images, can be invoked as follows:

$$FeatureMat.exe \quad face\_image\_1 \quad face\_image\_2$$

**Algorithm 1.** A face recognition algorithm on grid

**Input:** $X$, $N$, $K$. {$X$ is an face image for recognition. $N$ is the number of face views in the face database. Each face view has $K$ feature vectors}
**Output:** $R$. {$R$ is a face view in the database.}
1: Read face feature vectors from the face database
2: Extract $K$ feature vectors from $X$;
3: $Finished = 0$
4: **while** $Finished < N$ **do**
5: $Finished = Finished + 1$
6: **if** There exists a spare grid computation node **then**
7: Compare $X$ with a face view in the database
8: **else**
9: $Finished = Finished$ - 1, and wait a second
10: **end if**
11: Get a face view $R$ with the maximum similarity;
12: **end while**
13: Return $R$

We implemented each face recognition method and build 3 exe files respectively, then these files would be submitted to grid by GRAM Server and RSL (XML file) for recognition.

The uniform face recognition process on grid for different methods is shown in **Algorithm 1**. We have tested four face recognition methods on the same, but individually processed, face database. Moreover, as illustrated in and Fig. 4 (b), the average execution time of the improved line based face recognition algorithm can be shorten by increasing the number of grid computation nodes.

## 6 Conclusion

We propose the new concept and an architecture of BMG. The goal of the BMG system is to simplify both the resource management task, by making it easier for resource managers to make resource available to others, and the resource access task, by making biometric data as easy to access as Web pages via a Web browser. Also, BMG would support existing biometric applications with different QoS goals. Applications would be wrapped to Web Services based on WSRF and deployed into Web Services container.

BMG is designed for the contributions of saving processing time and improving resource management. However, the Web portal does not guarantee that QoS demands are meeting with the QoS goals, when defining QoS more broadly than bandwidth and capacity.

According to the analysis discussed in Section 4, the advent of BMG should be under the legal guidelines of governments. With the development of grid computing, the technical scheme of BMG will also be improved.

## References

1. M. Gamassi, Massimo Lazzaroni, et al.: Quality Assessment of Biometric Systems: A Comprehensive Perspective Based on Accuracy and Performance Measurement, IEEE Transactions on Instrumentation Measurement, Vol. 54, No. 4, (2005)
2. R. Bolle, S. Pankanti, and A.K. Jain: Guest editorial, IEEE Computer (Special Issue on Biometrics), vol. 33, No. 2, (2000) 46-49
3. J.D.M. Ashbourn: Biometrics: Advanced Identify VerificationThe Complete Guide, Springer-Verlag, Berlin (2000)
4. S. Waterman: Biometric borders coming, Times, Washington (2003)
5. General Accounting Office (GAO).: Technology assessment: Using biometrics for border security, GAO-03-174, Washington, (2002)
6. A.K. Jain, R. Bolle, et al.: Biometrics: Personal Identification in Networked Society, Kluwer Academic (1999)
7. D. Maltoni, D. Maio, A.K. Jain, and S. Prabhakar: Handbook of Fingerprint Recognition, Springer (2003)
8. PJouathou Phillips, Patrick Grother, et al.: Face recognitiou vendor test 2002: Evaluatiou report, Audio- and Video-Based Person Authentication (AVBPA) (2003)
9. Kai Meng, Guangda Su, et al.: A High Performence Face Recognition System Based on A Huge Face Database, Proceedings of the Fourth International Conference on Machine Learning and Cybernetics, Guangzhou (2005)

10. Satoshi Matsuoka, et al.: Japanese computational grid research project: NAREGI, Digital Object Identifier , Vol. 93, Issue 3, (2005) 522–533
11. David Bernholdt, et al.: The Earth System Grid: Supporting the Next Generation of Climate Modeling Research, Digital Object Identifier, Vol. 93, Issue 3 (2005) 485–495
12. Geoffrey Fox: Grid computing environments, Digital Object Identifier, Vol.5(2) (2003): 68-72
13. D.A. Menasc: Allocating applications in distributed computing, Digital Object Identifier, Vol. 9, Issue 1 (2005) 90-92
14. D.A. Menasc, V.A.F. Almeida, and L.W. Dowdy: Performance by Design: Capacity Planning by Example, Prentice Hall (2004)
15. Catherine J. Tilton, SafLink Corp: An Emerging Biometric API Industry Standard, Digital Object Identifier, Vol. 33(2) (2000) 130-132
16. C. Tilton: Status of Biometric APIs: Are We There Yet?, proceedings of confference on CardTech/SecurTech, Faulkner & Gray, Bethesda, Md., (1999) 403-410
17. O. de Vel and S. Aeberhard.: Line-based face recognition under varying pose, IEEE Trans. Pattern Analysis and Machine Intelligence, vol. 21 (1999) 1081-1088
18. Anlong Ming, Huadong Ma: An improved Approach to the line-based face recognition.pdf, Proceedings of the 2006 IEEE International Conference on Multimedia and Exposition (ICME), Toronto, 2006
19. Zhao WY, Chellappa R, Phillips PJ, Rosenfeld A: Face recognition: A literature survey, ACM Computing Surveys, Vol. 35 (2003) 399-458

# An Ontology Slicing Method Based on Ontology Definition Metamodel*

Longfei Jin and Lei Liu

College of Computer Science and Technology, Jilin University, Changchun, China
jinlongfei@vip.163.com

**Abstract.** Slicing is a method that can extract required segments from data according some special criteria. Program slicing and model slicing are two familiar slicing techniques. By introducing slicing technique into ontology engineering domain, an ontology slicing method is provided in this paper. In the method, an Ontology Dependency Graph (ODG) is derived from OMG's Ontology Definition Metamodel (ODM), and then ontology slices are generated automatically according slicing criteria. This method has many applications in which large scale ontology processing is needed.

## 1 Introduction

Slicing is a method that can extract required segments from data according some special criteria. Program slicing [1] and model slicing [2] are two familiar slicing techniques. Program slicing has many important applications in program test, debugging, parallelization, synthesizing, understanding, software security and software maintenance, etc. Model slicing technique is based on program slicing and has been applied to UML (Unified Modeling Language) [3] model analysis.

Currently, Semantic Web [4] is an active research domain of information technology, which takes ontology [5] as its kernel and basis. With the development and application of Semantic Web, ontologies become larger and larger. How to process these large scale ontologies is current a new issue of ontology engineering domain. Many research works have tight relation to it, such as ontology modularizing, ontology reuse, large scale ontology reasoning, ontology evolution, ontology mapping and contextual ontology, etc.

By introducing slicing technique into ontology engineering domain, we provide an ontology slicing method in this paper. In the method, an Ontology Dependency Graph (ODG) is derived from OMG's Ontology Definition Metamodel (ODM), and then ontology slices are generated automatically according slicing

* The present work has been supported by European Commission under grant No. TH/Asia Link/010 (111084), the Research Fund for the Doctoral Program of Higher Education of China under grant No. 20061083044 and Jilin Province Science Development Plan Project of China under grant No. 20050527.

W. Abramowicz (Ed.): BIS 2007, LNCS 4439, pp. 209–219, 2007.
© Springer-Verlag Berlin Heidelberg 2007

criteria. This method has may applications in which large scale ontology processing is needed.

The rest of this paper is organized as follows. Section 2 introduces Model Driven Semantic Web (MDSW) [6] and its kernel – Ontology Definition Metamodel [7]. Based on ODM, section 3 provides an ontology slicing method composed of construction of Ontology Dependency Graph, formal description of ontology slicing and semantic completeness processing. Section 4 simply analyzes applications of ontology slicing in some ontology engineering domains. In the last section, we discuss the results and mention some aspects for future work.

## 2 Ontology Definition Metamodel

Model Driven Semantic Web is a new direction of semantic web. It combines Model Driven Architecture (MDA)[8] of software engineering and ontology technique of Semantic Web, and provides a new idea for resolving problems in using large scale ontologies. The kernel of MDSW is Ontology Definition Metamodel, which is current undergoing standardization through the Object Management Group (OMG).

ODM contains a family of metamodels of knowledge representation languages and conceptual modeling languages, which are defined by MDA meta-meta model – Meta-Object Facility (MOF) [9]. Among these metamodels, RDF (Resource Description Framework) [10] and OWL (Web Ontology Language) [11] are two mainstream languages of current Semantic Web. We will investigate OWL ontology slicing in this paper. If not mentioned, ontologies are all OWL ontologies in the rest of the paper.

A segment of ODM is shown in Figure 1.

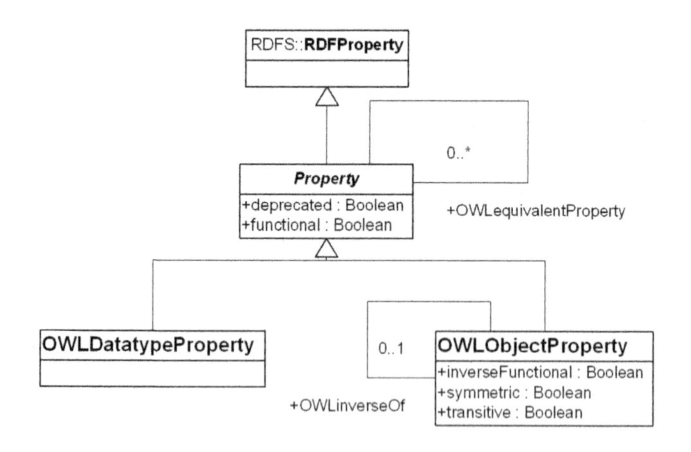


**Fig. 1.** A segment of ODM

Form the figure we can see that ODM uses several kinds of elements as follows.

- Hierarchical classes identified by class names (e.g. `OWLDatatypeProperty`);
- Attributes composed of attribute names (e.g. `functional`), attribute types (e.g. `Boolean` type for `functional` attribute), and attribute cardinalities (e.g. `[1..1]` for `functional` attribute);
- Associations composed of association names (e.g. `OWLinverseOf`), and association cardinalities (e.g. `[0..1]` for `OWLinverseof` association).

For an example, an ontology $\mathcal{O}$ represented by Description Logic [12] is as follows.

$$\mathcal{O} = \{Car \sqsubseteq Vehicle, Vehicle \sqsubseteq_{=1} (colorOfVehicle.Color), \\ Color \sqsubseteq_{=1} (colorName.string), string \sqsubseteq \top, Car(myCar), \\ colorOfVehicle(myCar, redColor), Color(redColor), \\ colorName(redColor, red), string(red)\}$$

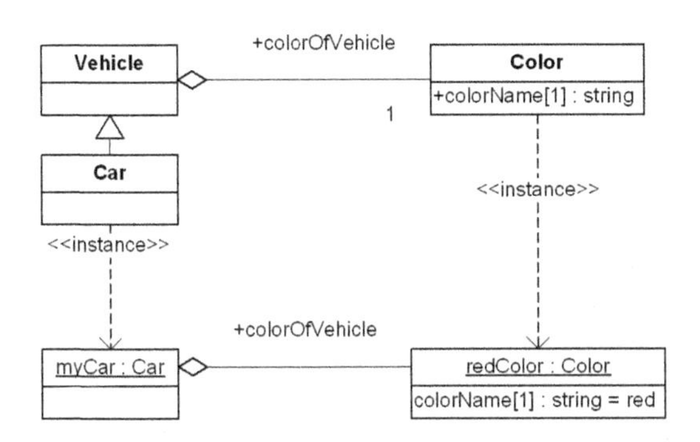


**Fig. 2.** An example of an ontology

Under ODM, it can also be represented in a UML-like graph as Figure 2. This kind of graph is simple, intuitionistic and easy understandable, and it suits for intercommunions between ontology engineers and domain engineers.

# 3 Ontology Slicing

Similar to the program slicing method based on Program Dependency Graphs (PDG) [13], our ontology slicing method is based on Ontology Dependency Graphs derived from ODM, which can represent dependent relations among ontology elements.

## 3.1 Ontology Dependency Graph

**Definition 1.** *An Ontology Dependency Graph (ODG) of an ontology $\mathcal{O}$ is a typed graph $G =< V, E >$, where $V$ is the vertex set and $E$ is the edge set, each element* *of vertex set* $v \in V \wedge v =< vt, A >$ *is an instance of ODM class, where the vertex type* $vt$ *is corresponding to ODM class name and* $A$ *is the attribute set, each attribute* $a \in A \wedge a =< name, value >$ *is a pair of name and value, the name is just ODM attribute name, and the value is an instance of corresponding ODM attribute type; each element of edge set* $e \in E \wedge e =< et, from, to >$ *is an instance of ODM association, where the edge type* $et$ *is just ODM association name, and* $from, to \in V$ *are start vertex and end vertex separately, the direction of an edge specifies dependent relations between start and end vertexes, there are two cases: (1) if* $et \in \{$*OWLequivalentClass, OWLdisjointWith, OWLequivalentProperty, OWLinverseOf, OWLdifferentFrom, OWLsameAs*$\}$*, for the sake of the symmetric feature of these associations, bidirectional edge is needed in the ODG, that is to say* $e =< et, from, to > \in E \wedge e' =< et, to, from > \in E$*; (2) otherwise,* $e$ *has a same direction as its corresponding ODM association.*

For an example, contents of ODG of the ontology shown in Figure 2 are shown in Table 1. Its graphical representation is shown in Figure 3, in which vertex types and attributes are omitted.

**Table 1.** Contents of the ODG

| | |
|---|---|
| $v_0 =< \texttt{CardinalityRestriction}, \{\} >$ | |
| $v_1 =< \texttt{TypedLiteral}, \{\texttt{lexicalForm} = "1", \texttt{datatypeURI} = "http://www.w3.org/2001/XMLSchema\#int"\} >$ | |
| $v_2 =< \texttt{OWLClass}, \{\texttt{localName} = "Car"\} >$ | |
| $v_3 =< \texttt{OWLClass}, \{\texttt{localName} = "Color"\} >$ | |
| $v_4 =< \texttt{OWLClass}, \{\texttt{localName} = "Vehicle"\} >$ | |
| $v_5 =< \texttt{OWLDatatypeProperty}, \{\texttt{localName} = "colorName", \texttt{functional} = "true"\} >$ | |
| $v_6 =< \texttt{OWLObjectProperty}, \{\texttt{localName} = "colorOfVehicle"\} >$ | |
| $v_7 =< \texttt{Individual}, \{\texttt{localName} = "myCar"\} >$ | |
| $v_8 =< \texttt{Individual}, \{\texttt{localName} = "redColor"\} >$ | |
| $v_9 =< \texttt{OWLClass}, \{\texttt{uri} = "http://www.w3.org/2001/XMLSchema\#string"\} >$ | |
| $v_{10} =< \texttt{OWLClass}, \{\texttt{uri} = "http://www.w3.org/2002/07/owl\#Thing"\} >$ | |
| $v_{11} =< \texttt{TypedLiteral}, \{\texttt{lexicalForm} = "red", \texttt{datatypeURI} = "http://www.w3.org/2001/XMLSchema\#string"\} >$ | |
| $v_{12} =< \texttt{ObjectSlot}, \{\} >$ | |
| $v_{13} =< \texttt{DatatypeSlot}, \{\} >$ | |
| $e_0 =< \texttt{OWLonProperty}, v_0, v_6 >$ | $e_1 =< \texttt{OWLcardinality}, v_0, v_1 >$ |
| $e_2 =< \texttt{RDFSsubClassOf}, v_2, v_4 >$ | $e_3 =< \texttt{RDFSsubClassOf}, v_4, v_0 >$ |
| $e_4 =< \texttt{RDFSsubClassOf}, v_4, v_{10} >$ | $e_5 =< \texttt{RDFSdomain}, v_5, v_3 >$ |
| $e_6 =< \texttt{RDFSrange}, v_5, v_9 >$ | $e_7 =< \texttt{RDFSdomain}, v_6, v_4 >$ |
| $e_8 =< \texttt{RDFSrange}, v_6, v_3 >$ | $e_9 =< \texttt{RDFType}, v_7, v_2 >$ |
| $e_{10} =< \texttt{objectSlot}, v_7, v_{12} >$ | $e_{11} =< \texttt{content}, v_{12}, v_8 >$ |
| $e_{12} =< \texttt{property}, v_{12}, v_6 >$ | $e_{13} =< \texttt{RDFType}, v_8, v_3 >$ |
| $e_{14} =< \texttt{datatypeSlot}, v_8, v_{13} >$ | $e_{15} =< \texttt{content}, v_{13}, v_{11} >$ |
| $e_{16} =< \texttt{property}, v_{13}, v_5 >$ | |

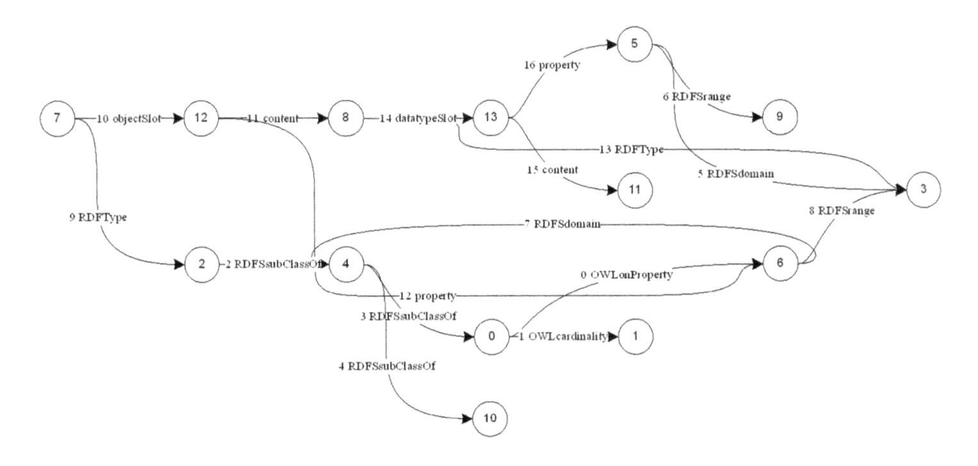


**Fig. 3.** Graphical representation of an ODG

## 3.2 Formal Description of Ontology Slicing

After construction of Ontology Dependency Graphs, ontology slicing becomes a process of graph reachable analysis.

**Definition 2.** *Ontology slicing is a mapping* $SL : G \times C \rightarrow G$*, where* $G =< V, E >$ *is an Ontology Dependency Graph;* $C =< I, S, D >$ *is an ontology slicing criterion; the initial-vertex set* $I = \{v | v \in V \wedge P_I(G)\}$ *specifies the initial vertexes of the slice, the predicate* $P_I(G)$ *is constructed to be satisfied for vertexes in the initial set, the selected-vertex set* $S = \{v | v \in V \wedge P_S(G)\}$ *specifies the vertexes selected for inclusion in the result slice, the predicate* $P_S(G)$ *is defined so that only vertexes of interest are selected; the dimension-set* $D = \{e | e \in E \wedge P_D(G) \wedge T(G) \wedge B(G)\}$ *specifies the edges of interest (a.k.a., dimensions) to be included in the slice and traversed in its computation; the predicate* $P_D(G)$ *defines which edges are included in the slice; the predicate* $T(G)$ *defines a terminating condition of the computation with respect to each of, or all, the edges; the bounding predicate* $B(G)$ *is the computational upper bound on the path length between vertexes with respect to each of, or all the edges, of the slice.*

For an example, as to the ontology shown in Figure 2, we let

$$P_I(G) := \{v_2\} \subseteq V, P_S(G) := True, P_D(G) := et \in ODM,$$
$$T(G) := False, B(G) := False$$

That is to say, the result is composed of all vertexes and edges that $v_2 =< \texttt{OWLClass}, \{\texttt{localName} = "Car"\} >$ of $G$ can reach. The result ontology slice is $G_s =< \{v_0, v_1, v_2, v_3, v_4, v_6, v_{10}\}, \{e_0, e_1, e_2, e_3, e_4, e_7, e_8\} >$, the OWL segment of this ontology slice is as follows.

```
<owl:Class rdf:ID="Color"/>
<owl:Class rdf:ID="Vehicle">
  <rdfs:subClassOf>
    <owl:Restriction>
      <owl:cardinality
        rdf:datatype="http://www.w3.org/2001/XMLSchema#int"
      >1</owl:cardinality>
      <owl:onProperty>
        <owl:ObjectProperty rdf:ID="colorOfVehicle"/>
      </owl:onProperty>
    </owl:Restriction>
  </rdfs:subClassOf>
  <rdfs:subClassOf
    rdf:resource="http://www.w3.org/2002/07/owl#Thing"/>
</owl:Class>
<owl:Class rdf:ID="Car">
  <rdfs:subClassOf rdf:resource="#Vehicle"/>
</owl:Class>
<owl:ObjectProperty rdf:about="#colorOfVehicle">
  <rdfs:range rdf:resource="#Color"/>
  <rdfs:domain rdf:resource="#Vehicle"/>
</owl:ObjectProperty>
```

### 3.3 Semantic Completeness

When using the ontology slicing method based on ODM, we have to consider two levels of semantic completeness: 1) ontology slices must satisfy the definition of ODM; 2) ontology slices must preserve connection information of original ontologies.

Our ontology slicing method is a general method. Users can provide arbitrary slicing criteria, such as criteria with restrictions on path length. So ontology slices may break semantic completeness. We give four semantic completeness processing strategies as follows.

1. Do nothing strategy
   In general cases, an ontology slice can not be looked as a complete ontology and can not be used by reasoners directly. But it can be used for ontology analysis. For an example, if users are only interested in inherit relations among ontology concepts, they can use $P_D(G) := et = \texttt{RDFSsubClassOf}$ as slicing criteria to generate corresponding ontology slices.
2. Minimal semantic completeness strategy
   As we have mentioned before, ODM uses attribute/association cardinalities as number semantics to restrict attributes/associations. When low bound of a cardinality is equal or great than 1, there must be at least one attribute or association in the ontology model. Such restriction can not be guaranteed by ontology slicing. In order to resolve this problem, we can analyze every vertex

during slicing process and add vertexes into the result slice even though they are not considered by slicing criteria. For an example, ODM requires just one `OWLClass` connecting to a `ComplementClass` through a `OWLComplementOf` association, then we must add the corresponding `OWLClass` vertex of the `ComplementClass` vertex into the result slice.

3. Simple semantic completeness strategy
For convenience, we give the concept of ontology slice boundary as follows.

**Definition 3.** *Given an Ontology Dependency Graph* $G =< V, E >$ *and an ontology slice of it* $G_s =< V_s, E_s >$, *we call* $V_b = \{v_b | v_b \in V \setminus V_s \wedge v_s \in V_s \wedge \forall e =< et, v_b, v_s > \in E \setminus E_s\}$ *the ontology slice boundary.*

That is to say, ontology slice boundary is a vertex set, of which each vertex connects to vertexes of the ontology slice directly.

**Definition 4.** *Binary relation* $R = (x, y)$ *is an equivalence relation on the ontology slice boundary* $V_b$ *of an Ontology Dependency Graph* $G$, *which satisfies* $x.vt \equiv y.vt$, *we call* $[x]_R = \{y | y \in V_b \wedge xRy\}$ *is an ontology slice boundary equivalence class of* $x$ *respect to* $R$, *written in* $[x]$, *types of every element of an equivalence class are the same, we call it the type of ontology slice boundary equivalence class, written in* $[x]^{vt}$.

New ontology slice $G_s \cup < V_n, E_n >$ is calculated under the simple semantic completeness strategy, where $V_n = \{v_n | v_n \notin V \wedge v_n.vt = [x]^{vt}\}$ is a set of new vertexes, which are used to replace every element of corresponding ontology slice boundary equivalence class, $E_n = \{e_n =< et, v_n, v_s > | e_n \notin E \wedge e' =< et, v_b, v_s > \in E \setminus E_s \wedge v_b \in V_b \wedge v_s \in V_s\}$ is a set of new edges constructed according to edges between the ontology slice boundary and the ontology slice. For an example, if there is an ontology slice boundary equivalence class $[x]$ that satisfies $[x]^{vt} = \texttt{OWLClass}$, we create a new vertex $v_n =< \texttt{OWLClass}, \{\texttt{localName} = "GeneralOWLClass0"\} >$ to replace all vertexes of $[x]$, and then create edges according to edges between vertexes of this equivalence class and original ontology slice, finally, we add $v_n$ and all these new edges to the ontology slice to preserve the connection semantics. During the processing of simple semantic completeness strategy, we must process minimal semantic completeness strategy to keep semantic completeness of these new added vertexes.

4. Maximal semantic completeness strategy
In simple semantic completeness strategy, we use a common vertex to replace an ontology slice boundary equivalence class, and only parts of connection semantics of the ontology slice is preserved though the result slices can be reasoned. We can also create different vertexes for every element of the ontology slice boundary, then we can get a more larger ontology slice that can preserve more information of connections.

To continue above example, because ontology slice $G_s$ satisfies ODM, process results of strategy 1 and strategy 2 are the same. Ontology slice boundary of

$G_s$ is $V_b = \{v_5, v_7, v_8, v_{12}\}$, equivalence classes of it are $\{v_5\}$, $\{v_7, v_8\}$ and $\{v_{12}\}$. Process result of strategy 3 is

$$\begin{aligned} G_s \cup \{\{v_{n5} &= < \texttt{OWLDatatypeProperty}, \ldots >, \\ v_{n78} &= < \texttt{Individual}, \ldots >, \\ v_{12} &= < \texttt{ObjectSlot}, \ldots >\}, \\ \{e_{n5} &= < \texttt{RDFSdomain}, v_{n5}, v_3 >, \\ e_{n9} &= < \texttt{RDFType}, v_{n78}, v_2 >, \\ e_{12} &= < \texttt{property}, v_{n12}, v_6 >, \\ e_{n13} &= < \texttt{RDFType}, v_{n78}, v_3 >\}\}, \end{aligned}$$

and process result of strategy 4 is

$$\begin{aligned} G_s \cup \{\{v_{n5} &= < \texttt{OWLDatatypeProperty}, \ldots >, \\ v_{n7} &= < \texttt{Individual}, \ldots >, \\ v_{n8} &= < \texttt{Individual}, \ldots >, \\ v_{12} &= < \texttt{ObjectSlot}, \ldots >\}, \\ \{e_{n5} &= < \texttt{RDFSdomain}, v_{n5}, v_3 >, \\ e_{n9} &= < \texttt{RDFType}, v_{n7}, v_2 >, \\ e_{12} &= < \texttt{property}, v_{n12}, v_6 >, \\ e_{n13} &= < \texttt{RDFType}, v_{n8}, v_3 >\}\}. \end{aligned}$$

## 4 Applications of Ontology Slicing

**Ontology Reasoning.** An ontology slice is a segment of original ontology, it has less contents and restrictions than original ontology, so reasoning on these slice can get coarser information and will be faster than original ontology.

**Ontology Debugging.** As an increasingly large number of OWL ontologies become available on the Semantic Web and the descriptions in the ontologies become more complicated, finding the cause of errors becomes an extremely hard task even for experts. Aditya Kalyanpur and co-authors provided a grass and a black box debugging approaches for OWL ontologies [14]. Because our ontology slicing method can determine dependent contents of special ontology elements and reduce the scope for finding errors, it can assist these two ontology debugging approaches and can improve efficiency of ontology debugging.

**Ontology Modularizing.** Modularity in ontologies is key both for large scale ontology development and for distributed ontology reuse on the Web [15]. Researchers have developed many ontology modularizing methods, among which ontology extraction method based on graphs is an important approach [16,17,18]. Our Ontology Dependency Graph has a strict theory basis and can preserve all information of original ontology, so it can replace current ontology graph model using in graph-based ontology modularizing method. And more, our ontology slicing method is a general method with flexible criteria, it can act as the basis of ontology modularizing implementation.

**Ontology Evolution.** Ripple-effect analysis is an important quantificational analysis method of ontology evolution [19]. Ontology Dependency Graph can determine dependent contents of special elements, which are just the changed ontology elements in ripple-effect analysis. At the same time, ontology slicing can reduce contents for analyzing, and improve analysis efficiency.

**Ontology Merging.** An important issue of ontology merging [20] is to calculate ontology similarity. We can take elements that may be similar among different ontologies as initial-sets, after ontology slicing, we can get several ontology slices that are simpler for compare than original ontologies. At the same time, to merge ontology slices may be easier than to merge original ontologies.

**Contextual Ontology.** Contextual ontology has been recognized as an important research domain of Semantic Web. Paolo Bouquet and co-authors provided Context OWL (C-OWL) [21] – an extension of the OWL, for describing contextual ontologies. If we look an ontology slice as a stand-alone ontology, then we can look the rest portion after slicing as its context. Thus ontology slicing can help to construct contextual ontologies.

## 5 Conclusion

By introducing slicing techniques into ontology engineering domain, we provide an ontology slicing method in this paper. In the method, an Ontology Dependency Graph is derived from OMG's Ontology Definition Metamodel (ODM), and then ontology slices can be generated automatically according slicing criteria. We also investigate the semantic completeness problem about ontology slicing.

As ontology slicing is a new research domain, we have many works to do in the future. Application domains of ontology slicing mentioned above is the result of simple analysis of features of ontology slices, and more works have to be done before ontology slicing becomes an important ontology engineering technique. Optimization of automatic ontology slicing algorithms, implementation of ontology slicing tools and ontology slice applications are the important works in the future.

## References

1. Weiser, M.: Program slicing. IEEE Transactions on Software Engineering **10** (1984) 352–357
2. Kagdi, H.H., Maletic, J.I., Sutton, A.: Context-Free Slicing of UML Class Models. In: 21st IEEE International Conference on Software Maintenance (ICSM'05). (2005) 635–638

3. Object Management Group, Inc.: UML 2.0 Infrastructure Specification. (2003) OMG Final Adopted Specification: ptc/03-09-15 http://www.omg.org/cgi-bin/apps/doc?ptc/03-09-15.pdf.
4. Berners-Lee, T., Hendler, J., Lassila, O.: The Semantic Web. Scientific American **284** (2001) 34–43
5. Gruber, T.R.: Toward principles for the design of ontologies used for knowledge sharing. International Journal of Human-Computer Studies **43** (1995) 907–928
6. Frankel, D., Hayes, P., Kendall, E., McGuinness, D.: The Model Driven Semantic Web. In: 1st International Workshop on the Model-Driven Semantic Web (MDSW2004), Monterey, Ca. (2004)
7. IBM, Sandpiper Software: Ontology Definition Metamodel – Third Revised Submission to OMG/ RFP ad/2003-03-40. Object Management Group, Inc. (2005) OMG Document: ad/05-08-01 http://www.omg.org/cgi-bin/apps/doc?ad/05-08-01.pdf.
8. Object Management Group, Inc.: MDA Guide Version 1.0.1. (2003) OMG Document: omg/03-06-01 http://www.omg.org/ cgi-bin/apps/doc?omg/03-06-01.pdf.
9. Object Management Group, Inc.: Meta Object Facility (MOF) 2.0 Core Specification. (2003) OMG Final Adopted Specification: ptc/03-10-04 http://www.omg.org/cgi-bin/apps/doc?ptc/03-10-04.pdf.
10. Klyne, G., Carroll, J.J.: Resource Description Framework (RDF): Concepts and Abstract Syntax. http://www.w3.org/TR/rdf-concepts/ (2004)
11. McGuinness, D.L., van Harmelen, F.: OWL Web Ontology Language Overview. http://www.w3.org/TR/owl-features/ (2004)
12. Baader, F., Calvanese, D., McGuinness, D., Nardi, D., Patel-Schneider, P., eds.: The Description Logic Handbook — Theory, Implementation and Applications. Cambridge University Press (2003)
13. Ballance, R.A., Maccabe, A.B., Ottenstein, K.J.: The program dependence web: a representation supporting control-, data-, and demand-driven interpretation of imperative languages. j-SIGPLAN **25** (1990) 257–271
14. Parsia, B., Sirin, E., Kalyanpur, A.: Debugging owl ontologies. In: The 14th International World Wide Web Conference (WWW2005), Chiba, Japan (2005)
15. Spaccapietra, S.: Report on Modularization of Ontologies. Technical report, Knowledge Web Consortium (2005) http://www.starlab.vub.ac.be/research/projects/knowledgeweb/KWebDel2131-Modularization.pdf.
16. Noy, N.F., Musen, M.A.: Specifying Ontology Views by Traversal. In McIlraith, S., Plexousakis, D., van Harmelen, F., eds.: The Semantic Web - ISWC 2004. Proceedings of the Third International Semantic Web Conference. Volume 3298 of Lecture Notes in Computer Science., Hiroshima, Japan, Springer-Verlag (2004)
17. Wouters, C., Dillon, T., Rahayu, W., Chang, E., Meersman, R.: Ontologies on the MOVE. In Lee, Y., Li, J., Whang, K.Y., Lee, D., eds.: Database Systems for Advanced Applications: 9th International Conference (DASFAA 2004). Volume 2937 of Lecture Notes in Computer Science., Jeju Island, Korea, Springer-Verlag (2004) 812–823
18. Bhatt, M., Wouters, C., Flahive, A., Rahayu, W., Taniar, D.: Semantic Completeness in Sub-ontology Extraction Using Distributed Methods. In Laganà, A., Gavrilova, M., Kumar, V., Mun, Y., Tan, C., Gervasi, O., eds.: Computational Science and Its Applications - ICCSA 2004. Volume 3045 of Lecture Notes in Computer Science., Assisi, Italy, Springer-Verlag (2004) 508–517
19. Jin, L., Liu, L.: A Graph Model based Ripple-Effect Analysis Method for Ontology Evolution. GESTS International Transactions on Computer Science and Engineering **24** (2005)

20. Noy, N.F., Musen, M.A.: PROMPT: Algorithm and Tool for Automated Ontology Merging and Alignment. In: Proceedings of the 17th National Conf. on Artificial Intelligence (AAAI'2000), Austin, Texas., MIT Press/AAAI Press (2000)
21. Bouquet, P., Giunchiglia, F., van Harmelen, F., Serafini, L., Stuckenschmidt, H.: C-owl – contextualizing ontologies. In Fensel, D., Sycara, K., Mylopoulos, J., eds.: The Semantic Web - ISWC 2003. Volume 2870 of Lecture Notes in Computer Science (LNCS)., Sanibel Island (FL, USA), Springer Verlag (2003) 164–179

# Facilitating Business Interoperability from the Semantic Web

Roberto García and Rosa Gil

Universitat de Lleida
Jaume II 69, E-25001 Lleida, Spain
rgarcia@diei.udl.es
rgil@diei.udl.es

**Abstract.** Most approaches to B2B interoperability are based on language syntax standardisation, usually by XML Schemas. However, due to XML expressivity limitations, they are difficult to put into practice because language semantics are not available for computerised means. Therefore, there are many attempts to use formal semantics for B2B based on ontologies. However, this is a difficult jump as there is already a huge XML-based B2B framework and ontology-based approaches lack momentum. Our approach to solve this impasse is based on a direct and transparent transfer of existing XML Schemas and XML data to the semantic world. This process is based on a XML Schema to web ontology mapping combined with an XML data to semantic web data one. Once in the semantic space, it is easier to integrate different business standards using ontology alignment tools and to develop business information systems thanks to semantics-aware tools.

**Keywords:** Semantic Web, ontology, B2B, integration, mapping.

## 1 Introduction

As more and more business is performed in the Internet and stronger inter-organisational links are established, enterprises needs for advanced information processing and data integration grow.

The objective is then to settle shared information spaces. The more mature initiatives come from standards bodies and enterprises organisations. They try to build common business model and languages and they usually root on standardised grammars based on XML technologies.

The previous approach finds difficulties due to the complexity of the business domain. Business languages grow more and more and it is difficult to manage them by computerised means if just their grammar is formalised. Moreover, different languages proliferate and their integration is almost impossible moving at just the syntactic level.

Therefore, recent approaches explore the possibilities of formal semantics through ontologies. They seem promising but, as they are recent, they lack momentum. Moreover, they find difficulties in getting it as they do not see support from the business world and stay as research issues.

W. Abramowicz (Ed.): BIS 2007, LNCS 4439, pp. 220–232, 2007.
© Springer-Verlag Berlin Heidelberg 2007

Our approach to solve this impasse situation is to take profit from the great efforts that have been done in the XML e-business world and transfer them to the semantics-oriented one. More concretely, our objective is to map them to Semantic Web ontologies as they are a way of formalising semantics in a way that integrates smoothly in the Web.

The key point is that, as B2B standards are developed by people with domain concept models in their minds, they partially formalise their semantics while defining the XML Schemas for the different business languages. However, as XML Schemas are not semantics aware, this implicit semantics remain hidden from the computational point of view. They can be made explicit by mapping XML Schema constructs to the ontology language ones that correspond to their implicit semantics.

The previous XML Schema to Web Ontology mapping is combined with a transformation from XML data to RDF semantic data. Altogether, both mappings allow a transparent transfer of existing XML-based business data to the Semantic Web. Although there are other attempts to connect the XML and Semantic Web worlds, they just concentrate on the XML Schemas or the XML data so they do not achieve this level completeness and transparency in the transfer.

Once in the Semantic Web, it is possible to use semantics tools that make it easier to integrate data coming from disparate sources and to develop business management applications. For instance, it is possible to perform intelligent retrieval by semantic queries.

The rest of this paper is organised as follows. First, Section 2 presents the state of the art in the B2B domain. Then, in Section 3, the proposed methodology for a transparent transfer from XML-based B2B data to the Semantic Web is detailed. The results for this methodology when applied to some of the main B2B standards are shown in Section 4, together with the benefits obtained for data integration and information processing. Finally, Section 5 concludes the paper and presents the future work.

## 2 State of the Art

There are many B2B initiatives from international standardisation bodies and organisations and most of them are based on XML Schemas. For instance Biztalk[1], RosettaNet[2], ebXML[3] or UBL[4].

As it has been introduced, the XML approach does not scale well when the standardised domains are as complex as the business one. In this case, the lack of formal semantics makes it very difficult to automate sophisticated information processing mechanisms like integrating data from different standards and making them interoperate. Moreover, there is a lot of domain knowledge that remains hidden to implementations due to XML Schema expressive limitations. It is written down in the standards specifications and must be read and interpreted each time an implementation of the standard is persuaded.

---

[1] http://www.microsoft.com/biztalk

[2] http://www.rosettanet.org

[3] Electronic Business using eXtensible Markup Language, http://www.ebxml.org

[4] Universal Business Language 1.0, http://docs.oasis-open.org/ubl/cd-UBL-1.0

Consequently, many research efforts have moved to the formal semantics domain and there are many business modelling initiatives based on ontologies, and more concretely Semantic Web ontologies [1]. Most of them take profit from the enhanced expressive power so they try to build a complete enterprise model on the basis of a formal conceptualisation [2,3,4].

There are also other initiatives that are based on ontologies but that start from existing conceptualisations, which is an approach nearer to the one proposed in this paper. For instance, there is the eClassOWL ontology based on the products and services categorization standard eCl@ss [5], and EDI Ontology[5] that formalises EDI X12 Transaction Sets or the RosettaNet Ontology[6], an OWL implementation of RosettaNet Partner Interface Processes (PIPs).

However, these attempts to move B2B standards to the Semantic Web are based on an ad-hoc mapping from the source documents. This effort has to be done each time a new standard is mapped. Moreover, they just deal with the schema part so any existing data based on these standards is not semantically enriched. Therefore, what we get is a quite empty semantic framework. It has good conceptualisations but it lacks the semantic data it can operate on.

# 3 XML Semantics Reuse Methodology

On the contrary to the previous initiatives to B2B standards semantics formalisation, our approach provides an automatic and transparent mapping from XML Schema-based standard conceptualisations to OWL ontologies complemented with an XML B2B data to semantic RDF data one. The objective is to get a functional semantic framework full of semantic data.

Section 3.1 presents the related work and its limitations for the objectives stated in the introduction. These limitations have motivated the development of the XML Semantics Reuse Methodology. There is an overview of the proposed methodology in Section 3.2 and its two main components, the XML Schema to OWL and XML to RDF mappings, are detailed in sections 3.3 and 3.4. The architecture of the system that implements the methodology is presented in Section 3.5.

## 3.1 Related Work

This has been already detected as a key issue in order to add momentum to the Semantic Web so there are many attempts to move conceptualisations and data from the XML domain to the Semantic Web.

Some of them just model the XML tree using the RDF primitives [6] so there are not formalised semantics. The same happens when the approach is to encode XML semantics integrating RDF into XML documents [7,8] as the result is not semantic data, just ad-hoc XML.

In order to get semantic data, other initiatives concentrate on modelling the knowledge implicit in XML languages definitions, i.e. DTDs or the XML Schemas, using web ontology languages [9,10,11]. However, they do not generate transparent semantic

[5] http://www.wsmo.org/TR/d27/v0.1

[6] http://lsdis.cs.uga.edu/projects/meteor-s/wsdl-s/ontologies/rosetta.owl

data as the formalised semantics root on RDF ad-hoc semantics, i.e. they rely on custom RDF constructs that should be interpreted in a specific way in order to capture the formalised semantics. Therefore, existing applications that are not aware of these interpretations have no access to the semantics.

The most transparent approach we have found is Gloze XML to RDF [12]. It provides a direct mapping from XML data to RDF semantic data, which is based on the XML Schema implicit semantics. However, it does not fix the XML Schema semantics into an ontology. Therefore, it is not possible to work with the semantic data using semantic tools that rely on ontologies formal semantics or to perform semantic data integration at the conceptual level by mapping ontologies.

As it has been shown, none of them facilitates an extensive transfer of XML data to the Semantic Web in a general and transparent way. Their main problem is that the XML Schema implicit semantics are not made explicit when XML data instantiating this schemas is mapped.

Consequently, they do not take profit from the XML semantics and produce RDF data almost as semantics-blind as the original XML. Or, on the other hand, they capture these semantics but they use additional ad-hoc semantic constructs that produce less transparent data.

### 3.2 Overview

In order to overcome the limitations detected in the previous section, the "XML Semantic Reuse Methodology" [13] has been developed and implemented in the ReDeFer[7] project. It combines an XML Schema to OWL web ontology mapping, called XSD2OWL, with a mapping from XML to RDF, XML2RDF. The ontologies generated by XSD2OWL are used during the XML to RDF step in order to generate RDF data enriched by the XML Schema semantics made explicit.

### 3.3 XSD2OWL Mapping

The XML Schema to OWL mapping is responsible for capturing the schema implicit semantics. This semantics are determined by the combination of XML Schema constructs. The mapping is based on translating this constructs to the OWL ones that best capture their semantics. These translations are detailed in Table 1.

Some mappings depend on the context where the XML Schema construct appears. Therefore, XPath pointers have been used in order to detail this context, e.g. "complexType//element" refers to an element definition that appears inside a complex type definition.

The XSD2OWL mapping is quite transparent and captures a great part XML Schema semantics. The same names used for XML constructs are used for OWL ones, although in the new namespace defined for the ontology. XSD and OWL constructs names are identical; this usually produces uppercase-named OWL properties because the corresponding element name is uppercase, although this is not the usual convention in OWL.

[7] ReDeFer, http://rhizomik.net/redefer

**Table 1.** XSD2OWL mappings and shared semantics

| XML Schema XPath | OWL | Shared informal semantics |
|---|---|---|
| element\|attribute | rdf:Property<br>owl:DatatypeProperty<br>owl:ObjectProperty | Named relation between nodes or nodes and values |
| element@substitutionGroup | rdfs:subPropertyOf | Relation can appear in place of a more general one |
| element@type | rdfs:range | The relation range kind |
| complexType\|group\|attributeGroup | owl:Class | Relations and contextual restrictions package |
| complexType//element | owl:Restriction | Contextualised restriction of a relation |
| extension@base\|restriction@base | rdfs:subClassOf | Package concretises the base package |
| @maxOccurs<br>@minOccurs | owl:maxCardinality<br>owl:minCardinality | Restrict the number of occurrences of a relation |
| sequence<br>choice | owl:intersectionOf<br>owl:unionOf | Combination of relations in a context |

One of the key points is the *xsd:element* and *xsd:attribute* mapping. As it can be observed in Table 1, there are three alternatives, which are selected depending on the kind of values they can take as specified in the schema. All attributes are mapped to *owl:DatatypeProperty* as all of them have simple type values. This is also the case for elements that have a simple type value. On the other hand, if the value is a complex type, they must be mapped to an *owl:ObjectProperty*.

The third option is necessary as *xsd:elements* can have both simple and complex type values. In these cases, the element must be mapped to an *rdf:Property*, which can deal with both data type (for *xsd:simpleType*) and class instance (for *xsd:complexType*) values.

From the combination of all these XML Schema constructs mappings, XSD2OWL produces OWL ontologies that make the semantics of the corresponding XML Schemas explicit. The only caveats are the implicit order conveyed by *xsd:sequence* and the exclusivity of *xsd:choice*.

For the first problem, *owl:intersectionOf* does not retain its operands order, there is no clear solution that retains the great level of transparency that has been achieved. The use of RDF Lists might impose order but introduces ad-hoc constructs not present in the original data. Moreover, as it has been demonstrated in practise, the element ordering does not contribute much from a semantic point of view. For the second problem, *owl:unionOf* is an inclusive union, the solution is to use the disjointness OWL construct, *owl:disjointWith*, between all union operands in order to make it exclusive.

The XSD2OWL mapping has been checked using OWL validators, e.g. Pellet[8], which have been used in order to test the semantic consistency of the resulting ontologies.

[8] Pellet OWL Consistency Checker, http://www.mindswap.org/2003/pellet/demo.shtml

### 3.4 XML2RDF Mapping

Once all the XML Schemas are available as OWL ontologies, it is time to map the XML data that instantiates them. The intention is to produce RDF data as transparently as possible. Therefore, a structure-mapping approach has been selected [6]. It is also possible to take a model-mapping approach [14], which is based on representing the XML information set using semantic tools. However, this approach reduces the transparency of the mapped data as it depends on the particular modelling decisions.

Transparency is achieved in structure-mapping models because they only try to represent the XML data structure, i.e. a tree, using RDF. The RDF model is based on the graph so it is easy to model a tree using it. Moreover, we do not need to worry about the semantics loose produced by structure-mapping. We have formalised the underlying semantics into the corresponding ontologies and we will attach them to RDF data using the instantiation relation *rdf:type*.

The structure-mapping is based on translating XML data instances to RDF ones that instantiate the corresponding constructs in OWL. The more basic translation is between relation instances, from *xsd:elements* and *xsd:attributes* to *rdf:Properties*. Concretely, *owl:ObjectProperties* for node to node relations and *owl:DatatypeProperties* for node to values relations.

However, in some cases, it would be necessary to use *rdf:Properties* for *xsd:elements* that have both data type and object type values. Values are kept during the translation as simple types and RDF blank nodes are introduced in the RDF model in order to serve as source and destination for properties. They will remain blank until they are enriched with type information from the corresponding ontology. The resulting RDF graph model contains all that can be obtained from the XML tree, as it is shown in Fig. 1.

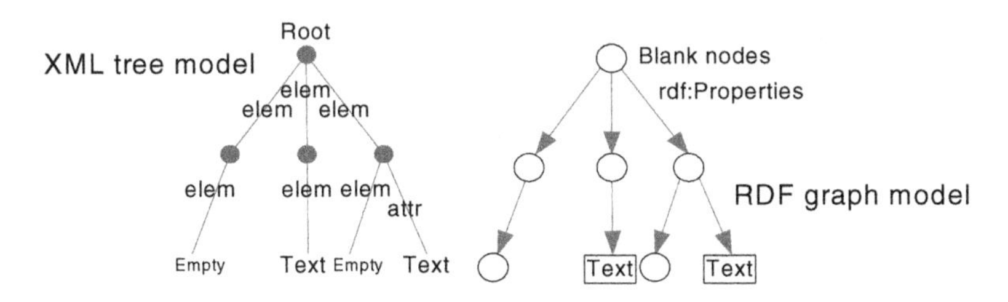


**Fig. 1.** RDF model for the XML tree

It is already semantically enriched thanks to the *rdf:type* relation that connects each RDF properties to the *owl:ObjectProperty* or *owl:DatatypeProperty* it instantiates. It can be enriched further if the blank nodes are related to the *owl:Class* that defines the package of properties and associated restrictions they contain, i.e. the corresponding *xsd:complexType*. This semantic decoration of the graph is formalised using *rdf:type* relations from blank nodes to the corresponding OWL classes.

At this point we have obtained a semantics-enabled representation of the input data. The instantiation relations can now be used to apply OWL semantics to data. Therefore, the semantics derived from further enrichments of the ontologies, e.g. integration links between different ontologies or semantic rules, are automatically propagated to instance data thanks to inference.

The XML2RDF mapping has been validated using some test XML instances. This test instances have been mapped to RDF and then back to XML. Then, it has been possible to compare the original and derived XML instances in order to detect mapping errors as, due to the mapping transparency, the underlying XML tree structure is preserved.

### 3.5 System Architecture

Based on the previous XML world to Semantic Web domain mappings, a system architecture that facilitates B2B data integration and retrieval has been developed. The architecture is sketched in Fig. 2.

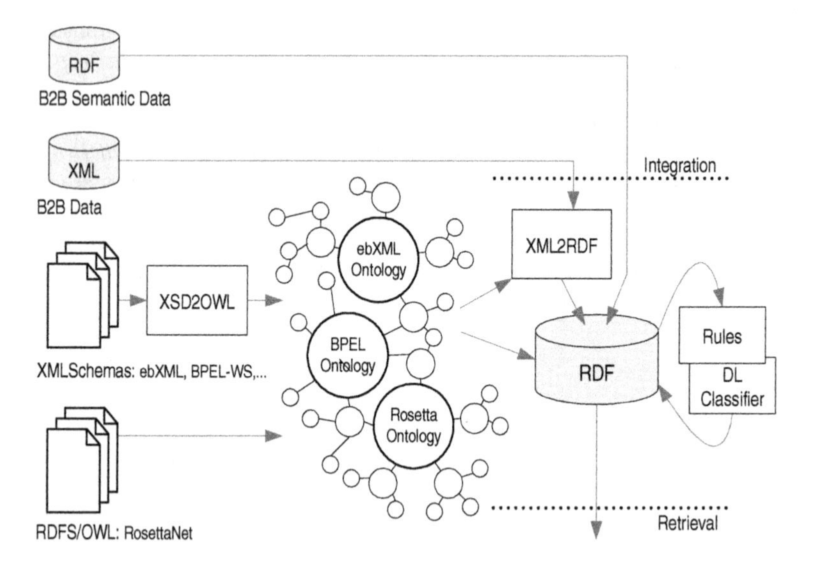


**Fig. 2.** B2B data integration and retrieval architecture

First of all, the architecture is fed with existing OWL ontologies, e.g. the RosettaNet Ontology, or those generated by XSD2OWL. For instance the ebXML or BPEL ontologies detailed in Section 4.1. These ontologies constitute the system semantic framework shown in the centre of Fig. 2.

Then, it is time to load B2B data. Semantic data can be directly fed into the system. On the other hand, B2B XML data must pass through the XML to RDF mapper. The mapping relies on the existing ontologies framework in order to produce semantically enriched RDF data.

All the data and ontologies are persisted in a RDF data store, where semantic tools like rule engines or semantic queries operate. Once all data has been put together, the semantic integration can take place. Finally, it is possible to perform intelligent data retrieval using semantic queries.

# 4 Preliminary Results

The XML Semantic Reuse Methodology has been tested with some XML-based B2B standards, ebXML and BPEL-WS. First, the XML Schemas for the standards selected parts have been mapped to OWL ontologies using the XSD2OWL mapping, as it is detailed in Section 4.1. Once the ontologies for these standards have been generated, automatic tools for ontology alignment have been used in order to derive some preliminary integration rules. This exercise shows how existing semantic web ontology alignment tools can be then applied in order to generate interoperability rules for B2B data based on these standards. Some of the integration rules are detailed in Section 4.2 while Section 4.3 shows the benefits of semantic queries compared to XML ones, fact exemplified with a piece of RDF mapped from XML B2B data.

## 4.1 Mapped Ontologies

Currently, the XML Semantics Reuse Methodology has been applied to some of the main B2B standards from the OASIS standardisation body. Their schemas are available online and they provide a good test bed for the XML Schema to OWL mapping and the subsequent semantic tools applications.

From the OASIS Electronic Business using eXtensible Markup Language (ebXML) initiative, the Business Process (ebBP) and Collaboration Protocol Profile and Agreement (ebCPPA) schemas have been considered. On the other hand, from the OASIS Web Services initiative, the Web Services Business Process Execution Language (BPEL-WS) schema has been selected.

This standards focus on business process modelling and on how collaborations among business parties are established in order to connect them. Although all of them come from the same standardisation body, there are not formal links among the corresponding schemas. Therefore, there are not integration means for them that facilitate interoperation.

All the resulting ontologies, together with some additional links and documentation, are available from the BizOntos web page[9]. Fig. 3 shows a portion of the ebBP class hierarchy. This hierarchy shows how the ontology makes the semantics in the XML extension relations among *xsd:ComplexTypes* explicit as a class hierarchy governed by inheritance relations.

## 4.2 Semantic Integration

Once the semantics of the B2B standards are formalised as OWL ontologies, they are easily integrated using OWL semantic relations for equivalence and inclusion: *subClassOf*, *subPropertyOf*, *equivalentClass*, *equivalentProperty*, *sameIndividualAs*, etc.

[9] BizOntos, http://rhizomik.net/ontologies/bizontos

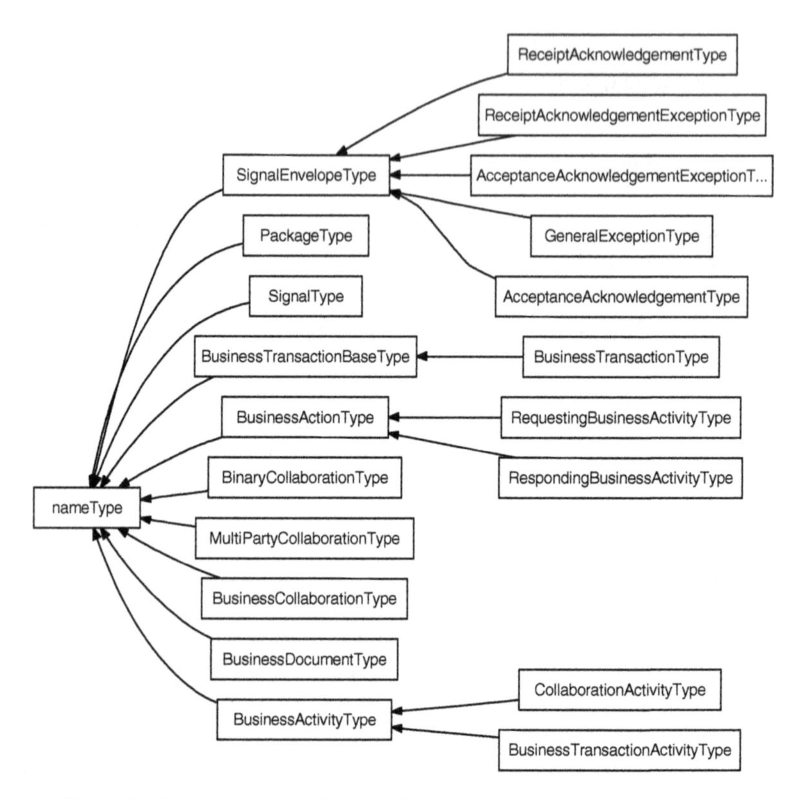


**Fig. 3.** Portion of the class hierarchy in the ebXML Business Process ontology

These relationships capture the semantics of the data integration. Moreover, this integration can be partially automated using ontology matching methods [15]. Some of these mappings generated by the OWL Ontology Aligner[10] are presented in the following subsections 4.2.1 and 4.2.2, which are devoted to the ebBP to ebCPPA mapping and the ebBP to BPEL-WS mappings respectively.

With all the relevant integration mappings, semantically-enriched B2B data coming from different standards is automatically integrated. The OWL relations formal semantic allow inference engines to derive the necessary mappings among instance data.

### 4.2.1 ebXML BP to ebXML CPPA Mappings

Some of the mappings derived by the OWL Ontology Aligner for the ebXML Business Process (ebBP) ontology and the ebXML Collaboration Protocol Profile and Agreement (CPPA) ontology are shown in Table 2. These are equivalence and inclusion relations for related ontology classes and properties.

### 4.2.2 ebXML BP to BPEL-WS

Some mappings for the ebXML Business Process (ebBP) ontology to the Business Process Execution Language for Web Services (BPEL-WS) ontology are shown in Table 3. They have been also generated by the OWL Ontology Aligner.

[10] OWL Ontology Aligner, http://align.deri.org

**Table 2.** ebXML BP to ebXML CPA mappings

| | | |
|---|---|---|
| ebbp:location | ≡ | cppa:location |
| ebbp:Role | ≡ | cppa:Role |
| ebbp:type | ≡ | cppa:type |
| ebbp:mimeType | ≡ | cppa:mimetype |
| ebbp:RoleType | ≡ | cppa:RoleType |
| ebbp:ProcessSpecificationType | ≡ | cppa:ProcessSpecificationType |
| ebbp:Start | ≡ | cppa:Start |
| ebbp:ProcessSpecification | ≡ | cppa:ProcessSpecification |
| ebbp:value | ≡ | cppa:value |
| ebbp:name | ≡ | cppa:name |
| ebbp:uri | ≡ | cppa:uri |
| ebbp:BusinessAction | ⊆ | cppa:action |
| ebbp:isAuthenticated | ≡ | cppa:authenticated |
| Specification | ⊆ | cppa:ProcessSpecification |
| ... | | ... |

**Table 3.** ebXML BP to BPEL-WS mappings

| | | |
|---|---|---|
| ebbp:expression | ≡ | bpel:expression |
| ebbp:expressionLanguage | ≡ | bpel:expressionLanguage |
| ebbp:name | ≡ | bpel:name |
| ebbp:pattern | ≡ | bpel:pattern |
| ebbp:Variable | ≡ | bpel:variable |
| ebbp:PreCondition | ⊆ | bpel:condition |
| ebbp:BusinessPartnerRole | ≡ | bpel:partnerRole |
| ... | | ... |

## 4.3 Semantic Applications

Once the B2B ontologies and the integration rules have been generated, it is time for the semantic enrichment of B2B data corresponding to the mapped standards. For instance, Table 4 shows a portion of the RDF data generated by the XML2RDF mapping from a CPPA XML example. As it can be observed, the input XML data tree is modelled using RDF properties and it is enriched with typed blank nodes. These types correspond to the XML Schema complex types mapped to OWL classes during the XSD2OWL step.

Thanks to this semantic enrichment, it is possible to take profit from the XML Schema implicit semantics using Semantic Web tools. Therefore, it is not even required to add more semantics to the resulting ontology; just with the semantics implicit in the original XML Schema, it is possible to provide new functionalities. For instance, the classes' hierarchy, which is formalised in the ontology and linked to the RDF data through the typed blank nodes, can be used by semantic query engines.

In order to illustrate this, a hypothetical ebBP application scenario can be considered. One of its functionalities is to process all the nodes referring to signal envelope types, which correspond to the *SignalEnvelopeType* complex type. If an XML-based

**Table 4.** Part of the RDF version of cpp-example.xml

```
<rdf:RDF xmlns:tp="...ontologies/2006/06/cpp-cpa-v1_0.owl#"
    xmlns:rdf="http://www.w3.org/1999/02/22-rdf-syntax-ns#">
<tp:CollaborationRoleType rdf:about=".../cpp-example.rdf#N00">
    <tp:ProcessSpecification>
      <tp:ProcessSpecificationType>
        <tp:name>buySell</tp:name>
        <tp:version>1.0</tp:version>
      </tp:ProcessSpecificationType>
    </tp:ProcessSpecification>
    <tp:Role>
      <tp:RoleType>
        <tp:name>buyer</tp:name>
      </tp:RoleType>
    </tp:Role>
    <tp:CertificateRef>
      <tp:CertificateRefType>
        <tp:certId>N03</tp:certId>
      </tp:CertificateRefType>
    </tp:CertificateRef>
    ...
  </tp:CollaborationRoleType>
  ...
</rdf:RDF>
```

tool like XQuery is used to do that, we must be aware of the implicit hierarchy of segment types and implement an XQuery that considers all possibilities, i.e. *ReceiptAcknowledgementType*, *GeneralExceptionType*, *AcceptanceAcknowledgementType*, etc.

On the other hand, once the hierarchy of segments types is available in OWL form, semantic queries benefit from the now explicit semantics. Therefore, a semantic query for *SignalEnvelopeType* will retrieve all subclasses without requiring additional developing efforts. Table 5 shows a query based on the SPARQL semantic query language [16] which demonstrates the simplicity of a query capable of dealing with all the kinds of *SignalEnvelopeType*.

This is so because, although XML Schemas capture some semantics of the domain they model, XML tools are based on syntax. The captured semantics remain implicit from XML processing tools point of view. Therefore, when an XQuery searches for a *SignalEnvelopeType*, the XQuery processor has no way to know that there are many other kinds of segment types that can appear in its place, i.e. they are more concrete kinds of segments.

Additionally, semantic queries use integration rules, like the ones presented in Section 4.2, in order to perform B2B data integration. For instance, when a semantic

**Table 5.** SPARQL semantic query for *SignalEnvelopeType*

```
PREFIX ebbp: <http://rhizomik.net/ontologies/2006/06/ebbp-2.0.3.owl#>
SELECT ?r
WHERE (?r <rdf:type> ebbp:SignalEnvelopeType)
```

query for "ebbp:location" elements is performed, the "ebbp:location ≡ bpel:location" is interpreted and due to the semantics it carries the query also retrieves "bpel:location" elements automatically. Therefore, ebPB and BPEL-WS data is integrated and processed seamlessly by existing ebPB or BPEL-WS semantic applications.

# 5 Conclusions and Future Work

As it has been shown, the increase of B2B interactions and their complexity make existing standards, most of them based on XML technologies, very difficult to manage and to make them interoperate. Consequently, there are many initiatives trying to profit from ontologies greater level of expressivity and the formal semantics they provide in order to make B2B management and interoperability easier.

However, these initiative lack momentum as most of the available B2B systems and data are based on XML tools. In order to mitigate this, this paper proposes to apply the XML Semantics Reuse Methodology, which contributes a complete and transparent transfer of B2B data from existing XML-based standards to the Semantic Web. It is based on a generic XML Schema to OWL mapping and complemented with an XML data to RDF mapping.

As it has been shown, XSD2OWL is used to map some of the main B2B standards XML Schemas and generate the corresponding OWL ontologies. These ontologies provide the anchor points for semantic integration and ontology matching tools are used to align them and derive integration rules.

These rules are used to integrate XML B2B data once it is mapped to RDF and semantically-enriched by the XML2RDF mapping. Then, semantic tools can be used to facilitate B2B applications development, e.g. semantic query engines. Semantic queries take profit from the complex types and elements hierarchies implicit in the XML Schema definitions.

All this process is transparent as it is based on a structure-mapping approach that preserves the original XML tree structure. Therefore, it is easy to go back to XML data if just the RDF graph properties are considered, which correspond to the XML data elements.

The future plans concentrate now on performing detailed alignments among the generated OWL ontologies in order to get a complete data integration framework. Another objective is to integrate the ontologies mapped from XML Schemas with existing ontologies like the Enterprise Ontology. These are rich ontologies with much more formal semantics as they are developed from the beginning using more expressive tools. If they are aligned with XSD2OWL ontologies, it will be possible to reuse the detailed semantics captured by this ontology in order to implement advanced B2B applications operating on existing XML-based standards.

# References


1. Singh, R; Iyer, L.S.; Salam, A.F. (ed.): The Semantic e-Business Vision. Communications of the ACM, 48(12):38-77
2. Albani, A.; Dietz, J. L. G.; Zaha, J. M.: Identifying Business Components on the basis of an Enterprise Ontology. In: Konstantas, D.; Bourrières, J.-P.; Léonard, M.; Boudjlida, N. (ed.): Interoperability of Enterprise Software and Applications. Springer, pp. 335-347, 2006

3. Uschold, M.: King, M.; Moralee, S.; Zorgios, Y.: The Enterprise Ontology. The Knowledge Engineering Review, 13(1):31-89, Cambridge University Press, 1998
4. Dietz, J. L. G.: Enterprise Ontology. In: Chen, C.-S.; Filipe, J.; Seruca, I.; Cordeiro, J. (ed.): ICEIS 2005, Proceedings of the Seventh International Conference on Enterprise Information Systems, 2005
5. Hepp, M.: Products and Services Ontologies: A Methodology for Deriving OWL Ontologies from Industrial Categorization Standards. Journal on Semantic Web and Information Systems, 2(1):72-99, 2006
6. Klein, M.C.A.: Interpreting XML Documents via an RDF Schema Ontology. ECAI Workshop on Semantic Authoring, Annotation & Knowledge Markup. Lyon, France, 2002
7. Lakshmanan, L.; Sadri, F.: Interoperability on XML Data. Lecture Notes in Computer Science Vol. 2870, pp. 146-163. Springer, 2003
8. Patel-Schneider, P.F.; Simeon, J.; The Yin/Yang web: XML syntax and RDF semantics. Proceedings of the 11th World Wide Web Conference, pp. 443-453, Hawaii, USA, 2002
9. Amann, B.; Beeri, C.; Fundulaki, I.; Scholl, M.: Querying XML Sources Using an Ontology-Based Mediator. In R. Meersman; Z. Tari (ed.): On the Move to Meaningful Internet Systems 2002, ODBASE'02. LNCS Vol. 2519, pp. 429-448, Springer, 2002
10. Cruz, I.; Xiao, H.; Hsu, F.: An Ontology-based Framework for XML Semantic Integration. 8$^{th}$ Database Engineering and Applications Symposium, Portugal, 2004
11. Halevy, A.Y.; Ives, Z.G.; Mork, P.; Tatarinov, I.: Piazza: Data Management Infrastructure for Semantic Web Applications. Proceedings of the 12th World Wide Web Conference, pp. 556-567, Budapest, Hungary, 2003
12. Battle, S.: Gloze: XML to RDF and back again. In Proceedings of the First Jena User Conference, Bristol, UK, 2006
13. García, R.: Chapter 7: XML Semantics Reuse. In: A Semantic Web Approach to Digital Rights Management. PhD Thesis, Technologies Department, Universitat Pompeu Fabra, Barcelona, Spain, 2006. Available at http://rhizomik.net/~roberto/thesis/html/Methodology.html#XMLSemanticsReuse
14. Tous, R.; García, R.; Rodríguez, E.; Delgado, J.: Architecture of a Semantic XPath Processor. In: E-Commerce and Web Technologies, 6th International Conference. Springer, Lecture Notes in Computer Science Vol. 3590, pp. 1-10, 2005
15. Noy, N.: Semantic Integration: A Survey of Ontology-based Approaches. Sigmod Record, Special Issue on Semantic Integration, 2004
16. Prud'hommeaux, E.; Seaborne, A.: SPARQL Query Language for RDF. W3C Candidate Recommendation 6 April 2006. Retrieved July 4, 2006, from http://www.w3.org/TR/rdf-sparql-query

# Efficient Automatic Selection of Semantically-Annotated Building Blocks for ERPs Customizing

Eufemia Tinelli[1,2], Tommaso Di Noia[1], Eugenio Di Sciascio[1], and Francesco di Cugno[1]

[1] SisInfLab, Politecnico di Bari, Via Re David, 200, Bari, Italy
{t.dinoia,disciascio,f.dicugno}@poliba.it
[2] Università di Bari, via Orabona 4, Bari, Italy
tinelli@di.uniba.it

**Abstract.** We present an approach for efficient semantic-based building-blocks selection in the context of ERPs fast customizing, introducing Enduring Oak, a framework that implements an optimized greedy concept covering algorithm, able to deal with thousands of building block descriptions with reasonable computational times. The proposed approach uses a Description Logics reasoning engine in conjunction with a RDBMS to reduce the computational burden. We motivate the approach, present the framework and algorithms and illustrate experiments confirming the validity of our setting.

## 1 Introduction

In [7,8] algorithms and a prototype system were proposed for the automated selection of semantically-annotated building blocks in ERPs business processes customization. The framework adopted a polynomial greedy concept-covering algorithm, exploiting non-standard inference services named *concept abduction* and *concept contraction* [9]. The system was specifically designed to ease SAP R/3 best-practices [11] re-usability during the so-called *customizing*. While effective and theoretically sound, the approach —though based on a greedy approach— was unable to efficiently scale-up to the thousands of building blocks descriptions that can be present in a real scenario deployment. In this paper we show that by smoothly combining inference services provided by a reasoning engine with efficient storage and retrieval of relational DBMS we are able to drastically reduce computational times obtaining the desired scalability, with only a limited reduction in the effectiveness of the semantic-based selection process. The remaining of the paper is as follows: next section summarizes the domain we tackle. Then we move on to basics of the formalisms and algorithms we use, in order to make the paper self-contained. Section 4 presents and motivates the approach we propose; In section 5 we briefly outline our semantic-based system and then we report on experiments carried out. Conclusions close the paper.

## 2 Application Scenario

Enterprise Resource Planning (ERPs) systems have become more and more common in a number of different enterprise and companies; nevertheless they are able to actually

W. Abramowicz (Ed.): BIS 2007, LNCS 4439, pp. 233–244, 2007.
© Springer-Verlag Berlin Heidelberg 2007

improve the quality and efficiency of companies business processes as long as they are properly tailored and tuned on the actual organization. To this aim they provide a huge number of parametric customizations in order to adapt the system to the various organizational contexts. This process is often much more expensive than the actual purchase of the ERP software [18]. To simplify this stage ERPs producers tend to offer support methodologies and tools to rapidly re-use solutions from previous well established implementations. In particular, for SAP R/3, such process is known as Customizing [2,13]. Customizing methodologies include: Global Accelerated SAP (GASAP), Accelerated SAP (ASAP) and Best Practices [13,11]. Here we focus on the Best Practices methodology, which has —at its core— the Building Block (BB) concept [11]. The basic idea is the modularization of a vertical solution, *i.e.*, in SAP terms a complete SAP R/3 solution developed for a well defined organization scenario, identifying and extracting all its client independent information. BB contents in SAP Best Practices are defined considering from the start the possibility of their reuse from an implementation point of view. Basically, the BB content is defined by the identification of which Business Process parts can be reused within a predefined solution. The BB Library [16] provided by SAP in fact aims at sharing SAP knowledge within the community of developers. It is also possible to develop specific BBs able to provide particular solutions within a company context. Nevertheless, because of the rapid growth of the BBs number, choosing the correct BB in order to satisfy part of a specific Business Process, is increasingly expensive in terms of time, as the selection is driven only by the developers experience. The need to automate such a process, and similar ones, providing support in the customization stage, is therefore increasingly acknowledged. Furthermore we note that, although the system we developed so far —which exploits knowledge representation techniques to automatically select annotated building blocks from available business processes to compose new ones satisfying a given need— has been designed for SAP R/3 BBs, it is obvious that algorithms and solutions devised are applicable to a number of different business processes composition scenarios.

## 3 Framework and Basic Algorithms

Our semantic-based approach relies on Description Logics (DLs) formalization [3]. Description Logics (DLs) are a family of logic formalisms for Knowledge Representation. We briefly present introductory notions of DLs. Basic syntax elements are: *concept* names, *role* names, *individuals*. Intuitively, concepts stand for sets of objects, and roles link objects in different concepts. Individuals are used for special named elements belonging to concepts. Depending on the expressivity of the language and on the allowed constructors *e.g.*, number restrictions, transitive roles, full negation, etc., different names have been used to identify different Description Logics. Many of them are built upon the simple $\mathcal{AL}$([3]); they all define two special concept names, namely $\top$ and $\bot$ representing respectively, all the objects within the domain and the empty set. Concept expressions can be used in *inclusion assertions*, and *definitions*, which impose restrictions on possible interpretations according to the knowledge elicited for a given domain. Definitions are useful to give a meaningful name to particular combinations. Sets of such inclusions are called TBox (Terminological Box), and amount to what is usually called an ontology. Individuals can be asserted to belong to a concept using membership assertions in an ABox. The semantics of inclusions and definitions is based on set containment: an interpretation $\mathcal{I}$ satisfies an inclusion $C \sqsubseteq D$ if $C^{\mathcal{I}} \subseteq D^{\mathcal{I}}$, and it satisfies a definition $C \equiv D$ when $C^{\mathcal{I}} = D^{\mathcal{I}}$. A *model* of a TBox $\mathcal{T}$ is an interpretation satisfying all inclusions and definitions of $\mathcal{T}$. DL-based systems are equipped with inference services: logical problems whose solution can make explicit knowledge that was implicit in the assertions. Basic inferences are:

*Concept Satisfiability* — $\mathcal{T} \models C \sqsubseteq \bot$ : given an ontology $\mathcal{T}$ and a concept $C$, does there exist at least one model of $\mathcal{T}$ assigning a non-empty extension to $C$?

*Subsumption* — $\mathcal{T} \models C \sqsubseteq D$ : given an ontology $\mathcal{T}$ and two concepts $C$ and $D$, is $C$ more general than $D$ in any model of $\mathcal{T}$?

Given an ontology $\mathcal{T}$ modeling the investigated knowledge domain, a request description $D$ and a resource description $BB$, in our setting a building block (BB), using subsumption it is possible to evaluate either (1)if $BB$completely fulfills the request – $\mathcal{T} \models BB \sqsubseteq D$, full match – or (2)if they are at least compatible with each other – $\mathcal{T} \not\models BB \sqcap D \sqsubseteq \bot$, potential match – or (3)not – $\mathcal{T} \models BB \sqcap D \sqsubseteq \bot$, partial match [10]. It is easy to see that in case of full match all the information specified on $BB$ is expressed, explicitly or by means of the ontology $\mathcal{T}$, also in $D$. Responses to calls for subsumption and satisfiability checks are obviously Boolean. When we need to deal with approximation, other inferences may help. Let us consider two concepts $C$ and $D$;

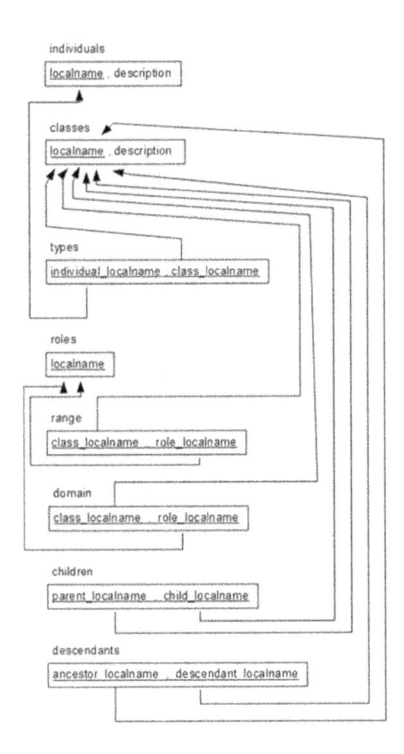


**Fig. 1.** Relational structure of the database

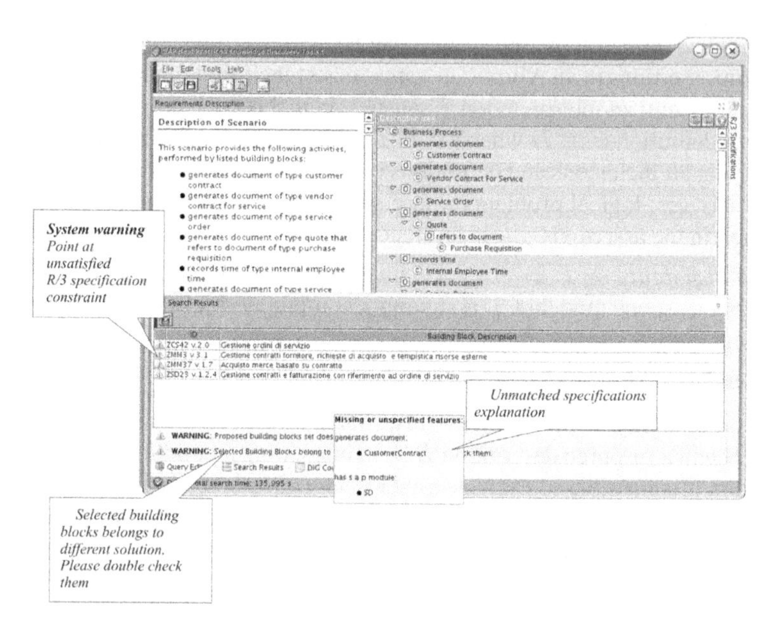


**Fig. 2.** An annotated snapshot of the User Interface

if their conjunction $C \sqcap D$ is unsatisfiable in the TBox $\mathcal{T}$ representing the ontology, *i.e.*, they are not compatible with each other, we may want, as in a belief revision process, to retract requirements in $D$, $G$ (for *Give up*), to obtain a concept $K$ (for *Keep*) such that $K \sqcap C$ is satisfiable in $\mathcal{T}$.

**Definition 1.** *Let $\mathcal{L}$ be a DL, $C$, $D$, be two concepts in $\mathcal{L}$, and $\mathcal{T}$ be a set of axioms in $\mathcal{L}$, where both $C$ and $D$ are satisfiable in $\mathcal{T}$. A* Concept Contraction Problem *(CCP), identified by $\langle \mathcal{L}, D, C, \mathcal{T} \rangle$, is finding a pair of concepts $\langle G, K \rangle \in \mathcal{L} \times \mathcal{L}$ such that $\mathcal{T} \models D \equiv G \sqcap K$, and $\mathcal{T} \not\models K \sqcap C \sqsubseteq \bot$. We call $K$ a* contraction *of $D$ according to $C$ and $\mathcal{T}$.*

$\mathcal{Q}$ is a symbol for a CCP, and $SOLCCP(\mathcal{Q})$ denotes the set of all solutions to $\mathcal{Q}$. Obviously, there is always the trivial solution $\langle G, K \rangle = \langle D, \top \rangle$ to whatever CCP, that is give up everything of $D$. When $C \sqcap D$ is satisfiable in $\mathcal{T}$, the "best" possible solution is $\langle \top, D \rangle$, that is, give up nothing — if possible. When subsumption does not hold *i.e.*, a full match is unavailable, one may want to hypothesize some explanation on which are the causes of this result.

**Definition 2.** *Let $C$, $D$, be two concepts in a Description Logic $\mathcal{L}$, and $\mathcal{T}$ be a set of axioms, where both $C$ and $D$ are satisfiable in $\mathcal{T}$. A* Concept Abduction Problem *(CAP), denoted as $\langle \mathcal{L}, C, D, \mathcal{T} \rangle$, is finding a concept $H$ such that $\mathcal{T} \not\models C \sqcap H \sqsubseteq \bot$, and $\mathcal{T} \models C \sqcap H \sqsubseteq D$.*

$\mathcal{P}$ is a symbol for a CAP, and $SOL(\mathcal{P})$ denotes the set of all solutions to $\mathcal{P}$. Given a CAP $\mathcal{P}$, if $H$ is a conjunction of concepts and no sub-conjunction of concepts in $H$ is a solution to $\mathcal{P}$, then $H$ is an **irreducible solution**.

The *rankPotential* algorithm [10] allows to numerically compute the *length* of $H$. Recalling the definition, *rankPotential* $(C, D)$ returns to a numerical measure of what is still missing in $C$ w.r.t. $D$. If $C \equiv \top$ the maximum value for *rankPotential* $(C, D)$ is computed, that is the maximum (potential) mismatch of $C$ from $D$. Hence, the value returned by *rankPotential*$(\top, D)$ amounts to how specific is a complex concept expression $D$ with respect to an ontology $\mathcal{T}$, what we call the *depth* of $D$: $\texttt{depth}(D)$.

Concept Covering, originally defined in [12] for a particular set of DLs with limited expressiveness, was later extended and generalized in terms of Concept Abduction in [5]. We recall here this definition:

**Definition 3.** *Let $D$ be a concept, $\mathcal{R} = \{S_1, .., S_k\}$ be a set of concepts in a Description Logic $\mathcal{L}$ and $\mathcal{T}$ be a set of axioms, where $D$ and $S_i, i = 1..k$ are satisfiable in $\mathcal{T}$.*

*1. A* Concept Covering Problem *(CCoP), denoted as* CCoP$(\langle \mathcal{L}, \mathcal{R}, D, \mathcal{T} \rangle)$*, is finding, if it exists, a set $\mathcal{R}_c \subseteq \mathcal{R}$, such that both for each $S_j \in \mathcal{R}_c$, $\mathcal{T} \not\models \sqcap S_j \equiv \bot$, and $H \in SOL(\langle \mathcal{L}, \sqcap S_j, D, \mathcal{T} \rangle)$ is such that $H \not\sqsubseteq D$.*
*2. We call $\langle \mathcal{R}_c, H \rangle$ a* solution *for the CCoP $\langle \mathcal{L}, \mathcal{R}, D, \mathcal{T} \rangle$.*

Intuitively, $\mathcal{R}_c$ is a set of concepts that completely or partially cover $D$ w.r.t. $\mathcal{T}$, while the abduced concept $H$ represents what is still in $D$ and is not covered by $\mathcal{R}_c$. A greedy approach is needed for performance reasons as also the basic set covering problem is NP-Hard. A Concept Covering Problem is similar, but has remarkable differences when compared to classical set covering [7]. There can be several solutions for a single CCoP, depending also on the strategy adopted for choosing candidate concepts in $\mathcal{R}_c$. In [7] the greedy Algorithm 1 was proposed, which we recall here for the sake of clearness. The algorithm *GsCCoP*$(\mathcal{R}, BP, \mathcal{T})$ ha as inputs $BP$, *i.e.*, the concept expression representing the required Business Process, $\mathcal{R} = \{BB_i\}, i = 1...n$ the set of available Building Blocks descriptions, the reference ontology $\mathcal{T}$.

The algorithm tries to cover $BP$ description "as much as possible", using the concepts $BB_i \in \mathcal{R}$. If a new building block $BB_i$ can be added to the already composed set $\mathcal{R}_c$, *i.e.*, $\mathcal{T} \not\models (\sqcap_{BB_k \in \mathcal{R}_c} BB_k) \sqcap BB_i \sqsubseteq \bot$, then an extended matchmaking process is performed (rows 8–21). If $BB_i$ is not consistent with the uncovered part of the business process $BP_u$, the latter is contracted and subsequently a CAP is performed between the contracted uncovered part and $BB_i$ (rows 8–10). If $BB_i$ is consistent with $BP_u$, only a CAP is solved (rows 11–15). Based on the previously computed concepts $G$, $K$ and $H$ a global score is computed as a metric to evaluate how good $BP_i$ is with respect to the covering set (rows 16–21). The score is determined through $\Psi$ function. It takes into account what has to be given up ($G$), kept ($K$) in the uncovered part ($BP_u$) of $BP$ in order to be compatible $BB_i$ and what has to be hypothesized ($H$) in order to fully satisfy the contracted $BP_u$, *i.e.*, $K$. For $\mathcal{ALN}$DL, that we adopt here the following formula is used [6]:

$$\Psi(G, K, H, BB_i, BP_u) = \left| 1 - \frac{N}{N - g} * (1 - \frac{h}{k}) \right| \tag{1}$$

where $N$, $k$, $g$, $h$ represent numerical evaluation, computed via the $rankPotential$ algorithm[10], for respectively:

```
    Input: BP, R, where BP and all BB_i ∈ R are satisfiable in T
    Output: R_c, \overline{BP}_u, K_BP, G_contraction
 1  BP_u = BP;
 2  H_min = BP;
 3  R_c = ∅;
 4  repeat
 5      BB_max = ⊤;
 6      d_min = ∞;
 7      foreach BB_i ∈ R such that R_c ∪ {BB_i} covers BP_u contraction according to
        Q = ⟨L, ⊓_{BB_k ∈ R_c} BB_k ⊓ BB_i, BP_u, T⟩ do
 8          if T ⊨ BP_u ⊓ BB_i ≡ ⊥ then
 9              ⟨G, K⟩ = contract(BB_i, BP_u, T);
10              H = abduce(BB_i, K, T);
11          else
12              H = abduce(BB_i, BP_u, T);
13              K = BP_u;
14              G = ⊤;
15          end
16          d = Ψ(G, K, H, BB_i, BP_u);
17          if d < d_min then
18              BB_max = BB_i;
19              H_min = H;
20              d_min = d;
21          end
22      end
23      if (BB_max ≢ ⊤) then
24          R = R \ {BB_i};
25          R_c = R_c ∪ {BB_i};
26          BP_u = H_min;
27      end
28  until (BB_max ≢ ⊤) ;
29  ⟨K_BP, G_BP⟩ = contract(⊓_{BB_k ∈ R_c} BB_k, BP, T);
30  \overline{BP}_u = abduce(⊓_{BB_k ∈ R_c} BB_k, K_BP, T);
31  return (R_c, \overline{BP}_u, K_BP, G_BP);
```

**Algorithm 1.** $GsCCoP(\mathcal{R}, BP, \mathcal{T})$ – Extended Concept Covering

-$BP_u$ the uncovered part of $BP$ in an intermediate step of the covering process: $N = rankPotential(\top, BP_u, \mathcal{T})$;
-$K$: belonging to a solution of $\mathcal{Q} = \langle \mathcal{ALN}, BB_i, BP_u, \mathcal{T} \rangle$: $k = rankPotential(\top, K, \mathcal{T})$;
-$G$: belonging to a solution of $\mathcal{Q} = \langle \mathcal{ALN}, BB_i, BP_u, \mathcal{T} \rangle$: $g = rankPotential(BP_u, K, \mathcal{T})$;
-$H$: a solution of $\mathcal{P} = \langle \mathcal{ALN}, BB_i, K, \mathcal{T} \rangle$: $h = rankPotential(BB_i, K, \mathcal{T})$.

The outputs of *GsCCoP* are:

-$\mathcal{R}_c$ : the set of building blocks selected to compose the business process;
-$\overline{BP}_u$ : the part of the business process description that has –in case no exact cover exists– not been covered;
-$K_{BP}$: the contracted BP;
-$G_{BP}$: the part of the business process description given-up at the end of the whole composition process.

We note that the proposed approach also allows to explicitly define mandatory requirements $M$ and preferences $P$ within their request as $BP \equiv M \sqcap P$ (as proposed in [6]).

## 4 Enduring Oak: An Efficient Concept Covering Framework

Several recent approaches try to merge DL-based reasoning with classical Relational DBMS. The objective is obvious: by pre-classifying large knowledge bases on RDBMS it is possible to reduce the burden on inference engines and deal efficiently with large datasets. Such approaches can be classified as either RDF-based, and include [4,17,19,20] or OWL-based [15,14]. In most cases they are limited to the persistent storing of ontologies on RDBMS, and sometimes have limited inference capabilities. We developed a novel Java-based framework, *Enduring Oak*, which was designed specifically to support persistence and the ability to efficiently solve Concept Covering problems. It is compliant with DIG specifications, uses Hypersonic SQL DB for storage and retrieval of persistent data, and calls a DIG-compliant DL reasoner for computing inferences. Currently we adopt MaMaS-tng [1], which implements *rankPotential* and algorithms to solve concept abduction and contraction problems. The database schema (see Figure 1) includes various tables, namely: Individuals, Concepts, Roles, Individual types, Children, Descendants, Roles range, and Roles domain, with the obvious meaning. The relational model we adopted is clearly redundant [1], as *e.g.*, , the *children* table could have been withdrawn by adding a Boolean attribute with a true value assigned whenever a descendant is also a child. Our design choices were guided by the will to improve retrieval efficiency, also at the expenses of normalization, with a greater storage area and a more complex insert mechanism. In the following we illustrate the modified Concept Covering algorithm implemented in the Enduring Oak framework (see Algorithm 2). Before starting to compute the actual Concept Covering, a preprocessing step is needed for the ontology $\mathcal{T}$. For each role $R$ in $\mathcal{T}$having as range a concept $C$, add the definition axiom $AUX \equiv \forall R.C$ to $\mathcal{T}$[2]. Notice, that since $AUX$ is a defined concept *i.e.*, it is introduced via an equivalence axiom, the structure of the domain knowledge modeled in the original $\mathcal{T}$is not modified.

It is easy noticing that the main difference w.r.t. the original algorithm are the calls to the `Candidate Retrieval` procedure (see Algorithm 3). Candidate Retrieval selects a subset of all available BB descriptions in order to avoid checking all the available dataset. Retrieval is carried out combining database queries and inference services calls to the reasoner. Differently from the basic version of the algorithm, using *GsCCoP* a "full" check for covering condition is not needed. In fact, $\mathcal{C}$ contains concepts

[1] http://dee227.poliba.it:8080/MAMAS-tng/

[2] We recall that a range relation can be modeled with the axiom $\top \sqsubseteq \forall R.C$ in $\mathcal{T}$.

**Input**: $BP, \mathcal{R}$, where $BP$ and all $BB_i \in \mathcal{R}$ are satisfiable in $\mathcal{T}$
**Output**: $\mathcal{R}_c, \overline{BP}_u, K_{BP}, G_{BP}$
1 $\mathcal{R}_C = \emptyset$;
2 $BP_u = BP$;
3 $H_{min} = BP$;
4 $\mathcal{C} = candidateRetrieval(D, \mathcal{R}, \mathcal{T})$;
5 **repeat**
6 $\quad BB_{MAX} = \top$;
7 $\quad d_{min} = \infty$;
8 $\quad$ **foreach** $BB_i \in \mathcal{C}$ **do**
9 $\quad\quad$ **if** $\mathcal{T} \models BP_u \sqcap BB_i \sqsubseteq \bot$ **then**
10 $\quad\quad\quad \langle G, K \rangle = contract(BB_i, BP_u, \mathcal{T})$;
11 $\quad\quad\quad H = abduce(BB_i, K, \mathcal{T})$;
12 $\quad\quad$ **else**
13 $\quad\quad\quad H = abduce(BB_i, BP_u, \mathcal{T})$;
14 $\quad\quad\quad K = BP_u$;
15 $\quad\quad\quad G = \top$;
16 $\quad\quad$ **end**
17 $\quad\quad d = \Psi(G, K, H, BB_i, BP_u)$;
18 $\quad\quad$ **if** $d < d_{min}$ **then**
19 $\quad\quad\quad BB_{MAX} = BB_i$;
20 $\quad\quad\quad H_{min} = H$;
21 $\quad\quad\quad d_{min} = d$;
22 $\quad\quad$ **end**
23 $\quad$ **end**
24 $\quad$ **if** $(BB_{MAX} \not\equiv \top)$ *and* $(\mathcal{R} \cup \{BB_i\} \cup \mathcal{T})$ *is satisfiable* **then**
25 $\quad\quad \mathcal{R}_c = \mathcal{R}_c \cup \{BB_i\}$;
26 $\quad\quad BP_u = H_{min}$;
27 $\quad\quad \mathcal{C} = candidateRetrieval(BP_u, \mathcal{R}, \mathcal{T})$;
28 $\quad$ **end**
29 **until** $S_{min} \equiv \top)$ ;
30 $\langle K_{BP}, G_{BP} \rangle = contract(\sqcap_{BB_k \in \mathcal{R}_c} BB_k, BP, \mathcal{T})$;
31 $\overline{BP}_u = abduce(\sqcap_{BB_k \in \mathcal{R}_c} BB_k, K_{BP}, \mathcal{T})$;
32 **return** $(\mathcal{R}_c, \overline{BP}_u, K_{BP}, G_{BP})$;

**Algorithm 2.** *GsCCoP+*$(\mathcal{R}, D, \mathcal{T})$ – Fast Concept Covering

which are surely consistent with $BP_u$. What is still to evaluate is if the best concept –from a covering point of view– in $\mathcal{C}$ is consistent with the ones in $\mathcal{R}_c$ (row 24 in Algorithm 2). Then, only a consistency check is needed instead of a full covering check. On the other hand, while computing a covering, the size of $\mathcal{C}$ cannot increase. In fact, $candidateRetrieval$ computes the candidates to be added to $\mathcal{R}_c$ considering only the ones which overlap $BP_u$. Since $BP_u$ decreases after each iteration, the same is for $\mathcal{C}$. Notice that, due to the preprocessing step, also $AUX$ concepts are taken into account and returned by $candidateRetrieval$, in this way also complex concepts involving roles are taken into account.

**Input**: $D, \mathcal{R}, \mathcal{T}$, where $D$ and all $C_i \in \mathcal{R}$ are satisfiable in $\mathcal{T}$
**Output**: $\mathcal{C}$
1 $\mathcal{CN} = \emptyset$;
2 $\mathcal{A} = parents(D, \mathcal{T})$;
3 $\mathcal{CN} = \mathcal{A}$;
4 **foreach** $CN_j \in \mathcal{A}$ **do**
5 $\quad \mathcal{CN} = \mathcal{CN} \cup descendants(CN_j, \mathcal{T})$;
6 **end**
7 **foreach** $C_i \in \mathcal{R}$ **do**
8 $\quad$ **foreach** $CN \in \mathcal{CN}$ **do**
9 $\quad\quad$ **if** $C_i \sqsubseteq_{\mathcal{T}} CN$ **then**
10 $\quad\quad\quad \mathcal{C} = \mathcal{C} \cup \{C_i\}$;
11 $\quad\quad$ **end**
12 $\quad$ **end**
13 **end**
14 **return** $\mathcal{C}$;

**Algorithm 3.** $candidateRetrieval(D, \mathcal{R}, \mathcal{T})$ — Candidates Retrieval

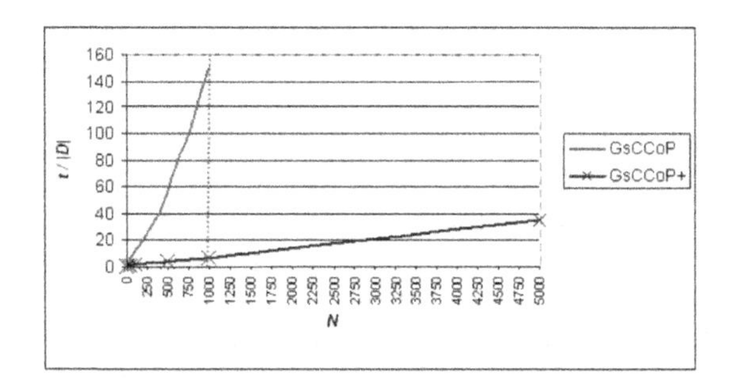


**Fig. 3.** Comparison of concept covering time performances (GsCCoP: basic *composer* algorithm, GsCCoP+: enhanced *composer* algorithm)

# 5 System and Performance Evaluation

The architecture of our system comprises MaMaS-tng (which is currently the only reasoner equipped with the above mentioned services), communicating via DIG HTTP and a Hypersonic SQL DBMS interfaced via JDBC. Building Blocks and Business Processes descriptions are all annotated in the $\mathcal{ALN}$ subset of OWL-DL (www.w3.org/TR/owl-features/) language. Obviously, interaction with the system is carried out using a GUI, both for query composition and retrieval of results. The GUI allows to compose a request for a single business process or a composed one. When a fully satisfying result cannot be found, the system returns the degree of completion together with a logical explanation of what remains missing ($BP_u$) and/or conflicting ($G_BP$) with the

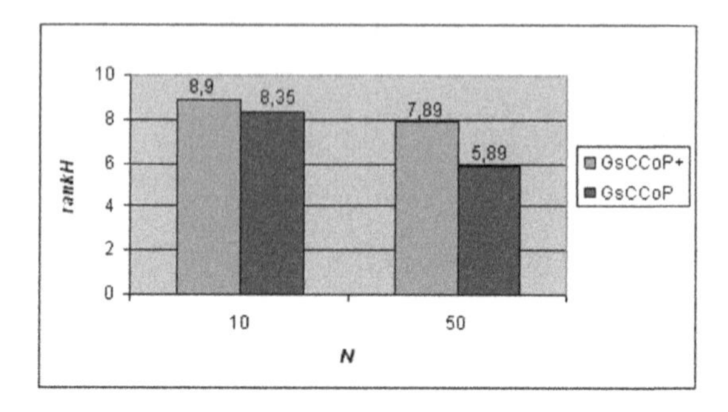


**Fig. 4.** Comparison w.r.t. the $BP_u$ (uncovered part $H$)(GsCCoP: basic *composer* algorithm, GsC-CoP+: enhanced *composer* algorithm

request, see Figure 2. Extensive experiments aimed at both quantitative and qualitative evaluation have been carried out, comparing the enhanced Enduring Oak approach with the existing system. We note that all experiments presented here have been carried out w.r.t. purely abductive versions of both algorithms, to ease comparison with existing data from previous tests. The test dataset is a randomly generated population of Building Blocks descriptions having a rank with a Gaussian distribution $\mu = 7, \sigma = 5$. Tests have been run using as sample requests descriptions randomly extracted from the dataset, having rank 5, 10, 15, *vs.* sets of 10, 50, 100, 500, 1000, 5000 descriptions randomly extracted from the same dataset at each iteration of the tests. Tests were run on PC platform endowed of a Pentium IV 3.06GHz, RAM 512MB, Java 2 Standard Edition 5.0. We point out that reported results include time elapsed for $\mathcal{T}$uploading and database connection, with a duration, on average of 5.81 and 0.26 secs, respectively. Obviously such bootstrap times should not be kept into account in the actual usage of the system. Figure 3 presents a comparison between the basic greedy algorithm and the one presented in this paper, clearly showing the computational benefits of the enhanced approach, also in the presence of huge datasets. In both cases, time is normalized w.r.t. to $\texttt{depth}(BP)$. To evaluate the effectiveness of the approach we carried out another set of tests using the same dataset. In this case requests were built as conjunction of a description randomly extracted from the selected dataset with a randomly generated description generated at run time. The rationale is that in this way we ensured that at least a part of the request would have to be satisfied by the description available in the dataset. Results refer to datasets of respectively 10 and 50 descriptions and are summarized in terms of $BP_u$ rank[3], see Figure 4, and covering percentage, see Figure 5, w.r.t. the basic *GsCCoP* and the enhanced *GsCCoP+* algorithms. The tests show that approximations introduced in *GsCCoP+* affect in a limited way the effectiveness of the covering procedure. We note that these are worst case tests, as the population is randomly generated; tests carried out with actual building block descriptions and requests provide results that are practically equivalent for both algorithms.

[3] Recall that $BP_u$ is the part that has to be hypothesized, *i.e.*, what remains uncovered.

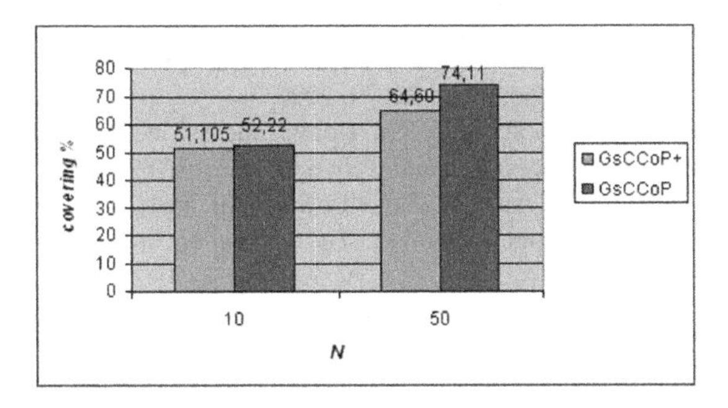


**Fig. 5.** Successful covering percentage(GsCCoP: basic *composer* algorithm, GsCCoP+: enhanced *composer+* algorithm

## 6 Conclusion

While semantic-based systems offer several benefits, they often become inapplicable on real world large datasets. In this paper we have faced such problem in the context of a framework for the automated selection of building-blocks for ERP customizing. Building on a pre-existing solution that implemented a greedy Concept Covering algorithm we have presented an enhanced approach that, partially relying on a relational DBMS, allows to scale our approach up to thousands of BB descriptions with negligible loss in terms of effectiveness.

## Acknowledgments

We wish to acknowledge partial support of Apulia regional project ACAB-C2, and EU-FP-6-IST STREP TOWL.

## References

1. S Abiteboul, R Hull, and V Vianu. *Foundations of Databases*. Addison Wesley Publ. Co., Reading, Massachussetts, 1995.
2. SAP AG. Contract-based Service Order with Third-Party Services and Time & Material Billing, SAP Best-Practices for Service Providers. help.sap.com/ bestpractices/ industry/serviceindustries/v346c_it/html/index.htm, 2003.
3. F. Baader, D. Calvanese, D. Mc Guinness, D. Nardi, and P. Patel-Schneider, editors. *The Description Logic Handbook*. Cambridge University Press, 2002.
4. J. Broekstra, A. Kampman, and F. Harmelen. Sesame: a generic architecture for storing and querying RDF and RDF-Schema. In *proc. of ISWC 2002*, 2002.
5. S. Colucci, T. Di Noia, E. Di Sciascio, F.M. Donini, G. Piscitelli, and S. Coppi. Knowledge Based Approach to Semantic Composition of Teams in an Organization. In *SAC-05*, pages 1314–1319, 2005.
6. S. Colucci, T. Di Noia, E. Di Sciascio, F.M. Donini, and M. Mongiello. Concept Abduction and Contraction for Semantic-based Discovery of Matches and Negotiation Spaces in an E-Marketplace. *Electronic Commerce Research and Applications*, 4(4):345–361, 2005.

7. F. di Cugno, T. Di Noia, E. Di Sciascio, F.M. Donini, and A. Ragone. Concept covering for automated building blocks selection based on business processes semantics. In *Joint 8th IEEE CEC'06 and 3rd EEE'06*, pages 72–79, 2006.
8. F. di Cugno, T. Di Noia, E. Di Sciascio, F.M. Donini, and E. Tinelli. Building-Blocks Composition based on Business Process Semantics for SAP R/3. In *SWCASE'05@ISWC2005*.2005.
9. T. Di Noia, E. Di Sciascio, and F.M. Donini. Semantic Matchmaking as Non-Monotonic Reasoning: A Description Logic Approach. *The Journal of Artificial Intelligence Research*, 2007. To appear.
10. T. Di Noia, E. Di Sciascio, F. M. Donini, and M.Mongiello. A system for Principled Matchmaking in an Electronic Marketplace. *Int. J. of Electronic Commerce*, 8(4):9–37, 2004.
11. R/3 Simplification Group. *Best Practices for mySAP All-in-One, Building Blocks Concept*. 2003.
12. M-S. Hacid, A. Leger, C. Rey, and F. Toumani. Computing Concept Covers: a Preliminary Report. In *DL'02*, volume 53 of *CEUR Workshop Proceedings*, 2002.
13. J. A. Hernández. *SAP R/3 Handbook*. McGraw Hill, 2nd edition, 2000.
14. I Horrocks, L Li, D Turi, and S Bechhofer. The instance store: DL reasoning with large numbers of individuals. In *DL 2004*, pages 31–40, 2004.
15. T. Kessel, M. Schlick, and O. Stern. Accessing configuration-databases by means of description logics. In *KI'95 workshop: knowledge Representation meets Databases*, 1995.
16. SAP Building Block Library. help.sap.com/bestpractices/BBLibrary/bblibrary_start.htm.
17. B. McBride. Jena:implementing the RDF model and syntax specifications. In *SemWeb'01*, 2001.
18. A. Scheer and F Habermann. Making ERP a success. *Comm. of the ACM*, 43(4):57–61, 2000.
19. R. Volz, D Oberle, S Staab, and B Motik. Kaon server - a semantic web management system. In *WWW 2003*, 2003.
20. D. Wood, P. Gearon, and T. Adams. Kowari: a platform for semantic web storage and analysis. In *Proc. of Xtech Conference*, 2005.

# Towards Ontology-Driven Information Systems: Redesign and Formalization of the REA Ontology

Frederik Gailly and Geert Poels

Faculty of Economics and Business Administration, Ghent University
{Frederik.Gailly,Geert.Poels}@UGent.be

**Abstract.** It is widely recognized that ontologies can be used to support the semantic integration and interoperability of heterogeneous information systems. Resource Event Agent (REA) is a well-known business ontology that was proposed for ontology-driven enterprise system development. However, the current specification is neither sufficiently explicit nor formal, and thus difficult to operationalize for use in ontology-driven business information systems. In this paper REA is redesigned and formalized following a methodology based on the reengineering extension of the METHONTOLOGY framework for ontology development. The redesign is focused on developing a UML representation of REA that improves upon existing representations and that can easily be transformed into a formal representation. The formal representation of REA is developed in OWL. The paper discusses the choices made in redesigning REA and in transforming REA's UML representation into a OWL representation.

**Keywords:** Business Ontologies, Interoperability, Ontology Engineering, UML, OWL.

## 1 Introduction

For many years now ontology research has been a growing research domain in Knowledge Engineering and Artificial Intelligence. The most cited definition of an ontology is the definition by Gruber: "an ontology is an explicit specification of a conceptualization" [1, p. 199]. A conceptualization is an intensional semantic structure which encodes the implicit rules constraining the structure of a piece of reality [2]. Ontologies can be used to represent explicitly the semantics of structured and semi-structured information enabling automatic support for maintaining and accessing information [3]. The Gruber definition was modified slightly by Borst [4] who added that the specification must be formal and the conceptualization should be shared. Formal means that a machine must be able to process the specification and shared indicates that the knowledge captured by the ontology is the consensus of a community of experts [5].

Domain ontologies specify a conceptualization of a selected part of reality (i.e. a domain) [6]. They describe the concepts that exist in a domain, the classification of the concepts, the relations between the concepts and their axioms (i.e. basic propositions

W. Abramowicz (Ed.): BIS 2007, LNCS 4439, pp. 245–259, 2007.
© Springer-Verlag Berlin Heidelberg 2007

assumed to be true). For instance, business ontologies (see [7] for an overview) have as 'Universe of Discourse' business, which is "the activity of providing goods and services involving financial, commercial and industrials aspects" (WordNet). Business ontologies have been proposed to support requirements elicitation, modeling and engineering of business applications, enterprise systems and e-collaboration systems [8-11]. For instance, in ontology-driven business modelling, they are used to constrain the contents and structure of business models, thereby helping to identify and organize relevant objects, relationships and other knowledge [12].

The use of ontologies at run-time (i.e. ontology-driven systems instead of ontology-driven development [12]) also offers great potential for business ontologies. Specifically in and between enterprises ontology-driven information systems can be used to create interoperability at different enterprise levels: shop-floor, intra-enterprise and inter-enterprise level. Recently there have been research efforts on how ontologies can be used for information integration [13, 14] and process integration [15-17]. However, real-world applications of business ontologies that explore these opportunities are still scarce.

The Resource Event Agent (REA) ontology [18] is an example business ontology that has been proposed for ontology-driven enterprise systems development. As an 'event ontology' [19], REA focuses on the events occurring within the realm of a company, their participating agents, affected resources, and regulating policies. REA can be used as a reference for modelling a single business cycle (e.g. sales-collection) or a chain of business cycles, connected through resource flows. Applications supported by REA-driven modelling include the design of accounting and operations management systems [20], auditing and internal control (e.g. SOX compliance checking) [21] and conceptual data modelling [19]. REA has been used in a number of international standardization efforts for e-collaboration systems. For instance, REA was the basis for the business process ontology in the UMM business process and information model construction methodology [22], the ECIMF system interoperability enabling methodology [23], and the Open-EDI business transaction ontology which is part of the ISO/IEC 15944-4 standard. REA has further been proposed as a theoretical basis for the reference models that underlie ERP systems [24].

In this paper we investigate REA's potential to be used at run-time (i.e. in ontology-driven information systems). The origin of REA is an accounting data model [25] that has been extended first into an enterprise information architecture and later into a full-scale enterprise ontology. This development followed an ad-hoc process rather than being guided by an Ontology Engineering methodology. The REA developers focused more on the theoretical background of the ontology (events accounting and Micro-Economic theories) than on the representation, formalization and computational correctness of the ontology (although they did perform in [18] an ontological analysis using Sowa's classification [26]). As a consequence, there is no formal representation of REA. Furthermore, the available literature sources on REA (e.g. [20, 21, 27, 28]) present different views on the REA conceptualization of an enterprise and use a variety of different formats (including text, tables and diagrams) to specify the ontology. The lack of a generally accepted conceptualization and a uniform, complete and formal representation of the ontology causes imprecision in its definition and ambiguity in its interpretation. For instance, in [29] we showed that the ontological concepts and the relations between the concepts are not strictly defined and that the ontological axioms, only informally defined in text, are confusing (mixing up types and instances of concepts). There have been attempts to formalize REA (e.g. [30, 31]), but the results obtained were highly dependent on the researchers' subjective interpretation of REA.

As currently REA is not formally represented, not sufficiently explicit, and not based on a demonstrable shared conceptualization, it needs improvement before it can be applied on a wide scale in ontology-driven systems. The aim of this paper is to facilitate the operationalization of REA (i.e. increasing its applicability to be used at run-time) by making it more explicit and formal. We first present the development of a new graphical UML representation of REA that presents a unified and consistent view on the ontology's concepts, relations and axioms. This new conceptual representation is subsequently used as the basis for a more formal representation of REA in OWL. Improving the conceptual and formal representations of REA makes the ontology more understandable, easier to compare to alternative business ontologies, easier to analyse (e.g. for consistency), and more operational (e.g. executable). The paper focuses on representation and formalization aspects, but not on the contents of REA. The new representations will not make REA more accepted per se (in the sense of a having a conceptualization that is widely shared in the business community). However, they do facilitate ontological analysis and the evaluation whether there is agreement upon the ontological definitions and axiomatization.

It is our position that the reengineering of business ontologies to improve their applicability should be guided by proven Ontology Engineering principles and techniques. The two REA reengineering activities presented in this paper are part of a more encompassing business ontology reengineering methodology that we have developed and are currently refining. The proposed reengineering methodology is presented briefly in the next section. Section three presents the new conceptual representation of REA (as a UML class diagram) based on our redesign and synthesis of the currently available representations. Section four then presents the OWL representation of REA and discusses the UML-OWL language mappings employed. Section five ends with conclusions and future work.

## 2 Reengineering Business Ontologies

An Ontology Engineering methodology can provide guidelines for different purposes. It can prescribe a stepwise ontology development process, provide decision rules to follow during the modelling of the ontology, and address various design, representation and management aspects of the ontology [32]. In [7] we proposed a reengineering methodology for business ontologies. The first part of this section outlines the process, phases and main activities of this methodology. The second part focuses on those activities that are relevant for the reengineering of REA as intended in this paper, i.e. developing an improved conceptual representation of the ontology and formalizing it.

## 2.1 Business Ontology Reengineering Methodology

The proposed reengineering methodology (see Figure 1) is based on the METHONTOLOGY framework for ontology development [33] which has its roots in Software Engineering and Knowledge Engineering methodologies. METHONTOLOGY is especially useful for our purpose because it has an ontology reengineering extension [34]. The METHONTOLOGY framework defines ontology reengineering as "the process of retrieving and transforming a conceptual model of an existing and implemented ontology into a new, more correct and complete conceptual model, which is reimplemented". Three phases are identified in this reengineering process: reverse engineering, restructuring and forward engineering. These phases provide the core structure of our business ontology reengineering methodology.

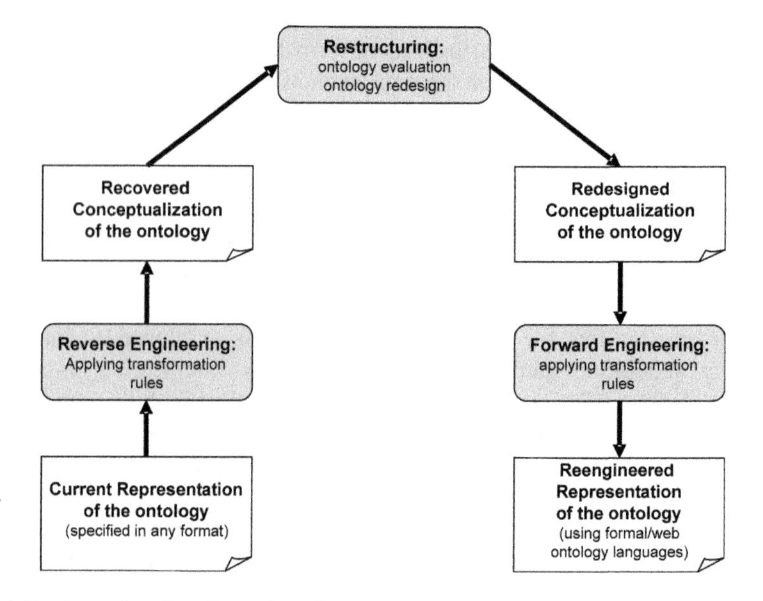


**Fig. 1.** Business Ontology Reengineering Methodology – Process, Activities and Artifacts

The *Reverse Engineering* phase starts from the currently available representation(s) of the ontology as specified in whatever language(s) (i.e. the *Current Representation* artifact). The goal is to recover the original conceptualization specified by the ontology. If only a formal representation of the ontology is available, then this formal representation should be transformed into a conceptual representation by using the right transformation rules for the chosen representation languages. For example, if the ontology is available in OWL, visualization rules that transform OWL into UML can be used to create a graphical representation of the ontology [35]. Sometimes there is neither a formal representation, nor a complete conceptual representation of the ontology (or there exists alternative specifications as with REA). In that case the reverse engineering will mainly be a 'unification' effort, gathering information on the conceptualization selected from the available sources, thereby focusing on the commonalities in the existing representations and underlying interpretations. The result of the reverse engineering activities is referred to as the *Recovered Conceptualization* of the ontology.

During the *Restructuring* phase the recovered conceptualization is evaluated and its specification is redesigned based on the shortcomings identified. The METHONTOLOGY restructuring process distinguishes two distinct sub-phases: analysis and synthesis. During the analysis phase the class hierarchy, the concept definitions and the axioms are evaluated. Studies that compare different ontology evaluation methods can be used for selecting the appropriate analysis method [36, 37]. The synthesis phase then uses the outcome of the analysis to re-specify the conceptualization. Also all changes made to the ontology are recorded.

The restructuring phase results in a *Redesigned Conceptualization* of the ontology. *Forward Engineering* activities transform the conceptual representation of this redesigned conceptualization into a machine readable representation, referred to as the *Reengineered Representation* of the ontology. The transformation rules used depend on the representation format of the redesigned ontology and the target ontology language.

## 2.2 From Lightweight Ontology to Formal Ontology

REA as it stands can be characterized as a lightweight ontology [38]. To formalize a lightweight ontology a specification is needed that provides a complete view of the conceptualization with as much explicit semantics as possible. Preferably such a specification is developed in a language that can easily be mapped onto an ontology language but at the same time allows for ontology modelling.

Recently, Ontology Engineering researchers have proposed the use of conceptual modelling languages (ER, UML, ORM, ...) for modelling ontologies [39-41]. Most conceptual modelling languages propose graphical representations with well-defined syntax and semantics that are close to how humans perceive the world [42]. Specifically in a business context the diagrammatic techniques offered by conceptual modelling languages are known (or can easily be learned) by business domain experts [43]. Conceptual modelling languages therefore provide a more suitable basis for the analysis and refinement of the content of an ontology than the more formal knowledge representation languages as the resulting representation of the ontology will be more natural and hence easier to understand for the domain experts.

REA uses a combination of informal text, table definitions and diagrams. These definitions and diagrams can be found in different sources by different authors. To facilitate the formalization of REA we first developed a representation of REA that unifies these partial (and often imprecise) definitions into a single coherent view with explicit semantics. In case of doubt we referred to the 'official' version of the ontology as described by the REA developers in their most recent papers (i.e. [20, 21]). We developed this new representation in UML (using a class diagram) and refer to it as the *conceptual* representation of REA because its intended users are humans and not machines. UML was chosen as ontology modelling language because it provides a standard and tool-supported notation. Furthermore, the Ontology Definition Metamodel (ODM) request for proposal of the Object Management Group (OMG) has resulted in the proposal of semantically correct mapping rules in both directions between UML and the ontology languages RDF and OWL[44].

The development of the new UML representation of REA corresponds roughly to the reverse engineering and restructuring steps of Figure 1. The currently available representations of REA had to be gathered, analysed and combined to recover to the best possible extent the original conceptualization (i.e. reverse engineering). Unifying the existing definitions, resolving inconsistencies and explicitly representing the recovered semantics in a UML class diagram can be seen as a restructuring activity. The result should however be seen as a redesigned representation of REA rather than a redesigned conceptualization as no fundamental changes to the intended content of the ontology were proposed. The proposed reengineering methodology is intended as an iterative process and in future iterations the new conceptual representation can be used for in-depth analysis of REA's conceptualization of business and subsequent discussion and refinement. In this first iteration, the UML class diagram was primarily meant as a starting point for the formalization of REA.

The purpose of the formalization activity is to map the conceptual representation into a *formal* representation, meaning a machine-readable representation (i.e. the intended users are machines and not humans). The ontology language chosen was OWL because of its wide acceptance as a web ontology language. Also, the availability of ontology tools that support OWL (e.g. Description Logic reasoners) will make it easier to experiment with REA-driven business applications. The OWL representation that results from this forward engineering step corresponds to the reengineered representation artefact in Figure 1.

## 3 Redesigning REA

REA specifies five basic concepts (Economic **R**esource, Economic **E**vent, Economic **A**gent, Commitment, Contract) in terms of which an enterprise is described. These five concepts (defined in Table 1) are the ontological primitives upon which the other ontological elements (e.g. specialized concepts, typified concepts, relations) are built and in terms of which domain axioms are defined. We also assume these definitions when redesigning the conceptual representation of REA.

**Table 1.** Definitions of the basic REA concepts

| Concept | Definition |
|---|---|
| Economic Resource | A thing that is scarce and has utility for economic agents and is something users of business applications want to plan, monitor and control. |
| Economic Agent | An individual or organization capable of having control over economic resources, and transferring or receiving the control to or from other individuals or organizations. |
| Economic Event | A change in the value of economic resources that are under control of the enterprise. |
| Commitment | A promise or obligation of economic agents to perform an economic event in the future |
| Contract | A collection of increment and decrement commitments and terms. |

REA further defines and names a number of relations between the basic concepts. For instance, economic resources are associated with the economic events that cause their *inflow* or *outflow* (*stockflow relations*). Economic events that result in resource inflows (e.g. purchases) are paired by economic events that result in resource outflows (e.g. cash disbursements) (*duality relations*). *Participation relations* identify the economic agents involved in economic events. A commitment will eventually be related to an economic event of the specified type by a *fulfilment relation*. *Reciprocity relations* are analogous to duality relations, but relate commitments instead of economic events.

These and other concept relations are defined informally in text or are depicted as relationships in ER diagrams, UML class diagrams or other graphical formalisms that provide a partial view on the ontology. Implicitly they introduce a number of derived concepts like specializations of basic concepts (e.g. economic events that cause an inflow (*increment* economic events) versus economic events that cause an outflow (*decrement* economic events)) and type images (e.g. economic event type), as well as constraints (e.g. if commitments A and B are linked by reciprocity, and economic events C and D fulfil respectively A and B, then C and D must be linked by duality). Apart from type images (described extensively in [21]) these inferred concepts and constraints are underspecified.

Also a minimal set of ontological axioms is defined, again informally. The definitions of axioms 1 and 2 are literally copied from [20]. Axiom 3 is based on [27] and is more precisely formulated than its equivalent in [20].

- ***Axiom1*** – At least one inflow event and one outflow event exist for each economic resource; conversely inflow and outflow events must affect identifiable resources.
- ***Axiom2*** – All events effecting an outflow must be eventually paired in duality relationships with events effecting an inflow and vice-versa.
- ***Axiom3*** – Each economic event needs to have at least one provide and one receive relationship with an economic agent.

The main problem with these definitions is that it not always clear whether the axioms apply to instances or types. For instance, an enterprise can own an economic resource that it has produced or acquired (e.g. a car) but not sold, used or consumed yet. Clearly the existence of a decrement event for every economic resource under the control of a company is not an axiom. However, following economic rationale (i.e. value creation), we could say that for every type of economic resource there exists at least one type of decrement economic event (e.g. a car manufacturer produces cars to sell them). Axiom 3 further introduces specializations of the participation relation: *provide* and *receive*.

The new conceptual representation that we propose for REA is shown in Figure 2. REA concepts are shown as classes, concept relations are shown as associations or association classes, and axioms are specified by means of multiplicities. For a more detailed and complete account of the shortcomings in the current REA representations and how we resolve them in our UML class diagram, we refer to [7]. The two main improvements are:

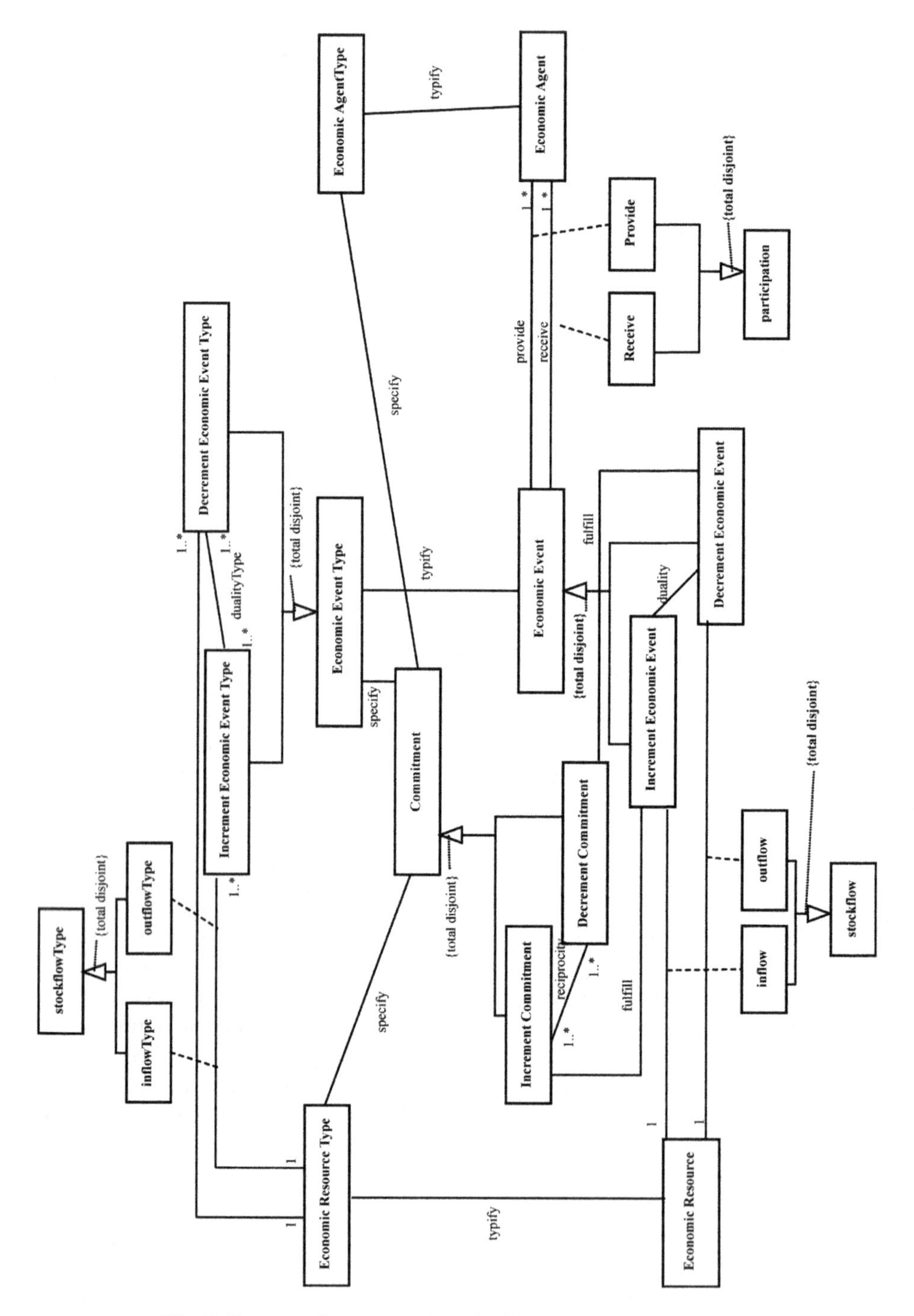


**Fig. 2.** Conceptual representation of REA as a UML class diagram

- The explicit specification of specializations of basic REA concepts and their type images: Increment Economic Event, Decrement Economic Event, Increment Commitment, Decrement Commitment, Increment Economic Event Type, and Decrement Economic Event Type. Less commonly used in UML is the specialization of association classes (that are used to represent concept relations): inflow and outflow as specializations of the stockflow relation and provide and receive as specializations of the participation relation. These new classifications add formerly implicit semantics to the conceptual representation. The diagram shows, for instance, that inflows relate increment events to resources, outflows relate decrement events to resources, increment events fulfil increment commitments, and decrement events fulfil decrement commitments. Note that Contract is not shown in the diagram, though it can be added by reifying the reciprocity relation. Further, both Contract and Commitment can be typified (again not shown in Figure 2).
- The extensions make it also possible to specify the REA axioms by means of multiplicities. For instance, the first part of axiom 1 can now be stated more precisely by enforcing the participation of every Economic Resource Type object in at least one inflowType link and at least one outflowType link. It is however not required that every Economic Resource object is related to at least one Economic Increment Event and one Economic Decrement Event. Other multiplicities than the ones shown in Figure 2 can be specified, but for the sake of clarity we only included the multiplicities that correspond to the three basic axioms. There is only one exception; the multiplicities shown for the reciprocity relation imply that every increment commitment must be paired with at least one decrement commitment, and vice versa (i.e. the economic reciprocity principle of capitalist market economies that underlies every written or unwritten contract). This example shows that additional domain axioms can easily be integrated into the conceptual representation.

## 4 Formalizing REA

Different authors have stipulated guidelines for transforming UML class diagrams into a formal representation in an ontology language [39, 40] and vice versa [35]. In the absence of a uniform approach, the recent adoptedOntology Definition Metamodel proposal [44] is used to guide the formalization of REA in OWL. All classes, relationships and constraints presented in the UML class diagram (Figure 2) are transformed into OWL by mapping them to OWL constructs. In most cases these transformations are straightforward but some of them require additional explanation. One of the problems with the ODM specification is that sometimes for the same UML construct different mapping rules can be used and that for some UML constructs no appropriate OWL construct exists. As a result the mapping from UML to OWL depends to some extent on the transformation choices made. The used UML to OWL transformations and corresponding examples are shown in tables 2, 3, 4 and 5. The complete OWL formalization of the conceptual REA representation developed in this paper can be downloaded from http://users.ugent.be/~fgailly/REAontology/.

In Table 2, the UML classes Economic Event (Type), Economic Agent (Type), Economic Resource (Type) and Commitment are mapped onto disjoint OWL classes (i.e. disjoint subclasses of `owl:Thing`). The binary associations and association classes (but not their sub-classes) in Figure 2 are represented by OWL properties (using the ObjectProperty construct). In OWL a property name has a global scope, while in UML the association name scope is limited to the class and subclasses of the class on which it is defined. As a result we have decided to give all associations a unique name. The transformation from UML binary association to OWL properties follows the approach taken in the ODM specification which states that every bidirectional binary association must be translated into a pair of properties where one is inverseOf the other. For an association class the same approach is followed.

**Table 2.** Transformations UML to OWL (UML class, UML association (class))

| **UML elements** | **OWL elements** |
|---|---|
| Class | Class, disjointWith |

```
<owl:Class rdf:ID="Economic_Agent">
   <owl:disjointWith rdf:resource="#Economic_Agent_Type"/>
   <owl:disjointWith rdf:resource="#Economic_Event"/>
   <owl:disjointWith rdf:resource="#Economic_Resource_Type"/>
   <owl:disjointWith rdf:resource="#Economic_Resource"/>
   <owl:disjointWith rdf:resource="#Economic_Event_Type"/>
   <owl:disjointWith rdf:resource="#Commitment"/>
</owl:Class>
```

| **UML elements** | **OWL elements** |
|---|---|
| Binary association | ObjectProperty, domain, range, inverseOf |

```
<owl:ObjectProperty rdf:ID="fulfill">
   <rdfs:domain rdf:resource="#Economic_Event"/>
   <rdfs:range rdf:resource="#Commitment"/>
   <owl:inverseOf rdf:resource="#inverse_of_fulfill"/>
</owl:ObjectProperty>
<owl:ObjectProperty rdf:ID="inverse_of_fulfill">
   <rdfs:domain rdf:resource="#Commitment"/>
   <rdfs:range rdf:resource="#Economic_Event"/>
   <owl:inverseOf rdf:resource="#fulfill"/>
</owl:ObjectProperty>
```

| **UML elements** | **OWL elements** |
|---|---|
| Association class | ObjectProperty, domain, range, inverseOf |

```
<owl:ObjectProperty rdf:ID="stockflow">
   <rdfs:domain rdf:resource="#Economic_Resource"/>
   <rdfs:range rdf:resource="#Economic_Event"/>
   <owl:inverseOf rdf:resource="#inverse_of_stockflow"/>
</owl:ObjectProperty>
<owl:ObjectProperty rdf:ID="inverse_of_stockflow">
   <rdfs:domain rdf:resource="#Economic_Event"/>
   <rdfs:range rdf:resource="#Economic_Resource"/>
   <owl:inverseOf rdf:resource="#stockflow"/>
</owl:ObjectProperty>
```

UML specializations can be very straightforward transformed into OWL by using the subclass construct for specialized classes (subClassOf) and the subproperty construct for properties (subPropertyOf) for specialized association classes (Table 3 and 4). In UML specialization structures can be declared as being disjoint, meaning each member of a superclass may be a member of no more than one subclass. In OWL this constraint is added by declaring that the subclass must be disjoint with the other subclasses.

**Table 3.** Transformations UML to OWL (total disjoint subclasses)

| UML elements | OWL elements |
|---|---|
| Total disjoint subclasses | disjointWith, unionOf, subClass |

```
<owl:Class rdf:ID="Economic_Event">
   <rdfs:subClassOf>
   <owl:Class>
   <owl:unionOf rdf:parseType="Collection">
   <owl:Class rdf:about="#Decrement_Economic_Event"/>
   <owl:Class rdf:about="#Increment_Economic_Event"/>
   </owl:unionOf>
   </owl:Class>
   </rdfs:subClassOf>
</owl:Class>
<owl:Class rdf:ID="Decrement_Economic_Event">
   <rdfs:subClassOf rdf:resource="#Economic_Event"/>
   <owl:disjointWith rdf:resource="#Increment_Economic_Event"/>
</owl:Class>
<owl:Class rdf:ID="Increment_Economic_Event">
   <rdfs:subClassOf rdf:resource="#Economic_Event"/>
   <owl:disjointWith rdf:resource="#Decrement_Economic_Event"/>
</owl:Class>
```

**Table 4.** Transformations UML to OWL (total disjoint subrelations)

| UML elements | OWL elements |
|---|---|
| Total disjoint 'subrelations' | subPropertyOf |

```
<owl:ObjectProperty rdf:ID="inflow">
   <rdfs:domain rdf:resource="#Economic_Resource"/>
   <rdfs:range rdf:resource="#Increment_Economic_Event"/>
   <owl:inverseOf rdf:resource="#inverse_of_inflow"/>
   <rdfs:subPropertyOf rdf:resource="#stockflow"/>
</owl:ObjectProperty>
<owl:ObjectProperty rdf:ID="inverse_of_inflow">
   <rdfs:domain rdf:resource="#Increment_Economic_Event"/>
   <rdfs:range rdf:resource="#Economic_Resource"/>
   <owl:inverseOf rdf:resource="#inflow"/>
   <rdfs:subPropertyOf rdf:resource="#inverse_of_stockflow"/>
</owl:ObjectProperty>
```

Notice that also the stockflow and participation relations are specialized in disjoint 'sub-relations'. The corresponding association classes in the UML class diagram are represented by means of the OWL ObjectProperty construct. The current specification of OWL does not allow declaring subproperties as being disjoint (this issue will probably be solved by the OWL 1.1 extensions). A solution could be reifying the stockflow and participate relations (i.e. declaring them as OWL classes and next declaring disjoint subclasses). A drawback of this approach is that additional OWL properties are needed to represent all associations between the reified associations and the original classes. These additional OWL properties will have no direct counterpart in the REA conceptual representation, so it might be hard to give them meaningful names. That is why we preferred not to reify them. The UML constraints on the specializations in Figure 2 are also total or complete, meaning that all members of a superclass must be members of at least one subclass. This can be formalized be defining a covering axiom on the superclass which states that the superclass is the union of the subclasses.

The formalization of the REA axioms is less straightforward and different approaches can be taken. Based on the ODM proposal we decided to formalize the multiplicities on the association ends by adding OWL minimum and maximum cardinality restrictions to the range of properties where necessary (see Table 5). This might not be the most appropriate solution because of semantic differences between cardinality restrictions in OWL and cardinality constraints in conceptual modeling languages [45]. An alternative for a maximum cardinality of one on the range of the property is to declare the OWL property as functional (see [7] for an example). A property is functional if for a given individual, there can be at most one individual that is related to the individual via the property.

**Table 5.** Transformations UML to OWL (multiplicities)

| UML elements | OWL elements |
|---|---|
| Multiplicities | maxCardinality, minCardinality |

```
<owl:Class rdf:ID="Increment_Economic_Event">
<rdfs:subClassOf>
   <owl:Restriction>
   <owl:onProperty rdf:resource="#inverse_of_inflow"/>
   <owl:minCardinality
       rdf:datatype="&xsd;int">1</owl:minCardinality>
   </owl:Restriction>
</rdfs:subClassOf>
<rdfs:subClassOf>
   <owl:Restriction>
   <owl:onProperty rdf:resource="#inverse_of_inflow"/>
   <owl:maxCardinality
       rdf:datatype="&xsd;int">1</owl:maxCardinality>
   </owl:Restriction>
</rdfs:subClassOf>
</owl:Class>
```

## 5 Conclusion and Future Research

This paper presents the development of two new representations of the Resource Event Agent (REA) business ontology. This development addresses shortcomings in the current specification that are likely to impede REA's use as a run-time ontology supporting semantic integration and interoperability of heterogeneous business applications and enterprise systems. The first representation is a UML class diagram that can be used by human users as an explicit, unified and uniformly represented specification of REA's conceptualization of business. The second representation is a formal, machine-readable specification obtained by applying UML to OWL mappings. The OWL formalization of REA, which would not be possible without first redesigning its conceptual representation, will facilitate its operationalization in ontology-driven systems as it makes REA executable.

A research contribution of the paper is the embedding of the REA redesign and formalization activities into a comprehensive business ontology reengineering methodology. Other activities suggested by this methodology, such as ontological evaluation and synthesis, can readily be performed based on the proposed conceptual representation. This new UML representation of REA may also serve as an improved basis to compare REA to other business ontologies and to evaluate the degree to which this business conceptualization is shared in the business community. The OWL representation of REA is of practical value for those wishing to explore the use of REA as a run-time ontology. It also allows for formal ontology evaluation, for instance verifying the consistency of the concept definitions using a Description Logic reasoner.

The limitation of this research is its focus on representation and formalization issues. Future research is needed to validate REA and to address ontological content rather than just form. An ontological analysis with respect to an upper-level ontology (e.g. SUMO, Dolce, GFO, BWW, SOWA, ...) is part of this validation. Future research may also investigate the use of other languages for business ontology modelling. In particular, enterprise modelling languages may offer the same advantages as UML (given the availability of UML profiles for these languages) whilst allowing semantically richer descriptions of business domain semantics (through domain-specific modelling constructs).

Currently, we are developing a 'proof of concept' REA-driven enterprise information integration application to demonstrate effectiveness (i.e. the use of the reengineered REA as an operational ontology) and feasibility (i.e. REA creating interoperability between heterogeneous business applications). Additionally we are also developing a revised graphical representation of the formalized REA-ontology specification by using the OWL UML profile which is part of the ODM specification. This graphical representation will be further discussed and evaluated in the REA research community.

## References


1. Gruber, T.R.: A translation approach to portable ontology specifications. Knowledge Acquisition 5 (1993) 199-220
2. Guarino, N., Giaretta, P.: Ontologies and Knowledge Bases: Towards a Terminological Clarification. In: Mars, N. (ed.): Towards Very Large Knowledge Bases: Knowledge Building and Knowledge Sharing. IOS press, Amsterdam (1995) 25-32
3. Fensel, D.: Ontologies: A Silver Bullet for Knowledge Management and Electronic Commerce. Springer-Verslag (2001)
4. Borst, W.N.: Construction of Engineering Ontologies. Centre for Telematica and Information Technology, Enschede, The Netherlands (1997)
5. Gómez-Pérez, A., Fernández-López, M., Corcho, O.: Ontological Engineering. Springer-Verslag (2004)
6. Guarino, N.: Formal Ontology and Information Systems. Proceedings of FOIS'98, Trento, Italy, IOS Press (1998) 3-15
7. Gailly, F., Poels, G.: Ontology-driven Business Modelling: Improving the Conceptual Representation of the REA-ontology. FEB Working paper series. Faculty of Economics and Business Administration, Ghent University (2007)

8. Assmann, U., Zchaler, S., Wagner, G.: Ontologies, Meta-Models, and the Model-Driven Paradigm. In: Calero, C., Ruiz, F., Piattini, M. (eds.): Ontologies for Software Engineering and Software Technology (2006)
9. Baida, Z., Gordijn, J., Saele, H., Akkermans, H., Morch, A.Z.: An ontological approach for eliciting and understanding needs in e-services. In: Pastor, O., Falcao e Cunha, J. (eds.): Advanced Information Systems Engineering (CAiSE 2005), LNCS 3520, Springer (2005) 400-414
10. Dietz, J.L.G.: System Ontology and its role in Software Development. In: Castro, J., Teniente, E. (eds.): Advanced Information Systems Engineering wokshops (CAiSE 2005), Porto, Portugal, 2, FEUP edicoes (2005) 273-284
11. Grunninger, M.: Enterprise Modelling. In: Bernus, P., Nemes, L., Schmidt, G. (eds.): Handbook on Enterprise Architecture. Springer (2003)
12. Guarino, N.: Understanding, building and using ontologies. International Journal of Human-Computer Studies 46 (1997) 293-310
13. Castano, S., Ferrara, A., Montanelli, S.: Ontology knowledge spaces for semantic collaboration in networked enterprises. Business Process Management Workshops 4103 (2006) 336-347
14. Vujasinovic, M., Marjanovic, Z.: Data level enterprise applications integration. Business Process Management Workshops, LNCS 3812, Springer (2006) 390-395
15. Izza, S., Vincent, L., Burlat, P.: A Framework for Semantic Enterprise Integration. In: Konstantas, D., Bourrières, J.-P., Léonard, M., Boudjlida, N. (eds.): Interoperability of Enterprise Software and Applications. Springer-Verslag (2006)
16. McIlraith, S.A., Son, T.C., Zeng, H.L.: Semantic Web services. IEEE Intelligent Systems & Their Applications 16 (2001) 46-53
17. W3C: Web Service Modeling Ontology (WSMO). (2006)
18. Geerts, G.L., McCarthy, W.E.: An Ontological Analysis of the Economic Primitives of the Extended-REA Enterprise Information Architecture. IJAIS 3 (2002) 1-16
19. Allen, G.N., March, S.T.: The effects of state-based and event-based data representation on user performance in query formulation tasks. MIS Quarterly 30 (2006) 269-290
20. Geerts, G., McCarthy, W.E.: The Ontological Foundation of REA Enterprise Information Systems. (2005)
21. Geerts, G., McCarthy, W.E.: Policy-Level Specification in REA Enterprise Information Systems. Journal of Information Systems Fall (2006)
22. UN/CEFACT: UN/CEFACT Modeling Methodology (UMM) User Guide. (2003)
23. ECIMF: E-Commerce Integration Meta-Framework. Final draft. ECIMF Project Group (2003)
24. O'Leary, D.: Different Firms, Different Ontologies, and No One Best Ontology. IEEE Intelligent Systems Sep/Oct (2000) 72-78
25. McCarthy, W.E.: The REA Accounting Model: A Generalized Framework for Accounting Systems in A Shared Data Environment. The Accounting Review july (1982) 554-578
26. Sowa, J.: Knowledge Representation: Logical, Philosophical, and Computational Foundations. Pacific Grove, Brooks/Cole (1999)
27. Hruby, P.: Model-driven design using business patterns. Springer, New York (2006)
28. Dunn, C.L., Cherrington, J.O., Hollander, A.S.: Enterprise Information Systems: A Pattern Based Approach. McGraw-Hill (2005)
29. Gailly, F., Poels, G.: Towards a Formal Representation of the Resource Event Agent Pattern International Conference on Enterprise Systems and Accounting (ICESAcc), Greece (2006)
30. Bialecki, A.: REA ontology. http://www.getopt.org/ecimf/contrib/onto/REA/ (2001)
31. Chou, C.-C.: Using ontological methodology in building the accounting knowledge model – REAP. 2006 AAA mid-year meeting - 2006 AI/ET Workshop (2006)

32. Jarrar, M.: Towards Methodological Principles for Ontology Engineering. STARLAB. Vrije Universiteit Brussel, Brussel (2005)
33. Fernández-López, M., Gómez-Pérez, A., Juristo, N.: METHONTOLOGY: From ontological art towards ontological engineering. Working Notes of the AAAI Spring Symposium on Ontological Engineering, Stanford, AAAI Press (1997)
34. Gómez-Pérez, A., Rojas, M.D.: Ontological Reengineering and Reuse. In: Fensel, D., Studer, R. (eds.): 11th European Workshop on Knowledge Acquisition, Modeling and Management, Dagstuhl Castle, germany, Springer-Verslag (1999) 139-156
35. Brockmans, S., Volz, R., Eberhart, A., Löffler, P.: Visual Modeling of OWL DL Ontologies Using UML. In: McIlraith, S.A., Plexousakis, D., van Harmelen, F. (eds.): The Semantic Web - ISWC 2004, Hiroshima, Japan, LNCS 3298, Springer (2004) 198-213
36. Hartmann, J., Spyns, P., Giboin, A., Maynard, D., Cuel, R., Suárez-Figueroa, M.C., Sure, Y.: Methods for ontology evaluation. Knowledge Web Deliverable D1.2.3, v. 1.3.1 (2005)
37. Gangemi, A., Catenacci, C., Ciaramita, M., Lehmann, J.: A theoretical framework for ontology evaluation and validation. Proceedings of SWAP 2005, the 2nd Italian Semantic Web Workshop, Trento, Italy, 166, CEUR Workshop Proceedings (2005)
38. Lassila, O., McGuiness, D.L.: The Role of Frame-Based Representations on the Semantic Web. Technical Report. Knowledge System Laboratory, Stanford University, Stanford, California (2001)
39. Kogut, P., Cranefield, S., Hart, L., Dutra, M., Baclawski, K., Kokar, M.K., Smith, J.: UML for ontology development. Knowledge Engineering Review 17 (2002) 61-64
40. Spaccapietra, S., Parent, C., Vangenot, C., Cullot, N.: On Using Conceptual Modeling for Ontologies. Proceedings of the Web Information Systems Workshops (WISE 2004 workshops), LNCS 3307. Springer-Verslag (2004) 22-23
41. Spyns, P.: Object Role Modelling for ontology engineering in the DOGMA framework. In: Robert Meersman et al. (ed.): On the Move to Meaningful Internet Systems 2005: OTM 2005 Workshops, Agia Napa, Cyprus, LNCS 3762, Springer (2005) 710-719
42. Mylopoulos, J.: Information modeling in the time of the revolution. Information Systems 23 (1998) 127-155
43. Davies, I., Green, P., Rosemann, M., Indulska, M., Gallo, S.: How do practioners use conceptual modeling in practice? Data & Knowledge Engineering 58 (2006) 358-380
44. OMG: Ontology Definition Metamodel: OMG Adopted Specification (ptc/06-10-11). Object Management Group (2006)
45. de Bruyn, j., Lara, R., Polleres, A., Fensel, D.: OWL DL vs. OWL Flight: Conceptual Modeling and Reasoning for the Semantic Web. World Wide Web Conference (WWW 2005), Chiba, Japan, ACM (2005)

# ProdLight: A Lightweight Ontology for Product Description Based on Datatype Properties

Martin Hepp

Digital Enterprise Research Institute (DERI), University of Innsbruck
mhepp@computer.org

**Abstract.** Web pages representing offerings of products and services are a major source of data for Semantic Web-based e-commerce. This data could be useful for numerous applications, e.g. (1) more precise product search engines and shopping bots, (2) aggregation or enrichment of multi-vendor catalogs using public product descriptions, or (3) the automated discovery of additional alternatives based on the combination of multiple items. While there are already some ontologies for products and services available, they are very large in size (20 – 70,000 classes), and thus not always suitable as ontology imports. In this paper, we take a different approach: We represent the semantics of offerings on the Web using a very lightweight ontology of datatype properties in combination with popular classifications like UNSPSC and eCl@ss. We then demonstrate how this representation can be mapped easily to comprehensive ontologies for products and services like eClassOWL[1]. Our approach provides a straightforward solution for annotating offerings on the Web while avoiding the overhead of importing fully-fledged products and services ontologies in every single annotation. We can show that our proposal has technical advantages and eliminates legal problems when reusing existing standards.

## 1 Introduction

Web pages representing offerings of products and services are a major source of data for Semantic Web-based e-commerce. This data covers technical and commercial aspects and could be useful for numerous future applications. Firstly, it could be used by novel product search engines and shopping bots that identify suitable alternatives for a given need and a given set of preferences. Secondly, the data could be used for assembling, augmenting, or maintaining multi-vendor catalogs for e-shops. For example, product descriptions and technical features from pages published by manufacturers of a product could be exploited for maintaining product descriptions in retail e-shops. Thirdly, it could support the discovery of additional, implicit purchasing options that are possible by the combination of multiple individual offers.

However, most of the product data on the Web is available only in a format suitable for rendering by browsers and interpretation by humans; and despite much research in using ontologies for supporting such scenarios, no comprehensive and practically useful framework exists so far for making Web offerings accessible at a semantic level.

[1] http://www.heppnetz.de/eclassowl/

W. Abramowicz (Ed.): BIS 2007, LNCS 4439, pp. 260–272, 2007.
© Springer-Verlag Berlin Heidelberg 2007

In B2B scenarios, quite comprehensive classification standards for products and services have evolved, namely UNSPSC [1] and eCl@ss [2]. Their reuse for e-commerce applications of the Semantic Web vision is, however, non-trivial, as has been shown in [3], [4], and [5].

With eClassOWL, there exists now a consistent OWL variant of the eCl@ss standard, containing ontology classes for more than 25,000 different goods categories plus more than 5,000 properties for representing product characteristics [6][2]. While this ontology is freely available and can be readily used for many applications, the idea of annotating existing Web resources that describe products and services still faces several problems.

a) Size
Due to their original design purpose of aggregating data on an enterprise-wide level, most available classifications are very large in size with usually more than 20,000 categories.

b) Intellectual Property Rights
It has recently been stressed that many standards are subject to copyright and other types of intellectual property rights [7]. Thus, creating and publishing an ontology by transforming an existing standard usually requires explicit agreements with the owners of the standard [8]. In the case of eCl@ss, such an agreement has been established. For UNSPSC, however, the issue has not yet been settled.

c) Unclear semantics of categories in standard classifications
Most informal classifications lack a clear notion of what it means to be an instance of a respective category. This is no problem when employing a classification schema in a well-defined context known to all users of the data, but it causes some problems when deriving ontologies from such classifications, since the latter requires a clear semantics. For example, UNSPSC codes are used both for classifying expenses and for describing actual products. Deriving an ontology from a classification thus includes important modeling decisions in this issue [4]. Otherwise, the meaning of an ontology class derived from a category must remain very broad, e.g. "anything that can, in any context, be subsumed under this type of good". As one direction, there has been work on adding a formal semantics to informal classifications by [9, 10].

From a practical perspective, we need to be able to describe and distinguish (1) actual products and services *instances*, (2) *models* of products and services, and (3) *entities just related to a certain type of goods* (e.g. invoices).

d) Type of relationship between a product category and a business entity
When representing the products and services domain, a lot of semantics is kept in the relation between a Web resource and a type of product or service. In fact, in most cases we do not just want to say that a specific Web resource is an instance of a specific product category. Much more, we want to represent a different type of relationship between a Web resource and a product definition, e.g. "This Web page contains an offer of product instances that meet the following specification" or "The company identified by this URI repairs products of the following kind". A valuable standard

[2] A similar ontology derived from UNSPSC called unspscOWL has been created but is waiting for copyright clearance.

providing common concepts for this problem is the UNSPSC Business Function Identifier (BFI) [11]. It is a simple two-digit value that reflects the kind of relationship between a business entity and a product category. Table 1 shows the currently supported values.

**Table 1.** UNSPSC Business Function Identifiers [11]

| UNSPSC BFI | Meaning |
|---|---|
| 10 | Rental or Lease |
| 11 | Maintenance or Repair |
| 12 | Manufacturer |
| 13 | Wholesale |
| 14 | Retail |
| 15 | Recycle |
| 16 | Installation |
| 17 | Engineered |
| 18 | Outsource |

In short, we assume that the creation and usage of true ontologies derived from products and services classifications is in general useful, for it provides a proper conceptual base for describing offerings, demands, and related items like invoices properly. However, it may not be realistic to annotate a majority of current Web shops by references to such ontologies in the near future, mainly for practical reasons. In particular, the scalability of current reasoners is still a bottleneck, which implies that drawing inferences about hundreds of thousands of offerings instances may not yet scale. Also, the amount of reasoning supported by such ontologies, e.g. eClassOWL, is rather limited. This is because popular product classifications have just four hierarchical levels with a limited amount of branching. Table 2 shows the median of branching between the levels of hierarchy in eCl@ss 5.1 and UNSPSC 7.0901. In here, the median tells us the maximum number of descendents for the lower half of the categories. We can see that from the third level to the fourth, half of all categories have not more than two descendents. This level of branching is particularly interesting, because the fourth level contains classes for actual *commodities*. UNSPSC is a bit more advanced in the branching at the lowest level with up to six descendents for the lower half of all categories. One may ask whether it always makes sense to import a very large ontology for all operations on product data if the amount of reasoning is that limited.

**Table 2.** Amount of branching in popular product classification standards [data taken from 12]

| | Median | |
|---|---|---|
| | **eCl@ss 5.1** | **UNSPSC 7,0901** |
| Top level → 2nd level | 18 | 5 |
| 2nd level → 3rd level | 6 | 4 |
| 3rd level → 4th level | 2 | 6 |

As a consequence, we are proposing a very lightweight mechanism that allows capturing the product semantics of Web offerings by referring to popular classification standards. This mechanism should avoid the burden of replicating the whole standard in the form of a huge ontology but still be upward compatible to such comprehensive ontologies. It should also be possible to work with respective annotations in a meaningful way even in the absence of scalable reasoning, e.g. on the level of plain RDF.

### 1.1 Related Work

The potential of using ontologies for e-commerce scenarios has been stressed e.g. by [13], [14], [15], and [16]. Gupta and Qasem argued that semantic e-commerce may reduce the price dispersion in markets [17]. In the information systems community, the benefits of product content standards have been discussed e.g. by [18].

A major focus of previous works on using ontologies in e-commerce scenarios was on the *integration* of product data, i.e. the challenge of harmonizing catalogue items or expenses referring to incompatible classifications. The complexity of this problem has been described e.g. in [19] and [20]. Respective tasks in B2B relationships have been detailed by [21]. The future role of ontologies on B2B marketplaces has been discussed in [22]. A prototype of catalogue data integration based on ontologies and ontological mappings was reported in [23]. An approach of identifying equivalences and other types of mappings between multiple product schemas is presented in [24].

Modeling aspects of product representation and respective ontology engineering tasks are covered by [25], [26], and [27].

The reuse of existing standards for e-commerce scenarios has been suggested e.g. in [28] and [29]. A domain-specific analysis of the problems faced when reusing product classifications for constructing products and services ontologies is given in [4]. A more generic analysis of the transformation of classifications into ontologies is presented in [10]. Recently, the W3C has proposed SKOS [30] as a very lightweight schema for representing classifications in the Semantic Web. While SKOS variants from standards can be created almost fully automatically, the resulting formalization cannot be directly used as a product and services ontology, like eClassOWL, while its size will still be very large.

Currently, there exist at least three ontologies for products and services ontologies, which were derived from eCl@ss or UNSPSC [31], [32], and [6]. While the last one is very current and also includes product attributes, the two former ones are somewhat dated snapshots of past UNSPSC versions. A comparison is given in [4].

### 1.2 Our Contribution

In this paper, we (1) describe how the semantics of offerings on the Web can be represented using a very lightweight ontology in combination with popular classifications like UNSPSC and eCl@ss. We then show how this representation can be mapped easily to comprehensive ontologies for products and services like eClassOWL. Our approach provides a straightforward solution for annotating offerings on the Web while avoiding the overhead of importing fully-fledged products and services ontologies in any single annotation.

## 2 The ProdLight Ontology

In the following, we describe a novel approach for using the class identifiers of common product and service classification schemas for the annotation of Web offerings. The main idea is to capture the class identifiers as literal values of simple datatype properties in the form "hasClassificationCode 1234" instead of making an offering an instance or a subclass of a fully-fledged products and services ontology like eClassOWL. This approach requires only a very small, hand-crafted ontology. At the same time, it is easy to create mappings to eClassOWL at any time.
Our approach requires the following steps:

1. We define four classes
   a. `ProductOrServiceOrRelatedInstance`
   b. `ProductOrService`
   c. `ProductOrServiceProxyInstance`
   d. `ProductOrServiceRelatedEntity`

   and make b, c, and d subclasses of class a.

The class `ProductOrServiceOrRelatedInstance` contains everything that is either an actual product or service, an entity that acts as a proxy for multiple actual product or services instances (which themselves are not exposed, i.e., unknown), or items that are related to product or services categories but which are no actual products or services (e.g. related invoices).

The class `ProductOrService` is a subclass of the former class and contains only actual product or services instances, i.e., individuals on which property rights can be established and transferred.

The class `ProductOrServiceProxyInstance` is a work-around for such products and services instances that are existentially quantified but not directly exposed. This is necessary, because in many situations, Web shops do not publish actual instances of a product or service, but just state that such exist. Since existential quantification, which would be the proper modeling for such situations, is computationally costly and makes a respective OWL ontology fall into OWL DL, we use instances of this class to represent existentially quantified product offerings.

The class `ProductOrServiceRelatedEntity` contains everything that may be characterized by a product or services category but is not a product or service instance (or "instance proxy") itself.
In OWL Abstract Syntax, the respective class definitions are as follows:

```
Class (ex:ProductOrServiceOrRelatedInstance partial)
Class (ex:ProductOrService partial
   ex:ProductOrServiceOrRelatedInstance)
Class (ex:ProductOrServiceProxyInstance partial
   ex:ProductOrServiceOrRelatedInstance)
Class (ex:ProductOrServiceRelatedEntity partial
   ex:ProductOrServiceOrRelatedInstance)
```

2. For the actual categories from the source classification standard themselves, we just define one single `owl:DatatypeProperty`, e.g. `relevantEclassCategory_v5.1` with the range of a string, and use *the primary keys of categories in eCl@ss version 5.1 as the literal value.*

In OWL Abstract Syntax, the respective property definition will be as follows:

```
DatatypeProperty(ex:relevantEclassCategory_v5.1
    range(xsd:string))
```

This can happen in parallel for multiple categorization schemes or versions from the same schema. For instance, we can define an additional datatype property for UNSPSC version 8:

```
DatatypeProperty(ex:relevantUnspscCategory_v8
    range(xsd:string))
```

If we allow the parallel usage of multiple classifications (e.g. UNSPSC and eCl@ss) or multiple values from the same standard, we need to be clear about the semantics of such assertions. The natural interpretation is that a resource annotated using multiple such statements reflects the *intersection* of both categories. For instance, if we use `relevantEclassCategory_v5.1` with the value "AAB29200202" ("Photo Cameras") plus `relevantUnspscCategory_v8` with the value "45121504" ("Digital Cameras"), then the actual meaning of this shall be that the resource is related to the intersection of both; in this case, that it is related to *digital* cameras only.

This aspect may be obvious for logicians; however, domain experts must be advised that a resource that contains two different offerings (e.g. cell phones and cell phone antennas) must be treated as two different resources (e.g. by adding URI fragment identifiers #phone and #antenna). Otherwise, a domain expert may add the UNSPSC codes for both categories to the same resource, which would be equivalent to the intersection of cell phones and cell phone antennas and thus, most likely, an empty set.

In addition to capturing the *type of good*, there is a need of expressing the *type of business function* that is supported for a particular product or service category, e.g., whether a Web page offers the sale, repair, or disposal of a particular type of goods. A first approach for capturing this aspect is by defining an additional datatype property that reflects the business function in the form of one or more UNSPSC Business Function Identifier values (see [11] for details).

In OWL Abstract Syntax, the respective property definition will be as follows:

```
DatatypeProperty(ex:supportedBusinessFunctionsUNSPSC-BFI
    range(xsd:int))
```

In here, we also need to agree upon the semantics of multiple property assertions for one resource. We suggest defining the semantics of this property such that multiple statements imply that the *union* of the respective business functions is supported. In our opinion, this is intuitive; it also means that we can express support for multiple business functions ("We sell and lease out boats") without duplicating much of the resource descriptions.

Instead of using `xsd:int` for the UNSPSC Business Function Identifier, we could also create instances in the ontology for any of the types as specified in Table 1. However, since we are aiming at the most lightweight solution, we decided against that. We are in parallel working on a fully-fledged ontology for business functions, into which the proposed lightweight approach can easily be grounded later. This can be achieved using similar patterns as for mapping categories to eClassOWL classes as shown in section 4.

# 3 Example

In the following, we give a brief example of our approach in plain OWL DLP. Even though we are using `owl:Class` for upward compatibility reasons, no reasoner support is necessary at this stage; i.e., the approach could also be used in a meaningful manner on the level of plain RDF.
We assume there are four entities:

1. `CameraInvoice`: An invoice related to a digital camera purchase.
2. `CameraWebsite:` A Web site on which cameras are *sold to end users (retail).*
3. `myCamera:` An actual camera.
4. `CameraModelSony123:` A Sony camera make and model, of which multiple actual instances exist, which will all share several technical features, e.g. weight, pixel resolution, etc. – however, the actual instances are only existentially quantified.

In the example, "AAB29200202" is the identifier of the category "Photo Camera" in eCl@ss 5.1. In UNSPSC v8, the closest category "Digital Cameras" has the identifier "45121504"[1]. The UNSPSC Business Function Identifier "14" means "Retail" [11].

Using the approach described in section 4.3.1, this leads to the following statements in OWL Abstract Syntax:

```
Individual(ex:CameraInvoice
   type(ex:ProductOrServiceRelatedEntity)
   value(ex:relevantEclassCategory_v5.1
   "AAB29200202"^^<http://www.w3.org/2001/XMLSchema#string>))

Individual(ex:CameraModelSony123
   type(ex:ProductOrServiceProxyInstance)
   value(ex:relevantEclassCategory_v5.1
   "AAB29200202"^^<http://www.w3.org/2001/XMLSchema#string>))

Individual(ex:CameraWebSite
   type(ex:ProductOrServiceOrRelatedInstance)
   value(ex:relevantEclassCategory_v5.1
   "AAB29200202"^^<http://www.w3.org/2001/XMLSchema#string>)
   value(ex:supportedBusinessFunctionsUNSPSC-BFI
   "14"^^<http://www.w3.org/2001/XMLSchema#int>))

Individual(ex:myCamera
   type(ex:ProductOrService)
   value(ex:relevantEclassCategory_v5.1
   "AAB29200202"^^<http://www.w3.org/2001/XMLSchema#string>))
```

We can also specify the product semantics by using eCl@ss **and** UNSPSC codes. However, we must keep in mind that this reflects the intersection of both product categories:

```
Individual(ex:CameraInvoice
   type(ex:ProductOrServiceRelatedEntity)
   value(ex:relevantEclassCategory_v5.1
   "AAB29200202"^^<http://www.w3.org/2001/XMLSchema#string>)
   value(ex:relevantUnspscCategory_v8
   "45121504"^^<http://www.w3.org/2001/XMLSchema#string>))
```

All those annotations can be queried in a meaningful manner even without reasoning support, e.g. by SPARQL queries against RDF data. At the same time, they can be integrated into eClassOWL or similar ontologies, as shown in the next section.

# 4 Upward Compatibility: Mapping to eClassOWL

In this section, we show how annotations based on the lightweight approach proposed in this paper can be mapped to eClassOWL or similar products and services ontologies. This allows taking into account the subsumption hierarchy of eClassOWL and other features when interpreting such data. It is important to note that this mapping does not need to be created individually for each annotation. Rather, one may create one big mapping document for each relevant release of eCl@ss. This document can then be imported as needed. Since the mapping within the same classification standard (e.g. between eCl@ss literal values and eClassOWL classes) follows a simple schema, such document can be created fully automatically. The next release of eClassOWL will include such a mapping document as a separate OWL file. The same holds for the mapping between UNSPSC categories and UNSPSC ontology variants. More difficult is a mapping that bridges two or more such standards, e.g. from UNSPSC to eCl@ss [see e.g. 19]. This challenge is a research topic in its own right and not considered further in this paper.

It is important to know that eClassOWL has two types of classes for each category in the original classification: Firstly, one *generic* class ("gen") that contains only *actual products* of that kind (i.e. an instance of "TV set" is a real product of that type), and secondly, a *taxonomic* class that contains "anything than can be subsumed under this label in any relevant context". Those taxonomic classes also cover *related* products. These characteristics of eClassOWL and the underlying rationales are explained in more detail in [4] and [5]. For the following example, it is sufficient to know the general semantics of the generic and the taxonomic classes.

The class identifiers in eClassOWL are derived from the category identifiers by adding a preceding "C_" for "class" and adding a trailing "-gen" (C_xxxxxxxxxxx-gen) for generic classes and "-tax" (C_xxxxxxxxxxx-tax) for the taxonomic classes. Thus, the two classes for the eCl@ss category "Photo Cameras" with the original identifier "AAB29200202" have the eClassOWL identifiers "C_AAB29200202-gen" (for actual cameras) and "C_AAB29200202-tax" (for anything that is camera-related in any relevant context).

The mapping itself can be implemented in OWL by value restrictions on the respective datatype properties.

Accordingly, the mapping definition for the *generic* classes will look as follows in OWL Abstract Syntax:

```
Class (eclass51:C_AAB29200202-gen complete
   intersectionOf(
      ex:ProductOrService
      restriction(ex:relevantEclassCategory_v5.1
       value ("AAB29200202"@EN)))
```

The mapping definition for the *taxonomic* classes will differ in that it is based on the intersection of `ProductOrServiceOrRelatedInstance`, i.e. the more general concept from our lightweight ontology:

```
Class (eclass51:C_AAB29200202-tax complete
   intersectionOf(
      ex:ProductOrServiceOrRelatedInstance
      restriction(ex:relevantEclassCategory_v5.1
       value ( "AAB29200202"@EN)))
```

Since value restrictions on datatype properties are outside of OWL DLP and OWL Lite, importing these mapping statements makes the resulting ontology reside in OWL DL. This is a bit of a disadvantage, since eClassOWL itself does not go beyond OWL DLP.

## 5 Evaluation

In the following, we compare the direct use of fully-fledged products and services ontologies with our lightweight approach and take into account practical and legal aspects. Our analysis considers the following dimensions:

1. **Size of the ontology:** How big is the ontology that needs to be imported when processing Web resources and their annotations?
2. **Upward Compatibility to eClassOWL:** Can the representation be mapped easily to fully-fledged products and services ontologies, namely eClassOWL?
3. **Ability to handle new types of products and services:** If additional products and services categories are needed, e.g. due to product innovation, how quickly and easily can they be agreed upon and used?
4. **Intellectual Property Issues:** Does the implementation of the approach require a legal agreement with the owners of existing e-commerce standards?
5. **Expressivity:** How much explicit and implicit knowledge about the products and services offerings can be gained from a respective annotation?

Table 3 gives a comparison of the two approaches.. One can see that our proposal of using category identifiers as datatype properties by means of the small ontology presented in section 2 has several appealing characteristics:

1. It does not require replicating the whole standard in the form of an ontology specification. This also eliminates the overhead for importing a large ontology when interpreting respective annotations.
2. It is easily upwards compatible with eClassOWL. In fact, the next release of eClassOWL will include respective mapping files.
3. It does not depend on an agreement with the standards owners prior to being allowed to publish the resulting ontology.
4. It can be used in a meaningful way even without the availability of a scalable reasoner, i.e., on the level of plain RDF data.

**Table 3.** Comparison of approaches for the representation of products and services

| Approach / Criterion | Ontologies derived from classification standards | Lightweight product description based on datatype properties |
|---|---|---|
| **Size of the ontology** | **Huge** (tens of megabytes) | **Very small** |
| **Upward compatibility to eClassOWL** | **Simple** (difficult may only be determining the correct relationship between product categories from *multiple* standards) | **Simple** (by value restrictions on the datatype properties) |
| **Ability to handle new types of products and services** | **Limited** (if new categories are needed, one must submit change requests to the respective standardization body; only after the standard has been updated, a new version of the ontology can be derived) | **Limited** (if new categories are needed, one must submit change requests to the respective standardization body; however, once the standard is updated, the new category identifiers are immediately usable – no delay caused by an ontology update).<br>**Note:** Major release changes in the standards require new datatype properties. |
| **Intellectual Property Issues** | **Problematic** (explicit agreements with IPR owners required) | **Simple** (no agreements with IPR owners required as long as the usage of the codes themselves is granted to the general public, which is mostly the case) |
| **Expressivity** | **Medium to High** (depending on the amount of effort invested in lifting the standard to an ontological level) | **Limited**, but grounding in rich ontologies is straightforward |

One might argue that the pure reference to flat concepts in external standards provides little help for reasoning about products in the Semantic Web. However, already a lot of gain in precision can be achieved by being able to search Web offerings by UNSPSC or eCl@ss category codes alone. Also, our approach eases the development of intelligent applications significantly, because it decouples the two spheres of (1) product data annotation and (2) the development of richly axiomatized vertical ontologies for small product domains. In other words, we can annotate offerings already now with respective category codes, while the development of richer ontology variants of eCl@ss or UNSPSC may be an ongoing research challenge. Also, we assume useful lightweight applications can be developed right away.

A possible extension of our approach is to combine it with just the large property library provided by eClassOWL. This part of eClassOWL defines more than 5,500 datatype and object properties for typical product characteristics, like "weight", "diameter", "screen size", etc. However, the property library of eClassOWL can currently not be separated from the class definitions for legal reasons.

One could also consider importing the vast amount of standardized properties from the DIN Properties Dictionary [33] in order to complement the lightweight approach with more specific product properties.

## 6 Conclusion

We have presented a straightforward solution for annotating offerings on the Web which avoids the overhead of importing fully-fledged products and services ontologies in every single annotation of a Web resource. The latter is unfeasible, since the respective ontologies are very large (20 – 70,000 classes), which makes them too large for frequent retrieval on-the-fly. Our approach does not require explicit legal agreements with the owners of popular classification schemas, while it can be easily mapped to fully-fledged ontology variants. The full ontology plus documentation will be released shortly at http://www.heppnetz.de/prodlight/. Also, the next release of eClassOWL will include mappings, as described in section 4, as an optional module, to be imported on demand. This will support interoperability between the ProdLight approach, meant for annotations embedded in Web resources and simple search, and heavyweight ontology applications that import eClassOWL and other large ontologies.

**Acknowledgements.** Parts of the work presented in this paper have been supported by the European Commission under the projects DIP (FP6-507483), SUPER (FP6-026850), and MUSING (FP6-027097), and by the Austrian BMVIT/FFG under the FIT-IT project myOntology (grant no. 812515/9284).

## References

1. United Nations Development Programme: United Nations Standard Products and Services Code (UNSPSC). Available at http://www.unspsc.org/ (retrieved May 22, 2006).
2. eClass e.V.: eCl@ss: Standardized Material and Service Classification. Available at http://www.eclass-online.com/ (retrieved December 20, 2006).
3. Hepp, M.: The True Complexity of Product Representation in the Semantic Web. Proceedings of the 14th European Conference on Information System (ECIS 2006), Gothenburg, Sweden (2006): 1-12.
4. Hepp, M.: Products and Services Ontologies: A Methodology for Deriving OWL Ontologies from Industrial Categorization Standards. Int'l Journal on Semantic Web & Information Systems (IJSWIS) 2 (1): 72-99, 2006.
5. Hepp, M.: Representing the Hierarchy of Industrial Taxonomies in OWL: The gen/tax Approach. Proceedings of the ISWC Workshop Semantic Web Case Studies and Best Practices for eBusiness (SWCASE05), Galway, Irland (2005).

6. Hepp, M.: eCl@ssOWL. The Products and Services Ontology. Available at http://www.heppnetz.de/eclassowl/ (retrieved December 20, 2006).
7. Samuelson, P.: Copyrighting Standards. Communications of the ACM 49 (6): 27-31, 2006.
8. Hepp, M.: Possible Ontologies: How Reality Constrains the Development of Relevant Ontologies. IEEE Internet Computing 11 (7): 96-102, 2007.
9. Giunchiglia, F., Marchese, M., Zaihrayeu, I.: Towards a Theory of Formal Classification. Proceedings of the AAAI-05 Workshop on Contexts and Ontologies: Theory, Practice and Applications (C&O-2005), Pittsburgh, Pennsylvania, USA (2005): 1-8.
10. Giunchiglia, F., Marchese, M., Zaihrayeu, I.: Encoding Classifications into Lightweight Ontologies. Proceedings of the 3rd European Semantic Web Conference (ESWC 2006). Springer, Budva, Montenegro (2006).
11. United Nations Development Programme: Business Function Identifiers (BFI). Available at http://www.un-spsc.org/AdminFolder/documents/BFI.doc (retrieved December 20, 2006).
12. Hepp, M., Leukel, J., Schmitz, V.: A Quantitative Analysis of Product Categorization Standards: Content, Coverage, and Maintenance of eCl@ss, UNSPSC, eOTD, and the RosettaNet Technical Dictionary. Knowledge and Information Systems 2006 (forthcoming).
13. Smith, H., Poulter, K.: Share the Ontology in XML-based Trading Architectures. Communications of the ACM 42 (3): 110-111, 1999.
14. Fensel, D.: Ontologies: A Silver Bullet for Knowledge Management and Electronic Commerce. Springer, Berlin etc. (2004).
15. Fensel, D., McGuinness, D.L., Schulten, E., Ng, W.K., Lim, E.-P., Yan, G.: Ontologies and Electronic Commerce. IEEE Intelligent Systems 16 (1): 8-14, 2001.
16. Zhao, Y., Sandahl, K.: Potential Advantages of Semantic Web for Internet Commerce. Proceedings of the International Conference on Enterprise Information Systems (ICEIS), Vol. 4, Angers, France (2003): 151-158.
17. Gupta, T., Qasem, A.: Reduction of price dispersion through Semantic E-commerce: A Position Paper. Proceedings of the Semantic Web Workshop 2002, Hawai, USA (2002): 1-2.
18. Fairchild, A.M., de Vuyst, B.: Coding Standards Benefiting Product and Service Information in E-Commerce. Proceedings of the 35th Annual Hawaii International Conference on System Sciences (HICSS-35) (2002): 258b.
19. Schulten, E., Akkermans, H., Botquin, G., Dörr, M., Guarino, N., Lopes, N., Sadeh, N.: The E-Commerce Product Classification Challenge. IEEE Intelligent Systems 16 (4): 86-89, 2001.
20. Fensel, D., Ding, Y., Omelayenko, B., Schulten, E., Botquin, G., Brown, M., Flett, A.: Product Data Integration in B2B E-Commerce. IEEE Intelligent Systems 16 (4): 54-59, 2001.
21. Omelayenko, B.: Ontology Integration Tasks in Business-to-Business E-commerce. In: Monostori, L., Vancza, J., Ali, M. (eds.): Proceedings of the Fourteenth International Conference on Industrial & Engineering Applications of Artificial Intelligence & Expert Systems. Springer, Budapest, Hungary (2001): 119-124.
22. Omelayenko, B.: Integration of Product Ontologies for B2B Marketplaces: A Preview. ACM SIGecom Exchanges archive 2 (1): 19-25, 2000.
23. Corcho, O., Gómez-Pérez, A.: Solving Integration Problems of E-commerce Standards and Initiatives through Ontological Mappings. Proceedings of the Workshop on E-Business and Intelligent Web at the Seventeenth International Joint Conference on Artificial Intelligence (IJCAI-2001), Seattle, USA (2001): 1-10.

24. Yan, G., Ng, W.K., Lim, E.-P.: Product Schema Integration for Electronic Commerce - A Synonym Comparison Approach. IEEE Transactions on Knowledge and Data Engineering 4 (3): 583-598, 2002.
25. Obrst, L., Wray, R.E., Liu, H.: Ontological Engineering for B2B E-Commerce. Proceedings of the International Conference on Formal Ontology in Information Systems (FOIS'01). ACM Press, Ogunquit, Maine, USA (2001): 117-126.
26. Omelayenko, B.: Preliminary Ontology Modelling for B2B Content Integration. In: Tjoa, A.M., Wagner, R.R. (eds.): Proceedings of the 12th International Workshop on Database and Expert Systems Applications. IEEE Computer Society, München (2001): 7-13.
27. Lee, H., Shim, J., Lee, S., Lee, S.-g.: Modeling Considerations for Product Ontology. In: Roddick, J.F., Benjamins, R.V., Cherfi, S.S.-S., Chiang, R., Claramunt, C., Elmasri, R., Grandi, F., Han, H., Hepp, M., Lytras, M., Misic, V.B., Poels, G., Song, I.-Y., Trujillo, J., Vangenot, C. (eds.): Proceedings of the First International Workshop on Ontologizing Industrial Standards, Vol. LNCS 4231. Springer, Tucson, Arizona, USA (2006): 291-300.
28. Zhao, Y., Lövdahl, J.: A Reuse-Based Method of Developing the Ontology for E-Procurement. Proceedings of the Nordic Conference on Web Services (NCWS), Växjö, Sweden (2003): 101-112.
29. Zhao, Y.: Develop the Ontology for Internet Commerce by Reusing Existing Standards. Proceedings of the International Workshop on Semantic Web Foundations and Application Technologies (SWFAT), Nara, Japan (2003): 51-57.
30. W3C: Simple Knowledge Organisation System (SKOS). Available at http://www.w3.org/2004/02/skos/ (retrieved December 20, 2006).
31. Klein, M.: DAML+OIL and RDF Schema representation of UNSPSC. Available at http://www.cs.vu.nl/~mcaklein/unspsc/ (retrieved December 20, 2006).
32. McGuinness, D.L.: UNSPSC Ontology in DAML+OIL. Available at http://www.ksl.stanford.edu/projects/DAML/UNSPSC.daml (retrieved December 20, 2006).
33. Deutsches Institut für Normung e. V.: DINsml.net - DIN-Merkmallexikon (DIN Properties Dictionary). Available at http://www.dinsml.net/ (retrieved December 20, 2006).

# A Context-Based Enterprise Ontology

Mauri Leppänen

Department of Computer Science and Information Systems
P.O. Box 35 (Agora), FI-40014 University of Jyväskylä, Finland
mauri@cs.jyu.fi

**Abstract.** The main purpose of an enterprise ontology is to promote the common understanding between people across enterprises, as well as to serve as a communication medium between people and applications, and between different applications. This paper outlines a top-level ontology, called the context-based enterprise ontology, which aims to advance the understanding of the nature, purposes and meanings of things in enterprises with providing basic concepts for conceiving, structuring and representing things within contexts and/or as contexts. The ontology is based on the contextual approach according to which a context involves seven domains: purpose, actor, action, object, facility, location, and time. The concepts in the ontology are defined in English and presented in meta models in a UML-based ontology engineering language.

**Keywords:** enterprise ontology, contextual approach, ontology engineering.

## 1 Introduction

A great number of applications are run in enterprises to provide information for, and to enable communication between, various stakeholders, inside and outside the organization. An increasingly large portion of enterprise knowledge is hold, processed and distributed by applications. Enterprise knowledge is "local knowledge" by its nature, in that its meaning and representation is agreed in relatively small, local contexts. A prerequisite for the successful use of the applications is, however, that the shared understanding about that knowledge is reached and maintained across the enterprise(s). Especially in modern inter- and intra-organizational applications the need to support the establishment of shared knowledge is crucial [5]. This implies that besides technical interoperability, the enterprises are facing with the challenge of achieving semantic and pragmatic interoperability among the applications.

For human beings to understand what individual things in reality mean necessitates that they know for what purposes the things are intended, by whom, when, and where, how the things are related to other things and the environment, how the things have emerged and evolved, etc. Shortly, people need to know about the contexts to which the things belong. Considering this, it is understandable how important the role of the notion of context is in many disciplines, such as in formal logic, knowledge representation and reasoning, machine learning, pragmatics, computational linguistics, sociolinguistics, organizational theory, sociology, and cognitive psychology. In the most of these fields, the notion is used, in particular, to specify, interpret, and infer from meanings of the things through the knowledge about the contexts they appear.

W. Abramowicz (Ed.): BIS 2007, LNCS 4439, pp. 273–286, 2007.
© Springer-Verlag Berlin Heidelberg 2007

In recent years a large number of enterprise and business ontologies and frameworks (e.g. TOVE [14, 15], EO [48], EKD [34], REA [36, 16], CEO [4], BMO [39]) have been proposed. Some of them are generic, whereas the others are aimed at specific business fields (e.g. UNSPC, NAICS, and OntoWeb for e-commerce). In addition, there are several languages (e.g. IDEF5, PIF, IEM, EEML, GRAI/Actigrams) intended for enterprise modeling. We focus here on enterprise ontologies. The main purpose of the enterprise ontologies is to promote the common understanding between people across different enterprises. They also serve as communication medium between people and applications, and between different applications, and refer, on a general level, to activities, products, resources and organizational structures in enterprises. Enterprise ontologies differ from business models (e.g. [17], [39], [35]), which contain concepts to express the business logic of a specific enterprise, and from business process models (e.g. [1]) which are more about how a business case is implemented in processes. Taking the significance of the role the sharing of meanings plays in communication in enterprises, and experience obtained from the use of context in capturing meanings in other disciplines, it is surprising how ignored a contextual view is in current enterprise ontologies. We propose that the semantic and pragmatic interoperability of applications in enterprises should be advanced by a more explicit use of context and other contextual concepts in the enterprise ontologies.

Our aim in this study is to present a context-based enterprise ontology. It is a top-level ontology [20], which provides a unified view of the enterprise as an aggregate of contexts. This ontology can be specialized into task ontologies or domain ontologies to meet particular needs of the enterprises, and still maintaining connections of the specialized things to their contexts. The concepts in the ontology are defined in English and presented in meta models in a UML-based ontology representation language. The UML language [6] has been applied because it has a large and rapidly expanding user community, it is supported by widely adopted engineering tools, and there are positive experience from the use of UML in presenting ontologies (e.g. [10], [49]). We apply a subset of the concepts of the class diagram.

The article is structured as follows. In Section 2 we will define the notions of context and contextual approach, and describe the overall structure of the context-based enterprise ontology. In Section 3 we will define the contextual concepts of the ontology and present them in meta models. In Section 4 we will end with some discussions and conclusions.

## 2 Context and Contextual Approach

Based on a large literature review about the notion of context in several disciplines, we conclude that a *context* is a whole composed of things that are connected to one another with contextual relationships. A thing gets its meaning through the relationships it has with the other things in that context. To define a proper set of contextual concepts and relationships we draw upon relevant theories about meanings. Based on the three topmost layers in the semiotic ladder [45], we identify semantics (e.g. case grammar [13]), pragmatics [31], and the activity theory [11] to be such theories. They concern sentence context, conversation context and action context, correspondingly.

Anchored on this groundwork, we define seven domains, which serve concepts for specifying and interpreting contextual phenomena. These contextual domains are: purpose, actor, action, object, facility, location, and time (Figure 1). Structuring the concepts within and between these domains is guided by the following scheme, called the seven S's scheme: *For **Some*** purpose, ***Somebody*** does ***Something*** for ***Someone,*** with ***Some*** means, ***Sometimes*** and ***Somewhere***.

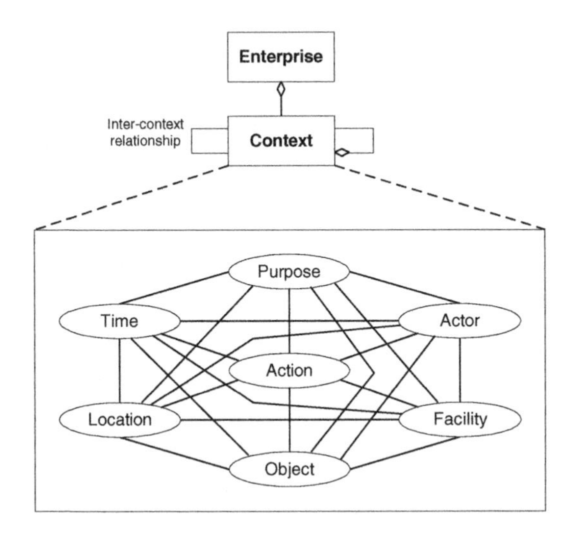


**Fig. 1.** An overall structure of the context-based enterprise ontology

We define the *contextual approach* to be the approach according to which individual things are seen to play certain contextual roles in a context, and/or to be contexts themselves. The contextual approach has been earlier applied to information systems development [30], method integration [29] and method engineering [28]. Here, we apply it to enterprises. We define an *enterprise* to be an aggregate of contexts that are composed of people, information and technologies, performing functions in the defined organizational structure, for agreed purposes, and responding to events, both internal and external, and needs of stakeholders. The contexts can be decomposed into more elementary contexts, and they are related to one another with inter-context relationships.

An ontology is an explicit specification of a conceptualization of some part of reality that is of interest [18]. The *context-based enterprise ontology* is an ontology which aims to promote the understanding of the nature, purposes, and meanings of the things in the enterprise with providing concepts and constructs for conceiving, structuring, and representing things within contexts, and/or as contexts. The ontology is intended to assist the acquisition, representation, and manipulation of enterprise knowledge.

The concepts and constructs in the enterprise ontology have been defined in a deductive and an inductive manner. Following an iterative procedure derived from

[48] and [12], we first determined the purpose, domain and scope of the ontology. Second, we searched for disciplines and theories that address social and organizational contexts and derived the basic categorization of concepts into seven contextual domains from them. Third, we defined the basic concepts and constructs for each contextual domain. Fourth, we analyzed current enterprise ontologies and modeling languages to find out whether they include parts that could be reused and integrated, as such or generalized, in our ontology. Results from this gradually evolving conceptualization were presented in a graphical form. The last step of the ontology engineering procedure was evaluation that was carried out in several stages, based on a set of quality criteria (e.g. [8], [19], [46]).

In the next section, we will first define five contextual domains (i.e. purpose, actor, action, object, facility) and the most essential concepts within them. Then we will shortly present relationships between the domains. Due to the limitation of space, the location and time domains are excluded.

# 3 Contextual Domains

## 3.1 Purpose Domain

The *purpose domain* embraces all those concepts and constructs that refer to goals, motives, or intentions of someone or something (Figure 2). The concepts are also used to express reasons for which something exists or is done, made, used etc. We use *purpose* as the general term in this domain.

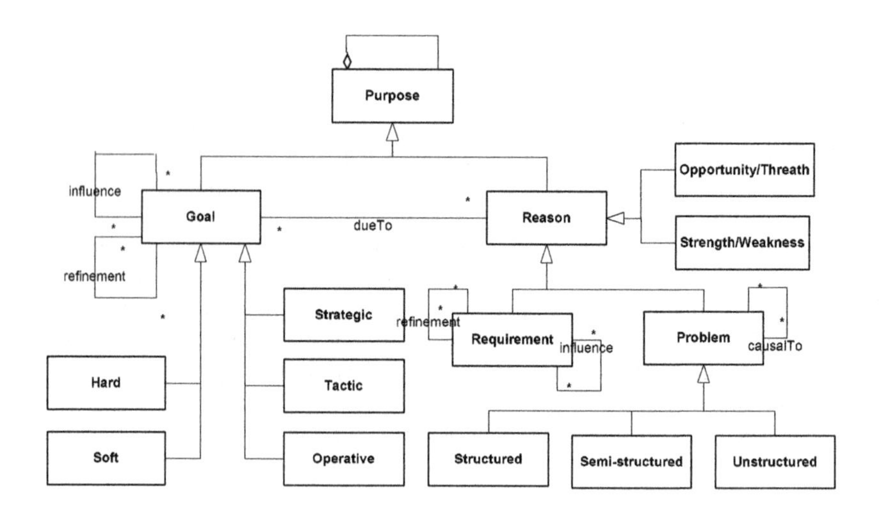


**Fig. 2.** Purpose domain

A *goal* means a desired state of affairs ([34], [26]). It can be related to other things (e.g. actors and actions) of the context. A *reason* is a basis or cause for some action, fact, event etc. [50]. It can be a requirement, a problem, a strength/weakness, or an opportunity/a threat. Between a goal and a reason there is the *dueTo relationship,* meaning that a reason gives an explanation, a justification or a basis for setting a goal. The goals can be specialized based on their lifespan. *Strategic goals* are kinds of missions, answering questions such as "What is the direction of an enterprise in the future". *Tactic goals* show how to attain strategic goals. *Operative goals* are generally determined as concrete requirements that are to be fulfilled by a specified point of time. The goals can also be categorized based on whether it is possible to define clear-cut criteria for the assessment of the fulfillment of goals. *Hard goals* have pre-specified criteria, and *soft goals* have not [32].

*Requirements* mean some things that are necessary and needed. They are statements about the future [38]. Actually, the goals and the requirements are two sides of a coin: some of the stated requirements can be accepted to be goals to which actors want to commit. Instead of directly referring to a desirable state, a purpose can also be expressed indirectly with a reference to problems that should be solved. A *problem* is the distance or a mismatch between the prevailing state and the state reflected by the goal [22]. To reach the goal, the distance should be eliminated, or at least reduced. Associating the problems to the goals expresses reasons, or rationale, for decisions or actions towards the goals [41]. The problems are commonly divided into structured, semi-structured and unstructured problems [43]. *Structured problems* are those that are routine, and can be solved using standard solution techniques. *Semi-structured* and *unstructured problems* do not usually fit a standard mold, and are normally solved by examining different scenarios, and asking "what if" type questions.

Other expressions for the reasons, of not so concrete kind, are strengths, weaknesses, opportunities and threats related to something for which goals are set (cf. SWOT-analysis, e.g. [23]). *Strength* means something in which one is good, something that is regarded as an advantage and thus increases the possibility to gain something better. *Weakness* means something in which one is poor, something that could or should be improved or avoided. *Opportunity* is a situation or condition favorable for attainment of a goal [50]. *Threat* is a situation or condition that is a risk for attainment of a goal.

A general goal is refined into more concrete ones. The *refinement relationships* between the goals establish a goal hierarchy, in which a goal can be reached when the goals below it (so-called sub-goals) in the hierarchy are fulfilled (cf. [25]). The *influence relationship* indicates that the achievement of a goal has some influence, positive or negative, on the achievement of another goal (cf. [34], [25]). The relationships between the requirements are similar to those between the goals. Consequently, a requirement can influence on another requirement, and a requirement can be a refinement of another requirement. The relationships between the problems manifest causality. The *causalTo relationship* between two problems means that the appearance of one problem is at least a partial reason for the occurrence of the other problem.

## 3.2 Actor Domain

The *actor domain* consists of all those concepts and constructs that refer to human and active parts in a context (Figure 3). Actors perform actions such as buy, pay, send, receive, communicate, plan etc. in the contexts. They are responsible for and/or responsive to triggering and causing changes in the states of objects in the same context, or in other contexts. Human actors are distinguished from non-human actors, often called agents, which are here regarded as tools (see Section 3.5).

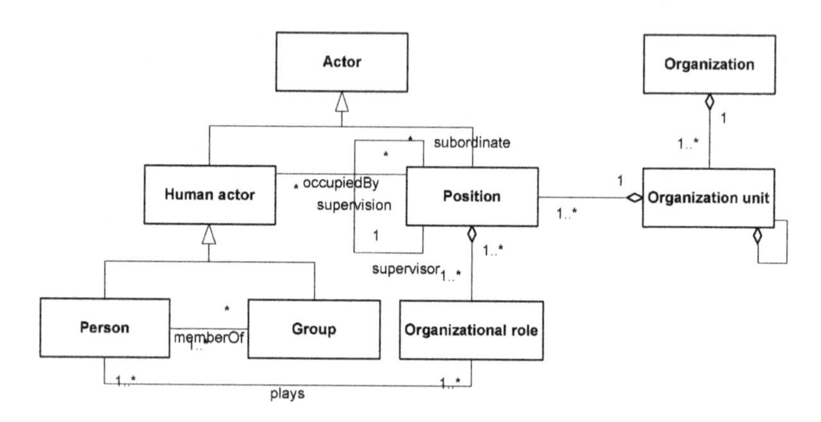


**Fig. 3.** Actor domain

An *actor* is a human actor or an administrative actor. A *human actor* is an individual person or a group of persons. A *person* is a human being, characterized by his/her desires, intentions, social relationships, and behavior patterns conditioned by his/her culture (cf. [7], [40]). A person may be a member of none or several *groups.* An administrative actor is a position or a set of positions. A *position* is a post of employment *occupied by* zero or many human actors. For each position, specific qualifications in terms of skills and demands on education and experience are specified.

An *organizational role*, shortly a role, is a collection of responsibilities, stipulated in an operational or structural manner. In the former case, a role is composed of tasks that a human actor occupying the position with that role has to perform. In the latter case, a role is charged with responsibilities for some objects. A role can be played by several persons, in or without the position(s) they hold. In business environment, common roles are manager, customer, vendor, partner, shareholder and competitor. The *supervision relationship* involves two positions in which one is a supervisor to another that is called a subordinate. A supervisor position has responsibility and authority to make decisions upon the positions subordinate to it, and those occupying the subordinate positions have responsibility for reporting on one's work and results to those occupying the supervisor position.

An *organization* is an administrative arrangement or structure established for some purposes, manifesting the division of labor into actions and the coordination of actions to accomplish the work. It can be permanent and formal, established with immutable regulations, procedures and rules. Or it may be temporally set up, like a project organization, for specific and often short-range purposes. An *organizational unit* is composed of positions with the established supervision relationships. An organization consists of organizational units.

### 3.3 Action Domain

The *action domain* comprises all those concepts and constructs that refer to deeds or events in a context. *Actions* can be autonomous or cooperative. They can mean highly abstract work such as strategic planning, or at the other extreme, physical execution of a step-by-step procedure with detailed routines. There are a large number of action structures, which an action is a part of. We distinguish between four orthogonal action structures: the decomposition structure, the control structure, the temporal structure, and the management – execution (Mgmt-Exec) structure (Figure 4).

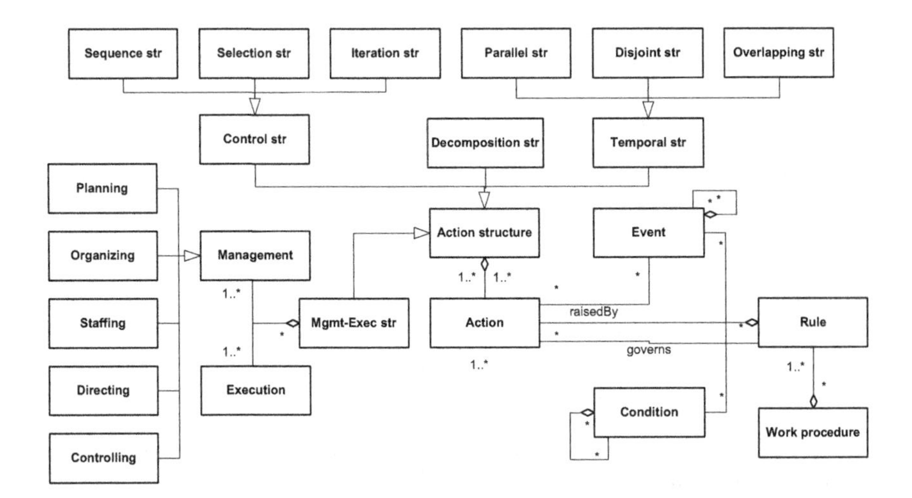


**Fig. 4.** Action domain

In the *decomposition structure,* actions are divided into sub-actions, these further into sub-sub-actions etc. Parts of actions are functions, activities, tasks or operations. Decomposition aims at reaching the level of elementary actions, where it is not possible or necessary to further decompose. The *control structure* indicates the way in which the actions are logically related to each other and the order in which they are to be executed. The control structures are: sequence, selection, and iteration. The *sequence relationship* between two actions $act_1$ and $act_2$ means that after selecting the action $act_1$ the action $act_2$ is next to be selected. The *selection relationship* means that after selecting the action $act_1$ there is a set of alternative actions $\{act_2,.., act_n\}$ from which one action (or a certain number of actions) is to be selected. The *iteration relationship* means that after selecting the action $act_1$ the same action is selected once more. The selection is repeated until the stated conditions become true. The *temporal structures* are like the control structures but with temporal conditions and events. They are specified using temporal constructs, such as during, starts, finishes, before, overlaps, meets, and equal [2]. With these constructs, overlapping, parallel and disjoint (non-parallel) executions of actions can be specified.

The *management – execution structure* (Mgmt-Exec) is composed of one or more management actions and those execution actions that put into action prescriptions got from the management actions (e.g. [37], [51]). *Management actions* mean *planning, organizing, staffing, directing and controlling*, in order to ensure the achievement of goals and constraints (cf. [9], [44]). *Execution actions* aim to implement plans and orders with the given resources. In business environment it is a commonplace to use specialized terms of execution actions such as order, transport, supply, invoice, and pay.

The action structures are enforced by rules. A *rule* is a principle or regulation governing a conduct, action, procedure, arrangement, etc [50]. It is composed of four parts event, condition, thenAction, and elseAction, structured in the ECAA structure [21]. An *event* is an instantaneous happening in the context, with no duration. A *condition* is a prerequisite for triggering an action. A *thenAction* is an action that is done when the event occurs and if the condition is true. An *elseAction* is an action that is done when the event occurs but the condition is not true. An aggregate of related rules constitutes a *work procedure*, which prescribes how the course of action should proceed. Depending on the knowledge of, and a variety of, actions, work procedures may be defined at different levels of detail.

### 3.4 Object Domain

The *object domain* contains all those concepts and constructs that refer to something, which actions are directed to (Figure 5). It can be a product, a message, a decision, an order, a service, a list of problems etc. We use *object* as the generic term to signify any concept in the object domain. The objects can be categorized into *goods* (e.g. car, CD-rom) and *services* (e.g. hair cut, consultation), on the one hand, and into *material objects* and *informational objects,* on the other hand. Business in many enterprises is, directly (e.g. in law firms, newspapers, and consultancies) or indirectly (e.g. in banks and insurance companies), related to information, its production and dissemination. Also for the other enterprises, such as in manufacturing industry, informational processing is crucial. Therefore, we next consider the informational objects more closely.

Informational objects can be classified based on the intentions by which they are provided and used (e.g. [27], [42], [45]). Informational objects can be descriptive or prescriptive. A descriptive object, called a *description,* is a representation about a slice of reality. An informational object can be descriptive in various ways. An *assertion* is a description, which asserts that a certain state has existed or exists, or a certain event has occurred or occurs. A *prediction* is a description of a future possible world with the assertion that the course of events in the actual world will eventually lead to this state (cf. [27]). A prescriptive object, called a *prescription,* is a representation of the established practice or an authoritative regulation for action. It is information that says what must or ought to be done. A prescription composed of at least two parts (i.e. (E or C) and A) of the ECAA structure is called a *rule*. A prescription with neither an event part nor a condition part is called a *command.* A *plan* is a description about what is intended. It can also be regarded as a kind of prediction, which is augmented with intentions of action. It is assumed that the future possible world described in the plan would not normally come out, except for the intended actions (cf. [27]). Informational objects can be *formal, semi-formal* or *informal* depending on a language used to represent them.

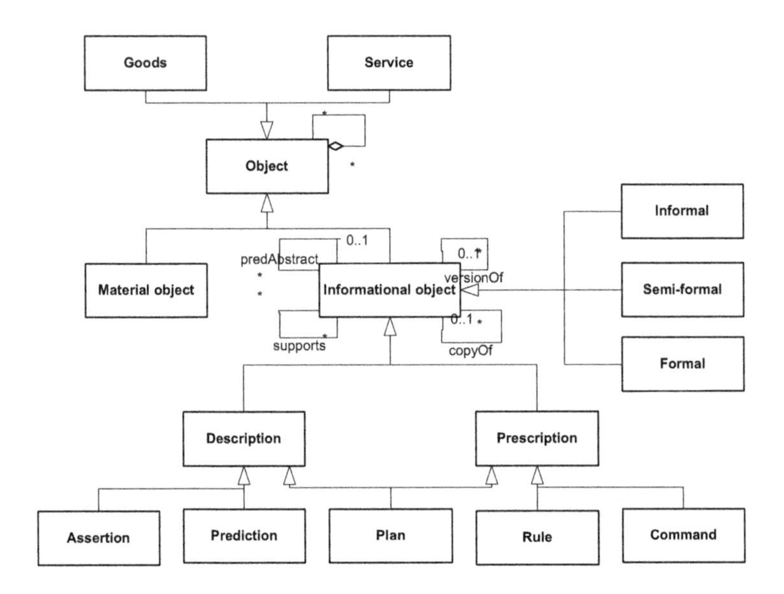


**Fig. 5.** Object domain

There are several relationships between the objects. The *versionOf relationship* holds between two objects $obj_1$ and $obj_2$, if properties of, and experience from, the object $obj_1$ have influenced the creation of another object $obj_2$ intended for the same purposes (cf. [24]). The *copyOf relationship* holds between two objects, the original object and a copy object, which are exactly, or to an acceptable extent, similar. The *supports relationship* involves two informational objects, $obj_1$ and $obj_2$, such that the information "carried" by the object $obj_1$ is needed to produce the object $obj_2$. The *predAbstract relationship* between two informational objects means that one object is more abstract than the other object in terms of predicate abstraction and both of the objects signify the same thing(s) in reality. The *partOf relationship* means that an object is composed of two or more other objects.

### 3.5 Facility Domain

The *facility domain* contains all those concepts and constructs that refer to the means by which something can be accomplished, in other words, to something which makes an action possible, more efficient or effective. We distinguish between two kinds of *facilities,* tools and resources (Figure 6).

A *tool* is a thing that is designed, built and installed to serve in a specific action affording a convenience, efficiency or effectiveness. A tool may be a simple and concrete instrument held in hand and used for cutting or hitting. Or, it may be a highly complicated computer system supporting, for instance, an engineer in his/her controlling a nuclear power station. Tools can be *manual, computer aided,* or *computerized.* A *resource* is a kind of source of supply, support, or aid. It can be

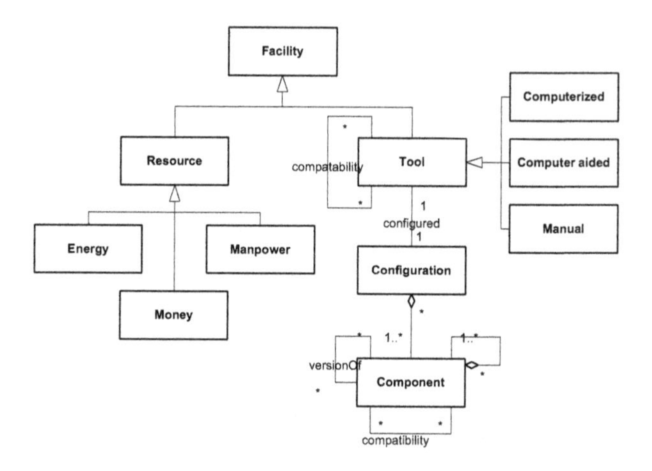


**Fig. 6.** Facility domain

money, energy, capital, goods, manpower etc. The resources are not interesting in terms of pieces, but rather in terms of amount. When a resource is used, it is consumed, and when consuming, the amount of the resource diminishes. Thus, a resource is a thing, about which the main concern is how much it is available (cf. [33].

There are a great number of relationships between the concepts within the facility domain, representing functional and structural connections, for instance. We only consider some of them. For being operative and useful, tools should be compatible. Two tools are *compatible* if their interfaces are structurally and functionally interoperable. Tools are composed of one or more components that develop through consecutive *versions*. Only some versions of components are compatible with the certain versions of the other components. A *configuration* is an assembly that is composed of the compatible versions of the components.

### 3.6 Inter-domain Relationships

Until now we have defined contextual relationships which associate concepts within the same contextual domain. There is, however, a large array of contextual relationships that relate concepts in different domains. For example, an actor carries out an action, an object is an input to an action, and a facility is situated in a location. We call these *inter-domain relationships.* Figure 7 presents an overview of inter-domain relationships. The space is divided into seven sub-areas corresponding to the seven contextual domains. In each of the sub-areas we present the concerned generic concepts and relate them with one another through the inter-domain relationships. It goes beyond the space available to define the relationships here.

In addition to the binary inter-domain relationships seen in Figure 7, there are multiple n-ary relationships. With these relationships it is possible to specify and interpret things in the enterprise in a way that discloses their contextual meanings. An

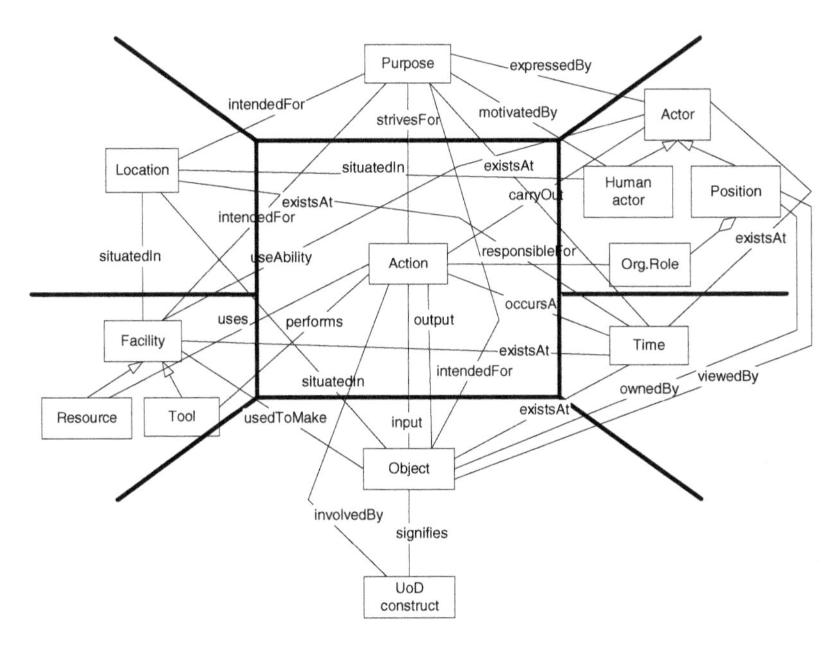


**Fig. 7.** Overview of inter-domain relationships

example of this kind of specification is: the customer c places the order o for the product p at time t, based on the offer o from the supplier s owned by the partners $\{p_1,\ldots p_n\}$, to be delivered by a truck tr to the address a by the date d. It depends on the situation at hand which contextual domains and concepts are seen to be relevant to be included in the specification.

## 4 Discussions and Conclusions

In this article we have presented the context-based enterprise ontology to promote the understanding of the nature, purposes and meanings of things about which information is stored and processed in, and transmitted between, various applications in enterprises. This ontology, grounded upon relevant theories [11, 13, 31], guides how to achieve a conceptualization of the structure and behavior of the enterprise in contextual terms.

The context-based enterprise ontology differs favorably from existing enterprise ontologies (cf. [4], [15], [48]). Based on our comparative analysis, the current ontologies mostly lack a theoretical background. Some of them do contain sub-ontologies that cover, to some extent, three of our contextual domains. For instance, the work area Organization in [48] and the ontology Organization in [15] correspond to the actor domain, Activity in [48] and Activity in [15] stand for the action domain, and Marketing::product in [48] and Requirements::part in [15] provide some concepts of the object domain. However, compared to the three domains in our ontology these ontologies are narrow-scoped and ignore many important contextual aspects of the enterprises.

Furthermore, they provide only a few of those concepts and constructs that are included in the purpose, facility and location domains in our enterprise ontology.

The context-based enterprise ontology can be deployed as a frame of reference to analyze and compare other enterprise ontologies in terms of contextual concepts. In addition, it can be specialized into a task ontology or a domain ontology for the needs of a particular business task or domain, respectively. An ontology resulted from this kind of specialization can be used, among other things, as a basis of business model design, enterprise architecture design, or legacy systems integration in enterprises.

In future research we continue to follow our top down approach to ontology engineering and focus on a more systematic derivation of specialized concepts and constructs, and use them in empirical studies on semantic and pragmatic interoperability of enterprise applications.

## References

1. Aguilar-Savén RS (2004) Business process modelling: review and framework. International Journal of Production Economics 90(2): 129-149
2. Allen J (1984) Towards a general theory of action and time. Artificial Intelligence 23(2): 123-154
3. Baclawski K, Kokar MM, Kogut PA, Hart L, Smith J, Holmes WS, Letkowski J, Aronson ML (2001) Extending UML to support ontology engineering for the Semantic Web. In: Gogolla M, Kobryn C (eds) Proc. of 4th Int. Conf. on the Unified Modeling Language. Springer-Verlag, Berlin, pp 342-360
4. Bertolazzi P, Krusisch C, Missikoff M (2001) An approach to the definition of a core enterprise ontology: CEO. In: Int. Workshop on Open Enterprise Solutions: Systems, Experiences, and Organizations (OES-SEO 2001), Rome, pp 14-15
5. Bianchini D, De Antonellis V, Melchiori M (2004). Ontology-based semantic infrastructure for service interoperability for interorganizational applications. In: Missikoff M (ed) Proc. of the Open InterOp Workshop on Enterprise Modelling and Ontologies for Interoperability, Riga, Latvia
6. Booch G, Rumbaugh J, Jacobson I (1999) The Unified Modeling Language – User Guide, Addison-Wesley
7. Bratman M (1987) Intentions, plans, and practical reason. Harvard University Press, Cambridge
8. Burton-Jones A, Storey V, Sugumaran V, Ahluwalia P (2005) A semiotic metric suite for assessing the quality of ontologies. Data & Knowledge Engineering 55(1): 84-102
9. Cleland D, King W (1972) Management: a systems approach. McGraw-Hill, New York
10. Cranefield S, Purvis M (1999) UML as an ontology modeling language. In: Proc. of the Workshop on Intelligent Information Integration
11. Engeström Y (1987) Learning by expanding: an activity theoretical approach to developmental research. Orienta-Konsultit, Helsinki, Finland
12. Fernandez-Lopez M, Gomez-Perez A, Pazos-Sierra A, Pazos-Sierra J (1999) Building a chemical ontology using METONTOLOGY and the ontology design environment. IEEE Intelligent Systems & Theory Applications 4(1): 37-46
13. Fillmore C (1968) The case for case. In: Bach E, Harms RT (eds) Universals in Linguistic Theory. Holt, Rinehart and Winston, New York, pp 1-88

14. Fox M (1992) The TOVE Project: A common-sense model of the enterprise. In: Belli F, Radermacher F (eds.) Industrial and Engineering Applications of Artificial Intelligence and Expert Systems. LNAI 604, Springer-Verlag, Berlin, pp 25-34
15. Fox MS, Gruninger M (1998) Enterprise modeling. AI Magazine 19(3): 109-121
16. Geert G, McCarthy W (2000) The ontological foundations of REA enterprise information systems. [online: htt://www.msu.edu/user/mccarh4/rea-ontology/]
17. Gordijn J (2002) Value-based requirements engineering – Exploring innovative e-commerce ideas. Dissertation Thesis, Vrije University, Amsterdam
18. Gruber T (1993) A translation approach to portable ontology specification", Knowledge Acquisition 5(2): 119-220
19. Gruber T (1995) Towards principles for the design of ontologies used for knowledge sharing. International Journal of Human-Computer Studies 43(5/6): 907-928.
20. Guarino N (1998) Formal ontology and information systems. In: Guarino N (ed) Proc. of Conf. on Formal Ontology in Information Systems (FOIS'98). IOS Press, Amsterdam, pp 3-15
21. Herbst H (1995) A meta-model for business rules in systems analysis. In: Iivari J, Lyytinen K, Rossi M (eds) Advanced Information Systems Engineering. LNCS 932, Springer, Berlin, pp 186-199
22. Jayaratna N (1994) Understanding and evaluating methodologies: NIMSAD – a systemic framework. McGraw-Hill, London
23. Johnson G, Scholes K, Sexty RW (1989) Exploring strategic management. Prentice-Hall, Englewood Cliffs
24. Katz R (1990) Toward a unified framework for version modeling in engineering databases. ACM Computing Surveys 22(4): 375-408.
25. Kavakli, V. & Loucopoulos, P. 1999. Goal-driven business process analysis application in electricity deregulation. Information Systems 24(3), 187-207.
26. Koubarakis M, Plexousakis D (2000) A formal model for business process modeling and design. In: Wangler B, Bergman L (eds) Proc. of 12th Int. Conf. on Advanced Information Systems Engineering (CAiSE 2000). Springer-Verlag, Berlin, pp 142-156
27. Lee R (1983). Epistemological aspects of knowledge-based decision support systems. In: Sol H (ed.) Proc. of Int. Conf. on Processes and Tools for Decision Support Systems. Elsevier Science Pub., Amsterdam, pp 25-36
28. Leppänen M (2006) Conceptual evaluation of methods for engineering situational ISD methods. Software Process: Improvement and Practice 11(5): 539-555
29. Leppänen M (2007) A contextual method integration. In: Proc. of the 15th Int. Conf. on Information Systems Development (ISD 2006). Springer-Verlag, Berlin (in print)
30. Leppänen M (2007) Towards an ontology for information systems development – A contextual approach. In: Siau K (ed.) Contemporary Issues in Database Design and Information Systems Development. Idea Group Inc. (in print)
31. Levinson S (1983) Pragmatics, London: Cambridge University Press
32. Lin C-Y, Ho C-S (1999) Generating domain-specific methodical knowledge for requirements analysis based on methodology ontology. Information Sciences 14(1-4): 127-164
33. Liu L, Yu E (2002) Designing web-based systems in social context: a goal and scenario based approach. In: Banks Pidduck A, Mylopoulos J, Woo C, Tamer Ozsu M (eds.) Proc. of 14th Int. Conf. on Advanced Information Systems Engineering (CAiSE'2002). LNCS 2348, Springer-Verlag, Berlin, pp 37-51

34. Loucopoulos P, Kavakli V, Prekas N, Rolland C, Grosz G, Nurcan S (1998) Using the EKD approach: the modelling component. ELEKTRA – Project No. 22927, ESPRIT Programme 7.1
35. Magretta J (2002) Why business models matter. Harvard Business Review 80(5): 86-92
36. McCarthy WE (1982) The REA Accounting model: A generalized framework for accounting systems in a shared data environment. The Accounting Review 58(3): 554-578.
37. Mesarovic M, Macko D, Takahara Y (1970) Theory of hierarchical, multilevel, systems. Academic Press, New York
38. NATURE Team (1996). Defining visions in context: models, processes and tools for requirements engineering. Information Systems 21(6): 515-547
39. Osterwalder A (2004) The Business Model Ontology - A proposition in a design science approach. Dissertation Thesis 173, University of Lausanne, Switzerland
40. Padgham L, Taylor G (1997) A system for modeling agents having emotion and personality. In: Cavedon L, Rao A, Wobcke W (eds.) Intelligent Agent Systems. LNAI 1209, Springer-Verlag, Berlin, pp 59-71
41. Ramesh R, Whinston A (1994) Claims, arguments, and decisions: formalism for representation, gaming, and coordination. Information Systems Research 5(3): 294-325.
42. Searle J, Vanderveken D (1985) Foundations of illocutionary logic. Cambridge University Press, New York
43. Simon H (1960) The new science of management decisions. Harper & Row, New York
44. Sisk H (1973) Management and organization. South Western Pub. Co., International Business and Management Series, Cincinnati
45. Stamper R (1975) Information science for systems analysis. In: Mumford E, Sackman H (eds.) Human Choice and Computers. Elsevier Science Pub., Amsterdam, pp 107-120
46. Uschold M (1996) Building ontologies: towards a unified methodology. In: Proc. of 16th Annual Conf. of the British Computer Society Specialist Group on Expert Systems. Cambridge, UK
47. Uschold M, King M (1995) Towards a methodology for building ontologies. In: Workshop on Basic Ontological Issues in Knowledge Sharing, held in conjunction with IJCAI'95, Montreal, Canada
48. Uschold M, King M, Moralee S, Zorgios Y (1998) The Enterprise Ontology. The Knowledge Engineer Review 13(1): 31-89
49. Wang X, Chan C (2001) Ontology modeling using UML. In: Wang Y, Patel S, Johnston R (eds.) Proc. of the 7th Int. Conf. on Object-Oriented Information Systems (OOIS'2001). Springer-Verlag, Berlin, pp 59-70
50. Webster (1989) Webster's Encyclopedic Unabridged Dictionary of the English Language. Gramercy Books, New York
51. Weick KE (1995) Sensemaking in organizations. Sage Publications, California

# Extracting Violent Events From On-Line News for Ontology Population

Jakub Piskorski, Hristo Tanev, and Pinar Oezden Wennerberg

Joint Research Center of the European Commission
Web and Language Technology Group of IPSC
T.P. 267, Via Fermi 1, 21020 Ispra (VA), Italy
{jakub.piskorski,hristo.tanev,pinar.oezden}@jrc.it

**Abstract.** This paper presents NEXUS, an event extraction system, developed at the Joint Research Center of the European Commission utilized for populating violent incident knowledge bases. It automatically extracts security-related facts from on-line news articles. In particular, the paper focuses on a novel bootstrapping algorithm for weakly supervised acquisition of extraction patterns from clustered news, cluster-level information fusion and pattern specification language. Finally, a preliminary evaluation of NEXUS on real-world data is given which revealed acceptable precision and a strong application potential.

**Keywords:** ontology population, event extraction, machine learning, security informatics.

## 1 Introduction

Nowadays, ontologies are widely used for organizing knowledge of a particular domain of interest. Since any knowledge evolves in time, a key part of the ontology engineering cycle is the ability to keep an existing ontology up to date. Recently, we have witnessed and ever-growing trend of utilizing natural language processing technologies for automatic ontology population.

This paper reports on on-going endeavour at the Joint Research Center (JRC) of the European Commission for populating violent incident knowledge base via automatically extracting event information from on-line news articles collected through the Internet with the Europe Media Monitor system [1]. Gathering information about violent events is an important task for better understanding conflicts and for developing early warning systems for automatic detection of precursors for threats in the fields of conflict and health.

In particular, we present NEXUS – an event extraction system. It deploys a novel bootstrapping algorithm for semi-automatically acquiring extraction patterns from clustered news. Exploiting clustered news for learning patterns intuitively guarantees better precision. Since information about events is scattered over different documents voting heuristics can be applied in order to aggregate information extracted locally within each cluster. Further, in order to find a trade-off between 'compact linguistic descriptions' and efficient processing, we

W. Abramowicz (Ed.): BIS 2007, LNCS 4439, pp. 287–300, 2007.
© Springer-Verlag Berlin Heidelberg 2007

have developed our own pattern matching engine. A preliminary evaluation of NEXUS on real-world data revealed acceptable precision and a strong application potential. Although our domain centers around security domain, the techniques deployed in NEXUS can be applied in other domains, e.g., tracking business-related events for risk assessment.

The rest of this paper is organized as follows. First, sections 2 and 3 briefly introduce the task of event extraction and our ontology model for violent events. The architecture of NEXUS is described in 4. Next, the pattern-matching engine is presented in 5. The issues concerning automatic pattern acquisition are addressed in section 6. Subsequently, section 7 elaborates on information fusion. Some evaluation figures are given in 8. Finally, we end up with some conclusions and an outlook in section 9.

## 2 Event Extraction

The task of event extraction is to automatically identify events in free text and to derive detailed information about them, ideally identifying *Who did what to whom, when, with what methods (instruments), where and eventually why?* Information about an event is usually represented in a so called *template* which can be seen as a attribute-value matrix. An example illustrating an event template is presented in figure 1. Automatically extracting events is a higher-level

| | |
|---|---|
| TYPE | killing |
| METHOD | shooting |
| ACTOR | Americans |
| VICTIM | Five Iraqi |
| LOCATION | Bagdad |
| TIME | 12.10.2003 |

**Fig. 1.** An example of an event template derived from the sentence *Five Iraqi were shot by the Americans in Bagdad on the 12th of October 2003*

IE task which is not trivial due to the complexity of natural language and due to the fact that a full event description is usually scattered over several sentences and documents. Further, event extraction relies on identifying named entities and relations holding among them. Since the latter tasks can be achieved with an accuracy varying from 80 to 90%, obtaining precision/recall figures oscilating around 60% for event extraction (usually involving several entities and relations) is considered to be a good result.

The research on automatic event extraction was pushed forward by the DARPA-initiated Message Understanding Conferences (1987-1998)[1], which organized several competitions for the research community in the area of IE and formally defined extraction tasks. A wide range of topics were covered, e.g., fleet

[1] MUC - http://www.itl.nist.gov/iaui/894.02/related/projects/muc

operation information, terrorist activities, joint ventures and leadership changes in business. The ACE (Automatic Content Extraction) Program[2] started in 1999 is a follow-up to MUC endeavor. It defines new extraction tasks focusing on entity, relation and event extraction which are significantly harder to tackle than MUC-defined tasks. In particular, varying source and quality of input data is considered. Further, ACE defines a wider spectrum of event types and introduced more complex template structures.

Although, a considerable amount of work on automatic extraction of events have been reported, it still appears to be a lesser studied area in comparison to the somewhat easier tasks of named-entity and relation extraction. Interestingly, most of the event extraction systems apply a blend of knowledge-based and machine-learning techniques. Two comprehensive examples of the current functionality and capabilities of event extraction technology dealing with identification of disease outbreaks and conflict incidents are given in [2] and [3] respectively. A most recent trends and developments in this area are reported in [4].

## 3 Ontology Model for Violent Events

The extracted information about violent events and related entities such as people and organisations is mapped to a domain ontology, namely the Ontology of Politically Motivated Violent Events (PMVE). It was constructed following a formal methodology presented in [5] and is described in more detail in [6]. The PMVE domain ontology consists of the concepts, the relationships and the properties that formally describe the PMVE as they are reported in the online news articles. The purpose of the ontology mapping process is to assign each violent event to its relevant concept in the ontology to obtain the instances for the knowledgebase. Currently there are 84 concepts covering events, places, people, groups and organisations in the ontology. 59 properties describe these concepts and 30 relationships connect them with each other. The figure 2 displays the higher and the mid-level concepts of the ontology hierarchy. Typically, a violent event has the following conceptualization:

```
Concept Name: ConflictEvent
Concept Properties: eventTitle (string)
                    damage (integer)
                          numberKilled (integer)
                          numberWounded (integer)

Concept Relations: affiliatedWith (Person|Organization)
                          hasActor (Person|Organization)
                          subeventOf (Event |Event)

Concept Name: Person
Concept Properties: ...
Concept Relations: affiliatedWith
                          collegeagueOf (Person|Person)
```

Note that the ontology adapts a property and relationship topology as discussed in [7], in that they can be hierarchical. This enables a convenient exploration of

[2] ACE - http://projects.ldc.upenn.edu/ace

the model. As seen above, `numberKilled` and `numberWounded` properties are subproperties of the `damage` property. Thus, finding the number of people `killedAt` a given violent event, we can also find the number of people `woundedAt` by first navigating from the `numberKilled` property up to `damage` and then down again to `woundedAt` property. Same applies to the relationships in the ontology, in that they can be not only hierarchial but also symmetric, reflexive, transitive and inverse. For example, the three relations `subeventOf`, `hasActor` and `collegeagueOf` are subrelationships of `affiliatedWith`. Moreover, `subeventOf` is reflexive-transitive and `colleagueOf` is symmetric-transitive. To provide semantic completeness for every relationship an inverse relationship is also defined. More concretely, if it is true that a relationship called `subeventOf` exists, it is true that a relationship `hasSubevent` also exists. These aspects particularly support the subsequent inference process.

```
Thing
   |-Agent
   |    |-Person
   |    |    | |-CivilianPerson
   |    |    | |-UnknownPesron
   |    |    | |-GovernmentalPerson
   |    |    |
   |    |    -------
   |    |          |
   |    |-Group    |
   |    |     |-PeopleGroup
   |    |            |-GovernmentalPeopleGroup
   |    |            |-CivilianPeopleGroup
   |    |            |-UnknownPeopleGroup
   |    |__ Organization
   |
   |-Location
   |-Event
        |-MilitaryEvent
        |-GovernmentalEvent
        |-CivilianEvent
        |-SocioReligiousEvent
        |-ConflictEvent
        |-CoupAttempt
        |-Sneaking
        |-Killing
        |-Arrest
        |-Kidnapping
        |-FailedTerrorPlot
        |-Uncovering
        |-Bombing
        |-Assasination
        |-HostageTaking
        |-Execution
```

**Fig. 2.** Ontology higher and mid-level structure

## 4 Nexus

This section describes our endeavor toward fully automatic event extraction. In order to tackle this task we created a prototype of an event extraction system called NEXUS (News cluster Event eXtraction Using language Structures). The system processes news articles on a daily basis and for each detected violent event it produces a frame, whose main slots are: date and location, number of killed and injured, kidnapped people, actors, and type of event. The architecture of NEXUS is depicted in figure 3.

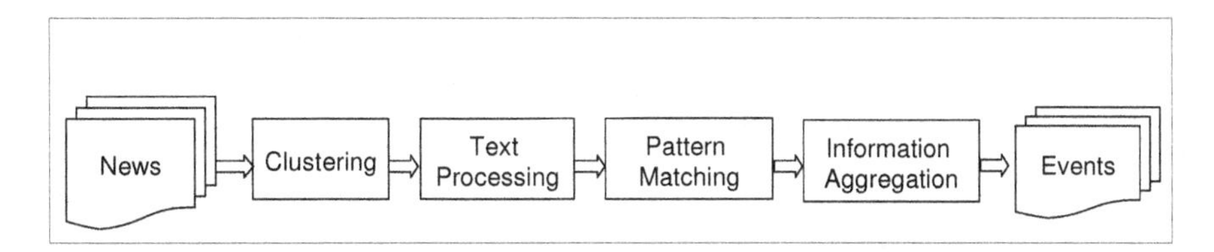


**Fig. 3.** The architecture of NEXUS

First, before the proper event extraction process can proceed, news articles are gathered by dedicated software for electronic media monitoring, namely the EMM system [1], which regularly checks for updates of headline across multiple sites. Secondly, the input data is grouped into news clusters ideally including documents on one topic. Further, clusters describing security-related events are selected via application of key-word based heuristics. For each such cluster the system tries to detect and extract only the main event via analyzing all documents in the cluster.

In the subsequent step, the documents in each cluster are further linguistically preprocessed in order to produce a more abstract representation of the texts. This encompasses following steps: sentence splitting, named-entities recognition (e.g., organizations, people, locations), simple chunking, labeling of key terms like action words (e.g. *kill*, *shoot*) and unnamed person groups (e.g. *five civilians*). In the text preprocessing phase, a geo-coding of documents in each cluster is performed, which is relevant for defining the place where the main event of the cluster took place.

Once texts are grouped into clusters and preprocessed, the pattern engine applies a set of extraction rules on each document within a cluster. For creating extraction patterns we apply a blend of machine learning and knowledge-based techniques. Firstly, we acquire patterns for recognition of event slots in a semi-automatic manner. Contrary to other approaches, the learning phase is done via exploiting clustered news, which intuitively guarantees better precision of the learned patterns. Secondly, we enhanced the set of automatically learned patterns by adding manually created multi-slot extraction patterns. The extraction patterns are matched against the first sentence and the title of each article from the cluster. By processing only the top sentence and the title, the system is more likely to capture facts about the most important event in the cluster.

Finally, since information about events is scattered over different documents, the last step consists of cross-document cluster-level information fusion, i.e., we aggregate and validate information extracted locally from each single document in the same cluster. For this purpose simple voting-like heuristics are deployed, e.g., among the phrases which appear as a filler of a given slot in the event templates extracted from all documents in the cluster, we select the most frequent one.

The more thorough description of our pattern engine, automatic pattern acquisition and information fusion follows in the subsequent sections. The data gathering, clustering and text preprocessing is addressed in [1] and [8].

## 5 Pattern Matching Engine

State-of-the art IE systems make use of patterns to recognize named entities, semantic relations and roles. In the context of event extraction patterns are used to find information relevant to events detected in a text: participants, places, dates, consequences, damages, etc. Consider the following sample patterns.

```
[1] [NUMBER] "people were killed"
[2] [ORGANIZATION] "launched an attack"
```

The first pattern(`[1]`) may be used to extract the number of the victims in a violent event, whereas the second extracts actors which initiate violent events. In general, IE patterns can be linear sequences or syntactic trees, they may contain NE classes or semantic elements. Usually, each pattern is a language construction which contains one or more slots and some context. In violent event extraction, patterns are intended to extract text entities which have different roles, e.g., affected dead, affected wounded, material damages, actors, places, dates, etc.

In the last decade, several high-level specification languages for creating extraction patterns have been developed and presented. The widely-known GATE platform comes with JAPE (Java Annotation Pattern Engine) [9]. A JAPE grammar consists of pattern-action rules. The left-hand side (LHS) of a rule is a regular expression over arbitrary atomic feature-value constraints (the recognition part), while the right-hand side (RHS) constitutes a so-called *annotation manipulation statement* which specifies the output structures. Additionally, the RHS may call native code, which on the one side provides a gateway to the outer world, but on the other side makes pattern writing difficult for non-programmers.

In order to find a trade-off between 'compact linguistic descriptions' and efficient processing we have developed our own extraction grammar formalism which is similar in spirit to JAPE, but also encompasses some features and syntax borrowed from XTDL, a significantly more declarative and linguistically elegant formalism used in SProUT platform [10]. In the new formalism, the LHS of a pattern is a regular expression over flat typed feature structures (TFS)[3], i.e., non-recursive TFSs without coreferencing, where types are not ordered in a hierarchy. Unlike JAPE, we allow for specifying variables tailored to string-valued attributes on the LHS of a pattern in order to facilitate information

[3] Typed feature structures are related to record structures in many programming languages. They are widely used as a data structure for natural language processing and their formalizations include multiple inheritance and subtyping, which allow for terser descriptions.

transport into the output structures. Further, like in XTDL, functional operators are allowed on the RHSs for forming and manipulating slot values (e.g., calling external components which post-process the slot values). In particular, the aforementioned features make our pattern formalism more amenable than JAPE since writing 'native code' is eliminated. Finally, we adapted the JAPEs feature of associating patterns with multiple actions, i.e., producing more than one annotation (eventually nested) for a given text fragment. The following pattern for matching partial information concerning violent events, where one person is killed by another, gives an idea of the syntax.

```
killing-event :> ((person & [FULL-NAME: #name1]):killed
                   key-phrase & [METHOD: #method, FORM: "passive"]
                  (person & [FULL-NAME: #name2]):killer):event
-> killed: victim & [NAME: #name1],
   killer: actor & [NAME: #name2],
   event: violence & [TYPE: "killing",
                       METHOD: #method,
                       ACTOR: #name2,
                       VICTIM: #name1,
                       ACTOR_IN_EVENTS: inHowManyEvents(#name2)]
```

The pattern matches a sequence consisting of: a structure of type `person` representing a person or group of persons who is (are) the victim of the event, followed by a key phrase in passive form, which triggers a *killing event*, and another structure of type `person` representing the actor. The symbol `&` links a name of the structure type with a list of constraints (in form of attribute-value pairs) which have to be fulfilled. The variables `#name1` and `#name2` establish variable bindings to the names of both humans involved in the event. Analogously, the variable `#method` establishes binding to the method of killing delivered by the `key-phrase` structure. Further, there are three labels on the LHS (`killed`, `killer`, and `event`) which specify the start/end position of the annotation actions specified on the RHS. The first two actions (triggered by the labels `killed` and `killer`) on the RHS produce structures of type `victim` and `actor` respectively, where the value of the `NAME` slot is created via accessing the variables `#name1` and `#name2`. Finally, the third action produces an output structure of type `violence`. The value of the `ACTOR_IN_EVENTS` attribute is computed via a call to a functional operator `inHowManyEvents()` which contacts some knowledge base to find out the number of all events the current actor was involved in the past (such information might be useful on higher strata). The pattern presented above matches the phrase *Five Iraqi were shot by the Americans.* It is worthwhile to note that grammars can be cascaded, i.e., output produced by one grammar can be used as input for the grammar on higher level.

## 6 Automatic Pattern Acquisition

While in the past, IE systems used patterns created manually by human experts, the modern approach is to use automatic or semi-automatic techniques for their acquisition [11,12]. However, since machine learning (ML) approaches are error-prone, building a high quality pattern library still requires some manual intervention. Therefore, we first automatically learn an initial set of patterns and subsequently, we manually filter out implausible ones. Due to the known data sparsity problem hard-to-learn patterns are added manually to the library.

### 6.1 Pattern Learning Approach

We adopted a combined iterative approach for pattern learning which involves both ML and manual validation. Currently, we run it separately for learning patterns for each event-specific semantic role (slot). The method uses clusters of news articles produced automatically by our European Media Monitoring Web mining infrastructure [13]. Each cluster includes articles from different sources about the same news story. Therefore, we assume that each entity appears in the same semantic role (actor, victim, injured) in the context of one cluster.

Here are the basic steps of the pattern acquisition algorithm:

1. Annotate a small corpus with event-specific information, e.g., date, place, actors, affected dead, etc. As an example consider the following two sentences:
   a. *<actor>Hezbollah</actor>claimed the responsibility for the kidnapping of the Israeli corporal.*
   b. *<actor>Al Qaida </actor>claimed the responsibility for the bombing which killed <affected_dead>five people</affected_dead>.*
2. Learn automatically single-slot extraction patterns (see 6.2), e.g., the pattern `[ORGANIZATION] "claimed the responsibility"` could be learned from both sentences, where the entity filling the slot `[ORGANIZATION]` is assigned the role `actor`.
3. Manually check, modify and filter out low quality patterns. Eventually add new patterns. If the size of the list is exceeds certain threshold (the desired coverage is reached)- terminate.
4. Match the patterns against the full corpus or part of it. Next, entities which fill the pattern slots and comply to the semantic constraints of the slot are taken as *anchor entities.* If an anchor entity $A$ (e.g., *five people*) is assigned a role $R$ (e.g., `affected_dead`) in the news cluster $C$, we assume with high confidence that in the cluster $C$ entity $A$ appears mostly in the same role $R$. Consequently, annotate automatically all the occurrences of $A$ in $C$ with the label $R$, e.g., in our example all the occurrences of *five people* in the cluster from which the second sentence originate will be labeled as `affected_dead`.
5. Go to step 2.

After single-slot patterns are acquired by this algorithm, we use some of them to manually create double slot patterns like `X shot down Y`. A similar kind of learning using the Web as a corpus has been reported previously [14,15]. However Web-based learning approaches exhibit following bottlenecks: (a) they dependent on the speed of public search engines, (b) the linguistic quality of the

Web documents is in general lower than the quality of a news corpus, and (c) Web documents are not clustered, consequently capturing the set of documents, where an anchor appears in the same semantic role, is hard. Some similar-in-spirit approaches to ours that exploit clustered data for learning extraction patterns were reported in [16,17].

### 6.2 Learning Linear Patterns from Anchor Contexts

In step 2 of the pattern acquisition schema single-slot patterns are learned from a corpus with labeled entities (anchors). We developed a ML approach which learns sequences of tokens which have a slot on its left (right) side. Therefore, the left and right contexts of all the equally labeled anchors is considered separately. Suppose we have to learn a pattern from the following right contexts of anchors labeled as `affected_wounded`, where the position of the anchor entity (i.e. the slot) is denoted by `P`:

```
P "was heavily wounded in a mortar attack"
P "was heavily wounded when a bomb exploded"
P "was heavily injured in an accident"
P "was heavily injured by a bomb"
P "was heavily injured when a bomb exploded"
```

There are different pattern candidates. For example `P "was heavily"` is the most general one, but it is not very informative. On the other hand, the pattern `P "was heavily wounded in a mortar attack"` is too specific and contains redundant words. Our criterion for pattern selection is based on the so called *local context entropy maximum.* For each candidate pattern we consider all its occurencies and their immediate context (i.e., adjacent words). In the above example, the pattern `P "was heavily"` has two context words: *wounded* and *injured*, each of which co-occurs twice with the pattern. Taking this into consideration, a *context entropy* for each pattern $t$ can be calculated using the following formula:

$$context\_entropy(t) = \sum_{w \in context(t)} p(w|t) \cdot ln(p(w|t)^{-1}),$$

where *context(t)* is the set of the words in the immediate context of the pattern $t$ and $p(w|t)$ is the probability that a word appears in this context.

Intuitively, the more words we add to a phrase, the lower its context entropy becomes. However, when a pattern is semantically consistent and complete, it may have higher context entropy than some of its sub-patterns. This is because a complete phrase is less dependent on the context and may appear in different linguistic constructions, while the incomplete phrases appear in a limited number of immediate contexts which complete it. For example, the phrase `P "was heavily injured"` has higher right-context entropy than `P "was heavily"` which can be completed by only two words in the above example.

In order to introduce formally our selection criterion, we have to consider a partial order of the pattern constructions. For simplicity we will explain this partial order for constructions in which the slot is on the left: We say that a phrase $t_1$ precedes $t_2$, when $t_2$ can be obtained from $t_1$ by adding one word on the right. For example `P "was heavily"` precedes `P "was heavily wounded"`. Considering this ordering, we may introduce our LOCAL CONTEXT ENTROPY MAXIMUM CRITERION: A pattern $t$ satisfies this criterion only when all the patterns which precede it have the same or lower context entropy and all the patterns it precedes have lower context entropy. Finally, we select only the patterns which satisfy this context entropy maximum criterion and do not contain other patterns which satisfy it. In the example above we will select only `P "was heavily wounded"` and `P "was heavily injured"`.

## 7 Information Aggregation

NEXUS firstly uses linear patterns in order to capture partial information concerning an event and secondly merges the single pieces into event descriptions via application of various heuristics. After the pattern matching phase, the system has a set of text entities with semantic roles assigned to them for each cluster in the collection. If one and the same entity has two roles assigned, a preference is given to the role assigned by the most reliable group of patterns. The double-slot patterns like `X shot down Y` which extract two entities at the same time are considered the most reliable. Regarding the one-slot constructions, the system considers the ones for detection of `affected_dead`, `affected_wounded`, and `affected_kidnapped` as more reliable than the ones for extraction of the `actor` (the latter one being more generic). All these preference rules are based on empirical observations.

Another ambiguity arises from the contradictory information which news sources give about the number of killed and wounded. NEXUS uses an ad-hoc algorithm for computing the most probable estimation for these numbers. After this estimation is computed, the system discards from each news cluster all the articles whose reported numbers of killed and wounded are significantly different from the estimated numbers. The entities which are extracted from the discarded articles are also filtered out. In this way, only the most important named entities of the news clusters are taken into consideration when merging the pieces of information and filling the slots of the final event description. To illustrate the approach, consider the following two sentences.

[1] *Two killed in U.S. hostage standoff.*

[2] *A gunman killed a female hostage and then himself in a Thanksgiving Day standoff in the northeastern U.S. city of Chicago, CBS News reported Friday.*

In sentence 1 NEXUS finds a match for the pattern `[NUMBER] "killed in"` for extracting the `affected-dead` slot (*two* is extracted as a candidate). The second candidate for this slot is *a female hostage*. NEXUS will extract it from the second sentence together with the actor (*a gunman*) via application of the 2-slot pattern

`[PERSON:ACTOR] "killed" [PERSON:VICTIM]`. Consequently, after the pattern extraction phase, the `affected-dead` slot of the event will contain two entities (*a female hostage* and *two*) which constitute candidate for the final value of this slot. Next, in the information-aggregation phase, the number of victims is estimated to be 2, referring to the second entity. The heuristic behind this estimation gives preference to the largest number of victims, since smaller numbers usually refer to victim sub-groups (In fact, it is the case in the example).

The sketched heuristics perform well on our news data. However, it has some limitations, the most important of which is that it considers only one main event per news cluster, ignoring events with smaller importance or incidents subsumed by the main event. In the security related domain it is often necessary to detect links between events. For example, a kidnapping typically includes capturing a hostage, a statement by the kidnappers in which they declare what they want to liberate the abducted person, police action to liberate the hostage, and finally his or her liberation. The current version of NEXUS detects these events separately, but it cannot aggregate them into one complex event since temporal reasoning is not applied, i.e. references to past events are captured as current events.

## 8 Evaluation

A preliminary evaluation of the NEXUS performance has been carried out on 26333 English-language news articles grouped into 826 clusters from the period 24-28 October 2006. We implemented in NEXUS an ad-hoc text classification algorithm for detection of security-related news clusters based on frequency of patterns matched, number of articles and keywords present. This algorithm classified as security-related 47 events; for 39 of them the classification was correct. Since only the texts returned by the system were taken into consideration, in general, only the precision was measured.

We observed that in the news articles the most important information about the main event appears typically in the first sentence and in the title. Therefore, for each news cluster estimated as security related, NEXUS considers only the title and the first sentence of each article in the cluster to find an estimate for the number of killed and wounded. Only the title and sentences, where the numbers of killed and wounded are close to the cluster estimate, are returned by the system. The aforementioned pieces of texts were used for evaluating NEXUS accuracy with respect to security-related events classification, date and place detection, and pattern-based slot filling for dead, wounded, kidnapped, and actors. Since NEXUS assumes that the date of the news cluster is the date of the extracted event, the date is considered correctly detected, only if the cluster reports a current event. In the contrary, references to past events are considered as incorrect date detection. The evaluation of the slot filling [4] was carried out only on the 39 events which were correctly classified as security-related. Table 1 summarizes the precision figures.

[4] The size of the annotated corpus for learning extraction patterns amounts to circa 4,5 MB.

**Table 1.** NEXUS performance

| Detection task (slot) | *Precision* | Detection task (slot) | *Precision* |
|---|---|---|---|
| security-related events classification | 83 | affected dead | 91 |
| date | 76 | affected wounded | 91 |
| country identification | 95 | affected kidnapped | 100 |
| city, town, village | 28 | actors | 69 |

Regarding the identification of places, our geo-location algorithm is precise at the level of country (95%), but due to some temporary technical problems relatively low accuracy (28%) at the level of city, town and village can be observed. Further, the precision of pattern-based slot filling (the last two columns in Table 1) oscialates between 69% and 100%. Additionally, for the slot `actor` the recall was measured (63%), but still considering only the texts which NEXUS returns. The evaluation revealed that there is still space for improvement, especially in case of date detection and security-related events classification. However, taking into account that NEXUS relies on superficial patterns and no syntactic information is exploited, the presented figures can be considered quite satisfactory at this stage.

## 9 Conclusions and Outlook

Automatic event extraction from free text is an extremely relevant application for many areas such as risk assessment and early warning. In this paper, we have presented results on event extraction from on-line news for populating violent event ontology. In particular, we introduced NEXUS – an event extraction system, which deploys a novel bootstrapping algorithm for semi-automatically acquiring extraction patterns and performs heuristic-based cluster-level information fusion in order to merge partial information into fully-fledged event descriptions. Considering the lightweight techniques which are applied, the results of the preliminary evaluation are satisfactory and show that NEXUS is already operational. Although our endavour is focused on violent events, tackling the task of extracting other types of events, e.g., company take overs, management successions, etc., can be done in a similar manner.

In order to improve on the semantics derived from event reports several extensions are envisaged. Firstly, some work on improving the coverage of preprocessing modules and manual creation of additional extraction patterns is is planed. Secondly, a long-term goal is to automatically discover hierarchical structure of events, i.e., discovering sub-events (or side events) of the main event and incidents which are part of it. Clearly such structure would provide a more meaningful information for discovering complex patterns by an inference apparatus on higher strata. Some of the problems encountered in the current version of the system were caused by the clustering module which occassionally returns a group of documents which do not only refer to the main event, but also to some related events or events happening in the same region (in particular for hot areas like Iraq and Afghanistan). Therefore, we intend to perform experiments with subclustering in order achieve a more fine-grained structure of the document collection.

## Acknowledgements

We are indebted to our colleagues without whom the presented work could not have been be possible. In particular, we thank Ralf Steinberger, Bruno Pouliquen, Clive Best, Andrea Heyl, Jenya Belyaeva and other EMM colleagues.

## References

1. Best, C., van der Goot, E., Blackler, K., Garcia, T., Horby, D.: Europe Media Monitor. Technical Report EUR 22173 EN, European Commission. (2005)
2. Grishman, R., Huttunen, S., Yangarber, R.: Real-time Event Extraction for Infectious Disease Outbreaks. In Proceedings of Human Language Technology Conference (HLT) 2002, San Diego, USA (2002)
3. King, G., Lowe, W.: An Automated Information Extraction Tool For International Conflict Data with Performance as Good as Human Coders: A Rare Events Evaluation Design. International Organization **57** (2003) 617–642
4. Ashish, N., Appelt, D., Freitag, D., Zelenko, D.: Proceedings of the workshop on Event Extraction and Synthesis, held in conjcnction with the AAAI 2006 conference. American Association for Artificial Intelligence, Menlo Park, California, USA (2006)
5. Sierra, J., Gomez Perez, A., Fernandez Lopez, M.: Building a Chemical Ontology Using METHONTOLOGY and the Ontology Design Environment. In: IEEE Intelligent Systems. (1999) 37–46
6. Oezden-Wennerberg, P., Piskorski, J., Tanev, H.: Ontology Based Analysis of Violent Events. (In: In Progress)
7. Sheth, A., Arpinar, I., Kahyap, V.: Relations at the heart of semantic web: Modeling, discovering, and exploiting complex relationships. In Azvin, B., Yager, R., Zadeh, L., eds.: Enhancing the Power of Internet Studies in Fuzziness and Soft Computing. (2003)
8. Steinberger, R., Pouliquen, B., Ignat, C.: Navigating multilingual news collections using automatically extracted information. Journal of Computing and Information Technology - CIT **13** (2005) 257–264
9. Cunningham, H., Maynard, D., Tablan, V.: Jape: a java annotation patterns engine (second edition). Technical Report, CS–00–10, University of Sheffield, Department of Computer Science (2000)
10. Drożdżyński, W., Krieger, H.U., Piskorski, J., Schäfer, U., Xu, F.: Shallow Processing with Unification and Typed Feature Structures — Foundations and Applications. Künstliche Intelligenz **2004(1)** (2004) 17–23
11. Jones, R., McCallum, A., Nigam, K., Riloff, E.: Bootstrapping for Text Learning Tasks. In: In Proceedings of IJCAI-99 Workshop on Text Mining: Foundations, Techniques, and Applications, Stockholm, Sweden. (1999)

12. Yangarber, R.: Counter-Training in Discovery of Semantic Patterns. In: Proceedings of the 41st Annual Meeting of the Association of Computational Linguistics. (2003)
13. Best, C., Pouliquen, B., Steinberger, R., van der Goot, E., Blackler, K., Fuart, F., Oellinger, T. Ignat, C.: Towards automatic event tracking. In: Intelligence and Security Informatics - Proceedings of IEEE International Conference on Intelligence and Security Informatics (ISI'2006), San Diego, California, USA. (2006) 26–34
14. Szpektor, I., Tanev, H., Dagan, I., Coppola, B.: Scaling Web-based acquisition of Entailment Relation. In: In Proceedings of EMNLP 2004, Barcelona, Spain. (2004)
15. Downey, D., Etzioni, O., Soderland, S. Weld, D.: Learning Text Patterns for Web Information Extraction and Assessment. In: Proceedings of the 13th international conference on World Wide Web. (2004)
16. Shinyama, Y., Sekine, S.: Preemptive information extraction using unrestricted relation discovery. In: Proceedings of Human Language Technology Conference (HLT)-NAACL. (2006)
17. Naughton, M., Kushmerick, N., Carthy, J.: Event Extraction from Heterogeneous News Sources. In: Proceedings of the Workshop Event Extraction and Synthesis. (2006)

# Improving the Accuracy of Job Search with Semantic Techniques

Malgorzata Mochol[1], Holger Wache[2], and Lyndon Nixon[1]

[1] Freie Universität Berlin, Institut für Informatik
Takustr. 9, D-14195 Berlin, Germany
mochol,nixon@inf.fu-berlin.de
[2] Vrije Universiteit Amsterdam
Artificial Intelligence Department de Boelelaan 1081a, NL-1081HV
Amsterdam The Netherlands
holger@cs.vu.nl

**Abstract.** In this paper we introduce a prototype job portal which uses semantically annotated job offers and applicants. In our opinion, using Semantic Web technologies substantially increase market transparency, lower transaction costs and speed up the procurement process. However adding semantics is not a panacea for everything. We identify some outstanding problems in job search using the system and outline how the technique of query approximation can be the basis for a solution. Through an Industry-Research co-operation we are extending the prototype with these semantic techniques to demonstrate a more accurate job search.

## 1 Introduction

Nowadays many business transactions are carried out via the Internet. Human resources management has also discovered the Internet as an effective communication medium. As reported in [7] 90% of human resource managers in Germany rated the Internet as an important recruitment channel. One reason for this high rate is that the Internet, in particular, reaches applicants who are young and highly qualified. Despite the fact that companies use more than one channel to publish their job openings, over half of all personnel recruitment is the result of online job postings [13]. Despite these achievements, the information flow in the online labour market is far from optimal.

As a result the German national project *Wissensnetze* (*Knowledge Nets*[1]) has worked together with German job portal provider WorldWideJobs GmbH[2] on developing an innovative semantic approach to searching for job offers or potential applicants on the Internet. In this paper we will present this approach, which has been demonstrated in a web-based prototype job portal. However, it is important to recognize that adding semantic annotation to the existing syntactic data is not a panacea for everything. On the other hand, the use of semantics

[1] http://wissensnetze.ag-nbi.de
[2] http://www.worldwidejobs.de/

W. Abramowicz (Ed.): BIS 2007, LNCS 4439, pp. 301–313, 2007.
© Springer-Verlag Berlin Heidelberg 2007

based on formal logic models such as the RDF and OWL specifications within a human resource domain [5,10] brings with it the possibility of new solutions to old problems, which would not be possible with pure syntactic data. In this paper, we introduce the approach of query approximation and show how it applies to the semantic job portal, further improving the results of job search. As a result we wish to demonstrate that while all aspects of a business problem will not be solved immediately by annotating semantically the existing business data, the introduction of semantics into business systems is the first step towards uncovering new IT solutions to typical business processes.

The rest of the paper is organized as follows: Section 2 gives a brief overview of a job procurement taking into account current (Sec. 2.1) as well as future (Semantic Web based) solutions (Sec. 2.2). Section 3 describes the semantic search approach used in the semantic job portal at the same time identifying some still outstanding problems. These issues are addressed in Section 4 which outlines how query approximation techniques can be the basis for the possible solution of such difficulties. We summarize the research on the application of query relaxation on the job procurement tasks with a brief status of the work and outlook considering future activities in Section 5.

## 2 Job Procurement: Old and New Ways

### 2.1 State of the Art

A large number of online job portals have sprung up, dividing the online labour market into information islands and making it close to impossible for a job seeker to get an overview of all relevant open positions. On the other side, the strong market position of the job portals as the prime starting point for job seekers allows them to charge employers high fees for publishing open positions which causes that in spite of a large number of portals employers still publish their openings on a rather small number of portals in order to keep costs and administrative effort down. The hiring organizations simply assume that a job seeker will visit multiple portals while searching for open positions.

Furthermore, the German Federal Employment Office (BA), for example, launched the platform "virtual employment market" in December 2003. This initiative is an effort to increase transparency of the job market and to decrease the duration of job procurement. In spite of high investments, these goals have not been reached yet. Some of the problems of the BA portal are, on the one hand, the necessity for all participants of the virtual employment market to use the proprietary data exchange format issued by the BA and some defects in quality of data and query results on the other hand.

Alternatively, companies can publish job postings on their own website [11]. This way of publishing, however, makes it difficult for job portals to gather and integrate job postings into their database. Thus search engines such as Google or Yahoo! have become vital in the job search. Furthermore, dedicated search engines, such as worldwidejobs.de[3], are entering into the market, allowing

[3] `http://www.wwj.de`

detailed queries as opposed to the keyword-based search of current search engines. The quality of search results depends not only on the search and index methods applied. Further influential factors include the processability of the used web technologies and the quality of the automatic interpretation of the company-specific terms occurring in the job descriptions. The deficiencies of a website's machine processability result from the inability of current web technologies, such as HTML, to semantically annotate the content of a given website. Therefore, computers can easily display the content of a HTML site, but they lack the ability to interpret the content properly.

## 2.2 Semantic-Based Solution

In our opinion, using Semantic Web technologies in the domain of online recruitment and skill management can cope with such problems and substantially increase market transparency, lower the transaction costs for employers and speed up the procurement process. For this reason, in the *Wissensnetze* project, which explores the potential of Semantic Web from a business and a technical viewpoint by examining the effects of the deployment of Semantic Web technologies for particular application scenarios and market sectors, we developed a job portal which is based on Semantic Web technologies. Every scenario includes a technological component which makes use of the prospected availability of semantic technologies in a perspective of several years and a deployment component assuming the availability of the required information in machine-readable form. The combination of these two projections allows us, on the one hand, to build e-business scenarios for analysis and experimentation and, on the other hand, to make statements about the implications of the new technology on the participants of the scenario in the current early stage of development.

In a Semantic Web-based recruitment scenario the data exchange between employers, applicants and job portals is based on a set of vocabularies which provide shared terms to describe occupations, industrial sectors and job skills [12]. In this context, the first step towards the realization of the Semantic Web e-Recruitment scenario was the creation of a *human resources ontology (HR-ontology)*. The ontology was intended to be used in a job portal, not only to allow for a uniform representation of job posting and job seeker profiles but also to support semantic matching (a technique that combines annotations using controlled vocabularies with background knowledge about a certain application domain) in job seeking and procurement tasks. Another important requirement was extending the Semantic Web-based job portal with user needs, considering this is already common practice in the industry. Accordingly to this specification we focused on how vocabularies can be derived from standards already in use within the recruitment domain and how the data integration infrastructure can be coupled with existing non-RDF human-resource systems.

In the process of ontology building we first identified the sub-domains of the application setting (skills, types of professions, etc.) and several useful knowledge sources covering them (approx. 25)[3]. As candidate ontologies we selected some of the most relevant classifications in the area, deployed by federal agencies or statistic organizations: German Profession Reference Number Classification (BKZ), Standard Occupational Classification (SOC), German Classification of Industrial Sector (WZ2003), North American Industry Classification System (NAISC), German version of the Human Resources XML (HR-BA-XML) and Skill Ontology developed by the KOWIEN Project[18]. Since these knowledge sources were defined in different languages (English/German) we first generated (depending on the language) lists of concept names. Except for the KOWIEN ontology, additional ontological primitives were not supported by the candidate sources. In order to reduce the computation effort required to compare and merge similar concept names we identified the sources which had to be completely integrated to the target ontology. For the remaining sources we identified several thematic clusters for further similarity computations. For instance BKZ was directly integrated into the final ontology, while the KOWIEN skill ontology was subject of additional customization. To have an appropriate vocabulary for a core skill ontology we compiled a small conceptual vocabulary (15 concepts) from various job portals and job procurement Web sites and matched them against the comprehensive KOWIEN vocabulary. Next, the relationships extracted from KOWIEN and various job portals were evaluated by HR experts and inserted into the target skill sub-ontology. The resulting conceptual model was translated mostly manually to OWL (since except for KOWIEN the knowledge sources were not formalized using a Semantic Web representation language) [15]. More information regardinR ontology can be found in [16,12].

The planned architecture for the Semantic Web-based job portal which uses our HR-ontology implies three basic roles [2] (cf. Fig.1) information providers, aggregators and consumers.

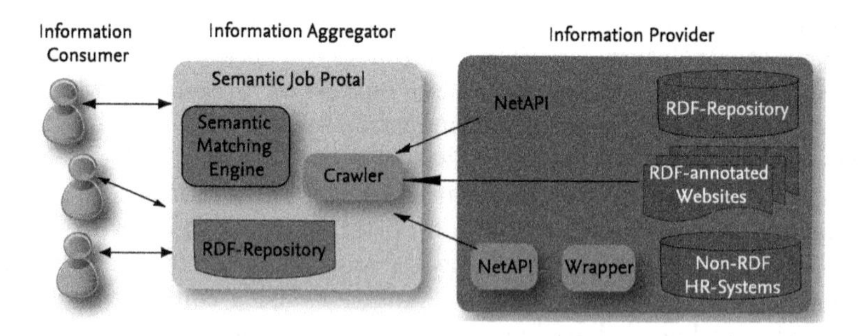


**Fig. 1.** Architecture of the Job Portal [2]

**Information Providers** who publish open positions and company background information in the RDF format using controlled vocabularies. They have two different approaches for publishing annotated job postings depending on their existing software infrastructure. If they use database standard software in the back-end, they can export related data directly into RDF using mapping tools like D2RQ [4]. If they do not use any enterprize software to manage open positions, they can annotate existing HTML versions of their postings using annotation tools and publish RDF version of their postings using, for example the RDF NetAPI or RAP.

**Information Aggregators** who crawl the published information and present it to the users in the desired format and level of aggregation. Combining job postings with background knowledge about the industrial sector allows the aggregators to offer semantic matching services. The usage of URIs as an employer identification mechanism allows the enrichment of postings with information about employers from different external information services.

**Information Consumers** who use one portal of their choice as a central access point to all published open positions and background information about employers, instead of collecting information fragments from different sites.

Having modelled the HR-ontology and prepared the RDF-Repository to store applicant profiles and job description, we developed the matching engine which as the core component of the system plays the crucial role in the procurement process (cf. Sec. 3).

## 3 Semantic Job Search

Semantic Matching is a technique which combines annotations using controlled vocabularies with background knowledge about a certain application domain. In our prototypical implementation, the domain specific knowledge is represented by concept hierarchies like skills, skill level classification, occupational classification, and a taxonomy of industrial sectors. Having this background knowledge of the recruitment domain (i.e. a formal definition of various concepts and specification of the relationships between these concepts) represented in a machine-understandable format allows us to compare job descriptions and applicant profiles based on their semantic similarity [17] instead of merely relying on the containment of keywords like most of the contemporary search engines do.

In our HR-scenario, our matching approach[4] utilizes metadata of job postings and candidate profiles and as the matching result, a ranked list of best candidates for a given job position (and vice versa) is generated. Inside both job postings and applicant profiles, we group pieces of information into "*thematic clusters*", e.g. information about skills, information regarding industry sector and occupation category, and finally job position details like salary information, travel requirement, etc. Each thematic cluster from a job posting is to be compared with the corresponding cluster from an applicant profile (and the other way round). The total similarity between a candidate profile and a job description is then calculated as the average of the cluster similarities. The *cluster similarity* itself is computed based on the similarities of semantically corresponding concepts from a job description and an applicant profile. The *taxonomic similarity* between two concepts is determined by the distance between them which reflects their respective positions in the concept hierarchy. Following this, the distance

[4] For further information about used matching framework SemMF see [14].

d between two given concepts in a hierarchy e.g. Java and C (cf. Fig. 2) represents the path from one concept to the other over the closest common parent. The semantic differences between upper level concepts are bigger than those between concepts on lower hierarchy levels (in other words, two general concepts like ObjectOrientedLanguages and ImperativeProceduralLanguages are less similar than two specialized like C++ and TURBOPASCAL) and the distance between siblings is greater than the distance between parent and child (d(Java,VisualBasic) > d(Java,PureObjectOrientedLanguages)).

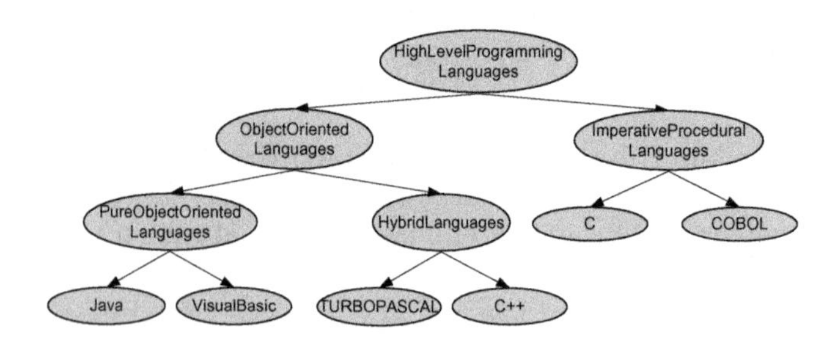


**Fig. 2.** Skills hierarchy

Since we also provide means for specifying *competence levels* (e.g. expert or beginner) in applicants' profiles as well as job postings we compare these levels in order to find the best match. Furthermore, our approach also gives employers the opportunity to specify the importance of different job requirements. The concept similarity is then justified by the indicated weight (i.e. the similarity between more important concepts) like the skills crucial for a given job position and will have greater influence on the similarity between a job position posting and an applicant profile.

### 3.1 Outstanding Problems

In the first version of the *HR-prototype* we concentrated on the modelling of the human resource ontology and development of a matching approach for comparisons of applicant profiles and job openings with a focus on skills and occupations as well as industry sector descriptions. The further specification of the job portal contains the comparison of applicants and vacancies not only under consideration of skills and their levels but also professional experience of a job seeker in relation to the requirements of the hiring company. We want to be able to express not only the duration of particular experience (e.g. 3 years experience in Java programming) but also to deliver those job applications which maybe do not fit 100% to the defined requirements but are still acceptable for the employer (e.g. 3 years instead of 5 years industry experience). Furthermore, to verify the consistency of the job opening descriptions we also have to avoid the definition of nonsensical requirements like job postings which demand only very young yet highly qualified people (e.g. under 25 with at least 10 years work experience).

Following this, we need an additional method which starts checking the data with the strongest possible query that is supposed to return the "best" answers satisfying most of the given conditions and then weaken the query if the returned result set is either empty or contains unsatisfactory results.

Since we have been working very close with one German job search engine we were able to define (in cooperation with the industry) several exemplary use cases which focuses on the definition of such requirements and issues. From the practical point of view the use-cases may represent the kind of queries which happen in the real world. However from the scientific point of view these use-cases are challenges to the techniques which we want to apply.

When implementing a (Semantic Web) job portal the requirements of the system depend on the meaningful use cases which are derived by the industrial project partner from its day to day business practices within the HR-domain. To clarify the still outstanding problems we will briefly present one such use case which (at first view) seems to be quite simple. However, if we look closer and try to represent the data in an ontology or satisfy the requirements in the semantic portal we will meet some difficulties, which at the same time show the complexity of such "simple" queries.
We are looking for a person which:

1. has experiences in CORBA, JavaScript, Java and Databases,
2. has worked for at least 5 years in industrial and 5 years in research projects,
3. should have experience as project or team manager,
4. should be not older then 25.

This example serves as a guideline and a thread in the rest of the article.

## 4 Query Relaxation

It lays in the nature of this application that instead of returning many exact answers to a request only a few or zero answers will be returned. Normally the description of the job requirements are too specific to be matched. A few or zero answers would immediately disappoint users and they would not continue to use a job portal. Therefore the job portal has to return not only the exact matches; it also has to return the most similar matches to a query.

Since *Semantic Web reasoning* is founded on logics which are perfect in returning exact answers but are inappropriate in order to return similar answers, it must be adopted or extended for the usage in such cases. The obvious approach uses a similarity measure which calculates the similarity between the job posting and request(cf. Sec. 3). Such a function $f(p, r) \mapsto [0..1]$ directly provide a ranking between the results because answers which are more similar can be higher ranked. However, each of such function does not explain how the job posting and the request differ since it only returns a value like 0.78. Furthermore, the difference between answers can not explored, i.e. the difference with another answer with ranking value 0.79 is not obvious and is not explained.

In order to ensure that the job portal ranks certain answers higher than others — and can ensure that the system will respect our preferences during the ranking

— similarity measures can be biased in that way that weights $w_i$ are attached to some parts of the calculation, i.e. $f(p,r) = \sum_{i=1}^{n} w_i * f_i(p,r)$. But again this give only the impression that the ranking function can be biased in the preferred way. Furthermore, it still does not explain the difference of one answer with the request or other answers. On the other hand, the user is directly able to specify how he wants to relax his request. The user may specify directly: "if nobody have 5 years industrial experience then I will also accept 3 years experience". Furthermore, the system can also explain how this set of returned answers is related to the original query, e.g. here comes now the answers not with 5 but with 3 years experiences (cf. Sec. 3.1).

In the following we describe an approach which uses *rewriting rules* to capture this knowledge explicitly and show how this knowledge is used to relax the original query into a set of *approximated queries*. We propose an approach for query rewriting based on conditional rewriting rules. This rewriting relaxes the over-constrained query based on rules in an order defined by some conditions. This has an advantage that we start with the *strongest possible query* that is supposed to return the "best" answers satisfying most of the conditions. If the returned result set is either empty or contains unsatisfactory results, the query is modified either by replacing or deleting further parts of the query, or in other words relaxed. The *relaxation* should be a continuous step by step, (semi-)automatic process, to provide a user with possibility to interrupt further relaxations.

In [6] the authors proposed a rule-based query rewriting framework for RDF queries independent of a particular query language. The framework is based on the notion of triple patterns[5] as the basic element of an RDF query and represents RDF queries in terms of three sets:

- triple patterns that must be matched by the result (mandatory patterns);
- triple patterns that may be matched by the results (optional triple patterns);
- conditions in terms of constraints on the possible assignment of variables in the query patterns.

Re-writings of such queries are described by transformation rules $Q \xrightarrow{R} Q'$ where $Q$ the original and $Q'$ the rewritten query generated by using $R$. Rewriting rules consist of three parts:

- a matching pattern represented by a RDF query in the sense of the description above;
- a replacement pattern also represented by an RDF query in the sense of the description above;
- a set of conditions in terms of special predicates that restrict the applicability of the rule by restricting possible assignments of variables in the matching and the replacement pattern.

A re-writing is now performed in the following way: If the predicates in the conditions are satisfied for some variable values in the matching and the replacement pattern and the matching pattern matches a given query $Q$ in the sense that the

[5] RDF statements that may contain variables.

mandatory and optional patterns as well as the conditions of the matching pattern are subsets of the corresponding parts of $Q$ then these subsets are removed from $Q$ and replaced by the corresponding parts of the replacement pattern.

To clarify the approach we take the example from the Section 3.1: someone who has experiences in CORBA, JavaScript, Java and Databases. Looking for such a person requires the system to translate this free text description into an instance retrieval problem[6]. The query must be translated into a concept expression. The retrieval process will return all job seekers which belong to that concept expression, i.e. fulfill all the requirement in the concept expression. The following OWL expression shows the concept expression for some person who has experience in some of CORBA, JavaScript, Java and Databases[7].

```
<owl:Class rdf:ID="Query">
  <rdfs:subClassOf>
    <owl:Class rdf:ID="Person"/>
  </rdfs:subClassOf>
  <rdfs:subClassOf>
    <owl:Restriction>
      <owl:someValuesFrom>
        <owl:Class>
          <owl:intersectionOf rdf:parseType="Collection">
            <owl:Class rdf:about="CORBA"/>
            <owl:Class rdf:about="JavaScript"/>
            <owl:Class rdf:about="Java"/>
            <owl:Class rdf:about="Databases"/>
          </owl:intersectionOf>
        </owl:Class>
      </owl:someValuesFrom>
      <owl:onProperty>
        <owl:ObjectProperty rdf:ID="hasExperience"/>
      </owl:onProperty>
    </owl:Restriction>
  </rdfs:subClassOf>
  ...
</owl:Class>
```

In the following we give some examples for the rewriting rules which use the aforementioned example as a basis.

Making use of the predicates we can use generic rewriting rules that are guided by information from the ontology. The predicate `subsumed` for example is satisfied when $X$ is more specific than $Y$. With the following rewriting rule we are able to consider the knowledge in the ontology.

[6] Users provide the appropriate description manually in the structured form over the web-interface.

[7] Originally we modelled these as nominals (enumerations like Week =Monday, Tuesday, ...). Nominals are instances and classes at the same time. However current DL systems have problems with nominals therefore we use classes in the current approach.

```
pattern(<owl:Class rdf:about="X"/>) ==>
 replace(<owl:Class rdf:about="Y"/>) && subsumed(X,Y).
```

Following the general rewriting rule a specific rule taking into account a required skill, e.g. Java, can be defined. The simplest approach relaxes some requirements in the applicant's experience, i.e. instead of `JAVA` the`PureObjectOriented Languages` or even the`ObjectOrientedLanguages` could be possible weakenings of the original query (cf. Fig. 2 ):[8]

```
pattern(<owl:Class rdf:about="Java"/>) ==>
 replace(<owl:Class rdf:about="PureObjectOrientedLanguages"/>) && true.
```

This means that if the term `<owl:Class rdf:about="Java"/>` appears in a query it can be replaced by:
`<owl:Class rdf:about="PureObjectOrientedLanguages"/>`. This rule is a very specific one as it only applies in one special case and does not make use of information stored in the experience ontology. In the same way some number restrictions can be applied. In our example the requirement that a person has five years experience in industrial projects is encoded with the help of the (artificial) class `FiveYearsOrMore`. This class represents all numbers representing years which are larger or equal to five. This class can be replaced by the class `TwoYearsOrMore` which obviously is more general (weaker) then the former. Furthermore we can restrict the replacement in that way that we only allow this for the restriction on property `hasDuration`. The corresponding rewriting rule look like:

```
pattern(<owl:Restriction>
         <owl:onProperty rdf:resource="#hasDuration"/>
          <owl:someValuesFrom>
           <owl:Class rdf:ID="FiveYearsOrMore"/>
          </owl:someValuesFrom>
         </owl:Restriction>)
 ==>
 replace(<owl:Restriction>
          <owl:onProperty rdf:resource="#hasDuration"/>
           <owl:someValuesFrom>
            <owl:Class rdf:ID="TwoYearsOrMore"/>
           </owl:someValuesFrom>
         </owl:Restriction>)
 && true.
```

The main problem of the rewriting approach to query relaxation is the definition of an appropriate control structure to determine in which order the individual rewriting rules are applied to generalize new queries. Different strategies can be applied to deal with the situation where multiple re-writings of a given query are possible. Example is a Divide and Conquer (i.e. Skylining) [8,9] strategy: The best results of each possible combinations of re-writings is returned. In the current version of the system we have implemented a simple version of skylining.

[8] For the sake of readability the examples are simplified.

In particular, we interpret the problem of finding relaxed queries as a classical search problem. The search space (Fig. 3) is defined by the set $Q$ of all possible queries. Each application of a rewriting rule $R$ on a query $Q$ is a possible action denoted as $Q \xrightarrow{R} Q'$. A query represents a goal state in the search space if it does have answers. In the current implementation we use breadth-first search for exploring this search space. Different from classical search, however, the method does not stop when a goal state is reached. Instead the results of the corresponding query are returned, goal state is removed from the set of states to be expanded and search is continued until there are no more states to be expanded. As each goal state represents the best solution to the relaxation problem with respect to a certain combination of re-writings, the goal states form a skyline for the rewriting problem and each of them is returned to the user together with the query answers.

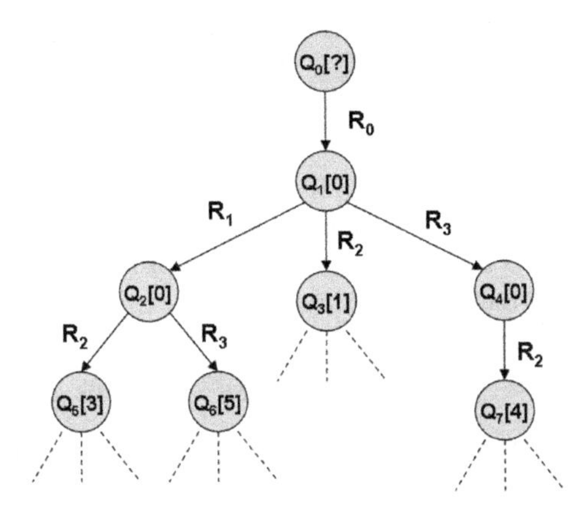


**Fig. 3.** Search space

The second difference to classical search is that we do not allow the same rule to be applied more than once with the same parameters in each branch of the search tree. The only kind of rules that can in principle be applied twice are rules that add something to the query[9]. Applying the same rule to extend the query twice leads to an unwanted duplication of conditions in the query that do not change the query result, but only increase the complexity of query answering.

## 5 Status and Future Work

The above mentioned co-operation between leading research in the area of semantic query approximation and the industry scenario of a semantic job portal has been enabled by the EU Network of Excellence KnowledgeWeb[10]. The aim

[9] Rules that delete or replace parts of the query disable themselves by removing parts of the query they need to match against.

[10] http://knowledgeweb.semanticweb.org/o2i

of KnowledgeWeb is to enable transfer of semantic technologies from academia to industry. To achieve this, one of the first steps taken in the network was to collect industrial use cases where semantic technologies could form a potential solution, as well as derive from those use cases industry requirements for Semantic Web research. The results of this use case collection and analysis showed a real potential for the Semantic Web to solve real world business problems [1].

Currently, Knowledge Web is promoting the co-operation of leading Semantic Web researchers with selected industry partners with a real world business problem to solve. The co-operation between VU Amsterdam, Free University Berlin and WorldWideJobs GmbH is just one of these co-operations. In the co-operation on the semantic job portal, the following tasks have already been achieved:

- Ca. 10-20 good examples of queries which will include the characteristic of experience have been defined,
- The HR ontology has been extended in order to be able to model the property of experience,
- Ca. 250 job seeker and 250 job position instances have been added to the HR knowledge base which include the extended properties of the ontology and can be used to test semantic queries such as the one given in Sec. 3.1,
- An interface to the rule rewriting tool has been specified (which is more general than the DIG interface in that it is not limited to Description Logics),
- The first concrete technical details of rule rewriting (e.g. abstract query syntax) have been defined,
- A first implementation of the rule rewriting tool has been completed.

It is planned by early 2007 to have a working extended version of the semantic job portal which will support queries which cover the types of sample query mentioned in Sec. 3.1. A benchmark will be used to test the extended prototype against the original prototype which does not support query approximation. As a result, we hope to acquire a clear view of the benefits of applying query approximation to enable more relevant responses to over-constrained or inconsistent queries, which is currently not possible in the present prototype.

The aim of query approximation is to allow more robust and efficient query response from knowledge bases which can scale to real world enterprise size. Furthermore, this approach is useful in areas such as eRecruitment to loosen queries that are too specific in order to allow users to find best matches rather than simply receive no results at all. Extending the HR-prototype provides us with a real life example to test the value of the query approximation approach. On top of the extended prototype we plan to continue our research in testing the scalability of the approach, as well as techniques for modelling and generating rewriting rules.

*Acknowledgement.* The work is supported by the EU Network of Excellence KnowledgeWeb (FP6-507482) and Wissensnetze Project which is a part of the InterVal-Berlin Research Centre for the Internet Economy funded by the German Ministry of Research.

## References

1. L. Nixon A. Leger, F. Paulus and P. Shvaiko. Towards a successful transfer of knowledge-based technology to European Industry. In *Proc. of the 1st Workshop on Formal Ontologies Meet Industry (FOMI 2005)*, 2005.
2. C. Bizer, R. Heese, M. Mochol, R. Oldakowski, R. Tolksdorf, and R. Eckstein. The Impact of Semantic Web Technologies on Job Recruitment Processes. In *International Conference Wirtschaftsinformatik (WI'05)*, 2005.
3. C. Bizer, M. Mochol, and D. Westphal. Recruitment, report, April 2004.
4. C. Bizer and A. Seaborne. D2RQ - Treating Non-RDF Databases as Virtual RDF Graphs. In *Proc. of the 3rd International Semantic Web Conference*, 2004.
5. M. Bourse, M. Leclère, E. Morin, and F. Trichet. Human Resource Management and Semantic Web Technologies. In *Proc. of the 1st International Conference on Information Communication Technologies: from Theory to Applications*, 2004.
6. Peter Dolog, Heiner Stuckenschmidt, and Holger Wache. Robust query processing for personalized information access on the semantic web. In *7th International Conference on Flexible Query Answering Systems (FQAS 2006)*, number 4027 in LNCS/LNAI, Milan, Italy, June 2006. Springer.
7. T. Keim et al. Recruiting Trends 2005. Working Paper No. 2005-22. efinance Institut. Johann-Wolfgang-Goethe-Universität Frankfurt am Main., 2005.
8. Werner Kießling and Gerhard Köstler. Preference sql - design, implementation, experiences. In *In Proc. of 28th International Conference on Very Large Data Bases*, pages 990–1001. Morgan Kaufmann, 2002.
9. M. Lacroix and Pierre Lavency. Preferences; putting more knowledge into queries. In Peter M. Stocker, William Kent, and Peter Hammersley, editors, *VLDB'87, Proceedings of 13th International Conference on Very Large Data Bases, September 1-4, 1987, Brighton, England*, pages 217–225. Morgan Kaufmann, 1987.
10. T. Lau and Y. Sure. Introducing ontology-based skills management at a large insurance company. In *Proc. of the of the Modellierung 2002*, pages 123–134, 2002.
11. W. Mülder. Personalinformationssysteme - Entwicklungsstand, Funktionalität und Trends. *Wirtschaftsinformatik. Special Issue IT Personal*, 42:98–106, 2000.
12. M. Mochol and E. Paslaru Bontas. Practical Guidelines for Building Semantic eRecruitment Applications. In *International Conference on Knowledge Management, Special Track: Advanced Semantic Technologies (AST' 06)*, 2006.
13. Monster. Monster Deutschland and TMP Worldwide: Recruiting Trends 2004. In *2. Fachsymposium für Personalverantwortliche.* Institut für Wirtschaftsinformatik der Johann Wolfgang Goethe-Universität Frankfurt am Main, 2003.
14. R. Oldakowski and C. Bizer. SemMF: A Framework for Calculating Semantic Similarity of Objects Represented as RDF Graphs. In *Poster at the 4th International Semantic Web Conference (ISWC 2005)*, 2005.
15. E. Paslaru Bontas and M. Mochol. Towards a reuse-oriented methodology for ontology engineering. In *Proc. of 7th International Conference on Terminology and Knowledge Engineering (TKE 2005)*, 2005.
16. E. Paslaru Bontas, M. Mochol, and R. Tolksdorf. Case Studies on Ontology Reuse. In *Proc. of the 5th International Conference on Knowledge Management (iKnow05)*, 2005.
17. J. Poole and J.A. Campbell. A Novel Algorithm for Matching Conceptual and Related Graphs. *Conceptual Structures: Applications, Implementation and Theory*, 954:293–307, 1995.
18. F. Sowa, A. Bremen, and S. Apke. Entwicklung der Kompetenz-Ontologie für die Deutsche Montan Technologie GmbH, 2003.

# Ontology-Based User Profiling

Carsten Felden[1] and Markus Linden[2]

[1] TU Bergakademie Freiberg, Fakultät für Wirtschaftswissenschaft
Professur ABWL, Informationswirtschaft/Wirtschaftsinformatik
Lessingstraße 45, 09599 Freiberg
carsten.felden@bwl.tu-freiberg.de
[2] Universität Duisburg-Essen, Campus Duisburg, Fachbereich Betriebswirtschaft
Lehrstuhl für Wirtschaftsinformatik und Operations Research
Lotharstraße 65, 47057 Duisburg
markus.linden@uni-due.de

**Abstract.** Profiles are the basis for individual communication, because they provide information about website users. Ontologies represent a possibility for modeling user profiles. The ontology development within the paper is based on a concept which shows firstly criteria of segmentation and secondly product programs of a retailer. A meta-ontology is built to enforce a mapping between the ontologies. Due to ontology-based recommendations it is obvious that they imply an additive character regarding conventional recommender systems. Therefore, the possibility arises to increase the turnover and to achieve customer satisfaction. But, the usage of ontology-based profiles is currently disputable relating to economic efficiency.

**Keywords:** Profiling, Ontologies, Recommender Systems, E-Commerce.

## 1 Introduction

A recommender system is a kind of software which presents a recommendation list to a user according to his/her preferences. Due to this, a website of an e-shop can be understood as an integration point, because the web presence is the semantic entanglement between the customers´ data and the product data. This integration point is origin of the necessary analytical tasks for recommendations. But a problem occurs due to the reason that there is often no semantic linkage between user profiles and product profiles. Therefore, we make the following case: The usage of ontologies enhances the usability of user profiles and product profiles and improves recommendation results and customer satisfaction.

This paper presents an approach to model ontologies in favor of a recommendation system. The ontologies link semantically harmonized heterogeneous data derived from different systems, for example shop systems and server protocols. In Section 3 we will show the ontology construction which enhances the idea of recommender systems. Firstly we build a product ontology and secondly a profile ontology. Based on this, we will use a meta-ontology as a linkage between e-shop-products and user information. Section 4 will analyze the realized ontology-models and section 5 will summarize the findings.

W. Abramowicz (Ed.): BIS 2007, LNCS 4439, pp. 314–327, 2007.
© Springer-Verlag Berlin Heidelberg 2007

## 2 Conceptual Framework

Ontologies represent knowledge within a system. Thus, they can be used to generate profiles. Ontology-based hierarchical classifications provide a clear illustration and structure of domains [4]. On this account, the following profile-, product- and meta-ontology are modelled in so called superclasses and subclasses. Based on this hierarchy, it is possible to connect relationships between classes of an ontology which cannot directly observed by customers´ behavior [14]. The used procedure to design ontologies is called *On-To-Knowledge*, introduced by Sure, Staab, and Studer [20].

We emphasize that a product program of a virtual retailer set out the bottleneck of a product-ontology. Reasons for using the program are quite simple: a product program already exists for planning and controlling methods. Moreover, those programs contain the property of a structured composition. Consequently, a transformation in a hierarchical classification is effortless to realize. A modification of the product-ontology is needed, if the retailer would provide new products to the customer.

The meta-ontology connects both individual-ontologies and maps them in case of modifications, for example a product launch. Due to this mapping, all concepts within the ontology are updated regularly. Moreover, the meta-ontology represents synonyms concerning the generated concepts, which support search functions as well as the profile development.

## 3 Ontology Construction

The user-oriented data analysis requires a user profile and the knowledge of valid problem-oriented variables of a specific problem domain. This chapter deals with the construction of a product-ontology and a profile-ontology.

### 3.1 Product- and Profile-Ontology

The breadth of a product program provides information about the amount of alternative products which are called product lines. The depth of a product program reflects the

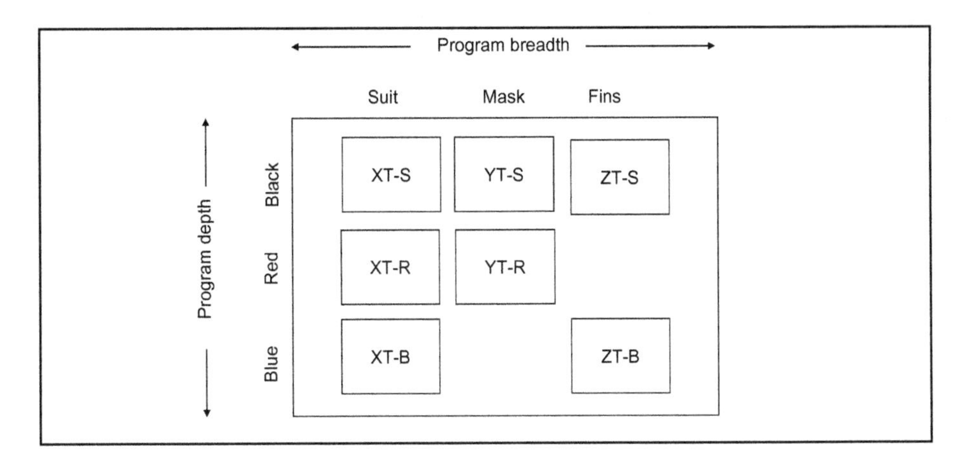


**Fig. 1.** Product program for diving area

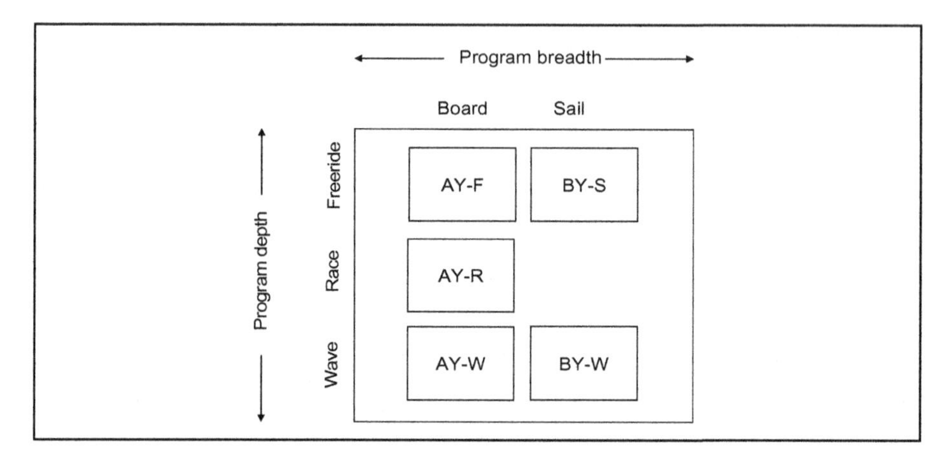


**Fig. 2.** Product program for windsurfing area

amount of products in a program line [12]. Figure 1 outlines a product program in a matrix and it contains various products in the area of diving sports. The product lines exhibit diving suits, diving masks, and diving fins.

Regarding the program depth, the product lines differ in various colors. Thus, product *XT-B* represents a blue diving suit. Another product program contains the area of windsurfing and can be interpreted in the same way as the program before (figure 2).

Finally, the last product program includes goods for snowboarding (figure 3). In this case, the breadth of the program gives information about different products. The depth of the program points out a gender specific manufacture.

The described programs are used to generate a product-ontology. Thereby, the first step affords the benefit of an indirect link to the profile-ontology. A problem can arise, if the retailer holds a large amount of products. Therefore, the hierarchical structure of concepts would be very complex. Beside the product notation, the ontology will be endorsed with heading terms of these products.

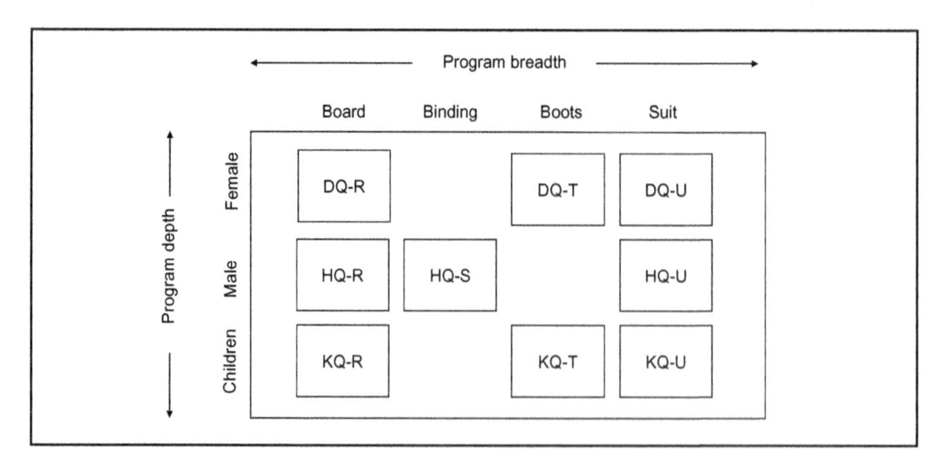


**Fig. 3.** Product program for snowboarding area

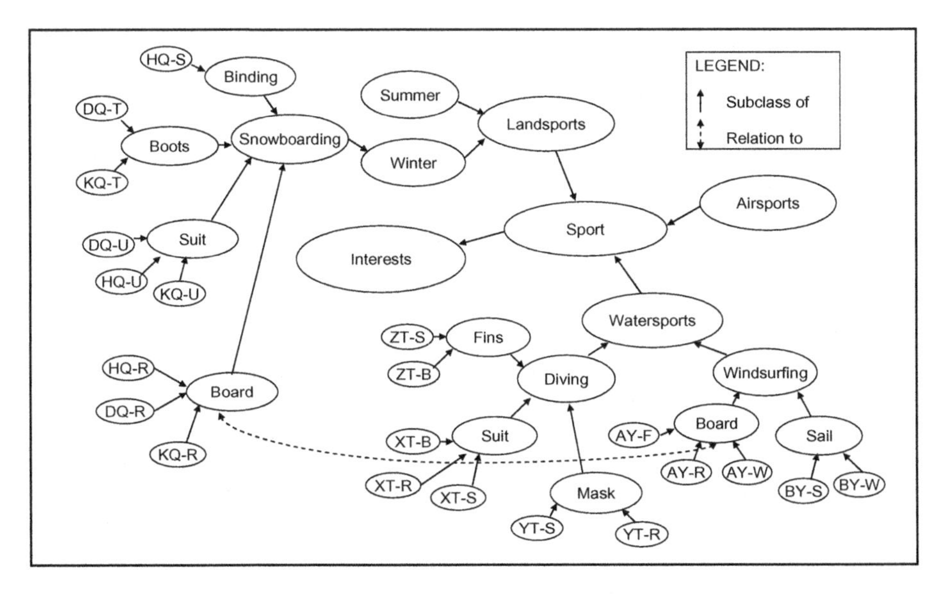


**Fig. 4.** Product-ontology for sport areas

To achieve a unique nomenclature of concepts, a determination of a technical agreement is needed [4]. As a result of the ontology engineering process, figure 4 shows a product-ontology based on three product programs which will be actualized in specific time intervals or after a product program modification.

From a technical point of view, a profile can be based on a tree structure. Thus, it is possible that every heading term, for example sport, assigns to a corresponding sub term like water sports [19]. The logical construction is reflected in a hierarchical classification of elements [1]. In this context, the concept *customer number* represents a root node and can be connected with concepts like interests or hobbies. There, a user profile contains descriptive information about the user and prescriptive information about determined and selectable interests and preferences [4].

## 3.2 Necessary Information Criteria

An online transaction is only feasible, if a customer is registered. During the registration, the user has to provide information about his/her personal data. Dispensable data should be avoided at the starting point, because a potential customer should not get annoyed by unnecessary questions and the volume of databases should not get increased without a reason. Segmentation criteria from marketing science provide an information basis to retailers in order to build a user profile. Those criteria contain important factors relating to a customer information basis and enable a segmentation of potential and de facto customers.

The geographic criteria show indications for data which cannot be acquired with observations of customers` behavior. Therefore, the retailer has to ask for those data within a registration form. Due to continuous actualizations of geographic data, a retailer has the ability to forecast buying probabilities for specific products [12].

As well as geographic data, sociodemographic data cannot be generated by keeping track of clickstreams. Even though a user would buy the product *DQ-R*, it is not ambiguous that the customer is in fact a woman, because the product may be a gift. Due to this reason, demographic data like age, sex, marital status and number of children as well as socioeconomic data like job, apprenticeship and revenue indicate social levels and should be acquired at the beginning of a business relation [12].

The psychographic criteria cover information about customers´ lifestyle, for instance activities, interests and attitudes. Product specific characteristics expose motives and cognitions relating to the benefit of a product. In this coherence, a segmentation of user types is possible regarding their interests in such a way as to enable for summing up similar users in a group. Even though such information can be discovered by observing the behavior of a user in the variation of time, a virtual retailer should collect information at the beginning of a business connection [19].

The behavior-oriented segmentation criteria refer to a customers´ price behavior and point out to that effect, if a customer buys frequently special offers. Furthermore, the choice of a brand or the buying volume can refine a user profile and improve the quality of recommendations. In return to the previous criteria, those data cannot be asked for, but allegorize a result of a logfile-analysis within the framework of dynamic user profile development.

### 3.3 Approach of Static Profile Development

Within the approach of static profile development, the illustrated geographic and socioeconomic criteria are used in a registration form. The explicit data acquisition in context of static profile development can be accomplished with a XML-document, because of an easy transformation from a typical tree structure of XML to a graphically modelled ontology. A representation of hierarchical dependencies will be arranged by utilizing nested elements [16]. The questions for psychographic criteria are integrated in a pull-down menu. In this context, the graphical representation orientates to the structure of the product-ontology. Thus, a website user is allowed to choose areas of sports, which fit to his/her interests. Accordingly, recommendation services provide a user who is interested in windsurfing, products of this area like *AY-R*. Afterwards, this product can be evaluated by the user via rating scale. Starting from this result, the retailer can estimate an individual product benefit.

### 3.4 Approach of Dynamic Profile Development

Behavior-oriented criteria figure within a dynamic profile development and can be identified on the basis of a logfile-analysis. This is a matter of implicit data acquisition, whereas for instance clickstreams and search keys provide information about product preferences of users. In the following description we analyze a users´ buying behavior (brand preference, price consciousness).

We act on the assumption that a user purchases the product *AY-R* after his/her registration. Thus, a preference regarding the brand *ProStyle* of *AY-R* can be discovered. The identification of buying brands plays a decisive role for an improvement of a user profile and will be put as a brand name and as a property in the profile-ontology. In this context, a recommender system can offer further products of this brand, in such a way continuative transactions verify or falsify the supposed disposition.

### 3.5 Static and Dynamic Ontology Engineering

A feasibility study contains an analysis of problems and possibilities in the engineering process. As a chance it is to state that a systematic attention handling of the most important segmentation criteria and a linked static and dynamic data acquisition generates a hierarchical structure. This structure can be used to infer individual recommendations during a business relation. However, a risk consists in a content of concepts, which results from a logfile-analysis. The analyzed information does not possess always an adequate goodness.

During the kick-off, a user profile represents the ontologies´ domain. As knowledge resources serve acquired information from the XML-form, the pull-down menu and the rating scale. Finally, user properties can be traced back to a logfile-analysis. For engineering a baseline-taxonomy, the top-down method is to be used and thus we start with the root node *customer number*. The tree structure within the XML-form can be consulted as a guideline regarding geographic and sociodemographic criteria. Psychographic criteria, which are specified by means of a rating result, orientate to a structure of an already determined pull-down menu. At the end, the behavior-oriented data can be added to the described user interests. A customers´ brand preference can be related to all product programs on the one hand. On the other hand it can be related to a single product.

The refinement shows connections between illustrated concepts of the baseline-taxonomy. These connections provide information about a relation without a hierarchical background.

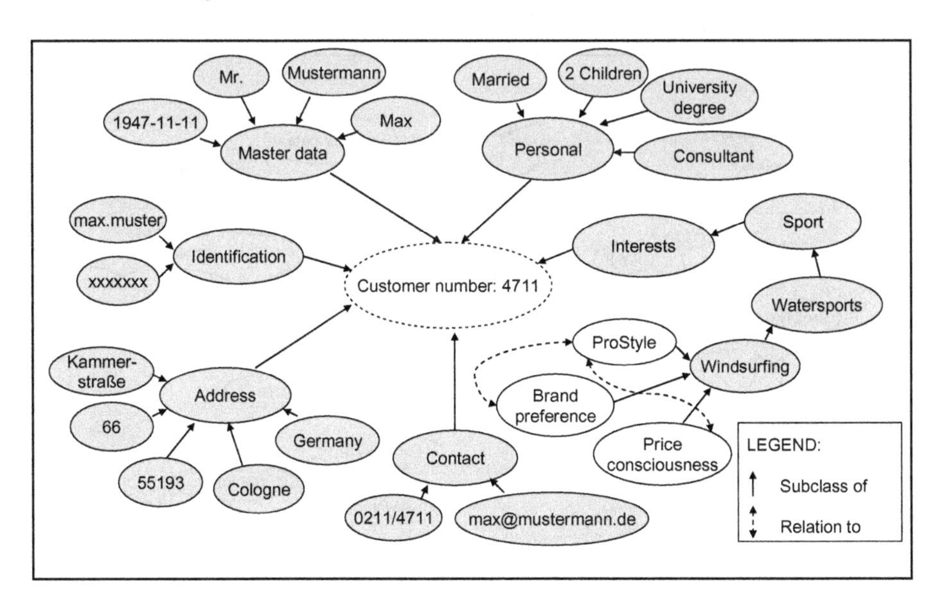


**Fig. 5.** Profile-ontology of a user

The grey fields of figure 5 represent all acquired data of the static approach, whereas the white fields illustrate the demonstrated results of the dynamic approach. An automatic generated customer number forms the root node of this profile-ontology.

### 3.6 Construction of a Meta-ontology

The objective of ontologies is to capture a common knowledge basis of a domain in a formal way. In this context, it can be inferred incorrectly that there is just a single ontology for one domain [7]. But this does not reflect the reality. The profile- and product-ontology will be mapped to a meta-ontology which specifies a common semantic of concepts [5]. There, a set of inference rules or relations must be generated for translating annotations from one ontology into another [17]. The meta-ontology shows a mapping of identical concepts of both individual ontologies and resolves redundancies. In addition, a meta-ontology presents, besides identical concepts, all other concepts of the individual-ontologies. Thus, against the background of mapping, new concepts can be integrated in the meta-ontology and can be transferred through its connections to the profile-ontology as well as the product-ontology.

Within our described example, a meta-ontology contains synonyms regarding determined terms of concept, comparable to a thesaurus. If a user feeds surfing into a website, the system interprets this term as an interest in windsurfing. Thus, on a semantic level both terms correspond to each other.

To construct a meta-ontology, we use the top-down method and start with a customer number. Afterwards, we add all identical concepts of the modelled ontologies. Finally, we put the residuals of the concepts to the meta-ontology. Figure 6 illustrates the result of the constructed meta-ontology as a joining element between the individual-ontologies.

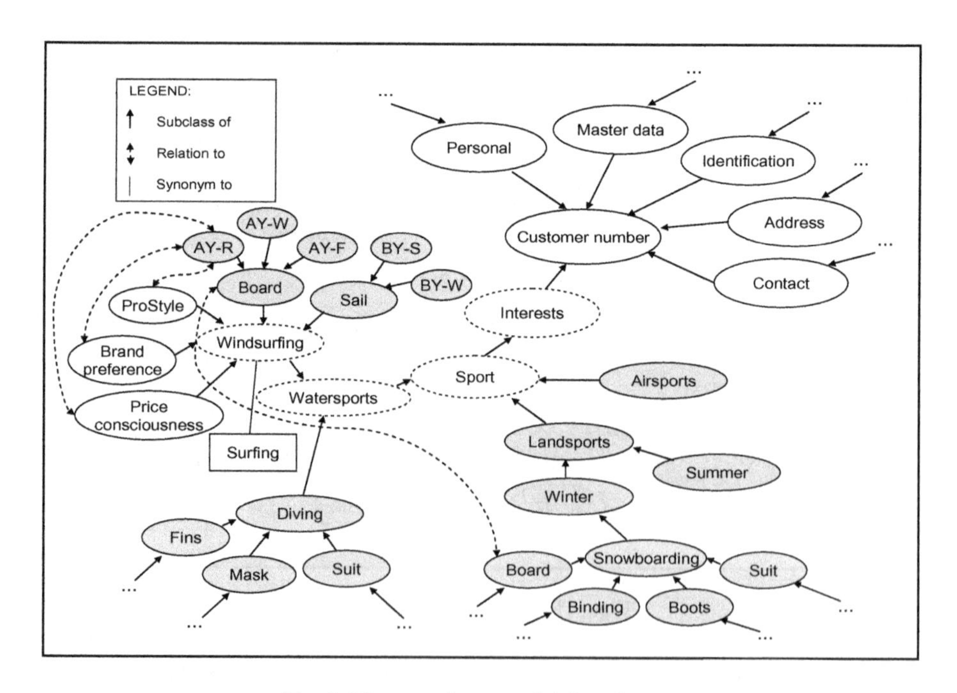


**Fig. 6.** Meta-ontology as a joining element

The grey fields represent constituents of the product-ontology, whereas the white fields reflect components of the profile-ontology. In the middle of the meta-ontology are dashed fields, which point out the identical concepts of the individual-ontologies.

### 3.7 Utilization of Ontology-Based User Profiles

Individualized recommender systems are divided in content-based and collaborative filtering relating to personalization methods. Within a content-based filtering, a classifier must be able to dispose objects in a database regarding positive and negative user interests. This can be enabled by analyzing a feedback or behavior of a user. On its basis a user profile is generated [14]. Afterwards, the product must be compared with user profiles and can be recommended, if the information fit his/her profile [21]. Thus, those objectives will be recommended, which consist the highest similarity with positive patterns and which are determined by a similarity measure of a classifier [14]. The comparison is called *filtering* in this context [21].

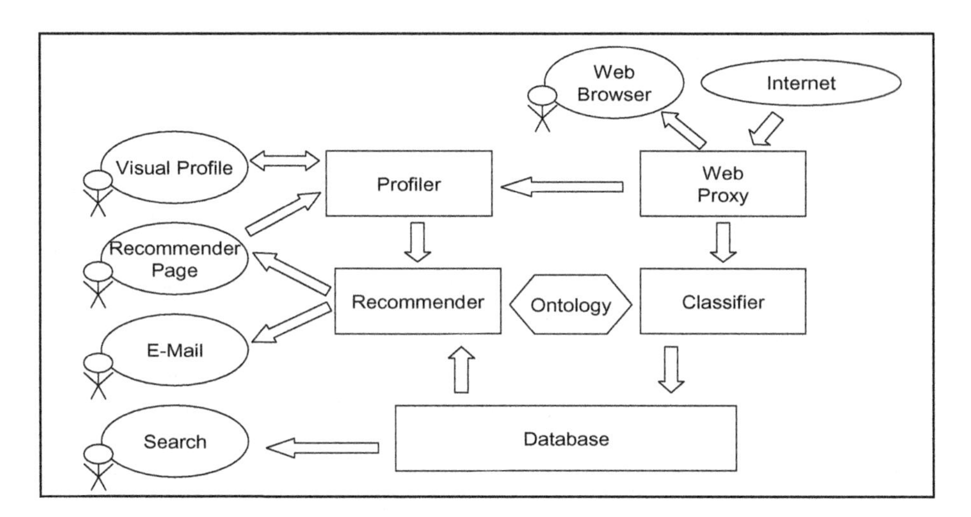


**Fig. 7.** Ontology-based recommender systems [14]

Software agents assume the accomplishment of deductions and presumptions and apply intelligent inference methods [21]. Such an agent can generate conclusions on basis of a connection between sensory cognitions of environment events and a combination of those events with knowledge [2]. Because of an agents´ learning aptitude, it can optimize and extend constantly user profiles [21].

The information about a user interest in windsurfing via pull-down menu results in an assimilation of a new concept in our profile-ontology. Due to a mapping of both individual-ontologies, all products can be found to this specific user interest within the meta-ontology. In our case, products like *AY-F*, *AY-W*, *BY-S*, *BY-W* are available to be recommended to potential customers. Product *AY-R* is excluded from this operation, because the customer already bought it. Now, all listed products can be filtered on the basis of observed characteristics, for instance brand preference, to generate individual recommendations.

This method is comparable to a product search. We assume that another user feeds *surfing* into a search tool of a website. Because of a synonym relation between *surfing* and *windsurfing* within meta-ontology, it can be inferred that this user is interested in windsurfing, too. This information was transferred through a mapping with the profile-ontology. After a filtering process with the user profile, search results or indirect recommendations of the product program for windsurfing area are presented on the website.

Besides the previous methods for generating product recommendations, there is a further possibility to offer profile adequate products due to a hierarchical structure. In this context, the fundament represents knowledge, which can be acquired by logical conclusions and is called *inference*. This knowledge within an ontology can be used to infer other interests which are not able to acquire with a user observation [14].

Ontology-based inference points out for example, that our above-mentioned customer of product *AY-R* has an interest in diving besides his/her well known enthusiasm in windsurfing. But this interest is to be inferred, because it was not emanated from observations and statements of the user. Due to an ontology structure with superclasses and subclasses we assume that a user, who is interested in windsurfing, has a weakness for further water sports like diving. In this context, *diving* represents as well as *windsurfing* a subclass of the superclass *water sports*. Thus, it appears that we climb up from a subclass to a superclass first and then we step down in another subclass. It is important to pay attention to potential preference areas of a user and to avoid supposing random classes. Against this background, we must calculate a likelihood which provides information whether products of other classes should be offered or not. For this purpose, we set up collaborative filtering to find similar profiles relating to the profile of our target user. Often a statistical function estimates such a correlation. Therefore, it can be provided evidence that 89 percent of all users who are interested in windsurfing have also a preference for diving. On the basis of this high value, our presumption is verified and the recommender system has to offer diving products, too.

In contrast, it is possible to generate recommendations by means of direct references to subclasses. We assume that a user offers only information about a preference for water sports. In this case, a user interest in water sports´ subclasses like *windsurfing* and *diving* can be inferred. As well as in the previous method, likelihoods support the verification process of stated presumptions. If a sufficient value is reached, the inferred user preference could be included in his/her profile and could be used for respective recommendations.

Finally, we state an interrelation of external concepts relating to user interests within an ontology. A reason for this presumption is a pretended similarity of products, which represents a tendency of a users´ hobby. Along the way it can be assumed that a user who is interested in windsurfing is also interested in snowboarding. This correlation must be verified again on the basis of a user clustering. If a high correlation value exists, the user will get product offers. It has to keep in mind that it remains unclear whether a user has a true interest in external concepts or not. Therefore, the user must be able to evaluate the recommendations due to a rating scale.

# 4 Analysis and Appraisal of Ontology-Based User Profiles

This chapter deals with an analysis of the constructed architecture and the included ontologies relating to a user profile development.

## 4.1 Analysis of the Ontology-Based Architecture

The architecture of an ontology-based user profile development points out its structure and constituents. In addition, the following remarks analyze the problems and the possibilities of ontologies in the framework of personalization.

As already mentioned, two specific domains are treated within the architecture and are represented in the individual-ontologies. A connection between these ontologies is reflected by a mapping of a meta-ontology, which allows a personalization based on ontology-based recommender systems. Moreover, a meta-ontology is easier to actualize in case of a product program modification and it transfers respective concepts to the individual-ontologies.

## 4.2 Appraisal of the Product-Ontology

If there would be a unique and correct ontology, this would lead to an existence of an implicit conformity of basic concepts of a domain. Thus, an ontology would be superfluously. We can see that ontologies differ for instance in a level of abstraction, a number of relations as well as a granularity of the taxonomy, even when consulting the same domains [10]. Against this background, all conventions about potential relations between concepts are determined by ontology-engineers and can pose problems relating to necessity and sense. The generated ontology does not claim to be universal, but represents one of many other potential solutions. *Non-universality* can be attenuated, if the existing product program provides an approach for the structure of product-ontologies.

Our described product program reflects a structure with two dimensions. Thus, it is illustrated in terms of a matrix and can be transferred easily into a baseline-taxonomy. In this context, problems arise in case of multidimensionality of a product, so that a cube must be utilized for the representation of product properties. Thus, the development of an ontology shows a higher complexity and reduces the benefit of clearly arranged concepts [4]. In conclusion, it makes sense to use and to extend existing ontologies [9].

## 4.3 Appraisal of the Profile-Ontology

The quality of the profile-ontology depends on segmentation criteria. These criteria are used in terms of milestones for necessary information, which are transferred into ontology concepts and are inspected regarding product- and company-specificity. The benefits of consulting segmentation criteria are an easy use as well as a holistic provision for all needed information.

However, problems can arise within an interpretation of geographic criteria, because it can lead to an incorrect appraisal of the user in some cases. But, the data reliability is to declare as correct, when a respective high value of likelihood was generated. Moreover, it can happen that users do not trust data acquiring services and feed no or false information [8].

Furthermore, a problem can occur within a clickstream-analysis, if an accidentally clicking on different websites generates incorrect data. In addition, some users deactivate so-called *cookies*, because they do not want that their surfing-behavior get analyzed. Such actions restrict the effectiveness of implicit data acquisition. Finally, we can state that analysis results can be incorrect sometimes, because it is just possible to identify a computer in general, but not a single person in a household [6]. Nevertheless, in principle clickstream results optimize a profile-ontology and improve recommendations.

In conclusion, the process contains high costs and thus it should be executed just one time or actualized after special events [4]. The representation of user interests in form of an ontology consists a deficit of a fine granularity, but in return it approves inference that supports the profile and builds communication to other ontologies [14].

### 4.4 Appraisal of the Meta-ontology

A final determination of equivalence relations must be accomplished by engineers, because term definitions of an ontology cannot be read automatically. As well as within the ontology engineering process, the ontology mapping costs time and money [17]. But at least costs of maintenance relating to individual-ontologies can be reduced, because new concepts must be included only in the meta-ontology. In addition, the benefit of a meta-ontology is a prevention of redundancies through a unification of concepts of individual-ontologies. The connected individual-ontologies provide a possibility of ontology-based recommendations, because relations between user interests and corresponding products as well as synonyms are represented.

### 4.5 Appraisal of Utilization

The described recommender systems are based on ontologies with potential inconsistencies and generate possibly fuzzy offers. In spite of using collaborative filtering to verify our presumptions about interests, the user can get bad recommendations. In this case, the recommender system bothers the user. This has to be avoided. Therefore, a user must have a possibility to evaluate products or to hide the recommendations. The most important thing in context of recommender systems and personalization is a realization of benefits from the perspective of a user.

Even though critical points, ontology-based recommender systems have a decisive advantage, because they can infer user interests on basis of a hierarchical structure. In addition, they can use relations between external concepts for recommendations. In this way, the *cold-start-problem* can be reduced significantly. Besides it can be supposed that ontology-based user profiles, which are developed by inference, generate better recommendations than profiles which do not use inference-rules. Due to the ontological inference, user profiles can be rounded off and can be matched better to the wide range of user interests [14]. There, we assume that a software agent consists of a large spectrum of intelligent properties. But an information interpretation based on software technology is very complex and currently hard to realize [3].

A general problem in user profile development can arise, if a profile is too sharp at its borders and if it becomes manifest in an unjustified direction [13]. So, it can happen that a customer gets a tendency to a product in the variation of time, even though he/she had an aversion for this product in a previous time. In this case, the recommendation system will never offer this product to the user because of the information within the profile.

### 4.6 Economic Appraisal

The benefit of ontologies is difficult to measure in a quantitative way from a company perspective. The utilization of ontologies in an information system raises questions about development time, consistency and maintenance [11]. Against this background, it is not possible to automate the process of maintenance. Thus, specialists are needed who allegorize an expense factor at least due to time intensity. In particular, a user profile development underlies dynamic changes and must be actualized immediately. Under these restrictions an adoption of ontologies seems to be inefficient, because the costs of modification are very high [17]. To capture costs relating to ontology engineering, appraisal methods from the discipline *software engineering* like *function point* can be used as a guideline, but must be customized to the specific requirements of ontologies.

Besides ontology development, further costs can arise for instance within an explicit data acquisition. Mostly, a user does not offer voluntarily data into a questionnaire and thus a vendor has to offer incentives. Based on individual and adequate recommendations, the customer satisfaction increases and thus the customer loyalty, too. Moreover, the potential of *cross-selling* emphasizes [18], which is supported by recommendations of external concepts.

From the perspective of a customer, providing personal data represent costs (time and effort) as well as the indirectly linked data misuse. The benefit for a customer consists in an individual consulting and in recommended products [18]. In the framework of such personalized services, those products are recommended which were mostly unknown before and are customized to his/her profile. Thus, an information overload can be avoided and the product choice is easier to handle for a customer. In addition, transaction costs can be reduced significantly regarding the time factor due to automated order transactions. However, the doubts relating to data security as well as administration efforts cannot balance the advantages of personalization in many cases [15].

## 5 Conclusion

Within this paper we pointed out a conceptual framework for constructing ontology-based user profiles. There, we used the explicit and implicit acquired data, which are represented as ontologies, to personalize recommendations in a virtual B2C-market. In this context, the utilization of ontology-based recommender systems showed innovative possibilities of inferences based on hierarchical structures and relations. A condition for inferring coherences is a semantic linkage between user profiles and product profiles.

In conclusion, we can verify our stated thesis according to an enhancement of the usability of user profiles and product profiles and its subsequent improvement of individual recommendations. Thus, good results of recommendations often increase customer satisfaction and turnover. In spite of an effective benefit regarding the usage of ontologies for individual offers, the efficiency in terms of engineering and maintenance costs is discussible in specific cases. Therefore, there is still a need to improve the engineering and the maintenance of ontologies.

## References


1. Böhnlein, M.; vom Ende, A.-U.: XML - Extensible Markup Language. In: WIRTSCHAFTSINFORMATIK 44 (1999) 3, p. 274 - 276. Wiesbaden (1999).
2. Clement, M.; Runte, M.: Intelligente Software-Agenten im Internet - Implikationen für das Marketing im eCommerce. http://www.runte.de/matthias/publications/ agents_clement_ runte_dermarkt.pdf#search=%22%22Intelligente%20SoftwareAgenten%20im%20Internet %22%22, last call 2006-10-06.
3. Erfurth, C.: Proaktive autonome Navigation für mobile Agenten. http://www.db-thueringen. de/servlets/DerivateServlet/Derivate4801/DISS.PDF#search=%22%22Proaktive%20autono me%20Navigation%20f%C3%BCr%20mobile%20Agenten%22%22, last call 2006-10-06.
4. Felden, C.: Personalisierung der Informationsversorgung in Unternehmen. Wiesbaden (2006).
5. Fensel, D.: Ontologies: A Silver Bullet for Knowledge Management and Electronic Commerce. 2. Ed., Berlin, Heidelberg, New York (2004).
6. Fritz, W.: Internet-Marketing und Electronic Commerce: Grundlagen - Rahmenbedingungen - Instrumente. 3. Ed., Wiesbaden (2004).
7. Gómez-Pérez, A.; Fernández-López, M.; Corcho, O.: Ontological Engineering. 3. Ed., London, Berlin, Heidelberg (2004).
8. Koch, M.; Möslein, K.: User Representation in E-Commerce and Collaboration Applications. http://www.communixx.de/files/Koch2003a.pdf, last call 2006-09-04.
9. Kuropka, D.: Modelle zur Repräsentation natürlichsprachlicher Dokumente: Ontologie-basiertes Information-Filtering und -Retrieval mit relationalen Datenbanken. Vol. 10. In: Becker, J.; Grob, H.-L.; Klein, S.; Kuchen, H.; Müller-Funk, U.; Vossen, G. (eds.): Advances in Information Systems and Management Science. Berlin (2004).
10. Leitner, J.: Extraktion von Ontologien aus natürlichsprachlichen Texten. http://www.inf. uni-konstanz.de/~leitner/docs/ontomining.pdf#search=%22%22 Extraktion%20von%20Ontologien%20aus%22%22, last call 2006-10-08.
11. Mädche, A.; Staab, S.: Ontology Learning. In: Staab, S.; Studer, R. (eds.): Handbook on Ontologies. Berlin, Heidelberg, New York (2004).
12. Meffert, H.: Marketing - Grundlagen marktorientierter Unternehmensführung: Konzepte - Instrumente - Praxisbeispiele. 9. ed., Wiesbaden (2000).
13. Mertens, P.; Höhl, M.: Wie lernt der Computer den Menschen kennen? Bestandsaufnahme und Experimente zur Benutzermodellierung in der Wirtschaftsinformatik. In: WIRTSCHAFTSINFORMATIK 41 (1999) 3, p. 201 - 209. Wiesbaden (1999).
14. Middleton, S.-E.; De Roure, D.; Shadbolt, N.-R.: Ontology-based Recommender Systems. In: Staab, S.; Studer, R. (eds.): Handbook on Ontologies. Berlin, Heidelberg, New York (2004).

15. Müller-Hagedorn, L.; Wierich, R.: Der Nutzen des Internets für den stationären Einzelhandel. In: Gabriel, R.; Hoppe, U. (eds.): Electronic Business - Theoretische Aspekte und Anwendungen in der betrieblichen Praxis. Heidelberg (2002).
16. Ollmert, C.: Extensible Markup Language. In: Thome, R. (ed.): Electronic Commerce - Anwendungsbereiche und Potential der digitalen Geschäftsabwicklung. 2. Ed., München (2000).
17. Schmaltz, R.: Semantic Web Technologien für das Wissensmanagement. http://www.wi2.wiso.uni-goettingen.de/getfile?DateiID=466, last call 2006-09-27.
18. Schwarze, J.; Schwarze, S.: Electronic Commerce - Grundlagen und praktische Umsetzung. Herne, Berlin (2002).
19. Stegmann, R.; Koch, M.; Wörndl, W.: Acquisition of Customer Profiles by means of Adaptive Text-Based Natural Language Dialog. http://www11.informatik.tu-muenchen.de/publications/pdf/Stegmann2004.pdf, last call 2006-09-04.
20. Sure, Y.; Staab, S.; Studer, R.: On-To-Knowledge Methodology (OTKM). In: Staab, S.; Studer, R. (eds.): Handbook on Ontologies, p. 117 - 132, Berlin, Heidelberg (2003).
21. Zarnekow, R.: Softwareagenten und elektronische Kaufprozesse - Referenzmodelle und Integration. Wiesbaden (1999).

# Xistree: Bottom-Up Method of XML Indexing

Xinyin Wang[1], Chenghong Zhang[2], Jingyuan Wang[1], and Yunfa Hu[1]

[1] Dept. of Computer and Info. Tech., Fudan University, Shanghai 200433, China
{041021050,042021137,yfhu}@fudan.edu.cn
[2] School of Management,Fudan University, Shanghai 200433, China
chzhang@fudan.edu.cn

**Abstract.** This article mainly proposes a bottom-up method to index XML document. Firstly we discuss the underlying properties of the method, architecture, creation algorithm and query algorithm, then conduct a set of experiments referring to the Timber and XIndice system. The demo system convinces that, this method can maintain excellent indexing and querying performance under given queries with normal PC on the DBLP XML test set of which the size is 315M, so it can be regarded as a prospective application with good performance. XML, indexing, Inter-relevant Successive Trees.

## 1 Introduction

With the wide acceptance of XML standard, XML format has become the major data exchange format inside and among enterprises, governments and research institutes. Along with the massive and autonomous generation of data, the problem of indexing and querying is emerging.

The problem of indexing and querying semi-structured data such as XML has been brought out at the end of last century, resulting in the birth of many demo systems. Nevertheless, Jagadish H.V. et al. pointed out that related presentation systems have good performance only when the data volume is relatively small but are incapable of processing huge storage [1]. To develop a new index method with high efficiency and low cost that can practically support huge storage is the aim of this paper.

This paper proposes a new XML index with bottom-up querying method. In Section 2, we briefly introduce the related researches. In the first part of Section 3, we present the coding scheme; in the second part we present the algorithms and some basic properties; and we analyze the performance at last. Section 4 introduces the location coding method and Section 5 describes the system architecture. In Section 5, We illuminate the organization of the experiments and report the results. We conclude our work and view the prospects in the end.

## 2 Related Work

At present, the methods of XML indexing can be categorized into three types: path coding, node coding and sequence coding. The core principle of path coding

W. Abramowicz (Ed.): BIS 2007, LNCS 4439, pp. 328–338, 2007.
© Springer-Verlag Berlin Heidelberg 2007

is to generate a certain path abstract from XML source data, and accomplish path query by join operator. Typical methods of this type include Dataguides [2], Index Fabric [3] etc.; the methods of node coding, such as XISS [4], mainly apply certain coding strategy to design codes for each node, in order that the relationship among nodes can be evaluated by computation; the methods of sequence coding transform the source data and query into sequences, leading to direct query of sequences. ViST [5] and D(k)-Index [6] are among the typical methods. From this point of view, the method proposed in this paper should be classified into path coding.

Generally speaking, every method has its own advantages, however, shortcomings do exist: path coding methods need large amount of join with elements; node coding method is very difficult to be applied to ever-changing data source; and sequence coding method is likely to generate approximate solutions, thus requiring a great deal of validation. Some kinds of index do need a mass of IO operations for simple query due to extremely separate storage, while some others can only be applied to query that begins from the tree root.

As far as bottom-up querying methods are concerned, Gottlob G. et al. explained its efficiency [7] and Ishikawa Y. et al. advised how to convert a XPath query to methods applying bottom-up query operator [8].

Catania, B. et al. systematically summarized the researches concerning XML indexing in recent years [9], in which systems such as dbXML, eXist, XIndice, Ipedo, Timber, Tamino are investigated. To check the performance of the method applied in this paper, we select XIndice and Timber as our reference systems.

The index of Timber is based on B-Tree, and supports value indices, element index, name index, and term-based inverted indices [1]. XIndice is an open source project on Apache Software Foundation, and it's the continuation of the project that used to be called the dbXML Core. XIndice supports Element and attribute values index [10].

# 3 Index Model

In this section, we mainly introduce how to develop the index. In the first subsection, we propose some concepts and properties to lay the ground of the index model, in the second part the creation and query algorithms are discussed, and the complexity of the algorithms are analyzed at last. A sample XML data of DBLP in Figure 1 is quoted from [11].

## 3.1 General Introduction

Every tree is composed of nodes and edges. Firstly we define the node set and edge set.

**Definition 1.** If we assemble all edges with the same name $C \rightarrow P$, of which C is the child of P, in XML tree as a vector, we can call it **edge vector** $V_{cp}$.

**Definition 2.** If we assemble all edge vectors $V_{cx}$, of which X is all possible parents of C, with the same node name C according to the sequence of X, we can regard it as **node vector** $V_c$.

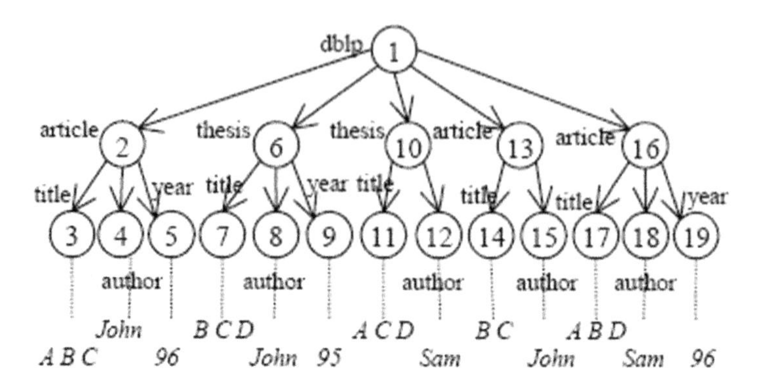


**Fig. 1.** A Sample XML

Now we can use a node name and a logic position in node vector to address an edge, for example, if one edge with name $C \rightarrow P$ has a position i in node vector C, its address is $C_i$, which is short for (C, i).

**Definition 3.** The **upper edge** $P \rightarrow G$ of $C \rightarrow P$ is defined when there is an edge $P \rightarrow G$, of which G is the parent node of P and the grandpa node of C.

To preserve the relationship between nodes and edges, for every edge $C \rightarrow P$, we should store the address $P_i$ of upper edge $P \rightarrow G$ in corresponding position of $C \rightarrow P$.

**Definition 4.** All the Node Vectors represented by logic addresses of their Upper Edge compose a set of vectors of edges, and we call them XML Inter-Relevant Successive Trees(Xistree).

Xistree is the applications of Inter-Relevant Successive Trees (IRST) [13], [14], [15] in the domain of XML indexing. The Xistree model of Figure 1 is presented in Figure 2.

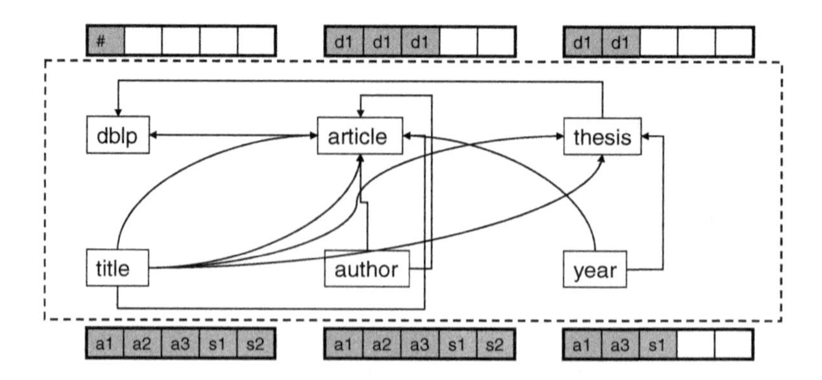


**Fig. 2.** The Index Structure of Figure 1

Note that in Figure 2, inside a node vector we use different colors to represent different edge names, and the same color to represent the same edge names. The node name is substituted by a character, for example, "d" for dblp, "a" for article, and "#" for a document which is the parent of the root node, etc.. If an element stores "d1", its parent is the first element of dblp node vector. It is easy to discover that the same node name is always contiguous in a single edge vector, therefore not all elements are required to store node names, and what should be done is to collectively label the vector section corresponding to each node name.

**Example 1.** Take path /1/10/12 for example, whose name is /dblp /thesis /author. After decomposition, we get $author \rightarrow thesis, thesis \rightarrow dblp, dblp \rightarrow null$, whose position in Xistree is (s, 2), (d, 1), (#, 0). Note that the position of (s, 1) has been occupied by /1/6/8 now.

**Definition 5.** Every edge vector $V_{cp}$ of tag pair $C \rightarrow P$ has a count denoted as $Count_{cp}$, and $Count_{cp} \geq 0$. So vector $V_c$ of node C corresponds to a vector $Count(Count_{c1}, Count_{c2}, ..., Count_{cn})$. Set $Sum_{cm} = \sum Count_{ci} (0 < i \leq m)$, here $(Sum_{c1}, Sum_{c2}, ..., Sum_{cn})$ is called **section count vector**.

**Table 1.** Section Count Vector

| | dblp | article | thesis | title | author | year | # |
|---|---|---|---|---|---|---|---|
| dblp | $+\infty$ | $+\infty$ | $+\infty$ | $+\infty$ | $+\infty$ | $+\infty$ | 1 |
| article | 1 | 4 | 0 | 0 | 0 | 0 | 0 |
| thesis | 1 | 3 | 0 | 0 | 0 | 0 | 0 |
| title | $+\infty$ | 1 | 4 | 6 | 0 | 0 | 0 |
| author | $+\infty$ | 1 | 4 | 6 | 0 | 0 | 0 |
| year | $+\infty$ | 1 | 3 | 4 | 0 | 0 | 0 |

**Example 2.** Suppose a query "/dblp/thesis/author" in Figure 2. Firstly, it should be decomposed into "$author \rightarrow thesis, thesis \rightarrow dblp, dblp \rightarrow null$". Then the set (s1, s2)(whose original code is (8, 12)) is fetched from the vector; From section count vector we get the count section of thesis/dblp, (1, 2), because 1 and 2 both fall into the section (1, 2), so s1, s2 are the solution of the join; a further step can get the content of s1 and s2, which is (d1, d1)(original code is (6, 10)), and the count section of dblp/null should be (1, 1), so the final solution is (u4, u5).

### 3.2 Algorithms and Properties

The following is the creating algorithm in Figure 3 and querying algorithms in Figure 4.

**Theorem 1.** All the path patterns can be composed by edge vectors and join operations, so Xistree can express the structure of the source xml document correctly.

```
Algorithm 1: Xistree building algorithm
Input: XML documents set xmlDocs;
Output: Section count vector V_cs; indices xistree;

1 traverse xmlDocs and generate the section count vector V_cs;
2 for each doc_i in xmlDocs do
3   add all edges of the root to xistree;
4   for each edge C → P in doc_i do
5     arrange the position of the current edge pos_c according to V_cs;
6     record the upper edge at P → R at pos_c of xistree;
7     add the value of the node C to xistree according to its type;
8   end for
9 end for
```

**Fig. 3.** Creating Algorithm of Xistree

Some good properties of Xistree are presented as follows.

**Property 1.** In Xistree, each node tag corresponds exclusively to a node vector, and each node in source tree per se corresponds exclusively to a vector element. In this way, the information of nodes path is not compact, so the path information will not be ill retrieved.

**Property 2.** Xistree can search out all the results of path sequence initially from any node name in the source tree. This property guarantees that Xistree can retrieve arbitrary path segment queries, not only queries from the root of XML trees.

**Property 3.** Xistree adopts the "node name: relative position" logic address pair, resulting in a simple structure. The coding scheme is independent of device. Once the storage is transferred, the addresses may not be changed.

```
Algorihm 2: Query processing based on Xistree
Input: Xpath query expression exp; indices xistree;
Output: Query results set resultset;

1  assign the lowest edge set to resultset;
2  translate exp into elementary bottom-up operations connected by logic ops;
3  for each layer of query path C → P and its upper edge P → G do
4      solve the value query ops separately and store the results in newset;
5      for each value addr in resultset do
6        compare addr with Sum_P,G and Sum_P,G+1 in V_cs;
7        if addr falls in section [Sum_P,G, Sum_P,G+1] then add the value of G_addr to pathset;
8      end for
9      compute resultset with newset and pathset;
10 end for
```

**Fig. 4.** Query processing algorithm based on Xistree

In Xistree model, we can compute operators such as "and", "or", "not", etc. on the basis of edge vector, and we can convert the operator "child of" to "parent of". So according to the method mentioned in [8], Xistree can satisfy most of Xpath operators. As far as operators such as "//" and "*" are concerned, the method that we apply now is to create an edge index for every node tag name. From the index we can get all the edge name occurred for a given node tag name, and then we can compose all the possible path patterns that can be used to accelerate the query.

Now let's take a look at time complexity. In order to reserve position for edges to facilitate the subsequent infilling, the system firstly needs to visit once to calculate the section count vector. Then normally the index can be established after one visit of each document, therefore the time complexity is $O(n)$.

In algorithm 2, the lowest edge set is the initial set, which may be the vector of bottom edge in the query. If the size of the upper edge is m, then the time complexity of path join operation is basically linear(that is $O(m)$). For the similar structure with invert files, the first step of path selection can be done without any join cost and reduce to a candidate set of $1/n^2$ in average, therefore the initiatory candidate edge set is relatively small and the actual query complexity is very low.

In Figure 2, according to Property 1, in order to code each node in the entire document set, the coding space should be the count of all node n in the whole system. So the bit count to represent each element should be $log_2 n$. It has been mentioned above that not every element will take storage space to save the node name since the same edges all have the same node name. The requirement is only to store the relative position. Therefore, if the number of node names is z, each element takes approximately $(log_2 n - log_2 z) bit$. Thus the entire system has a storage cost of approximately $(n * (log_2 n - log_2 z)) bit$.

Updates on index can be accomplished with deletion and insertion operations. The method of deletion is, we should query the object edge and set a flag to represent the deletion. As for insertion, firstly some empty address spaces are left between edge vectors and edge vectors are stored separately in physical storage, so insert operations just means appending at the end of edge vector.

## 4 Location Coding

Today, locating query results precisely becomes very important. Catering to this need, we propose a method to find the nodes in original XML documents based on Xistree. Inspired by [4], all children of each node can be numbered from 1 respectively, as is shown in Figure 5. If we find node t4 in the index, i.e. node 7 in Figure 1, we may locate its position in original documents precisely by its relative position number (1, 2, 1).

It can be noticed, in Xistree model, each edge corresponds to an edge vector exactly and each node corresponds to an element in the edge vector uniquely. As indicated in Figure 6, to denote the position information of a node in the index, we only need to mark its relative position number. For example, node 7 referred before, has a relative number "1" in its parent node, while node 18 has a relative number "2" in its parent node.

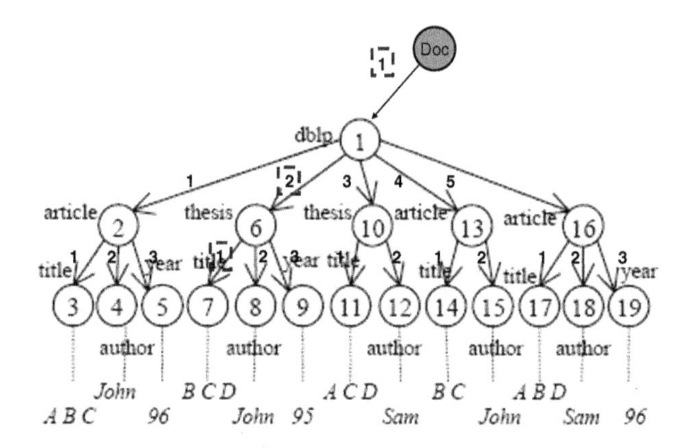


**Fig. 5.** The Location Code of Fig. 1

The relative numbers corresponding to Xistree model in Figure 2 are indicated by row 3 of each edge vector in Figure 6, which coincides with their original nodes and Xistree nodes exactly. So, to give the exact position in query result set, we need to write down relative position number when query is executed. It is likely to locate result set precisely through relative position number. Moreover, the relative position number has many advantages, such as fixed length, simple and short, and easy to locate, etc. Therefore it has great application prospect.

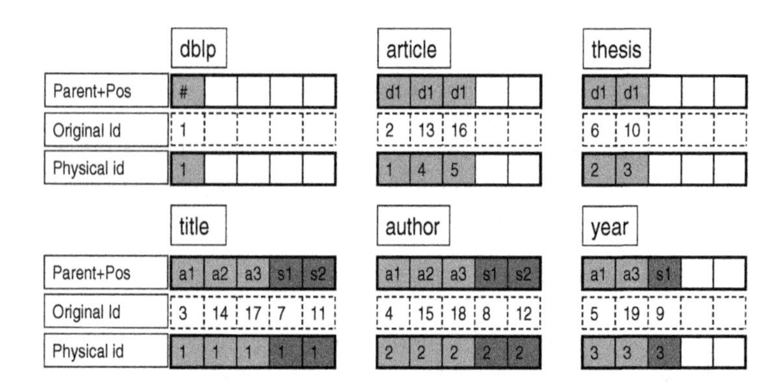


**Fig. 6.** The Location Code of Xistree

## 5 System Architecture

In this section, we'll depict architecture of Xistree model. The framework includes source data, which is composed of the XML Data and DTD, the command interpreting interface, the index engine, the query engine, index supervision agent and index data. The command interpreting interface complies with international XML standards of Xpath, Xquery Language or a subset of Xquery Language, which realizes conversion between system functions and users or applications. The index engine will create indices when files are put into the system or a user asks to create indices.

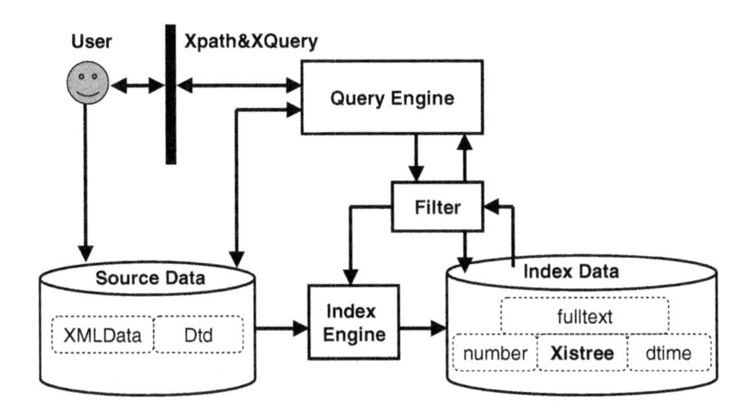


**Fig. 7.** The System Architecture

As shown in Figure 7, query engine is the key connection component between user and indexed data. It dominating task is to perform specific functions and operations, including supervision of index model, optimization of query execution, and etc. Sometimes the query engine may query source data directly.

The index supervision proxy is responsible for operations concerning specific index structure. It may query, update, maintain indices, or coordinate multiple concurrent threads running, or even operating in parallel. The proxy likely implements query task independent of specific index structure, localizes index structure according to its characteristics and assures that the alteration of index structure has no impact on the running of the whole system. In a multi-computers, multi-indices system, the index supervision proxy may help to fulfill the task efficiently.

## 6 Experimental Results

We have implemented path query function and full text query function of every node in the demo system. We finally select Timber [17] and XIndice [10] as reference systems for two reasons: on one hand, they are prominent systems in this field, on the other hand, they share similarity with Xistree as for the category of indexing methods.

The experiment is conducted under the environment of AMD2500+ CPU, 448M memory and 160G hard disk, and the systems all run on windows 2003. The test set selected is the standard data set of 315M DBLP [18]. To show the variation of the performance on different sizes of data set, we create data file of 0.5M, 1M, 3M, 5M, 11M, 20M, 30M, 50M, 113M, 315M separately. After

**Table 2.** Sample Queries

| | |
|---|---|
| Q1 | /dblp//journal[contains(.,"Advances in Computers")] |
| Q2 | /dblp//url[contains(.,"http://www.yahoo.com/")] |
| Q3 | /dblp//author[contains(.,"Tim Berners-Lee")] |
| Q4 | //*[contains(mdate,"2002-01-03")] |
| Q5 | //year[contains(.,"1994")] |

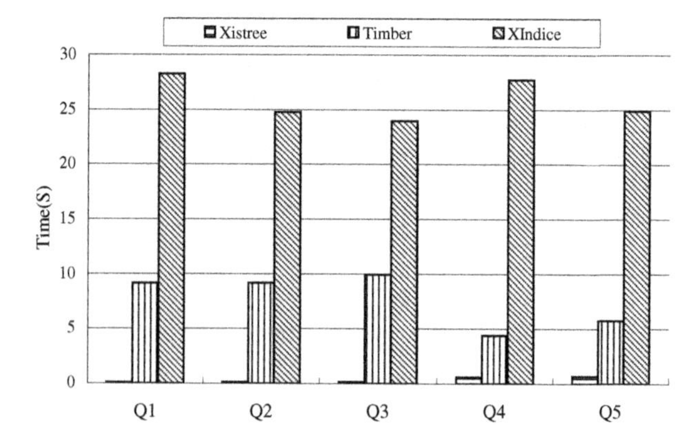


**Fig. 8.** The Query Time on 5MB Set

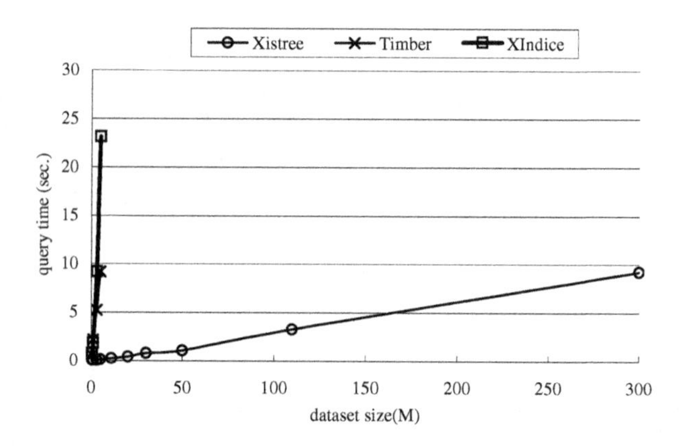


**Fig. 9.** The Query Time of Q2

removing the biggest and the smallest figures, all the following data is the average result of 5 groups of experiments.

The query cases are selected and presented in Table 2.

Figure 8 shows the query time cost of Timber, XIndice and Xistree on 5MB data set. The average time cost of Xistree with Q1, Q2, Q3, Q4, Q5 is 0.4s, while that of Timber is 7.7s, and that of XIndice is 25.9s.

Figure 9 shows the experiment result of Timber, XIndice and Xistree on all data set under Q2. What we should mention is that the system crash on 11M data set occurs due to memory deficiency during experiments of XIndice and Timber, therefore no more result data can be collected. And Xistree runs smoothly on a single file of 315M.

The time of index creating consumed on 11MB data set by Timber is 402s, higher than 401.9s, which is the cost on 110MB data set by Xistree.

At present, the index data of Xistree is nearly 1.85 times of the original size, while that of Timber is 2.64 times. It should be noted that even though the index occupies a great deal of space, the ratio of path index to original XML data is only 0.11, that is, the path index part of 315M data set is 35.1M.

## 7 Conclusion

The demo system is developed from scratch, and the index structure is simple, consistent and easy to maintain. It can support queries started at any point in the path with high efficiency, and it is very suitable for queries with a known path pattern. Experiments demonstrate that this method is practically applicable to query of huge XML data storage.

In the subsequent research, we will realize the optimizing method mentioned above and go into the problem of XML tags overlapping [12]. Meanwhile, we will carry out more experiments in order to develop this method into an all-round and excellent practical system.

**Acknowledgments.** This work is supported by National Natural Science Foundation of China under grants no. 60473070 and no. 70471011.

## References

1. H. V. Jagadish, S. Al-Khalifa, A. Chapman, et al. Timber: A native XML database. The VLDB Journal, 11: 274-291, 2002.
2. R. Goldman and J. Widom. Dataguides:enabling query formulation and optimization in semistructured databases. Proc. of VLDB'97, pp. 436-445, 1997.
3. B. Cooper, N. Sample, M. Franklin, G. Hjaltason, and M. Shadmon. A fast index for semistructured data. Proc. of VLDB'01, pp. 341-350, 2001.
4. Q. Li, and B. Moon. Indexing and querying xml data for regular path expressions. Proc. of VLDB'01, pp. 361-370, 2001.
5. H. Wang, S. Park, W. Fan and P. Yu. Vist: A dynamic index method for querying xml data by tree structures. Proc. of ACM SIGMOD'03, pp. 110-121, 2003.
6. Q. H. Chen,A. Lim and K. Ong. D(k)-index: An adaptive structural summary for graph-structured data. Proc. of ACM SIGMOD'03, pp. 134-144, 2003.
7. G. Gottlob, C. Koch and R. Pichler. Efficient algorithms for processing xpath queries. Proc. of VLDB'02, pp. 95-106, 2002.
8. Y. Ishikawa, T. Nagai and H. Kitagawa. Transforming XPath Queries for Bottom-Up Query Processing. Proc. of ISDB'02, pp. 210-215, 2002.
9. B. Catania, A. Maddalena and A. Vakali. XML document indexes: A classification. IEEE Internet Ccomputing, 64-71, 2005.

10. XIndice: available at http://xml.apache.org/xindice/index.html, 2006.
11. Q. Zou, S. Liu, and W. Chu. Ctree: a compact tree for indexing XML data. Proc. of WIDM'04, pp. 39-46, 2004.
12. G. Kazai, M. Lalmas, and A. Vries. The Overlapping problem in Content-Oriented XML Retrieval Evaluation. Proc. of SIGIR'04, pp. 72-79, 2004.
13. S. Zhou, Y. Hu, and J. Guan. Adjacency matrix based full-text indexing models. Journal of Software, 13(10): 1933-1942, 2000.
14. Y. Hu. Inter-relevant successive trees – a new mathematical model for full-text database. Technical Report no. TR022031, Department of Computer and Information Technology, Fudan University, 2002.
15. H. Ma, C. Zhang, C., et al. Mining frequent patterns based on is+-tree model. Journal of Computer Research and Development, 42(4): 588-593, 2005.
16. C. Chung, J. Min and K. Shim. Apex: An adaptive path index for xml data. Proc. of ACM SIGMOD'02, pp. 121-132, 2002.
17. Timber: available at http://www.eecs.umich.edu/db/timber, 2004.
18. DBLP: available at http://dblp.uni-trier.de/xml/, 2006.

# Efficient Algorithms for Spatial Configuration Information Retrieval

Haibin Sun[1] and Xin Chen[2]

[1] College of Information Science and Engineering,
Shandong University of Science and Technology,
Qingdao 266510, China
Offer_sun@hotmail.com

[2] Foreign languages college, Shandong University of Science
and Technology, Qingdao 266510, China
ciccychen1@hotmail.com

**Abstract.** The problem of spatial configuration information retrieval is a Constraint Satisfaction Problem (CSP), which can be solved using traditional CSP algorithms. But the spatial data can be reorganized using index techniques like R-tree and the spatial data are approximated by their Minimum Bounding Rectangles (MBRs), so the spatial configuration information retrieval is actually based on the MBRs and some special techniques can be studied. This paper studies the mapping relationships among the spatial relations for real spatial objects, the corresponding spatial relations for their MBRs and the corresponding spatial relations between the intermediate nodes and the MBRs in R-tree. Three algorithms are designed and studied, and their performances are compared.

## 1 Introduction

Spatial configuration retrieval is an important research topic of content-based image retrieval in Geographic Information System (GIS), computer vision, and VLSI design, etc. A user of a GIS system usually searches for configurations of spatial objects on a map that match some ideal configuration or are bound by a number of constraints. For example, a user may be looking for a place to build a house. He wishes to have a house A north of the town that he works, in a distance no greater than 10km from his child's school B and next to a park C. Moreover, he would like to have a supermarket D on his way to work. Under some circumstances, the query conditions cannot be fully satisfied at all. The users may need only several optional answers according to the degree of configuration similarity. Of the configuration similarity query problem, the representation strategies and search algorithms have been studied in several papers[1,3,7,16,17,21,25,26].

A configuration similarity query can be formally described as a standard binary constraint satisfaction problem which consists of: (1) a set of $n$ variables, $v_0, v_1, \cdots v_{n-1}$ that appear in the query, (2) for each variable $v_i$, a finite domain $D_i = \{u_0, \cdots, u_{m-1}\}$ of m values, (3) for each pair of variables $(v_i, v_j)$, a constraint $C_{ij}$ which can be a simple spatial relation, a spatio-temporal relation or a disjunction of relations. In addition, unary constraints such as physical and semantical features can be added to the variables. The goal of query processing is to find instantiations of variables to image objects so that the input constraints

W. Abramowicz (Ed.): BIS 2007, LNCS 4439, pp. 339–352, 2007.
© Springer-Verlag Berlin Heidelberg 2007

are satisfied to a maximum degree. The *dissimilarity degree* $d_{ij}$ of a binary instantiation $\{v_i \leftarrow u_k, v_j \leftarrow u_l\}$ is defined as the dissimilarity between the relation $R(u_k, u_l)$ (between objects $u_k$ and $u_l$ in the image to be searched) and the constraint $C_{ij}$(between $v_i$ and $v_j$ in the query). The inconsistency degree can be calculated according to the principles such as conceptual neighborhood[25] or binary string encoding[26]. Given the inconsistency degrees of binary constraints, the inconsistency degree $d(S)$ of a complete solution $S = \{v_0 \leftarrow u_p, \cdots v_{n-1} \leftarrow u_l\}$ can be defined as:

$$d(S) = \sum_{\forall i,j,i \neq j,0 \leq i,j<n} d_{ij}(C_{ij}, R(u_k, u_l)), \{v_i \leftarrow u_k, v_j \leftarrow u_l\} \quad (1.1)$$

Given the defined dissimilarity degree $d(S)$, the similarity degree $sim(S)$, which is not affected by the problem scale and is within the range [0,1], can be defined as:

$$sim(S) = \frac{n(n-1) \cdot D - d(S)}{n(n-1) \cdot D} \quad (1.2)$$

Where $d(S)$ is the dissimilarity degree of the solution $S$ for a query, $n$ is the number of variables in a query, $n(n\text{-}1)$ is the set of constraints between distinct variable pairs (including inverse and unspecified constraints), and $D$ is the maximum dissimilarity degree between two constraint relations. Setting an appropriate minimum value MIN for $sim(S)$ can help to obtain the balance between the approximation degree of the solutions to query conditions and processing cost. The smaller the MIN, the more the solutions obtained, while the processing cost increases too.

In the real world, spatial data often have complex geometry shapes. It will be very costly if we directly to calculate the spatial relationships between them, while much invalid time may be spent. If $N$ is the number of spatial objects, and $n$ the number of query variables, the total number of possible solutions is equal to the number of $n$-permutations of the $N$ objects: $N!/(N-n)!$ . Using Minimum Bounding Rectangles (MBRs) to approximate the geometry shapes of spatial objects and calculating the relations between rectangles will reduce the calculation greatly. So we can divide the spatial configuration retrieval into two steps: firstly the rectangle combinations for which it is impossible to satisfy the query conditions will be eliminated, and then the real spatial objects corresponding to the remaining rectangle combinations will be calculated using computational geometry techniques. To improve the retrieval efficiency, the index data structure which is called R-tree[4]or the variants R+-tree[5] and R*-tree[6] can be adopted.

The next section takes topological and directional relations as examples to study the mapping relationships between the spatial relationships for MBRs and the corresponding relationships for real spatial objects; section 3 studies three spatial configuration retrieval algorithms; section 4 presents the experimental system for comparing the three algorithms, designs the experiments, analyzes the experimental results and make a conclusion; the last section concludes this paper.

## 2 Spatial Mapping Relationships

This paper mainly concerns the topological and directional relations for MBRs and the corresponding spatial relationships for real spatial objects. The ideas in this paper can be applied to other relationships such as distance and spatio-temporal relations, etc.

## 2.1 Topological Mapping Relationships

This paper focuses on RCC8[9] (see Fig.1) relations and studies the mapping relationship between the RCC8 relations for real spatial objects and the RCC8 relations for the corresponding MBRs. Let p and q be two real spatial objects, p' and q' be their corresponding MBRs. If the spatial relation between p and q is PO (Partly Overlap), then the possible spatial relation between p' and q' is PO(Partly Overlap) or TPP (Tangential Proper Part) or NTPP (Non-Tangential Proper Part) or EQ (Equal) or TPPi (inverse of Tangential Proper Part) or NTPPi (inverse of Non-Tangential Proper Part) which can be denoted by the disjunction form PO(p', q') TPP(p', q') NTPP(p', q') EQ(p', q') TPPi(p', q') NTPPi(p', q'). To use R-tree to improve the efficiency of the spatial configuration retrieval, the topological relations in the query condition should first be transformed to the corresponding topological relations for the MBRs, which can be used to eliminate the rectangle combinations that cannot fulfill the constraints from the leaf nodes in the R-tree. The intermediate nodes in the R-tree can also be used to fast the retrieval process. Let p" be the rectangle that enclose p', i.e. the parent node of leaf node p' in the R-tree, which is called intermediate node. Given the spatial relation between p' and q', the spatial relation between p" and q' can be derived. For example, from the spatial relation TPP(p', q'), the spatial relation PO(p",q') TPP(p",q') EQ(p",q') TPPi(p",q') NTPPi(p",q') can be obtained. It is very interesting that the parents of the intermediate nodes also have the same property. Table 1 presents the spatial relations between two real spatial objects, the possible spatial relations that their MBRs satisfy and the possible spatial relations between the corresponding intermediate node and the MBR.

Based on the above mapping relationship and the R-tree, the candidate MBR combinations can be retrieved efficiently, and then a refinement step is needed

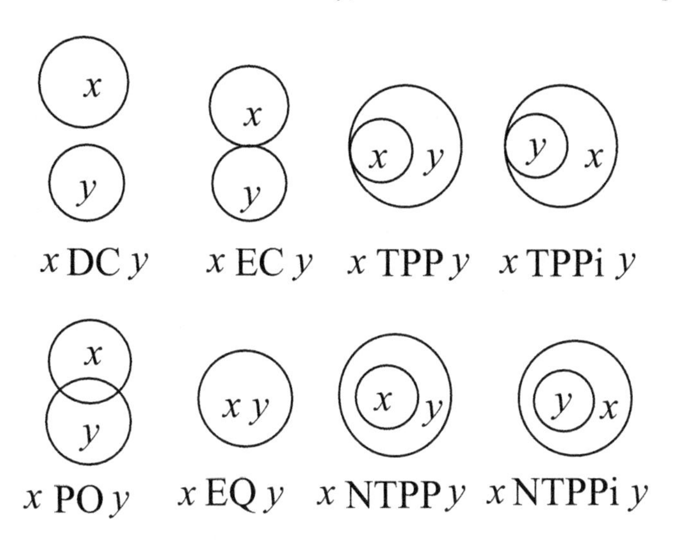


**Fig. 1.** Two-dimensional examples for the eight basic relations of RCC8

**Table 1.** The spatial relations between two real spatial objects, the possible spatial relations that their MBRs satisfy and the possible spatial relations between the corresponding intermediate node and the MBR

| RCC8 relation between p and q | RCC8 relation between MBRs p' and q' | RCC8 relation between p" and q' |
|---|---|---|
| DC(p,q) | DC(p',q') ∨ EC(p',q') ∨ PO(p',q') ∨ TPP(p',q') ∨ NTPP(p',q') ∨ EQ(p',q') ∨ TPPi(p',q') ∨ NTPPi(p',q') | PO(p",q') ∨TPP(p",q') ∨ NTPP(p",q') ∨ EQ(p",q') ∨TPPi(p",q')∨ NTPPi(p",q') ∨ EC(p",q') ∨ DC(p",q') |
| EC(p,q) | EC(p',q') ∨ PO(p',q') ∨ TPP(p',q') ∨ NTPP(p',q') ∨ EQ(p',q') ∨ TPPi(p',q') ∨ NTPPi(p',q') | EC(p",q') ∨ PO(p",q') ∨ TPP(p",q') ∨ NTPP(p",q') ∨ EQ(p",q') ∨ TPPi(p",q') ∨ NTPPi(p",q') |
| PO(p,q) | PO(p', q') ∨ TPP(p', q') ∨ NTPP(p',q') ∨ EQ(p',q') ∨ TPPi(p', q') ∨ NTPPi(p', q') | PO(p", q') ∨ TPP(p", q') ∨ NTPP(p", q') ∨ EQ(p", q') ∨ TPPi(p", q') ∨ NTPPi(p", q') |
| TPP (p,q) | TPP(p', q') ∨ NTPP(p', q') ∨ EQ(p', q') | PO(p", q') ∨ TPP(p", q') ∨ NTPP(p", q') ∨ EQ(p", q') ∨ TPPi(p", q') ∨ NTPPi(p", q') |
| NTPP (p,q) | NTPP(p', q') | PO(p", q') ∨ TPP(p", q') ∨ NTPP(p", q') ∨ EQ(p", q') ∨ TPPi(p", q') ∨ NTPPi(p", q') |
| TPPi (p,q) | EQ(p', q') ∨ TPPi(p', q') ∨ NTPPi(p', q') | EQ(p", q') ∨ TPPi(p", q') ∨ NTPPi(p", q') |
| NTPPi (p,q) | NTPPi(p', q') | NTPPi(p", q') |
| EQ(p,q) | EQ(p', q') | EQ(p", q') ∨ TPPi(p", q') ∨ NTPPi(p", q') |

to derive the spatial relations among the real spatial objects that the MBRs enclose, which means that the spatial relation between p and q should be derived from the spatial relation between p' and q'. From the spatial relation between two MBRs, we can derive several possible spatial relations or only one definite spatial relation between two real spatial objects that the MBRs enclose. In the former case the complex geometry computation will be applied whereas it will be omitted in the latter case. For example, given the spatial relation NTPPi(p', q'), we can derive DC(p, q)∨ EC(p, q)∨ PO(p, q)∨ NTPPi (p, q)∨ TPPi (p, q), the geometry computation must be adopted to ascertain the spatial relation between p and q. But if we know the spatial relation DC(p', q'), then spatial relation DC(p, q) can be derived directly.

## 2.2 Direction Mapping Relationships

According to Goyal and Egenhofer's cardinal direction model[10], there are 9 atomic cardinal direction relations(O, S, SW, W, NW, N, NE, E, SE) (see Fig.2) and totally 218 cardinal direction relations for non-empty connected regions in the Euclidean space $\Re^2$ (illustrated by 3 × 3 matrix, see Fig.3) [11].

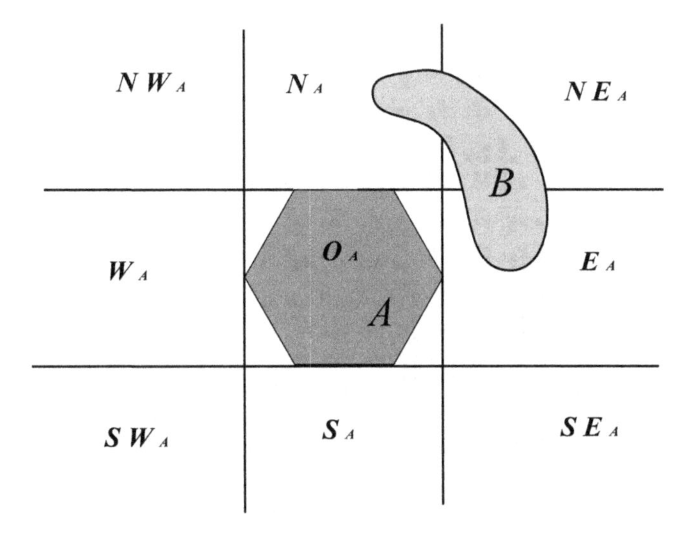


**Fig. 2.** Capturing the cardinal direction relation between two polygons, A and B, through the projection-based partitions around A as the reference object

There are 36 cardinal direction relations for the non-empty and connected regions'MBRs: O, S, SW, W, NW, N, NE, E, SE, S:SW, O:W, NW:N, N:NE, O:E, S:SE, SW:W, O:S, E:SE, W:NW, O:N, NE:E, S:SW:SE, NW:N:NE, O:W:E, O:S:N,SW:W:NW,NE:E:SE, O:S:SW:W, O:W:NW:N, O:S:E:SE, O:N:NE:E, O:S:SW:W:NW:N, O:S:N:NE:E:SE, O:S:SW:W:E:SE, O:W:NW:N:NE:E, O:S:SW:W:NW:N:NE:E:SE(see Fig.4). This kind of cardinal direction relation has the rectangle shape, so it is also named *rectangle direction relation*, otherwise it is called *non-rectangle direction relation.*

In the following, we study the mapping relationships between the cardinal direction relations for real spatial objects and the cardinal direction relations for the corresponding MBRs. First of all, we give a definition as follows.

**Definition 1.** *a cardinal direction relation R contains another cardinal direction relation R', if all the atomic relations in R' also exist in R.*

The mapping relationships from the cardinal direction relations for real spatial objects to the ones for their MBRs can be described using the following theorems.

**Theorem 1.** *if the cardinal direction relation between the real spatial objects p and q is rectangle direction relation R(see Fig.4), the cardinal direction relation between their MBRs p' and q' is also R; if the cardinal direction relation between the real spatial objects p and q is non-rectangle direction relation R, the cardinal direction relation between their MBRs p' and q' is the rectangle direction relation R' in Fig.4 which contains relation R and has the minimum area.*

Theorem 1 can be derived by combining Fig.3 and Fig.4. Assume that the cardinal direction relation between two real spatial objects p and q is N:NW:W which obviously is not rectangle direction relation, from Fig.4 the rectangle

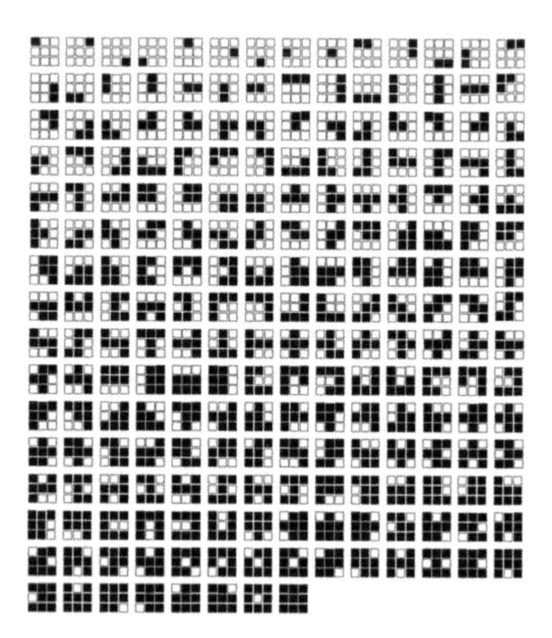

**Fig. 3.** 218 cardinal direction relations between two non-empty and connected regions[11]

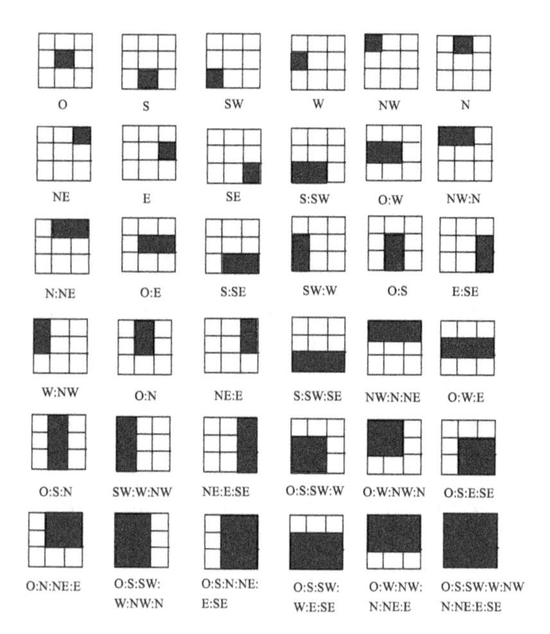


**Fig. 4.** 36 cardinal direction relations for MBRs

direction relation that *contains* N:NW:W and has the minimum rectangle area is O:W:NW:N, so the cardinal direction relation between two MBRs p' and q' is O:W:NW:N.

Similarly the mapping relationships from the cardinal direction relations for MBRs to the ones for the possible real spatial objects can be described as follows.

**Theorem 2.** *if the cardinal direction relation R between two MBRs p' and q' contains no more than 3 atomic cardinal direction relations (including 3), the corresponding cardinal direction relation between the real spatial objects p and q is also R; otherwise, the possible cardinal direction relations between p and q will be the subsets of relation R which can be transformed to relation R when p and q are approximated by p' and q'.*

For example, if the cardinal direction relation between two MBRs is S:SW:SE(including three atomic relations:S,SW,SE), then the cardinal direction relation between the corresponding two real spatial objects definitely is S:SW:SE. If the cardinal direction relation between two MBRs is O:S:SW:W, the possible cardinal direction relations between two real spatial objects include O:W:SW, W:O:S,SW:S:O,SW:S:W and O:S:SW:W.

Given the cardinal direction relation between the MBRs p' and q', the cardinal direction relation between p", which is the parent node of p' in R-tree, and q' can be described using the following theorem.

**Theorem 3.** *if the cardinal direction relation between MBRs p' and q' is R, the possible cardinal direction relations between p" and q' are the rectangle direction relations containing R.*

For example, if the cardinal direction relation between p' and q' is O:S:SW:W, the possible cardinal direction relations between p" and q' will be O:S:SW:W, O:S:SW:W:NW:N, O:S:SW:W:E:SE and O:S:SW:W:NW:N:NE:E:SE.

# 3 Algorithms for Spatial Configuration Information Retrieval

As mentioned above, topological relations and directional relations can all be transformed into spatial relations for MBRs. Given a spatial configuration query, we can first transform the spatial relations constraints among spatial objects into spatial relations constraints among their MBRs, then some efficient retrieval algorithm and the R-tree that organizes the spatial data can be applied to pick out the MBR combinations that will later be checked using computational geometry technique(some real spatial relations can be derived directly from MBR combinations according to spatial relation properties, e.g., theorem 2). This paper concentrates on systematic algorithms in spatial configuration information retrieval for spatial relations among MBRs, which are transformed from the spatial relations in the query constraints according to table 1 and theorem 1. Heuristic algorithms are not in the scope of this paper. Systematic algorithms use techniques for constraint satisfaction problems(e.g., forward checking), dynamic variable ordering and R-tree, etc. to process the domain values of variables efficiently. This paper studies and implements three algorithms.

## 3.1 SFC-DVOSolver Algorithm

The spatial configuration information retrieval problem is essentially a binary constraint satisfaction problem[12]. Many efficient algorithms have been put forward to solve constraint satisfaction problems. One of these efficient methods is Forward Checking (FC), which has been proved to be better than other methods on many problems[13,14]. Dynamic Variable Ordering (DVO) is a strategy that some algorithms use to improve efficiency for solving constraint satisfaction problems[15]. The main idea of FC-DVO is to sort the uninstantiated variables according to the domain size and select the variable having the smallest domain as the next one to be instantiated after each forward checking process. The algorithm that adopts the FC-DVO strategy to solve spatial configuration information retrieval problem is named SFC-DVOSolver.

The R-tree and its variants have been used to improve the efficiency of spatial information retrieval in many applications[16,17,18]. There are two ways to combine SFC-DVOSolver algorithm and R-tree: one is to use R-tree to fast pruning in the forward checking process, e.g., RSFC-DVOSolver and RSFC-DVOSolver which will be introduced in the following section; another is to apply SFC-DVOSolver to R-tree from the top level to the bottom level, which means all the nodes in the same level of R-tree compose the domain of every variable. The latter way has been proved to be very costly, so we particularly present the former way.

### 3.2 RSFC-DVOSolver and HRSFC-DVOSolver Algorithms

The basic idea for the RSFC-DVOSolver algorithm is that when the current variable $V_i$ is assigned a value $U_k$, the domain of any uninstantiated variable $V_j$ is rebuilt by searching all the leaf nodes in the R-tree satisfying the query constraint $C_{ji}$ between $V_i$ and $V_j$ using $U_k$ as the search window. We present the algorithm WindowSearch (see Fig.5) that utilizes R-tree to search values that satisfy the query constraint $C_{ji}$ with $U_k$, and the other parts of RSFC-DVOSolver algorithm are basically the same as SFC-DVOSolver algorithm. Given the query constraint $C_{ji}$ between two real spatial objects p and q, we can obtain the relation constraint $R_{ji}$ for their MBRs p' and q' and the relation constraint $R^{'}_{ji}$ between the intermediate node p" and q' according to table 1, theorem 1 and theorem 3. Algorithm WindowSearch checks whether the parameter root is a leaf node or not (when the function is first called, it is the root node of the R-tree). If it is a leaf node, for any MBR $U_l$ that the parameter root contains the algorithm checks if there exists the spatial relation $R_{ji}(U_l, U_k)$ . if the relation $R_{ji}(U_l, U_k)$ exists and $U_l$ also belongs to the old domain of $V_j$ , then $U_l$ is added to the new domain of $V_j$ ; if root is not a leaf node, for any intermediate node $U_l$ that the parameter root contains the algorithm checks if there exists the spatial relation $R^{'}_{ji}(U_l, U_k)$. if the relation $R^{'}_{ji}(U_l, U_k)$ exists, for any childnode $U_l$ of the root it is passed to the parameter root and the algorithm WindowSearch is called recursively. In the above algorithms the MBRs that satisfy the query constraints are not directly added to the new domain but have the set intersection operation with the old domain, because the old domain might have been pruned. The set intersection operation includes the float computation of checking if two rectangles are equal, which is time-consuming. To improve the efficiency of rectangle matching, we use the hash index technique [18] for reference to reorganize the MBRs data (i.e. the rectangle data that the leaf nodes in R-tree contain). We use a number of buckets to accommodate the MBRs, and through calculating the index value

```
WindowSearch(Variable: Vi,Vj, Rtree_Node root, Value
Uk)
  IF root.isLeaf() THEN
    FOR all Ulroot DO
      IF Rji (Ul, Uk) THEN
        IF Ul is in the domain of Vj then Add Ul  to the
newdomain of Vj;
      END IF
    END FOR
  ELSE
    FOR all Ulroot DO
     IF Rji' (Ul, Uk) THEN
       WindowSearch(Vi,Vj, root.getChild(), Uk);
     END IF
    END FOR
```

**Fig. 5.** WindowSearch function in the RSFC-DVOSolver algorithm

according to an MBR's coordinates the algorithm decides to put it into which bucket. The set intersection operation will be replaced by calculating the index value i of an MBR that satisfies the query constraint and comparing it with all the MBRs in the bucket with the index value i. The RSFC-DVOSolver algorithm using the hash index technique is named HRSFC-DVOSolver.

## 4 Experiments

The algorithms SFC-DVOSolver, RSFC-DVOSolver and HRSFC-DVOSolver are implemented in this paper and their performance is analyzed. The programmes are developed in Java language with the free Eclipse development environment [27], which have been run on a PC(1.7G) with 256MB of RAM. The traditional CSP algorithms are implemented using the JCL (Java Constraints Library) algorithm library[19] developed by Bhattacharjee etc.. The R-tree algorithms by Guttman[4]are implemented using Hadjieleftheriou's source code[20]. Based on these open source codes, the algorithms mentioned above in this paper are realized. To evaluate these algorithms, this paper only considers topological relations (adding the other spatial relations can make the algorithms run faster, but it is unnecessary) and requires that the query condition must be completely satisfied, i.e. the similarity degree is 1. The system implemented in this paper can be applied to the spatial information system [23] based on qualitative spatial reasoning[22].

The experiment data in this paper include MBR datasets that are randomly produced with different quantities and different data densities (the data density is the sum of all the areas of the MBRs divided by the area of the whole space) and the real German street MBR dataset[24]. For every dataset, we design queries respectively with 3, 5, 7, 10 and 15 variables and 10 query instances for every query. The results of 10 query instances are averaged for every query. The algorithms SFC-DVOSolver, RSFC-DVOSolver and HRSFC-DVOSolver are run with a certain dataset and a query.

To use R-tree there are two parameters that should be decided. One is the maximum number of entries that will fit in one node of the R-tree (we call it *capacity*); the other is the worst-case space utilization for all nodes, i.e. the rate of the minimum number of entries that will fit in one node of the R-tree to capacity (we call it *fillfactor*). According to the previous experiences and our experiments, the capacity and the fillfactor are set to 40 and 0.5, respectively.

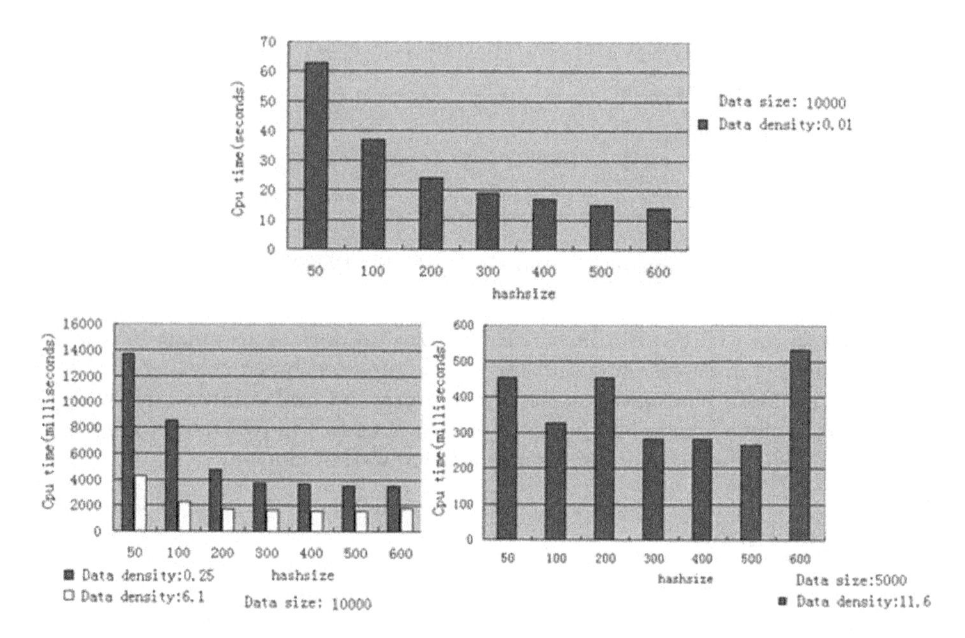


**Fig. 6.** The illustration of the relationship between the number of buckets and the running time of algorithm HRSFC-DVOSolver (R-tree fillfactor=0.5, R-tree capacity=40, and the number of query variables=5) on four datasets

The number of the hash buckets (hashsize) should also be decided first. We use three datasets, each of which has 10,000 MBRs, with data densities 0.01, 0.25 and 6.1, respectively, and one dataset, which has 5,000 MBRs, with data density 11.6. For every dataset, the algorithm HRSFC-DVOSolver with 5 query variables and R-tree (fillfactor=0.5, capacity=40) is run to test the performance when the hashsize is assigned the value 50, 100, 200, 300, 400, 500, 600, respectively. The results are illustrated in Fig.6. From the figure with data size 10000 and data density 0.01, we can see that the efficiency of this algorithm improves sharply when hashsize changes from 50 to 500, which gives us the intuition that the efficiency of the algorithm should become better and better as the hashsize increases. But from the figure with data size 10000 and data densities 0.25 and 6.1, we can see that the efficiency decreases as the figure with data size 5000 and data density 11.6 also shows. The reason for this phenomenon is that when the data density increases, the data will be assigned to each hash bucket more uniformly, which results in that the performance of this algorithm will not improve much. The performance of this algorithm maybe decrease because when the number of hash buckets increases to an extent the stack space of the Java Running Environment will be occupied so much that the efficiency of the running programme will be affected negatively. So for the experiments in this paper, the hashsize is set to 500.

The algorithms SFC-DVOSolver, RSFC-DVOSolver and HRSFC-DVOSolver are tested using query sets with 3, 5, 7, 10 and 15 variables (10 queries per set) over the dataset with data size 5000 and data density 1.99. The parameters fillfactor and capacity of the R-tree are set to 0.5 and 40, respectively, and the hashisize is set to 500. The experiment results are displayed in table 2. The first column of table 2 shows the number of variables, the second column shows the topological relations and their proportions in the queries, the third column shows the cpu time in milliseconds required for the algorithm SFC-DVOSolver to find a solution for different queries, the fourth column shows the cpu time in milliseconds required for the algorithm RSFC-DVOSolver to find a solution for different queries, and the fifth column shows the cpu time in milliseconds required for the algorithm HRSFC-DVOSolver to find a solution for different queries. From table 2, we can see that with all the query conditions except for the one with 3 variables and topological relations property PO:NTPP=1:2 (i.e. in the topological relations of the query the proportion of relation PO is 1/3, and the proportion of relation NTPP is 2/3) the algorithm SFC-DVOSolver outperforms the algorithm RSFC-DVOSolver. For the query with 3 variables and topological relations property PO:NTPP=1:2, both the algorithm RSFC-DVOSolver and the algorithm HRSFC-DVOSolver outperform the algorithm SFC-DVOSolver, and especially the algorithm HRSFC-DVOSolver outperforms the algorithm SFC-DVOSolver by an order of magnitude. In all the cases the algorithm HRSFC-DVOSolver outperforms, or has almost the same performance as, the algorithm SFC-DVOSolver. In some cases (especially 5, 7, and 15) the algorithm RSFC-DVOSolver performs the worst.

In the following we analyze the results presented in table 2. From table 1 we can see that the corresponding topological relation for MBRs of topological relation DC is the disjunction of all RCC8 relations, which adds the processing of many intermediate nodes to the algorithms using the R-tree. So for the

**Table 2.** the efficiency comparison of algorithms SFC-DVOSolver, RSFC-DVOSolver and HRSFC-DVOSolver tested with different query conditions

| The number of query variables | Topological relations and their proportions in the query | SFC-DVO Solver(ms) | RSFC-DVO Solver(ms) | HRSFC-DVO Solver(ms) |
|---|---|---|---|---|
| 3 | PO:NTPP= 1:2 | 2994 | 661 | 301 |
| 5 | PO:DC= 2:3 | 3175 | 110249 | 2233 |
| 7 | PO:NTPP:DC= 7:3:11 | 9373 | 339538 | 10745 |
| 10 | PO:NTPP:DC= 28:5:12 | 23945 | 67487 | 10055 |
| 15 | PO:NTPP:DC= 53:9:43 | 64503 | 2167257 | 45415 |

queries that contain the topological relation DC the algorithm RSFC-DVOSolver performs worse than the algorithm SFC-DVOSolver, by orders of magnitude in some cases (e.g., 5, 7, 15). According to table 1 the corresponding topological relations for MBRs and intermediate nodes of PO, TPP, NTPP, TPPi, NTPPi and EQ are much less than the corresponding topological relations of DC, which helps the algorithms RSFC-DVOSolver and HRSFC-DVOSolver using R-tree to improve the efficiency of pruning the search space. For example, for the query with 3 variables and property PO:NTPP=1:2 the algorithms RSFC-DVOSolver and HRSFC-DVOSolver outperform the algorithm SFC-DVOSolver. For the algorithm HRSFC-DVOSolver combines the R-tree technique with the hash index strategy that improves the efficiency of MBRs matching, it outperforms, or has almost the same performance as, the algorithm SFC-DVOSolver even if the query contains the topological relation DC. We get the same conclusion when the datasets with different size and density are used to test the performances of the three algorithms.

Based on the above experiment results, we can see that the algorithms SFC-DVOSolver and HRSFC-DVOSolver are feasible ones that solve the spatial configuration information retrieval problems, and they should be applied according to the query property. If the query constraint is weak (e.g. DC, that the query constraint is weak or strong can be judged by the knowledge base like table 1), the algorithm SFC-DVOSolver can be applied; if an ordinary or strong query constraint is given, we should use the algorithm HRSFC-DVOSolver. But it should be known that when the number of query variables is big (e.g., bigger than 20) or the dataset is huge (e.g., the number of MBRs it contains is above one hundred thousand), the systematic algorithms (like SFC-DVOSolver and HRSFC-DVOSolver) are not recommended, and the heuristic algorithms should be considered.

## 5 Conclusion

This paper has studied the spatial configuration information retrieval problem which includes 1) the mapping relationship among the spatial relations (topological and directional relations) for real spatial objects, the corresponding spatial relations for the corresponding MBRs and the corresponding spatial relations between intermediate nodes and the MBRs in R-tree, and 2) three systematic search algorithms. Through experiments two of the three algorithms are shown to be feasible, i.e. SFC-DVOSolver and HRSFC-DVOSolver. This paper has worked out the parameter values the algorithms need and the query conditions under which we can choose one of the two algorithms to be applied. The research work of this paper is valuable for the information retrieval system related to spatial data.

In the future, the strategies (i.e. R-tree and hash index) used in this paper can be applied to the heuristic search algorithms, and the research results in this paper can be generalized to the spatio-temporal information retrieval.

## References


1. Bergman, L. Castelli, V., Li C-S. Progressive Content-Based Retrieval from Satellite Image Archives. D-Lib Magazine, October 1997. http://www.dlib.org/dlib/october97/ibm/10li.html.
2. Chang S.F, Smith J.R., Meng H.J, Wang H., Zhong D. Finding Images/Video in Large Archives. CNRI Digital Library Magazine, Feb.1997. http://www.dlib.org/dlib/december02/marchionini/12marchionini.html.
3. Gupta A., Jain R. Visual Information Retrieval. Communications of ACM, May 1997, 40(5): 70-79.
4. Guttman, A. R-trees: A Dynamic Index Structure for Spatial Searching. In: Proc. Of ACM SIGMOD, 1984, pages 47-57.
5. Timos K. Sellis, Nick Roussopoulos, Christos Faloutsos. The R+-Tree: A Dynamic Index for Multi-Dimensional Objects. In: Proceedings of 13th International Conference on Very Large Data Bases, September 1-4, 1987, Brighton, England, pages 507-518.
6. Norbert Beckmann, Hans-Peter Kriegel, Ralf Schneider, Bernhard Seeger. The R*-tree: An Efficient and Robust Access Method for Points and Rectangles. In: Proceedings of the ACM SIGMOD, 1990, pages 322-331.
7. Orenstein, J. A. Spatial Query Processing in an Object-Oriented Database System. In: Proc. Of the 1986 ACM SIGMOD international conference on Management of data, 1986, pages 326-336.
8. James F. Allen. Maintaining knowledge about temporal intervals. Communications of the ACM, November 1983, 26(11): 832-843.
9. D. A. Randell, Z. Cui and A. G. Cohn. A Spatial Logic Based on Regions and Connection. In: Proc. 3rd Int. Conf. on Knowledge Representation and Reasoning, Morgan Kaufmann, San Mateo, 1992, pages 165-176.
10. R. Goyal and M. Egenhofer. Cardinal Directions between Extended Spatial Objects. IEEE Transactions on Knowledge and Data Engineering, 2000(in press).
11. S.Cicerone and P. Di Felice. Cardinal directions between spatial objects: the pairwise-consistency problem. Information Sciences. 2004, 164(1-4): 165-188.
12. Nadel, B. Constraint Satisfaction Algorithms. Computational Intelligence, 1989, 5: 188-224.
13. Haralick, R.M., Elliott, G.L. Increasing tree search efficiency for constraint satisfaction problems. Artificial Intelligence, 1980, 14: 263-313.
14. Bacchus, F., Grove, A. On the Forward Checking Algorithm. In Proceedings the First International Conference on Principle and Practice of Constraint Programming, 1995,pages 292-309.

15. Bacchus, F., van Run, P. Dynamic Variable Ordering in CSPs. In Proceedings the First International Conference on Principle and Practice of Constraint Programming, 1995, pages 258-275.
16. Xiao Yu-Qin, Zhang Ju, Jing Ning; Li Jun. Direction relation query processing using R-Trees. Journal of Software 2004,15(1): 103-111.
17. Dimitris Papadias, Nikos Mamoulis, Vasilis Delis. Algorithms for Querying by Spatial Structure. In: Proceedings of 24rd International Conference on Very Large Data Bases, August 24-27, 1998, New York, USA, pages 546-557.
18. Jiang xiajun, Wu huizhong, and Li Weiqing. R-tree method of matching algorithm for data distribution management. Journal of Computer Research and Development,2006,43(2): 362-367.
19. Rajat Bhattacharjee, et al.. JCL: Java Constraints Library, http://liawww.epfl.ch/JCL/index.htm.
20. Marios Hadjieleftheriou. SIL: Spatial Index Library. http://u-foria.org/marioh/spatialindex/index.html.
21. M. Egenhofer. Query Processing in Spatial-Query-by-Sketch. Journal of Visual Languages and Computing, 1997, 8(4): 403-424.
22. Haibin Sun, Wenhui Li. Spatial Reasoning Combining Topological and Cardinal Directional Relation Information. Journal of computer research and development2006,43(2):253-259.
23. Yao, X., Thill, J.C. Spatial Queries With Qualitative Locations In Spatial Information Systems. Computers, Environment and Urban Systems. 2005 (In Press). http://www.ggy.uga.edu/people/faculty/xyao/QL05.pdf
24. MBRs of streets (polylines) of Germany. http://isl.cs.unipi.gr/db/projects/rtreeportal/spatial.html
25. Nabil M., Ngu A., Shepherd J. Picture Similarity Retrieval using 2d Projection Interval Representation. IEEE TKDE, 8(4), 1996.
26. Papadias D., Mantzourogiannis M., Kalnis P., Mamoulis N., Ahmad I. Content-Based Retrieval Using Heuristic Search. ACM SIGIR, 1999.
27. Eclipse. http://www.eclipse.org/

# String Distance Metrics for Reference Matching and Search Query Correction

Jakub Piskorski[1] and Marcin Sydow[2]

[1] Joint Research Center of the European Commission
Web and Language Technology Group of IPSC
T.P. 267, Via Fermi 1, 21020 Ispra (VA), Italy
Jakub.Piskorski@jrc.it

[2] Polish-Japanese Institute of Information Technology (PJIIT)
Department of Intelligent Systems
Koszykowa 86, 02-008 Warsaw, Poland
msyd@pjwstk.edu.pl

**Abstract.** String distance metrics have been widely used in various applications concerning processing of textual data. This paper reports on the exploration of their usability for tackling the reference matching task and for the automatic correction of misspelled search engine queries, in the context of highly inflective languages, in particular focusing on Polish. The results of numerous experiments in different scenarios are presented and they revealed some preferred metrics. Surprisingly good results were observed for correcting misspelled search engine queries. Nevertheless, a more in-depth analysis is necessary to achieve improvements. The work reported here constitutes a good point of departure for further research on this topic.

**Keywords:** string distance metrics, reference matching, search engine query correction, information retrieval, inflective languages.

## 1 Introduction

In many knowledge discovery applications the major issue is to combine information originating from different sources. The crucial task is to identify equivalent data, since data often lacks a unique, global identifier. This is in particular the case of natural language textual data, where one can refer to the same thing in many different ways. Therefore, a frequently appearing problem in the context of text processing technologies involves making a decision whether two distinct strings refer to the same object. Since the emergence of computer science various communities, e.g., information retrieval, database, artificial intelligence, statistics, have been heavily researching string distance metrics for tackling this problem.

In this paper, we focus on exploring the usability of the well-established string distance metrics for reference matching and automatic spelling correction of Web queries in the context of processing textual data in highly inflected languages.

W. Abramowicz (Ed.): BIS 2007, LNCS 4439, pp. 353–365, 2007.
© Springer-Verlag Berlin Heidelberg 2007

In particular, we present some results of our experiments carried out on a Polish proper-name dataset and on a collection of Web logs from a leading search engine in Poland.

The problem of reference matching centers around identifying co-referring entities in a text, i.e., finding text fragments which refer to the same real-world object, e.g., person, organization, etc. For instance, the text phrases *Dr. Jan Kowalski*, *Dr Kowalski*, *Kowalski* and *Kowlski* (typographical error) might constitute different textual mentions of the same person. Reference matching has been studied thoroughly in the past and approaches ranging from linguistically oriented ones [1] to very lightweight approximate-string matching techniques have been proposed. A comparison of string distance metrics for name matching tasks has been given recently in [2,3]. The main motivation for our work is the fact that processing highly inflective languages adds another complication to reference matching. The intuitive way of combating the inflection problem would be to lemmatize names, and then to apply techniques which turned out to work fine for inflection-poor languages like English. However, the lemmatization of proper names in Polish (and similar languages) is both knowledge and time intensive. Accuracy figures of more than 80% have not been reported [4]. To illustrate the difficulty of the task, let us consider the declension of full names consisting of a first name and a surname. Its lemmatization might depend on several factors, including: (a) the gender of the first name, (b) the part-of-speech information and gender of the word which constitutes the surname, and (c) origin (country) and pronunciation of the name. Clearly, even obtaining some of such information might not be possible without contextual knowledge. Therefore, we decided to choose the string similarity lane to cope with reference matching. However, it is important to note at this stage that this paper explores only the performance of the string distance metrics, but it does not address a fully-fledged solution to the reference matching task.

Our second line of experiments deals with applying string distance metrics to the problem of an automated correction of spelling errors in search engine queries in the same context of highly inflected languages, in particular Polish. Automatic spelling correction in search engine queries poses a problem of high practical importance, since a substantial portion of users' queries are misspelled. There exist special techniques based on statistical modeling and statistical learning for solving the problem [5,6], but application of the simple string-similarity techniques for reference matching turned out, as our experiments show, to be quite successful for 'guessing' intended queries. Actually, automatic spelling correction can be seen as a special variant of reference matching problem, since for an erroneous query one tries to find a correct one from a Web log collection, which refers to the same concept.

The remaining part of this paper is organized as follows. First, in section 2 we introduce the standard string distance metrics used in the experiments. Next, sections 3 and 4 describe in detail the experiments and evaluation. Finally, we end up with some conclusions and an outlook in section 5.

## 2 String Distance Metrics

Today, quite a number of string distance metrics[1] exist which are applied in different fields of science. In our work, we explore the ones used mainly by the database community for record linkage and matching. The point of departure constitutes the well-known *Levenshtein* edit distance metric given by the minimum number of character-level operations (insertion, deletion, or substitution) needed to transform one string into the other [7]. In case of the *Levenshtein* metric all operations are assigned a unit cost. The basic algorithm for computing Levenshtein distance between two strings $s$ and $t$ runs in $O(|s| \cdot |t|)$ time. However, detecting whether two strings have Levenshtein distance less than $k$ can be done in $O(max\{|s|, |t|\} \cdot k)$ time [8].

There are several extensions to the basic Levenshtein metric. The *Needleman-Wunsch* [9] metric modifies the original one in that it allows for variable cost adjustment to the cost of a gap, i.e., insert/deletion operation and variable cost of substitutions (e.g., the cost of replacing $A$ with $a$ might be smaller than the cost of replacing $k$ with $z$). Another variant considered here is the *Smith-Waterman* metric [10], which additionally uses an alphabet mapping to costs. A further variant of the latter one introduces two extra edit operations, *open gap* and *end gap*. The cost of extending the gap is usually smaller than the cost of opening a gap, and this results in small cost penalties for gap mismatches than the equivalent cost under the standard edit distance metrics. We will refer to the aforesaid metric as *Smith-Waterman-With-Affine-Gaps*. Computation of the latter metric requires $O(|s| \cdot |t| \cdot g)$ time, where $g$ denotes the maximum length of the gap, for $g \leq \min\{|s|, |t|\}$. All other aforementioned variants of the basic edit distance metric can be computed in $O(|s| \cdot |t|)$ steps.

Good results in the context of name-matching tasks [2] have been reported using variants of the *Jaro* metric [11], which is not based on the edit-distance model. It considers the number and the order of the common characters between two strings. Given two strings $s = a_1 \ldots a_K$ and $t = b_1 \ldots b_L$, we say that $a_i$ in $s$ is *common* with $t$ if there is a $b_j = a_i$ in $t$ such that $i - R \leq j \leq i + R$, where $R = \min(|s|, |t|)/2)$. Further, let $s' = a'_1 \ldots a'_K$ be the characters in $s$ which are common with $t$ (with preserved order of appearance in $s$) and let $t' = b'_1 \ldots b'_L$ be defined analogously. A *transposition* for $s'$ and $t'$ is defined as the position $i$ such that $a'_i \neq b'_l$. Let us further denote the number of transposition for $s'$ and $t'$ as $T_{s',t'}$. The Jaro similarity for strings $s$ and $t$ is then defined as follows.

$$Jaro(s,t) = \frac{1}{3} \cdot (\frac{|s'|}{|s|} + \frac{|t'|}{|t|} + \frac{|s'| - (T_{s',t'}/2)}{|s'|}) \tag{1}$$

A *Winkler* variant of this measure takes into account the length $P$ of the longest common prefix of $s$ and $t$ and is defined as follows.

[1] Distance metrics map a pair of strings $s$ and $t$ to a real number $r$, where a smaller value of $r$ indicates greater similarity. In this paper we also refer to *similarity metrics* which are defined in an analogous way, except the difference that larger values of $r$ indicate greater similarity.

$$Jaro\text{-}Winkler(s,t) = Jaro(s,t) + \frac{\max(P,4)}{10} \cdot (1 - Jaro(s,t)) \quad (2)$$

Jaro family of metrics are intended primarily for short strings. Their time complexity is $O(|s| \cdot |t|)$.

The *q-gram* metric [12], heavily exploited for approximate string matching, is based on the intuition that two strings are similar if they share a large number of character-level q-grams. Let $G(s)$ denote the set of all q-grams of a string $s$ obtained by sliding a window of length $q$ over the characters of $s$. The formal definition of q-gram metric is given below

$$q\text{-}gram\text{-}distance(s,t) = 1 - \frac{|G(s) \cap G(t)|}{|G(s) \cup G(t)|} \quad (3)$$

Since q-grams at the beginning and the end of the string can have fewer than $q$ characters, the strings are conceptually extended by *padding* the beginning and the end of the string, i.e., adding $q-1$ unique initial and trailing characters to a string. Most frequently literature reports on using 3-grams for name-matching tasks (our default setting). Clearly, q-gram distance can be computed in $O(\max\{|s|,|t|\})$ steps.

Character-level similarity metrics focus on string-based representation, but strings may be phonetically similar even if their string representation is not similar. One of the phonetic-based encodings of strings, namely *Soundex* [13], turned out to be useful for matching surnames. The idea is to assign each string a code, consisting of a character and three digits. The character corresponds to the first character of the string, whereas the digits are computed via: (1) assigning each consonant (except the first character in the string) a code digit, so that phonetically similar consonants are mapped to the same digit (vowels are not coded), (2) consolidating the sequence constructed in (1) by replacing sequences of identical digits into one digit, and (3) by eventually dropping digits or padding zeros depending on the length of the code after step (2). Intuitively, names referring to the same person have identical or similar Soundex code. Consequently, the Soundex codes are compared for their similarity using an arbitrary string similarity metric. Although Soundex coding can be parametrized w.r.t. encoding of single characters, we used the default settings in our experiments, which was reported to work satisfactorily for many languages. Computing Soundex code is linear in the length of the string.

Finally, for multi-token strings we tested the recursive matching schema, known also as *Monge-Elkan* distance [14]. Let us assume that the strings $s$ and $t$ are broken into substrings (tokens), i.e., $s = s_1 \ldots s_K$ and $t = t_1 \ldots t_L$. The intuition behind Monge-Elkan measure is the assumption that $s_i$ in $s$ corresponds to a $t_j$ with which it has highest similarity. The similarity between $s$ and $t$ equals the mean of these maximum scores. Formally, the Monge-Elkan metric is defined as follows, where *sim* denotes some secondary similarity function.

$$Monge\text{-}Elkan(s,t) = \frac{1}{K} \cdot \sum_{i}^{K} \max_{j=1 \ldots L} sim(s_i, t_j) \quad (4)$$

We experimented with different secondary similarity functions in this context, including Jaro, Jaro-Winkler, Smith-Waterman, Levenshtein, and q-gram.

# 3 Matching Entities

This section describes our experiments on using different metrics for the entity matching task. There are several ways of formulating this problem. In this context, we define it as follows. Let $A$, $B$ and $C$ be three sets of strings over some alphabet $\Sigma$, with $B \subseteq C$. Further, let $f : A \rightarrow B$ be a function representing a mapping of inflected forms into their corresponding base forms. Given, $A$ and $C$ (the latter representing the search space), the task is to construct an approximation of $f$, namely $\widehat{f} : A \rightarrow C$. If $\widehat{f}(a) = f(a)$ for $a \in A$, we say that $\widehat{f}$ returns a correct answer for $a$. Otherwise, i.e., if $\widehat{f}(a) \neq f(a)$, we say that $\widehat{f}$ returns an incorrect answer. Since obtaining a set of answers, of which at least one is correct, might be useful, we defined an additional task of constructing another approximation of $f$, namely function $f^* : A \rightarrow 2^C$, where $f^*$ is said to return a correct answer for $a \in A$ if $f(a) \in f^*(a)$.

We could also define the reference matching as a clustering problem. Given a set of strings, group them into clusters of strings, each referring to the same object.

In subsection 3.1 our test data is described. Further, in 3.2 the evaluation methodology is presented. The experiment set-up and the results thereof are given in 3.3.

## 3.1 Data

For the experiments on name matching we have mainly used two resources: (a) a lexicon of the most frequent Polish first names (PL-FNAMES) consisting of pairs $(in, base)$, where $in$ is an inflected form and $base$ stands for the corresponding base form, and (b) an analogous lexicon of inflected forms of country names in Polish (PL-COUNTRIES). The sizes of the aforementioned lexica are 5941 and 1765 respectively. It is important to note that the second resource contains multi-words since full names of some countries might consist of more than one token (e.g., *Republika Demokratyczna Kongo* vs. *Kongo*). Further, in our experiments on string distance metrics, we did not consider the lexicon entries $(in, base)$, where $in = base$ since in such a case finding an answer is straightforward.

Secondly, we utilized the data in the PL-FNAMES lexicon and an additional list of 58038 uninflected foreign first names as anchors for extracting full person names (first + last name) from a corpus of 15,724 on-line news articles from *Rzeczpospolita*, one of the leading Polish newspapers. This resulted in a list of 22485 full person-name candidates (some of them are clearly not person names due to the known problem of first names being also valid word forms in a given language). We will refer to this resource as PL-PNAMES.

### 3.2 Evaluation Methodology

Since for a given string more than one answer can be returned (identical value of the similarity metric), we measured the accuracy in three ways. Firstly, we calculated the accuracy with the assumption that a multi-result answer is not correct and we defined an accuracy measure (*all-answer accuracy*) which penalizes the accuracy for multi-result answers. Secondly, we measured the accuracy of single-result answers (*single-result accuracy*) disregarding the multiple-result answers. Finally, we used a somewhat weaker measure which treats a multi-result answer as correct if one of the results in the answer is correct (*relaxed-all-answer accuracy*). Formal definitions follow.

Let $S$ denote the number of strings, for which a single result (base form) was returned. Analogously, let $M$ be the number of strings for which more than one result was returned. Further, let $SC$ denote the number of correct single-result answers returned. Finally, $MC$ stands for the number of multi-result answers containing at least one correct result. The definitions of the accuracy metrics are given below.

$$all{-}answer{-}accuracy = \frac{SC}{S+M} \tag{5}$$

$$single{-}result{-}accuracy = \frac{SC}{S} \tag{6}$$

$$relaxed{-}all{-}answer{-}accuracy = \frac{SC+MC}{S+M} \tag{7}$$

### 3.3 Experiments

We started our experiments with the PL-FNAME lexicon and applied all but the Monge-Elkan metric since it is rather dedicated to matching multi-token strings. Although Soundex was designed to match surnames, we believed that the nature of first names is somewhat similar and included this metric in the tests. Further, 5 different metrics have been applied as a metric for comparing Soundex codes.

Additionally, taking a look at the data, we have also defined three simple similarity measures which can be computed in linear time, to compare with the standard metrics. Let us first denote the longest common prefix of two strings $s$ and $t$ with $lcp(s,t)$. The additional measures are defined as follows.

$$common{-}prefix(s,t) = |lcp(s,t)| \tag{8}$$

$$common{-}prefix{-}sq(s,t) = \frac{|lcp(s,t)|^2}{|s|\cdot|t|} \tag{9}$$

$$common{-}prefix{-}sq2(s,t) = \frac{|lcp(s,t)+\delta|^2}{|s|\cdot|t|} \tag{10}$$

The symbol $\delta$ in the last metric is an additional parameter for favoring certain suffix pairs in $s$ ($t$). For the task of matching the first names we have set $\delta$ to 1 if $s$ ends in one of the characters: $o$,$y$,$ą$,$ę$, and $t$ ends in an $a$. Otherwise $\delta$ is set to 0. This setting is a result of our empirical study of the data and the declension paradigm.

The results of the accuracy evaluation are given in table 1. The acronyms **AA**, **SR** and **RAA** refer respectively to all-answer accuracy, single-result accuracy and relaxed-all-answer accuracy. Additionally, in order to get a better picture of the performance of the measures, the columns labeled with **AV** and **MAX** give the average and maximum number of results returned in an answer.

**Table 1.** Results for first name matching

| Metric | AA | SR | RAA | AV | MAX |
|---|---|---|---|---|---|
| Levenshtein | 0.708 | 0.971 | 0.976 | 2.08 | 8 |
| Needleman-Wunsch | 0.728 | 0.833 | 0.826 | 3 | 20 |
| Smith-Waterman | 0.625 | 0.763 | 0.786 | 3.47 | 74 |
| Smith-Waterman with Affine Gaps | 0.603 | 0.728 | 0.749 | 3.36 | 74 |
| Jaro | 0.776 | 0.822 | 0.828 | 2.05 | 3 |
| Jaro-Winkler | 0.805 | 0.830 | 0.834 | 2 | 2 |
| q-grams | 0.812 | 0.829 | 0.831 | 2.06 | 3 |
| Soundex with Smith-Waterman | 0.498 | 0.942 | 0.953 | 2.58 | 7 |
| Soundex with Jaro | 0.455 | 0.947 | 0.958 | 3.41 | 34 |
| Soundex with Jaro-Winkler | 0.498 | 0.942 | 0.953 | 2.58 | 7 |
| Soundex with Levenshtein | 0.474 | 0.971 | 0.975 | 3.21 | 9 |
| Soundex with q-grams | 0.476 | 0.971 | 0.974 | 3.11 | 28 |
| common-prefix | 0.660 | **0.979** | **0.983** | 2.48 | 25 |
| common-prefix-sq | 0.829 | 0.843 | 0.844 | 2.11 | 3 |
| common-prefix-sq2 | **0.947** | 0.956 | 0.955 | 2.18 | 3 |

Interestingly, the simplest common prefix-based measures turned out to work best in all the three accuracy categories. Even when we disregard the common-prefix-sq2 metric which integrates some knowledge on first-name declension, the basic variant, namely common-prefix-sq2 performs better than q-grams and Jaro metrics in the **AA** category, which is the most relevant one. From the computational point of view common-prefix family and q-grams metric are the ones which can be computed in linear time, which makes them the most attractive in this category. Further, Soundex and Levenshtein scored surprisingly well w.r.t. **SR** and **RAA**, but again the 'cheap' common-prefix metric beats the other ones. It is important to note at this stage, that circa 10% of the inflected first name forms in Polish may either refer to a male or female (e.g., *Stanisława* - genitive form of the male name *Stanisław* vs. nominative form of the female name *Stanisława*), which illustrates additional complexity.

The next test was carried out on the PL-COUNTRIES lexicon, which contains many multi-token strings. We considered the 'best' metrics from the previous experiment and included also Monge-Elkan metric, originally designed for matching multi-field records in databases, to better cope with multi-token strings. For the latter one, we used 8 different settings for the internal metric. Although Soundex is not designed to perform best on country names, we included one result on applying it on PL-COUNTRIES for comparison. The results of the

**Table 2.** Results for country name matching

| Metric | AA | SR | RAA | AV | MAX |
|---|---|---|---|---|---|
| Levenshtein | 0.564 | 0.590 | 0.586 | 2.94 | 12 |
| Needleman-Wunsch | 0.714 | 0.777 | 0.757 | 2.93 | 11 |
| Smith-Waterman | **0.903** | **0.937** | **0.926** | 3.22 | 10 |
| Smith-Waterman with Affine Gaps | 0.870 | 0.899 | 0.889 | 2.35 | 5 |
| Jaro | 0.432 | 0.437 | 0.436 | 2 | 2 |
| Jaro-Winkler | 0.433 | 0.434 | 0.435 | 2 | 2 |
| q-Grams | 0.700 | 0.707 | 0.707 | 2 | 2 |
| Soundex with Smith-Waterman | 0.373 | 0.661 | 0.426 | 2.04 | 5 |
| common-prefix-sq | 0.416 | 0.421 | 0.420 | 2.35 | 3 |

**Table 3.** Results for country name matching with Monge-Elkan variants

| Metric | AA | SR | RAA | AV | MAX |
|---|---|---|---|---|---|
| ME with Smith-Waterman | 0.206 | 0.484 | 0.291 | 4.94 | 10 |
| ME with Smith-Waterman with affine Gaps | 0.172 | **0.881** | 0.261 | 5.10 | 6 |
| ME with Jaro | 0.520 | 0.576 | 0.532 | 2.97 | 5 |
| ME with Jaro-Winkler | 0.521 | 0.565 | 0.526 | 3 | 4 |
| ME with Levenshtein | 0.573 | 0.639 | 0.593 | 2.79 | 4 |
| ME with Needleman-Winch | 0.530 | 0.658 | 0.574 | 3.08 | 11 |
| ME with q-grams | 0.644 | 0.766 | 0.665 | 3.08 | 4 |
| ME with common-prefix | 0.579 | 0.840 | 0.632 | 3.11 | 21 |
| ME with common-prefix-sq | **0.694** | 0.868 | **0.716** | 3.08 | 4 |
| ME with common-prefix-sq2 | 0.631 | 0.845 | 0.669 | 3.13 | 4 |

accuracy evaluation are given in table 2 and table 3, where the second one contains only the results for the various settings of the Monge-Elkan metric.

Surprisingly, the best results were achieved by the Smith-Waterman metrics. On the contrary, Monge-Elkan performed rather badly. Using each of the common-prefix measures as internal metric yielded the best results. However, even simple q-grams turned out to perform better in the **AA** category than the best Monge-Elkan setting. Clearly, the Monge-Elkan metric needs some fine-tuning regarding the order of the tokens in multi-token strings and comparing strings which consist of a different number of tokens.

We have carried out some preliminary tests on the PL-PNAMES dataset too, but due to some erroneous entries we encountered in this data, matching full names has been postponed until PL-PNAMES has been cleaned accordingly.

## 4 Search Query Correction

Another potentially promising field of application of the string distance metrics is automated correction of spelling errors in search engine queries. In particular, an evaluation thereof for highly inflective languages such as Polish has not been reported. To explore this, we applied an extensive experiment on a properly preprocessed sample from query logs of a major Polish search engine [15]. We formulated the problem in an analogous way as in the case of entity matching, i.e., match a misspelled query with the 'intended' query. Consequently, we used the same evaluation methodology as described in 3.2.

### 4.1 Data

In order to perform an evaluation of different metrics, we had to identify in a huge collection of query logs of search-engine pairs of subsequent user queries, such that the second query was the correction of a misspelled one (ground-truth) - rather a non-trivial data-mining task [16].

Obtaining this kind of data was done as follows; from an excerpt of search engine Web query logs permitted by the major Polish search engine Netsprint, we extracted time, IP number and the HTTP request part. This initial log data was filtered via removing: queries containing non-letter characters, foreign-language queries and multi-word queries. Next, for each pair of neighbouring queries (same IP and similar time stamp), only those 'differing little' were identified and recorded as potential *misspelling-correction* pairs. With 'little difference' we mean: similar length, similar sum of characters' codes, and Levenshtein distance between the queries not exceeding 2.

Subsequently, to distinguish between real query correction and other minor query modifications (such as *query refinement* or following a misspelled query with another misspelled one) the following simple technique was applied. The search engine's index was consulted, where for each given word, the number of the Polish Web documents containing this word is specified. The pairs of queries for which the terms from the second query were present in less Web documents than for the first one were rejected as potential examples of so called *query refinement*, in which case a user slightly modifies the query to obtain *more specific* results (what may be viewed as the opposite to the query correction case, in some way). Additionally, all the pairs for which the second query was contained in less than 1000 Polish Web documents (as being unlikely to be spelled correctly) were rejected. From the remaining data a small dataset of about 400 pairs, being real misspelling-correction pairs, was obtained. We will refer to it as PL-LOGS. Additionally, we created another dataset via filtering some misspelling-correction pairs from PL-LOGS. Each pair, for which the original query term is in the lexicon of known words, was removed. The rationale behind this is that our lightweight technique is not capable of 'correcting' such typographical errors without considering the context. Secondly, pairs, where both query terms differ in two non-adjacent positions were removed too, i.e, only the pairs representing transposition-like errors (of character distance 2) remained. The latter dataset will be referred to as PL-LOGS-2. Intuitively, PL-LOGS-2 dataset represents errors which are slightly easier to be corrected.

### 4.2 Experiments

We applied 5 different string distance (similarity) metrics, namely Jaro, Jaro-Winkler, Levenshtein, q-grams and Smith-Waterman with affine gaps on the PL-LOGS and PL-LOGS-2 datasets. For each 'misspelled' query term, the correct answer was searched in a lexicon containing circa 130,000 words. The latter was obtained via consulting Netsprint's index consisting of circa 2 million entries for filtering out the most frequent/rare words and removing non-alphanumeric strings and foreign words. The results of the experiments are shown in tables 4 and 5.

**Table 4.** Results for search query correction on PL-LOGS

| Metric | AA | SR | RAA | AV | MAX |
|---|---|---|---|---|---|
| Levenshtein | 0.538 | **0.835** | **0.813** | 2.65 | 19 |
| Jaro | 0.521 | 0.679 | 0.673 | 2.83 | 10 |
| Jaro-Winkler | 0.555 | 0.643 | 0.652 | 2.86 | 7 |
| Smith-Waterman With Affine Gaps | 0.396 | 0.572 | 0.600 | 3.39 | 22 |
| q-grams | **0.578** | 0.737 | 0.718 | 3.16 | 15 |

**Table 5.** Results for search query correction on PL-LOGS-2

| Metric | AA | SR | RAA | AV | MAX |
|---|---|---|---|---|---|
| Levenshtein | 0.652 | **0.898** | **0.873** | 3.58 | 19 |
| Smith-Waterman with Affine Gaps | 0.457 | 0.628 | 0.643 | 3.58 | 22 |
| Jaro | 0.608 | 0.752 | 0.743 | 2.8 | 10 |
| Jaro-Winkler | 0.643 | 0.719 | 0.723 | 3 | 7 |
| q-grams | **0.690** | 0.830 | 0.805 | 3.21 | 15 |

They revealed that via using the time-efficient q-gram metric on PL-LOGS, in almost 60% of the cases it is possible to successfully *guess* the exact next 'correct' query of the user. Furthermore, in circa 81% of the cases (Levenshtein metric) it is possible to include the 'correct' query suggestion for a misspelled query in a small set of possible candidates (2-3 candidates on average). Analogously, for PL-LOGS-2, representing a subclass of misspelling errors, q-grams and Levenshtein metrics obtained the top results (0.69% and 0.87% resp.) in the same categories, namely **AA** and **RAA**. q-gram metric seems to be the most universal one when applied on queries of unknown type (we do not know what the query refers to). An intuitive next step would be to classify the query terms, e.g. person names, locations, etc., and to explore how different metrics cope with different query (entity) types.

## 5 Conclusions and Outlook

String distance metrics have been successfully deployed in various text processing applications and scenarios. This paper reports on the preliminary exploration of the usability of some well-known string distance metrics for tackling the reference matching task and for the automatic correction of misspelled search engine queries in the Polish Web.

Our experiments revealed that in the case of matching single-token names even declension-unaware common-prefix-like distance metrics seem to outperform other time-costly ones. A combination of common-prefix-like distance metrics with the q-gram metric might constitute a reasonable alternative to be explored in future. For multi-word tokens, we found that a Smith-Waterman metric performs best. However, this metric has quadratic complexity. Clearly, some fine-tuning of metrics that were expected to deal better with multi-token strings (Monge-Elkan) is necessary. Experiments considering more entity types and experiments with mixed data, i.e., data consisting of names of different types, are necessary to get a better picture of the behaviour of different metrics. On average, the q-gram metric seems to be the most universal and effective measure due to its performance and linear-time complexity. In the future, an extension of q-grams, namely *positional q-grams* [17], which also records the position of q-grams in the string, should be investigated. Although the default cost settings for the Smith-Waterman metrics explored in this work proved to be quite useful in one of the scenarios, in a proximate step we will investigate employment of training methods for automatically learning optimal cost functions for edit-distance metrics [18].

The second line of tests proved that simple techniques based on q-grams and Levenshtein metrics for 'guessing' the proper version of a misspelled search query are so successful (automatic suggestion of a correct spelling with circa 80-90% accuracy), so that it is interesting to investigate the application of more elaborated, Web-query-specific methods in future work. The experiments described here are, to the knowledge of the authors, the first published ones which concern automatic spelling correction of Polish search engine queries. In the next step, we intend to apply similar methods for multi-word queries, and to classify the queries w.r.t. the type of entity they refer to. Potentially, each string distance metric would perform differently on a different type of query.

In the summary, we believe that, although we do not present here fully-fledged solutions to the problems of reference matching and a very special case thereof, namely the search-engine query correction, the presented results constitute kind of handy guidelines for further research in this area, in the context of highly inflected languages like Polish.

## Acknowledgments

The work reported here was partially supported by the Polish Ministry of Education and Informatization under grant no. 3 T11C 007 27 (*Ekstrakcja informacji z języka polskiego*) and by the Polish-Japanese Institute Of Information Technology internal research grant ST/SI/06/2006 (*Analysis of search engine query logs*). Further, some contribution is an additional non-financed personal effort of the authors.

We would like to thank Netsprint - the leading Polish search engine - for giving access to part of its data, which was used for the experiment with automatic correction of misspelled queries. Some code from Open Source Java library, called SimMetric [19], was used in the experiments concerning comparing strings as a basis for further extensions. All the rest of the experimentation was done with an intensive use of the (wonderful) standard GNU/Linux toolset.

## References

1. Morton, T.: Coreference for NLP Applications. In: Proceedings of ACL 1997. (1997)
2. Cohen, W., Ravikumar, P., Fienberg, S.: A comparison of string metrics for matching names and records. In: Proceedings of the KDD2003. (2003)
3. Cohen, W., Ravikumar, P., Fienberg, S.: A comparison of string distance metrics for name-matching tasks. In: Proceedings of the IJCAI-2003. (2003)
4. Piskorski, J.: Named Entity Recognition for Polish with SProUT. In: Proceedings of Intelligent Media Technology for Communicative Intelligence: Second International Workshop, IMTCI 2004, Warsaw. Lecture Notes in Computer Science Vol 3490. (2005) 122–132
5. Linden, K.: Multilingual modeling of cross-lingual spelling variants. Information Retrieval **9** (2006) 295–310
6. Kukich, K.: Techniques for automatically correcting words in text. In: CSC '93: Proceedings of the 1993 ACM conference on Computer science, New York, NY, USA, ACM Press (1993)
7. Levenshtein, V.: Binary Codes for Correcting Deletions, Insertions, and Reversals. Doklady Akademii Nauk SSSR **163** (1965) 845–848
8. Landau, G., Vishkin, U.: Fast Parallel and Serial Approximate String Matching. Journal of Algorithms **10** (1989) 157–169
9. Needleman, S., Wunsch, C.: A General Method Applicable to Search for Similarities in the Amino Acid Sequence of Two Proteins. Journal of Molecular Biology **48** (1970) 443–453
10. Smith, T., Waterman, M.: Identification of Common Molecular Subsequences. Journal of Molecular Biology **147** (1981) 195–197
11. Winkle, W.: The state of record linkage and current research problems. Technical report, Statistical Research Division, U.S. Bureau of the Census, Wachington, DC (1999)
12. Ukkonen, E.: Approximate String Matching with q-grams and Maximal Matches. Theoretical Computer Science **92** (1992) 191–211
13. Russell, R.: U.S. Patent 1,261,167. http://pattft.uspto.gov/netahtml/srchnum.htm (1918)
14. Monge, A., Elkan, C.: The Field Matching Problem: Algorithms and Applications. In: Proceedings of Knowledge Discovery and Data Mining 1996. (1996) 267–270
15. NetSprint. (http://www.netsprint.pl)
16. Cucerzan, S., Brill, E.: Spelling correction as an iterative process that exploits the collective knowledge of web users. In: Proceedings of EMNLP 2004. (2004) 293–300

17. Gravano, L., Ipeirotis, P., Jagadish, H., Koudas, N. Muthukrishnan, S., Pietarinen, L., Srivastava, D.: Using q-grams in a DBMS for Approximate String Processing. IEEE Data Engineering Bulletin **24** (2001) 28–34
18. Bilenko, M., Mooney, R.: Employing trainable string similarity metrics for information integration. In: Proceedings of the IJCAI Workshop on Information Integration on the Web, Acapulco, Mexico, August 2003. (2003)
19. SimMetric. (http://www.dcs.shef.ac.uk/ sam/stringmetrics.html)

# Natural Language Technology for Information Integration in Business Intelligence

Diana Maynard[1], Horacio Saggion[1], Milena Yankova[2,1], Kalina Bontcheva[1], and Wim Peters[1]

[1] Department of Computer Science, University of Sheffield
Regent Court, 211 Portobello Street,
Sheffield, S1 4DP
United Kingdom
{diana,saggion,wim,milena,kalina}@dcs.shef.ac.uk
[2] Onotext Lab, Sirma Group Corp.
135 Tazrigradsko Chassee, Fl.5
1784 Sofia, Bulgaria
milena@sirma.bg

**Abstract.** Business intelligence requires the collecting and merging of information from many different sources, both structured and unstructured, in order to analyse for example financial risk, operational risk factors, follow trends and perform credit risk management. While traditional data mining tools make use of numerical data and cannot easily be applied to knowledge extracted from free text, traditional information extraction is either not adapted for the financial domain, or does not address the issue of information integration: the merging of information from different kinds of sources. We describe here the development of a system for content mining using *domain ontologies*, which enables the extraction of relevant information to be fed into models for analysis of *financial and operational risk* and other business intelligence applications such as *company intelligence*, by means of the XBRL standard. The results so far are of extremely high quality, due to the implementation of primarily high-precision rules.

**Keywords:** Information Extraction, Ontology, Business Intelligence, Natural Language Processing, Information Fusion.

## 1 Introduction

Business intelligence requires the collecting and merging of information from many different sources, both structured and unstructured, in order to analyse for example financial risk, operational risk factors, follow trends and perform credit risk management. The information is published either by the companies themselves on their web sites (e.g. balance sheets, company reports), by financial newspapers, specialised directories (e.g. Yahoo! Company and Fund Index[1]),

[1] http://biz.yahoo.com/i/

W. Abramowicz (Ed.): BIS 2007, LNCS 4439, pp. 366–380, 2007.
© Springer-Verlag Berlin Heidelberg 2007

governamental bodies, etc. The analytical techniques frequently applied in business intelligence, however, have been largely developed for dealing with numerical data so, unsurprisingly, the industry has started to struggle with making use of this distributed and unstructured information. For example, Ellingsworth and Sullivan [8] found that traditional analytic techniques to understand trends in insurance claims could not help because the information was not fully described by structured data.

One solution to this problem is to apply text processing and Natural Language Processing (NLP) techniques to unstructured sources in order to transform them into structured representations suitable for such analysis. Information Extraction (IE) is a key NLP technology which automatically extracts specific types of information from text to create records in a database or populate knowledge bases, for example. One typical scenario for information extraction in the business domain is the case of insurance companies tracking information about ship sinkings around the globe [21]. Without an IE system, company analysts would have to read hundreds of textual reports and manually dig out that information. Another typical IE scenario is the extraction of information about joint ventures or other types of commercial company agreements from unstructured documents [2,12]. This kind of information can help identify not only information about who is doing business with whom, but also market trends, such as which world regions or markets are being targeted by companies.

One additional problem with business information is that even in cases where the information is structured (e.g. balance sheets), it is not necessarily represented in a way machines can understand - and this is particularly true with legacy systems and documentation. One response to this problem has been the development of the emerging standard XBRL (eXtensible Business Reporting Language)[2]. XBRL is a universal XML-based specification for business information, providing both public and private companies with an effective way to prepare and distribute various business reports using the Internet in a cost effective and universal manner [18]. Structured data such as that from company balance sheets and tabular reports can be mapped into XBRL using automatic processes [10]. But when the information is unstructured, then NLP and text mining techniques are of paramount importance.

In this paper, we report on our work on information extraction for business intelligence in the context of the EU Musing project[3]. We are working with domain ontologies which represent our understanding of the domain of application and which capture the experts' knowledge. Ontologies contain concepts arranged in class/sub-class hierarchies (e.g. a bank is a financial institution), relations between concepts (e.g. a bank has a manager), and properties (e.g. a company has only one CEO). We have developed different applications in the business domain targeting real business scenarios defined by real users in the areas of financial risk management, internationalisation, and IT operational risk - ontologies are being developed for each of the scenarios. We focus here on applications

[2] http://www.xbrl.org

[3] IST-2004-027097 http://www.musing.eu

for extracting information from company profiles and country/region for the developing internationalisation applications but we will also briefly describe techniques used in other scenarios. One key aspect of our work is the development of *ontology-based information extraction systems*[4] which are being developed using robust and adaptable tools from the GATE architecture [5]. A second key aspect of our work is a framework for *merging* information across different sources which also uses a domain ontology. The ontology acts as bridge between the text and a knowledge base, which in turn feeds reasoning systems or provides information to end users.

The following section describes the approach to text analysis we have adopted, while Section 3 describes the information extraction system in more detail. Section 4 describes the evaluation, and in Section 5 we compare our approach with previous work in the fields. Finally we discuss some related work and future directions in the last section.

## 2 Information Extraction

Information extraction (IE) is a technology for automatically extracting specific types of information from text [11]. The information to be extracted or the concepts to be targeted by the IE system are predefined in knowledge resources such as a domain ontology or templates. These concepts are elucidated by domain experts or can be automatically learnt (at least partially) from domain-specific texts. In the business domain, an information extraction template for joint ventures might be made up of the following key variables or concepts which need to be instantiated from text: partners (e.g. companies), nationalities, type of contractual form (e.g. alliance or joint venture), name of the contractual form, business sector, date of constitution of the alliance, etc.

Once target concepts, relations, and attributes have been defined for each domain, the information extraction system can be developed so that new documents can be semantically annotated by identifying instances of those concepts, attributes and relations. For example, company names can be identified in a number of ways such as gazetteer lookup, regular expression matching, or a combination of techniques. Relations between entities in text can be identified from syntactic relations found in parse trees or predicate-argument structures obtained from semantic analysis. The instances identified in text can then be mapped to the domain ontology, stored in a database, or used as semantic indexes for further processing (e.g. searching, reasoning). Some instances in the text may be already known to the system, while others may never have been encountered before: this is one of the key features of the IE technology.

We focus on an information extraction task which targets different domain ontologies. Ontology-based information extraction is a task which consists of finding in a text instances of concepts and relations between them as expressed in an ontology. This process is domain-specific and is carried out with a domain

---

[4] Musing ontologies extend the Proton Ontology http://proton.semanticweb.org

ontology over texts which belong to that domain. Figure 1 shows our development environment and a text which has been automatically annotated with respect to an ontology for company information.

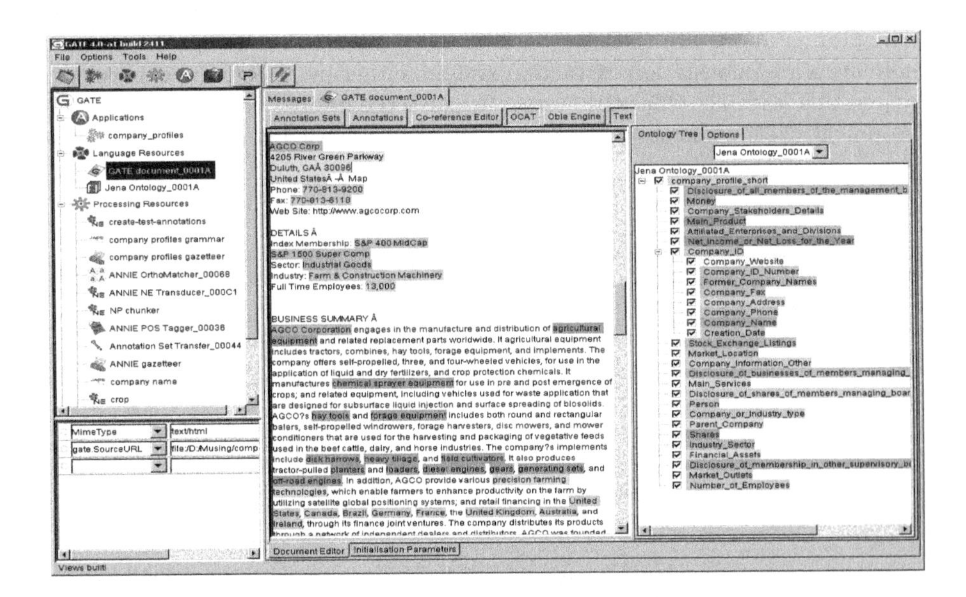


**Fig. 1.** GATE Development Environment and Text Automatically Annotated with Ontological Classes

## 2.1 Data Sources and Ontology-Based Annotation Tool

When developing an information extraction system, it is essential to have textual documents where the key domain concepts have been identified, so that a language engineer can create accurate information extraction rules. In addition to data provided by different partners in the project[5], a number of on-line data sources for business intelligence (e.g. Yahoo! Finance, World Bank, CIA Fact Book) have been targeted in order to boost system accuracy. We rely on the Ontology-based Corpus Annotation Tool (OCAT), a GATE plugin which uses one or more ontologies for annotation of concepts/classes. The required ontology can be selected from a pull-down list of available ontologies which are loaded into the system. GATE currently provides support for ontologies in both OWL and RDF. The current version of the tool supports only annotation with information about the ontology class, however future work will include the annotation of relations from the ontology. Ontology-based annotations in the text can be viewed by selecting the desired classes in the ontology tree.

[5] The European Business Register (EBR), Belgium and Verband der Vereine Creditreform e.V. (VVC), Germany are members of the Musing consortium.

We have developed a Web service which allows the user to *annotate* texts with ontological information over the Web (Figure 2). First, a set of documents (corpus) is annotated with key information using an initial information extraction system. This information may only be partially correct, so the user uses a corpus annotation tool to edit the annotations proposed by the system. The human annotations are then fed back to the system and developer to create a more accurate information extraction system, either by re-defining new rules or by machine learning. Once the system achieves the desired performance, the development cycle comes to an end and the system can be deployed by the final user.

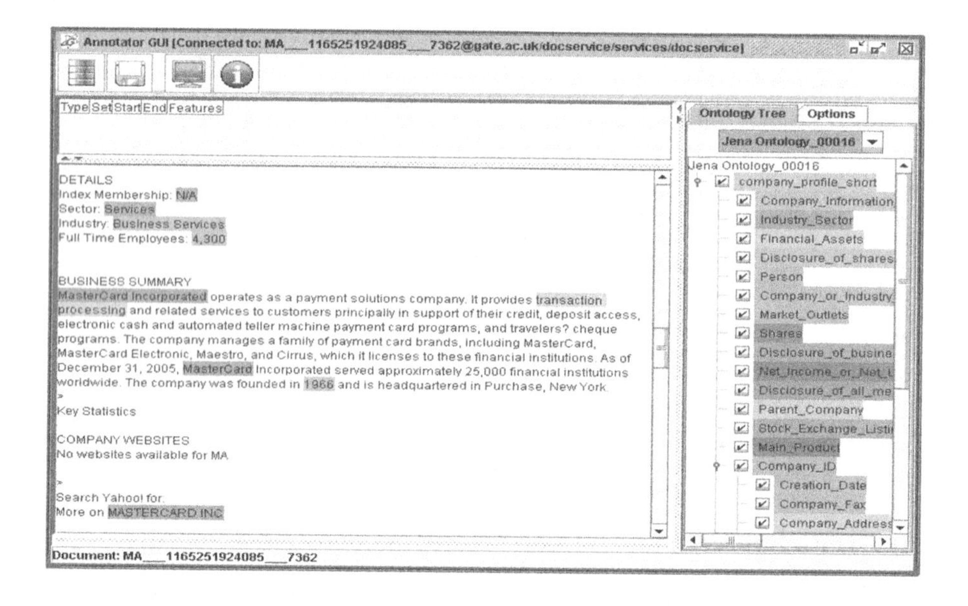


**Fig. 2.** Document Service for Ontology-based Annotation

## 2.2 Natural Language Processing Tools

We have developed our information extraction system using GATE. While GATE comes with a default information extraction system called ANNIE [16], it is only partially relevant to the business domain. The ANNIE system identifies generic concepts such as persons, locations, organization, dates, etc., so we had to develop new rules or adapt rules for our applications. The tools available in GATE to perform text analysis consist of: a document structure analyser which parses different input files into GATE documents; a tokeniser which identifies different types of words; a sentence splitter which segments the document into sentences; a part-of-speech tagger which associates POS tags to words and symbols; a morphological analyser which produces a root and affix for each word in the document; a named entity recognition sub-system composed of a gazetteer lookup component and a rule-based pattern matching engine; and a coreference resolution algorithm. Other components which are sometimes necessary, depending on the text and task, are parsers which associate syntactic and semantic structures with sentences. For the work reported here, we have mainly adapted the named entity recognition components and developed a conceptual mapping to map concepts identified by our system into the ontologies of the application domains. The named entity system in GATE is a rule-based system developed using a pattern-matching engine called JAPE [6] which is ontologically aware; making the mapping of entities into ontological classes possible during entity recognition.

The ease of adaptation of the core ANNIE system to new applications depends on many factors: language, annotation types to be recognised, document type, level of structure in the text, and level of accuracy required (tradeoff between precision and recall). ANNIE does not deal with ontologies, however, so an ontology-based IE application requires a lot more initial adaptation than just the recognition of new entity types, for example. For more information about the adaptation process in general, we refer the reader to [13,15]; for another example of adaptation to ontologies, see [17].

### 2.3 Merging Information Across Different Sources

One of the fundamental problems one has to address with the proliferation of information is the identification and merging of ontological instances extracted from multiple sources. In the Semantic Web community, this problem is known as ontology population. An example of this is presented in Figure 3, where three texts refer to the same company Alcoa, using different expressions "ALCOA", "Alcoa Inc." and "Alcoa". It is important to identify the three instances as the same company because of the complementary information they bring (note that the interlinking or coreference between entities in the same text is solved by our coreference resolution mechanism). While one text provides information about the company profile (e.g. address, management), a second text provides information about where the company has business (e.g. 8 plants in the UK), the third one provides relevant financial information (e.g. share prices). The merging of these complementary sources provide a clearer picture about the company for BI purposes.

In the work presented here, merging and interlinking between pieces of information are carried out in an identity resolution framework which provides a generic solution to the merging problem. The framework is based on an ontology of the domain and a knowledge base containing known instances. For each new ontological instance discovered by the extraction process, the resolution process operates in four stages. First, a set of possible candidates is retrieved from the knowledge base (e.g. instances with the same class information). Second, evidence is collected from each of the candidate instances (e.g. attributes and values stored in the knowledge base). Third, a decision is made based on the similarity between the new instance and the instances retrieved from the database. The decision is based on a set of rules defined by the domain expert which are used to compute a similarity score between the new instance and each candidate (these rules may for example check name aliases; or similarity between values for similar attributes). Finally, the new instances and their attributes are asserted in the knowledge base. The framework uses the KIM [19] semantic repository implemented in OWLIM/Sesame.

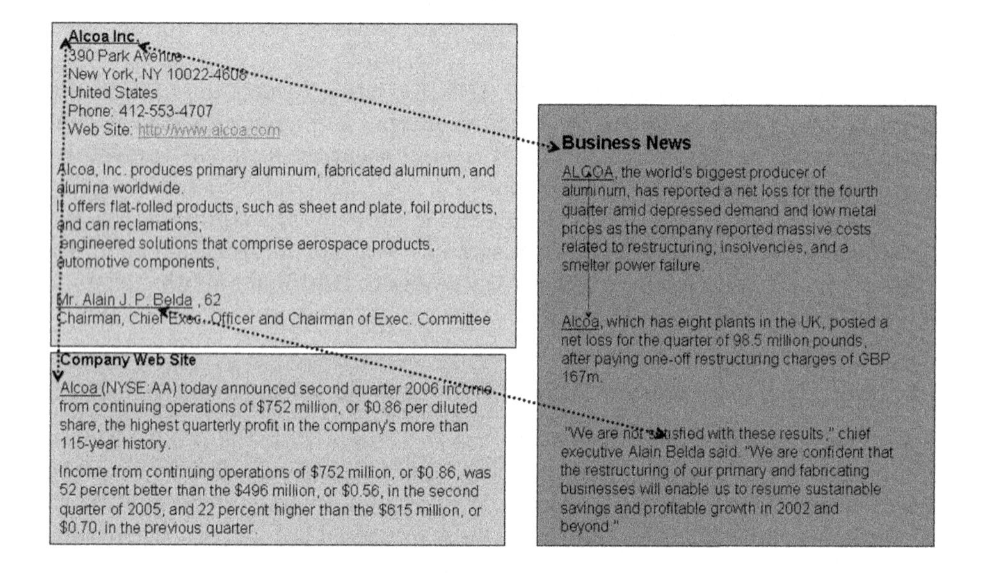


**Fig. 3.** Related information from multiple sources

## 3 MUSING Information Extraction System

In our framework, the documents to be analysed are first loaded into GATE and undergo document format analysis, which enables the documents to be processed by the application. Document structure analysis is then carried out in order to identify the layout. This consists of pre-processing modules such as tokenisation and sentence splitting. For example, a special splitting module is run in order to identify each row in a table in documents such as balance sheets. Then the information extraction system is run and the information is identified as annotations on the document. Finally, this information is mapped into XBRL and the appropriate ontology.

Because the system needs to take into account information from different kinds of sources, different applications are needed which may use slightly different sets of components. Not only do gazetteer lists and grammar rules differ for different kinds of concepts, but also pre-processing may differ to take into account different structural information. For example, some web pages may contain a lot of extraneous information that should not be processed. Web pages in particular often contain information which is useful to the human user looking for other sources of information, such as information about other countries when looking at information for a specific country. These are often in the form of tables or drop-down boxes. Such information is very useful to a visual user but can be very misleading to a system which cannot distinguish the relative importance of information in different kinds of formats. We use some of GATE's processing resources to help us detect such information and ignore what is unimportant.

In the following sections we describe 3 applications for identifying and extracting information relevant for business intelligence from 3 kinds of domain-specific unstructured text: company profiles, country profiles, and balance sheets.

### 3.1 Information Extraction from Company Profiles

Structured information from company profiles needs to be extracted in order to be able to feed this data into statistical models of financial risk assessment or investment, e.g. assessment of the creditworthiness of a company. In addition, such information is necessary for providing services to companies who are looking for commercial partners working in the same sector in a different country, e.g. all software companies in Russia. The information from country profiles is therefore also needed as input. For example, if the system extracts the fact that Russia's investment Fitch rating is BBB+, increased from BBB, then the risk assessment model can take this into account and correspondingly revise risk downwards. One prototype we are developing is an International Enterprise Intelligence application whose objective is to provide customers with up-to-date and correct information about companies, mined from many different sources such as web pages, financial news, and structured data sources. A set of company profiles has been downloaded from Yahoo! and the most relevant concepts to extract have been identified in the ontology. Each concept is extracted along with the relevant information, for example the concept "number of employees" is associated with a feature and a value, such as "Number=2000".

**Table 1.** Relevant Concepts for Company Information

| Address Data | Company Data | Financial Data |
|---|---|---|
| Name of Company | Branch | Turnover |
| Telephone | Main Activities | Number of Employees |
| Postcode | Import/Export Activities | Turnover per Employee |
| Country | Legal Form | Shareholders |
| E-Mail | Managerial head | Related persons |
| ... | ... | ... |

Table 1 presents some examples of key concepts which, according to our users, need to be extracted from text for each company. The company domain specific ontology which extends the Proton model contains at the present time 24 concepts and 38 properties.

### 3.2 Country and Region Information Extraction

Our country/region profiles application enables us to extract general information about countries/regions from unstructured text. A set of country profiles has been downloaded from the CIA World Factbook[6], and a list of concepts to be extracted has been identified from the domain ontology. The following concepts have been extracted so far: country name; population; surface area; official language; currency; exchange rate; foreign debt; unemployment rate; GDP; and foreign investments. Each concept is extracted with features and values depicting the information associated with it. In the case where we wish to extract information for multiple years (for example if we want to extract the exchange rate for the last 3 years), we extract separate features and values for each year.

There still remain some further concepts wich require a deeper level of analysis such as ratings, sustainability and vulnerability, which can be quite vague and hard to define in free text. The extracted concepts will be used in a Musing specific internationalisation application which will help companies or businesses searching for appropriate regions for internationalisation of their businesses.

### 3.3 Extracting Information from Financial Statements

While balance sheets and other financial statements contain both structured (tables) and unstructured information (explanatory notes), these statements are only currently available in documents in pdf, tiff, or similar binary formats which are difficult to process automatically. When a bank needs financial information about a company, a balance sheet would be requested and then analysed by a human analyst, who would typically re-enter all the information of the balance sheet in the bank system to produce a structured file before credit rating can be performed. This is a very tedious and error-prone practice. As an additional disadvantage, it is currently impossible for a bank to automatically obtain key information (relevant for our users) from a balance sheet such as *what were the net assets of the company in the 31 December 2001?* or *what is the purchase plan of the company?*: the analyst has to dig into the files in order to find the appropriate answers to these key questions. Some answers are found in free text descriptions in balance sheets, but this information is currently inaccessible to models of risk or the company's creditworthiness. The latter is required, for example, by the Basel II accord which lays down guidelines for matters such as how much capital a bank or financial institution needs to keep in reserve to

[6] https://www.cia.gov/cia/publications/factbook/index.html

cover all its current lending. There are various methods of calculating the bank's expected loss/unexpected loss with differing degrees of complexity.

Our information extraction application over balance sheets aims to identify all specific financial information such as details of fixed assets, profits, goodwill (reflecting good relationship of a business enterprise with its customers) etc. from the files. The application identifies the structure of the balance sheets using patterns developed in JAPE, and maps each line of the balance sheet into the appropriate XBRL concept - as specified by an FRM expert. Another important aspect of this work is the identification of explanatory notes in the balance sheets as well as any concepts related to the financial risk management described by the domain experts in the ontology which curretly contains 45 concepts.

We have developed an application in GATE that extracts such information from company balance sheets in PDF format – some balance sheets are also available in other formats such as TIFF files or HTML pages. One of the problems of PDF files is that it is very difficult to extract information that is in tabular form. One solution is to first convert the PDF directly into a more easily processable format such as HTML, XML or XBRL. Alternatively, we can process the application directly as a PDF file in GATE, making use of GATE's language processing capabilities and the JAPE pattern-matching language [6] to identify things like column headings and separate rows. It is important to note that because the original documents are in PDF, the spatial/graphical structure of the document is not fully preserved and this will have consequences for extraction. For example, the numbers in each line are associated with particular dates which are given once at the top of the balance sheet. Some numbers appear to be totals but this is not explicitly mentioned, so analysis has to be performed on such figures based on positional information, and the meaning made clear.

Once the PDF file is loaded into GATE, the Balance Sheet application identifies each row in the table, using a specially modified version of the ANNIE sentence splitter which identifies each row as a separate sentence. Usually in balance sheets each column is headed by a date (usually a year), i.e. information in each column represents the information for that date. A JAPE grammar first identifies a line of date information in the table, e.g. 2001, 2002 etc., and then stores this information as annotations on the document as a whole (e.g. that the first column represents 2001, the second column represents 2002, etc.). Then various grammars look for the row entries in the table, for example identifying labels such as "Fixed Asset". For each concept, features and values are added to the annotation representing the amount and year. One annotation is thus produced for each row in the table, with the following information:

- year (e.g. year=2005)
- amount value (e.g. value=73,000)
- positive or negative (e.g. type=negative)
- string of the asset (e.g. string=Total Current Liabilities)

Negative values are sometimes displayed by a number in round brackets. A special grammar rule identifies these as negative. Our current work is looking at extensions to work with other document formats. Next stages in the process are to link the concepts denoting the entries in the table with concepts in the ontology, and to transform the final annotations (currently in XML) into XBRL, performed in collaboration with our financial partners.

## 4 Evaluation

Evaluation is an essential component of any information extraction application. Our quantitative evaluation compares annotations produced by the automatic system with annotations produced by human experts (known as key or gold standard annotations). We make use of traditional metrics used in information extraction [4]: precision, recall, and F-measure. Precision measures the number of correctly identified items as a percentage of the number of items identified. It measures how many of the items that the system identified were actually correct, regardless of whether it also failed to retrieve correct items. The higher the precision, the better the system is at ensuring that what is identified is correct. Recall measures the number of correctly identified items as a percentage of the total number of correct items measuring how many of the items that should have been identified actually were identified. The higher the recall rate, the better the system is at not missing correct items. The F-measure [20] is often used in conjunction with Precision and Recall, as a weighted average of the two – usually an application requires a balance between Precision and Recall. For the application on extraction of company information from different textual sources, we have obtained very encouraging results. An expert manually annotated the texts (using the tool described in Section 2.1) and we compared the results of the system annotations against this gold standard set. The results for each type as well as the totals are shown in Table 2.

**Table 2.** Evaluation of company profiles application

| Concept | Precision | Recall | F-Measure |
|---|---|---|---|
| Company Address | 100.00 | 66.70 | 80.00 |
| Company Fax | 100.00 | 100.00 | 100.00 |
| Company Name | 88.90 | 80.00 | 84.20 |
| Company Phone | 100.90 | 100.00 | 100.00 |
| Company Website | 50.00 | 70.00 | 58.30 |
| Company or Industry Type | 60.00 | 75.00 | 66.70 |
| Creation Date | 100.00 | 100.00 | 100.00 |
| Industry Sector | 60.00 | 100.00 | 75.00 |
| Market Outlets | 85.00 | 94.40 | 89.50 |
| Market Location | 69.60 | 94.10 | 80.00 |
| Number Of Employees | 100.00 | 100.00 | 100.00 |
| Stock Exchange Listings | 100.00 | 100.00 | 100.00 |
| Total | 85.60 | 93.60 | 84.00 |

For comparison purposes, our generic IE system ANNIE which identifies classical types of information such as People, Location, Organization, etc. has levels of precision of 93.5%, recall of 92.3%, and F-measure of 92.9% on general news texts.

The other two applications also show very encouraging results, although they require more work to complete the extraction of all relevant concepts.

# 5 Related Work

In a pure information extraction context in the business domain, JV-FASTUS [2] developed for the Message Understanding Conferences performed shallow and robust text analysis using a set of finite state transducers. For joint ventures the system achieved recall levels of 34%, precision levels of 62%, and combined F-score of around 45%. As with other systems in the MUC context, FASTUS targeted a template and not a domain ontology. Our work is different from traditional approaches to extraction not only because of the complexity of the domain but also because we address the problem of merging information extracted from different sources.

h-TechSight [17] is a system which also uses GATE (amongst other tools) to detect changes and trends in business information and to monitor markets. It uses semantically-enhanced information extraction and information retrieval tools to identify important concepts with respect to an ontology, and to track changes over time. This enables companies to keep an eye on competitors' products in the news and in company reports etc., and enables job seekers and job providers to monitor changes in the employment market (for example, required skills, salaries payable, locations of jobs, trends in company hiring policies, etc.). This system differs from MUSING in that the information acquired is only related to a quite shallow and simple ontology with a few fairly fixed concepts. The information discovery module realised in GATE is part of a much larger knowledge portal combining a number of different tools. It acts as a very good starting point or baseline from which to continue.

Information extraction is also used in the MBOI tool [9] for discovering business opportunities on the internet. The main aim is to help users to decide about which company tenders require further investigation. This enables the user to perform precise querying over named entities recognised by the system. Similarly the LIXTO tool is used for web data extraction for business intelligence [3], for example to acquire sales price information from online sales sites. However, this requires a semi-structured data source which is not always available or sufficient for the kind of financial information we are concerned with.

Ahmad et al. [1] have developed a system for analysing sentiment in business and financial news streams, using term recognition and collocation extraction techniques. The idea behind this is that positive and negative sentiments expressed in news can often make or break people, companies and even governments, creating effects such as economic bubbles through the power of financial journalism. While this work does not directly address the problem we have in mind, the sentiment research supports the underlying theory about the importance of extracting such information from free text.

None of the systems above deals specifically with extracting information useful for financial business intelligence, and although there are systems which do so [7], they do not deal adequately with gathering information from unstructured text and the problem of merging information from different data sources or using an ontology to assist these processes.

## 6 Conclusions and Further Work

We have described the design and implementation of a system for knowledge extraction in business intelligence. The aim is to extract relevant information from a number of sources including the Web in order to build up a financial picture of a particular company for applications in financial risk management and internationalisation. Our system targets an ontology of the application domains containing the most relevant domain concepts and relations. The system produces annotations which will be used to populate a knowledge base or semantic repository with the assistance of a multi-source merging mechanism. The identification and extraction of such information has been largely implemented, and this paper describes the design approach to these tasks. Work will continue on refining this work and on the merging process which will follow. So far the actual extraction is of extremely high quality and there are few errors. Our future work on extraction will concentrate on different text types such as business reports and company web sites. As a continuation of our work on evaluation, we shall be looking at an evaluation metric specifically adapted to ontology-based information extraction, such as [14], since this will give us a more informed and practical result (giving credit for answers which are closely linked in the ontology to the correct answer).

## Acknowledgements

This work is partially supported by the EU-funded MUSING project (IST-2004-027097).

## References

1. Khurshid Ahmad, Lee Gillam, and David Cheng. Sentiments on a grid: Analysis of streaming news and views. In *5th Language Resources and Evaluation Conference*, 2006.
2. D.E. Appelt, J.R. Hobbs, J. Bear, D. Israel, M. Kameyama, and M. Tyson. Description of the JV-FASTUS system as used for MUC-5. In *Proceedings of the Fourth Message Understanding Conference MUC-5*, pages 221–235. Morgan Kaufmann, California, 1993.

3. R. Baumgartner, O. Frlich, G. Gottlob, P. Harz, M. Herzog, and P. Lehmann. Web data extraction for business intelligence: the lixto approach. In *Proc. of BTW 2005*, 2005.
4. Nancy Chinchor. Muc-4 evaluation metrics. In *Proceedings of the Fourth Message Understanding Conference*, pages 22–29, 1992.
5. H. Cunningham, D. Maynard, K. Bontcheva, and V. Tablan. GATE: A Framework and Graphical Development Environment for Robust NLP Tools and Applications. In *Proceedings of the 40th Anniversary Meeting of the Association for Computational Linguistics (ACL'02)*, 2002.
6. H. Cunningham, D. Maynard, and V. Tablan. JAPE: a Java Annotation Patterns Engine (Second Edition). Research Memorandum CS–00–10, Department of Computer Science, University of Sheffield, November 2000.
7. T. Declerck and H. Krieger. Translating XBRL into Description Logic: an approach using Protege, Sesame and OWL. In *Proceedings of Business Information Systems (BIS)*, Klagenfurt, Germany, 2006.
8. Marty Ellingsworth and Dan Sullivan. Text mining improves business intelligence and predictive modeling in insurance. *DM Review Magazine*, 2003.
9. J.-Y. Nie F. Paradis and A. Tajarobi. Discovery of business opportunities on the internet with information extraction. In *Workshop on Multi-Agent Information Retrieval and Recommender Systems (IJCAI)*, pages 47–54, Edinburgh, Scotland, 2005.
10. Franseco Fornasari, Alessandro Tommasi, Cesare Zavattari, Roberto Gagliardi, Thierry Declerck, and Michele Nannipieri. Xbrl web-based business intelligence services. In Paul Cunningham and Miriam Cunningham, editors, *Innovation and the Knowledge Economy: Issues, Applications, Case Studies. Proceedings of eChallenge 2005*. IOS Press, 2005.
11. R. Gaizauskas and Y. Wilks. Information Extraction: Beyond Document Retrieval. *Journal of Documentation*, 54(1):70–105, 1998.
12. P.S. Jacobs and L.F. Rau. Scisor: Extracting information from on-line news. *Communications of the ACM*, 33(11):88–97, 1990.
13. D. Maynard, K. Bontcheva, and H. Cunningham. Towards a semantic extraction of Named Entities. In *Recent Advances in Natural Language Processing*, Bulgaria, 2003.
14. D. Maynard, W. Peters, and Y. Li. Metrics for evaluation of ontology-based information extraction. In *WWW 2006 Workshop on "Evaluation of Ontologies for the Web" (EON)*, Edinburgh, Scotland, 2006.
15. D. Maynard, V. Tablan, K. Bontcheva, and H. Cunningham. Rapid customisation of an Information Extraction system for surprise languages. *Special issue of ACM Transactions on Asian Language Information Processing: Rapid Development of Language Capabilities: The Surprise Languages*, 2003.
16. D. Maynard, V. Tablan, C. Ursu, H. Cunningham, and Y. Wilks. Named Entity Recognition from Diverse Text Types. In *Recent Advances in Natural Language Processing 2001 Conference*, pages 257–274, Tzigov Chark, Bulgaria, 2001. `http://gate.ac.uk/sale/ranlp2001/maynard-etal.pdf`.
17. D. Maynard, M. Yankova, A. Kourakis, and A. Kokossis. Ontology-based information extraction for market monitoring and technology watch. In *ESWC Workshop "End User Apects of the Semantic Web")*, Heraklion, Crete, 2005.
18. J. Montes. Consumer entertainment software - industry trends. In Brian Stanford-Smith and Enrica Chozza, editors, *E-Work and E-Commerce*, pages –7. IOS Press, Amsterdam, 2001.

19. B. Popov, A. Kiryakov, A. Kirilov, D. Manov, D. Ognyanoff, and M. Goranov. KIM – Semantic Annotation Platform. *Natural Language Engineering*, 2004.
20. C.J. van Rijsbergen. *Information Retrieval.* Butterworths, London, 1979.
21. Yorick Wilks and Roberta Catizone. Can We Make Information Extraction More Adaptive? In *M. Pazienza (ed.) Proceedings of the SCIE99 Workshop*, pages 1–16, Rome, Italy, 1999.

# Semantic Similarity Measure of Polish Nouns Based on Linguistic Features

Maciej Piasecki and Bartosz Broda

Institute of Applied Informatics, Wrocław University of Technology,
Wybrzeże Wyspiańskiego 27, Wrocław, Poland
maciej.piasecki@pwr.wroc.pl

**Abstract.** A word-to-word similarity function automatically extracted from a corpus of texts can be a very helpful tool in automatic extraction of lexical semantic relations. There are many approaches for English, but only a few for inflective languages with almost free word order. In the paper a method for the construction of a similarity function for Polish nouns is proposed. The method uses only simple tools for language processing (e.g. it does need the application of a parser). The core is the construction of a matrix of co-occurrences of nouns and adjectives on the basis of application of morpho-syntactic constraints testing agreement between an adjective and a noun. Several methods of transformation of the matrix and calculation of the similarity function are presented. The achieved accuracy of 81.15% in WordNet-based Synonymy Test (for 4 611 Polish nouns, using the current version of Polish WordNet) seems to be comparable with the best results reported for English (e.g. 75.8% [5]).

**Keywords:** semantic similarity function, Polish, automatic extraction, nouns, LSA.

## 1 Introduction

WordNet [4] is a large electronic thesaurus applied to many Natural Language Processing tasks, cited in many papers, often criticised, but it is always a reference point. The worst thing about WordNet is that its creation for a new language is a laborious and costly process. The question whether a WordNet-like thesaurus (or one of richer structure) can be created automatically is still open, but a lot of methods have been proposed, e.g. [24,7,14]. The prise to be won is the reduction of time and money, e.g. by a semi-automatic method of creation.

A first step in automatic creation of a thesaurus is often the construction of a *semantic similarity function* (SSF) of words: $W \times W \rightarrow R$, where $W$ is a set of words, or, more generally, a set of *lexical units* and $R$ — set of real numbers. A *lexical unit* (LU) is a set of words or short phrases (multiword) possessing the same meaning (with accuracy to different values of morphological categories like number, gender, etc.) and differing only by the values of morphological categories. LU is represented by its basic morphological form, i.e. in our approach SSF: $LU \times LU \rightarrow R$.

W. Abramowicz (Ed.): BIS 2007, LNCS 4439, pp. 381–390, 2007.
© Springer-Verlag Berlin Heidelberg 2007

The goal of this work is to construct a SSF for Polish nouns on the basis of a corpus of Polish using as small set of language tools as possible (as simple tools as possible) and demanding a limited manual effort in construction (e.g. measured in the number of working days). However, the method should cope with rich inflection and free word order (i.e. can not be based on simple positional patterns, has to be resistant to permutations of components of phrases that often happens in languages like Polish). We decide to limit the method to the domain of nouns because of the existence of a trustful measure of accuracy based on the preliminary version of Polish WordNet (plWN) in which the noun-part is already developed in enough detailed way [17]. Moreover, we will consider only one word noun LUs and we assume that the corpus includes initially no other information than sequences of words. Contrary to works utilising patterns and directly extracting semantic relations, e.g. [7,14], we want our method to assign some value of similarity to any pair of nouns.

The works on construction of SSF (and then a similarity thesaurus built on its basis) started in the sixties, but the early approaches used monolingual dictionaries in a machine readable form as the main resource. The first works on corpora date from the beginning of nineties. The general idea for the construction of SSF is constantly the same: the more often two LUs occur in similar text contexts, the more similar are their meanings. The context can be a document or a part of it, e.g. a text window of the fixed size moved across a document. Grefenstette [8] in the *Sextant* system enriched context performing morphological analysis and a limited shallow parsing. Next, for each occurrence of LU $l_0$ a vector of attributes — LUs appearing in the context and the type of syntactic relation connecting them to $l_0$, were collected. LUs were next compared by their vectors of attributes and Jacard measure. The evaluation was performed in relation to some existing electronic thesauruses. The context described by syntactic relations was also used in [19]. Landauer and Dumais [11] introduced a technique called *Latent Semantic Analysis* in which a context is the whole document but the created occurrence matrix is next transformed by Singular Value Decomposition [1] to a matrix of reduced dimensions. The comparison of LUs is done by cosine measure on the reduced matrix. They introduced the comparison with the *Test Of English as a Foreign Language* (TOEFL) as the measure of the quality of SSF, achieving 64.4% of hits. The main problem of LSA is the limmitation on the size of the corpus, SVD works only for a limited dimensions of a matrix.

Shütze [21] proposed a method called *Word Space* in which a text window is moved accross documents and in each position of the window statistics of co-occurrence of a word in the centre of the context with a number of *meaning bearers* (some selected general words) are collected. Turney [22] used searching for co-occurences of LUs in millions of documents in Internet as SSF. Two tested LUs were delivered in one question to Altavista search engine. He achieved 73.75% of proper hits in TOEFL, recently increased to 97.5% by combining several methods [23]. A new measure for SSF, called *WordNet-based Synonymy Test* (WBST), was proposed in [5]. In WBST WordNet is used to generate "a large set of questions identical in format to those in the TOEFL". The best results of 71.6% [3] and 75.8% [5] are reported for nouns according to WBST.

Most of the works were done for English, Polish as an inflective language poses several additional problems. The rich morphology increases the number of words very much, if they are not reduced to morphological base forms (e.g. up to 14 forms of a noun lexeme and 119 of a verb one). However, the reduction is not trivial and needs contextual disambiguation among many potential ones. The almost free word order blocks possibilities of simple shallow parsing. Moreover, there is no resources for Polish comparable to those used in [11,21] and the size of the Polish Internet resources is much smaller than the resources used in [23]. Thus, our aim is to construct SSF on the basis of the largest corpus of Polish [18], including different genres of text, but any encyclopaedia, and to apply as the only more sophisticated tool the general purpose morpho-syntactic tagger called TaKIPI [16]. It is worth to emphasise that the accuracy of TaKIPI (93.44% for all words, 86.3% for ambiguous words) is significantly lower than the accuracy of a typical English tagger (about 97%). The interesting question is how this low accuracy can influence the accuracy of statistical extraction of SSF?

## 2 LSA Applied to Short News

As there is little information about application of LSA [11] to Polish (and to other inflective languages, too) we started with the direct application of LSA method to a corpus of Polish. The largest existing corpus of Polish is The IPI PAN Corpus [18] (IPIC). It is too big ($\tilde{3}50$ 000 documents, 260 millions of words) to process a matrix: documents×words created on its basis by the application of SVD. Moreover, the documents in IPIC are of very different sizes. Thus, for the LSA experiment[1] we selected a subcorpus of relatively short news from a daily newspaper (i.e. *Dziennik Polski*). The subcorpus includes 185 066 documents from the years 1998–2001.

The main goal of our LSA experiment was to define the base line for further experiments with more sophisticated construction of SSF. Using the well known method of LSA, we wanted also to analyse the influence of TaKIPI errors on the erroneous associations in the constructed SSF.

In [11] words from documents were directly used in the construction of the matrix. There are many morphological words derived for the same lexeme in a inflective language. It can be assumed that the meaning of different morphological words is identical with the corresponding lexeme. We use *morphological base forms* (shortly *base forms*) as representing lexemes. As many Polish words are ambiguous according to their possible base forms, a morpho-syntactic tagger must be applied to contextually determine the appropriate base form for each word. Thus the process of the construction of SSF can be dived into four main phases: *Initial Preparation*, *Matrix Construction*, *Matrix Transformation* and *Evaluation of SSF*. In the Initial Preparation phase, we applied TakIPI tagger to disambiguate words, including base form, in spite of the fact that TaKIPI

[1] The preliminary version of this experiment was presented in [15].

expresses (the state on 09.2006) the accuracy of 93.44% for all words (the accuracy of base form disambiguation is higher, but not tested).

On the basis of documents annotated and disambiguated by TaKIPI a matrix $\mathbf{M}$ of the size $B \times D$ is created, where $D$ is the number of documents, $B$ — the total number of base forms, and $\mathbf{M}[b, d]$ — the number of occurrences of the $b$-th base form in the $d$-th document (calculated in the *Matrix Construction* phase). We limited the set of base forms analysed only to the nouns ($\mathbf{N_{WN}}$) included in the preliminary version of plWN (state 12.2006) [17,15], i.e. the 4 611 most frequent nouns from IPIC v.1.0. It was motivated by the following: the intended evaluation procedure, (discussed further in this section) based on WordNet, and the technique proposed in Sec. 3 limited to nouns. As result, all other base forms were eliminated from the counting of occurrences in $\mathbf{M}$.

During the transformation phase $\mathbf{M}$ we perform (following [11]):

- the *logarithmic scaling* of cells — for each $d, b$: $\mathbf{M}[b, d] = ln(\mathbf{M}[b, d] + 1)$,
- the *entropy normalisation*: $\mathbf{M}[b, d] = \mathbf{M}[b, d] / \sum_{1 \leq i \leq D} -p(b, i) \; log \; p(b, i)$, where $p(b, d) = \mathbf{M}[b, d] / \sum_{1 \leq i \leq D} \mathbf{M}[b, i]$,
- and SVD transformation of the whole $\mathbf{M}$ to the $k$ dimensions — $\mathbf{M}_k$: $B \times k$.

In order to achieve an objective measure, we applied the WBST test of [5] on the basis of plWN. In WBST, for each $b \in \mathbf{N_{WN}}$ we choose randomly its synonym from plWN, and we create all ordered question/answer pairs for every two words in the same synset. As in the fine grained structure of plWN synsets quite often include only a single LU (in such cases the meaning of a given LU is defined by the other lexical semantic relations), we have had to extend the basic WBST procedure by extending singleton synsets with their direct hyperonyms. After choosing a pair of synonyms, we add to every Q/A pair 3 other words that are not in the same synset as the pair is. The task is to choose the answer for the question, i.e. the synonym of $b$ from the 4 possible answers on the basis of SSF. The accuracy is defined as the number of proper answers. The baseline for WBST is 25% — random selection. For better analysis of the results, WBST has been applied to 7 native speakers of Polish. 79 Q/A sets randomly generated were tested with the average score of 86%. An example of a Q/A pair is:

Q: produkt (*product*)
A: bóg (*god*) model (*model*) niesprawiedliwość (*injustice*) **wyrób** ( *wares*)

The results achieved in the LSA experiment, presented in Tab. 1, are all below the result of [11] in TOEFL, i.e. 64.4%, in spite of the six time larger number of documents used in our experiment. However, it must be emphasised that WBST is claimed to be more difficult for SSFs than TOEFL [5]. Moreover, the manual analysis of the 20 most similar base forms shows a strong topic bias in the corpus, i.e. some topics like sport or local culture dominate in the corpus. In [11], a very good corpus of encyclopaedia definitions was applied. Decreasing of the bias by the use of a larger corpus was hardly possible because of technical limitations of SVD. Errors of TaKIPI were not visible in the errors of the constructed SSF.

**Table 1.** The LSA experiment on news documents

| No. of Singular Values | 50 | 300 | 500 | 750 | 1000 |
|---|---|---|---|---|---|
| Accuracy | 53.74 | 57.76 | 58.07 | 58.06 | 58 |

Tab. 1 shows the accuracy of SSF in relation to the different numbers of singular values of SVD (the different row dimensions) which gets the highest value around 500–750 singular values.

## 3 Geometrical Measures

The limitation of SVD on the size of the corpus used was overcome in [21] by the usage of a text window and contextual features. Looking for good indicators of the semantic properties of nouns on the level of syntax one can take into account: noun modifiers and predicates (especially verbs) taking nouns as their arguments. However, identification of predicate-argument structure without a robust parser or at least a dictionary of verb subcategorisation is impossible. The proper identification of many types of noun modifiers is also non-trivial task, e.g. in the case of prepositional phrases we need to determine the boundaries of noun phrases and in the case of modification by other noun in genitive case we need to distinguish between many different reasons of using the genitive case (the modifier can take any position and genitive case can be expected e.g. by a verb or numeral). Finally, we decided to start from using only adjectival–modifiers as the only indicators. Adjectives are claimed in [6] to express properties (or attributes) of entities represented by nouns and their syntactic association with nouns should be relatively easy to be checked in the text, even without an efficient shallow parser for Polish — free word order makes it difficult to construct (especially in contrast to the positional syntactic chunking of English), e.g. objects of a verb can appear in the text typically in any order.

Initially, we limited ourselves to the set of the 4 157 adjectives[2] included in plWN — $\mathbf{A_W}$, next the extended set of the 15 768 adjectives collected from the whole IPIC v.2.0 (tagged by TaKIPI) was used — $\mathbf{A_I}$.

In the construction phase, the window of $\pm k$ words was iteratively moved across documents with the centre positioned each time on the nouns from $\mathbf{N_{WN}}$. For each position of the window, occurrences of subsequent adjectives were counted and added to $\mathbf{M}[b, a]$, where $b$ is the base form of the noun in the centre and $a$ is the base form of an adjective from the window. The calculation was sentence borders wise, i.e. the position outside the sentence were marked with '*an empty base form*' `none`. As we wanted to count only the adjectives potentially syntactically associated with the nouns from $\mathbf{N_{WN}}$, only small values of $k$ were tested from the range: 2–6 plus 10.

In the transformation phase row vectors were transformed by entropy normalisation (see Sec. 2) [11] but we tested the lack of transformation, too.

[2] Some functional words classified as adjectives in IPIC (e.g. determiners or quantifiers) and ordinal numbers have been removed from each set of adjectives.

**Table 2.** Accuracy [%] for the nouns×adjective matrices created according to the different windows of co-occurency

| | $\mathbf{A_W}(k=3)$ | $\mathbf{A_I}(2)$ | $\mathbf{A_I}(3)$ | $\mathbf{A_I}(4)$ | $\mathbf{A_I}(5)$ | $\mathbf{A_I}(6)$ | $\mathbf{A_I}(10)$ |
|---|---|---|---|---|---|---|---|
| cosine — entropy | 68.22 | 74.16 | 68.66 | 65.77 | 64.37 | 63.07 | 60.79 |
| cosine — no transformation | 50.23 | 61.04 | 57.99 | 56.99 | 56.82 | 56.72 | 55.94 |

The results achieved in WBST for both sets of adjectives, different values of $k$ and both transformations for *non-reduced matrices* are presented in Tab. 2 — only the results for nouns occurring $\geq 1000$ (the threshold used in [5]) are included there, the results in relation to the frequency of nouns are given in Tab. 5. The larger set of adjectives (i.e. $\mathbf{A_I}$) gave better results in both cases (for $\mathbf{A_W}$ only the result for $k = 3$ is given). The application of SVD did not bring improvement, the result was increasing with the increasing number of singular values up to the results for the non-reduced matrices. It seems that the matrices describe attributes and any reduction of dimensions causes some loss of information. The results are decreasing with increasing value of $k$. It is caused by the increasing number of erroneous associations adjective-noun as the size of the window is increasing. The result of entropy transformation is better than in [11] but still lower than 75.8% reported for nouns (WBST) in [5].

Counting every occurrence of an adjective in the $\pm k$ window as associated with a noun in the centre is very naive from the linguistic point of view (it is visible in the decreasing result in Tab. 2). The obvious improvement is to introduce a mechanism of recognition of the adjectives really modifying the tested noun. Instead of a non-existing robust parser of Polish, the potential morpho-syntactic agreement between words (e.g. on gender, number, case, etc.) can be tested by constraints written in the language JOSKIPI developed for TaKIPI [16], i.e. a part of TaKIPI engine was applied to the processing of the corpus during the construction phase. A very simplified version of a constraint testing the existence of an particular adjective which agrees on the appropriate attributes with a noun in the window centre is shown below:

```
or( llook(-1,-5,$A,and(inter(pos[$A],{adj}),
                      inter(base[$A],{"base_form_of_an_adjective"}),
                      agrpp(0,$A,{nmb,gnd,cas},3) )),
    rlook(1,5,$A,and( ... ) ) )
```

The operator `llook` looks for a word fulfilling the complex condition in the range of the 5 words to the left of the window centre (the position of a noun). In the complex condition, the part of speech is compared to an adjective (for the sake of efficiency). Next the base form must be equal to the one defined, and finally, the agreement of the found adjective on number, gender and case with the noun is checked. The second symmetric part of searching to the right has been removed from the example. The matrix cell $\mathbf{M}[b, a]$ is increased when the condition with $a$ inside is fulfilled for the context with the $b$ noun in the centre. There is one condition (with some adjective) for each column of the matrix.

**Table 3.** Accuracy [%] for the nouns×adjective matrices based on the application of syntactic constraints

| | JOSKIPI+$\mathbf{A_W}$ range=5 | JOSKIPI+$\mathbf{A_I}$ range=5 |
|---|---|---|
| cosine — entropy | 80.71 | 81.15 |
| cosine — no transformation | 62.34 | 62.84 |

The final form of the condition has been developed on the basis of several tests performed manually on IPIC and it has been extended with several additional constraints taking the following scheme (symmetric for left and right context):

- search for a particular adjective or adjectival participle agreed on number, gender and case,
- tests for the presence of allowed modifiers (e.g. adverbs or numerals) between the found adjective (participle) and noun,
- tests for the non-presence of several other words or phrases.

The detailed linguistic condition was developed in two working days.

The sophisticated constraint eliminated a lot of false associations of adjectives and nouns which resulted in the large increase of the accuracy — up to 81.15% for entropy normalisation and the larger set of adjectives, see Tab. 3.

The used range of 5 words for the search operators does not influence the result as the linguistic constraints eliminate the false associations with adjectives in larger distance from the noun in the centre. The tests performed for the range reduced to 2 have not given any significant change, only a small increase in the results. It means that the constraints successfully eliminate false associations.

# 4 Probabilistic Measures

Following the claims of [13, pp 303] that "The Euclidean distance is appropriate for normally distributed quantities, [...]", as the words in the corpus don't have a normal distribution, two measures of similarity based on probability distributions were tested, too [2,13]:

- *information radius* (*IRad*): $D(p||\frac{p+q}{2}) + D(q||\frac{p+q}{2})$,
  and similarity: $Sim_{IRad} = 10^{-\beta IRad(p||q)}$,
- $L_1$ *norm* ($L_1$): $\sum_i |p_i - q_i|$, and similarity $Sim_{L_1} = (2 - L_1(p,q))^\beta$

During the transformation, the rows of $\mathbf{M}$ were converted into probability distributions of modification of nouns by adjectives — the two types of distribution were tested:

- $P(Adj|Noun)$ — $\mathbf{M}[b,a] = \mathbf{M}[b,a]/\sum_i \mathbf{M}[b,i]$
- $P(Adj, Noun)$ — $\mathbf{M}[b,a] = \mathbf{M}[b,a]/(\sum_i \mathbf{M}[b,i] + e(b)$,
  and $p(e_b) = e(b)/(\sum_i \mathbf{M}[b,i] + e(b)$, where $e()$ returns the number of occurrences of the $b$ noun without a modifier from the set of adjectives.

The second type of distribution was intended to take into account the total number of occurrences of a noun and how strongly it is described by the attributes. However, as the results in Tab. 4 show (all test were performed on the smaller set of adjectives, i.e. $\mathbf{A_W}$ and with the matrix created with JOSKIPI constraints, where the range of search operators is 5), the introduction of a negative event i.e. lack of a modifier, decreased significantly the accuracy. The number of occurrences of an attribute, i.e. some adjective, is the most important for semantic similarity. Several values of the $\beta$ parameter for $Sim_{IRad}$ and $Sim_{L_1}$ were tested, achieving the best results for: 10 and 4, respectively. However, there were no significant changes of the result for the different values of $\beta$ in large range. A smoothing based on the Laplace's law was introduced, too, but the accuracy was significantly lower — the difference among the attributes was lost.

**Table 4.** Accuracy [%] of probabilistic SSF in relation to different distributions ($A$ means Adj, $N$ — Noun)

| | $P(A\|N)$ range=5 | $P(A,N)(5)$ | smoothed $P(A\|N)(5)$ | smoothed $P(A,N)(5)$ |
|---|---|---|---|---|
| IRad | 80.16 | 46.23 | 41.73 | 34.69 |
| L1 | 76.98 | 47.70 | 46.76 | 30.89 |

The best accuracy of *IRad*-based SSF (80.16%) is slightly lower than the best score of the cosine-based SSF (81.15%). It seems that the attributes expressed by adjectives form a kind of meaning space in which differences can be better expressed by geometrical notions than probabilistic ones. Mannual analysis of the $n = 20$ most similar nouns generated on the basis of *IRad*-based SSF revealed that in the case of probabilistic measures nouns are grouped more on the basis of similar syntactic behaviour than semantic similarity.

## 5 Conclusion

Until now, the results were given only for nouns occurring $\geq 1000$ times in IPIC. However, this set of nouns includes many very general words like *kobieta* (*a woman*) that are used in many different contexts and create corpus-dependent associations with many different words that was clearly visible in the experiments in [15]. Theise general words are relatively more difficult for the synonymy test than more specific words.

In Tab. 5 we compared the accuracies achieved for different types of nouns in relation to their frequencies in IPIC (206 millions of words). The result for the threshold 1000, identical to the one used in [5], shows the potential of our method based on precise analysis of syntactic associations: 81.15% in our method vs. 75.8% for nouns in [5], achieved by the application of the combinations of several measures. Obviously, in [5] SSF was constructed for more than 15 000 words including 9 887 nouns, when in our approach SSF is defined only for 4 611 nouns, and was tested on the preliminary version of plWN, in which some synstets are still quite large and include e.g. hyperonyms. In order, to check the influence of the state of plWN on the results, we tested also the method on the very initial version of plWN: including only large, broad synsets and no hyperonymy links at all. The achieved result in this test, which is possibly easier, is almost identical to the present one (a little lower, but without statistical significance). Anyway, this good result is very promising for the method of modeling the context of occurrence by precise exploration of morpho-syntactic associations. The method is able to cope with rich morphology of Polish and its free word order, utilises relatively simple tools, has been applied to a general corpus and can be easily transferred to other inflective languages like Czech.

**Table 5.** Accuracy [%] in relation to different thresholds on noun frequecies and different types of similarity functions

| type of SSF \ threshold | 0 | 1 | 10 | 100 | 500 | 1000 |
|---|---|---|---|---|---|---|
| *Cosine* | 64.03 | 68.33 | 70.08 | 75.90 | 79,46 | 81.15 |
| *IRad* for $P(Adj\|Noun)$ | 63.39 | 66,81 | 69.36 | 74.63 | 78.85 | 80.16 |
| *L1* for $P(Adj\|Noun)$ | 60.00 | 63.67 | 66.17 | 71.03 | 76.27 | 76.98 |
| *IRad* for $P(Adj, Noun)$ | 36.46 | 38.28 | 39.03 | 42.04 | 43.56 | 46.23 |
| $L_1$ for $P(Adj, Noun)$ | 40.09 | 41.89 | 43.04 | 45,31 | 45.69 | 47.70 |

Manual analysis showed that the method based exclusively on attributes (adjectives) has some problems with the proper distinction of some nouns, e.g. some names of large buildings are grouped together, one needs to analyse their functions by analysing associations with verbs or participles to differentiate them. In further research we plan to explore other types of modifiers e.g. modification of nouns by nouns. Next we plan to extend the method to other Parts of Speech.

**Acknowledgement.** This work was financed by the Ministry of Education and Science project No 3 T11C 018 29.

## References

1. Berry M.: Large Scale Singular Value Computations. International Journal of Supercomputer Applications, 6:1, pp. 13–49, (1992).
2. Dagan I., Pereira F., Lee L.: Similarity-based estimation of Word Co-occurrence Probabilities. In ACL 32, pp. 272-278, (1997).
3. Ehlert B.: Making Accurate Lexical Semantic Similarity Judgments Using Word-context Co-occurrence Statistics. Masters thesis, University of California, San Diego(2003).
4. Fellbaum Ch. (ed.): WordNet An Electronic Lexical Database. MIT Press, (1998).
5. Freitag D., Blume M., Byrnes J., Chow E., Kapadia S., Rohwer R., Wang Z.: New Experiments in Distributional Representations of Synonymy. Proceedings of the 9th Conference on Computational Natural Language Learning, pp.25-32, ACL, (2005).
6. Gärdenfors P. Conceptual Spaces — The Geometry of Thought. The MIT Press, 2000.

7. Girju R., Badulescu A., Moldovan D.: Automatic Discovery of Part-Whole Relations. Computational Linguistics, 32(1): pp. 83–135, 2006.
8. Grefenstette G.: Evaluation Techniques for Automatic Semantic Extraction: Comparing Syntactic and Window Based Approaches. In Proceedings of The Workshop on Acquisition of Lexical Knowledge from Text, Columbus. SIGLEX/ACL, (1993)
9. Harris Z.: Mathematical Structures of Language. Interscience Publishers, New York, (1968).
10. Kłopotek, M.A., Wierzchoń, S.T., Trojanowski, K., eds.: Intelligent Information Processing and Web Mining — Proceedings of the International IIS: IIPWM'06 Conference, Zakopane, Poland. Advances in Soft Computing. Springer, Berlin (2006)
11. Landauer T., Dumais S.: A Solution to Plato's Problem: The Latent Semantic Analysis Theory of Acquisition. Psychological Review, 104(2):211-240 (1997)
12. Lin D., Pantel P.: Induction of Semantic Classes from Natural Language Text. ACM, (2001)
13. Manning Ch.D., Schütze H.: Foundations of Statistical Natural Language Processing. The MIT Press, (1999).
14. Pantel P., Pennacchiotti M.: Esspresso: Leveraging Generic Patterns for Automatically Harvesting Semantic Relations. Proceedings of 21st International Conference on Computational Linguistics (COLING-06), Sydney 2006, pp. 113–120, ACL, 2006.
15. Piasecki M.: LSA Based Extractionof Semantic Similarity for Polish. In Proceedings of Multimedia and Network Information Systems. Wrocław University of Technology, (2006)
16. Piasecki M., Godlewski G.: Effective Architecture of the Polish Tagger. In Proceedings of the Text, Speech and Dialogue Conference, Brno, 2006. LNCS, Springer, Berlin, (2006).
17. Polish WordNet — Homepage of The Project. `http://www.plwordnet.pwr.wroc.pl/main/`, State: the December 2006.
18. Przepiórkowski, A.: The IPI PAN Corpus Preliminary Version. Institute of Computer Science PAS (2004).
19. Ruge G.: Experiments on Linguistically-based Term Associations. Information Processing and Management, Vol. 28, No. 3, pp. 317–332, (1992).
20. Ruge G.: Automatic Detection of Thesaurus Relations for Information Retrieval Applications. Lecture Notes in Computer Science, Vol. 1337, pp. 499-506, (1997).
21. Shütze H.: Automatic Word Sense Discrimination. Computational Linguistics, 24(1), 97-123, (1998).
22. Turney P.T. Mining the Web for Synonyms: PMI-IR versus LSA on TOEFL. In Proceedings of the Twelfth European Conference on Machine Learning, Springer-Verlag, Berlin (pp. 491502) (2001).
23. Turney P.T., Littman M.L., Bigham J., Shnayder V.: Combining Independent Modules to Solve Multiple-choice Synonym and Analogy Problems. In Proceedings of the International Conference on Recent Advances in Natural Language Processing. (2003).
24. Widdows D.: Geometry and Meaning, CSLI Publications, (2004).
25. Woliński, M.: Morfeusz — a practical tool for the morphological analysis of polish. [10]

# Automatic Document Structure Detection for Data Integration

Radek Burget

Faculty of Information Technology, Brno University of Technology
Bozetechova 2, 612 66 Brno, Czech Republic

**Abstract.** A great amount of information is still being stored in loosely structured documents in several widely used formats. Due to the lack of data description in these documents, their integration to the existing information systems requires sophisticated pre-processing techniques to be developed. To the document reader, the content structure is mostly presented by visual means. Therefore, we propose a technique for the discovery of the logical document structure based on the analysis of various visual properties of the document such as the page layout or text properties. This technique is currently being tested and some promising preliminary results are available.

**Keywords:** semistructured data, document analysis, visual segmentation, logical structure.

## 1 Introduction

Although there exist several formats for representing the documents containing structured data, as for example XML, a great amount of information is still being stored in documents that don't allow the explicit specification of the structure and semantics of the content. Most frequently, this is the case of on-line web resources in the HTML language, however, many important documents are often exchanged via e-mail in some commonly used office formats. Due to the lack of data description in these documents, the integration of this information source to the existing information systems requires sophisticated pre-processing techniques to be developed.

To the document reader, the content structure is mostly presented by visual means. The document may be divided into sections with headings, paragraphs, tables and other structures. Therefore, analyzing the visual part of documents may bring an useful infomation about their structure.

In this paper, we introduce a technique for the document structure discovery that is currently being tested. This technique is based on an analysis of a range of visual features of the document and its content and therefore, it doesn't rely on any particular document format. As the result, we obtain a hierarchical structure of the document as it is expressed by the visual means, which can be used for identifying important parts of the documents.

W. Abramowicz (Ed.): BIS 2007, LNCS 4439, pp. 391–397, 2007.
© Springer-Verlag Berlin Heidelberg 2007

## 2 Related Work

The visual organization of the documents has been mostly investigated for HTML documents. Most approaches work with a tree representation of the HTML code [1,2,3,4]. However, the direct analysis of the HTML code must be always based on certain assumptions on the HTML usage that are often not fulfilled in reality.

Advanced methods of page segmentation work with the visual representation of the rendered document [5,6]. This approach is more general and a similar approach can be used other types of documents as for example PDF documents [7].

Some approaches try to detect page objects with certain semantics and to model the semantic relations among these objects by creating a logical schema of the document [8,9].

## 3 Document Structure Discovery

The proposed approach consists of two basic steps:

1. **Page segmentation** – we detect the basic visual organization of the page represented by various visually separated areas that are laid out on the page and possibly nested. The result of this phase is a tree of visual areas discovered in the document.
2. **Logical structure analysis** – in this phase, we consider additional visual features of the discovered visual areas. Based on these features, we try to guess the mutual relations among the visual areas at the same level.

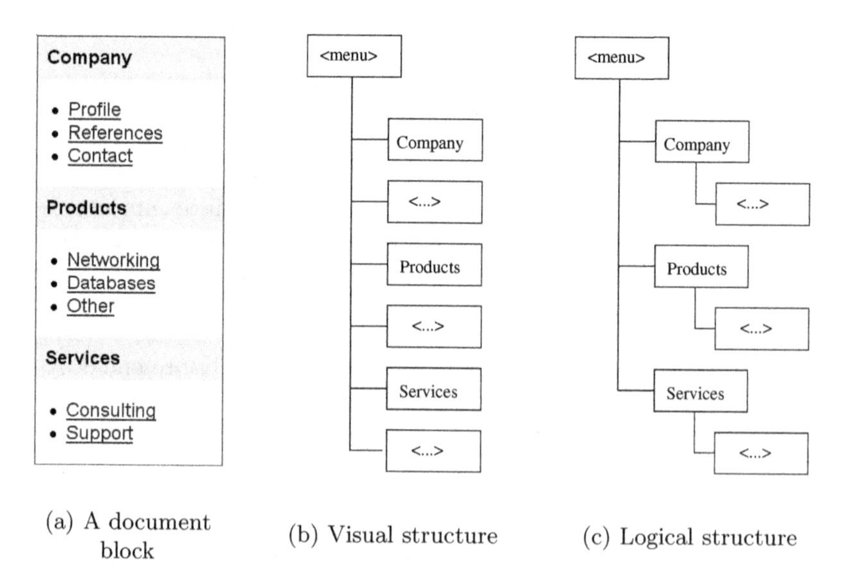


(a) A document block (b) Visual structure (c) Logical structure

**Fig. 1.** The difference between the visual and logical structure

These steps are demonstrated in Fig. 1. When analyzing a particular visual block of a document (a), the visual segmentation discovers that this block is divided in six subblocks. This corresponds to the tree structure (b). When analyzing the logical structure, the blocks “Company”, “Products” and “Services” are apparently used as labels that denote the meaning of the following block. Therefore, from the logical point of view, there exist other relations among these blocks as shown in (c).

## 4 Page Segmentation

The purpose of the page segmentation is to discover *visual areas* in the document. We define a visual area as a rectangular area in the page that is visually separated from the remaining content by any combination of the following means:

- **Visual properties of the area itself** – a frame around the whole area or a different background color of the area
- **Surrounding objects** – an area can be delimited by other visual areas placed in the page
- **Whitespace borders** delimiting the area

Since the areas can be nested, the result of the page segmentation is generally a tree of visual areas.

### 4.1 Source Document Representation

The input to the segmentation algorithm is a set of *boxes* produced by a layout engine specific to the document format. Currently, the layout engines are freely available at least for HTML, RTF, OpenDocument and PDF formats.

A box is a rectangular area in the page that contains a part of the document content – a text portion, an image, graphics and other. Each box is determined by its position and size in the page. For describing the box positions in the page, we define a rectangular coordinate system of the page with the beginning in the top left corner of the page. Additionally, most of the document formats define the behavior in case that the boxes are overlapping. For example in HTML/CSS, the box defined later in the document code is placed in front of the boxes defined before, if not specified otherwise. We represent this behavior by a virtual $z$ coordinate, where the boxes with a greater value of $z$ are placed in front of the boxes with a lower value of $z$.

Depending on the format, additional properties may be specified for the box such the background color or a frame around the box. If the box is formed by text, the text and font properties can be examined too.

### 4.2 The Segmentation Algorithm

The proposed algorithm is based on a bottom-up approach to page segmentation. It consists of following steps:

- **A box tree creation.** We create a tree from the input set of boxes. The root of this tree is a virtual box $b_0$ that covers the whole page. The tree itself models the box nesting. Any box $b_1$ is represented as a child node of another box $b_2$ if $b_1$ is enclosed in $b_2$. If the boxes partially or fully overlap, we say, that the box with a greater value of $z$ is enclosed in the box with a lower value of $z$.
- **Standalone area detection.** We detect all the boxes, that form standalone visually separated areas. This happens when the box is visually separated from its neighborhood by a frame around or by a different background color or it directly contains a text. The remaining boxes are omitted. As the result, we obtain a basic tree of visual areas which is processed further.
- **Continuous area detection.** For each area, we find the directly adjoining areas that are not visually separated. These areas are then replaced by a single area. This is often the case of the text paragraphs built from several line boxes.
- **Detection of covering areas.** For each area, we detect the largest possible rectangular area containing this area that is limited by other, visually separated areas. We start with the examined area and we try to expand to all the directions it while maintaining its rectangular shape until some visually separated area is encountered. As a result, we obtain a virtual area covering several subareas such as for example columns in the document.

By applying these steps, we obtain a tree of visual areas corresponding to Fig. 1b, which we will call a *visual structure tree*.

## 5 Logical Structure Detection

The visual layout forms only a part of the information necessary to obtain the logical structure. The remaining part of the information is expressed by other means. In our technique, we analyze following visual features:

- **Font properties** that may be used for creating a labeling system in the document. Usually, several levels of headings and labels are used that denote particular parts of the document. The most important properties are the *font weight* and *size*, in some cases the *text color* and *underlining* too.
- **Punctuation**, which is a frequently used mean of expressing the relation between two parts of the text. A good example is using colons or dashes.

Based on the above features, we determine a value of *weight* $w$ that directly depends on the font size and weight for each part of the text. The remaining visual features mentioned above influence this value via various heuristics:

1. Underlining and using a color that is not common for the remaining text increases the weight
2. A colon at the end of a part of text increases the weight of the text

Then, for each visual area, we can compute an *average weight* as the average value of the weights of all the nested areas.

Subsequently, we convert the previously created visual structure tree to the logical structure by considering the weights of the areas. For each visual area in the visual structure tree, we go through the list of its child areas. If the weight of an area is lower than the weight of its preceding area, we make this area the child area of the preceding area. This step is repeated recursively. As the result, we obtain a logical structure tree corresponding to the Fig. 1c.

## 5.1 Mutual Block Positions

Mutual positions of the blocks play an important role in tables, where the particular position in a row and column allows the data interpretation. A combination with the above mentioned analysis of the font properties allows to identify the header cells in the table. Fig. 2 shows an example of a table and a possible representation of its logical structure. Several spatial reasoning applications have been proposed for obtaining the structure based on the analysis of the table layout [10,11]. More generally, relations among the visual blocks that can be observed in a table, can be found in the overall page organization too.

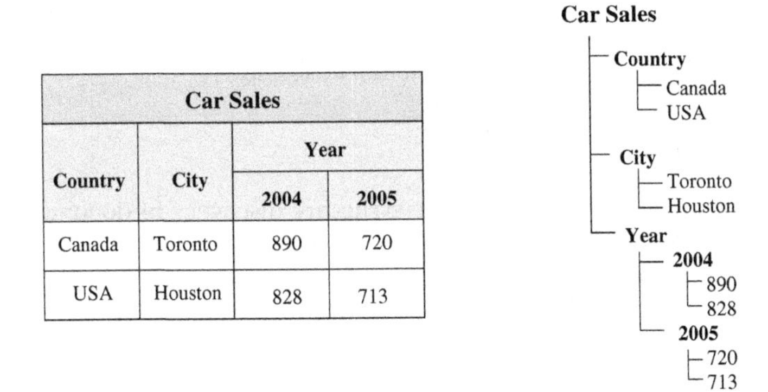

| Car Sales | | | |
|---|---|---|---|
| Country | City | Year | |
| | | 2004 | 2005 |
| Canada | Toronto | 890 | 720 |
| USA | Houston | 828 | 713 |



**Fig. 2.** Logical structure of a table

# 6 Expected Applications

Once the logical structure of the document is known, the document can be represented in a structured form, for example by a single XML file. This allows adopting various already existing methods of XML document querying, classification, indexing or retrieval. By supplying an additional ontological information, the semantic of the discovered visual areas can be guessed, as we have shown in [12]. Other approach to detecting the semantics of individual areas is also presented in [8].

The main expected application is information extraction from electronic documents on the web (e.g. prices, stock quotes) or exchanged through the e-mail (e.g. orders, invoices) and their integration to the information systems.

## 7 Development Status

The described technique is currently being implemented and tested. Current implementation allows testing on HTML documents. It includes a HTML/CSS layout engine, the page segmentation and the logical structure detection as described above. The resulting logical structure is represented by an XML file.

We have tested the proposed techniques on 30 home pages of various news servers such as `cnn.com`, `nytimes.com`, `elmundo.es`, etc. with the aim of the story extraction. These pages usually contain a large amount of visual areas. The preliminary tests have shown following issues that are being solved:

- In some cases, the nested areas are incorrectly recognized as area separators. This leads to detecting a greater number of visual areas in the document.
- Not all the visual areas have the same importance. A more sophisticated area classification algorithms could be used for distinguishing the less important areas such as advertisement as discussed for example in [13].

Although the resulting tree contains noisy parts that result mainly from advertisements and other auxiliary content, the subtrees that correspond to the main content of the page in most cases describe the actual structure of the content as expected.

Concurrently, we are implementing the PDF and RTF document parsers based on the available open-source libraries in order to test the applicability of the proposed techniques to non-HTML document formats.

## 8 Conclusion

We have presented a technique for the structure discovery in documents, where the structure is not explicitly defined. In contrast to already existing methods, our technique is general, it doesn't rely on the properties of a particular markup language and thus, it is not limited to a single document format. It is based on the analysis of various types of visual information present in the document such as the page layout, font properties or colors. The resulting structure can be used for further processing of the document. Currently the proposed technique is in the phase of implementation and testing; the preliminary results indicate that the resulting information can allow better integration of loosely structured documents in the information systems.

## Acknowledgements

This research was partially supported by the Research Plan No. MSM 0021630528 – Security-Oriented Research in Information Technology.

## References

1. Chen, J., Zhou, B., Shi, J., Zhang, H., Fengwu, Q.: Function-based object model towards website adaptation. In: Proceedings of the 10th International Wold Wide Web Converence. (2001)

2. Gupta, S., Kaiser, G., Neistadt, D., Grimm, P.: Dom-based content extraction of html documents. In: WWW2003 proceedings of the 12 Web Conference. (2003) 207–214
3. Kovacevic, M., Diligenti, M., Gori, M., Maggini, M., Milutinovic, V.: Recognition of common areas in a web page using visual information: a possible application in a page classification. In: Proceedings of 2002 IEEE International Conference on Data. (2002)
4. Mukherjee, S., Yang, G., Tan, W., Ramakrishnan, I.: Automatic discovery of semantic structures in html documents. In: International Conference on Document Analysis and Recognition, IEEE Computer Society (2003)
5. Cai, D., Yu, S., Wen, J.R., Ma, W.Y.: VIPS: a Vision-based Page Segmentation Algorithm. Microsoft Research (2003)
6. Gu, X.D., Chen, J., Ma, W.Y., Chen, G.L.: Visual based content understanding towards web adaptation. In: Proc. Adaptive Hypermedia and Adaptive Web-Based Systems. (2002) 164–173
7. Hassan, T., Baumgartner, R.: Intelligent wrapping from pdf documents with lixto. In: RAWS 2005, FEI VB (2005) 17–24
8. Chung, C.Y., Gertz, M., Sundaresan, N.: Reverse engineering for web data: From visual to semantic structures. In: 18th International Conference on Data Engineering, IEEE Computer Society (2002)
9. Yang, Y., Zhang, H.: HTML page analysis based on visual cues. In: ICDAR '01: Proceedings of the Sixth International Conference on Document Analysis and Recognition, Seattle, USA, IEEE Computer Society (2001) 859
10. Gatterbauer, W., Bohunsky, P.: Table extraction using spatial reasoning on the CSS2 visual box model. In: Proceedings of the 21st National Conference on Artificial Intelligence (AAAI 2006), AAAI, MIT Press (July 2006)
11. Kruepl, B., Herzog, M.: Visually guided bottom-up table detection and segmentation in web documents. In: WWW '06: Proceedings of the 15th international conference on World Wide Web, New York, NY, USA, ACM Press (2006) 933–934
12. Burget, R.: Hierarchies in html documents: Linking text to concepts. In: 15th International Workshop on Database and Expert Systems Applications, IEEE Computer Society (2004) 186–190
13. Song, R., Liu, H., Wen, J.R., Ma, W.Y.: Learning block importance models for web pages. In: WWW '04: Proceedings of the 13th international conference on World Wide Web, New York, NY, USA, ACM Press (2004) 203–211

# Bottom-Up Discovery of Clusters of Maximal Ranges in HTML Trees for Search Engines Results Extraction

Dominik Flejter and Roman Hryniewiecki

Poznan University of Economics, Department of Information Systems, al. Niepodleglosci 10, 60-967, Poznan, Poland
D.Flejter@kie.ae.poznan.pl, Roman@Hryniewiecki.net

**Abstract.** Unsupervised HTML records detection is an important step in many Web content mining applications.

In this paper we propose a method of bottom-up discovery of clusters of maximal, non-agglomerative similar HTML ranges in nested set HTML tree representation. Afterward we demonstrate its applicability to records detection in search engines results. For performance measurement several distance assessment strategies were evaluated and two test collections were prepared containing results pages from almost 60 global and country-specific search engines and almost 100 methodically generated complex HTML trees with pre-set properties respectively.

Empirical study shows that our method performs well and can detect successfully most of search results ranges clusters.

## 1 Introduction

As the Web matures, the number of resources accessible in dynamically generated Web sites resulting from underlying databases grows dynamically, including the huge amount of information available from the Deep Web sites [5]. Therefore information extraction from structured or semi-structured Web pages becomes an important task in contemporary Internet information processing.

Several approaches to the problem can be found in existing publications. In wrapper construction task wrappers are build for specific Web sites manually in visual or programmatic way. In wrapper induction task extraction rules are learnt from annotated sample of expected results. In automatic extraction data is extracted basing on several documents (less ambitious) or one document (more ambitious) without preceding learning; automatic extraction results may be subsequently used for fully-automatic wrapper induction.

In information extraction from semi-structured Web sources two steps are usually performed: 1) records and 2) columns detection with diverse sets of used techniques. In this paper we focus on one-document based automatic extraction of records from HTML documents. We demonstrate our approach on text search engines results pages; however, as no domain-level heuristics were adopted, the method should perform well on other type of Web sites such as e-commerce and Deep Web sites.

Several techniques were reported to be applicable to records extraction. Document text similarity measurement [16], punctuation and HTML separators use [16], mining of repeating or similar tag tree subtrees [6, 15, 20, 22, 23, 24], use of extraction

W. Abramowicz (Ed.): BIS 2007, LNCS 4439, pp. 398–410, 2007.
© Springer-Verlag Berlin Heidelberg 2007

ontologies [9] and most recently visual perception-enhanced analysis [19, 22, 24] as well as multiple "experts" based techniques combining several sources of information [11] were proposed.

In this paper we propose a tag-tree-structure-based bottom-up method of discovery of clusters of maximal similar ranges of HTML document and demonstrate its application in the domain of search results records detection on empirical and theoretical collections.

The structure of this paper is as follows: Section 2 introduces the basic terminology, Section 3 presents our algorithm for finding clusters of maximal similar HTML ranges, Section 4 describes algorithm implementation, evaluation methodology and test datasets, Section 5 contains evaluation results, Section 6 lists possible further development and Section 7 concludes our contribution.

## 2 Background

In this section we define the most important basic terms used throughout this paper. Section 2.1 elaborates on tree-related terms and conventions and Section 2.2 introduces preliminary notions of edit distances of ranges.

### 2.1 Trees

A rooted tree is defined a set $T$ of one or more nodes such that

- there is one specially designated node called the root of the tree, *root (T)*; and
- the remaining nodes (excluding the root) are partitioned into $m \geq 0$ disjoint sets $T_1, \ldots, T_m$ and each of these sets in turn is a tree [14].

The trees $T_1, \ldots, T_m$ are called the subtrees of the tree $T$. This relation will be designated $subtree(T, T_1)$ in the remaining of this paper and $|nodes(T)|$ will be used for the number of nodes in the tree.

The one-to-many relation of parenthood that places a hierarchical structure on $T$ is defined [2]: every node $n \in T$, $n \neq root(T)$ has exactly one parent node $parent(n)$ and $|children(n)|$ of children nodes. All tree nodes with no children are called leaves.

In this paper our interest is in rooted, ordered, labelled trees. Tree is ordered if given a node $n$ with $|children(n)|$ children we can uniquely identify $i^{th}$ child for $i \in <1, |children(n)|>$ [7]; in the remaining of this paper $i^{th}$ child of node $n$ will be identified by $child(n, i)$. Tree is labelled if any label $label(n)$ is assigned to each node $n$. In this paper nodes' labels consist of HTML tag name, attributes name/value pairs.

In this paper $ancestor(n_A, n_B)$ will be used to denote that $n_A$ is an ancestor of $n_B$ and $descendant(n_B, n_A)$ to denote that $n_B$ is a descendant of $n_A$. A set $subtree(n_S) = \{n_S\} \cap \{n \in T: descendant(n, n_s)\}$ is a subtree generated by node $n_S$.

Embedded subtree $ET$ of $T$ is such a tree that [8]

$$ET \subset T \text{ and}$$
$$\forall n_A, n_B \in ET : ancestor(n_A, n_B) \text{ in } T \Rightarrow ancestor(n_A, n_B) \text{ in } ET$$

For given node $n_R$ (called range parent, *parent(R)*) of an ordered tree and integers $1 \leq a \leq b \leq |children(n_R)|$ we define a range R as

$$R = \bigcup subtree(n) \ for\ n = child(n_R, i), a \leq i \leq b$$

Less formally, a range consists of any sequence of consecutive children of the same parent node $n_R$ and all their descendants. As range parent is not a part of the range, a range is not necessarily a tree.

Out of a number of possible representations of rooted ordered trees [2] nested set model [4] is the most convenient for tasks requiring continuous verification of inclusion of one subtree in another. In nested set model each node is considered a set that is nested in its parent set, each parent node contains a union of its children sets and root node contains all other sets as its direct or indirect subsets.

To facilitate the verification of sets inclusion sets are described as intervals with two integers corresponding to left and right node boundary *<left(node), right(node)>*; $subtree(n_A) \subset subtree(n_B)$ iff $left(n_A) > left(n_B)$ *and* $right(n_A) < left(n_B)$. Typically consecutive integers in depth first (preorder) tree transversal are used for right and left boundaries assignment. In that setting interval corresponding to root node equals to $<1, 2*|nodes(T)|>$ and for each leaf node $n$: $right(n) = left(n) + 1$. Figure 1 presents an exemplary tree with calculated corresponding intervals.

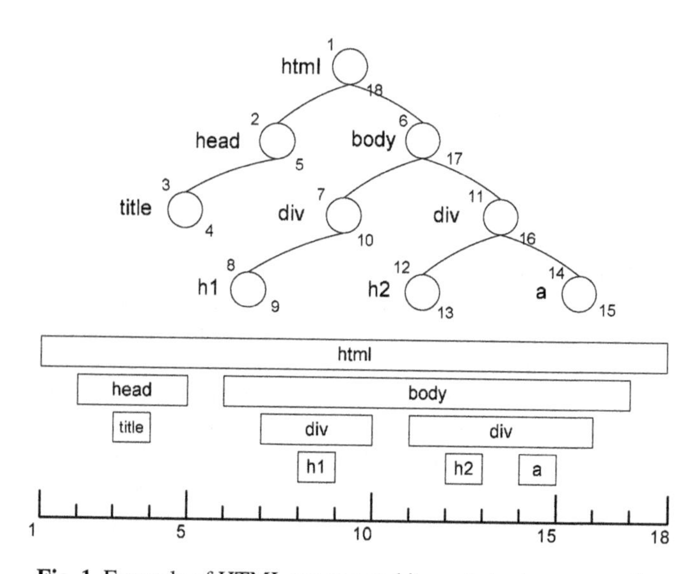


**Fig. 1.** Example of HTML tag tree and its nested set representation

Nested set model allows not only to verify inclusion or find intersection of two subtrees but also of two ranges with $left(R) = min(left(n))$, and $right(R) = max(right(n))$, $n \in R$.

This paper focuses on finding clusters of similar HTML ranges; for simplicity an informal definition of cluster as a set of objects similar to each other was adopted; similarity measurement will be described more profoundly in Section 4.1. For each cluster C the number of ranges it contains will be denoted $|ranges(C)|$. For easier algorithm implementation each cluster will be treated as ordered by *left(R)* and $i^{th}$ range in the cluster will be accessed by *range(C, i)*. For convenience:

$$|ranges(C_1)| = |ranges(C_2)| \Rightarrow C_1 \Theta C_2 = \\ = \{R_i : R_i = range(C_1, i)\ \Theta\ range(C_2, i)\ for\ i = 1, \ldots, |ranges(C_1)|\}$$

for each operator Θ unless stated otherwise.

### 2.2 Edit Distance

Edit distance measures for two sequences are measures based on finding minimal-cost edit script transforming one sequence to the another by means of insertion, deletion and update operations. To find the edit distance of two sequences $A = (A[1]\ A[2]\ \ldots\ A[m])$ and $B = (B[1]\ B[2]\ \ldots\ B[n])$ an $(m+1)\times(n+1)$ edit graph is constructed with point $(x, y)$ corresponding to pair $(A[x], B[y])$ and directed edges corresponding to edit operations: $((x\text{-}1, y), (x, y))$ – to removing $A[x]$, $((x, y\text{-}1), (x, y))$ – to inserting $B[y]$ and $((x\text{-}1, y\text{-}1), (x, y))$ – to updating $A[x]$ with $B[y]$ value (if value is not equal) with operation costs assigned to edit graph edges [7]. The task is therefore transformed to finding minimal-cost path in the edit graph between vertices $(0, 0)$ and $(m, n)$.

Distance measure with all aforementioned operations allowed is called Levenshtein distance and can used for two strings. In case of measuring distance between two trees, some of the operations are disallowed – e.g. it is impossible to remove parent node without removing all its descendants. As a result some edges in edit graph does not exists. In our research we used edit graph calculation based on tree nodes depth information proposed in [7]. The same paper provides more information about edit distance measures and edit graphs.

## 3 Ranges Clusters Finding

This section describes a bottom-up algorithm for finding similar ranges in HTML trees: the goal of the algorithm, used tree representation, method of finding maximal clusters and method for agglomerative clusters removal.

### 3.1 Goal

The goal of our algorithm described in the next sections is to find all clusters conforming to the following constraints:

O1. Intersection of any pair of two ranges in one cluster is empty set.
O2. The cluster is maximal in its clusters family.
O3. The cluster is not an agglomeration of any other clusters.
O4. Cluster internal incoherence is not larger than pre-set parameter.

First of the objective is understandable *per se*; however O2, O3 and O4 require a more detailed explanation.

Family of clusters is defined as a maximal set of clusters CS having the same number of ranges such that

$$\forall c \in CS \; \exists c_1 \neq c \in CS : c \cap c_1 \in \{c, c_1\}$$

Less formally, it is a set of clusters with equal number of ranges such that each cluster is included in or includes at least one another cluster. Cluster is maximal within its cluster family if it has minimal internal incoherence out of all clusters with maximal total number of nodes in all clusters ranges. Discussion of cluster internal incoherence measurement can be found in Section 4.1.

We call a cluster an agglomeration of another cluster iff

$$a)\ ranges(c) = n * ranges(c_1), n \in N$$

$$b)\ \forall i = 1,\dots, ranges(c_i) \forall j = 1,\dots,n : range(c_1,(i-1)*n + j) \subset range(c,i)$$

### 3.2 HTML Tree Representation

HTML document can be considered a tree with <HTML> element acting as root and elements with no nested tags corresponding to leaves. The representation of this tree for further processing is built in one preorder transversal of the original HTML document and consists of two outputs:

- Transformed tree constructed with respect to tag filtering rules and containing at each node additional properties necessary for the following stages of the algorithm.
- Set of initials clusters.

Transformed tree is constructed by copying only nodes corresponding to allowed tags (tag filtering approaches are discussed in Section 4.1); it is an embedded subtree of original HTML document tree. For each of the nodes of transformed tree there are several properties calculated at this step of the algorithm:

- ***depth(n)*** – depth of the node,
- ***left(n)*** – left boundary of the node in nested set representation,
- ***nodeid(n)*** – node number in preorder (depth-first) tree traversal,
- ***right(n)*** – right boundary of the node in nested set representation,
- ***tagid(n)*** – id of used tag equivalence class from the document tags vocabulary (see Section 4.1 for details).

During the same transversal **array representation of the tree** *T* (*array(T)*) is build with element at index *left(n)* corresponding to *tagid(n)* and element at index *right(n)* corresponding to *–tagid(n)* of node *n*. Initial set of clusters is also constructed with every cluster containing one-node ranges identical in terms of tags equivalence classes. These clusters are being expanded in following steps in a bottom-up way to find maximal clusters in families they belong to.

### 3.3 Maximal Ranges Finding

The first part of our algorithm is responsible for finding clusters of ranges fulfilling objectives O1, O2 and O4. It is based on expansion of the initial clusters to the parent and to the left / right sibling.

Expansion to the parent of a range *R* is a procedure consisting in constructing new range (expanded range) *ER* = *subtree(parent(R))*. Range *R* is expandable if *R* does not contain root and *ER* has small enough internal incoherence.

Expansion to the parent of a cluster (expanding cluster) consists in construction of a new cluster (expanded cluster) containing deduplicated expanding cluster's ranges expanded to the parent. For a cluster to be expandable to the parent at least one of the cluster's ranges must be expandable to the parent.

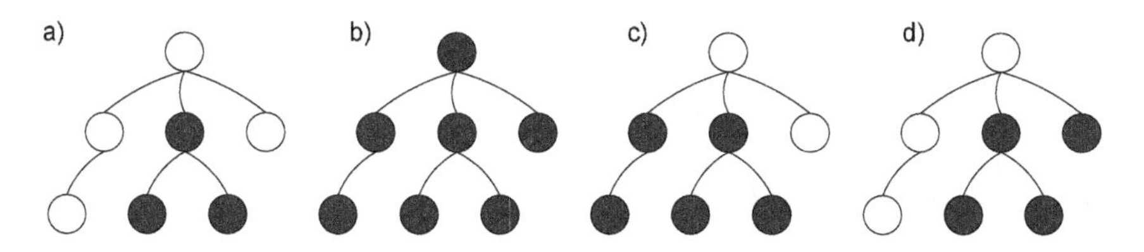


**Fig. 2.** Example of range expansions: a) original range b) range expanded to the parent c) range expanded to the left d) range expanded to the right

The procedure of expansion of an initial cluster to the parent is repeated as long as expansion is possible. Each time when expansion is impossible or expanded cluster contains less ranges than expanding cluster, expanding cluster is recorder as cluster candidate for further processing.

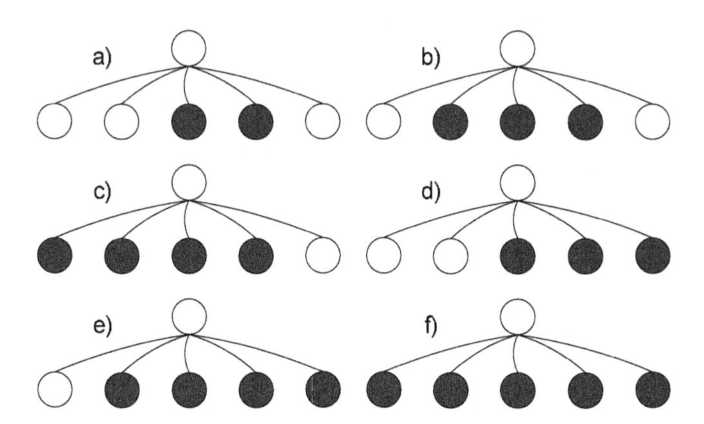


**Fig. 3.** Range and its all allowed expansions to the siblings: a) original range b) 1 step left c) 2 steps left d) 1 step right e) 1 step left, 1 step right f) 2 steps left, 1 step right

After all clusters are found that can not be expanded to the parent without falling out of their families, all possible expansions to the siblings for each of the candidates are performed and maximal clusters are discovered.

Expansion of a range (expanding range) to the left / right consists in constructing new range (expanded range) equal to union of expanding range and subtree generated by next left / right sibling of expanding ranges leftmost / rightmost child. Range is expandable if such sibling exists, it is not already included in another range of the same cluster and expanded range has acceptable internal incoherence.

Expansion of a cluster (expanding cluster) to the left / right consists in construction of a new cluster (expanded cluster) containing deduplicated expanding cluster's ranges expanded to the left / right. For a cluster to be expandable to the left / right at least one of the cluster's ranges must be expandable to the left / right.

An expansion of original candidate cluster is done for every allowed combination of left and right steps and a separate maximal cluster is found for each ranges count in expanded clusters. Set of all found clusters is designated *fclusters*. For further processing the collection *counts* is constructed with *counts(i)* defined as follows:

$$counts(i) = \{c \in fclusters: |ranges(c)| = i\}$$

### 3.4 Aggregated Clusters Removal

The second part of our algorithm is responsible for filtering out excessive clusters resulting from minor irregularities in HTML structure but embodying exactly the same content although arranged differently in ranges.

Simple heuristic proved to be useful. For each cluster $c_1$, starting from the least numerous, our algorithm checks if there exists any other cluster $c_2$ that satisfies following conditions (objective O3):

- $c_2$ has the number of ranges being multiplicity of $c_1$'s count,
- $|left(c_2) - left(c_1)| \leq b$ and $|right(c_2) - right(c_1)| \leq b$,

and if found – $c_1$ is removed from clusters collection;

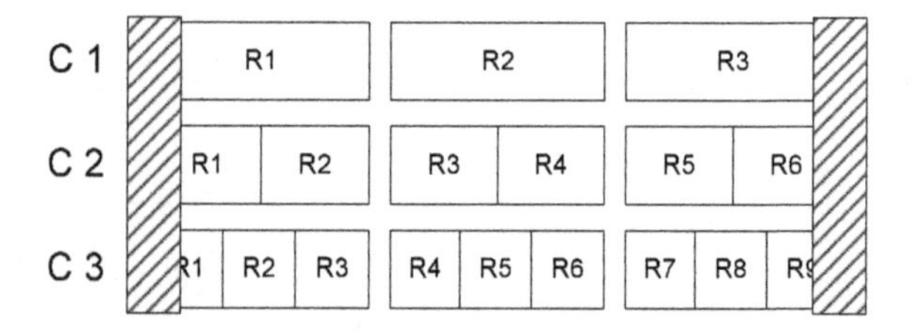


**Fig. 4.** Examples of agglomerative clusters of ranges (C1 and C2)

Parameter $b$ stands for the acceptable distance between left and right boundaries of clusters.

## 4 Experimental Setup

The algorithm described in section 3 was implemented and evaluated for several scenarios applying different tag filtering approaches, tag equivalence strategies and cluster coherence measurement methods on two test collections. The following section gives details on this experimental procedure and datasets.

### 4.1 Algorithm Implementation

The algorithm was implemented in C# language in Visual Studio.NET 2005 environment. For HTML cleaning and processing Tidy [12] and Chilkat XML [9] libraries were used respectively.

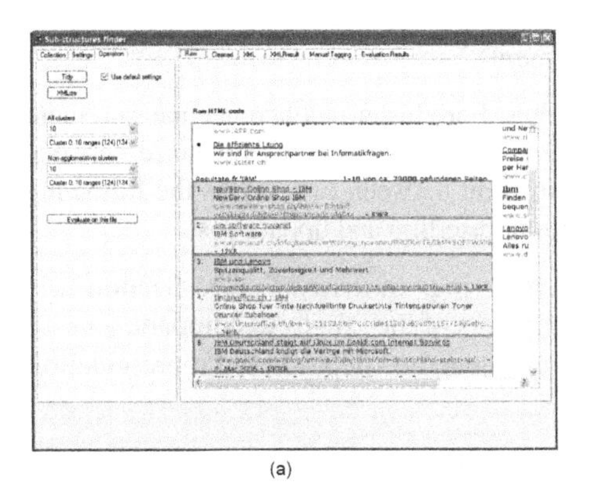

(a)

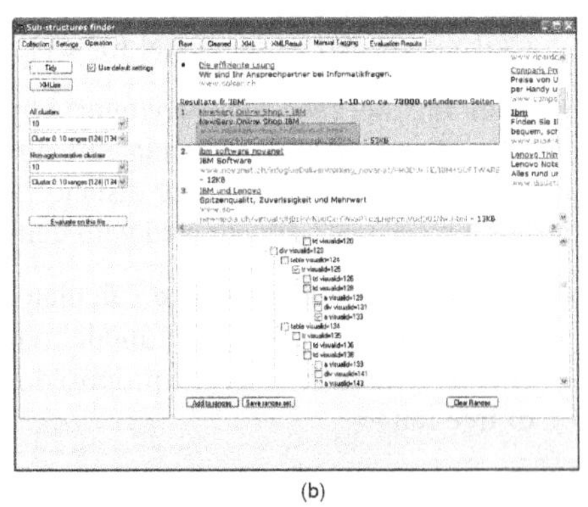

(b)

**Fig. 5.** Screenshots of (a) found clusters visualization (b) manual tagging mode

The application consists of modules responsible for manual tagging of reference clusters (see Fig. 5a.), generating theoretical collections, finding and visualizing clusters (see Fig. 5b.) as well as performing and evaluating clusters detection on a collection for a set of evaluation scenarios. For work with individual files complex GUI was developed including view of original Web page, Web page processed by Tidy, XML representation of the pages and found clusters as well as manual tagging interface with preview on the original Web page. Settings adjustable in evaluation scenarios are described below.

**Tags Filtering Approaches.** Tag filtering approach is used in tree construction phase and decides if given tag is to be included in further analysis. All tags filtering implementations share a common interface. Interface input consists in full information on HTML nodes (including tag name and attributes) and output corresponding to decision if the tag is allowed.

Three different tags filtering approaches were implemented:

- **only block tags** – only block tags as defined in [21] are allowed,
- **block tags, images, spans and links** – block tags and link (<A>), image (<IMG>) and text range (<SPAN>) elements are allowed,
- **all tags** – all tags (conforming or not to [21]) are allowed.

**Tags Equivalence Strategies.** Tag equivalence strategy is responsible for translation of HTML nodes (tag name and all attributes) to simplified text expressions for better tags comparisons. Several tag equivalence strategies were used in this research:

- **tag text** – original, complete tag text with all attributes found in source code,
- **normalized tag text** – similar to tag text but with attributes ordered alphabetically,

- **tag name** – only the tag name is used,
- **tag name, CSS class** – tag name and CSS class name (if present) are used,
- **tag name, CSS class, CSS styles** – tag name, CSS class name (if present) and CSS styles (if present) with selectors and values ordered alphabetically by selector,
- **tag-specific approach** – tags group name for similar tags (e.g. "cells" for <TD> and <TH> elements) and a tag-dependant list of attributes are used (e.g. COLSPAN and ROWSPAN for <TD> element).

**Cluster Incoherence Measurement Approaches.** For cluster incoherence measurement distance matrix *dist(c)* is calculated with *dist(c)[i, j]* corresponding to distance between *range(c, i)* and *range(c, j)*. For distance measurement between two ranges the following distance measures were adopted with unit deletion, insertion and update cost:

- **Levenshtein edit distance** calculated on substrings of *array(T)* (see Section 3.1) corresponding to nested set model representations of the ranges to compare,
- **edit tree distance** method proposed in [7] calculated on depth-based representation of tree range.

Given *distance(c)* matrix overall internal cluster incoherence is calculated using one of the three adopted approaches:

- **maximum of all distances** defined as:

$$incohm(c) = \max_{1 \prec i,\, j \leq |ranges(c)|} dist(c)[i, j],$$

- **average of all distances** defined as:

$$incoha(c) = \frac{2 * \sum_{2 \leq i \leq |ranges(c)|} \sum_{1 \leq j \leq i} dist(c)[i, j]}{|ranges(c)| * (|ranges(c)| + 1)},$$

- **maximum of average distances of a range** to all others defined as:

$$incohma(c) = \max_{1 \leq i \leq |ranges(c)|} \frac{\sum_{j \neq i} dist(c)[i, j]}{|ranges(c)| - 1}.$$

**Other Parameters.** Other parameters that can be set in the application include maximal allowed internal cluster incoherence, minimal and maximal allowed size of the range, minimal allowed number of ranges in a cluster and limit of left / right expansion steps in expansion to the siblings part of the algorithm and *b* parameter of agglomerative clusters removal (see Section 3.4).

### 4.2 Test Collections

For evaluation purposes two test collections were prepared:

- empirical collection containing results pages from a number of search engines,
- theoretical collection containing methodically generated HTML documents.

**World-Wide Search Engines Collection.** The first test collection contains manually tagged results pages from most important global and European country-specific search engines in several languages. The initial list of global and country-specific popular search engines was constructed based on [1, 3, 18]. Out of 67 initial search engines, 9 were removed due to their irresponsiveness, problems with query processing, parsing problems at Tidy or Chilkat libraries or no required minimum of 10 results per page met. The final list contains 58 search engines.

Global companies names "IBM" and "Microsoft" were used as query terms due to their language-independence and large number of returned results. For each search engine the first page of results was recorded using HTTrack application [13] and Opera Web browser [17] for GET and POST based search forms respectively.

The cluster of the largest ranges corresponding to search results was manually tagged for each of the pages with respect to their visual similarity. Sponsored links were included in the cluster if they were judged visually similar to other results.

**Generated Theoretical Collection.** The theoretical collection was generated with aim to cover any reasonable combination of the following binary properties:

- **block per result** – if true, each result range is embedded in one tag; if false – result is composed of several tags sharing common parent (non-tree range),
- **same depth** – if true, all ranges' parents have the same depth; otherwise some ranges are more embedded than others,
- **regular ranges** – if true, all result ranges has identical structure; otherwise some ranges have elements (including "More from this site" and "Copy" links) not present in others,
- **code between ranges** – if true, there is some varying HTML code (some advertisement code in this case) between some or all pairs of consecutive ranges,
- **code around cluster** – if true, there are some HTML elements before and after the cluster of results ranges; otherwise only cluster is embedded in <BODY> element,
- **grouped ranges** – if true, pairs of consecutive ranges are grouped in common <DIV> element,
- **code per group** – true if there are some non-range HTML elements repeating in all results ranges groups; true is allowed only if grouped ranges is true.

The theoretical collection contains 96 HTML documents.

### 4.3 Experimental Procedure

This section gives details on how our algorithm's performance was evaluated on test collections including description of evaluation scenarios as well as used success measures.

**Evaluation Scenarios.** For both test collections described in Section 4.2 several evaluation scenarios were prepared corresponding to different algorithm settings.

For World-Wide Search Engines Collection a separate evaluation scenario was created for any combination of strategies of tag filtering, tag equivalence and cluster incoherence measurement. A total of 108 scenarios were implemented and performed for all documents.

For theoretical collection only incoherence measurement strategies were taken into consideration while tag filtering strategy was to allow all tags and both tag name and CSS class were used as tags equivalence approach. As a result a total of 6 evaluation scenarios were used.

**Success Measures.** For each document, for each cluster following measures are calculated, based on the ranges offset as compared to manually-tagged reference cluster. For each range $r$ the best fitting reference range $R$ is chosen by comparing $r$ and $R$ boundaries. If more ranges than one fit, algorithm additionally takes in count range order in calculated and reference cluster. Ranges distant from all reference ranges add severe penalty to overall cluster measure so the exact clusters are preferred. Range variation RV for range $r_i$ in cluster $c$ is defined as follows:

$$rv_i = \min_{j1..|ranges(c_{ref})|}(|left(r_i) - left(r_{ref,j})| + |right(r_i) - right(r_{ref,j})|),$$

$$r_i \in c, r_{ref,j} \in c_{ref}, i = 1..|ranges(c)|$$

Then average range variation and range variation standard deviation for cluster are calculated:

$$arv_c = \frac{\sum_{i=1}^{|ranges(c)|} rv_i}{|ranges(c)|}, \quad rvsd_c = \frac{\sum_{i=1}^{|ranges(c)|} (rv_i - arv_c)^2}{|ranges(c)|}.$$

The former serves as a cluster quality measure, the latter provides additional information about how much ranges vary among them. Considerable *arv* and small values of *rvsd* indicate rather poor pattern fitting but not necessary poor cluster quality i.e. because of the non-visual tags not included in the pattern but influencing the HTML document structure and thus clustering algorithm results.

Many WWW pages contain other repetitive structures like pagination results, advertisement panels or menus, which are valid result of our method, although are not represented in pre-tagged reference clusters, hence only general measure for entire collection can be calculated. As a success measure number of successfully found anticipated clusters comparing to collection count is adopted, and then average *arv* and *rsvd* for best clusters collection measure overall method efficacy.

## 5 Experimental Results

**Empirical Collection Results.** As said before our success measurement procedure rates clusters and scenarios in two steps. Firstly, it is verified if algorithm has been able to find clusters with the exact number of ranges, secondly, clusters with best values of *arv* are being chosen as representative for given combination of scenario and document. Based on these results the best scenario has been chosen as well as some general conclusions have been formulated.

44% of scenarios can handle at least 50% of documents. These are well-formed documents in which many scenarios were able to find correct clusters. The best scenario has found exact clusters in 37 of 57 cases witch constitutes 65% with *arv* of 2.2 and good quality cluster in 9 further document which gives in total 81% of successful cases with the *arv* of 4.4. The standard deviation of arv are respectively 4.6 and 7.8 which means that even if ranges differ from the sample pattern, the algorithm produces rather regular results. Optimal parameters are "all tags" for filtering, "Tag name, CSS class, CSS styles" for tag equivalence, "Levenshtein edit distance" for distance measure and "Average of all distances" for cluster incoherence measure (see Section 4.1);

Our analysis reveals also existence of several HTML documents classes. For each class another subset of scenarios seems to work best. This approach to document categorization constitutes promising field for further research.

Further analysis of most troubling documents reveals that in most cases algorithm failures are due to unusual document structure, which puts additional emphasis on scenario and algorithm vulnerability to malformed HTML.

**Theoretical Collection.** Three out of six evaluation scenarios (where distance measurement is based on Levenshtein measure) have performed exceptionally well. The best scenario handles entire collection with the *arv* of 1.0 and found all ranges in every document with excellent precision (no excessive ranges have been found).

As collection is composed of 96 generated documents which reflect different types of HTML regular structures, we assume that considerable part of problems with empirical collection comes from poor HTML quality. In that case the most important characteristic of particular scenario is its tolerance to structure noise (random as well as intended by the author i.e. advertisement). This result suggests also that some important live HTML documents structure properties were not included in generation algorithm.

## 6 Future Work

Some of our plans for future work include:

- In-depth analysis of cases when no clusters were discovered for any scenario.
- Experiments with another tags equivalence strategies (including visual analysis based on tag, class and inherited styles as well as size and location of tags).
- Better distance measurement strategies (including use of non-unit edition cost depending on how much two tags differ visually or functionally).
- Better evaluation of results (e.g. based on comparison of visual block or effective content corresponding to reference and found range).
- Evaluation on other test collections (e.g. Internet auctions or Web-based stores).
- Creation of domain specific heuristics for cluster selection.
- Use of clusters extracted from multiple-items pages (e.g. first page of search results) for ranges discovery on pages with low number of items (e.g. last page of search results).
- Evaluation of influence of additional parameters listed in Section 4.1 on the algorithm output.

## 7 Conclusion

In this paper we proposed a simple tag-tree-based bottom-up method for discovery of maximal non-agglomerative HTML ranges clusters and evaluated its usefulness for detection of search engines results records. Experiments for several tag filtering, tag equivalence and clusters incoherence measurements strategies were performed on two test collections. Our results demonstrate that the method performs well on simple and complex standard-compliant generated theoretical results Web pages and on most of gathered search engines results pages. However for a group of search engines our algorithm was unable to discover any similar ranges clusters for any evaluation scenario; we plan in-depth analysis of this cases for future work.

## References

[1] Age Of .Com, http://www.ageof.com
[2] Aho, A.V., Hopcroft, J.E., Ullman, J.D.: Data Structures and Algorithms, Addison-Wesley (1983).
[3] Big Search Engine Index, http://www.search-engine-index.co.uk
[4] Celko, J.: Trees and Hierarchies in SQL for Smarties (2004).
[5] K.C.C. Chang, B. He. Structured databases on the web: observations and implications. SIGMOD Record, 2004.
[6] C. Chang, S. Lui. IEPAD: Information Extraction based on Pattern Discovery. WWW 2001.
[7] S.S. Chawathe. Comparing Hierarchical Data in External Memory. VLDB, 1999.
[8] Y. Chi, Y. Yang, R.R. Muntz. Canonical forms for labelled trees and their applications in frequent subtree mining. Knowledge and Information Systems 2005.
[9] Chilkat XML .NET, http://www.chilkatsoft.com/xml-dotnet.asp
[10] D.W. Embley and Tao C. Automating the Extraction of Data from HTML Tables with Unknown Structure. Data & Knowledge Engineering, 2005.
[11] B. Gazen and S. Minton. AutoFeed: an unsupervised learning system for generating webfeeds. K-CAP, 2005.
[12] HTML Tidy Library, http://tidy.sourceforge.net/
[13] HTTrack Website Copier, http://www.httrack.com/
[14] D.E. Knuth: The Art of Computer Programming, Addison-Wesley (1968).
[15] Liu, R. Grossman and Y. Zhai. Mining Data Records in Web Pages. SIGKDD, 2003
[16] S. Minton, C.A. Knoblock and K. Lerman. Automatic data extraction from lists and tables in web sources. IJCAI, 2001.
[17] Opera Web Browser, http://www.opera.com/
[18] Pandia Powersearch, http://www.pandia.com/powersearch
[19] K. Simon, G. Lausen. ViPER: Augmenting Automatic Information Extraction with Visual Perceptions. CIKM, 2005.
[20] J. Wang, F. Lochovsky. Data Extraction and Label Assignment for Web Databases. WWW, 2003.
[21] World Wide Web Consortium. HTML 4.01 Specification, 1999.
[22] Y. Zhai, B. Liu. Web Data Extraction Based on Partial Tree Alignment. WWW 2005.
[23] H. Zhao, W. Meng, Z. Wu, V. Raghavan, C. Yu. Fully Automatic Wrapper Generation for Search Engines. WWW, 2005.
[24] H. Zhao, W. Meng and C. Yu. Automatic Extraction of Dynamic Record Sections From Search Engine Result Pages. VLDB, 2006.

# Usability of GeoWeb Sites: Case Study of Czech Regional Authorities Web Sites

Jitka Komarkova, Martin Novak, Renata Bilkova, Ondrej Visek, and Zdenek Valenta

USII, University of Pardubice, Studentska 95,
532 10 Pardubice, Czech Republic
{Jitka.Komarkova,Martin.Novak,Renata.Bilkova}@upce.cz,
ondrej.visek@seznam.cz, valic@seznam.cz

**Abstract.** Today, many of the solved problems are spatially oriented. It means that more and more people need to use spatial information. They need to be able to use it quickly, without any special software tool and without any special training. So they need an easy-to-use solution. GeoWeb sites can perfectly fit the demands but they must be properly designed. The article describes usability testing of GeoWeb sites of all Czech regional authorities and proposes some recommendations how user interface should be designed. Recommendations can be generalized because there are no significant differences between private and public sector. The main differences are between target groups of the users: regular or casual users.

**Keywords:** GeoWeb, Internet GIS, Usability, Usability Testing.

## 1 Introduction

The Internet and Web have significantly influenced our lives because they allow easy, remote, and fast access to information for everyone. Information itself has become goods and it belongs to competitive advantages to have information earlier than the others. On the other side, amount of information stored on the Internet is really vast so it is more and more difficult to quickly find relevant information. Information overload has become a serious problem not only for regular users [38] but for example even for intelligence services [37] so various data mining techniques [17], [23], [30], [37] and content extraction and integration information systems [14] must be used to help users quickly obtain demanded information.

Spatial information (or geographic information, sometimes called geoinformation) is a special kind of information – it is located in the space, e.g. on the Earth. An increasing frequency of solving spatially oriented problems and then making decisions in both public and private sector is one of the reasons why accessibility of spatial information is so important today. Along with development of information society and increasing utilization of information for supporting decision-making processes, spatial information is more often required by the users as well. Because of the special nature of spatial information special software tools are required to its treatment and analyses – geographic information systems (GIS) [13], [18], [22]. The complexity of spatially oriented problems, like crisis management, land management, urban planning, and

W. Abramowicz (Ed.): BIS 2007, LNCS 4439, pp. 411–423, 2007.
© Springer-Verlag Berlin Heidelberg 2007

public administration, leads even to integrated systems approach, such as a spatial decision support system [28]. In the case of spatial data there is a significant problem how to handle imprecision and uncertainty in data. This is why possibility of utilization of the fuzzy set techniques for spatial data treatment and querying has been studied [12], [24]. In some cases, there can be a significant problem with some languages and their special letters and symbols too. Czech language can be given as an example [10]. But regardless of the previously shown problems, GIS have become an indispensable part of a variety of information systems used for supporting decision-making processes in business, public administration, and personal matters so an easy access of all potential users to spatial information and services is strongly required [18], [21], [22].

Unfortunately, classic GIS software packages (desktop or professional GIS) are too sophisticated so they cannot meet today's demands of all the users who need to use spatial information. They limit users at least for the following reasons [18], [21], [22]:

- Every user has to buy a full license even if only a small part of available functions will be used.
- Desktop software is accessible only from the computer on which it is installed.
- Complicated user interface requires training and disallows a fast solving of simple tasks.
- Desktop GIS is still a proprietary technology.

The above stated problems along with spreading of the Internet and increasing demand for spatial information have driven a rapid process of geo-enabling the Web and a rapid development of Internet GIS applications.

Nowadays, many various technologies which can geo-enable the Web exist. The technologies allow remote and easy access of end-users without any special education in the field of geoinformatics to spatial information by means of variety devices, e.g. computers, notebooks, pocket PCs, mobile phones, etc. Many various terms can be found which are used while talking about GIS applications on the Internet/intranet. The most common are: GIS on-line, distributed geographic information [25], Web-based GIS [21], Internet GIS, mobile GIS [18], [21] interactive mapping [4], [6], distributed GIservice [21], geo-enabled Web [20] GeoWeb, Internet map servers, and many others. These terms are sometimes understood as synonyms but it is not the best way of their understanding. For example, Internet provides not only WWW service, so the term Internet GIS has a different meaning from GeoWeb or Web-based GIS. Mobile GIS is not the same as an Internet GIS because mobile devices like PDA, and mobile phones use different protocols and technologies, and so on [21]. In the framework of this paper the term Internet GIS will be used too although attention will be paid only to the Web-based solutions (GeoWeb).

## 2 Internet Geographic Information Systems

Six main types of GIS software are usually recognized: professional GIS, desktop GIS, mobile GIS, component GIS, viewers, and Internet GIS. Internet geographic information systems represent a quite new branch of information and communication technologies – they have risen in the end of the 20th century and they undergo a rapid development. Today, they are used by the largest number of the users for the lowest costs per user. They, of course, provide only limited number of functions but in fact they are expected to provide only the required functionality [18], [21].

## 2.1 System Architecture

Internet GIS are now considered to be a category of information systems, very often the necessary one. Their architecture follows architecture of the other information systems and it is usually based on the n-tier client/server architecture. At least the following parts can be usually recognized [1], [21]:

- *Data layer* – data management system which is able to store and provide both spatial and non-spatial data.
- *Application layer* (business logic) – processing and analytical functionality (at minimum map server and Web server must be available).
- *Presentation layer* – users interface.

Today, Internet GIS are used as a tool for fast providing terabytes of data from various sources including data from remote sensing and various kinds of maps from various sources, e.g. in libraries [29]. Due to this demand and thanks to the existence of interoperability standards which are accepted by a wide community, Internet GIS were found as a suitable domain for application of ideas of parallel and distributed computing too [9], [21].

## 2.2 Users Groups

Internet GIS now have the highest number of the users in comparison with the other types of GIS applications (including desktop GIS). Users classifications vary from author to author and they are dependent on the purpose of classification too. Anyway, at least the following basic types of users are usually distinguished [18], [21], [27]:

- *High-end users*, usually GIS specialists who treat data, run spatial analyses, and provide the results of their work to the other users by means of Internet GIS.
- *Regular users*, e.g. civil servants, managers, controllers, regular customers, cooperating partners, etc. Regular, everyday use of Internet GIS is typical for this group of the users. They usually need only several functions. All needed functions are known in advance and they are used repeatedly. The users can be trained in advance if it is necessary. It can be supposed that they access Internet GIS application by means of appointed Web browser or other defined client software. It means their working environment is known in advance and can be influenced in the case of necessity.
- *Casual users*, e.g. tourists, residents, businessmen, some managers, etc. They use Internet GIS application irregularly and casually. Their digital literacy may be very low. On the other side, only a few functions are required by these users. They usually need to set region of interest, select appropriate data layers, display geographic information, change scale, run very simple queries, and print outputs or save result maps. It is supposed that they can use various platforms including various Web browsers. They may not be able to install any software, and their Internet connection can be slow.

- *Mobile users*, i.e. people who use mobile devices like PDAs or mobile phones and wireless technologies to connect to a server and access spatial information and/or GIS functionality provided by the server. The users can vary from casual low-end users to high-end users. The set of demanded functions is limited but some special function, like disconnected editing of data, may be required. Utilities management (e.g. water, electrical utilities) can be given as an example of a branch where mobile GIS solutions are very often used as a regular tool by non-GIS specialists.

# 3 Quality and Usability of GeoWeb

As far as functioning of information systems including Internet GIS application can be critical for businesses and public administration, some quality requirements are laid on them. Today, it is preferred to "make the design to fit the users" to the previous attempt: "make the user to fit the design" [26]. The "fitness for use" definition takes customer's requirements and expectations into account, which involve whether the products or services fit their needs. Quality of design and quality of conformance are two most important parameters [11]. Today, applications are required to be intuitive because the digital literacy of many users is not very high. This requirement is even more strict in the case of Web applications. Their designers should remember that users do not read precisely the entire Web page and they are very often in a hurry [15].

Today, there are several models for measuring quality and customer satisfaction available. Every specialist can propose his/her own quality model which will meet the needs of the given situation [2]. Anyway, there is one quality model standardized by ISO/IEC: ISO/IEC 9126 - Information technology - Software Product Quality. ISO/IEC 9126 defines a quality model which is applicable to every kind of software and which uses six main characteristics to evaluate software quality. Each characteristic is further sub-divided into sub-characteristics. The main characteristics of this approach are shown on the Fig. 1. This quality model is widely used for both measuring architectures and intranet applications [16], [19]. This model uses functionality, reliability, efficiency, usability, maintainability, and portability as main software quality characteristics [16].

Customer's satisfaction is usually measured by percent satisfied or dissatisfied from customer satisfaction surveys. In addition to overall customer satisfaction with the software product, satisfaction toward specific attributes is also assessed. For instance, IBM monitors satisfaction with its software products in levels of CUPRIMDSO model (capability, usability, performance, reliability, installability, maintainability, documentation, service, and overall). Hewlett-Packard focuses on FURPS model (functionality, usability, reliability, performance, and serviceability). Other companies use similar dimensions of software customer satisfaction [11].

In many of the quality models usability is one of the characteristics of software quality. In according to ISO/IEC 9126 usability is divided into the following sub-characteristics: understandability, learnability, operability, attractiveness, and compliance. The aim of usability testing is to collect empirical data about the tested application. In the framework of human-computer interaction research usability is studied by changing interface variables and measuring user performance. It means representative end users are observed

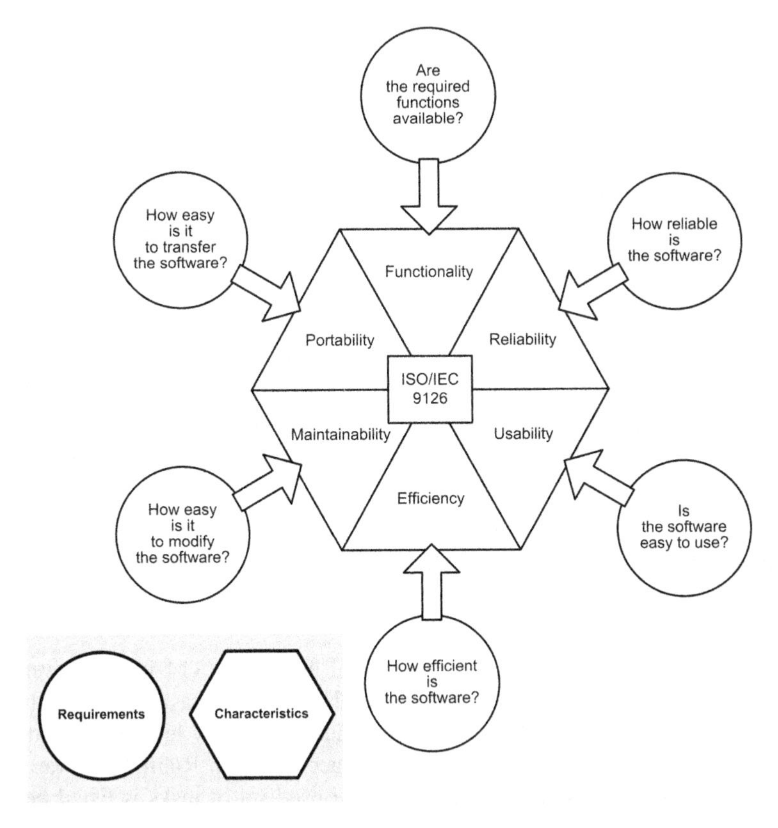


**Fig. 1.** Explanation of ISO/IEC 9126 quality model and quality characteristics used by this model [7]. Requirements on software are represented by the circles, quality characteristics by the parts of hexagon.

when they use the application to perform selected typical tasks. Usually, efficiency, effectiveness, and satisfaction are used as indicators of interface usability and user satisfaction [2], [3], [25].

Usability testing has been used to improve many Web-based systems. Usability testing and re-design of one US state e-government portal resulted in a more efficient, enjoyable, and successful experience for visitors to the site. Study showed that mean task time was even significantly shorter on the new site for some tasks (time was reduces by 62%). The failure rate of the old portal was found to be 28%, while the new site was 5% [36]. Usability testing was used as a one of techniques used to improve Web-based learning environment for the high school students. The aim of the project was to support and increase students' motivation to learn science [33], [34]. Usability testing was successfully used to evaluate and improve two interactive systems for monitoring and improving the home self-care of patients suffering with asthma and dyslipidemia [5]. The United States Computer Emergency Readiness Team (part of the Department of Homeland Security of the USA) changed significantly its Web site after the first usability test. Then, the second usability test of the new Web site showed that technical users' success was improved by 24%, their satisfaction was improved by 16%. The non-technical (home) users' success was improved by 20%, and their satisfaction was improved by 93% [31]. The usability of the encryption program PGP 5.0 was studied in the framework of the theoretical study [35] to allow unskilled users use security concepts (send encrypted emails) perfectly in the future.

A significant problem is that the users of the same level of digital literacy still individually differ so usability cannot be ensured by training or education. Users' diversity must be respected by design of the application as well [3]. Danish project Public Participation GIS can be used as an example of project targeted to the all citizens but resulting in a participation of only limited group (middle-age well-educated males with income above average) [8].

Concerning user interface of GeoWeb applications, it should be: visually balanced, enough contrast, typographically correct, readable, enough contrast, using familiar presentations, done in according to the target user group. Tools should be clearly presented and the interface should support and guide the user [32].

## 4 Case Study: Usability Testing of GeoWeb Sites of the Czech Regional Authorities

Internet GIS applications of all Czech regional authorities (14 in total) were tested. Target user's group of these applications is the public – casual users without any special GIS knowledge and skills and sometimes with a low level of digital literacy. For the testing purpose a usability method in according to Rubin [26] was used. At first, set of the testing tasks was proposed. The used set of tasks is listed below. For the first testing, a user with higher practical skills was selected.

### 4.1 Experimental Conditions

*Tested subject*

University student, male, 23 years old. Study branch: informatics in public administration. High level of digital literacy and good practical skills including intermediate knowledge of GIS.

*Used information and communication technologies*

Configuration of used computer: PC x86, processor: Intel Pentium 4 1.4 GHz, RAM: 760 MB. Display resolution: 1024x728. Operating system: Microsoft Windows XP Home Edition, version: 5.1.2600 Service Pack 2. Web browser: Internet Explorer 6.0, SP2 (pop-up windows and all other technologies were allowed).

Internet connection: connection speed: 1024 kbit/s.

*Set of testing tasks*

1. Accessibility of Internet GIS applications of the region
   - Finding a list of GIS applications on web pages of the region
   - Finding a given kind of map
   - Opening a map output

2. Orientation within the map frame
   - Overview map
   - The name of the map
   - Menu and tools
   - Scale
   - Map window
   - Displayed layers and legend
3. Changes of the map and its appearance
   - Zoom in and zoom out (change of scale)
   - Pan (moving) of the map by hand tool
   - Movement of the map by means of keyboard or map field arrows
   - Printing of the map
   - Adding and removing data layers
4. Searching (orientation) in the map
   - Finding the county (regional) town
   - Finding the given city or territory
   - Measurement of the distance between given features
   - Cancelling the selection of features after measuring the distance
   - Finding given road
   - Finding given object
5. Orientation in the GIS application in general (variety of available services and searching within them)
   - Return to the list of all available services (maps)
   - Finding maps with the given topic: biking paths, environment, monuments, disasters (e.g. flooding), land use

## 4.2 Results

All results of the usability testing of GeoWeb sites of all Czech regional authorities are given in the Appendix.

At first, there is the time (in minutes) provided which was needed to complete each particular task (see the list of the set of the typical tasks in chapter 4.1) or find out that the functionality is not available. Next, the degree of tasks completion (in percents) is stated. Lower percentage share means that some tasks could not be completed because the functionality was not working or it was missing at all. For example, overview maps were missing, it was impossible to select and deselect data layers, it was impossible to change the scale of the map or find it at all, it was impossible to measure distances, it was impossible to print the map, etc. Then the absolute numbers of found problems which caused that some tasks could not be completed are given. After each part of the results ranking of the regions is specified (the lower number means the better result). Final order of the regions is then calculated as an average value of all partial results.

### 4.2.1 General Remarks

Design of Internet GIS applications is often not very user-friendly from the point of view of the target group of the users. Common users do not have knowledge and skills as GIS specialists. It is supposed the number of the users will be high and there will be great differences between their knowledge and skills. The most important task for all users is to work with data layers but common users may not understand the layered approach of GIS at all.

Next problem is connected to the design of HTML form: more than one HTML form item, namely checkbox and radio button, are very often situated next to the name of one data layer. This attempt again comes from GIS software logic: it is necessary to select which layers will be displayed, and at the same time it is necessary to set active layers which will be treated (for example queried). Unfortunately, common users do not know the difference between them so they could be frustrated.

A 'hand tool' is a very common control tool – it is used for panning/moving. But again, only users of GIS or Adobe software know this function. Another problem with this function is slow server response because server has to remake the whole map.

Many Web sites demand special extensions/plug-ins (the most frequent one is Java) which common user needn't to have installed on his computer so the application then does not work. Not all users have a right to install new software on the computer.

The same problem appears if user does not use "the right browser" which is selected by the authors of the Web application. The most common "the only one" is Microsoft Internet Explorer. Some of GeoWeb sites suffer with this problem - users with any other Web browser cannot get to the application. This restriction is not acceptable on the Internet nowadays.

#### 4.2.2 Recommendations – Target Group: Casual Users

Technologies/extensions which need to be installed to the user's computer/browser (for example Java, Macromedia Flash) should not be used in this case. Only common and native technologies (e.g. JavaScript) should be used. Installation of plug-ins often needs a higher user rights to the operating system then the user usually has (for example firm computers, in the Internet coffees, etc.). Even if user can install a plug-in, this operation is often very frustrating for the user due to the difficult questions during installation, necessity of restarting Web browser or the whole computer.

Very useful technology for Web-based GIS is AJAX - Asynchronous JavaScript and XML. With AJAX there is no need to reload of complete Web pages. Loading new data proceeds on the background. If application server has a short response time, there are no visible delays during movement of the map with a hand tool – the nearest neighborhood is loaded in browser cache. This solution has a great effect on efficiency of work with GeoWeb. Movement (panning) of map is very often used function so optimization of this task has a great impact on all work.

Possibility of continual map movement (panning) with hand tool seems to be the most useful type of movement. Often used type of movement on the map is utilization of eight arrows which are situated on the each corner and between them. Every movement on the map again has to wait to the response of the server. Movement with preloading the nearest neighborhood into the server cache can have a great positive impact on productivity of user.

The most important part of each GIS application is, of course, a map. Therefore, it should occupy the biggest area of viewport. If navigation or some other informational boxes need a lot of space, it should be preferably solved by hiding menus via CSS + JavaScript. When user needs it, this item can be shown via 'mouse over' on icon, text or some other item which is enough expressive. After utilization of pop-up box or menu, it should be again hidden to keep low occupied space of viewport. But this solution is not always useful. For example in the case of search user should be able to see the entered query in the box with result to remember what exactly he was searching for so he can refine the query. What should never be forgotten: map and basic navigation are the most important parts. Other items which users do not need so often should be hidden or minimized to maximize space for the map.

Another useful feature of GeoWeb applications is resizing map window (box for map) after resizing viewport (for example when user resize or maximize Web browser window). This can be done by means of JavaScript. Dynamical resizing is useful for example when users want to get a large map or a map of a larger area so they maximize Web browser window.

A special page or view should be provided for printing the output maps. Only a map and desired information such as results of a query, scale, displayed layers, etc should be on the print output. Users do not want to print icons.

Opening a GIS application into a new Web browser window without menus and icons of the browser is next very often used feature. This is a good solution which allows maximization of viewport. But it is not so good for user interaction with the browser. Thus, this kind of feature should be used very carefully because it can prevent users to use functions which are accessible only via menus of Web browser.

Next useful feature is to allow users to save or get URL of displayed map and searched items to let users to share the results or save them for their future need.

Only common and/or very expressive icons should be used in the application, i.e. icons which are similar to the icons used by the most common applications like Web browsers, text editors, and viewers of images or documents. The icons should be used for the same actions (e.g. print, save, open document, etc.). It cannot be expected that the users know special icons used by desktop GIS software. Every icon must have a short description for the case when user does not recognize its function. One of the best solutions is to use parameter 'title' with <img> tag of the icon in HTML. Then, 'mouse over' the icon shows the description written down to the parameter 'title'. Another solution is to write down a very short description under icons. But this solution must be used very carefully. This solution is often redundant so it wastes space because icons are always more expressive than text.

Load of application server should be balanced because response time can be a critical parameter of the usability of GeoWeb. In the case of necessity, dedicated server should be used.

Utilization of introduction page for entering the real Internet GIS application is at least arguable. This page means that user has to click through to enter the demanded application so it costs him time.

#### 4.2.3 Recommendations – Target Group: Regular Users

A completely different approach should be applied to this target group of the users. Probably the only similar thing to the previous target group is necessity of user-friendly interface which is easy to use. In this case it can save money and time needed for training of the users.

In many cases homogenous environment on the computers of the users (e.g. a given type and version of Web browser) can be expected so some special plug-ins and programs like Java can be used.

These users often solve some specific tasks which may require specific functions and consequently a specific user interface. This interface may be specific per user or users group (group of the users who solve similar tasks, e.g. from the same department of a company).

## 5 Conclusions and Future Work

Internet GIS applications have become an indispensable part of business and public information systems. Internet GIS applications of the Czech Regional Authorities have been studied in the framework of this research but the results can be applied to business systems like facility management because there are no significant differences between requirements of the users from private sector and public administration on usability of Internet GIS application.

User-friendly and easy-to-use interface is today the most important thing. Users should not need any long-lasting training how to use the application, they only need to run application and work with it.

Usability testing of GeoWeb sites of all Czech regional authorities was done in according to the method recommended by Rubin [26]. Set of the typical tasks for the testing purpose (usability testing of GeoWeb) was proposed in the framework of this study. As it was found out, many of the GeoWeb sites are not properly designed because they do not respect users' needs, knowledge, skills, and abilities. Some of them could use more modern Web technologies and should respect general principles how to create a good Web site.

For the future work, usability testing will be done with casual users who have no GIS knowledge and skills. Next, heuristic evaluation can be conducted. After assessment of obtained results a user interface can be proposed with utilization of available modern technologies and general principles of creating Web sites because GeoWeb sites are still web sites. At the same time the proposed user-interface have to take into account users' needs and skills. It means that design of user interface should respect target group of the users because casual users have completely different needs in comparison with regular users.

**Acknowledgments.** This research was kindly supported by the University of Pardubice, grant project Nr. FG462029.

## References

1. Alter, S.: Information Systems: Foundation of E-Business. 4th edn. Prentice-Hall, Upper Saddle River (2002)
2. Azuma, M.: Software Products Evaluation System: Quality Models, Metrics and Processes - International Standards and Japanese Practice. Information and Software Technology 38 (1996) 145-154

3. Dillon, A.: Spatial-Semantics: How Users Derive Shape from Information Space. J. of Am. Soc. for Information Science 51 (2000) 521-528
4. Doyle, S., Dodge, M., Smith, A.: The potential of Web-based mapping and virtual reality technologies for modelling urban environments. Computers, Environment and Urban Systems 22 (1998) 137-155
5. Farzanfar, R., Finkelstein, J., Friedman, R.H.: Testing the Usability of Two Automated Home-Based Patient-Management Systems. Journal of Medical Systems 28 (2004) 143-153
6. Friedl, M.A., McGwire, K.C., Star, J.L.: MAPWD: An interactive mapping tool for accessing geo-referenced data sets. Computers & Geosciences 15 (1989) 1203 – 1219
7. Gross, H.: Guidance for validation of software – Proposal [online]. [cit. 2007-02-10]. Available from: <www.amctm.org/getres.asp?id=81>
8. Hansen, H.S., Reinau, K.H.: The Citizens in E-Participation. Lecture Notes in Computer Science, Vol. 4084. Springer-Verlag, Berlin Heidelberg New York (2006)
9. Hawick, A.K., Coddington, P.D., James, H. A.: Distributed frameworks and parallel algorithms for processing large-scale geographic data. Parallel Computing 29 (2003) 1297–1333
10. Janakova, H.: Text categorization with feature dictionary problem of Czech language. WSEAS Transactions on Information Science and Applications 1 (2004) 368-372
11. Kan, S.H.: Metrics and Models in Software Quality Engineering. 2nd edn. Pearson Education, Upper Saddle River (2003)
12. Kollias V.J., Voliotis A.: Fuzzy reasoning in the development of geographical information systems. FRSIS : a prototype soil information system with fuzzy retrieval capabilities. Int. J. of Geographical Information Systems 5 (1991) 209-223
13. Konecny, G.: Geoinformation: Remote Sensing, Photogrammetry and Geographic Information Systems. 1st publ. Taylor & Francis, London (2003)
14. Kowalkiewicz, M., Orlowska, M.E., Kaczmarek, T., Abramowicz, W.: Towards more personalized Web: Extraction and integration of dynamic content from the Web [online]. In Proceedings of the 8th Asia Pacific Web Conference APWeb 2006. Springer Verlag, Harbin China (2006). [cit. 2006-12-18]. Available from: <http://hermes.kie.ae.poznan.pl/~marek/me/Publications/apweb2006%20Final.pdf >
15. Krug, S.: Don´t Make Me Think: A Common Sense Approach to Web Usability. 2nd edn. New Riders Press (2005)
16. Leung, H.K.N.: Quality Metrics for Intranet Applications. Information & Management 38 (2001) 137-152
17. Lin, Ch.-Ch.: Optimal Web site reorganization considering information overload and search depth. European Journal of Operational Research 173 (2006) 839-848
18. Longley, P.A.: Geographic Information Systems and Science. 1st edn. John Wiley & Sons, Chichester (2001)
19. Losavio, F., Chirinos, L., Matteo, A., Lévy, N., Ramdane-Cherif, A.: ISO Quality Standards for Measuring Architectures. Journal of Systems and Software 72 (2004) 209-223
20. Open Geospatial Consortium [online]. [cit. 2006-12-07]. Available from <http://www.opengeospatial.org/about/>
21. Peng, Z.-R., Tsou, M.-H.: Internet GIS: Distributed Geographic Information Services for the Internet and Wireless Networks. John Wiley & Sons, Hoboken (2003)
22. Peng, Z.-R., Zhang, Ch.: The roles of geography markup language (GML), scalable vector graphics (SVG), and Web feature service (WFS) specifications in the development of Internet geographic information systems (GIS). J. Geograph. Syst. 6 (2004) 95-116

23. Perkowitz, M., Etzioni, O.: Towards adaptive Web sites: Conceptual framework and case study. Artificial Intelligence 118 (2000) 245-275
24. Petry, F.E., Cobb, M.A., Wen, L., Yang, H.: Design of system for managing fuzzy relationships for integration of spatial data in querying. Fuzzy Sets and Systems 140 (2003) 51-73
25. Plewe, B. S.: GIS Online: Information Retrieval, Mapping, and the Internet. OnWord Press, Santa Fe (1997)
26. Rubin, J.: Handbook of Usability Testing: How to Plan, Design, and Conduct Effective Tests. John Wiley & Sons, New York (1994)
27. Schaller, J.: GIS on the Internet and environmental information and planning. In 13th ESRI European User Conference, 7. – 9. 10. 1998, Firenze, Italy, [online]. [cit. 2006-07-01]. ESRI (2002). Available from <http://gis.esri.com/library/userconf/europroc98/proc/idp27.html>
28. Sharma, D.K., Sharma, R.K., Ghosh, D.: A Spatial Decision Support System for Land Management. Int. J. of Computers & Applications 28 (2006) 50-58
29. Shawa, T.W.: Building a System to Disseminate Digital Map and Geospatial Data Online. Library Trends 55 (2006) 254-263
30. Smith, K.A., Ng, A.: Web page clustering using a self-organizing map of user navigation patterns. Decision Support Systems 35 (2003) 245-256
31. US-CERT Usability Lessons Learned [online]. [cit. 2007-02-10]. Available from <http://www.us-cert.gov/usability/>
32. Voženílek, V.: Cartography for GIS: Geovisualization and Map Communication. 1st edn. Univerzita Palackeho, Olomouc (2005)
33. Wang, S.-K., Reeves, T.C.: The Effects of a Web-Based Learning Environment on Student Motivation in a High School Earth Science Course. Educational Technology, Research and Development 54 (2006) 597-621
34. Wang, S.-K.,Yang, Ch.: The Interface Design and the Usability Testing of a Fossilization Web-Based Learning Environment. Journal of Science Education and Technology 14 (2005) 305-313
35. Whitten, A., Tygar, J.D.: Usability of Security: A Case Study [online]. [cit. 2007-02-10]. Available from <http://reports-archive.adm.cs.cmu.edu/anon/1998/CMU-CS-98-155.pdf>
36. Withrow, J., Brinck, T., Speredelozzi, A.: Comparative Usability Evaluation for an e-Government Portal. Diamond Bullet Design Report #U1-00-2. Diamond Bullet Design, Ann Arbor, MI (2000)
37. Wu, H., Gordon, M., DeMaagd, K., Fan, W.: Mining web navigations for intelligence. Decision Support Systems 41 (2006) 574-591
38. Yang, Ch.C., Yen, J., Chen, H.: Intelligent internet searching agent based on hybrid simulated annealing. Decision Support Systems 28 (2000) 269-277

# Supporting Use-Case Reviews*

Alicja Ciemniewska, Jakub Jurkiewicz, Łukasz Olek, and Jerzy Nawrocki

Poznań University of Technology, Institute of Computing Science,
ul. Piotrowo 3A, 60-965 Poznań, Poland
Alicja.Ciemniewska@gmail.com,
{Jakub.Jurkiewicz,Lukasz.Olek,Jerzy.Nawrocki}@cs.put.poznan.pl

**Abstract.** Use cases are a popular way of specifying functional requirements of computer-based systems. Each use case contains a sequence of steps which are described with a natural language. Use cases, as any other description of functional requirements, must go through a review process to check their quality. The problem is that such reviews are time consuming. Moreover, effectiveness of a review depends on quality of the submitted document - if a document contains many easy-to-detect defects, then reviewers tend to find those simple defects and they feel exempted from working hard to detect difficult defects. To solve the problem it is proposed to augment a requirements management tool with a detector that would find easy-to-detect defects automatically.

## 1 Introduction

Use cases have been invented by Ivar Jacobson [13]. They are used to describe functional requirements of information systems in a natural language ([1], [4], [12], [5]). The technique is getting more and more popular. Use cases are extensively used in various software development methodologies, including Rational Unified Process [16] and XPrince [19].

Quality of software requirements, described as use cases or in any other form, is very important. The later the defect is detected, the more money it will cost. According to Pressman [22], correcting a defect in requirements at the time of coding costs 10 times more than correcting the same defect immediately, i.e. at the time of requirements specification. Thus, one needs quality assurance. As requirements cannot be tested, the only method is requirements review. During review a software requirements document is presented to interested parties (including users and members of the project team) for comment or approval [11]. The defects detected during review can be minor (not so important and usually easy to detect) or major (important but usually difficult to find). It has been noticed that if a document contains many easy-to-detect defects then review is less effective in detecting major defects. Thus, some authors proposed to split reviews into two stages (Knight calls them "phases" [15], Adolph uses term "tier"

* This work has been financially supported by the Ministry of Scientific Research and Information Technology grant N516 001 31/0269.

W. Abramowicz (Ed.): BIS 2007, LNCS 4439, pp. 424–437, 2007.
© Springer-Verlag Berlin Heidelberg 2007

[1]): the first stage would concentrate on finding easy-to-detect defects (e.g. compliance of a document with the required format, spelling, grammar etc.) and the aim of the second stage would be to find major defects. The first stage could be performed by a junior engineer or even by a secretary, while the second stage would require experts.

In the paper a mechanism for supporting use-case reviews, based on natural language processing (NLP) tools, is presented. Its aim is to find easy-to-detect defects automatically, including use-case duplication in the document, inconsistent style of naming use cases, too complex sentence structure in use cases etc. That will save time and effort required to perform this task by a human. Moreover, when integrating this mechanism with a requirements management tool, such as UC Workbench developed at the Poznan University of Technology ([20], [21]), one can get instant feedback on simple defects. That can help to learn good practices in requirements engineering and it can help to obtain a more 'homogeneous' document (that is important when requirements are collected by many analysts in a short time).

The idea of applying NLP tools to automate analysis of requirements is not new. First attempts were aiming at building semi-formal models ([3], [10], [24]) and detecting ambiguity ([17], [14], [18]) in "traditional" requirements. Requirements specified as use cases were subject of research done by Fantechi and his colleagues [8]. They have used three NLP tools (QuARS [7], ARM [23] and SyTwo) to automatically detect lexical and semantical ambiguity as well as too long and too complicated sentences in a requirements document specifying Nokia's FM radio player. Our work is oriented towards use-case patterns proposed by Adolph and others [1] and our aim is to extend UC Workbench with automatic detection of easy-to-dected defects.

The next section describes two main concepts used in the paper: use case and bad smell (a bad smell is a probable defect). Capabilities of natural language processing tools are presented in Section 3. Section 4 describes defects in use cases that can be detected automatically. There are three categories of such defects: concerning the whole document (e.g. use-case duplication), injected into a single use case (e.g. an actor appearing in a use case is not described), and concerning a single step in a use case (e.g. too complex structure of a sentence describing a step). A case study showing results of application of the proposed method to use cases written by 4th year students is presented in Section 5. The last section contains conclusions.

## 2 Use Cases and Bad Smells

Use cases are getting more and more popular way of describing functional requirements ([1], [4], [12]). In this approach, scenarios pointing up interaction between users and the system are presented using a natural language. Use cases can be written in various forms. The most popular are 'structured' use cases. An example of structured use case is presented in Figure 1. It consists of the main scenario and a number of extensions. The main scenario is a sequence of steps. Each step describes an action performed by a user or by a system. Each

**UC1. Search a product**
**Main Actor: Customer**
**Main Scenario:**
1. Customer chooses search option.
2. System shows search box.
3. Customer enters search criteria, and asks for results.
4. System shows a list of found products.
5. Customer chooses one of the products.

**Extensions:**
3.A. Search criteria are invalid.
    3.A.1. System marks invalid fields, and asks for correction.
    3.A.2. Go back to step 3.

**Fig. 1.** An example of a use case in a structured form

extension contains an event that can appear in a given step (for instance, event 3.A can happen in Step 3 of the main scenario), and it presents an alternative sequence of steps (actions) that are performed when the event appears.

The term 'bad smell' was introduced by Kent Beck and it was related to source code ([9]). A bad smell is a surface indication that usually corresponds to a deeper problem in the code. Detection of a bad smell does not have to mean, that the code is incorrect; it should be rather consider as a symptom of low readability which may lead to a serious problem in the future.

In this paper, a bad smell is a probable defect in requirements. Since we are talking about requirements written in a natural language, in many situations it is difficult (or even impossible) to say definitely if a suspected defect is present or not in the document. For instance, it is very difficult to say if a given use case is a duplication of another use case, especially if the two use cases have been written by two people (such use cases could contain unimportant differences). Another example is complexity of sentence structure. If a bad smell is detected, a system displays a warning and the final decision is up to a human.

## 3 Natural Language Processing Tools for English

Among many other natural language processing tools, the Stanford parser [2] is the most powerful and useful tool from our point of view. The parser has a lot of capabilities and generates three lexical structures:

– probabilistic context free grammar (PCFG) structure - is a context-free grammar in which each production is augmented with a probability
– dependency structure - represents dependencies between words
– combined structure - is a lexicalized phrase-structure, which carries both category and (part-of-speech tagged) head word information at each node (Figure 2)

The combined structure is the most informative and useful. To tag words it uses Penn Treebank set. As an example in Figure 2 we present structure of a sentence "User enters input data". In the example the following notation is used:

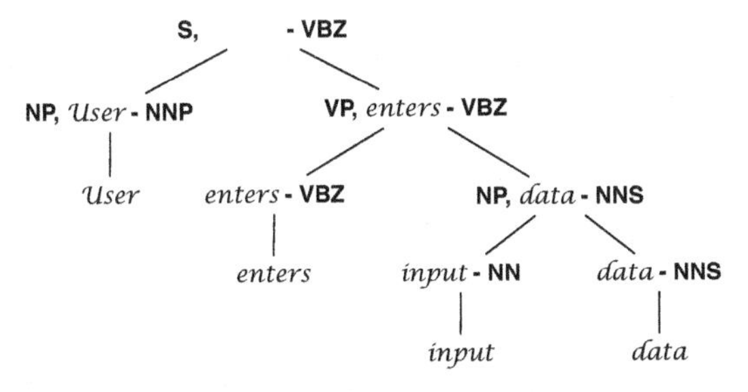


**Fig. 2.** An example of a Combined Structure generated by the Stanford parser

*User enters input data*
1 2 3 4

**nsubj(*enters* - 2, *User* - 1) - nominal subject**
**nn(*data* - 4, *input* - 3) - noun compound modifier**
**dobj(*enters* - 2, *data* - 4) - directobject**

**Fig. 3.** An example of a typed dependencies generated by the Stanford parser

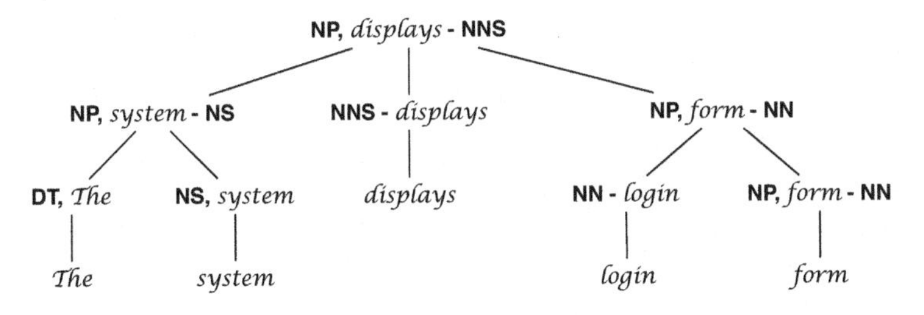


**Fig. 4.** An example of an incorrect sentence decomposition

S - sentence, NP - noun phrase , VP - verb phrase, NN - noun, NNP - singular proper noun, NNS - plural common non, VBZ - verb in third person singular.

Moreover, the Stanford parser generates grammatical structures that represents relations between individual pairs of words. Below in Figure 3 we present typed dependencies for the same example sentence, as in the above example. Notation used to describe grammatical relations are presented in [6].

The ambiguity of natural language and probabilistic approach used by the Stanford parser cause problems with automatic language analysis. During our research we have encountered the following problem. One of the features of the English language is that there are some words which can be both verbs and nouns. Additionally the third person singular form is composed by adding "s" to the end of the verb base form. In a similar way the plural form of a noun is built. This leads to the situation when the verb is confused with the noun. For example word "displays" can be tagged as a verb or a noun. Such confusion may have great influence on the further analysis. An example of such situation is presented in Figure 4.

In our approach we want to give some suggestions to the analyst about how the quality of use cases can be improved. When the analyst gets the information about the bad smells, he can decide, whether these suggestions are reasonable or not. However, this problem does not seem to be crucial. In our research, which involved 40 use cases, we have found only three sentences in which this problem occurred.

# 4 Defects in Use Cases and Their Detection

Adolph [1] and Cockburn [4] presented a set of guidelines and good practices about how to write effective use cases. In this context "effective" means clear, cohesive, easy to understand and maintain. Reading the guidelines one can distinguish several types of defects in use cases. In this Section we present those defects that can be automatically detected. Each defect discussed in the paper contains a description of automatic detection method. They have been split into three groups presented in separate subsections: specification-level bad smells (those are defect indicators concerning a set of use cases), use-case level bad smells (defect indicators concerning a single use case), and step-level bad smells (they concern a step - a use cases consists of many steps).

## 4.1 Specification-Level Bad Smells

At the level of requirements specification, where there are many use cases, a quite common defect which we have observed is use-case duplication. Surprisingly, we have found such defects even in specifications prepared by quite established Polish software houses. The most frequent is duplication by information object. If there are two different information objects (e.g. an invoice and a bill), analysts have a tendency to describe the processes which manipulate them as separate use cases, even if they are processed in the same way. That leads to unnecessary thick documentation (the thicker the document, the longer the time necessary to read it). Moreover, it is dangerous. When someone finds and fixes a defect in one of the use cases, the other will remain unchanged, what can be a source of problems in the future. There are two sources of duplicated use cases:

- **Intentional duplication.** An analyst prefers that style and/or he wants to have a thick document (many customers still prefer thick documents - for them they look more serious and dependable, which is of course a myth). Some of such analysts perhaps will ignore that kind of warning, but some other - more proactive - may start to change their style of writing specification.
- **Unintentional duplication.** There are several analysts, each of them is writing his own part of the specification and before the review process no one is aware of the duplications. If this is the case, the ability to find duplicates in an automatic way will be perceived as very attractive.

**Detection method** is two-phased. In the first stage a signature (finger print) of each use case is computed. It can be a combination of a main actor identifier (e.g. its number) and a number of steps a use case contains. Usually a number of steps in a use case and a number of actors in the specification are rather small (far less than 256), thus a signature can be a number of the integer type. If two use cases have the same signature, they go through the second stage of analysis during which they are examined step by step. Step number j in one use case is compared against step number j in the second use case and so-called step similarity factor, s, is computed. Those similarity factors are combined into use-case similarity factor, u. If u is greater than a threshold then two use cases are considered similar and a duplication warning is generated.

A very simple implementation of the above strategy can be the following. We know that two similar use cases can differ in an information object (a noun). Moreover, we assume that most important information about processing an information object is contained in verbs and nouns. Thus, step similarity factor, si, for steps number i in the two compared use cases can be computed in the following way:

- If *all the corresponding verbs and nouns* appearing in the two compared steps *are the same*, then $s_i = 1$.
- If all the corresponding verbs are the same and *all but one corresponding nouns are the same* and a difference in the two corresponding nouns has been observed for the first time, then $s_i = 1$ and InfObject1 is set to one of the "conflicting" nouns and InfObject2 to the other (InfObject1 describes an information object manipulated with the first use case and InfObject2 is manipulated with the second use case).
- If all the corresponding verbs are the same and *all but one corresponding nouns are the same* and the conflicting nouns are InfObject1 in the first use case and InfObject2 in the second, then $s_i = 1$.
- In all other cases, $s_i$ for the two analyzed steps is 0.

Use-case similarity factor, $u$, can be computed as a product of step similarity factors: $s_1 * s_2 ... * s_n$.

The described detection method is oriented towards *intentional duplication*. To make it effective in the case of *unintentional duplication* one would need a dictionary of synonyms. Unfortunately, so far we do not have any.

### 4.2 Use-Case Level Bad Smells

Bad smells presented in this section are connected with the structure of a single use case. The following bad smells have been selected to detect them automatically with UC Workbench, a use-case management tool developed at the Poznan University of Technology:

- **Too long or too short use cases.** It is strongly recommended [1] to keep use cases 3-9 steps long. Too long use cases are difficult to read and understand. Too short use cases, consisting of one or two steps, distract a reader from the context and, as well, make the specification more difficult to understand. To detect that bad smell it is enough to count the number of steps in each use case.

- **Complicated extension.** An extension is designed to be used when an alternative course of action interrupts the main scenario. Such an exception usually can be handed by a simple action and then it can come back to the main scenario or finish the use case. When the interruption causes the execution of a repeatable, consistent sequence of steps, then this sequence should be extracted to a separate use case (Figure 5). Detection can be based on counting steps within each extension. A warning is generated for each extension with too many steps.
- **Repeated actions in neighboring steps.**
- **Inconsistent style of naming**

The last two bad smells will be described in the subsequent subsections.

**Repeated Actions in Neighboring Steps.** Every step of a use case should represent one particular action. The action may consist of one or more moves which can be taken as an integrity. Every step should contain significant information which rather reflect user intent then a single move. Splitting these movements into separate steps may lead to long use cases, bothersome to read and hard to maintain.

**Detection method:** Check whether several consecutive steps have the same actor (subject) and action (predicate). Extraction of subject and predicate from the sentence is done by the Stanford parser. The analyst can be informed that such sequence of steps can be combined to a single step.

**Main Scenario:**
1. System switches to on-line mode and displays summary information about data that have to be uploaded and downloaded.
2. User confirms action.
3. System executes action.

**Extensions:**
1.A. TMS in unreachable.
  1.A.1. System shows information that there is no connection to TMS.
1.B. There is no data to synchronize.
  1.B.1. System shows information that no data have to be synchronized.
  1.B.2. End of use case.
**2.A. TMS does not recognize user's login and password.**
  **2.A.1. System displays information about the problem and shows the login form.**
  **2.A.2. User fills the form.**
  **2.A.3. System saves new data.**
  **2.A.4. Go to step 2.**

**Fig. 5.** Example of a use case with too complicated extension (bolded)

**Wrong:**
~~*1. Administrator fills in his user name*~~
~~*2. Administrator fills in his telephone number*~~
~~*3. Administrator fills in his email*~~
**Correct:**
*Administrator fills in his user name, telephone number and email*

**Fig. 6.** Example of repeated actions in neighboring steps

**Wrong:** ~~*Title: Main Use Case*~~
**Correct:** *Title: Buy a book*

**Fig. 7.** Example of inconsistent style of naming

**Inconsistent Style of Naming.** Every use case should have a descriptive name. The title of each use case presents a goal that aprimary actor wants to achieve. There is a few conventions of naming use cases, but it is preferable to use active verb phrase in the use case name. Furthermore, chosen convention should be used consistently in all use cases.

**Detection method:** Approximated method of bad smell detection in use case names, is to check whether use case name satisfies the following constraints:

- The title contains a verb in infinitive (base) form
- The title contains an object

This can be done using the Stanford parser. If the title does not fulfill these constraints, a warning is attached to the name.

### 4.3 Step-Level Bad Smells

Bad smells presented in this section are connected with use-case steps. Steps occur not only in the main scenario, but also in extensions. The following bad smells are described here:

**Too Complex Sentence Structure.** The structure of a sentence used for describing each step of use case should be as simple as possible. It means that it should generally consists of a subject, a verb, an object and a prepositional phrase ([1]). With such a simple structure one can be sure that the step is grammatically correct and unambiguous. Because of the simplicity, use cases are easy to read and understand by readers. Such a simple structure helps a lot when using natural language tools. It is essential to notice that the simpler the sentence is, the more precisely other bad smells can be discovered. An example of a too complex sentence and its corrected version is presented below.

**Detection method:** Looking at the use cases that were available to us we have decided to build a simple heuristic that checks whether a step contains more than one sentence, subject or predicate, coordinate clause, or subordinate clause. If

**Wrong:** ~~*The system switches to on-line mode and displays summary information about data that have to be uploaded and downloaded*~~
**Correct:** *The system displays list of user's tasks*

**Fig. 8.** Examples of too complex sentence structure

the answer is YES, then a warning is generated. Obviously, it is just a heuristic. It can happen that a step contains one of the mentioned defect indicator, but it is still readable and easy to understand. Providing a clear and correct answer in all possible cases is impossible - even two humans can differ in their opinion what is simple and readable. In our research we have encountered some examples of this problem. However, we have distinguished some rules, which can be used to verify, whether a sentence is too complex and unreadable. Below we present the verification rules. The numbers in the brackets show applicability of a rule (in how many use cases a given rule could be applied) and its effectiveness (in how many use cases it has found a real defect approved by a human).

- step contains more than one sentence (2 / 2)
- step contains more than one subject or predicate (2 / 2)
- step contains more than one coordinate clause (8 / 4)
- step contains more than one subordinate clause (7 / 5)

**Lack of the Actor.** According to [1] it should be always clearly specified who performs the action. The reader should know which step is performed by which actor. Thus, every step in a use case should be an action that is performed by one particular actor. Actor's name ought to be the subject of the sentence. Thus, in most cases the name should appear as the first or second word in the sentence. Omission of actor's name from the step may lead to a situation in which the reader does not know who is the executor of the action.

**Wrong:** ~~*The form is filled in*~~
**Correct:** *Student fills in the form*

**Fig. 9.** Example of lack of the actor

**Detection method:** In the case of UC Workbench, every actor must be defined and each definition is kept within the system. Therefore it can be easily verified whether the subject of a sentence describing a step is defined as an actor.

**Misusing Tenses and Verb Forms.** A frequent mistake is to write use cases from the system point of view. This is easier for the analyst to write in such a manner, but the customer may be confused during reading. Use cases should be written in a way which is highly readable for everyone. Therefore the action ought to be described from the user point of view. In order to ensure this approach, the present simple tense and active form of a verb should be used. Moreover, the present simple tense imply that described action is constant and the system should always respond in a determined way.

**Wrong:** ~~*Send the message*~~
**Wrong:** ~~*System is sending an emai*~~
**Wrong:** ~~*The email is sent by System*~~
**Correct:** *System sends an email*

**Fig. 10.** Example of misusing tenses and verb forms

**Wrong:** ~~*User chooses the second tab and marks the checkboxes*~~
**Correct:** *User chooses appropriate options*

**Fig. 11.** Example of using technical jargon

**Wrong:**
~~*1. Administrator types in his user name and password*~~
~~*2. System checks if the user name and the password are correct*~~
~~*3. If the given data is correct the system logs Administrator in*~~
**Correct:**
*1. Administrator types in his user name and password*
*2. System finds that the typed-in data are correct and logs Administrator in*
**Extensions:**
*2.A. The typed-in data are incorrect*
*2.A.1. System presents an error message*

**Fig. 12.** Example of conditional steps

**Detection method:** Using combined structure from the Stanford parser, the nsubj relation [6] can be determined. It can be checked if the subject is an actor and the verb is in the third person singular active form.

**Using Technical Jargon.** Use case is a way of describing essential system behavior, so it should focus on thefunctional requirements. Technical details should be kept outside of the functional requirements specification. Using technical terminology might guide developers towards specific design decisions. Graphical user interface (GUI) details clutter the story, make reading more difficult and the requirements more brittle.

**Detection method:** We should create a dictionary containing terminology typical for specific technologies and user interface (e.g. button, web page, database, edit box). Then it is easy to check whether the step description contains them or not.

**Conditional Steps.** A sequence of actions in the use case depends on some conditions. It is natural to describe such situation using conditionals "if condition then action ...". This approach is preferred by computer scientists, but it can confuse the customer. Especially it can be difficult to read when nested "if" statement is used in a use case step. Use cases should be as readable as possible. Such a style of writing makes it complex, hard to understand and follow.

It is preferable to use the optimistic scenario first (main scenario) and to write alternative paths separately in the extension section.

**Detection method:** The easiest way to detect this bad smell is to look for specific keywords, such as if, whether, when. Additionally, using the Stanford parser it can be checked that the found keyword is a subordinating conjunction. In such a case the analyst can be notified that the step should be corrected.

## 5 Case Study

In this section we would like to present an example of applying our method to a set of real use cases. These use cases were written by 4th year students participating in the Software Development Studio (SDS) which is a part of the Master Degree Program in Software Engineering at the Poznan University of Technology. SDS gives the students a chance to apply their knowledge to real-life projects during which they develop software for real customers.

Let us consider a use case presented in Figure 14. Using the methods presented in Section 4 the following bad smells can be detected:

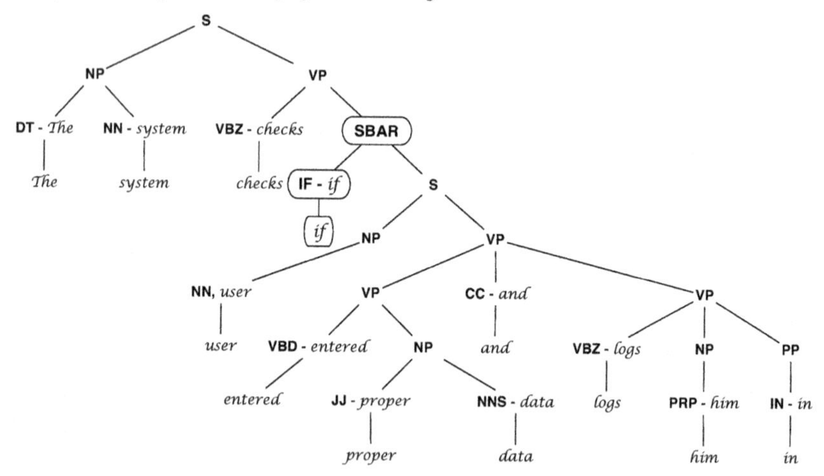


**Fig. 13.** Example of a use case that contains a condition

### Misusing Tenses and Verb Forms

- **Step:** *3. User fill the form*
- **Tagging** (from the Stanford parser): *User*/NNP *fill*/VBP *the*/DT *form*/NN
  NNP - singular proper noun
  VBP - base form of auxiliary verb
  DT - determiner
  NN - singular common noun
- **Typed dependencies** (from the Stanford parser):
  nsubj(*fill*-2, *User*-1) - nominal subject
  det(*form*-4, *the*-3) - determiner
  dobj(*fill*-2, *form*-4) - direct object

- **Conclusion:**
  From the typed dependencies we can determine the *nsubj* relation between *User* and *fill*. From the tagging it can be observed that *fill* is used in wrong form (proper form would be VBZ - verb in third person singular form).

**UC2: Log in**

**Main Scenario:**
1. User runs TMS Mobile.
2. The system presents login form.
3. User fill the form.
4. The system checks if user entered proper data and loges him in.

**Extensions:**
4.A. Entered data is invalid.
    4.A.1. The system shows information about problem.
    4.A.2. Go to step 2.
4.B. User enters login data for the first time.
    4.B.1. The system ask user to confirm his password.
    4.B.2. User enters password one more time.
    4.B.3. The system saves data, switch to on-line mode and downloads auxiliary data.

**Fig. 14.** A use case describing how to log in to the TMS Mobile system

### Conditional Step

- **Step:** *4. The system checks if user entered proper data and loges him in*
- **Tagging** (from the Stanford parser): *The*/DT *system*/NN *checks*/VBZ *if*/IN *user*/NN *entered*/VBD *proper*/JJ *data*/NNS *and*/CC *loges*/VBZ *him*/PRP *in*/IN
  VBZ - verb in third person singular form
  IN - subordinating conjunction
  VBD - verb in past tense
  JJ - adjective
  NNS - plural common noun
  PRP - personal pronoun
- **Combined structure** (from the Stanford parser): Presented in Figure 13
- **Conclusion:**
  As it can be observed the step contains the word *if*. Moreover from the combined structure we can conclude that the word *if* is subordinating conjunction.

### Complicated Extensions

- **Extension:** *4.b User enters login data for the first time*
- **Symptom:** The extension contains three steps.
- **Conclusion:** The extension scenario should be extracted to a separate use case.

## 6 Conclusion

So far about 40 use cases have been examined using our methods of detecting bad smells. Almost every use case from the examined set, contained a bad smell. Most common bad smells were: Conditional Step, Misusing Tenses and Verb Forms and Lack of the Actor. Thus, this type of research can contribute to higher quality of requirements specification.

In the future it is planned to extend the presented approach to other languages, especially to Polish which is mother tongue to the authors. Unfortunately, Polish is much more difficult for automated processing and there is lack of appropriate tools for advanced analysis.

## Acknowledgements

First of all we would like to thank the students involved in the UC Workbench project. We would like to thank the IBM company for awarding Eclipse Innovation Grant to UC Workbench project. It allowed students focus on the development work. This research has been financially supported by the Ministry of Scientific Research and Information Technology grant N516 001 31/0269.

## References

1. Steve Adolph, Paul Bramble, Alistair Cockburn, and Andy Pols. Patterns for Effective Use Cases. Addison-Wesley, 2002.
2. Advances in Neural Information Processing Systems 15. Fast Exact Inference with a Factored Model for Natural Language Parsing, 2003.
3. Vincenzo Ambriola and Vincenzo Gervasi. Processing natural language requirements. In Automated Software Engineering, pages 36–45. IEEE Press, 1997.
4. Alistair Cockburn. Writing Effective Use Cases. Addison-Wesley, 2001.
5. Larry L. Constantine and Lucy A. D. Lockwood. Software for use: a practical guide to the models and methods of usage-centered design. ACM Press/Addison-Wesley Publishing Co., New York, NY, USA, 1999.
6. Marie-Catherine de Marneffe, Bill MacCartney, and Christopher D. Manning. Generating typed dependency parses from phrase structure parses. In LREC, 2006.
7. F. Fabbrini, M. Fusani, S. Gnesi, and G. Lami. The linguistic approach to the natural language requirements quality: benefit of the use of an automatic tool. In Software Engineering Workshop. Proceedings. 26th Annual NASA Goddard, pages 97–105, 2001.
8. Alessandro Fantechi, Stefania Gnesi, G. Lami, and A. Maccari. Application of linguistic techniques for use case analysis. In RE '02: Proceedings of the 10th Anniversary IEEE Joint International Conference on Requirements Engineering, pages 157–164, Washington, DC, USA, 2002. IEEE Computer Society.
9. Martin Fowler, Kent Beck, John Brant, William Opdyke, and Don Roberts. Refactoring: Improving the Design of Existing Code. Addison-Wesley, 1999.
10. Zbigniew Huzar and Marek Łabuzek. A tool assisting creation of business models. Foundations of Computing and Decision Sciences, 27(4):227–238, 2002.
11. IEEE. Ieee standard for software reviews (ieee std 1028-1997), 1997.

12. Ivar Jacobson. Use cases - yesterday, today, and tomorrow. Technical report, Rational Software, 2002.
13. Ivar Jacobson. Object-Oriented Software Engineering: A Use Case Driven Approach. Addison-Wesley, 2004.
14. E. Kamsties and B. Peach. Taming ambiguity in natural language requirements. In ICSSEA, Paris, December 2000.
15. John C. Knight and E. Ann Myers. An improved inspection technique. Commun. ACM, 36(11):51–61, 1993.
16. Per Kroll and Philippe Kruchten. The rational unified process made easy: a practitioner's guide to the RUP. Addison-Wesley Longman Publishing Co., Inc., Boston, MA, USA, 2003.
17. B. Macias and S. G. Pulman. Natural language processing for requirement specifications. In Safety Critical Systems. Chapman and Hall, 1993.
18. L. Mich and R. Garigliano. Ambiguity measures in requirement engineering. In Int. Conf. On Software Theory and Practice, Beijing, China, August 2000.
19. Jerzy Nawrocki, Michał Jasiński, Bartosz Paliświat, Łukasz Olek, Bartosz Walter, Błażej Pietrzak, and Piotr Godek. Balancing agility and discipline with xprince. In Proceedings of RISE 2005 Conference (in print), volume 3943 of LNCS, pages 266 – 277. Springer Verlag, Jan 2006.
20. Jerzy Nawrocki and Łukasz Olek. Uc workbench - a tool for writing use cases. In 6th International Conference on Extreme Programming and Agile Processes, volume 3556 of LNCS, pages 230–234. Springer Verlag, Jun 2005.
21. Jerzy Nawrocki and Łukasz Olek. Use-cases engineering with uc workbench. In Krzysztof Zieliński and Tomasz Szmuc, editors, Software Engineering: Evolution and Emerging Technologies, volume 130, pages 319–329. IOS Press, oct 2005.
22. R. Pressman. Software Engineering - A Practitioners Approach. McGraw-Hill, 2001.
23. W. M. Wilson, L. H. Rosenberg, and L. E. Hyatt. Automated analysis of requirement specifications. In Proceedings of the 1997 (19th) International Conference on Software Engineering, pages 161–171, 1997.
24. Marek Łabuzek. Modelling the meaning of descriptions of reality to improve consistency between them and business models. Foundations of Computing and Decision Sciences, 29(1-2):89–101, 2004.

# Investigation of Application Specific Metrics to Data Quality Assessment

Dariusz Król[1], Tadeusz Lasota[2], Maciej Siarkowski[3], and Bogdan Trawiński[1]

[1] Wrocław University of Technology, Institute of Applied Informatics,
Wybrzeże S. Wyspiańskiego 27, 50-370 Wrocław, Poland
[2] Wrocław University of Environmental and Life Sciences,
Faculty of Environmental Engineering and Geodesy
C.K. Norwida 25/27, 50-375 Wroclaw, Poland
[3] Computer Association of Information BOGART Ltd.
Rejtana 9-11, 50-015 Wroclaw, Poland
{dariusz.krol,trawinski}@pwr.wroc.pl,
tadeusz.lasota@wp.pl, siarkowski@bogart.wroc.biz

**Abstract.** Databases have risen to be one of the most important corporate assets, but usually their data quality is poor or even not manageable at all. Several metrics of data quality have been designed and implemented to monitor a database of an information system. The primary goal of data quality metrics design was to provide the managers of information centres the tools for monitoring of their databases and for alerting that the amount of errors crossed a given threshold value and it is necessary to undertake activities aimed at data cleansing. The metrics should also be useful for the software producers to enable them to improve their applications. The proposed metrics have been evaluated using databases of several cadastral information systems. The investigation of data quality changes during the period of last three years is presented in the paper. Several metrics applying to domain and referential defects have been used. The study has revealed that simple metrics based on the number of defects detected in a database are not sufficient. The metrics calculating the cost of defect removal have been proposed and evaluated. The experiments covered also metrics specific for cadastral data.

## 1 Introduction

Data quality is related to the satisfaction of the intended use and data must be accurate, timely, relevant, and complete [4]. The concepts of valid versus invalid, inconsistencies in representation, object-level inconsistency, representation of values not known and missing information are all part of defining quality. There are two methods of determining the quality of data: reverification (modernizations) and data analysis [11]. In our approach both are used.

The primary job of the data quality investigation is to identify specific instances of wrong values [1]. The most often used definitions of software quality are defect density rate (DDR) and mean time to failure (MTTF) [3]. Examples of metrics of data quality that can be gathered are: key violations (duplicates, orphans), rows with at least

W. Abramowicz (Ed.): BIS 2007, LNCS 4439, pp. 438–448, 2007.
© Springer-Verlag Berlin Heidelberg 2007

one wrong value, percent of records never returned as answers, object fan-in, and fan-out, number of used attributes per object. Examples of table related measures and schema oriented metrics for databases quality are the following [9]: depth of referential tree (DRT), referential degree (RD), percentage of complex columns (PCC), table size (TS), schema size (SS), number of foreign keys (NFK). Another type of quality metrics used by majority of systems indicates the customer problems when using database. The metric is usually expressed in terms of problems per user month (PUM) and is calculated for each month after the new software is released. However, metrics can show improvements after modifications, but they do not solve problems [10].

We focus on the solution of one problem: the lack of domain-specific metrics for evaluating the quality of cadastral databases. To illustrate the difficulty of cadastral data quality, we first introduce what is known about these data sets. First, there is no universal record key to establish their identity. Second, there are several differences between the records coming from various data sources. Third, some data can be inconsistent such as cadastral subjects i.e. the owners or users of parcels, buildings and apartments. Finally, data can be wrong due to errors introduced during the data input and therefore records may hold different information, e.g. details missing.

Many companies developed to assess data quality the following dimensions [8]: accessibility, believability, completeness, integrity, flexibility, free-of-error, timeliness, simplicity, and understandability. By these dimensions three functional forms help in practice: simple ratio, min or max operators, and weighted average. In our approach we implement integrity dimension and free-of-error dimension.

## 2 Data Quality Metrics Used in the Study

Three groups of data quality metrics have been designed and implemented, because there is no single metrics that would be capable to fulfil all monitoring goals. The first group comprises simple metrics based on the number of defects detected in database. These metrics can be used very easily and their values can be interpreted clearly. However they do not provide any complete image of data quality in a system, because they do not distinguish the significance of defects and the costs of their removal. Therefore the second group of metrics taking into account significance and costs has been proposed, which seem to be a good extension of the simple metrics. These metrics in turn require the estimation of significance and the cost of removing each kind of defects, which is not a trivial task and in most cases it is necessary to engage experts.

Due to this fact we developed some metrics specific for cadastral data. These metrics concern complex objects called registration units which are main units being exposed to modifications or transferred to other systems. For example a land registration unit comprises cadastral parcels, owners and users of those parcels and their land shares. In this respect individual objects are not so much important as the groups of relationally bound objects constituting registration units.

Moreover we distinguish two main types of defects in a database: domain and referential. Domain defects apply to lacking or invalid values and values incompatible with regulations or with dictionaries, i.e. a female name for a man, a birth date for living person before year 1850. In turn referential defects disturb referential integrity of data, i.e. a person record without an address record, land share without person object.

All the metrics enumerated below are based on the results of elementary tests detecting defects in a database. The following denotation has been used to describe data quality metrics: $d_{dom}(i)$ is the number of domain defects indicated by one elementary *i-th* test executed on one table and $d_{ref}(j)$ is the number of referential defects detected by one elementary *j-th* test run for the records of one table. In turn $f(t_{dom}(k))$ is a function, which returns 0 or 1 depending on the result of the *k-th* elementary test $t_{dom}(k)$ detecting domain defects. The value of 1 is returned when the test $t_{dom}(k)$ has detected at least one domain defect and 0 when no defect has been found. In turn, $g(t_{ref}(l))$ is a similar function to the latter, but its argument is the *l-th* elementary test $t_{ref}(l)$ detecting referential defects. REC is the total number of records tested and kREC is equal to this number divided by portion, i.e. one thousand, in turn RU means the total number of register units tested. NIR denotes the number of invalid records, that means the number of records where at least one defect was detected.

## 2.1 Simple Metrics Based on the Number of Defects Detected

**Number of domain defects.** It is the number of domain defects detected in a database which can be expressed by the following formula:

$$ND_{dom} = \sum_{i=1}^{m} d_{dom}(i) \tag{1}$$

**Number of referential defects.** It is the number of referential defects detected in a database which can be expressed by the following formula:

$$ND_{ref} = \sum_{j=1}^{n} d_{ref}(j) \tag{2}$$

**Total number of defects.** It is the sum of the number of domain and referential defects detected in a database which can be expressed by the following formula:

$$ND_{tot} = ND_{dom} + ND_{ref} = \sum_{i=1}^{m} d_{dom}(i) + \sum_{j=1}^{n} d_{ref}(j) \tag{3}$$

**Number of domain defects per portion.** It is equal to the number of domain defects detected falling on portion tested and is expressed as:

$$NDP_{dom} = \frac{\sum_{i=1}^{m} d_{dom}(i)}{kREC} \tag{4}$$

**Number of referential defects per portion.** It is equal to the number of referential defects detected falling on portion tested and is expressed as:

$$NDP_{ref} = \frac{\sum_{j=1}^{n} d_{ref}(j)}{kREC} \tag{5}$$

**Total number of defects per portion.** It is equal to the sum of number of domain and referential defects detected falling on portion tested and is expressed as:

$$NDP_{tot} = NDP_{dom} + NDP_{ref} = \frac{\sum_{i=1}^{m} d_{dom}(i) + \sum_{j=1}^{n} d_{ref}(j)}{kREC} \tag{6}$$

**Percentage of invalid records in records tested.** It is the ratio of the number of records where at least one defect was detected to the number of all records tested, expressed by the following formula:

$$PIR = \frac{NIR}{REC} * 100\% \tag{7}$$

## 2.2 Metrics Calculating Costs of Defect Removal

**Cost of domain defects removal.** It is the total cost of removing all domain defects detected in a database, which can be expressed by the following formula:

$$C_{dom} = \sum_{i=1}^{m} c_{dom}(i) * d_{dom}(i) + \sum_{k=1}^{o} c_{dom}(k) * f(t_{dom}(k)) \tag{8}$$

The first part of the formula concerns those defects, which have to be removed one by one and $c_{dom}(i)$ denotes the cost of removing *i-th* domain defect. The second part applies in turn to such kind of defects which can be all removed by means of one procedure, so that the cost $c_{dom}(k)$ regards to the occurrence of *k-th* domain defect not to the number of these defects.

**Cost of referential defects removal.** It is the total cost of removing all referential defects detected in a database, which can be expressed by the following formula:

$$C_{ref} = \sum_{j=1}^{n} c_{ref}(j) * d_{ref}(j) + \sum_{l=1}^{p} c_{ref}(l) * g(t_{ref}(l)) \tag{9}$$

The first part of the formula concerns those defects, which have to be removed one by one and $c_{ref}(j)$ denotes the cost of removing *j-th* referential defect. The second part applies in turn to such kind of defects, which can be all removed by means of one procedure, so that the cost $c_{ref}(l)$ regards to the occurrence of *l-th* referential defect not to the number of these defects.

**Total cost of defects removal.** It is the sum of the costs of removing both domain and referential defects and which can be expressed by the following formula:

$$C_{tot} = C_{dom} + C_{ref} \tag{10}$$

In the same manner we define **cost of domain defects removal per portion** ($CP_{dom}$), **cost of referential defects removal per portion** ($CP_{ref}$), **total cost of defects removal per portion** ($CP_{tot}$).

### 2.3 Metrics Specific for a Cadastral System

**Number of invalid register units.** It is the number of register units where at least one defect was detected in their objects. It can be denoted by NIRU.

**Percentage of error-free register units.** It is the ratio of the number of register units where no one defect was detected in their objects to the number of all register units tested, expressed by the following formula:

$$PRU_{errf} = \frac{NRU_{errf}}{RU} * 100\% \tag{11}$$

**Percentage of acceptable good register units.** It is the ratio of the number of register units where no one critical defect, which makes it impossible to process data correctly, was found in their objects to the number of all register units tested, expressed by the following formula:

$$PRU_{accg} = \frac{NRU_{crif}}{RU} * 100\% \tag{12}$$

## 3 Investigation of Quality of Cadastral Databases

Cadastral systems are mission critical systems designed for the registration of parcels, buildings and apartments as well as their owners and users. In Poland there are above 400 information centres located in district local self-governments as well as in the municipalities of bigger towns which exploit their local cadastral systems. Data quality of databases taken from four cadastral information centres of different size was monitored. Two of information centres were located at district self-governments (Centre 3 and 4) and two at municipalities of towns with the population of about 200000 (Centre 1 and 2). In order to examine how the size of databases and data quality changed in the course of time seven backup copies made in each centre in the period from 2003 to 2006, one copy per one semester, were investigated. In each database only current records were taken into account. We do not evaluate the quality of historical records because of administration rules. Only the current state of data is important. So that altogether about 18 million records were tested. 232 elementary tests have been designed and implemented in the form of scripts containing SQL expressions. Some scripts were designed to detect defective or lacking values in determined fields of records containing data of main cadastral objects whereas the other to examine incorrect or lacking references between records of different objects. So we could distinguish two groups of scripts, namely the first comprising scripts to detect domain defects and the second containing scripts to indicate referential defects.

In order to estimate costs of defect removal four experts were engaged. They used a cost unit having its values in the form of real numbers in the range of [0, 1]. They assigned values of cost units to each of 232 elementary tests. Due to substantial divergence of experts' assessments the maximum and minimum values for each defect were discarded and the final estimations were calculated as the average of two remaining values.

The defects occurring in databases are of different importance, some of them can be neglected or removed in any time, because they have minor influence on proper functioning of a cadastral system, but some of defects are critical and should be immediately fixed. Due to this respect, each defect has been assigned to one of three significance categories: **critical defects** which disable proper operation of a system and which cannot be removed using any function available in system application, **significant defects** which hamper users' work or cause some data to be illegible or even useless and **non-significant** defects which occur in data not used in everyday work or have no effect on system operation.

### 3.1 Investigation of Changes in the Course of Time

The goal of first series of experiments was to investigate how usable are individual metrics to study changes of data quality in the course of time. At the beginning we examined what the size of cadastre databases monitored was and how it changed in the course of time. As it can be seen in Fig. 1 the number of records in databases ranged from about 200 000 to 1 400 000 depending on the area and population of a district covered by the cadastre system. In each database the number of records has increased in the course of time because information centres were systematically inputting the data of buildings and apartments and their owners.

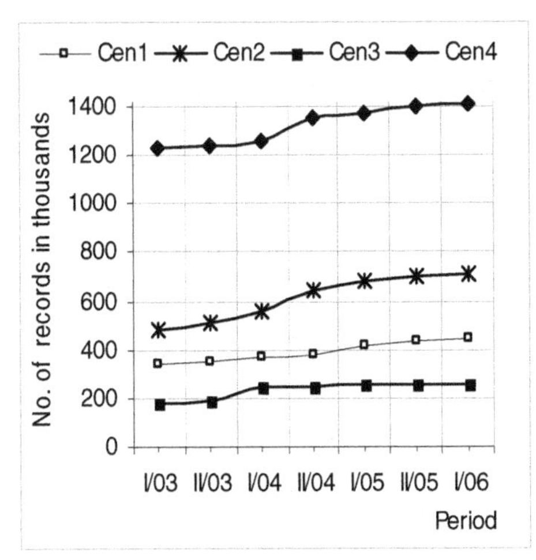


**Fig. 1.** Size of cadastral databases tested

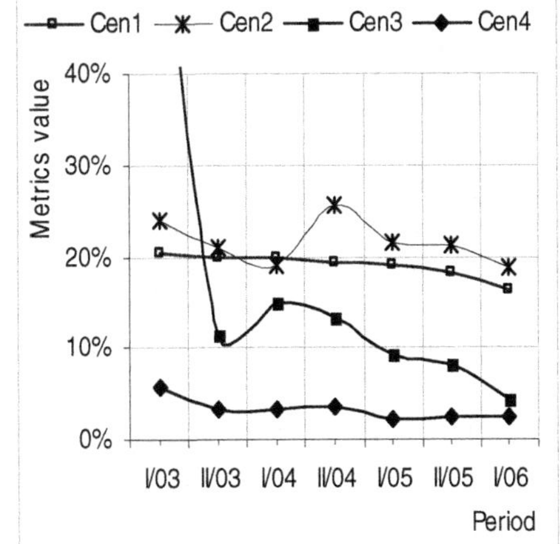


**Fig. 2.** Percentage of invalid records (PIR)

In order to compare data quality of the databases the simple metrics determining percentage of invalid records has been applied. The results presented in Fig. 2 revealed that this number decreased in each database in the course of time because the information centres have systematically been performing data cleansing and also the operators have gained bigger and bigger experience in maintaining building and apartment registers. The percentage of invalid records in the Centre 3 reached 61 per cent in the first semester of 2003 due to the modernization of cadastre system application and after conversion from old database model. The rise of the value of metrics for Centre 2 and Centre 3 in 2004 was caused by loading data of buildings and apartments provided by geodesic companies. This proves that the surveyors' work is not faultless and the tools enabling the transfer of data from surveyors' database are not perfect yet.

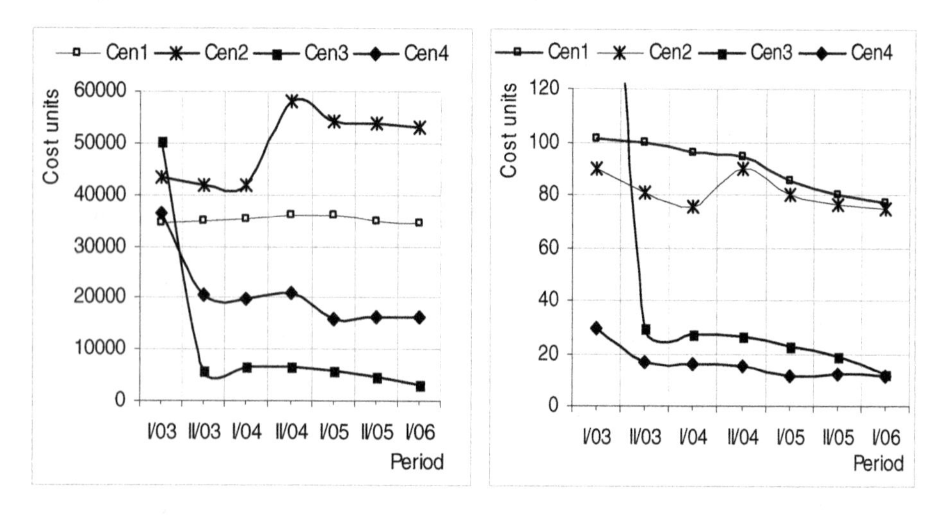


**Fig. 3.** Total costs of defect removal ($C_{tot}$)

**Fig. 4.** Costs of defect removal per 1000 records ($CP_{tot}$)

Similar phenomena could be observed when analyzing the results for costs of defect removal (Fig. 3 and 4). The metrics calculating total values are useful for the managers of information centres because they allow them to estimate costs of database repair and trace the changes of the costs. On the other hand the metrics figuring out the values falling on 1000 records enable the supervisors to analyze the state of the quality of cadastral databases in a province and to compare the efficiency of cleansing activities undertaken by individual information centres. It can be clearly seen that the best data are maintained in the Centre 4, despite the greatest size of the database and the good work has been also done by the Centre 3.

### 3.2 Investigation of the Contribution of Domain and Referential Defects

The absolute and relative number of defects, absolute and relative costs of defect removal is presented in Fig. 5, 6, 7, 8 respectively. The general conclusion is that costs strongly depend on the number of defects detected in a given database. It can be clearly seen that the number of domain defects is much greater than the number of referential ones. In the case of costs the figures concerning referential defects compared to domain ones are ignorable, because most referential defects can be easily removed using adequate single procedures or scripts.

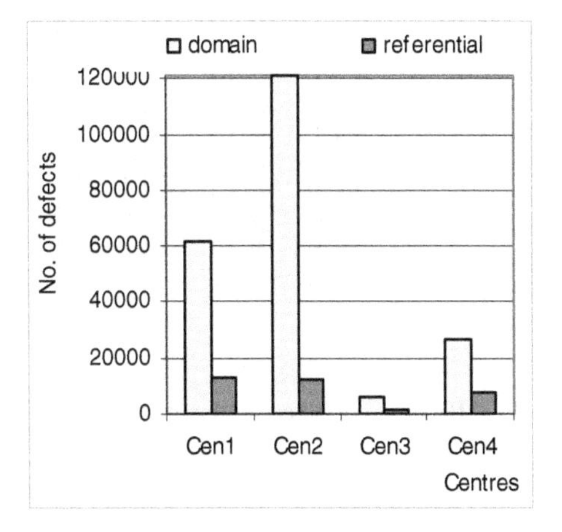


**Fig. 5.** Number of defects (ND)

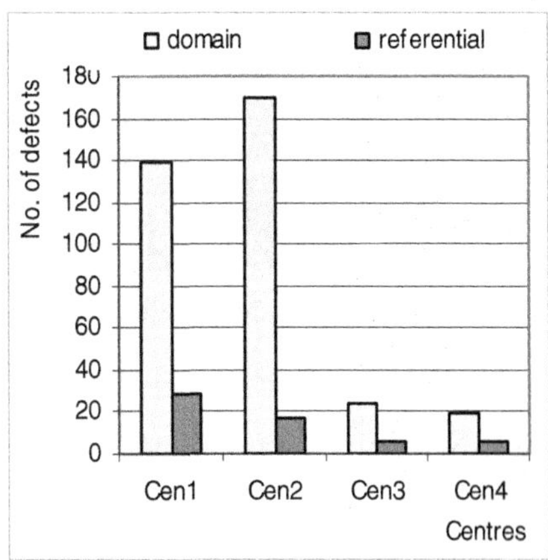


**Fig. 6.** Number of defects per 1000 records (NDP)

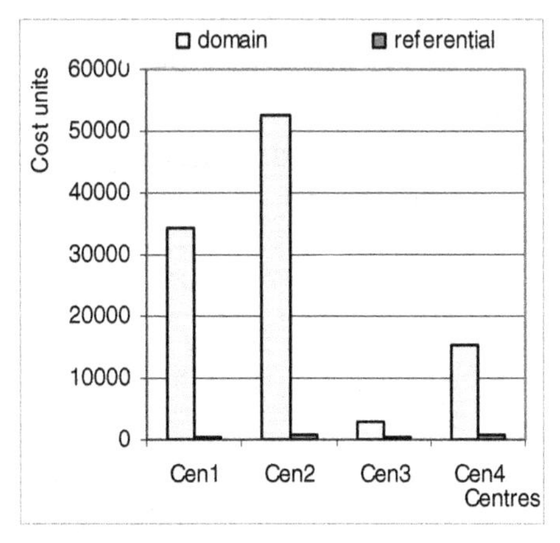


**Fig. 7.** Costs of defect removal (C)

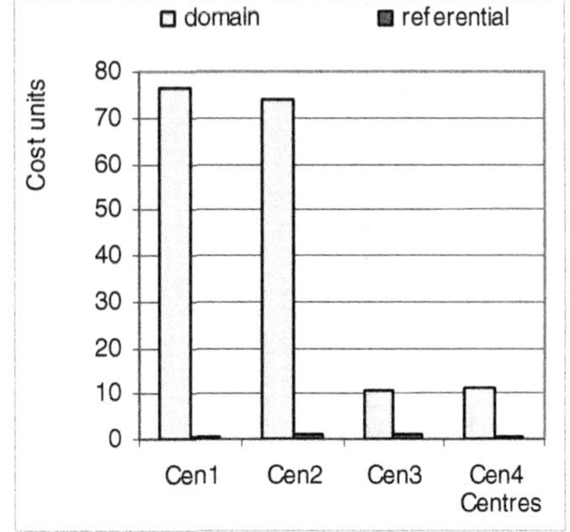


**Fig. 8.** Cost of defect removal per 1000 records (CP)

### 3.3 Investigation of Defect Significance

The number of defects falling into each significance category is shown in Fig. 9 and the total cost of removing all defects from each category is presented in Fig. 10. Due to the great differences between figures logarithmic scale was used in both graphs.

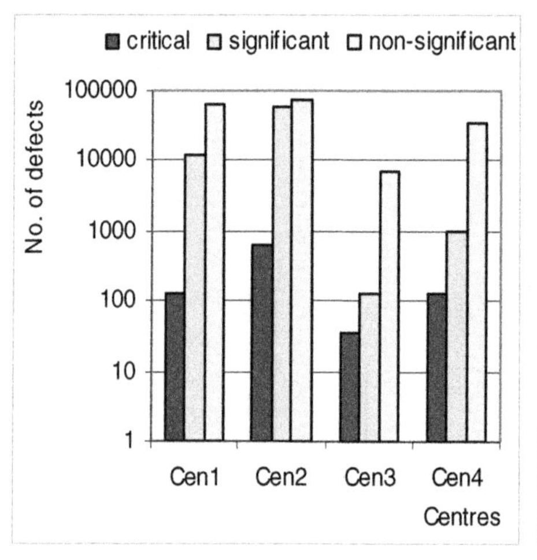


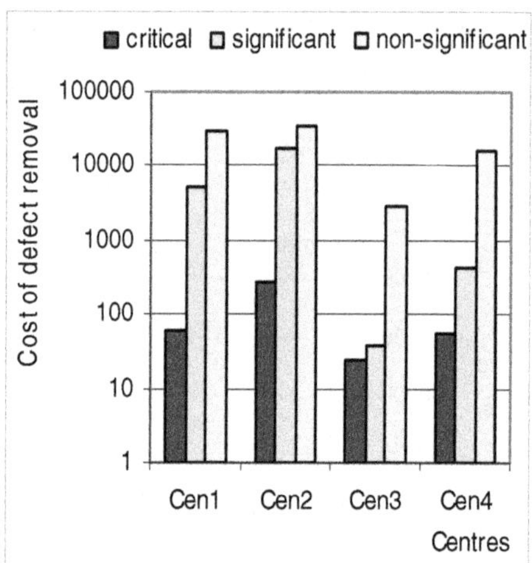


**Fig. 9.** Number of defects in significance groups (ND)

**Fig. 10.** Costs of defect removal in significance groups (C)

## 3.4 Investigation of Data Quality of Registration Units

The last series of investigations concerned registration units which are especially important for the cadastral system, since just the registration units are the main processing units in the system. Almost all modifications may be input to the database only in the window of a registration unit, where all objects comprised by one registration unit, i.e. land parcels, land uses, land shares, subjects and attributes of a registration unit itself, are available.

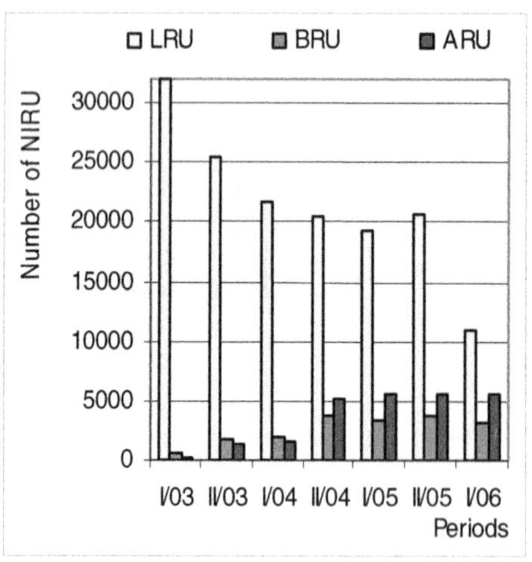


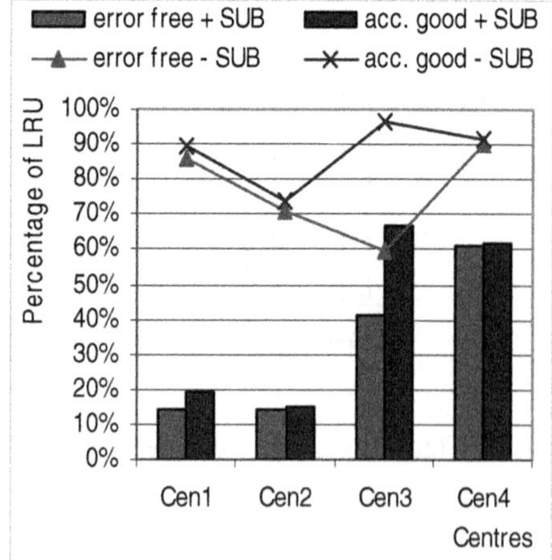


**Fig. 11.** Number of invalid registration units in Centre 2 (NIRU)

**Fig. 12.** Percentage of correct land registration units (PRU)

So the number of invalid registration units approximates well the effort needed to repair the database. The graph representing the number of invalid registration units in the course of time in the Centre 2 is presented in Fig. 11, where LRU, BRU and ARU denote land, building and apartment registration units respectively.

One of the important functions of the cadastral system is to provide data for other systems like IACS (Integrated Administration and Control System) – the system of payments to agricultural land or real estate management or financial-accounting systems. The most adequate data quality in this case is the percentage of correct land register units. In Fig. 12 values of two metrics: percentage of error-free register units and percentage of acceptable good register units calculated for information centres in 2006 are shown. In order to reveal the impact of defects occurring in subject data the results regarding data of persons and institutions (denoted by +SUB) and disregarding these data (denoted –SUB) are presented. The conclusion is unambiguous: data transferred to external systems are of low quality. The best result for Centre 4 reached only 60 per cent of error-free land register units.

# 4 Conclusion

The improvement the quality of data is a complex task, the definition of which is not straightforward. In the paper, we have illustrated a series of methods for quality evaluation. It is important that data were evaluated for quality characteristics using widely accepted metrics. The metrics presented in the paper can be used to alert the information centre management that the amount of defects crossed a given threshold value and it is necessary to undertake activities aimed at data cleansing. Moreover they may be also helpful in estimating resources needed to accomplish this task. They are also useful for the supervisors to analyze the state of the quality of databases and to compare the efficiency of cleansing activities undertaken by information centres.

With the growing availability of knowledge, methodologies, and software tools, high-quality database systems will become the norm, and there will be no excuse for not having them. Experiments performed on cadastral databases hint that it is strongly desired to institute corrective actions on erroneous data. This can be employed including transformation mapping tables, error-correction routines, custom SQL code and manual correction using the original source documentation. It should be also noted that correcting all of today's errors is impossible.

We have done some experiments, but more others are being developed at this moment. Metrics concerning the number of defects in a whole database (only current records) are not sufficient; because they do not provide any hint which objects are the most susceptible to errors. Using elementary tests it is possible to determine the number of defects occurring in individual objects. Also the examination of quality of data stored in numeric map and the consistency of spatial data in numeric map with descriptive data in the registers can be the subject of further study. Moreover, on the basis of ideas presented in [7] and [12], the project for fuzzy expert system to capture the relationship between the metrics and the data quality level is under way.

## References

1. Bobrowski, M., Marre, M., Yankelevich, D.: Measuring Data Quality. Universidad de Buenos Aires. Report 99-002, Buenos Aires, Argentina (1999)
2. Calero, C., Piattini, M., Pascual, C., Serrano, M.: Towards Data Warehouse Quality Metrics. Proceedings of the International Workshop on Design and Management of Data Warehouses, Interlaken, Switzerland (2001)
3. ISO 9126. Software Product Evaluation-Quality Characteristics and Guidelines for their Use. ISO/IEC Standard 9126, Geneva (1994)
4. Hinrichs, H., Aden, T.: An ISO 9001:2000 Compliant Quality Management System for Data Integration in Data Warehouse Systems. Proceedings of the International Workshop on Design and Management of Data Warehouses, Interlaken, Switzerland (2001)
5. Martn, M.: Measuring and Improving Data Quality. Part 1: The Importance of Data Quality. www.aphis.usda.gov/vs/ceah/ncahs/nsu/outlook (2005)
6. Motro, A., Rakov, I.: Estimating the Quality of Databases. LNAI 1495 (1998) 298-307
7. Moura-Pires, F., Ribeiro, R.A., Pereira, A., Varas, J., Mantovani, G., Donati, A.: Data quality fuzzy expert system. Proceedings of the 10th Mediterranean Conference on Control and Automation – MED2002, Lisbon, Portugal (2002)
8. Pipino, L., Lee, Y., Wang, R.: Data Quality Assessment. Communications of the ACM, Vol. 45, No 4 (2002) 211-218
9. Scannapieco, M., Virgillito, A., Marchetti, C., Mecella, M., Baldoni, R.: The DaQuinCIS architecture: a platform for exchanging and improving data quality in cooperative information systems. Information Systems, 29 (2004) 551-582
10. Sieniawski, P., Trawiński, B.: An Open Platform of Data Quality Monitoring for ERP Information Systems. IFIP Working Conference on Software Engineering Techniques - SET 2006, Warsaw, Poland (2006)
11. Signore, O.: Towards a Quality Model for Web Sites. CMG Annual Conference, Warsaw, Poland (2005)
12. Weis, M.: Fuzzy Duplicate Detection on XML Data. Proceedings of the 31st VLDB Conference, Trondheim, Norway (2005)

# Tool-Supported Method for the Extraction of OCL from ORM Models

Sergejus Sosunovas and Olegas Vasilecas

Vilnius Gediminas Technical University, Sauletekio al. 11, Vilnius, Lithuania
{sergejus,olegas}@isl.vtu.lt

**Abstract.** It is recognized that conceptual models expressed in ORM are more suitable for analysis stage and the relational database design, and because of natural verbalization are better tolerated by stakeholders, whereas UML models are more often used in the design of object oriented systems. If the system requires both of these characteristics in particular relational database and object oriented implementation of business logic, the problem of transformation from ORM to UML arises. This paper propose the approach to transform between two well-known modelling techniques: ORM models are transformed into UML models constrained by Object Constraint Language (OCL). The approach precisely describes properties of the transformation. This opens the approach for seamless refining of resulted models using UML tools and transformation to executable code. The transformation of ORM to UML/OCL transformation is validated for correctness by means of a widely used UML tools. Paper illustrates the proposed approach by a number of representative examples.

**Keywords:** ORM, metamodel, UML, OCL.

## 1 Introduction

Within the concept modelling community the object role modelling (ORM) [1] models have been studied and used for decades. These models are subject to introductory courses in database and software engineering education. A typical course will introduce the main concepts in an informal way, explain how to transform ORM schemas into Relational database schemas and will deepen the subject by practical exercises using a design tool and a database system. Conceptual modelling intends to support the quality checks needed before building physical systems by aiming at the representation of data at a high level of abstraction, and therefore acquire a high degree of, often implicit, semantics.

Within the software engineering community, Unified Modelling Language (UML) [2] has gained much attention, in particular in connection with the Model Driven Architecture (MDA) [3]. This paper proposes approach to transform ORM models to UML and OCL [4] using transformation languages and tools that satisfy MDA requirements. Making transformation specification design decisions we will use only such UML and OCL features that are implemented in the popular UML and OCL tools [5, 6, 7, 8, 9]. In contrast to known ORM – UML transformation approaches, this paper however describes with its transformation specification not only the basic

W. Abramowicz (Ed.): BIS 2007, LNCS 4439, pp. 449–463, 2007.
© Springer-Verlag Berlin Heidelberg 2007

ORM concepts but also, an important ORM part, ORM constraints that vaguely can be presented in pure UML. The paper formally connects ORM constraints to OCL constraints. Furthermore, the transformation between models is also described in formal executable language ATL [10]. Resulted UML models and OCL constraints are validated by before mentioned tools. We are not aware of another approach handling these two classical models with respect to practical applicability and their transformation in a rigorous and uniform way. In particular, we are not aware of an approach being able to express the ultimate goal of the model transformation process, namely the equivalence between the constraints for the different models, in a formal and explicit way. Although both models are well-known, the paper gives new insights into the models and their transformation by making usually implicit assumptions explicit which is particular useful for non-experts, beginners and practitioners of MDA.

The rest of this paper is structured as follows. Section 2 will give an overview on the related works. Section 3 concentrates on transformation rule specification. Section 4 presents transformation example of ORM model. Section 5 discusses typical questions and section 6 finishes the paper with a conclusion. The formal transformation rules are not given in form of ATL expressions in this paper but are explained informally when discussing transformation and are available at BRidgeIT site [11].

## 2 Related Works

Our work has connections to some related approaches that we will describe. Author of [12] analyzes UML data models from ORM perspective and identifies ORM constructs that can be transformed to UML. He also compares UML associations and related multiplicity constraints with ORM relationship types and related uniqueness, mandatory role and frequency constraints, discusses exclusion constraints, and summarizes how the two methods compare with respect to terms and notations for data structures and instances. Finally authors of [12, 13] draw to the conclusion that ORM set constraints are lost when transformed to UML. It is presented in the paper [13] how to compensate these defects by augmenting UML with concepts and techniques from the Object Role Modelling (ORM) approach. In general, set constraints in UML would normally be specified as textual constraints (in braced comments) or OCL should be used in more complicated cases.

The author of [15] provides a way to map ORM facts to UML constructs, leaving out the rest elements of the model. Although several papers [14, 13, 15] show how fragments of ORM model can be potentially encoded as fragments of UML models, a formal procedure for mapping onto logical schemas [14] that specifies how a target UML class diagram and OCL constraints can be created for any given 'source' ORM model is lacking. Both papers [15] and [12] propose to map ORM n-ary fact type to ternary associations in UML which is rarely supported in UML, tools in general and is not supported by our target tools.

Although ORM has been used for three decades and now has industrial modelling tool support, it has no official, standard meta-model necessary for the MDA transformations. Authors of [16, 17] discusses in their recent research to pave the way for a standard ORM metamodel. Our approach may be understood as one specific variant of metamodel proposed in [16]. The speciality of our presented approach is that differently from suggested in [16], where the ORM metamodel extends UML metamodel, we use independent ORM metamodel implemented in open source tool [11].

Technically, alternative approach to transform ORM models is based on ORM-ML XML-based ORM mark-up language as it is proposed in [18]. It is possible to create style sheets to transform ORM models presented in ORM-ML into another different syntax, e.g. pseudo natural language. Additionally there are several tools available with built in transformation capabilities from ORM to UML [18] and to relational database schema [19, 20]. In our approach we employ MDA higher order transformation specification language because of its explicit rule based transformation specification simplifies understanding and maintainability of transformation rules.

Our approach is based on wellformedness rules for the ORM [22]. However, no complete satisfiability checker is known for ORM. That complicates transformation task of ORM model. It is necessary to provide such transformation rules that result in correct UML model and are resistible for incorrect ORM models.

# 3 Transformation Rules

## 3.1 Object Type and Value Type Transformation Rules

According to UML metamodel [2] each class should belong to the package and the package should be in the model namespace. Therefore the first rule in the transformation specification creates UML package and appropriate UML model. ORM model is composed from entity types and value types. These are the first ORM model elements that should be transformed. ORM entity types are proposed to map to UML classes within the namespace of the created model. Reference schemas of the entity types are transformed to the attributes of the appropriate classes. Value types are transformed to UML attributes if they are connected to one fact type and to the classes otherwise. We argue that it is expedient to transform ORM value type to class

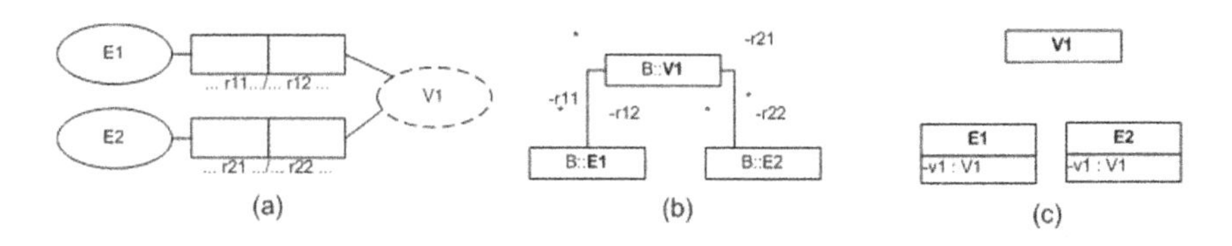


**Fig. 1.** (a) Value type of source ORM model can be transformed to several types of UML class diagrams (b,c)

(Fig. 1,b) in case of participation in several fact types (Fig. 1,a) than to attribute (Fig. 1,c) because of the existence of explicit associations between value type UML class and object type UML classes, besides connection names to both directions are preserved. The overview of transformation approach is presented in table 1.

**Table 1.** Overview of proposed approach for ORM transformation to UML/OCL

| ORM model elements | UML/OCL model elements |
|---|---|
| Entity Type | Class |
| Value Type | Class, Attribute |
| Fact Type | Class, Attribute, Association |
| Objectified Fact Type | Class |
| Subtype | Generalization |
| Mandatory constraint | Association end multiplicity range lower value |
| Uniqueness constraint | Association end multiplicity range upper value, OCL constraint |
| Frequency constraint | Association end multiplicity range lower value, Association end multiplicity range upper value |
| Set constraint | OCL constraint |
| Value constraint | OCL constraint |
| Ring constraints | OCL constraint (limited) |

## 3.2 Fact Type Transformation Rules

Transformation rules for fact types can be divided to three groups based on the cardinality of fact types: unary fact-types, binary-fact types and n-ary fact-types. Unary fact types attached to entity types are transformed to binary attributes. Unary fact types attached to value type that was not transformed to class results in an exception, it is treated as illogical model.

Binary fact types attached to the entity types results to binary association. Association end names are provided based on the first ORM phrase with the first appropriate role. Binary fact types with one value type, as it is stated earlier, are transformed to attribute or to the association if the value type is connected to several fact types. Binary fact types with two value types are transformed to the association or to the attribute based on the rules provided earlier.

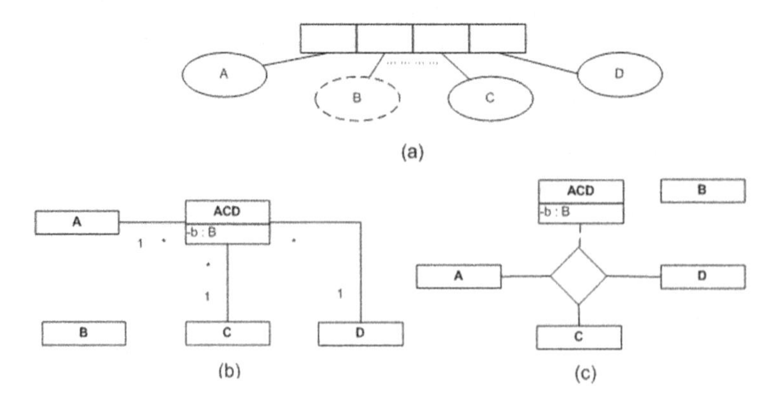


**Fig. 2.** Transformation of n-ary fact(a) type to combination of association and class(b) and to UML ternary association (c)

N-ary fact types (Fig. 2,a) differently from the proposed in [12][15] ternary association(Fig. 2,c) are transformed to UML class that have 1 multiplicity connections to participating entity types and value types (Fig. 2,b). The main reason for such transformation is that ternary associations are rarely supported by the UML tools. Objectified fact-types of any arity are transformed to UML classes as well.

## 3.3 Constraint Transformations

### 3.3.1 Uniqueness, Frequency and Mandatory Constraints

Internal uniqueness constraints are depicted as arrow tipped bars, and are placed over one or more roles in a fact type to declare that instances for that role (combination) in the relationship type population must be unique. For the transformation purposes we have identified three cases internal uniqueness constraints: one-role, two role on binary fact-type and n-ary role on n-ary fact-type.

One role internal uniqueness constraint is transformed to the multiplicity range upper value 1 of the appropriate association end for binary and n-ary fact types that was transformed to association. If it is applied on unary fact-type or on the fact type that was transformed to attribute then the multiplicity range upper value 1 is applied to attribute. If the constraint's binary or n-ary fact type was transformed to attribute and internal uniqueness constraint was applied to value type's role it constraints the following OCL constraint is generated for the UML model presented in Fig. 3,b:

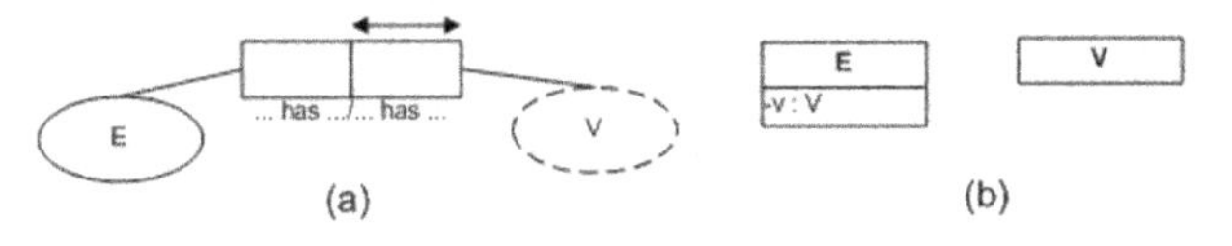


**Fig. 3.** (a) Internal uniqueness constraint on one role value type role of binary fact type, (b) resulted UML model

```
Context E
inv: let a: Set(E) = E.allInstances in
not a->exists(b|b.v=self.v)
```

Two role internal uniqueness constraints (Fig. 4.a) is transformed to following OCL statement for UML model in Fig. 4,b:

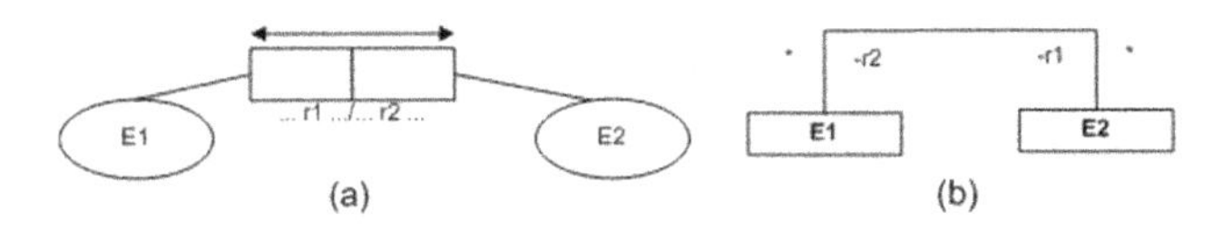


**Fig. 4.** (a) two role internal uniqueness constraint on binary fact type, (b) resulting UML model

```
Context A
inv: not (self.r1->exists(b|b.r2->includes(self)))
```

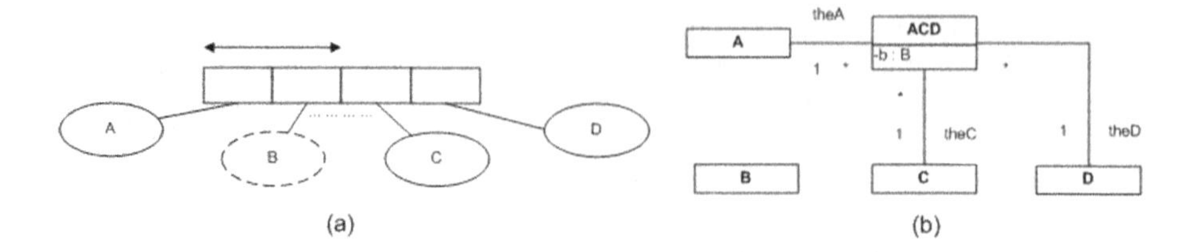


**Fig. 5.** (a) N-ary role internal uniqueness constraint, (b) resulting UML model

N-ary role internal uniqueness constraints (Fig. 5.a) is transformed to following OCL statement for UML model in Fig. 5,b:

```
context ACD
inv: let a:Set(ACD)=ACD.allInstances in
not (a->exists(it|it.theA=self.theA and it.b=self.b))
```

An external uniqueness constraint (Fig. 6,a) shown as a circled "u" may be applied to two or more roles from different fact types by connecting to them with dotted lines. This indicates that instances of the combination of those roles in the join of those fact types are unique. In order to efficiently implement this constraint we have had to introduce ORM model wellformedness constraint on scope of the external uniqueness constraint. It constrains external uniqueness constrain to be put only on roles of the fact types connected to the same value or entity types. The necessity of introducing such wellformedness constraint arises because of inability of OCL to iterate through the model and find joins that external uniqueness constraint requires. The OCL constraint's context in this case is any class that is attached to all fact types

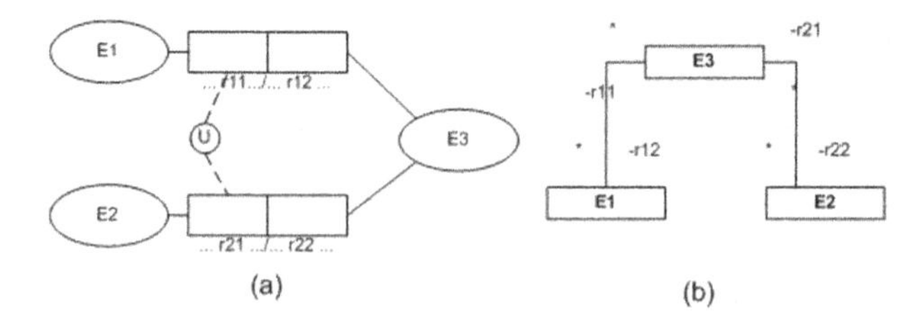


**Fig. 6.** (a) External uniqueness constraint on the binary fact type, (b) resulting UML model

constrained by the ORM external uniqueness constraint. OCL constraint on UML model in Fig. 6,b is following:

```
context E3 inv: let a:Set(E3)=E3.allInstances in (not
a->exists(b|b.r12=self.r12 and b.r22=self.r22))
```

A mandatory role constraint declares that every instance in the population of the role's object type must play that role. Mandatory constraint is transformed to association's other's end multiplicity range lower value. Default value is 0 if the role does not have mandatory constraint [12, 15].

Frequency constraint applied to a sequence of one or more roles, these indicate that instances that play those roles must do so exactly n times, between n and m times, or at least n times. This type of constraints is transformed to appropriate multiplicity range lower and upper value of the association end or attribute.

### 3.3.2 Set Constraints

A dotted arrow (Fig. 7,a) from one role sequence to another is a subset constraint, restricting the population of the first sequence to be a subset of the second. Resulting OCL constraints for UML model in Fig. 7,d:

```
Context E2 inv: self.r11->includesAll(self.r21)
Context E1 inv: self.r12->includesAll(self.r22)
```

Equality constraint (A double-tipped arrow Fig. 7,b) indicate the populations must be equal. Resulting OCL constraints for UML model Fig. 7,d:

```
Context E2 inv: self.r11=self.r12
Context E1 inv: self.r12=self.r22
```

A circled "X" (Fig. 7,c) is an exclusion constraint, indicating the populations are mutually exclusive. Exclusion constraints may be applied between two or more sequences. Resulting OCL constraints for UML model in Fig. 7,d:

```
Context E2
inv: self.r11->isEmpty() or self.r12->isEmpty()
```

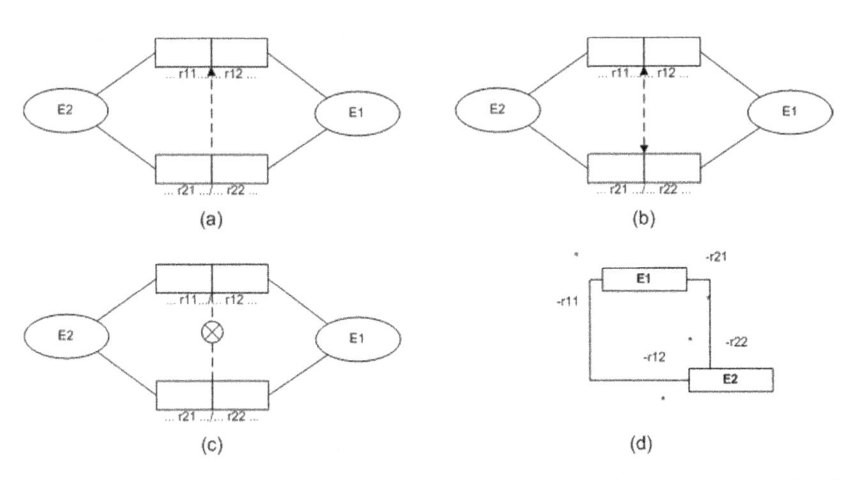


**Fig. 7.** (a) subset, (b) equality, (c) exclusion constraint on binary fact type, (d) resulting UML model

```
Context E1
inv: self.r12->isEmpty() or self.r22->isEmpty()
```

### 3.3.3 Value Constraint

To restrict an object type's population to a given list, the relevant values may be listed in braces (Fig. 8,a). If the values are ordered, a range may be declared separating the first and last values by ".." (Fig. 8,b). OCL constraint for range value constraints for UML model in Fig. 8,c:

```
context A inv: self.code>=a1 and self.code<=a2
```

OCL constraint for list value constraint for UML model in Fig. 8,d):

```
context B
inv: self.code='b1' or self.code='b2' or self.code='b3'
```

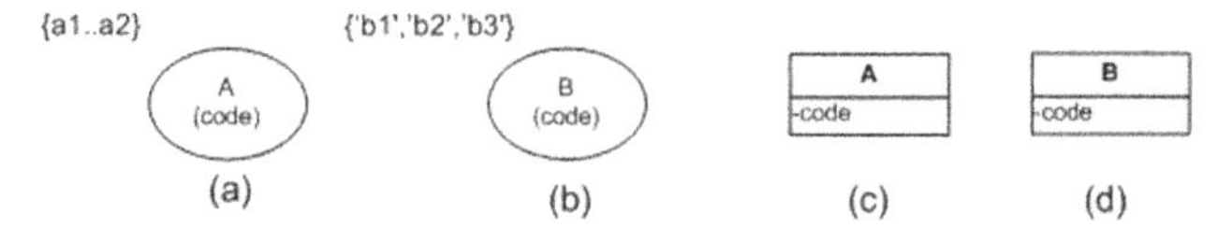


**Fig. 8.** Entity type with (a)value range constraint and (b) value list constraint, (c,d) resulting UML model

### 3.4 Ring Constraints

Ring constraint, that may be applied to a pair of roles played by the same host type. These indicate that the binary relation formed by the role population must be irreflexive (ir), intransitive (it), acyclic (ac), asymmetric (as), antisymmetric (ans) or symmetric (sym). We will illustrate OCL constraints for the ORM ring constraints using UML model presented in Fig. 9,b. Ring constraints can be put on roles that can

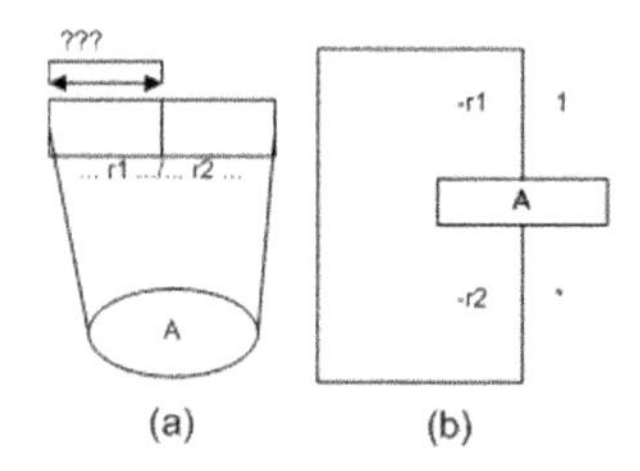


**Fig. 9.** (a) Role with undefined ring constraint, (b) resulting UML model

be transformed to association end of different multiplicity. Therefore we are presenting OCL constrains with navigation statements for one to many multiplicity case (r2 association end in Fig. 9,b) and constraint for single value for many to one case (r1 association end in Fig. 9,b).

Irreflexive means the object cannot bear the relationship to itself. OCL constraint for navigation to set:

```
Context A inv: self.r2->excludes(self)
```

OCL constraint for single value:

```
Context A inv: not (self.r1=self)
```

Intransitive means that if the first bears the relationship to the second, and the second to the third, then the first cannot bear the relationship to the third.

Intransitive OCL constraint for navigation to set:

```
context A
inv: self.r2->collect(b|b.r2)->excludesAll(self.r2)
```

Intransitive OCL constraint for single value:

```
context A inv: not(self.r1.r1=self.r1)
```

Asymmetric means that if the first bears the relationship to the second, then the second cannot bear that relationship to the first

Asymmetric OCL constraint for navigation to set:

```
context A inv: self.r2->collect(b|b.r2)->excludes(self)
```

Asymmetric OCL constraint for single value:

```
context A inv: not (self.r1.r1=self)
```

Anti-symmetric means that if the objects are different, then if the first bears the relationship to the second, then the second cannot bear that relationship to the first.

Anti-symmetric OCL constraint for navigation to set:

```
context A
inv: self.r2->select(a|not(a=self))->collect(a| a.r2)-
>excludes(self)
```

Anti-symmetric OCL constraint for single value for:

```
context A
inv: not (self.r1=self) implies (self.r1=self.r1)
```

Symmetric means that if the first bears the relationship to the second, then the second bears that relationship to the first.

Symmetric OCL constraint for navigation to set:

```
context A inv: self.r2->collect(a|a.r2)->includes(self)
```

Symmetric OCL constraint for single value:

```
context A inv: self.r1.r1=self
```

Acyclic means that a chain of one or more instances of that relationship cannot form a cycle (loop). It is the only type of ORM constraint that cannot be fully implemented in OCL. This constraint requires recursive OCL statement; however recursion is still unsolved issue of OCL [23]. But it is possible to generate through transformation specification OCL constraint of practically unlimited depth. We have shown in bold repeatable part of OCL constraints.

Acyclic OCL constraint for navigation to set:

```
context A
inv: inv: (self.r2->collect(a|a.r2)->excludes(self))
```

Acyclic OCL constraint for single value:

```
context A
inv: not (self.r1.r1=self)
```

Acyclic deeper OCL constraint for navigation to set:

```
context A
inv: (self.r2->collect(a|a.r2)->
collect(a|a.r2)->excludes(self))
```

Acyclic deeper OCL constraint for single value:

```
context A inv: not (self.r1.r1.r1=self)
```

Acyclic even deeper OCL constraint for navigation to set:

```
context A inv: (self.r2->collect(a|a.r2)->
collect(a|a.r2)-> collect(a|a.r2)->excludes(self))
```

Acyclic even deeper OCL constraint for single value:

```
context A inv: not (self.r1.r1.r1.r1=self)
```

## 4 An Example of ORM-UML/OCL Transformation

We present a case study of the use of the transformation specification in ATL to create UML model constrained by OCL statements from ORM model.

For our case study, we consider a fragment of scientific conference management domain ORM model (Fig. 10). It is information system used by a conference programme committee chair to maintain details about submitted papers, reviewers and assigned reviews.

The source ORM model was encoded to XMI format according to ORM metamodel and transformed to UML model in appropriate XMI format using ATL language execution environment. OCL statements constraining resulted UML model were generated as textual strings.

We have mapped example ORM model constraints to the OCL statements. In the following part of the chapter we will provide ORM constraint textual description and appropriate OCL statement resulted from the transformation.

Uniqueness role constraint on “Author has written Paper”:

```
context Paper inv:not self.iswrittenby -> exists
(a|a.haswritten->includes(self))
```

Uniqueness role constraint on n-ary fact type “Paper review evaluation according Evaluation Criteria is equal to Evaluation Value” is transformed to:

```
Context PaperreviewevaluationaccordingEvaluationCriteri
aisequaltoEvaluationValue
inv:let a: Set(PaperreviewevaluationaccordingEvaluation
CriteriaisequaltoEvaluationValue)
=PaperreviewevaluationaccordingEvaluationCriteriaisequa

ltoEvaluationValue.allInstances in not a->exists(a|
a.thePaperreview =  self.thePaperreview and
a.theEvaluationCriteria =  self.theEvaluationCriteria)
```

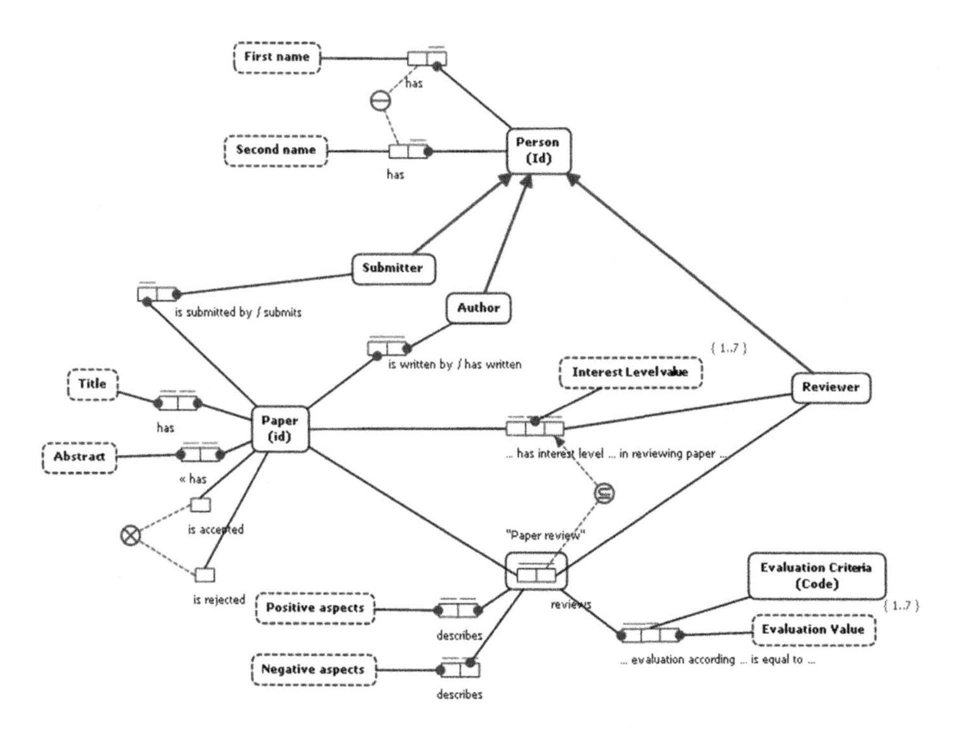


**Fig. 10.** Source ORM model for the transformation example

Uniqueness role constraint on objectified binary fact type "Reviewer reviews Paper" is transformed to:

```
context Paperreview
inv:let a: Set(Paperreview) =Paperreview.allInstances
in not a->exists(a|  a.theReviewer =  self.theReviewer
and a.thePaper = self.thePaper)
```

Uniqueness role constraint on n-ary fact type "Reviewer has interest level Interest Level value in reviewing Paper" is transformed to:

```
context ReviewerhasInterestLevelvalueinreviewingPaper
inv: let a:
Set(ReviewerhasInterestLevelvalueinreviewingPaper) =
ReviewerhasInterestLevelvalueinreviewingPaper.
allInstances in not a->exists(a| a.theReviewer =
self.theReviewer and a.thePaper = self.thePaper)
```

External uniqueness role constraint on fact types "Person has First name" and "Person has Second name" is transformed to:

```
context Person inv:
let a: Set(Person) =Person.allInstances in
not a->exists(a| a.Firstname = self.Firstname and
a.Secondname = self.Secondname)
```

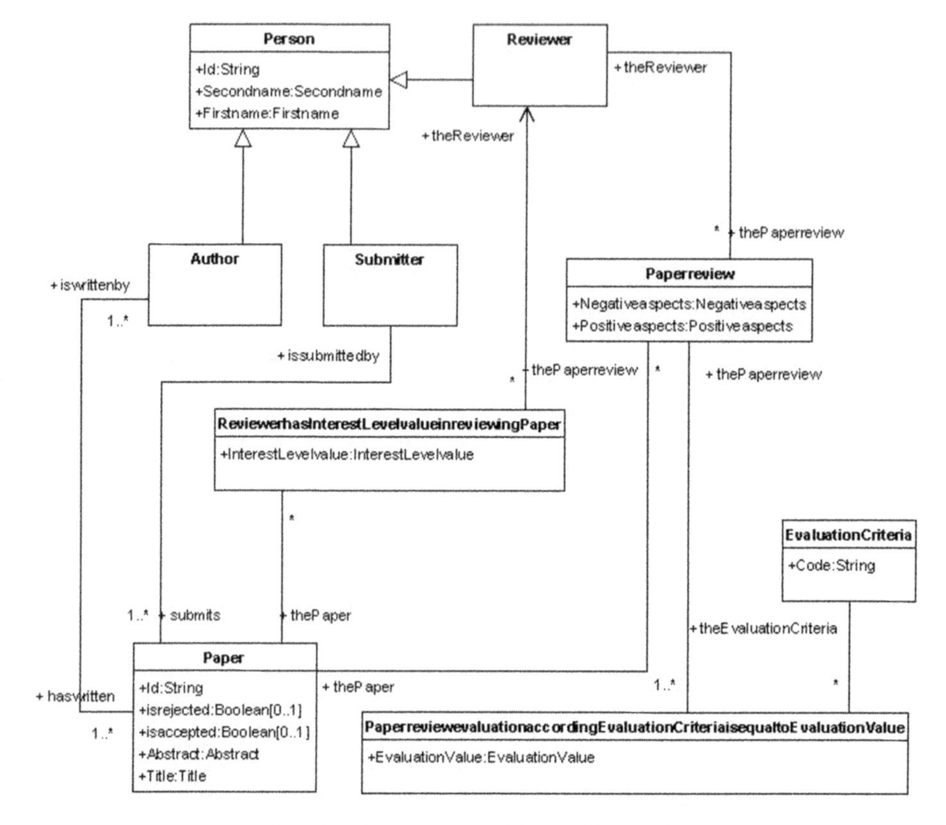


**Fig. 11.** Example of resulting UML model

Subset constraint on fact types "Reviewer has interest level Interest Level value in reviewing paper Paper" and "Reviewer reviews Paper" is transformed to:

```
context Paper inv:
self.theReviewerhasInterestLevelvalueinreviewingPaper->
collect ( a|a.theReviewer)->
includesAll(self.thePaperreview->
collect ( b|b.theReviewer))
```

Exclusion constraint on fact types "Paper is accepted" and "Paper is rejected" is transformed to:

```
context Paper
inv: self.isaccepted or self.isrejected
```

We have checked all presented constrains for the syntactic and semantic correctness using OCL tool OCLE and Dresden OCL toolkit. Additionally in order to verify that the OCL constraint semantics fully represent ORM constraint semantics we used approach described in [24] and implemented in USE tool. The principle for the approach is to define properties that should be verified on the model. Then the

USE tool checks whether it is possible to generate snapshots from the model that verify the property. Appropriate UML model snapshots were generated for the each OCL constraint.

## 5 Discussion

In this section we want to debate typical questions that may showup during discussions about the subject of this paper.

What are the business cases of the approach? A 'business case' for our approach could be tuning of the general database model, developed by ORM, and application, developed using UML to handle that database. Constraints provided in ORM should be preserved in both of them.

What role plays tool support in the approach? Transformation rules and resulting UML and the OCL constraints are quite complex. Our experience shows that this complexity requires tool support in order to understand the consequences of design decisions, for example, the consequences of a particular constraint. We use OCLE and Dresden OCL for constraint validation and Poseidon for target UML model validation.

Is transformation extensible? Transformation specification is provided as a fully executable ATL file containing transformation rules. One can change the transformation specification and adopt it for its own needs.

Is transformation fully reversible? At the moment transformation is not fully reversible. In case of reverse transformation of UML model to ORM objectified and n-ary fact types would be not recreated. Transformation of OCL constraint to ORM constraints is hardly possible at the moment. The alternative is to transform OCL to ConQuer language proposed in [25]. During reverse transformation only the basic phrases and sentences will be recreated.

## 6 Conclusion

In this paper we have employed MDA as a framework for the transformation of ORM models to UML class models constrained constraints represented in OCL. We have proposed and formally specified the transformation rules and transformation decisions for the resulted model to be accessible for the widely used UML tools (e.g. Poseidon for UML, Rational Rose, Eclipse UML). Differently from already existing approaches have covered ORM constraints with our transformation specification ORM constraints in addition to transformation of structural ORM elements. However, due to the limitations of OCL, in particular there still exists unresolved unlimited iteration issue. Therefore, we have had to limit transformation specification to the predefined iteration depth of resulting OCL constraints in case of transformation ORM set and acyclic ring constraints.

The proposed approach has proven to be very effective for generating UML and OCL constraints from ORM by presentation of representative transformations examples. Each presented OCL statement was validated to be correct syntactically and semantically using OCLE tool. In order to prove transformation of semantics of

ORM constraints we used snapshots of resulting UML model generated using USE tool for each OCL constraint.

Our approach enables software system engineers to focus on the application domain and architectural design decisions issues without being limited by the used tools, because MDA insures exchangeability of models. It is especially important if conceptual models were developed by separate teams and brought together for the creation of enterprise wide system.

## References

1. Halpin, T.: Object-Role Modeling (ORM/NIAM). In: Bernus, P., Mertins K., and Schmidt G. (eds.): Handbook on Architectures of Information Systems, Springer-Verlag, Berlin, (1998), (March, 2006) URL: http://www.orm.net/pdf/springer.pdf
2. OMG: UML 2.0 Superstructure Final Adopted specification, OMG document: ptc/03-08-02,OMG, (2003), (August, 2006) URL: http://www.omg.org/cgi-bin/doc?ptc/03-08-02.pdf
3. Siegel, J.: Developing in OMG's Model-Driven Architecture, OMG document: 01-12-01, OMG, (2001), (May, 2006) URL: http://www.omg.org/docs/omg/01-12-01.pdf
4. OMG UML 2.0 OCL Draft Adopted Specification, OMG document: ptc/03-08-08, OMG, (2003), (June, 2006) URL: http://www.omg.org/docs/ptc/03-10-14.pdf
5. IBM: Rational Rose tool (2006) (June, 2006) URL: http://www-306.ibm.com/software/rational/
6. Eclipse Project: Eclipse UML2 (October, 2006) URL: http://www.eclipse.org/uml2
7. LCI team: Object constraint language environment. Computer Science Re-search Laboratory, "BABES–BOLYAI" University, Romania, (2005) (November, 2006) URL: http://lci.cs.ubbcluj.ro/ocle/
8. Gentleware: Poseidon UML tool (May, 2006) URL: http://www.gentleware.com/
9. Demuth B.: The Dresden OCL Toolkit and its Role in Information Systems Development. In O. Vasilecas at al (Eds). Proc. of Thirteenth International Conference on Information Systems Development. Advances in Theory, Practice and Education. Vilnius, Technika, 2004, (February, 2006) URL: http://dresden-ocl.sourceforge.net/
10. Jouault, F., Kurtev, I.: Transforming Models with ATL. In: Proceedings of the Model Transformations in Practice Workshop at MoDELS 2005, Montego Bay, Jamaica, (June, 2006) URL: http://www.sciences.univ-nantes.fr/lina/atl/bibliography/MTIP05
11. Sosunovas S.: Open source tool Bridge IT, (March, 2006) URL:http://isl.vtu.lt/BRidgeIT/
12. Halpin, T.: UML data models from an ORM perspective: Parts 1 – 10, In: Journal of Conceptual Modeling, Inconcept, (1998-2001) (July, 2006) URL: http://www.orm.net/
13. Halpin, T., Augmenting UML with Fact-orientation, In: workshop proceedings: UML: a critical evaluation and suggested future, HICCS-34 conference, (2001), (September, 2006) URL: http://citeseer.ist.psu.edu/halpin00augmenting.html
14. Halpin, T., Information Modeling and Relational Databases 3rd edn., Morgan Kaufmann Publishers, (2001)
15. Bollen, P., A Formal ORM-to -UML Mapping Algorithm. (June, 2006) URL: http://arno.unimaas.nl/show.cgi?fid=465
16. Krogstie, J., Halpin, T., Siau, K.: Two Meta-Models for Object-Role Modeling. In: Krogstie, J., Halpin, T., Siau, K. (Eds.): Information Modeling Methods and Methodologies, Idea Group Publishing (2005) 17-42
17. Cuyler, D., Halpin, T., Metamodels for Object-Role Modeling, (2003), (August, 2006) URL: http://www.emmsad.org/2003/Final%20Copy/26.pdf

18. BCP Software: CaseTalk tool (October, 2006) URL: http://www.casetalk.com/php/index.php?FCO-IM%20Bridge
19. Sourcefourge team Object-Role Modeling team: The Object-Role Modeling (ORM) standard version 2, associated schemas and generation tools, and a reference implementa tion in the form of Neumont ORM Architect for Visual Studio (August, 2006) URL: http://sourceforge.net/projects/orm
20. Microsoft corporation: Visio tool (June, 2006) URL: http://office.microsoft.com/en-us/visio/default.aspx
21. Bird, L., Goodchild, A., Halpin, T.A.: Object Role Modelling and XML-Schema. In: Laender, A., Liddle, S., Storey, V. (Eds.): Proc. of the 19th International Conference on Conceptual Modeling (ER'00). Lecture Notes in Computer Science, Vol. 1920. Springer-Verlag, Berlin Heidelberg New York (1999) 309-322
22. Jarrar, M., Unsatisfiability Reasoning in ORM Conceptual Schemes. In: Illarramendi, A., Srivastava, D.: Proceeeding of International Conference on Semantics of a Networked World, LNCS 4254, Springer-Verlag, (2006) in press
23. Brucker A.D., Doser J., Wolff B.: Semantic Issues of OCL: Past, Present, and Future. In: OCL for (Meta-)Models in MultipleApplication Domains, Available as Technical Report, University Dresden, number TUD-FI06-04-Sept, (2006) 213–228 (November 2006) http://www.brucker.ch/bibliography/download/2006/brucker.ea-semantic-2006.pdf
24. Gogolla M., Bohling, J., Richters, M.: Validation of UML and OCL Models by Automatic Snapshot Generation. In: Booch, G., Stevens, P., Whittle, J., (Eds.): Proc. 6th Int. Conf. Unified Modeling Language (UML'2003). Springer, Berlin, LNCS 2863, (2003) 265-279
25. Bloesch, A., Halpin, T.: ConQuer: a conceptual query language. In: Thalheim B. (Eds.):Proc. 15th International Conference on Conceptual Modeling ER'96 (Cottbus, Germany), LNCS 1157, Springer-Verlag, (1996) 121-133

# Model-Driven Architecture for Mobile Applications*

Jürgen Dunkel and Ralf Bruns

Hannover University of Applied Sciences and Arts, Department of Computer Science,
Ricklinger Stadtweg 120, 30459 Hannover, Germany
{dunkel,bruns}@fh-hannover.de

**Abstract.** Although significant improvements in the development of business applications for mobile devices have been made in recent years, the software development in this area is still not as mature as it is for desktop computers. Therefore, declarative and code generation approaches should be preferred instead of manually coding. In the BAMOS project an architecture has been designed and implemented for the generic and flexible development of mobile applications. The architecture is based on the declarative description of the available services. In this paper we present a model-driven approach for generating almost the complete source code of mobile services. By applying model-driven development within the proposed approach, a new service can be conveniently modeled with a graphical modeling tool and the graphical models are then used to generate the corresponding XML descriptions of the mobile user interface and the workflow specification. In order to use such a service no specific source code has to be implemented on the mobile device.

**Keywords:** model-driven architecture (MDA), mobile applications, XForms, meta models, code generation.

## 1 Introduction

Nowadays mobile devices, e.g. mobile phones, personal digital assistants (PDA) or smart phones, are ubiquitous and accompany theirs users almost every time and everywhere. Their capability of connecting to local area networks via Bluetooth or Wireless LAN potentially enables new types of mobile applications expanding the limits of present ones. So far, mobile devices do not fully exploit the whole potential of these networks. They are mostly employed only for communication or personal information management purposes.

While moving with a mobile device, the user enters a large number of different local networks; each might offer different localization-specific services. Examples for such location-based services [HaRo04] are the timetable and location plan of the next bus stop, the recent programs of the local cinemas, or a car reservation service of the car rental agencies nearby.

Today, the software development for mobile devices is cumbersome and not as mature as for desktop computers. Therefore, declarative and code generation approaches

* This work was supported by AGIP (Arbeitsgruppe Innovative Projekte beim Ministerium für Wissenschaft und Kultur des Landes Niedersachsen) under Research Grant F.A.-Nr.2004.612.

W. Abramowicz (Ed.): BIS 2007, LNCS 4439, pp. 464–477, 2007.
© Springer-Verlag Berlin Heidelberg 2007

should be preferred instead of manually coding. In the BAMOS project (**B**ase **A**rchitecture for **MO**bile applications in **S**pontaneous networks) [SPBD05], an architecture has been designed and implemented for the flexible development of mobile applications. The BAMOS architecture can serve as a powerful base for code generation approaches. Using the BAMOS platform a mobile device can dynamically connect to a local network and use all the available services offered there. To make this approach successful the development of mobile services should be as easy as possible. In this paper we present a model-driven approach for generating nearly the complete source code of mobile BAMOS services. Furthermore, on the mobile devices no line of code has to be implemented when the BAMOS platform is used.

The paper is organized as follows: in section 2 we outline the architectural approach of the BAMOS platform which provides the destination platform for our model-driven development. Subsequently, section 3 motivates the usage of a model-driven architecture and derives a meta model for a domain specific language (DSL). An example is presented that illustrates the proposed approach. Finally, section 4 summaries the most significant features of the approach and provides some directions of future research.

# 2 Architectural Approach

An indispensable prerequisite for applying model-driven development is a powerful architectural base providing the destination platform for code generation. The BAMOS platform enables the development of mobile applications by providing two software components. The first component is an Adhoc Client that – similar to a Web browser – enables the mobile device to access information services in spontaneous networks. The second component is a Service Broker that – similar to a Web Server – serves as an interface between the Adhoc Client and the services available in the network.

With the BAMOS platform a mobile device can use different services in diverse local networks. The Adhoc Client is a generic software component that does not require any information about the specific services. It loads the declarative descriptions of the services at run-time and generates a service-specific graphical user interface. The core concept underlying this generic approach is the declarative description of the process flow as well as of the graphical user interface.

## 2.1 BAMOS Components

The BAMOS platform serves as the implementation base for the generation of mobile applications. It consists of three main components. Figure 1 illustrates the architecture and the relationship between the different architectural components.

The **Service Provider** offers services to other systems. To access these services on a mobile device some prerequisites have to be fulfilled:

- The implemented services must be accessible for remote programs. For example they may be implemented as a Web Service that can be invoked over the Internet.

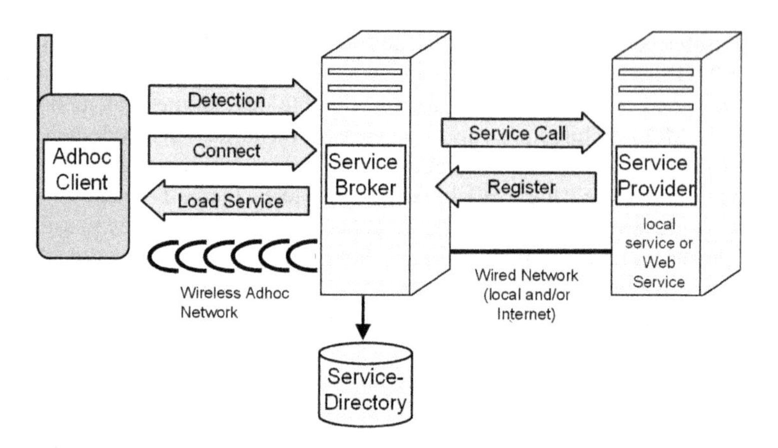


**Fig. 1.** BAMOS components

- In BAMOS all services must be described in a declarative manner to permit their usage on a mobile device. Each service description defines the mobile user interfaces and the corresponding control flow (more details are discussed in section 2.2.).
- The service providers have to register their service descriptions at the Service Broker.

The **Service Broker** mainly acts as a mediator between Service Providers and Adhoc Clients. It can be described by the following characteristics:

- It is integrated into two different networks: on the one hand in a local wireless network (e.g. Bluetooth) for connecting with the mobile devices, on the other hand in a wired network (local area network or Internet) for accessing the services provided by the Service Providers.
- It delegates the client service requests to the appropriate Service Provider and forwards the response to the Adhoc Client.
- It holds a service directory where all available services must have been registered. The directory contains the declarative descriptions of all available services. Services can be published and searched in the directory.

The **Adhoc Client** is a software component that is running on a mobile device. The device can enter and leave a local wireless network. In this case, the Adhoc Client acts as part of a wireless adhoc network. It provides the following features:

- When entering a local wireless network, it connects spontaneously to the Service Broker (using Bluetooth or WLAN).
- It loads the declarative service descriptions from the Service Broker for generating a user interface on the mobile device. Afterwards the user can enter data on the mobile device that is sent as a service request to the Service Broker. The Broker delegates the request to the Service Provider offering the requested service.

The presented BAMOS architecture allows an Adhoc Client to use different kinds of services, e.g. services that are generally available like Web Services. This architetural concept is independent of particular data transfer technologies in adhoc networks. The communication between Service Broker and Service Provider exploits common network technologies; this aspect is not further considered in this paper. The origin of the services is transparent to the Adhoc Client because the Service Broker is its only communication partner.

## 2.2 Service Descriptions

The description of a service must specify the mobile user interface of the service as well as the sequence of steps necessary in order to perform the complete service. To use a service on a mobile device normally a sequence of different screens is necessary: for selecting the desired service, for entering the input data and for presenting the output information returned by the service. The mobile user interface can be characterized by two different aspects:

(a) The layout of each screen on the mobile device.
(b) The workflow determining the sequence of screens on the mobile device.

**(a) Specification of Mobile User Interfaces by XForms**
Although the Adhoc Client is domain-independent, it should be able to interact with domain-specific services. Thus, in order to cooperate with such a service the client requires a description of the mobile user interface. This user interface description is provided by the Service Broker and can be accessed by the client. With XForms a W3C standard has been chosen as the mobile user interface description language. The main advantage of XForms is its close correlation to MIDP, the core technology used to implement graphical user interfaces on mobile clients. MIDP has been chosen as the implementation technology for the Adhoc Client.

The original intention of XForms was to build the next generation of forms in the World Wide Web. XForms is a XML-based language, issued as an open standard by W3C, with several improvements compared to traditional HTML forms [XFor06].

MIDP (Mobile Information Device Profile) is a J2ME profile suitable for the development of simple, but structured mobile user interfaces [Sun06]. MIDP user interfaces show significant similarities to traditional HTML forms: elements like input fields, radio buttons or lists offer very similar input capabilities. In addition, the possibilities for human-computer interaction are very often restricted to the submission of the entered input data – also similar to HTML forms. Thus, MIDP elements can be directly mapped to XForms and, consequently, XForms has been chosen for describing the mobile user interface.

**Structure of XForms**
One of the main concepts of XForms is the clear separation of model and view. An XForms document consists of two parts: The model part contains the data of the form, which can be displayed and altered. In the example shown in figure 2, the model contains the first name and surname of a person. The submission element holds the information about the action that shall be executed on the model data. In addition to the model part, the second part of an XForms document specifies the visual presentation of the model data as well as the possibilities for user interaction. Similar to HTML forms, XForms offers several input and output elements. Every presentation element

refers to a model element by means of a ref attribute. In the example, two input fields enable the presentation and alteration of the first name and surname of the person. The submit element specifies the presentation component for presenting the submission specified in the model, typically a button. If the submit button is pressed, all input fields are mapped to the elements in the model and the model is processed according to the action specified in the submission element. In the context of the WWW, the browser would transfer the model to a web server. In the context of BAMOS, the model is transferred to the Service Broker, e.g. via Bluetooth.

```
<?xml version="1.0" encoding="UTF-8"?>
<xf:xforms xmlns:xf="http://www.w3.org/2002/xforms">

  <xf:model>
    <xf:instance>
      <person>
        <first_name/>
        <surname/>
      </person>
    </xf:instance>
    <xf:submission id="names" action="names.jsp"
                   method="get"/>
  </xf:model>                                         } model

  <xf:input ref="first_name">
    <xf:label>First Name</xf:label>
  </xf:input>
  <xf:input ref="surname">
    <xf:label>Surname</xf:label>
  </xf:input>
  <xf:submit submission="names">
    <xf:label>Submit</xf:label>
  </xf:submit>                                        } view

</xf:xforms>
```

**Fig. 2.** XForms example

Figure 3 displays the message exchanges between the Adhoc Client on the mobile device and the Service Broker. XForms serves as the message format in every BAMOS interaction. (1) The Service Broker holds the XForms description of the mobile user interface and instantiates a specific XForms document for the requested interaction. (2) This XForms document is transferred via a wireless network to the mobile device where (3) the Adhoc Client renders the mobile user interface according to the information in the received XForms. (4) After an XForms submit the corresponding XForms model data is transferred back to the Service Broker where it is processed, usually by invoking an offered service. (5) The XForms for the subsequent interaction step is determined, an instance of it is generated and the output data of the service call is included in it. (6) This new XForms document is send to the mobile device and the interaction process can proceed.

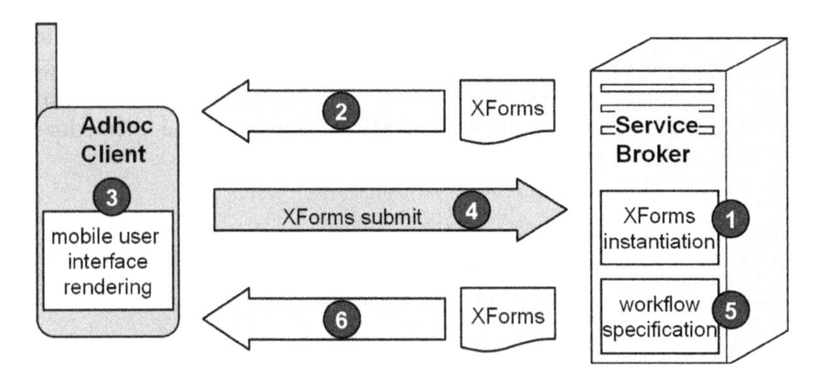


**Fig. 3.** Message exchanges between Adhoc Client and Service Broker

**(b) Workflow Specification**

In addition to the declarative description of the mobile user interface by XForms, the Service Broker has to determine the sequence of the dialogue steps necessary to execute a service, i.e. which XForms document has to be displayed next in response to a submit, and to invoke the requested service operation on the Service Provider. The core concept of the Service Broker is a process control component that can interpret the declarative service descriptions stored in the service registry.

In order to describe the process flow a simple XML-based workflow language has been designed. The concept of this language is based on the concept of a service in BAMOS and its parts. Figure 4 illustrates in detail the components of a service in BAMOS.

A BAMOS service consists of the sequence of interactions, which must all be performed in a predefined order to complete the service. A service can be a very simple one with a limited scope, e.g. querying information. Yet, it can also be a complex, composite service that is implemented by invoking other services, e.g. weather forecast service and timetable service for the public transport.

Every interaction step is implemented by a screen (a XForms template) displayed to the user on the mobile device. This template usually contains information reflecting the actual processing status of the previous interactions. The user can enter or alter data in the template and can issue a submission. Through a submission the input data is transferred as parameter to the service operation that is invoked. The output of this service operation serves as the initial input for the subsequent interaction, i.e. the next

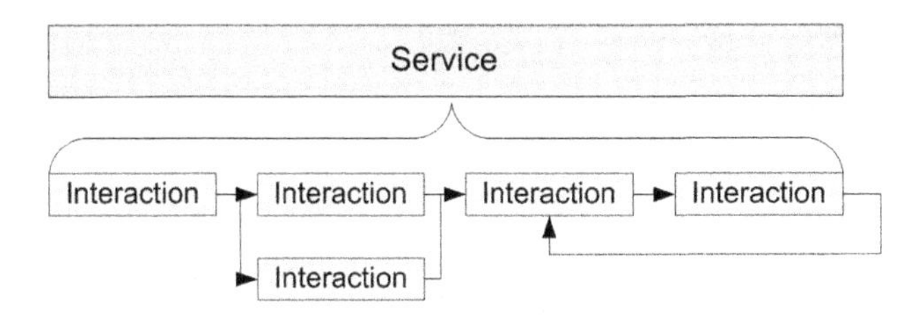


**Fig. 4.** Service composition and interactions

screen. Within the current interaction the next interaction step has to be determined. Due to the user input different conditional interactions can follow a finished one.

The presented concept of a BAMOS service is modeled by an XML workflow language. Figure 5 shows an example of the XML description of the interactions constituting a service.

```
<service id="MyService" caption="MyService">
  <class name="de.fhhannover.impl.MyService"/>

  <!-- Step1 -->
  <interaction id="Step1" start="yes">
    <xform src="Step1.xml"></xform>
    <method name="performStep1" />
    <next decisionpath="/MyService/choice">
      <case content="weatherService"   target="Step2"/>
      <case content="transportService" target="Step3"/>
    </next>
  </interaction>

  <!-- Step2 -->
  <interaction id="Step2">
    <xform src="Step2.xml"></xform>
    <method name="getStep2">
      <param id="0" path="/Step2/input"/>
      <outparam name="result" id="0"/>
    </method>
    <nextdefault target="Step3"/>
  </interaction>
  ...
</service>
```

**Fig. 5.** Service interactions specified in XML workflow language

The element service is the root element and contains several interaction elements. Every interaction element contains information about the corresponding XForms template (in the xform element) and about the service operation to be called with the mapping of the input and output attributes (specified in the element method).

The elements next and nextdefault define the subsequent interaction and determine the sequence of the interactions constituting a service.

## 3 Model-Driven Development

As discussed above, an XML-based specification of the mobile user interface and the workflow description enables the development of mobile applications in a very generic and flexible manner. In order to make a new service available in BAMOS, the descriptions of its mobile user interface as well as its workflow specification have to be defined in XML format. Only the service itself has to be implemented by the Service Provider in any programming language and wrapped by a Web Service.

To make the BAMOS concept successful the development of services should be as easy as possible. The entire effort for implementing services is already moderate but writing XForms screen descriptions and especially XML workflow descriptions for a service is too error-prone and time consuming. XML code is intended for the usage of software programs and cumbersome for humans as the short examples in the previous section illustrate. Model-driven development [Schm06], [AtKu03], [SVBH06] provides an approach to cope with these problems. It is based on the systematical use of models as primary artifacts in the software engineering process. By applying model-driven development within the BAMOS approach, a new service can be conveniently modeled with a graphical modeling tool and the graphical models are then used to generate the corresponding XML descriptions of the mobile user interface and the workflow specification.

The prerequisite for code generation is a semantically rich model. Because general-purpose modeling languages like UML [UML03] do not provide enough information for code generation a Domain Specific Language (DSL) is required that contains the necessary details for the automatic code generation. A DSL describes the elements of a certain modeling language and therefore can be considered as a meta model. Transformation engines and generators allow to analyze DSL models and to create artifacts such as other models, source code or XML deployment descriptors.

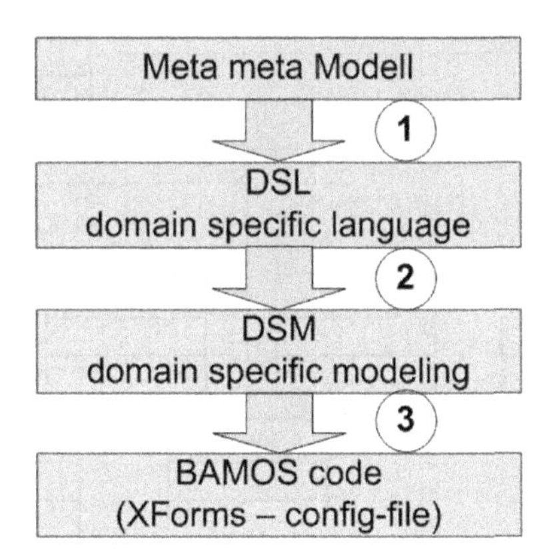


**Fig. 6.** Model-Driven Development steps

Figure 6 outlines the approach. First the DSL (or meta model) must be formally defined. This can be achieved by means of a model describing a meta model, i.e. a meta metamodel (step 1). Then the DSL can be used for domain-specific modeling (DSM); in this stage the mobile services are being modeled (step 2). Then the DSM model is finally used for generating the code; in this stage the XForms and the workflow description of the BAMOS platform are created (step 3). The four levels of description correspond to the four meta levels of the OMG Model-Driven Architecture: M0 - data, M1 - models, M2 - metamodels, M3 - meta metamodels [MOF04]. Each step is described in some more detail in the following subsections.

### 3.1 Specifying a Domain Specific Language (DSL)

General-purpose modeling languages for designing, specifying and visualizing software systems are not sufficient for code generation. They lack domain-specific model elements and concepts that specify the details required for generating the code. In our case the domain is about mobile applications based on the BAMOS framework. For example, a domain-specific model element could specify that a class attribute should appear as a choice box on a mobile device screen.

To let the code generators make use of the domain-specific model elements they must be defined in a consistent and formal way. Modeling languages can be formally defined by meta-modeling languages as Meta Object Facility (MOF) of OMG [MOF04] or Eclipse Encore [BSME03]. In a meta-modeling language the key concepts in a domain, their corresponding relationships, semantics and constraints can be precisely specified.

Domain-specific languages are mostly based on extensions of the Unified Modeling Language (UML) the de-facto standard for modeling languages. A pragmatic approach for defining a DSL are UML profiles [UML03] which use the built-in extensibility mechanisms of UML: stereotypes, tagged values and constraints. Figure 7 shows the meta model of a DSL for mobile applications based on BAMOS. The introduced new modeling elements inherit from the MOF-concepts MOF::Class and MOF::Attribute.

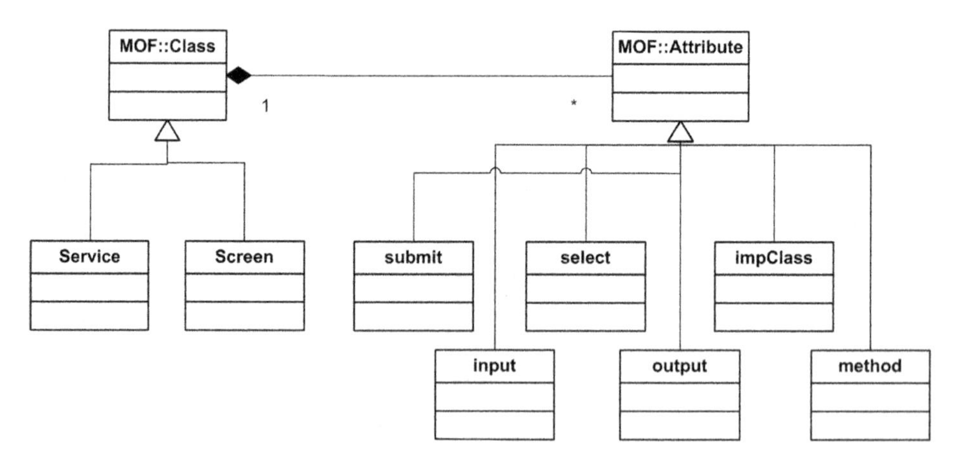


**Fig. 7.** The meta model of the BAMOS DSL (in parts)

The DSL defines two meta classes (inheriting from MOF::Class): a Service class describes mobile services and a Screen class describes screens on a mobile device. Furthermore, different types of attributes are defined for these classes (inheriting from MOF::Attribute). Special meta attributes for the Screen classes are used to define elements of the graphical mobile user interface: e.g. input defines an input field, select a radio button and submit a command button. For the Service class the implClass and the method attributes define the name of the method and the class of the Service

Broker which delegates a service invocation. The precise dependencies between the new model elements are defined by OCL constraints.

The sequence of interactions, i.e. the workflow between different mobile device screens can be specified by UML activity diagrams. Depending on the user input the Service Broker selects different interactions with corresponding XForms screens. Activity diagram guards are used for annotating transitions and specifying the appropriate interaction. An example is given in figure 8: depending on the value of attribute selectedService in screen1 the control flow will be directed to screen2 or to screen3.

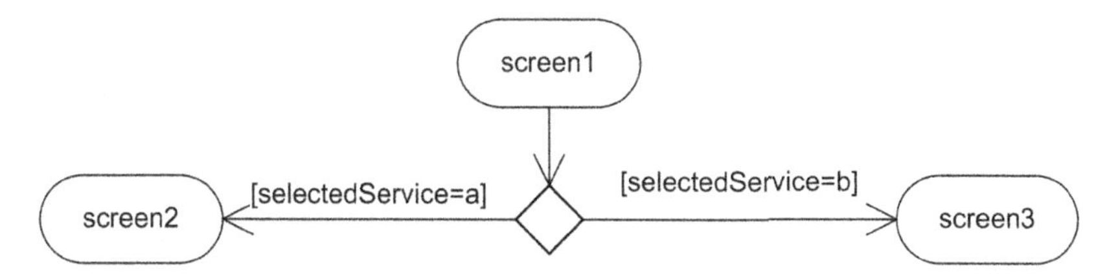


**Fig. 8.** UML activity diagrams for specifying the control flow

### 3.2 Domain-Specific Modeling (DSM)

Mobile services can be easily modeled using the DSL defined in the previous subsection. First the screens on the mobile device and the corresponding services must be specified. This can be achieved by a class diagram using stereotypes defined by the BAMOS DSL of figure 7. An example is presented in figure 9: the Service class MyService defines a service, which contains all the methods that are needed in the interactions of MyService. The Screen class weatherService defines an XForms screen,

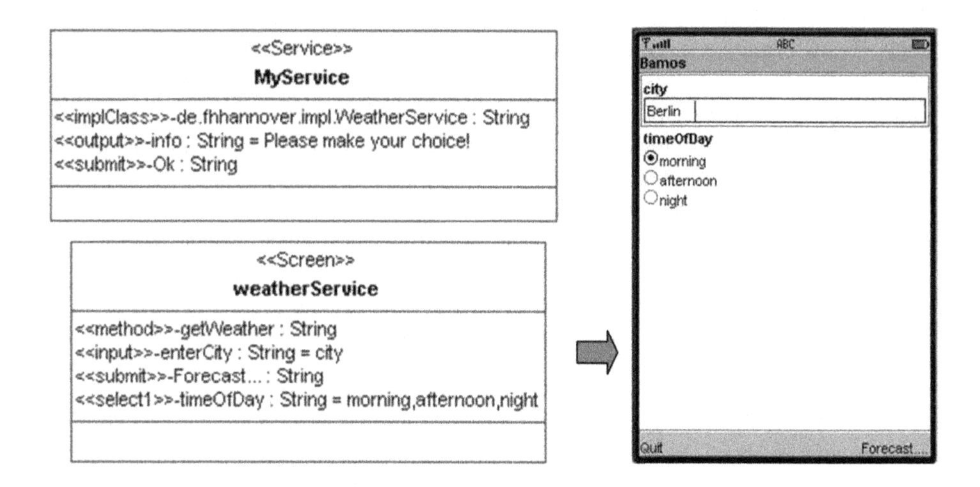


**Fig. 9.** Service and Screen class in the DSM

which is presented on a mobile phone as illustrated in Figure 9 to let the user enter a city and a time for the weather forecast. The attribute enterCity with stereotype <<input>> defines an input field to specify the city a weather forecast is requested for. The attribute timeOfDay with stereotype <<select1>> is presented as a radio button for choosing one of the initial attribute values (morning, afternoon, night).

In a second step, the control flow must be defined by an activity diagram as shown in figure 10. Depending on the choice made in the mobile user interface an appropriate sequence of screens is sent to the mobile device. The right side of figure 10 shows the corresponding initial screen to choose one of the two available services.

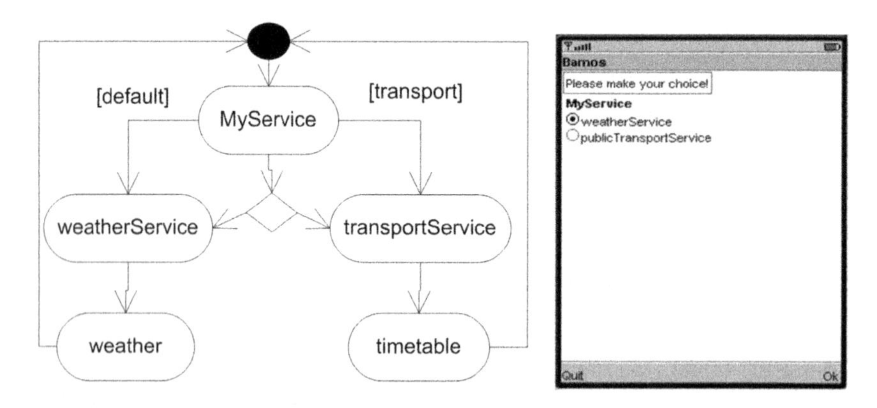


**Fig. 10.** Activity diagram specifying the control flow

### 3.3 Code Generation

In a final code generating step the DSM model must be transformed into code artifacts. Transformation engines and generators first analyze a certain DSM model and then synthesize various types of artifacts, such as source code, simulation inputs, XML deployment descriptions, or alternative model representations. To make DSM models processable by code generators the OMG standard interchange format XMI (XML Metadata Interchange) [XMI06] can be used. A XMI model representation can be imported by transformation engines. Each engine usually provides its own proprietary transformation languages for code generation, e.g. Java ServerPages, XPand. Currently the OMG is working on a standard called QVT.

In our domain the XForms and the workflow specification documents of the BAMOS platform are generated. Figure 11 shows the XForms code of the weatherService screen according to the model described in figure 9.

The corresponding part of the generated workflow description document is presented in figure 12. It contains the definition of the MyService service specifying its implementation class MyService and a corresponding XForms file MyService.xml. The next decisionpath element determines the sequence of interactions: if a user selects weatherService in the radio button choice of the MyService XForms screen, the cotrol flow is directed to the weatherService interaction, otherwise the publicTransportService interaction is chosen.

The weatherService interaction refers to the file weatherService.xml containing the XForms code of figure 11 and to the method getWeather of the Service class MyService. The two parameters of the method (id=0 and id=1) correspond to the values of the elements in the XForms screen (i.e. input fields city and time, see figure 11 and figure 9).

```
<?xml version="1.0" encoding="UTF-8"?>
<xf:xforms xmlns:xf="http://www.w3.org/2002/xforms">

<xf:model>                                                    
  <xf:instance>
    <weatherService>
      <city/>
      <timeOfDay/>
    </weatherService>                                          model
  </xf:instance>
  <xf:submission id="weatherServicesubmission"
                 action="do_bamos" method="get"/>
</xf:model>

<input ref="city">
  <xf:label>city</xf:label>
</input>
<select1 ref="timeOfDay">
  <xf:label>timeOfDay</xf:label>
  <xf:item>
    <xf:label>morning</xf:label>
    <xf:value>morning</xf:value>
  </xf:item>
  <xf:item>
    <xf:label>afternoon</xf:label>
    <xf:value>afternoon</xf:value>                            view
  </xf:item>
  <xf:item>
    <xf:label>night</xf:label>
    <xf:value>night</xf:value>
  </xf:item>
</select1>
<xf:submit submission="weatherServicesubmission">
  <xf:label>weatherForecast</xf:label>
</xf:submit>

</xf:xforms>
```

**Fig. 11.** Example of the generated XForms code for the weatherService screen

```
<!--MyService-->
<service id="MyService" caption="MyService">
  <class name="de.fhhannover.impl.MyService"/>

  <interaction id="MyService" start="yes">
    <xform src="MyService.xml"></xform>
    <nextdefault target="weatherService"/>
    <next decisionpath="/MyService/choice">
      <case content="weatherService" target="weatherService"/>
      <case content="publicTransportService"
                                target="publicTransportService"/>
    </next>
  </interaction>

  <!-- weatherService -->
  <interaction id="weatherService">
    <xform src="weatherService.xml"></xform>
    <method name="getWeather">
      <param id="0" path="/weatherService/city"/>
      <param id="1" path="/weatherService/time"/>
      <outparam name="result" id="0"/>
    </method>
    <nextdefault target="weatherResult"/>
  </interaction>

  <!--publicTransportService-->
  <interaction id="publicTransportService">
    <xform src="publicTransportService.xml">
  ...

</service>
```

**Fig. 12.** Part of the generated workflow description file

# 4 Conclusion

In this paper, we have described a model-driven approach to generate applications for mobile devices. Model-driven architecture (MDA) provides a higher level of abstrtion for developing software: it allows modeling software systems instead of programming. Only a few domain-specific functionalities remain for manual implementation. The indispensable prerequisite for an MDA approach is a powerful architectural base providing the destination platform for code generation and the development of a Domain Specific Language (DSL).

We introduced the BAMOS platform which allows the specification of complex mobile application using XML files to generate XForms screens for mobile devices.

The presented model-driven approach avoids the error-prone coding of XML files. Altogether, MDA shortens significantly the development time and improves software quality. Because UML models are the main artifacts instead of XML code, maintenance and reuse of model elements is made easier.

One drawback of the approach is that many different tools must be used and integrated. Actually, most UML modeling tools do not satisfactorily support meta modeling and the code generating tools are still proprietary and not yet stable.

## References


[AtKu03] Atkinson, C., Kuhne, T., Model-driven development: a metamodelling foundation, IEEE Software, IEEE Computer Society, vol. 20, pp. 36-41, 2003.

[BSME03] Budinsky, F., Steinberg, D., Merks, E., Ellersick, R., Grose, T. J., Eclipse Modeling Framework: A Developer's Guide: Addison Wesley, 2003.

[DeKV00] Deursen, A. V., Klint, P., Visser, J., "Domain-specific languages: An annotated bibliography," ACM SIGPLAN Notices, vol. 35, pp. 26-36, 2000.

[HaRo04] Hadig, T.,Roth, J.: Accessing Location and Proximity Information in a Decentralized Environment, International Conference on E-Business und Telecommunication Networks, Setúbal, Portugal, 2004, pp. 88-95

[Herr03 ] Herrington Jack. Code generation in action. Manning Ed. 2003.

[MDA06] OMG. MDA. http://www.omg.org/mda

[Sun06] Sun: Mobile Information Device Profile MIDP) http://java.sun.com/products/midp/

[MOF04] MOF, "Meta Object Facility 2.0 Core Specification" 2004, DocId: ptc/03-10-04.

[oAW06] openArchitectueWare: http://www.openarchitectureware.org/

[Schm06] Schmidt, D.C. (February 2006). Model-Driven Engineering. IEEE Computer 39 (2). Retrieved on 2006-05-16.

[SVBH06] Stahl, T, Völter, M., Bettin, J., Haase, A., Helsen S., Model-Driven Software Development: Technology, Engineering, Management: Wiley, 2006.

[SPBD05] Schmiedel, M., Pawlowski, O., Bruns, R., Dunkel, J., Nitze, F., Mobile Services in Adhoc Networks, in: Proc. of Net.ObjectDays 2005, Erfurt, Germany, 2005, pp. 167-178.

[UML03] UML2.0, "UML 2.0 Superstructure Specification, Final Adopted Specification, available at www.omg.org," 2003

[Wile01] Wile, D. S., Supporting the DSL Spectrum, Journal of Computing and Information Technology, vol. 9, pp. 263-287, 2001.

[XFor06] W3C, The Forms Working Group: http://www.w3.org/MarkUp/Forms/

[XMI06] OMG/XMI XML Model Interchange (XMI) 2.0. Adopted Specification. Formal/03-05-02, 2003.

# Automated Integration Tests for Mobile Applications in Java 2 Micro Edition

Dawid Weiss and Marcin Zduniak

Poznan University of Technology, Piotrowo 2, 60-965 Poznań, Poland
`dawid.weiss@cs.put.poznan.pl`, `mzduniak@j2me.pl`

**Abstract.** Applications written for mobile devices have become more and more complex, adjusting to the constantly improving computational power of hardware. With the growing application size comes the need for automated testing frameworks, particularly frameworks for automated testing of user interaction and graphical user interface. While such testing (also called *capture-replay*) has been thoroughly discussed in literature with respect to desktop applications, mobile development limits the possibilities significantly. To our best knowledge only a few solutions for creating automated tests of mobile applications exist and their functionality is very limited in general or constrained to only proprietary devices. In this paper we demonstrate preliminary results of our attempt to design and implement a framework for capturing and replaying user interaction in applications written for the Java 2 Micro Edition environment. Our evaluation test bed is a complex commercial mobile navigation system and the outcomes so far are very promising.

**Keywords:** Software Testing, Agile Development, Quality Assurance, Mobile Development.

## 1 Introduction

Software testing is a process of verifying the quality of computer programs to make sure they are doing what was expected in a consistent, error-free manner. But software testing in practice depends on *how* and *when* it takes place in the development process. We can distinguish several types of tests [1,2]. *Unit testing* concentrates on low-level pieces of software, such as classes and methods. These tests are typically a responsibility of the programmer transforming the design into implementation. *Acceptance tests* occur at the end of the development process – when the software is confronted with its initial requirements specification and expectations of target users. *Integration tests* (also called *regression tests*), which we focus on in this paper, happen in between unit and acceptance tests and cover larger blocks of the program, often the entire product. By running integration tests frequently, we ensure all the modules work together as a whole and provide results consistent with previously released versions of the software. Regression tests are thus a quality control aspect – they allow early detection of the program's suspicious behavior caused by changes (refactorings or new features) introduced to the product. This kind of constant software testing in

W. Abramowicz (Ed.): BIS 2007, LNCS 4439, pp. 478–487, 2007.
© Springer-Verlag Berlin Heidelberg 2007

anticipation of potential errors is part of most modern software development methodologies and is called the *continuous integration* principle [3].

While theoretically appealing, writing integration tests for applications with a rich graphical user interface (GUI) presents a generally complex technical problem. Since the human-computer interaction is quite unpredictable, GUI applications resist rigorous testing. A common solution is to *record* real scenarios of user interaction with the program (directly off the application's screen) and then try to reproduce the same stimuli at the testing phase, validating program's response accordingly. This kind of procedure is made possible with various GUI automation tools and programming interfaces; programs for recording GUI events are called *robots* and the technique is dubbed *capture-replay testing methodology.*

Java 2 Micro Edition environment (J2ME) lacks most of the above facilities for implementing GUI automation. All existing products (research or commercial) for testing mobile applications in J2ME are very simple and lack capture-replay testing support. This fact raises the following questions:

- In spite of technical difficulties, is it possible to devise a cross-platform architecture facilitating unit and integration testing of mobile applications? How much overhead (code, time) is required for running such a solution?
- Is there an industry need for integration tests aimed for mobile applications?
- How much time and resources can we save by implementing semi- or fully automatic integration tests in J2ME?

The first question is very technical in nature, but poses great technical difficulties because of the limited functionality available in the J2ME environment. We believe overcoming such major obstacles, although definitely with a technical in nature, qualifies as a research activity. In this paper we demonstrate an architecture that allows capture and replay of GUI events in the J2ME environment by means of dynamic code injection. This is a significant improvement over all the products available in the literature and on the market. We also estimate the overhead of this solution in terms of space and time needed for its execution at runtime.

To answer the second question we present some preliminary results and feedback from the evaluation of our proposal in a leading commercial company developing mobile navigation systems in Java.

As for the last question, there seems to be no direct answer to how using regression tests translates into economic value. While we could try a controlled user-study to assess the time or effort savings gained from using regression tests, this kind of experiment is always subjective and lacks the real-life constraints of a commercial company's environment. This problem is actually omnipresent with respect to software testing in general – common sense suggests tests provide certain measurable value, but hard estimation of this value is very difficult.

## 2 Related Works

We can distinguish two different types of related works: research about GUI testing principles (theory) and programs allowing automated GUI tests in practice.

The former topic has been broadly covered in research literature [1,2,4] and due to space limits we omit an in-depth background here in favor of surveying testing tools available at the moment. We reviewed the existing products (commercial and open source) that somehow tackle the problem of testing mobile applications in order to see to what extent they allow automated integration tests.

An open source project J2MEUNIT [A] can run simple unit tests. It does not allow testing application as a whole and the test cases are hard to maintain. It is also not possible to integrate J2MEUNIT into an automatic build process because results of performed tests must be verified by the programmer (which excludes its use for integration testing). Sony Ericsson's MOBILE JUNIT [B] is a more advanced framework, allowing unit testing on the device and collecting code coverage statistics automatically while running tests. MOBILE JUNIT is bound exclusively to the Microsoft Windows operating systems and on-device testing is limited to Sony Ericsson's telephones. Moreover, the tool's configuration and launching is quite complex and involves a pipeline of different tools which cannot be separated. Recently a few other toolsets similar to Sony's emerged: MOTOROLA GATLING [E] and CLDCUNIT [F] for example. The functionality they offer is close to that of Sony's.

So far we have only mentioned unit testing frameworks. One solution going beyond that point, towards GUI testing, is IBM's Rational Test RT, in short TESTRT [C]. TESTRT is a commercial package with a custom implementation of unit tests. The program allows GUI testing, but only on so-called *emulators* (software substitutes of real devices), not on the devices themselves. The simulation script knows nothing about the emulator or about the mobile environment – it merely replays the operating system's events such as keyboard actions or mouse clicks at certain positions over the emulator window. This implies that the product is testing a software emulation of a real device rather than the program running on that device. Unfortunately, TEST RT also lacks an automated test verification mechanism, the programmer is responsible for checking whether the replayed test passed or not.

A more sophisticated testing solution comes from Research In Motion and is bundled with development tools for this company's flagship device BlackBerry. The software emulator of a BlackBerry device (called FLEDGE [D]) is equipped with a controller tool that can interpret predefined event scripts. These scripts can contain events such as: starting and pausing the application, changing the readouts of GPS location API for devices supporting GPS positioning, generating keypad and other input device events, generating various phone events such as remote phone calls or changing battery level. BlackBerry's controller has several limitations: it runs only with the simulator, not with real devices, it lacks an automated test verification mechanism (assertions) and, most of all, the developers are unable to record test scenarios – all scripts must be written by hand prior to testing.

The conclusion from the list above is that in spite of the evolving theory of GUI testing, practical implementations for testing mobile applications remain within the domain of the simplest unit and limited GUI tests.

# 3 Writing and Testing Applications in Java 2 Micro Edition

Java applications written for mobile devices (mobile phones in vast majority) are simpler and smaller compared to their desktop cousins. The environment provides a simple virtual machine (JVM) for executing the program's code and a set of generic application programming interfaces (API) for accessing hardware layer – the device's display, network or communication ports.

Both programming and particularly testing are much more difficult in such a constrained environment compared to writing programs for the desktop. Each mobile phone, for example, has a different hardware configuration: display size and capabilities (number of colors), size of memory and varying computational power. Application interfaces defined by the J2ME specification and considered a 'standard', are implemented by different vendors and often contain differences that must be taken into account, increasing the complexity of the program. The same application looks, but often also *behaves* a bit different depending on the target device it was installed on. We summarized these key differences between mobile and traditional software development in Table 1.

**Table 1.** Differences between development and testing of mobile and traditional (desktop and server) applications

| Element | Mobile | Traditional |
|---|---|---|
| Test recording | Lack of programmatic access to recording GUI events. Emulation of user interaction impossible. | Standard `java.awt.Robot` class for recording GUI events. |
| Deployment automation | Tedious (manual) routine of on-device deployment and testing. | Deployment usually fully automatic. Testing and harvesting test results automatic and relatively easy. |
| Test environment differences | Differences across devices (different virtual machines, varying memory and resource availability). Requirement to run tests on all possible configurations. | Virtually identical development and deployment/ testing environment. In very rare cases operating-system specific. |
| Programming interfaces | A number of non-standard APIs and proprietary solutions (playing sounds, access to external ports, access to the current display). | More mature and standard APIs, portable across JVMs from different vendors. |

Because of the differences in hardware and software, software for mobile devices should be tested on each individual piece of equipment separately. Knowing that the deployment process takes some time, testing quickly becomes a tedious routine software developers grow to hate in no time. Writing a *capture-once, replay-on-all* testing framework seems like a natural answer addressing the problem, even if the experiences with this type of tests in desktop applications are not always rosy (contrary to the desktop, mobile applications are much simpler, so test scenarios should retain manageable size). Unfortunately, the J2ME environment does not offer any system-level support with respect to handling GUI events and any other events for that matter. In the following section we show

how to substitute this required and missing functionality with automatic preprocessing of the binary code of the tested program (a process generally known as *bytecode-level instrumentation* or *code injection*).

## 4 Automating Tests with Code Injection

We divided the problem into fairly independent goals. The first goal was to design a mechanism that would allow us to intercept and record the events resulting from user's interaction with the program (running on an emulator or a real device). This is called the *recording phase.* The second goal was to programmatically simulate the previously recorded events (user actions) – this is called the *replay phase.* Finally, we compare the initial recording with stimuli resulting from the replayed events; certain *assertions* are checked to ensure the program followed identical sequence of state transitions (this implies a correct outcome of the entire test). Note that states can be fairly low-level, such as action selection, but also high-level, perhaps even explicitly hardcoded in the program by the developer. We took extra care to facilitate future maintenance of the recorded scripts. Unlike with desktop applications, where an event is typically described by a mouse position or some obscure component identifier, our events are described with identifiers meaningful to the programmer (an action's label for example). Our goal is to make the recorded script comprehensible and comparable to a typical (unstructured) use-case scenario used in requirements engineering.

### 4.1 Recording Phase

Java 2 Micro Edition does not expose any standard system hooks for intercepting GUI events. To overcome this problem we instruct (dynamically rewrite) the tested program's bytecode before it is deployed, injecting our custom proxies anywhere on the border between the program and the J2ME environment. We identified several such *injection points*, trying to capture events related to application lifecycle, changes made to the active display and alternations of form fields. For intercepting the injection points we first considered aspect-oriented programming but this turned out very hard due to their different placement and handling. The details of how major injection points have been implemented are given in subsections below.

– *MIDlet Lifecycle.* A mobile application in J2ME must have at least one class that extends `javax.microedition.midlet.MIDlet` class.[1] The midlet is an entry point to the application and receives events connected to its lifecycle (start, pause, destroy and resume). We intercept these events by locating classes extending the `MIDlet` class, adding our own event methods in place of existing event handlers and moving the code from original implementations to private methods in the same class. A simple example of this operation is shown in Figure 1 on the facing page.

[1] In the remaining part of this paper we will omit package names for brevity.

```
public final class MyMidlet extends MIDlet {
    protected void startApp()
        throws MIDletStateChangeException {
        // original code
    }
...
```

```
public final class MyMidlet extends MIDlet {
    protected void startApp()
        throws MIDletStateChangeException {
        // record: before-start-event
        try {
            this.orig$startApp();
            // record: after-start-event
        } catch (Throwable t) {
            // record: start-exception-event
        }
    }

    private void orig$startApp()
        throws MIDletStateChangeException {
        // original code
    }
...
```

**Fig. 1.** Rewriting `MIDlet` classes to intercept lifecycle events. Original code on the left, modified (instructed) code on the right. We denote event recording blocks with comments for brevity.

– *Display changes.* A mobile application changes the display by setting a selected subclass of the `Displayable` class on the `Display`. From the point of view of a testing framework, switching one screen to another is a change of state. This event is useful because a sequence of display changes should typically be identical during the replay phase and can be considered an assertion. We intercept every change of the active display by locating (and generating an event upon) all invocations of `setCurrent(Displayable d)` method on the `Display` class.

– *Intercepting command actions.* Commands (instances of `Command` class) are issued when the user selects an option on an active display, which is a subclass of the `Screen` class (any type of screen with selectable buttons). Every command must be registered with the screen before it is visible. When the user presses a button on the mobile device, a listener (instance of `CommandListener` interface) is informed about such an event. To intercept all command events we must locate all implementations of the `CommandListener` interface and wrap the event-receiving method `commandAction(Command, Displayable)`. Commands do not carry any special identifier (which could be recorded for use in assertions at replay time), so we decided to use their visual representation (labels) instead. It is vital to separate and identify commands uniquely because in the replay phase the framework must know exactly which command to simulate.

An additional problem to solve in the recording phase was related to storage of the recorded events. Saving all events directly on the device was inconvenient because we could collide with the application's data or exceed the device's limited capacity (memory or persistent storage). Eventually we decided to transmit all events directly over the wire during the simulation and have an option to save them locally in case network protocol is unavailable on the device.

After the recording phase is over, some additional work can be done by a person responsible for recording the tests to make the recorded script more robust. A raw scenario recorded off the device is typically too verbose and could lead

```
<scenario>
  <event timestamp="1000">
      <displayable-changed title="Hello screen" type="TEXTBOX" />
  </event>

  <event timestamp="2000">
     <command cmdLabel="Start app" displayableTitle="Hello screen" />
  </event>

  <event timestamp="3000">
     <textbox-modification assertion="true" strongAssertion="true" string="I like testing" />
  </event>
</scenario>
```

**Fig. 2.** A sample fragment of a test script written in the XML language

**Fig. 3.** Screenshots from test recording session. Server console (left) and emulator window (right).

to maintenance problems, failing in response to the smallest change in the user interface (see [5] for example). The person designing the test case should review the recorded scenario and add or remove assertions or events as appropriate (as we already mentioned, the goal was to make test scripts as comprehensible to a human as possible). The script is originally in a compact, binary format to save network bandwidth and storage on the device. To modify the script we translate it to an XML file and, after changes, compile it back to the binary format. An example of a test script is shown in Figure 2. A screenshot from test recording session (server console, emulator window) is shown in Figure 3.

## 4.2 Replay Phase

In the replay phase our 'robot' class wraps the original midlet and manages its entire lifecycle. The robot continuously reads the binary test script stimulating corresponding events or checking for assertion violations.

There are two types of assertions: strong and weak ones. A failure report is made when any type of assertion fails, but with strong assertions the simulation is terminated and with weak ones the program is continued and the robot tries to execute the remaining part of the test script. By default all assertions are weak, the test designer may alter their type manually.

Hardware events recorded in the script are generated by the robot back to the tested application in different ways, depending on their original injection point.

- *MIDlet Lifecycle.* Midlet lifecycle events are handled much like during recording. The only difference is that the injected code is instructed to stimulate events (call corresponding application methods) rather then capture them. One important method is the midlet's startup call (constructor). The testing framework is interested in intercepting this special call because it is a sign that the replay phase should begin and the framework should start stimulating events for the application under test.
- *Mapping commands to test command identifiers.* During the recording phase every event about a command the user invoked was recorded. At replay, we simulate the same commands by invoking the current display listener's `commandAction` method. To do so, we must associate previously recorded commands (or rather their labels) with real objects created during the current test execution (instances of `Command` class). We intercept every instance of a command by locating invocations of the `addCommand(Command)` method on subclasses of the `Displayable` class. Once the mapping between the command objects and their labels is known, generation of corresponding test events only requires the knowledge of a listener where events should be proxied.
- *Intercepting command listener registration.* Listeners receiving commands from the application screens register directly with subclasses of the `Displayable` class. We intercept these registrations by locating all invocations of the `setCommand-Listener(CommandListener)` method. Once we know which listener registered on the current display, simulation of (previously mapped) command events becomes trivial.
- *Display changes.* This type of event is tracked during replay (to allow state-change tracking) in an identical way as in the recording phase.

Putting the described code injection procedures together, the testing framework is able to fully reproduce the original behavior recorded in the test script. The framework performs the simulation by spawning a background thread that continuously reads events and assertions from the test script and invokes corresponding event-generation routines, at the same time tracking objects to which the events should be delegated. While it may seem a bit complex at first glance, the replay phase is actually quite simple and efficient at runtime.

## 5 Preliminary Results

At the moment of writing, the test framework introduces an overhead of about 30 kilobytes of bytecode (unobfuscated bytecode). Our estimate is that the overhead will reach about 50 kilobytes in the final version of the framework. Comparing this figure to storage constraints of present mobile phones (handling up to a few megabytes) this seems not to be an issue. Runtime memory consumption increased only about 30 kilobytes (roughly identical to the size of the code), so it should not be an issue.

The testing framework has different performance overhead depending if it is in the capture or in replay mode. In the capture mode the overhead is mostly bound to network traffic (sending events over the wire), which can be easily neglected by using some engineering tricks (asynchronous queue of events waiting to be sent). In the replay mode the overhead is connected to the background thread reading and stimulating events. We found this overhead negligible as well.

The framework has been put to use at NaviExpert (www.naviexpert.pl), a Polish company offering complex navigation software for mobile phones, written entirely in Java. The initial feedback was very positive and we plan to collect some usage statistics to determine the value gained from using regression tests in production use.

We should emphasize that this paper reports on preliminary results from an initial implementation of the presented concepts. The prototype is fully functional with respect to a large slice of the J2ME specification, but does not cover all the possibilities (for example, `Canvas` class events are a matter of future work).

## 6 Summary and Future Directions

We have presented a proof of concept demonstrating that fully-fledged capture-replay testing framework is feasible in Java 2 Micro Edition environment. The prototype implementation has been well received and deployed in a commercial software house.

Our biggest challenge at the moment is to provide some objective means to assess the value gained from using the framework. What common sense states as obvious is quite difficult to express with hard numbers. We considered a controlled user experiment where participants would be given the same application and a set of tasks to implement (refactorings and new features). Half of the group would have access to the results of integration tests, the other half would just work with the code. We hoped this could demonstrate certain gains (number of early detected bugs, for example) that eventually translate into economic value for a company. Unfortunately, this kind of experiment is quite difficult to perform and its results are always disputable (i.e., due to ranging skills between programmers), so we decided to temporarily postpone it. Other possible research and technical directions are:

– Design a flexible architecture adding support for events that are outside the scope of the J2ME specification, but are commonly used in mobile development. These events include, for example, vendor-specific APIs such as vibration or backlight by Nokia.
– Implement alternative event serialization channels – through serial cables or Bluetooth connections.
– Consider evaluation schemes for the presented solution. A real feedback from developers translates into a proof of utilitarian value of the concept – does the testing framework help? How much time/ work does it save? What is the ratio of time spent on recording/ correcting test scripts compared to running them manually? We should emphasize that these questions are just as important as they are difficult to answer in a real production environment.

- Integrate the framework with popular integrated development environments. This goal is very important because developers must be comfortable with the tool to use it and must feel the benefits it provides. Instant hands-on testing toolkit would certainly assimilate faster in the community than an obscure tool (such as Sony's).
- We also think about extending the concepts presented in this paper to other Java-based platforms for building mobile applications, such as NTT DoCoMo Java, BlackBerry RIM API or Qualcomm Brew. They may not be as popular as J2ME, but the concepts we have presented should be applicable in their case as well.

## References


1. Kaner, C., Falk, J.L., Nguyen, H.Q.: Testing Computer Software, Second Edition. John Wiley & Sons, Inc., New York, NY, USA (1999)
2. Jeff Tian: Software Quality Engineering. Testing, Quality Assurance, and Quantifiable Improvement. John Wiley & Sons, Inc. (2005)
3. Fowler, M., Foemmel, M.: Continuous integration. Available on-line (2007): http://martinfowler.com/articles/continuousIntegration.html
4. William E. Perry: Effective Methods of Software Testing (second edition). John Wiley & Sons, Inc. (1999)
5. Mark Fewster and Dorothy Graham: Software Test Automation. Addison-Wesley (1999)


## Projects Surveyed

A. J2MEUnit, http://j2meunit.sourceforge.net.
B. Mobile JUnit, Sony Ericsson, http://developer.sonyericsson.com/getDocument.do?docId=87520.
C. TestRT (Rational Test RealTime), IBM, http://www.ibm.com/software/awdtools/test/realtime/.
D. Fledge (Java Development Environment and Fledge simulator), Research in Motion, http://na.blackberry.com/eng/developers/downloads/jde.jsp.
E. Gatling, Motorola, https://opensource.motorola.com/sf/sfmain/do/viewProject/projects.gatling.
F. CLDCUnit, Pyx4me, http://pyx4me.com/snapshot/pyx4me/pyx4me-cldcunit/.

# Pitfalls of Agent System Development on the Basis of a Travel Support System

Maciej Gawinecki[1], Mateusz Kruszyk[2,3],
Marcin Paprzycki[1], and Maria Ganzha[1]

[1] Polish Academy of Sciences, Systems Research Institute,
Warsaw, Poland
{Maciej.Gawinecki,Marcin.Paprzycki,Maria.Ganzha}@ibspan.waw.pl
[2] Adam Mickiewicz University, Department of Mathematics and Computer Sciences,
Poznań, Poland
[3] *Content Forces*, Content Management Services provider,
Amsterdam, Netherlands and Poznań, Poland
www.contentforces.pl

**Abstract.** Belief that a particular software engineering paradigm is universal for all domains is an illusion and agent-oriented engineering is not an exception. This we have learned during the development of an agent-based Travel Support System. The system was developed as a distributed environment to provide user with personalized content helping in travel planning. In this article we focus on these issues of our systems, where agents fit and give practical alternatives, where they do not. We believe that lessons learned in our project generalize to other project involving utilization of agent technology.

**Keywords:** multi-agent system, development methodology, content management, personalization.

## 1 Motivation

Nowadays a software architect, challenged to develop an application solving certain problems does not have to start building it from the scratch. Being a supporter of *re-use-what-available* philosophy, she can rather select relevant software development paradigm and existing off-the-shelf technologies. Obviously, each existing paradigm provides different abstraction for conceptualizing a given problem. The role of an architect is to know limitations and possibilities of different abstractions and choose the most intuitive and efficient one(s).[1] Therefore, believing the a particular paradigm is universal for all domains is an illusion and agent-oriented engineering is not an exception [1]. The same way as in all other software engineering tasks, a number of factors must be considered when selecting an agent-based approach [2,3]. Let us list some of the more prominent ones.

[1] In business practice the choice of right approaches is of course much more complicated and depends not only on software requirements, but also on various costs of using specific technology, skills of available programmers, predicted long-term support for existing technologies, etc.

W. Abramowicz (Ed.): BIS 2007, LNCS 4439, pp. 488–499, 2007.
© Springer-Verlag Berlin Heidelberg 2007

*An environment that is distributed, highly dynamic, uncertain and complex.* Such an environment requires distribution of data, control or expertise and these objectives can be naturally supported by agents. For example consider production system in a factory; where points of control behave in both autonomous and cooperative way, and can adapt to local environment changes in order to realize a global goal [4]. Moreover, in a distributed environment access to remote resources can be improved by providing a light agent with mobility: the agent representing certain point of control can move to the target location where data necessary for computation is stored—instead of transferring large quantities of data over the network (as is the case in a traditional remote procedure call).

*Agents as a natural metaphor.* Organizations and societies consisting of cooperative or competitive entities can be naturally modeled by agent teams. Agent-oriented engineering allows classical methods of building complex systems (*decomposition*, *abstraction* and *organization*, as in object-oriented paradigm [5]) to be applied in distributed dynamical environments [6].

*Dealing with legacy systems.* Genesereth and Ketchpel suggests using agents as wrappers for legacy software, which in such a way can be reused by other components in heterogeneous system [7].

At the same time it is important to acknowledge that agent paradigm is relatively novel and may fail in cases in which traditional approaches (client-server architecture, object-oriented paradigm etc.) and technologies (Web Services, Java RMI, Content Managements Systems etc.) have taken their deserved place, confirmed by business practice. This is also the lesson that we have learned, building our Travel Support System and thus we would like to share our experiences in this article. This knowledge may also be helpful, in the case when someone may naively may claim that agents are a "silver bullet" for software development, while these arguments is still largely untested in practice [8].

In the next section we briefly summarize the main design characteristics of our agent-based Travel Support System. We follow with a description of major problems that we have run into. We complete the paper by description of proposed solutions to these problems.

## 2 Background

Travel Support System (TSS) is an academic project aiming, among others, at convincing agent-idea skeptics that building an agent-based system for planning a travel is nowadays both reasonable and possible with use of on-the-shelf technologies [9]. Our work was inspired by the following scenario. *Hungry foreign tourist arrives to an unknown city and seeks a nice restaurant serving cuisine that she likes. Internet, contacted for advice about restaurants in the neighborhood, recommends mainly establishments serving steaks, not knowing that the tourist is a fanatic vegetarian.* This scenario determines the following functionalities of the system:

- *Content delivery.* Content should be delivered to the user in browser-processable form, i.e. HTML, WML etc. and match the user query.
- *Content personalization.* Delivered content should be personalized according to the user-model to avoid situations like the one presented in the scenario.
- *Adaptation of personalization.* Habits of the user can change, therefore her model should be adapted on the basis of her activities recorded by the system.

In fact, these functionalities are realized only by a part of the Travel Support System, called *Content Delivery Subsystem.* In what follows we focus our attention only on this particular subsystem (hereafter called system). The remaining parts of the TSS, responsible for data management and collection have been depicted on figure 1 and described in detail in [9]. This latter reference (and references collected there to our earlier work) should be consulted for all remaining details concerning the TSS. As far as the technologies utilized in the TSS, the RDF language has been applied to demarcate data (to allow machines process semantically rich data and meet requirements of Semantic Web applications [10,11]). Jena framework has been used to manage RDF graphs [12] (RDF graphs are persisted as Jena models in traditional relational databases). When conceptualizing the system, the Model-View-Controller design pattern [13] has been applied for clear separation between pure data (model) and its visual representation (view).

Let us now list the most important agents that have been designed and implemented in our system

- *Proxy Agent (PrA)* integrates non-agent user environment with the agent-based system (precisely described in [14]). It is able to receive HTTP requests from a user browser (since it wraps a simple "home-made" HTTP server), and forward them to the system and return an answer from the system in the form of an HTTP response.
- *Session Handling Agent (SHA)* is responsible for realizing user requests. It plays the role of controller in the MVC pattern. Specifically, it (1) receives user request from the *PrA*, (2) creates model responding the request

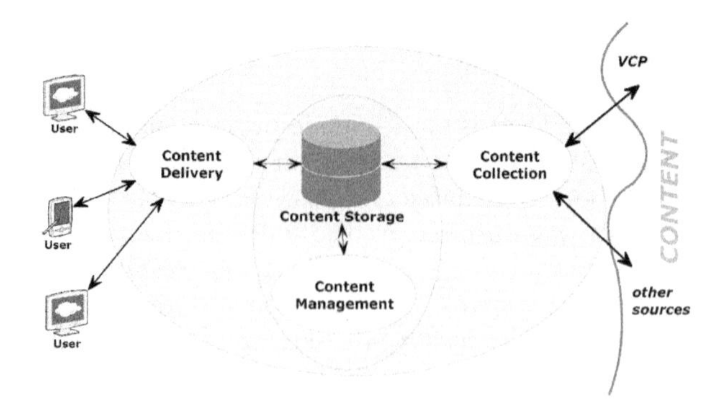


**Fig. 1.** Travel Support System general architecture

or delegates the *PA* to do it, (3) requests the *VTA* to transform the model into the browser-readable view, and (4) passes the response to the *PrA*. Additional responsibility of the *SHA* is to track user feedback and log it in the *History* database.
- *Profile Managing Agent (PMA)* is responsible for initializing and learning user profile on the basis of user feedback (see [15], for more details about learning algorithm used in the system). It provides a user profile to the *PA*.
- *View Transforming Agent (VTA)* is a response to the need of providing content to various user devices, which can render documents described in markup language (e.g. HTML) as well as simple TXT messaging. The *VTA* generates a view in terms of a HTML/WML/TXT document matching a given model. It wraps and utilizes Python-based Raccoon server, which applies pointed XSL stylesheet to a given XML document [16].
- *Personal Agent (PA)* acts on behalf of its user, personalizing recommendations restaurants with respect to the user profile (see [15], for more details about filtering algorithm used in the system). It is created only for a duration of the session, for a user who is logged in. Notice that user can log-in and log-out during a session and while user is logged out the *PA* can orchestrate work that is preparing a response set. When the response set is delivered and user logs-out, the *PA* is "killed."
- *Restaurant Service Agent (RSA)* Wraps Jena model with data of Polish restaurants.

## 3 Problems Encountered During the Development of the System

In this section we present four problems, we met during agentifing our system. We tried to present them as general issues with aids, so other developers could utilize our experience in their work.

### 3.1 Agents as Wrappers for Legacy Software

Utilization of agent as wrappers for legacy software, was proposed as aid for systems with heterogeneous software [7]. This is representation of a general move toward message-oriented communication, which increases the interoperability, portability, and flexibility of a distributed applications [17]. However, we found such utilization of agents justified only in one of the following situation: (a) where there is no other middleware solution, which would connect heterogeneous parts of an application, or (b) usage of agents brings additional functionality to wrapped software (as e.g. as Observer or Adaptor design pattern).

Let us illustrate this situation by an example from our system. JADE agents use two semantically rich languages: SL and ACL [18]. For example, in the following message the *RSA* agent informs the *PA* about requested restaurants:

```
(request
  :sender ...
  :receiver ...
  :ontology tss-ontology
  :language fipa-sl
  :content ''
        (result
           (action (find-restaurants :query .... ) )
           (''<?xml version=''1.0''?>
      <rdf:RDF>
         <res:Restaurant rdf:ID=''Poland_LD_Lodz_
             _Kuchnia_Polska_Obiady_domowe996614020''>
          <loc:streetAddress> ul. Lutomierska 8
          </loc:streetAddress>
        </res:Restaurant>
        ...
      </rdf:RDF>")
    )
  ''
)
```

It can be seen that the message contains also RDF/XML serialized data describing restaurants. The process of creating an ACL message by the *RSA* and reading it by the *PA* has been depicted in Figure 2. RDF data describing restaurants and persisted in the Jena model must be serialized to RDF/XML. The rest of the message content is constructed with use of Java beans representing certain concepts in communication ontology and then encoded in Lisp-like strings. All resulting message-parts are combined into a single ACL message and communicated to the *PA* with uses the RMI technology (standard technology that JADE uses to transfer messages). The *PA* reads the content of the message in exactly the reverse way. This process is definitely time- and resource-consuming and it is not justified in our system, where agents do not take advantage of SL's features. The SL language was developed to provide agents with ability of communicating their beliefs, desires and uncertain beliefs; this takes place particularly in the case of, so called, BDI agents. However, our agents are not BDI agents and do not utilize semantically rich communication. Moreover, as it was described above,

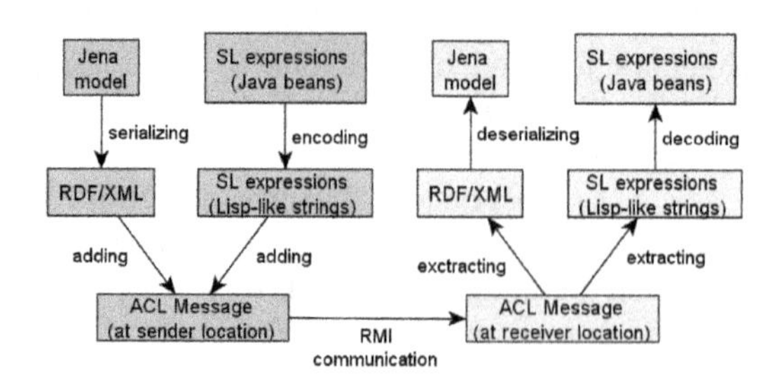


**Fig. 2.** Communication costs in Travel Support System

remote Jena models persisted in the database can be reached with use of simple database connection, without time-consuming serialization of Jena models and putting them inside of ACL messages. In this context, one should also remember about additional effort of a developer, who needs to design communication ontology (in the SL language).

An alternative could be (1) to use of simple database connections, in cases where data sources where interfaced by agent wrappers, (2) introducing traditional technologies, such as Java RMI, for requesting remote services (such as view transformation), or (3) if possible, integrating interacting agents within a single host. Summarizing, interoperability among parts of an application is simply warranted by Java-based interfaces of an application.

### 3.2 Replacing Traditional Technologies with Agents

*You see agents everywhere.* Many young developers narrowly follows the vision of Nwana and Ndumu to *agentify* all software functionalities [19]. However, this is very common mistake to design whole system in an agent-oriented architecture, while most of the work can be done with use of traditional approaches and technologies [1]. We have made this mistake for a purpose. The main **objective** of the original design was to utilize agents in all possible functions. Let us now look into some more details as to what we have found.

The main scenario of the system is content delivery, which is realized in a client-server architecture, where the system plays a passive role of server. This client-server architecture has been naturally transformed into the *FIPA Request Interaction Protocol* [18], in which the *Initiator* plays the client role, and the *Responder* plays the server role. Specifically, the *SHA* (Initiator) requests that the *VTA* (Responder) generates a view from the model. Separate functionality has been created as the *PrA*, which wraps the HTTP server. All these agents plays vital roles in our adaptation of the Model-View-Controller pattern in agent-like environment.

Summarizing, in general the MVC pattern utilizing the HTTP protocol can be characterized as:

- *stateless*—each user request is independent to others, so the results of response to a user request have no influence on results of another one, by an analogy to the HTTP;[2]
- *reactive*—MVC components stay inactive between user requests, so they react only to external requests, simply like *active objects* [20];
- *synchronous*— as process of realizing a single user request is a sequence of steps, where each next step cannot be realized until the previous one has been finished: receiving HTTP request, preparing model, preparing view and returning HTTP response;
- *parallel*, but not *concurrent*—parallelism is utilized to decrease interleaving in I/O operations.

[2] With an exception to a term session, which—however—has been successfully handled by traditional CMS frameworks.

Therefore, in this case, the well known properties of agents defined as *proactiveness*,[3] *asynchronous* communication, *statefulness* and *concurrency*[4] cannot be utilized.

Previously, the MVC pattern has been successfully incorporated and tested in business practice by use of traditional technologies, such as the Spring Framework [21]. In our case utilization of agents for this pattern resulted in the following disadvantages of the system:

- difficulty of integration of the proposed solution with traditional Content Management Systems, due to use of niche technologies (Raccoon, agents); this seems more reasonable in the situation where content presented to a user is composed also from fragments not delivered by agents.
- forcing a potential developer to learn designing web content from scratch,
- introduction of the solution that can be less stable and efficient that thoroughly tested solutions, for example (1) not properly managed concurrency in the HTTP server can be a bottleneck in the system and (2) above mentioned communication overhead can slow down data flow in the system.

Overall, upon reflection we can say that we have modeled a part of the system with higher abstraction than naturally necessary, which resulted in difficulties of verifying and reasoning about such solution (i.e. the simpler the model the easier it is to think about it, to verify its correctness and to remove errors).

### 3.3 Solving Conflicts in Functionalities Acquaintance

Creating too many agents, each one realizing separate functionality, is yet another pitfall in agent-oriented development [1]. The main problem is a potential communication overload caused by exchange of messages between separated functionalities [22]. Guidelines for solving such problems can be found in the Prometheus methodology, which proposes to analyze, in the given order, the following factors, while specifying particular agents: (1) data and knowledge acquaintance, (2) relationships among agents and (3) interactions frequency. Functionalities which use the same data or interacts with each other often are suggested to be integrated into a single agent [23]. However this approach often conflicts with a situation, where related resources, data etc. must be located on separate hosts (e.g. due to performance issue or because they belong to different owners). However, in this case communication can be limited by introducing mobile agents. Let us see this approach on the example.

At first, we applied the Prometheus methodology to the Travel Support System; the results can be seen in Figure 3. Red lines demarcate access of agents to particular databases and ontologies (directions of arrows point read and/or write access). Black lines describe dependency of a pointed agent on a given functionality. For instance, *PMA* reads data from the *Stereotypes* database, and both writes

[3] Agents can react also changes of their internal state (being usually results of reasoning), what can be perceived as proactive behaviour.

[4] Programming technique, in which two or more, often cooperative, threads are making progress, often used to model real world entities.

to and reads data from *Profiles* and *Statistics* databases. Moreover, the learning process requires access to (1) *History* database (and thus communicating with the *SHA* to obtain the learning data) and *Restaurants* database (wrapped by the *RSA*). Let us consider this example further. *Content-based learning* adapted in the system requires great amount of information: both description of recommended objects (*Restaurants*) and history of user feedback about those objects (*History*). According to the integration rule we should have integrated all related functionalities of the *PMA*, the *SHA* and the *RSA* agents together with access to the necessary data sources in a single agent. But we have made an exception to this rule for the following reasons: (1) restaurant data could be provided by the company not belonging to the system and thus external to it, (2) the *PMA* and the *SHA* should be hosted on separate machines, because learning process requires large amount of CPU resources, while the *SHA* is obligated to respond to many users' requests in a timely fashion. An alternative would be to designate an agent that learns single user profile and moves to the host where appropriate data for a particular learning phase is stored. Mobility of a such an agent has two advantages: (1) *performance boost*—it decreases communication overload while accessing remote data, and (2) *design metaphor*—it provides the developer with possibility to realize certain computations from a single point of control which can move itself, while releasing her from necessity of passing control along various remote hosts (as in standard *remote procedure calling*). Example of such a solution can be seen in Figure 4, where Mateusz's *PA* travels across remote hosts.

### 3.4 Placing Personal Agent in Real Environment

Generally speaking, automated personal assistants are one of the ways that agents are viewed [24]. This perspective says, that each user is represented in a

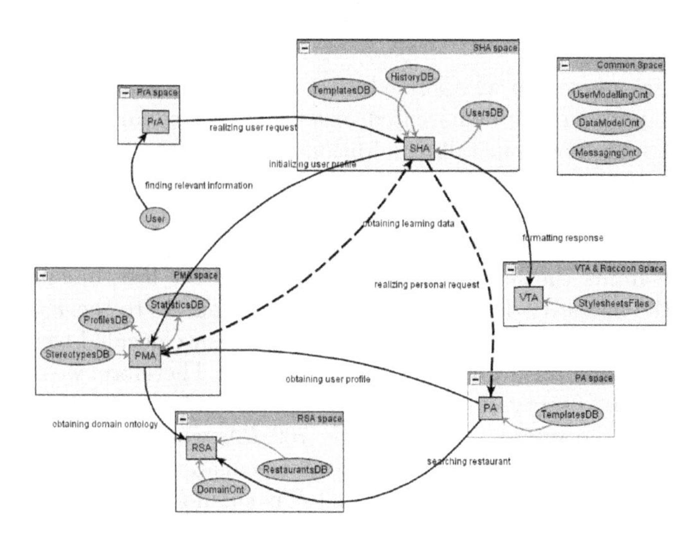


**Fig. 3.** Dependencies in the Travel Support System

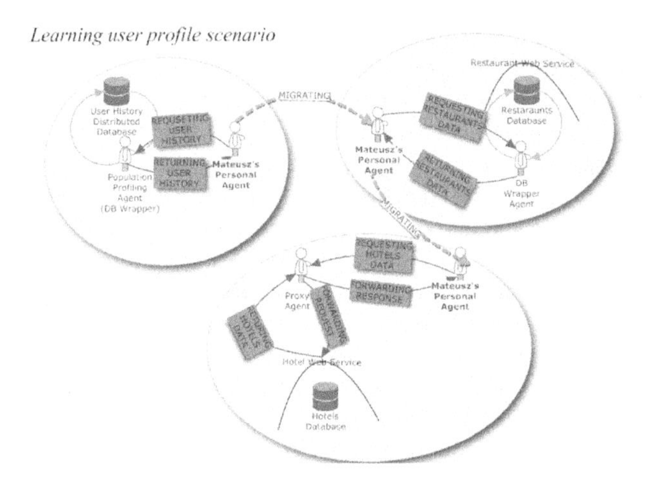


**Fig. 4.** Personal Agent migrating during learning process

system by a personal agent. We have learned that this approach can be justified mostly in situations, where such an agent resides on a user machine. Let us see reasons for this conclusion.

In our system we decided to utilize personal agent that is responsible for filtering and personalizing data delivered to the user [25]. In its original design, such an agent (also called *intelligent interface agent*) was supposed to exist on a user machine or mobile phone. The main reasons for this proposed design were:

- *security*—user profile is not explicitly accessible to the system,
- *mobility*— an agent can accompany the user in her travel, moving together with her mobile device,
- *resource separation*— an agent utilizes resources belonging to its user, not to the system.

However, in our system we assumed, that user devices are light (allowing only for visualizing documents demarcated in a markup language) and that user is unable or prevented from installing additional applications (e.g. for security reasons in a corporate environment). This decision has automatically swept away the expected advantages. Currently it is the system that stores user profiles. Moreover, a *PA* can be accessed by its user only through the *PrA* and this in an obvious ways limits the sense of *PA*'s mobility. And the last, but not least consequence is the necessity of providing additional resources to the *PA*, which now utilizes system hardware. To avoid wasting resources we decided to instantiate a *Personal Agent* only for duration of its user session. Therefore, we could have—for example—1000 users registered in the system, but only 100 of them actively interacting with the system, and thus only 100 *PA*s would exists. However, this leads to an interesting open question related to collaborative filtering. It is typically assumed that in this case all trusted *PA*s should be accessible every time another *PA* is about to ask them for their opinion about (restaurant) recommendations (to provide maximally relevant response to its user). Since not all *PA*s are "alive" all the time, they cannot provide a response to such a query. Therefore, usage of collaborative filtering, or even usage of *PA*s must be re-considered.

# 4 How the TSS Should Be Re-designed

Having analyzed problems that have arisen during the design and implementation of the original system, we have found that utilization of traditional technologies for some of its parts would be beneficial for the overall architecture. Therefore let us briefly discuss which functionalities of the system should be realized as agents and which—by utilizing traditional technologies, and how they can cooperate with each other.

## 4.1 Utilization of Agents

Agents can be placed in the following system scenarios:

- *profile learning*—because of their ability to move across different locations; this approach (1) gives also ability to organize agents in teams of agents, each team realizing different algorithm of learning, and (2) introduces possibility of selecting agents to realize a particular task by use of negotiations, depending on agent's current location and current access to the resource;
- *content filtering*—creation of a single *Personal Agent* for each user should replaced by utilization of agents representing groups of users and using different filtering algorithms;

Traditional technologies should be utilized in the following cases:

- MVC-based framework, for example Spring, can be used for content delivery, replacing functionality of the *SHA* and the *VTA* (together with the need of utilization of the Raccoon server),
- *PrA* wrapping home-made HTTP server can be replaced by a traditional servlet container, such as an Apache Tomcat, on which the Spring will be hosted,
- access to remote data sources can be "unwrapped" and provided directly via a simple database connection; concurrent access issue must also be considered here (e.g. by use of *multiply-read-single-write* policy).

## 4.2 Integration of Agents and Traditional Technologies

It has been shown that in the TSS there exists a demand for co-existence of both *traditional* and *agent* technologies. Therefore appropriate *middleware* must be created allowing heterogeneous parts of the system to communicate with each other.

Servlets executed by the CMS need to issue commands to JADE agents, for example through the so called *gateway*, which can be featured by the `Gateway Agent` class from `jade.wrapper.gateway` package.

Agent requiring functionality of a certain Web Service (speaking the WSDL language) can send its request to the *Gateway Agent* (from Web Services Integration Gateway add-on [26]), which translates ACL messages into SOAP messages in both directions and forwards them between the Web Service and the requesting agent. See figure 4 for example of such a situation, where the *PA* requires data about hotels provided by an appropriate Web Service.

# 5 Conclusion

In this paper we have discussed lessons learned from the design and initial implementation of the agent-based Travel Support System. We have shown, that the initial assumption—everything should be an agent has only educational value. However, in a realistic system both agent and traditional technologies have to coexist and be utilized in a judicious way. Following the critical analysis of pitfalls of our design, we have outlined a solution that would make our system more realistic, flexible and efficient. Finally, specific middleware solution was proposed for integration of non-JADE parts with JADE-based parts of the system.

# Acknowledgements

The authors would like to thank for fruitful discussions: Minor Gordon from Computer Laboratory of University of Cambridge in United Kingdom, Pawel Kobzdej from the Systems Research Institute of the Polish Academy of Sciences and Pawel Kaczmarek from Hewlett-Packard Poland. Many thanks also to Juan A. Bota Blaya from Information and Telecommunicatinos Engineering Department of the Murcia University in Spain for suggestion about the semantic overload issue.

# References

1. Wooldridge, M., Jennings, N.R.: Pitfalls of agent-oriented development. In Sycara, K.P., Wooldridge, M., eds.: Proceedings of the 2nd International Conference on Autonomous Agents (Agents'98), New York, ACM Press (1998) 385–391
2. Bond, A.H., Gasser, L., eds.: Readings in Distributed Artificial Intelligence. Morgan Kaufmann, San Mateo, CA (1988)
3. Jennings, N.R., Wooldridge, M.: Applications of intelligent agents. In: Agent Technology: Foundations, Applications and Markets. Springer, Berlin (1998)
4. Bussmann, S., Schild, K.: An Agent-based Approach to the Control of Flexible Production Systems. In: Procedding of the 8th IEEE International Conference of Emergent Technologies and Factory Automation (EFTA 2001), Abtibes Juan-les-pins, France (2001) 481–488
5. Booch, G.: Object-oriented analysis and design with applications (2nd ed.). Benjamin-Cummings Publishing Co., Inc., Redwood City, CA, USA (1994)
6. Jennings, N.R.: Agent-oriented software engineering. In: Proceedings of the 12th international conference on Industrial and engineering applications of artificial intelligence and expert systems : multiple approaches to intelligent systems, Secaucus, NJ, USA, Springer-Verlag New York, Inc. (1999) 4–10

7. Genesereth, M.R., Ketchpel, S.: Software agents. Communications of the ACM **37** (1994) 48–53
8. Wooldridge, M.: An introduction to multiagent systems. John Wiley & Sons (2002)
9. Ganzha, M., Gawinecki, M., Paprzycki, M., Gąsiorowski, R., Pisarek, S., Hyska, W.: Utilizing Semantic Web and Software Agents in a Travel Support System. In: Semantic Web Technologies and eBusiness: Virtual Organization and Business Process Automation. Idea Publishing Group (2006)
10. Resource Description Framework (RDF). `http://www.w3.org/RDF/` (2005)
11. Semantic Web Activity Statement. `http://www.w3.org/2001/sw/Activity` (2001)
12. Jena a semantic web framework for java. `http://jena.sourceforge.net/` (2005)
13. Ramachandran, V.: Design Patterns for Building Flexible and Maintainable J2EE Applications. `http://java.sun.com/developer/technicalArticles/J2EE/despat/` (2002)
14. Gawinecki, M., Gordon, M., Kaczmarek, P., Paprzycki, M.: The Problem of Agent-Client Communication on the Internet. Parallel and Distributed Computing Practices **6** (2003) 111–123
15. Gawinecki, M.: User modelling on a base of interaction with WWW system. Master's thesis, Deparment of Mathematics and Computer Science, Adam Mickiewicz University, Poznan (2005)
16. Raccoon. `http://rx4rdf.liminalzone.org/Racoon` (2005)
17. Rao, B.R.: Making the most of middleware. **12** (1995) 89–06
18. Foundation for intelligent physical agents. `http://www.fipa.org` (2007)
19. Nwana, H.S., Ndumu, D.T.: A perspective on software agents research. The Knowledge Engineering Review **14** (1999) 1–18
20. Guessoum, Z., Briot, J.P.P.: From active objects to autonomous agents. IEEE Concurrency **7** (1999) 68–76 `citeseer.ist.psu.edu/guessoum99from.html`.
21. Spring Application Framework. `http://www.springframework.org` (2006)
22. Tusiewicz, M.: System wieloagentowy: teoria, projekt, implementacja oraz przyklady zastosowań. Master's thesis, Department of Mathematics, Physics and Computer Science, Jaggielonian University, Kraków (2003)
23. Padgham, L., Winikoff, M.: Prometheus: a methodology for developing intelligent agents. In: AAMAS '02: Proceedings of the first international joint conference on Autonomous agents and multiagent systems, New York, NY, USA, ACM Press (2002) 37–38
24. Laufmann, S.C.: Agent software for near-term success in distributed applications. (1998) 49–69
25. Nesbitt, S.: Collaborative Filtering on the Web: An agent-based Approach (Literature Review) (1997)
26. JADE Board, Whitestein Technologies AG: JADE Web Services Integration Gateway Guide. `http://jade.tilab.com` (2006)

# Anticipative Agent Based System Synchronization Example

Andrej Škraba, Miroljub Kljajić,
Davorin Kofjač, Blaž Rodič, and Matevž Bren

University of Maribor, Faculty of Organizational Sciences,
Kidričeva cesta 55a, 4000 Kranj, Slovenia
{andrej.skraba,miroljub.kljajic,davorin.kofjac,blaz.rodic,matevz.bren}
@fov.uni-mb.si
http://www.uni-mb.si

**Abstract.** The concept of *feedback-anticipative* control as the extension of classical Wiener paradigm is considered in the context of multi agent systems. The behavior of complex real world agents is based on the consideration of feedback information as well as on the anticipation. A linear model of the agents with a nonlinear interaction rule is proposed as the mean for the methodological conception. The results of the developed system display a periodic response. An analytical determination of periodicity conditions for individual agents was performed by the application of *z-transform*. Proof of system stability for the case of two interacting agents has been provided. The hyperincursivity paradigm is presented as an interesting methodological platform for further investigation of multi agent systems.

**Keywords:** Agent, interaction, synchronization, periodicity, stability.

## 1 Introduction

In the present paper the introduction of Dubois feedback-anticipative principle [1] into agent based systems will be analyzed. The findings presented are of general importance not only for conventional agent based systems but also for the multi agent systems where a) each agent has incomplete information or capabilities for solving the problem and, thus, has a limited viewpoint; b) there is no global system control; c) data are decentralized; and d) computation is asynchronous [2].

First, let us consider the classical Wiener [3] control theory where the feedback mechanisms play a major role. One could formulate the respected idea as a recursive discrete system, which computes the successive time states as a function of its past and present states:

$$x_{t+1} = \mathbf{R}(\ldots, \mathbf{x}_{t-2}, \mathbf{x}_{t-1}, \boldsymbol{x}_t; \mathbf{p}) \tag{1}$$

where $\mathbf{x}_t$ are the vector states at time $t$, $\mathbf{R}$ the recursive vector function and $\mathbf{p}$ a set of parameters. Knowing the function $\mathbf{R}$, the values of parameters $\mathbf{p}$ and

W. Abramowicz (Ed.): BIS 2007, LNCS 4439, pp. 500–509, 2007.
© Springer-Verlag Berlin Heidelberg 2007

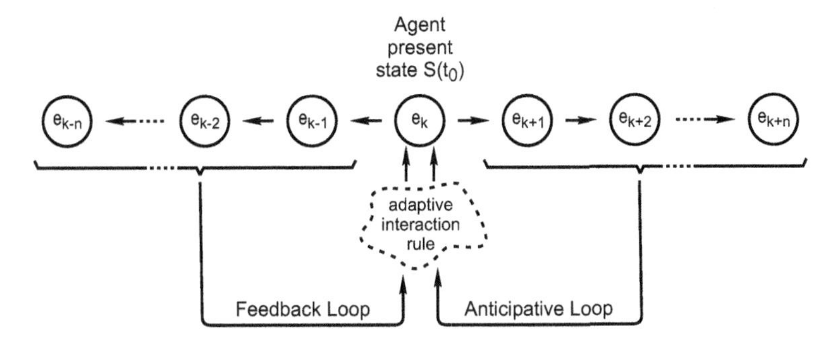


**Fig. 1.** Agent's present state depends on the feedback and anticipative event chain as well as adaptive interaction rule which determines the interaction of $n$ agents

the initial conditions $\mathbf{x}_{-2}, \mathbf{x}_{-1}, \mathbf{x}_0$, the successive states $x_1, x_2, x_3, \ldots$ can be recursively computed where the time interval $\Delta t = 1$ is fixed.

The interaction of individual feedback-loop controlled systems is the root of complexity, which is a hallmark of agent based systems, where such complex terms apply as e.g. Simon's Bounded Rationality [4]. One could question why should we emulate the human behavior. Well, agents by their definition simulate the real world and in the case of multi agent systems the impact of the society as well as an individuum should be thoroughly considered. It is only a question whether such a society is composed of rather technical entities such as software agents.

If we consider an individuum as well as the agents' "techno-society" it is important to notice that such systems are not controlled only by the means of a feedback loop. Their inevitable property is an *Anticipation* which is immanent in complex systems. Such a concept might be considered as a recursive differential system where the forward derivative as well as the backward derivative are considered as the means for feedback-anticipative control modelling. We are therefore dealing with a self-referential anticipatory system [1] characterized as an incursive system from the contraction of "inclusive" or "implicit" recursion [5]. Applicability of the feedback–anticipative principle is important since the presented principles determine the agents interaction which has an impact on the information systems development and performance. In practice, one could consider agents determining the stock exchange dynamics and supplementary IS support as demonstrating the key attributes indicated in this paper.

**Definition 1.** *The incursive system is defined by:*

$$x(t+1) = F[\ldots, x(t-1), x(t), x(t+1), \ldots] \tag{2}$$

*where the value of a variable $x(t+1)$ at time $t+1$ is a function of this variable at past, present and future times.*

Proposed incursivity paradigm should be considered as the concept for the examination of the system structure and its relation to the time component. Def. 1

states, that complex agent-based systems include feedback as well as a feedforward loop. Stated differently, one could not govern the reality only by knowing the past; one ought to consider the future as well. In Fig. 1 the Agent's state at time $t_0$ depends not only on the past events marked by $e_{k-1}, e_{k-2}, \ldots$ but also on the future event chain marked by $e_{k+1}, e_{k+2}, \ldots$. Both feedback as well as anticipative loop determine the particular agent realization in present time $t_0$ which determines the response by adaptive interactive rule.

## 2 Anticipative Agent Based System

In order to analyze the interactions between several entities modelled as agents the following agent based model is proposed. In our case agent interaction represents an alternative control mechanism, which should provide standing oscillations and global equilibrium-seeking behaviour found in real world cases [6]. The initial *feedback-anticipative* agent based model oscillates only for a certain parameter set, which represents a thin borderline in parameter space. If the system parameters are not on this border, the model either decays or produces explosive behaviour [7]. There are several approaches to fixing this problem. One of the methods for solving this problem is the application of the *floor-roof* principle, which should limit the values considered [7]. Note that an agent based system working in a stable environment either oscillates in its response or, on the other hand, has found itself in an exceptional situation (deadlock, communication loss, etc.). It will be shown that several agent based entities in the phase of standing oscillatory behaviour could be sustained in this working condition by their interaction. The proposition of a stable response for a system of two agents will be proved to show that the concept provides the stability of standing oscillations as time goes to infinity; $t \to \infty$.

Consider the following agent-based anticipative system where the dynamics is denoted by the variable $P$ as the function of control parameters $f(a, b, c, d)$:

$$\mathbf{P}_k = \frac{b\,\mathbf{P}_{k-1} + a - c}{\mathbf{d}_k} + \frac{b}{\mathbf{d}_k}\mathbf{P}_{k+1} \tag{3}$$

Equation 3 captures the general feedback-anticipative mechanism of system control where the present state at time $k$ is dependent on the state at time $k-1$ as well as on the state at time $k+1$. Such model has many possible applications in the field of complex dynamics modelling. In the above equation the matrix annotation represents column vectors, which have the same arbitrary dimension $n$ determined by the number of agents. Initial conditions for Eq. (3) should be stated in matrix form as:

$$\mathbf{P}_{k+1} = \frac{p-a}{b} \tag{4}$$

$$\mathbf{P}_k = \frac{bP_{k+1} + a - c}{\mathbf{d}} \tag{5}$$

For the computation of the new values of $P$, shift operator $\rho$ on sequence $P$ is applied, which shifts sequence $p \in P$ one step to the left:

$$\rho(\langle \mathbf{P} \rangle) = \langle \mathbf{p}_{n+1} \rangle \tag{6}$$

providing the forward shifted values for $\mathbf{P}_{k-1}$ and $\mathbf{P}_k$ in Eq. (3). The decision of change in parameter $d$ will be dependant on the sum of two values of variable $P$ at time $k+1$ and time $k-1$. Here, the relative value of $P$ by taking the range of system response in the denominator will be considered:

$$\mathbf{e} = \frac{\xi_{k+1} + \xi_{k-1}}{\left| \lceil \xi_k \rceil - \lfloor \xi_k \rfloor \right|} \tag{7}$$

In Eq. (7) $\xi$ represents the estimation chain for $r$ time steps computed in a similar manner to $\mathbf{P}$ in Eq. (3) except for the initial conditions, which are stated in matrix form as in Eqs. (4, 5) for time 0, while for $\xi(0)$ the shift operator $\rho$ is applied forcing the anticipation principle as $\xi_k = f(\mathbf{P}_{k+1})$. Besides the notation for absolute value in the denominator, the *roof* and *floor* operators are applied. In order to perform the control by variation of parameter $d$, where $n$ agents are present, the following state equation with the adaptive rule for $\Delta \mathbf{d}_k$ is introduced:

$$\mathbf{d}_{k+1} = \mathbf{d}_k + \Delta \mathbf{d}_k \tag{8}$$

where $\Delta \boldsymbol{d}$ determines the change in control parameter $d$:

$$\Delta \mathbf{d}_k = \begin{cases} \beta & \text{if} \quad e = \lceil \mathbf{e} \rceil \\ -\beta & \text{if} \quad e = \lfloor \mathbf{e} \rfloor \end{cases} \tag{9}$$

In the above definition of the agent's rule, the *floor* and *ceiling* functions over a vector of relative values $\mathbf{e}$ consider only a finite number of lags. One should notice that the mentioned *floor-roof* operators are applied on vector $\mathbf{d}$ rather than on vector $\mathbf{P}$, which would mean the strict, conventional implementation of the *floor-roof* principle [7]. Parameter $\beta$ is the *intensity* of agents reaction to the system disequilibrium; $\beta \in (0,1)$. Initialization of vector $\mathbf{d}$ is determined by random value $r_i \in [-2, 2]$, which falls within the interval of periodic solutions [8,9,10,11] for the anticipative agent based system. Certainly, one could also assign an arbitrary value for $\mathbf{d}$ as this will also be considered.

The idea captured in the above definition considers a situation where an agent based system where the state space values in the past and estimated future are at their peak, should be controlled by increasing the value of control parameter $d$, thus changing the frequency response of the system [12]. The case at the lower end of the system response is inverse.

Fig. 2 represents the interaction of eight agents defined by Eqs. (3)-(9) as an example of the system response. Here the values of the parameters are: $a = 1$, $b = 1$, $c = 1$, $p = \frac{1}{2}$. On $x$-axis the time step $k$ is represented and on $y$-axis value of parameter $d$ is shown. Each line of the graph represents the variation

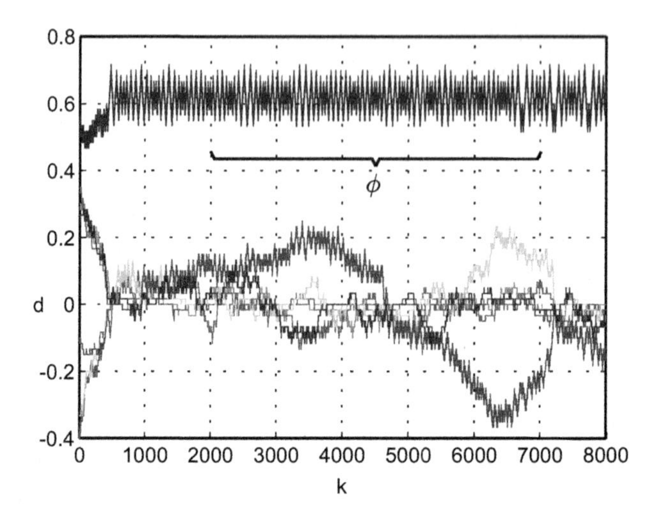


**Fig. 2.** Agents (8) interacting with synchronization plateaux at $\phi$; each curve represents the adjustment dynamics of particular agent

of parameter $d$ for particular agent $A$. From the Fig. 2 synchronization plateaux could be observed, which are marked as $\phi$. Synchronization is indicated as the plateaux in the system response, where dynamical equilibrium occurs. One of the main properties of the system of $n$ agents is that equilibrium could occur only as a trivial solution of the system, meaning that the system is not active, i.e. there is no interaction with the environment. This is the property of an agent based system which should be considered by the proposed agent-based implementation. Therefore, in the case of equilibrium, system $S$ does not exist hence $\nexists S$; the system is closed. Another important property which should be considered by the proposed agent-based implementation is dynamical equilibrium synchronization of agent response. Analyzing real time series of interacting agents, one could observe vivid synchronization plateaux [13,14].

## 3 System Periodicity

Periodicity that determines the agents interaction should be determined in order to provide the system with control. The analytical determination of periodicity conditions is provided by the application of $z$-transform, which is the basis of an effective method for the solution of linear constant-coefficient difference equations [15]. The application of $z$-transform on Eq. (3) with initial conditions stated by Eq. (4) for particular agent gives:

$$Y(z) = \frac{-y_1 z + y_0 d z - y_0 z^2}{-1 + dz - z^2} \tag{10}$$

Inverse $z$-transform yields the following solution:

$$Y^{-1}(z) = 2^{-1-n}\, y_0 \left(d - \sqrt{-4+d^2}\right)^n - \frac{y_1 \left(d - \sqrt{-4+d^2}\right)^n}{2^n \sqrt{-4+d^2}} + \\ + \frac{2^{-1-n}\, y_0\, d \left(d - \sqrt{-4+d^2}\right)^n}{\sqrt{-4+d^2}} + 2^{-1-n}\, y_0 \left(d + \sqrt{-4+d^2}\right)^n + \\ + \frac{y_1 \left(d + \sqrt{-4+d^2}\right)^n}{2^n \sqrt{-4+d^2}} - \frac{2^{-1-n}\, y_0\, d \left(d + \sqrt{-4+d^2}\right)^n}{\sqrt{-4+d^2}} \tag{11}$$

In order to gain conditions for the periodic response of the system the following equation should be solved:

$$Y^{-1}(z) = y_0 \tag{12}$$

Considering the equation for the roots of complex numbers [16]:

$$\sqrt[n]{z} = \sqrt[n]{r}\left(\cos\frac{\theta + 2k\pi}{n} + i \sin\frac{\theta + 2k\pi}{n}\right) \tag{13}$$

the general form of the solution for parameter $d$ could therefore be defined as:

$$d = 2\cos\frac{2\pi m}{n} \tag{14}$$

where $n$ is the period and $m = 1, 2, 3, ..., n-1$.

The value that emerges in Fig. 2 as $\phi$ is in our case present at period $n = 5$, where the value for parameter $d = \frac{\sqrt{5}-1}{2} = 0.61803...$, often called the *Golden Ratio Conjugate* (usually denoted with $\phi$).

## 4 System Stability

In our case, system stability of the anticipative agent-based model is dependent on the value of agent reaction $\beta \in (0,1)$. Higher values of $\beta$ result in a higher volatility of system response. Another important variable is the number of anticipative agents $n$ considered. Note, that the value for parameter $\mathbf{d}$ is not limited, i.e. by applying $\Delta d$; the values could range from $-\infty$ to $+\infty$, while the standing oscillation response of the agent could only be possible in the interval $d \in [-2, 2]$. Therefore, let us formulate the following proposition:

**Proposition 1.** *The anticipative agent-based system defined by Equation 3 is stable if* $n = 2$, $\Delta d = 1 - \phi$, *initial conditions* $d_1 = 1$, $d_2 = -1$ *when* $k \to \infty$; *and* $\mathbf{d} \in [-1, 1]$.

*Proof.* Due to the feasibility of proof the interaction of two agents will be analyzed here. One of the agents will be marked with the $[+]$ sign as $A_{[+]}$ and the other with the $[-]$ sign as $A_{[-]}$ due to the initial conditions for the value of parameter $d$. The observed parameters are: $a = 1$, $b = 1$, $c = 1$, $d_{[+]}(0) = 1$, $d_{[-]}(0) = -1$,

$p = \frac{1}{2}$. Let us consider the value for $\Delta d = 1 - \phi$. For the first step the value for $\mathbf{d}$ is $d_{[+]}(1) = d_{[+]}(0) - \Delta d$ and $d_{[-]}(1) = d_{[-]}(0) + \Delta d$.

$$\mathbf{d}(1) = \begin{bmatrix} d_{[+]}(0) - \Delta d \\ d_{[-]}(0) + \Delta d \end{bmatrix} \tag{15}$$

By examining the system response, one would expect that further values for $\mathbf{d}$ would vary in the interval $[-1, +1]$. According to this proposition, the following should hold:

$$\mathbf{d}_{k+1} = \begin{bmatrix} 1 - \Delta d \\ -1 + \Delta d \end{bmatrix} \;\; if \;\; \mathbf{d}_k = \begin{bmatrix} 1 \\ -1 \end{bmatrix} \quad \forall \, \mathbf{P} \in \Re / \{0\} \tag{16}$$

This would limit the value for parameter $\mathbf{d}$ in the prescribed interval $[-1, +1]$. There is an exception at critical point 0. In a further investigation of system stability one should consider the following condition for parameter $\mathbf{d}$:

$$(|P_{[+]k+1}| = |P_{[-]k+1}|) \wedge (|P_{[+]k}| = |P_{[-]k}|) \wedge (|P_{[+]k-1}| = |P_{[-]k-1}|) \tag{17}$$

Eq. (17) determines the parameter space in which the solution of the system could exist.

For the condition $(P_{[-]k-1} = P_{[+]k-1}) \wedge (P_{[-]k} = -P_{[+]k})$ at $d_{[+]}(0) = 1$ and $d_{[-]}(0) = -1$, the condition $e_{[+]} \leq e_{[-]}$ from Eq. (9) is determined by two planes:

$$\alpha = \frac{P_{k-1}}{|\max(s_1)| + |\min(s_2)|} + \frac{P_k - P_{k-1}}{|\max(s_3)| + |\min(s_4)|} \tag{18}$$

$$\beta = \frac{P_{k-1}}{|\max(s_5)| + |\min(s_6)|} + \frac{P_k - P_{k-1}}{|\max(s_7)| + |\min(s_8)|} \tag{19}$$

In Eq. (18) and Eq. (19) $s_n$ represents the sequence from $\mathbf{P}$ and $\xi$. On account of the periodicity condition, which is met at $d \in \{-1, -\phi, \phi, 1\}$, the minimum number of sequence values are taken for determining the planes. Planes $\alpha(P_{k-1}, P_k)$ and $\beta(P_{k-1}, P_k)$ cross symmetrically with respect to the origin, which could be proven by a reduction of Eq. (18) and Eq. (19):

$$\frac{P_k}{|\max(\varsigma_a)| + |\min(\varsigma_b)|} < \frac{P_k}{|\max(\varsigma_c)| + |\min(\varsigma_d)|} \tag{20}$$

Inequality defined by Eq. (20), where $\varsigma$ represents proper system sequence, holds except for the critical point 0 and limit values as $P \to \pm\infty$. The condition $e_{[+]} \leq e_{[-]}$ for the postitive combination of signs is met, meaning that for such values of $\mathbf{P}$ the direction of parameter $d$ change is correct. The procedure for the negative set of signs is performed respectively.

Above procedure does not provide an answer to what will happen in the limit. The answer is provided by the following four limits:

$$\lim_{P_k \to -\infty} \frac{P_k(\sqrt{5} - 1)}{2(|\max(\psi_1)| + |\min(\psi_2)|)} = \frac{1 - \sqrt{5}}{2(\sqrt{5} + 1)} \tag{21}$$

$$\lim_{P_k \to \infty} \frac{P_k(\sqrt{5}-1)}{2(|\max(\psi_1)| + |\min(\psi_2)|)} = \frac{\sqrt{5}-1}{2(\sqrt{5}+1)} \tag{22}$$

$$\lim_{P_k \to -\infty} \frac{P_k(\sqrt{5}-1)}{2(|\max(\psi_3)| + |\min(\psi_4)|)} = \frac{1}{4}(\sqrt{5}-3) \tag{23}$$

$$\lim_{P_k \to \infty} \frac{P_k(\sqrt{5}-1)}{2(|\max(\psi_3)| + |\min(\psi_4)|)} = \frac{1}{4}(3-\sqrt{5}) \tag{24}$$

The results of limits in Eqs. (21, 22, 23, 24) with fulfilled conditions for $-\infty < P_k < +\infty$ including critical point 0 confirm that the critical stability condition is not met for $P_k \in [-\infty, +\infty]$, thus providing a proper change of parameter $\mathbf{d}$ in a critical step before $\mathbf{d}$ takes the values $d_{[+]k+1}(0) = 1$ and $d_{[-]k+1}(0) = -1$ □

## 5 Discussion

Feedback–anticipative principle is an important concept in the modelling of multi agent systems. Dubois anticipative paradigm [1,17] could be further extended to the field of Hyperincursive systems (see *Appendix* for short description). It is important to know that complex systems such as multi agent systems incorporate two loops: a) a feedback loop and b) an anticipative loop. These two loops inevitably produce oscillatory behavior of the system which is the main property of real world complex agents. By the proposition of agent based system stated in the form of linear system with nonlinear rule of interaction the periodic response was determined with significant $\phi$ value in the example of system response. Gained periodicity results are applicative in further analysis of interacting agent-based systems [18,19,20].

The analysis of the agent-based model provides proof of system stability, which is one of the key conditions that should be meet by agent-based models simulating complex systems. The provided proof of system stability for the case of two agents, which also provides promising results for the *n-agent* case, confirms that the model could be set in the global equilibrium mode. All the stated characteristics of the agent-based model as well as the response of the system for eight agents provides a promising methodological platform for the study of the interaction between several agent-based systems which incorporate feedback-anticipative principles. The proposed model provides the means for analyzing interaction, feedback, anticipation, frequency response, synchronization, standing oscillations and system equilibrium. An introduction of *feedback-anticipative* systems interconnection and control by varying the parameter, which influences system frequency response, represents a new perspective for the analysis of complex evolutionary agent-based systems. Findings presented here provide interaction rules of program agents with potential business applications in the field of informational systems.

**Acknowledgment.** This research was supported by the Ministry of Higher Education, Science and Technology of the Republic of Slovenia (Programme No. UNI-MB-0586-P5-0018).

## References

1. Dubois, D., Resconi, G.: Hyperincursivity: a new mathematical theory. Presses Universitaires de Liège, Liège (1992)
2. Sycara, K.: Multiagent systems. AI Magazine **19** (1998) 79–92
3. Wiener, N.: Cybernetics or Control and Communication in the Animal and the Machine. Wiley, New York, NY (1948)
4. Simon, H.: Models of Man: Social and Rational – Mathematical Essays on Rational Human Behavior in a Social Setting. Wiley, New York, NY (1957)
5. Dubois, D.: Orbital stability and chaos with incursive algorithms for the nonlinear pendulum. International Journal of Computing Anticipatory Systems **14** (2004) 3–18
6. Rosenblum, M., Pikovsky, A.: Synchronization: from pendulum clocks to chaotic lasers and chemical oscillators. Contemporary Physics **44** (2003) 401–416
7. Puu, T., Sushko, I.: A business cycle model with cubic nonlinearity. Chaos, Solitons and Fractals **19** (2004) 597–612
8. Agarwal, R., Bohner, M., Grace, S., O'Regan, D.: Discrete Oscillation Theory. Hindawi Publishing Corp., New York, NY (2005)
9. Sonis, M.: Critical bifurcation surfaces of 3d discrete dynamics. Discrete Dynamics in Nature and Society **4** (1999) 333–343
10. Sonis, M.: Linear bifurcation analysis with application to relative socio–spatial dynamics. Discrete Dynamics in Nature and Society **1** (1996) 45–56
11. Strogatz, S.: Nonlinear dynamics and chaos: with applications to physics, biology, chemistry and engineering. Addison–Wesley Co., Reading, MA (1994)
12. Hogg, T., Huberman, B.: Controlling chaos in distributed systems. IEEE Transactions on Systems Man and Cybernetics **21** (1991) 1325–1332
13. Kociuba, G., Heckenberg, N., White, A.: Transforming chaos to periodic oscillations. Physical Review **E 64** (2001) 056220–1–056220–8
14. Matsumoto, A.: Ergodic cobweb chaos. Discrete Dynamics in Nature and Society **1** (1997) 135–146
15. Luenberger, D.: Introduction to Dynamics Systems: Theory, Models and Applications. John & Wiley Sons Inc., USA (1979)
16. Kreyszig, E.: Advanced Engineering Mathematics. John Wiley & Sons, Hoboken, NJ (1993)
17. Kljajić, M.: Contribution to the meaning and understanding of anticipatory systems. In Dubois, D., ed.: Computing anticipatory systems, American Institute of Physics (2001) 400–411
18. Škraba, A., Kljajić, M., Kofjač, D., Bren, M., Mrkaić, M.: Periodic cycles in discrete cobweb model. WSEAS Transactions on Mathematics **3** (2005) 196–203
19. Škraba, A., Kljajić, M., Kofjač, D., Bren, M., Mrkaić, M.: Anticipative cobweb oscillatory agents and determination of stability regions by lyapunov exponents. WSEAS Transactions on Mathematics **12** (2006) 1282–1289
20. Pažek, K., Rozman, Č., Turk, J., Bavec, M., Pavlovič, M.: Ein simulationsmodell für investitionsanalyse der nahrungsmittelverarbeitung auf ökologischen betrieben in Slowenien. Bodenkultur **56** (2005) 121–131

## Appendix: Definition and Example of Hyperincursive System

Hyperincursion is an incursion with multiple solutions which is an interesting principle applicative to the field of analysis of multi agent systems. The following definition characterizes the principle of Hyperincursion as the possible extension for the dynamical systems analysis [1]:

**Definition 2.** *Discrete hyperincursive anticipatory system is an incursive discrete anticipatory system generating multiple iterates at each time step. Discrete hyperincursive anticipatory system is defined by the structural set of equations:*

$$\Gamma_i[q_1(t+1),\ldots,q_n(t+1),q_1(t),\ldots,q_n(t)]=0 \tag{25}$$

*where* $i=1,2,\ldots,n$.

Hyperincursivity Def. 2 presents the possibility for extension of ordinary system structures to the nonlinear systems of higher complexities. Example of such system is presented in the Appendix. Present paradigm should be considered as the idea for the "*out of the box*" examination of the system structure and its relation to the time component.

The following equation represent an example of hyperincursive [1] system where:

$$x(t)=ax(t+1)(1-x(t+1)) \tag{26}$$

defines an hyperincursive anticipatory system. Hyperincursion is an incursion with multiple solutions. With $a=4$, mathematically $x(t+1)$ can be defined as a function of $x(t)$

$$x(t+1)=\frac{1}{2}\pm\frac{1}{2}\sqrt{1-x(t)} \tag{27}$$

where each iterate $x(t)$ generates at each time step two different iterates $x(t+1)$ depending of the plus minus sign. The number of future values of $x(t)$ increases with the power of 2. This system is unpredictable in the sense that it is not possible to compute its future states just by knowing the initial conditions. It is necessary to define the successive final conditions at each time step. As the system can only take one value at each time step, something new must be added to allow solving. Thus, the following decision function $u(t)$ can be added for making a choice at each time step:

$$u(t)=2d(t)-1 \tag{28}$$

where $u=+1$ for the decision $d=1$ (true) and $u=-1$ for the decision $d=0$ (false). By introducing Eq. 27 and Eq. 28, the following equation is obtained:

$$x(t+1)=\frac{1}{2}+\left(d(t)-\frac{1}{2}\right)\sqrt{1-x(t)} \tag{29}$$

The decision process could be explicitly related to objectives to be reached by the state variable $x$ of this system.

# Ubiquitous Commerce Business Models Based on Ubiquitous Media

Kyoung Jun Lee and Jeong-In Ju

School of Business, Kyung Hee University
Hoegi-dong, Dongdaemun-gu, Seoul, 130-701, Korea
{klee,jji99}@khu.ac.kr

**Abstract.** Conventional media, such as newspapers, radio, TV and Internet appeal human cognitive and perceptual organisms such as brain, eyes and ears. The producers of text, image, and video use their cognitive and perception processes and their consumers also receive and interpret the messages using the same two kinds of processes. However, the media in ubiquitous environment not only takes advantage of human biological systems, but also the digital systems of human beings while conventional media appeals only to people's bio-systems. Ubiquitous media creates and consumes content through not only human cognitive and perceptual processes but also through the interactions between surrounding digital systems. U-Media(Ubiquitous media) provides information by generating, collecting, and attaching the content itself and the related information based on the interaction of the bio-systems incorporating digital information and devices embedded in humans, and surrounding objects including external digital devices. This paper investigates the concept of media in ubiquitous environments and proposes a commerce business model based on U-Media.

## 1 Introduction

Media is affected by the technology that surrounds it during its content production and consumption process. It is evident that advancements in traditional media, such as newspapers and TV were developed alongside printing, editing, and photo and image technologies. The Internet affects many things not only in the content production and consumption processes, but also in the process where producers (senders) transmit content to their consumers (receivers). The Internet creates a prosumer that illustrates the obscure boundary between producers and consumers making it possible to generate and share new content, such as replies to BLOGs and videos on mini-home pages. Furthermore, the World Wide Web and especially the so called Web 2.0 have produced a new concept 'seamlessness' in the production and consumption process of content through links that connect between content and the new content that is derived from the content. One example is Flicker.com, a photo-sharing site that makes it possible to seamlessly link other content to a certain part of a photo. In addition, Beedeo.com provides a Cut & Tag function that opens a seamless link to a still image of a video and supports new content when a user clicks on a certain scene.

The seamless link between content is an important factor in U-Commerce (Ubiquitous Commerce). U-Commerce is commercial activity that creates seamless communication between provider, consumer, product, and service in which seamlessness

W. Abramowicz (Ed.): BIS 2007, LNCS 4439, pp. 510–521, 2007.
© Springer-Verlag Berlin Heidelberg 2007

means the continuous transmission of information between product, service, space, and economic entities in a commercial process (Lee and Ju, 2005). Several U-Commerce business models, such as U-Comparison Shopping (Lee and Seo, 2006), U-Referral Marketing (Lee and Lee, 2006), U-Recommendation (Kim, Lee, and Kim, 2006) and U-Payment, U-Payment & Receipt (Lee, Jeong, and Ju, 2006; Lee, Ju, and Jeong, 2006) were studied in this paradigm. These business models have the examples of seamless links that make it possible to have links between objects in the actual world and online information. According to the ubiquitous technology and business models, media will be affected in its content generation, consumption, distribution, sharing, and derivation processes thus it is possible to expect the widespread emergence of seamless business models in the media. A new network produces new media, and new media creates new industries and commerce. Ubiquitous environments make it possible to realize U-Media, and a new commerce model is anticipated based on U-Media. This paper investigates the concept of media in ubiquitous environment and proposes a commerce business model based on the U-Media.

## 2 Definition and Characteristics of U-Media

Before the existence of ubiquitous environments, conventional media, such as newspapers, radio, TV and Internet appeal human cognitive and perceptual organisms such as brain, eyes and ears. The producers of text, image, and video use their cognitive and perception processes and their consumers also receive and interpret the messages using the same two kinds of processes. However, the media in ubiquitous environment not only takes advantage of human biological systems, but also the digital systems of human beings while conventional media appeals only to people's bio-systems. Ubiquitous media is defined as a media where human creates and consumes content through not only human cognitive and perceptual processes but also through the interactions between surrounding digital systems. Hypermedia is a term created by Ted Nelson, and used in his 1965 article (Nelson 1965). It is used as a logical extension of the term hypertext, in which graphics, audio, video, plain text and hyperlinks intertwine to create a generally non-linear medium of information. U-Media is a hypermedia where the hyperlinks are automatically and systematically generated and the hyperlinks can be connected to any objects in real or virtual world.

The definition of U-Media can be easily understood with U-Camera, an example of U-Media, 'a camera that takes photographs of digital links as well as real world image'. Assume that a man takes a picture using a U-Camera. Through wireless networking such as Bluetooth and RFID, he receives digital information or links of objects and the people at the angle that the picture was taken. When it is possible to recognize the digital link in a picture and connects to another content that the link connects to, we can call it as U-Media since people can seamlessly obtain the information using the automatically generated links of the objects or people in the Picture The acquisition and usage processes of the digital links and the related information is far more automatic and systematized than those of Flicker.com and Beedeo.com where users themselves produces tags for every single image or specified section of a video. Fig. 1 illustrates the production and consuming process of conventional media and U-Media.

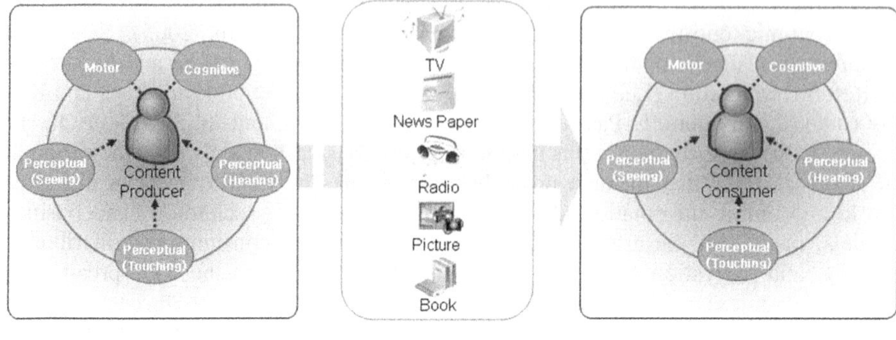


Conventional Media

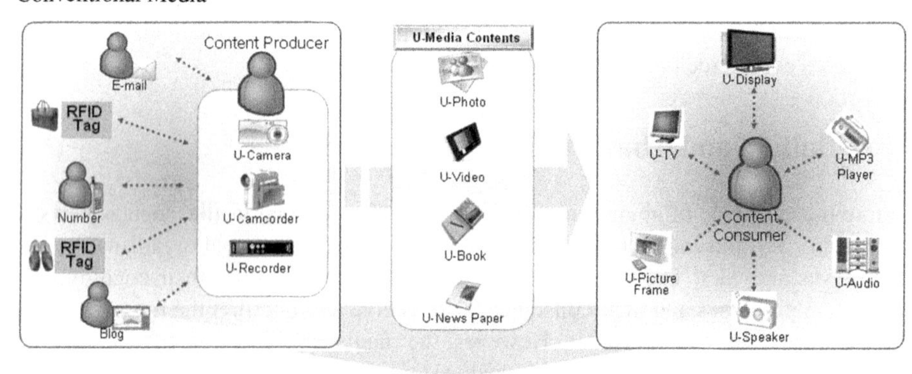


U-Media

**Fig. 1.** U-Media

The characteristics of U-Media can be analyzed as follows.

U-Media reduces information production. Although it is currently impossible to automatically annotate content-related links to content in traditional media, U-Media aides in the automatic and systematic annotation of surrounding digital information into media files. This is due to the improvement in wireless communication infrastructure, such as RFID and Bluetooth, and the computing power of user devices. As previously mentioned, a method that links to the object existing in image/video of Web 2.0 sites such as Flicker.com or Beedeo.com requiring work forces for every single link is entirely different from that of U-Media. Thus, it is evident that U-Media reduces content production costs. In addition, the automatic and systematic annotation of information and links affects the reduction in commerce costs. It is possible to reduce the costs of searching, sharing, advertising and advertisement effects assessment using the systematic and serialized annotation functionality. Search costs can be reduced using the link annotated in image files. Sharing costs can be reduced when users want to share image files just after taking a picture using the U-Camera. A receiver is able to receive image files using predefined methods such as e-mail and a cell phone with contact points of collected persons when u-photo are generated.

Advertisements and advertisement effect assessment costs can be reduced due to the use of the new media and new advertisement methods.

U-Media brings a new level of content. At present, it is difficult to produce content that includes links regardless of manual or automatic forms except for the text built into web pages. Furthermore, it is impossible to obtain information on objects or people from images or videos. However, U-Media enables high integrity regarding information and open media in information collecting compared with existing media. Moreover, U-Media supports meanings that are more precise by annotating the object links or tag information. U-Media is able to support the precise transmission of meanings due to the annotation of related information because media plays a role in the precise transmission of messages to receivers by producing it as text, picture, and video. Various information annotated in content may be included that introduces the spontaneous and creative involvement of the participant surrounding media.

Using U-Media, producers can track media consumption patterns using a link applied to the real world. In addition, it is possible to present a new advertising business model using U-Media. If businesses want to have links to the same object in U-photos, an advertising business model is possible. Furthermore, U-Media can provide economic incentives to purchasers and providers. Incentive mechanism enabled by the concept of seamlessness is the most important attribute of U-Commerce (Lee & Ju 2006). In U-Media, a business model can be designed that shares the benefits of content producers and senders by providing links to objects and people.

# 3 U-Media Business Model

This section offers an easier understanding of U-Media using a U-Media scenario from the viewpoint of users and analyzes U-Media business models in detail. Though U-Media scenarios can be produced in a variety of forms according to classification criteria, this paper considers images and videos as basic scenarios for U-Media.

## 3.1 Scenario 1: U-Camera

*Daniel takes some pictures of his friends in front of a gallery using a U-Camera. His friends gather to check the picture using the U-Camera display. Contact information for each friend is included and linked to each picture, and some links are connected to surrounding objects. All friends predefine their contact information to their own UDA in order to send it when requested. In addition, the information of surrounding objects is annotated to the picture by receiving the information from the RFID Tag embedded in surrounding objects. In order to send the picture to his friends, Daniel selects the picture sending function in the U-Camera. Then, the pictures are sent to friends via e-mail or UDA.*

*That evening, James, who appeared in the pictures and is a friend of Daniel, likes the background image that appears on Daniel's blog. James wants to hang the picture on his wall. Thus, he clicks on the image to obtain detailed information. The web page of the gallery is loaded and displays additional information regarding the price and painter.*

### 3.2 Scenario2: U-Camcorder and U-TV

*Daniel watches a drama while waiting for his train at a station. The drama is famous and shown by "A" broadcasting company. In addition to the popularity of this drama, fashion products, such as accessories, dresses, and hairstyles are also popular. Daniel especially loves the bag used by the male actor in the drama. He selects the scene that appeared on TV using his UDA and a list of linked objects appears on the screen. From the list, he clicks on the bag and the product information is fully displayed. Daniel checks the price and selects the bag from the purchasing menu. He completes the purchase.*

### 3.3 Process Analysis

***Information included in U-Photo and U-Video if allowed to be open***
Location information: GPS information and URL for further information on the location
People/Actor information: contact information such as e-mail address or cell phone number, and URLs for further information on the people or actors
Objects in photos: basic information and URLs for further information on the objects

***Major economic subject in commerce***
Content producer: Producing images and movies, expecting future incentives
Content distributor: Diffusing uploaded content to their own blogs or other sites
Content consumer: Enjoying the produced images and videos; In the case of commerce, they are able to easily obtain detailed information by linking provided links.

#### 3.3.1 U-Camera System Architecture
**Process of U-Camera**

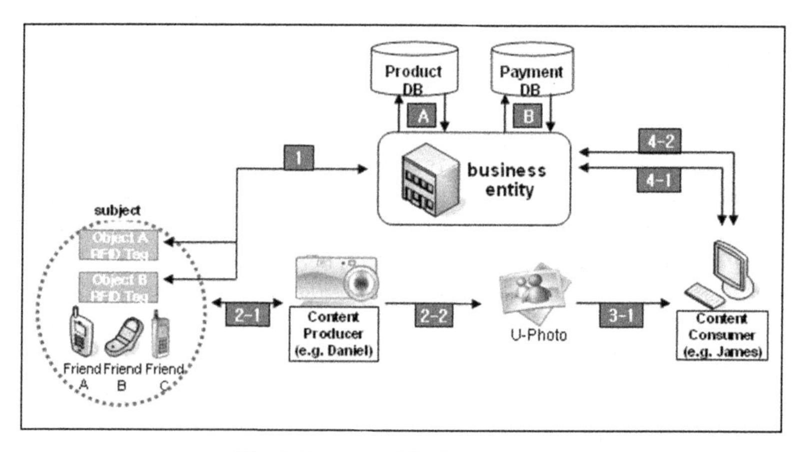


**Fig. 2.** Process of U-Camera Scenario

0. Specify the level of openness of contact information
1. RFID Tag: Insert digital information
2. Taking pictures: 2-1) Collecting digital information of people and objects, 2-2) Completing image files (U-Photo)

3. Image file transmission: 3-1) Receiving image files
4. Clicking images: 4-1) Loading additional information, 4-2) Clicking additional information
5. Commerce

In Step 0, each member predefined their contact information before taking pictures. Members who first consider privacy protection configure their e-mail as contact information, and members that are more open-minded configure their blog or mini-homepy page. In addition, members who are not concerned about the exposure of their cell phone number can link their number to the picture. In Step 1, providers and owners input digital information to the RFID Tag of various objects. Then, they link it to their own homepage or detailed information in order to use it for commercial purposes. In Step 2, contact information is collected from the RFID Tag of objects in the background of a picture and annotated to the picture in order to complete the picture (U-Photo). In Step 3, Daniel who took the picture transmits the image file to his friends. In the future, instant transmission of a picture will be possible using the WiBro or Bluetooth or WCDMA module in a camera then sent to a blog, e-mail, messenger, and other forms of communication. In Step 4, James who received the picture, clicks on a certain object in the picture to obtain additional information from the picture. In Step 5, a commerce process is achieved from the information embedded in the picture.

### 3.3.2 U-Camcorder System Architecture

**Process of U-Camcorder and U-TV Scenario**

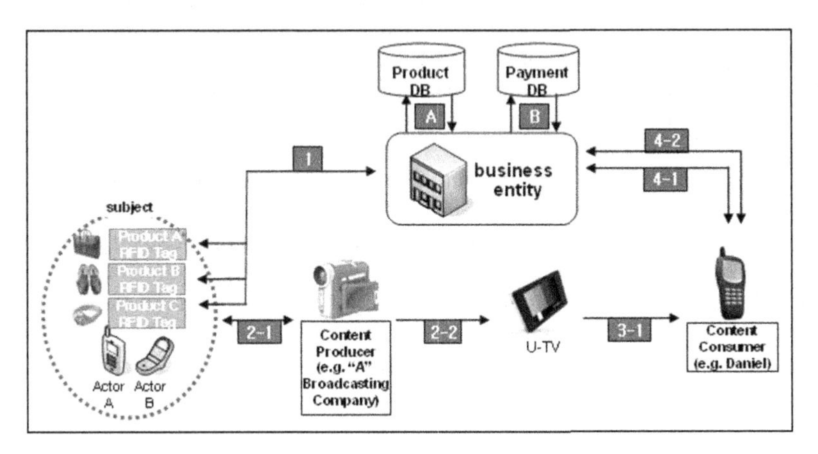


**Fig. 3.** Process of U-Camcorder and U-TV Scenario

0. Specify the level of openness of contact information
1. RFID Tag: Insert digital information
2. Taking movies: 2-1) Collecting digital information of people and objects, 2-2) Completing image files (U-Movie)
3. Video file transmission: 3-1) Receiving movie files, 3-2) Loading the scene in UDA screen

4. Listing of product: 4-1) Loading the additional information 4-2) clicking on the additional information
5. Commerce

In Step 0, each actor in the drama appears with predefined contact information. New actors can link their information to their agency's home page and famous stars may link it to their own home page. Steps 1 and 2 are the same process as the U-Camera. In Step 1, providers and owners insert digital information into the RFID tag of various objects. In Step 2, a U-Camcorder used to shoot the drama collects digital information of people and objects and completes U-Video files. In Step 3, Daniel loads a certain scene into his UDA while he watches the drama in order to know more about the bag that appeared in the drama. The linked product list is displayed on his UDA screen. In Step 4, Daniel clicks on the bag to obtain additional information about the product. In Step 5, a commerce process is achieved from the information embedded in the picture.

### 3.4 Analysis of a U-Media Business Model

Timmers (1998) defines business model as 1) an architecture for the product, service and information flows, including a description of the various business actors and their roles, 2) A description of the potential benefits for the various business actors, 3) a description of the sources of revenues. Applying the definition to U-Media business model, its participants can be classified as a content creator, distributor, consumer, and advertiser. A content creator first produces images, movies, and other objects, distributor diffuses the produced content to the distributor's blog or other sites, and content consumer enjoys the produced content in web sites or on mobile devices in which consumers are able to obtain more detailed information of content by clicking images in content. An advertiser publishes advertisements which connect the objects in content to the advertiser's site.

Regarding the potential benefits for such participants, Content producers and distributors will receive a proper incentive because they contribute to a commercial transaction to increase connecting points between the advertiser and the consumer. In the case of the content consumer, it is possible to easily obtain detailed information by simply clicking objects in content. An advertiser is able to overcome a ruptured process that has not been connected to a commercial transaction due to the fact that consumers couldn't obtain the information of the object published in photos. Table 1 denotes the role and potential benefit of the participants in this business model.

**Table 1.** The Values and Incentives of the Participants in U-Media

| Subject | Value | Incentive |
|---|---|---|
| Content producer | Guaranteeing connecting points between advertiser and consumer by producing and distributing content for consuming the produced content | Self-satisfaction, entertainment, and forthcoming incentives |
| Content distributor | | |
| Content consumer | - | Easy and convenient information access |
| Advertiser | Detailed information of the object appeared in the advertisement | Contact points with consumers, increasing sales |

The U-Media business model represents a sort of ubiquitous UCC (User Created Content) models. The reason that end users produce their own images or movies is to remember their joyful memories in the life and is due to the easiness of using the recording device. Since the content of U-Media includes location information (e.g. GPS), entity information, and object information that couldn't be included in traditional photos or movies and it can promote various types of participation. Thus, it can be regarded as an open type of content in the production and consumption of information.

The previous two scenarios are significant in that they support seamless commerce that could not be achieved by conventional media. The existing picture file has difficulty in obtaining additional information because objects in the picture have no links making it impossible to lead to commerce. In addition, the dresses and accessories worn by actors in the TV drama have not been easily connected to commerce. Thus, the two scenarios propose greater benefits for the customer using an advanced commerce process compared to conventional methods and provide clues to similar models used in other media.

# 4 Classification and Design of U-Media

This section investigates the classification of U-Media and analyzes user requirements regarding U-Camera and U-Photo.

## 4.1 Classification of U-Media for Input/Output Device and Content

Scenarios in U-Media can be classified according to devices and content. U-Camera, U-Camcorder, and U-Recorders are classified as input devices, and U-Display (e.g. U-TV, U-Picture Frame, etc.) and U-Speaker (e.g. U-Audio, U-MP3 Player, etc.) can be regarded as output devices. Furthermore, U-content for these devices can be referred to as U-Photo, U-Image, U-Video, U-Sound, U-Book, and U-News.

Characteristics in each device can be described as follows; U-Camera and U-Camcorder mean an intelligent camera or camcorder not only taking a picture but also annotating digital information (e.g. Hyperlink or Tag) of objects. A U-Voice Recorder records voices including surrounding digital information and provides the collected information through interaction with user or use device when the voice is reproduced. U-TV reproduces video content, including hyperlinks and the U-Picture Frame display pictures including links. U-Audio, U-MP3, U-Speaker, can be used as a sound player that reproduces sound signals including annotated information. U-Photo and U-Image include pictures and photos with links, U-Video includes video with links, U-Sound includes voice signal with links, and U-Book and U-Newspaper include books and newspapers with links. These are all classified as U-Content. Table 2 shows the classification of U-Media.

## 4.2 U-Media Design

This section demonstrates basic conditions applied to U-Media. These conditions are illustrated using U-Photo and U-Camcorder.

### 4.2.1 User Requirements of U-Photo

U-Camera should include the information of people and objects within the angle along the direction of the camera when the U-Camera creates U-Photos. If certain objects that are not allocated along the direction of the camera are included in the picture, unwanted information will be included. If the orientation of the camera is neglected, the information of passersby or objects in the background will be included in the picture. This is not desirable in the use of this U-Camera.

**Table 2.** Classification of U-Media

<table>
<tr><td colspan="2" rowspan="2">Input Device</td><td>U-Camera/U-Camcorder: an intelligent camera/camcorder that takes pictures of objects including digital information of objects (e.g. Hyperlink or Tag) and annotates it to the picture/video</td></tr>
<tr><td>U-Recorder: a recorder that records voices including surrounding digital information</td></tr>
<tr><td rowspan="3">Output Device</td><td rowspan="2">U-Display</td><td>U-TV: a reproduction device of broadcasting content that include hyperlinks</td></tr>
<tr><td>U-Picture Frame: a picture frame that includes pictures with links</td></tr>
<tr><td>U-Speaker</td><td>U-Audio/U-MP3 Player: a player that reproduces music and voice signals and additional information</td></tr>
<tr><td colspan="2" rowspan="4">U-Content</td><td>U-Photo/Image: objects in pictures and photos with links</td></tr>
<tr><td>U-Video: objects in videos with links</td></tr>
<tr><td>U-Sound: sound streaming files with links</td></tr>
<tr><td>U-Book/U-Newspaper: books and newspapers with links</td></tr>
</table>

It is recommended that a user decide whether the annotation of digital information should be limited to the object that is correctly focused on the pictures. In a picture with a foggy in the background and focused on a specific object, it is difficult to determine an efficient link between objects that are correctly focused and objects that include both clear and foggy focus. Only the photographer can recognize this situation. Thus, the U-Camera should support a function that enables a photographer to decide on links for the object according to the state of the focus.

The number of objects and people in a picture should be the same as that of the annotated information. If there are four people in the picture, the number of links related to the people should be four. Likewise, 10 objects in a picture should have 10 links. This means that N Tags have N links related to these Tags. However, digital information that includes non-embedded objects or people who do not open their contact information should be excluded.

There is an issue of privacy protection. Privacy that includes information regarding people and objects is the most sensitive when this information is annotated to files using U-Media. Thus, 1) the people appearing in a picture should have the right to determine the level of information and exposure, 2) the people appearing in a picture should have the right to annotate only the necessary information to a media file, and 3) a function is required for deleting information that should not be accessible after completing media files. In U-Camera, a method that annotates the information of people in a picture may be classified as follows: i) unconditional notification, ii) unconditional non-notification, iii) notification only when requested, and iv) knowing the photographer and whose picture is being taken (e.g. each name is listed in their cell phone). Because privacy is the most sensitive element regarding U-Media, it should be designed precisely.

The interface used in U-Media should be a simple and intuitive design. In addition, U-Media should configure options in the production of media files. In addition, it is necessary that certain settings be in place for the object and people appearing in pictures. Thus, an interface is required that minimizes complexity occurring in the configuration of options. Furthermore, applications are required to edit the information annotated in a file after completing media files.

## 5 Related Works

Wilhelm et al. (2004) described the system of annotation, such as CellID, User Name, Date, and Time, using a cell phone camera. Annotation of digital information to the image produced by a cell phone camera was attempted by considering a cell phone as a proper platform producing photos that can be applied to a network. Annotated digital information designated MMM (Mobile Media Metadata) was configured into four different steps. In Step 1, a user takes pictures using his/her cell phone. In Step 2, the Cell ID of GSM captures its location, user name, and production time and date of pictures. Step 3 is the transmitting process in a cell phone that transmits pictures and annotated information to the Metadata Repository of a server in which the matching algorithm of the server transmits a list that is most frequently selected by users by comparing the annotated information and metadata stored in the server. In Step 4, a user confirms the appropriate item from the list that is most similar to the pictures.

The difference between our approach and the research of Wilhelm et al. can be summarized in several ways. Our approach collects information using a communication process in the RFID Tag of subjects located along the angle of U-Camera and digital devices. Then, links are applied to the object in pictures. However, they provide the opportunity for users to confirm the most appropriate information by comparing the produced metadata using a server. Pictures produced using the system designed by them does not include individual links for all objects in the picture, but single metadata that is created based on the previous situation of the user using a mobile phone is annotated to a single picture file. They investigated user interfaces and systems in order to design an effective service design for the limited display size of a mobile terminal.

Sarvas (2004) demonstrated some improvements though the main idea was regarded similar to the study performed by Wilhelm et al. MobShare, a type of blog, is a system that immediately publishes pictures on the Internet using a cell phone, and provides album and comment functions. Although MobShare can produce and share picture content simultaneously, it cannot provide the information of objects appearing in the picture. Furthermore, it does not provide commerce functions.

Finally, Kohtake et al. (2004) employs a similar approach in which U-Photo is produced using the direct interaction between objects located in the real world and a camera. They propose the use of U-Photo as an interface for the remote control of electric home appliances that included a network, computer, and sensors. The U-Photo can be produced using several processes. First, a PDA including a CCD camera applied to a wireless LAN is used to take pictures by attaching an LED transmitter to the object that is being taken. Next, a U-Photo creator recognizes the ID of objects using the color of the LED and detailed information by searching the information applied in the ID from database. Following this process, a user can remotely control objects using the icon on the LED in the U-Photo of a PDA. This research was conducted on remotely controlled objects using the content of photos and demonstrated the differences compared to the goals of our approach, which were to design a commerce model based on U-Media. In addition, the research recognized objects using the color of the attached LED showed a difference in the annotation process of digital information compared to our study. Our approach applied several links to the objects appearing in pictures using the interaction between the RFID and the U-Camera.

## 6 Conclusion

This paper investigates the concept of media in ubiquitous environments. The media in ubiquitous environment not only takes advantage of human biological systems, but also the digital systems of human beings while conventional media appeals only to people's bio-systems. Ubiquitous media creates and consumes content through not only human cognitive and perceptual processes but also through the interactions between surrounding digital systems. Thus, media consumers can use the content linked to other content to obtain information seamlessly using the link.

We also proposed a commerce model based on U-Media by demonstrating scenarios and processes of a U-Camera/U-Camcorder. The classification of input/output devices and content used in U-Media help understand U-Media scenarios and recognize which models could be applied for commercial purposes according to device and content. In addition, we analyzed the business model using the Timmers' definition to understand the model as a perspective of business entity.

Future research topics which should be investigated include intellectual property rights issues, legal constraints of U-Media, and cost-benefit analysis model of auto-identification infrastructure etc.

## Acknowledgments

This research is supported by the Ubiquitous Autonomic Computing and Network Project, the Ministry of Information and Communication (MIC) 21st Century Frontier R&D Program in Korea.

## References


1. Lee, K., Ju, J.: Research Trend and Approach of Ubiquitous Commerce, Kyung Hee Business Review, Vol.2, No.2, (2005) 443–458
2. Lee, K., Seo, Y.: Design of a RFID-Based Ubiquitous Comparison Shopping System, Lecture Notes in Artificial Intelligence, vol. 4251. Springer-Verlag (2006) 1251–1267
3. Lee, K., Lee, J.: Design of Ubiquitous Referral Marketing: A Business Model and Method, Lecture Notes in Computer Science, vol. 4082. Springer-Verlag (2006) 103–112
4. Kim, H., Lee, K., Kim, J.: A Peer-to-Peer CF-Recommendation for Ubiquitous Environment, Lecture Notes in Computer Science, vol. 4088. Springer-Verlag (2006) 678–683
5. Lee, K., Jeong, M., Ju, J.: Seamlessness & Privacy Enhanced Ubiquitous Payment, Lecture Notes in Computer Science, vol. 4082. Springer-Verlag (2006) 143–152
6. Lee, K., Ju, J., Jeong, M.: A Payment & Receipt Business Model in U-Commerce Environment, Proceedings of the 8th International Conference on Electronic Commerce, Fredericton, Canada (2006) 319-324
7. Nelson, T. H.: Complex information processing: a file structure for the complex, the changing and the indeterminate, In Proceedings of the 20th ACM National Conference (1965) 84–100
8. Timmers, P.: Business Model for Electronic Markets, Electronic Markets, Vol. 8 No.2 (1998) 3–8
9. Wilhelm, A., Takhteyev, Y., Sarvas, R., House, N.V., Davis, M.: Photo Annotation on a Camera Phone, CHI 2004, Vienna, Austria (2004)
10. Sarvas, R.: Media Content Metadata and Mobile Picture Sharing, Proceedings of the 11th Finnish Artificial Intelligence Conference 2004, September 1–3, Vantaa, Finland (2004)
11. Kohtake, N., Iwamoto, T., Suzuki, G., Aoki, S., Maruyama D., Kouda, T., Takashio, K., Tokuda, H.: u-Photo: A Snapshot-based Interaction Technique for Ubiquitous Embedded Information, Second International Conference on Pervasive Computing (PERVASIVE2004), Advances in Pervasive Computing, Linz/Wienna Austria, (2004) 389–392

# Context-Awareness in Mobile Service Meets Fine Location Estimate

Tomohisa Yamashita, Daisuke Takaoka, Noriaki Izumi,
Akio Sashima, Koichi Kurumatani, and Koiti Hasida

National Institute of Advanced Industrial Science and Technology (AIST)
CREST, Japan Science and Technology Agency (JST)
Sotokanda 1-18-13, Chiyoda-ku, Tokyo, 101-0021 Japan
{tomohisa.yamashita,daisuke.takaoka,n.izumi,
sashima-akio,k.kurumatani,hasida.k}@aisit.go.jp
http://itri.aist.go.jp/en/

**Abstract.** In this paper, to tackle with uncertainty in the real world, the light-weight ontology drive approach is proposed for the realization of context dependent services. We concentrate on position information and an operation history, as a user's context, and develop our location-aware content delivery system. The evaluation experiment of our location estimate engine is performed in Akihabara Software Showcase at Information Technology Research Institute. Furthermore, through the proofing experiment in Expo 2005 Aichi, our proposed architecture is confirmed to enables us to realize the real world application of context dependency. Finally, we compare our location-aware content delivery system and related researches, and discus the advantage of our system.

**Keywords:** Active RFID system, Context-awareness, Mobile service, Content delivery, Location estimate.

## 1 Introduction

Because of the rapid development of ubiquitous computing technology, various mobile services have been realized using personal devices, cellular phones and IC cards, including i-mode, EZweb, Edy, Suica, and so on. These services emphasize information access during movement, but it is necessary to deal with rich context of users, such as operation histories as position information, to also provide context dependency. In fact, development of weak radio and positioning technologies make it difficult to realize context sensitive services in the real world. Therefore, research issues of uncertainty remain among various theories and applications.

To tackle the issues described above, two main approaches have been proposed. The first approach is realization of pull-type services based on passive IC-tag technologies. Although this approach can achieve extremely robust service to the uncertainty of device operation, the data utility in context dependency has been left in the operation of a user. Another approach is based on the logic of

W. Abramowicz (Ed.): BIS 2007, LNCS 4439, pp. 522–535, 2007.
© Springer-Verlag Berlin Heidelberg 2007

probability, including Bayesian networks. Although the introduction of various parameters enables detailed tuning, time and cost issues pertain in adjustment because of embedded logic.

From the standpoint of tackling uncertainty in the real world, we propose a lightweight-ontology drive approach to achieve context-dependent services. In this study, we specifically examine position information and an operation history as a user's context. To distinguish the observed objects and the observing subject in a real-world application, ontologies of areas and services were developed. A mechanism of adjustment for the real-world application is developed in mapping between the proposed ontologies. Through the proofing experiment in Expo 2005 Aichi, we have confirmed that our proposed architecture enables realization of the real-world application of context dependency.

## 2 Designing Context Awareness in Mobile Service

### 2.1 Location-Aware Service

In many kinds of information, position information and the operation history are useful clues to grasp the user context. They can be acquired through observation with sensors in an environment. For example, if a user stands in front of an exhibit at a museum and pushes a button of a user device to play its explanation, it is possible to guess that the user is interested in the exhibition and that they will listen to the explanation. In this paper, the concept of "Context Awareness" is realized through development of location-dependent content delivery.

### 2.2 Bridging by Linearizing Simplification

By supposing that we can obtain an ideal environment in which radio wave conditions are extremely stable in the real world, the distance from an RFID tag to an RFID receiver is calculable based on the Received Signal Strength (RSS) of an RFID receiver. However, it is difficult in the real world to observe the RSS precisely for the reason that the received RSS by RFID receivers is extremely unstable even though an RFID tag remains in a single position. Moreover, the instability of radio wave conditions is also reported as a result of reflection and phasing phenomena. Furthermore, because the human body absorbs radio waves, user crowding might give unexpected results in practice. Therefore, it is impossible to exclude the above factors, which all affect the electronic stability of environments. In actual situations where RSS is used for location estimation, it is difficult for any computational algorithm to adjust parameters and estimate locations precisely because of unstable environments.

To tackle real-world instability, we employ the number of detections of a tag ID as the key parameter for estimating its location, instead of the RSS. Regarding the robustness of computation, we propose a linearizing simplification to the relationship between the number of detections of a tag ID by an RFID receiver (antenna) and the distance from an RFID tag to an RFID receiver, we then regard this relationship as "the closer an RFID tag is located to an

RFID receiver, the higher the number of detections of a tag ID by an RFID receiver." Although introduction of various parameters enables precise tuning, a complicated model, e.g., a stochastic model, presents the problem of time cost in adjustment. This linearizing simplification decreases the complexity of the location estimation algorithm and enables adjustment of the parameters in a practical period.

### 2.3 Lightweight Ontology

To realize a method of location estimation with the above approximation as a computational algorithm, we must clarify a distinction between a representation of how we recognize location and a parameter that indicates how a computation is adjusted to the real world. Furthermore, from the standpoint of practical use, it is necessary for the real-world application to complete both estimating location and adjusting parameters in a very short period.

From the viewpoint of ontologies, to maintain an adjustment of a practical application, we must distinguish an object that is observed and a subject that is observing: we must also devise a means to recognize the real world and a method to infer a location. Furthermore, both of the above ontological aspects of the location estimation must be compatible to realize a service that is effective in the real world. The above discussion underscores the necessity of simplifying a model of a location estimation and a strategy of a service that is provided. Therefore, to retain robustness for instability of the real-world environment, we employ a hierarchical representation of areas that are recognized and contents that are serviced. At the same time, strong restrictions and rigorous constraints are unnecessary from the computational aspect. Moreover, correspondence between areas and contents is considered as a parameter for adjustment. In this paper, the above architecture of a hierarchical structure including estimated locations and provided services with fewer constraints and parameters is designated as a lightweight ontology.

## 3 Implementation of a Mobile Content Delivery Service

### 3.1 Outline of Our System

Aimulet GH+ was developed as a users' mobile device for our location-aware content delivery service in Global House, Expo 2005 Aichi. In our service with Aimulet GH+, a content list displayed on Aimulet GH+ is updated automatically as a user moves within the Global House. Based on the user's location, the system updates the content list containing some items of explanations about exhibits that are near the user's location. The user chooses one item from the content list and touches it on the display to play an explanation with sound, text, and graphics.

The components and data flow of location-aware content delivery service with Aimulet GH+ in Figure 2 is as follows. An active RFID tag on Aimulet GH+ transmits its tag ID; then RFID receivers detect the tag ID through RFID antennas and send it to the RFID server. The RFID server stores RFID data (RFID

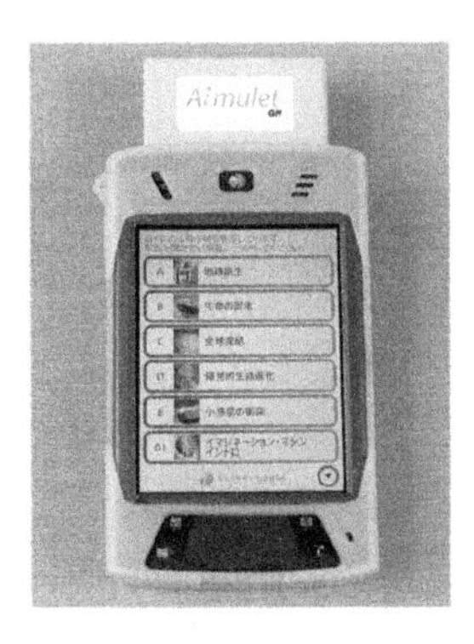


Fig. 1. Aimulet GH+

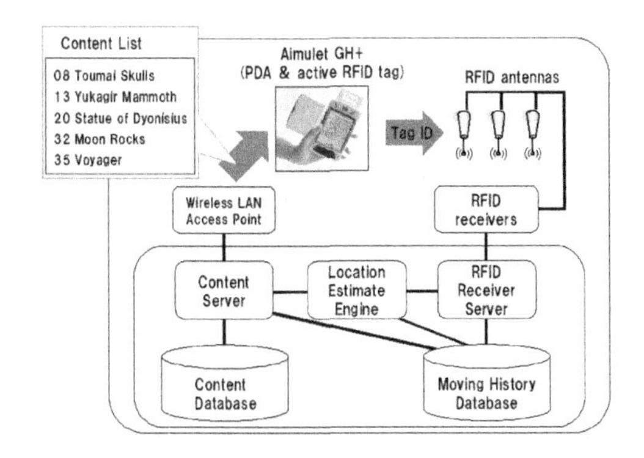


**Fig. 2.** Components and data flow in location-aware content delivery system

receiver IDs detecting a tag ID and their respective time stamps). The location estimation engine uses current and past RFID data and estimates the locations of Aimulet GH+ users. The content server sends the user's subarea to the content database, and requests a reply including the most optimal content list. The content database then submits a reply with a prearranged content list for that subarea. Subsequently, the content server sends it to Aimulet GH+ through a wireless LAN. Aimulet GH+ then receives a content list and displays it.

We use the active RFID system manufactured by Totoku Electric Co. Ltd. The product name of this system is MEGRAS. This system uses an On-Off Keying (OOK) modulation scheme; its frequency is 315 MHz or 303 MHz, depending on noise in the environment. The RFID receiver is equipped with an omnidirectional helical antenna, which detects a maximum 60 tag IDs per second. The RFID tag battery life is 1.5 years for a 1.0 s transmission interval.

### 3.2 Location Estimation

The whole area is divided conceptually into subareas. The division method is conceptualization as layers; some kinds of layers are prepared. Subareas in the same layer have similar size. The sizes of subareas in the lowest layer are smallest of all layers. Higher-layer subareas are larger.

The location estimation engine selects one subarea in a layer where the user with Aimulet GH+ is considered, with the highest probability, to be located. First, the location estimation engine selects a subarea in the lowest layer containing subareas of the smallest size. However, according to transmission of a tag ID, i.e. many RFID receivers detect the tag ID very well, the location estimation engine sometimes cannot select one subarea in a lower layer. Consequently, the location estimation engine chooses a higher layer containing larger subareas; it then tries to select one subarea within that layer. Based on the number of detections of tag IDs by RFID receivers, the selection is processed as follows.

First, a set of layers $L$ is defined as

$$L = \{l_1, l_2, \ldots, l_i, \ldots, l_n\}. \tag{1}$$

Second, set of subareas $S_i$ in layer $l_i$ is defined as

$$S_i = \{s_{1,i}, s_{2,i}, \ldots, s_{j,i}, \ldots, s_{m_i,i}\}. \tag{2}$$

A set of RFID receivers $R$ is defined as

$$R = \{r_1, r_2, \ldots, r_k, \ldots, r_l\}. \tag{3}$$

Each RFID receiver has a receiver point. Here, the receiver point $rp_{r_k}(id, T)$ for a tag ID $id$ at time $T$ is the number of detections of tag ID $id$ by RFID receiver $r_k$ for one second at time $T$.

To consider past RFID data of $t$ seconds ago, a time weight of $w_{time}(t)$ is applied. With time weight $w_{time}(t)$, we define the total receiver point $trp_{r_k}(id, T)$ for a tag ID $id$ at time $T$ as

$$trp_{r_k}(id, T) = \sum_{t=0} w_{time}(t) rp_{r_k}(id, T - t). \tag{4}$$

Here, the time weight $w_{time}(t)$ is a monotonically decreasing function.

Each subarea has a subarea point. Here, subarea point $sp_{s_j}(id, T)$ of subarea $s_{j,i}$ in layer $l_i$ for tag ID $id$ at time $T$ indicates how a user with Aimulet GH+ transmitting tag ID $id$ is considered to exist in subarea $s_{j,i}$ based on receiver points around subarea $s_{j,i}$.

To calculate subarea point $sp_{s_{j,i}}(id, T)$, a contribution ratio is defined. The contribution ratio $c_{r_k}(s_{j,i})$ of receiver $r_k$ to subarea $s_{j,i}$ in layer $l_i$ indicates how the detection of RFID receiver $r_k$ contributes to inferring that an RFID tag transmitting tag ID $id$ exists in subarea $s_{j,i}$ when RFID receiver $r_k$ detects the tag ID $id$. The value of the contribution ratio is determined based on this supposition: "The closer an RFID tag is located to an RFID receiver, the higher the number of detections of an RFID tag by an RFID receiver." Therefore, the closer an RFID receiver is to a subarea (or included into a subarea), the greater the contribution ratio of an RFID receiver to a subarea. Each contribution ratio is set as real number in the range of $[0, 1.0]$. With contribution ratio $c_{r_k}(s_{j,i})$, subarea point $sp_{s_{j,i}}(id, T)$ is defined as

$$sp_{s_{j,i}}(id, T) = \sum_{p=0}^{m_i} c_{r_p}(s_{j,i}) trp_{r_p}(id, T). \tag{5}$$

After calculation of all subarea points in a layer, the location estimation engine selects one subarea with the highest subarea point in all subareas in a layer. However, if little difference exists between the highest subarea point $sp_1$ and the second highest subarea point $sp_2$, the location estimation engine select no subarea in layer $i$; instead, it selects one subarea in the next-highest layer $i + 1$. The condition by which the location estimation engine rises from layer $i$ to layer $i + 1$ is defined as

$$min\_ratio_{i,i+1} \leq sp_2 / sp_1. \tag{6}$$

Here, $min\_ratio_{i,i+1}$ indicates the minimum ratio by which the location estimation engine rises from layer $i$ to layer $i+1$. Otherwise, the location estimation engine considers that a user with Aimulet GH+ transmitting tag ID $id$ exists in the subarea with the highest subarea point.

The subarea selection process is repeated every second.

### 3.3 Content Delivery

The content server sends the user's subarea to a content database, and requests a reply that includes the most optimal content list. The content database has a hierarchical structure of content lists and replies with a prearranged content list.

For example, in a museum containing some exhibition rooms and passages with many exhibits, in the case that the content database receives a larger subarea in a higher layer, e.g. an exhibition room, the content database responds with a prearranged content list that includes (an explanation of) the concept of the room above and (explanations of) the main exhibits in the room below. In contrast, when the content server receives a smaller subarea in a lower layer, e.g., the place in front of a specific exhibit, the contents database replies with a prearranged contents list that has the exhibit above and other exhibits around it, along with the concept of the room below. However, it is difficult to assign a contents list a priori to each subarea because of the large number of subareas of all layers.

In light of the problems posed by these issues, lightweight ontology is introduced to connect the hierarchical structures of the subarea and the content list. The hierarchical structures are not equivalent. If the content lists are assigned to characteristic subareas and the hierarchical structure of subareas is defined, then, based on the subarea's hierarchical structure, the content database assigns the content list of the parent subarea to the subarea to which a content list was not assigned previously. For example, in the case of a subarea in a middle layer, e.g., the half of a room to which a content list was not assigned previously, the content database responds with the same content list as its parent subarea.

### 3.4 Evaluation Experiment

The evaluation experiment of our location estimation engine was performed in the Akihabara Software Showcase (SSC) at the Information Technology Research Institute (ITRI).

In the Akihabara SSC, 30 RFID antennas and receivers are set on the ceiling, as shown in Fig. 4. The whole area of the Akihabara SSC contains six parts shown in Fig. 3, which are named based on their facilities: lounge space, seminar room, closed meeting room, open meeting room, reception, and living room. Because the living room was under construction, it was impossible to enter it for our evaluation experiment.

Five layers were prepared for the evaluation experiment, as shown in Fig. 5. Layer 5 contained the whole area of the Akihabara SSC. Layer 4 had six subareas.

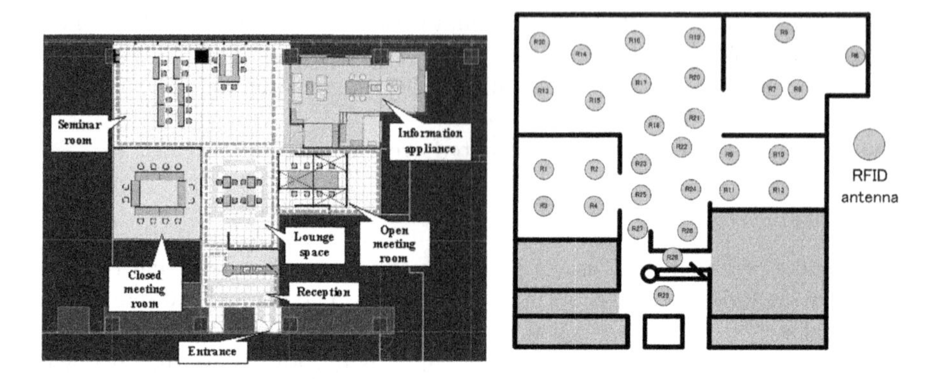


**Fig. 3.** Ground plan of Akihabara SSC

**Fig. 4.** Arrangement of RFID antennas in Akihabara SSC

**Table 1.** Minimum ratio in each layer

| $min\_ratio_{5,4}$ | $min\_ratio_{4,3}$ | $min_{r}atio_{3,2}$ | $min\_ratio_{2,1}$ |
|---|---|---|---|
| 0.1 | 0.05 | 0.05 | 0.001 |

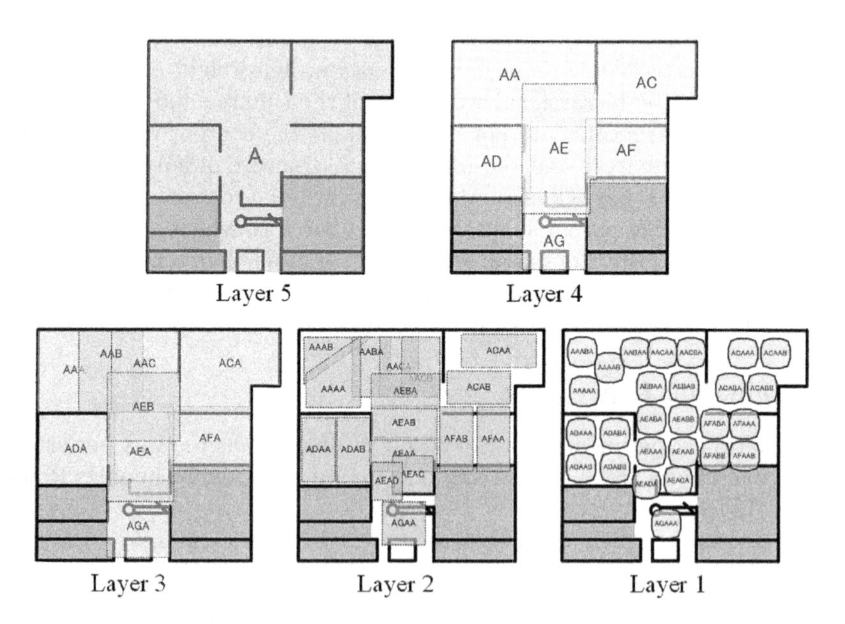


**Fig. 5.** Division of subareas in each layer

Each subarea corresponded to a room of the Akihabara SSC in Fig. 3. Layer 1 had 27 subareas, which were the smallest of all layers. The tree structure of the subarea multi-layer system is portrayed in Fig. 6. The tree structure of the content list is also shown in Fig. 6. Here, we show the case where the location

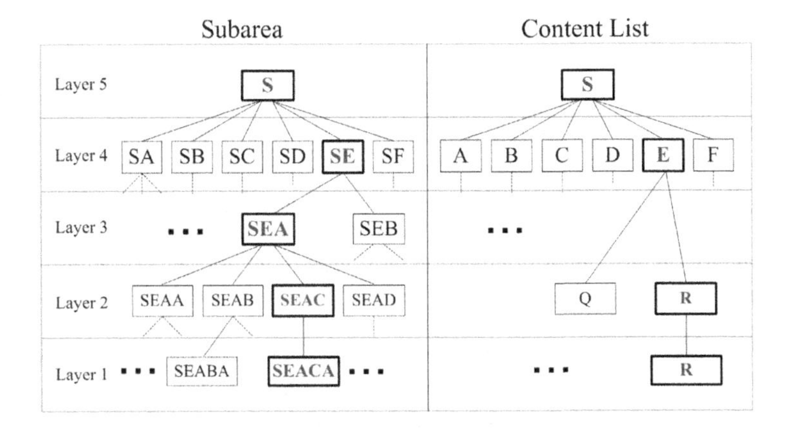


**Fig. 6.** Tree structure of subareas and content lists in multi-layer system

**Table 2.** Accuracy rate of location estimation engine in layer 4

| | trial | correct answers | accuracy rate |
|---|---|---|---|
| lounge space | 449 | 406 | 0.904 |
| seminar room | 459 | 346 | 0.754 |
| closed meeting room | 270 | 239 | 0.885 |
| open meeting room | 280 | 227 | 0.811 |
| reception | 80 | 55 | 0.688 |
| total | 1538 | 1273 | 0.828 |

estimation engine output subarea $SEAC$ and $SEACA$, content database assigns a content list $R$ based on the content list tree structure. The minimum ratios used in the evaluation experiment are shown in Table 1.

In our evaluation experiment in the SSC, a subject with Aimulet GH+ moved around the subareas in layer 5, and produced an actual subarea record (the subarea in layer 5 and time in which the subject existed actually). The active RFID data (RFID receiver IDs detecting a tag ID and their time stamps) were stored by the system. Subsequently, we confirmed that the subarea in the actual subarea record corresponded with the subarea estimated by the location estimation engine based on RFID data. This one comparison is defined as a single trial.

As results of the evaluation experiment, the number of trials, the number of correct answers, and the accuracy rate are listed in Table 2. From Table 2, the accuracy rate of the location estimation engine in layer 4 was greater than 80% overall. In the Akihabara SSC, trial subjects were satisfied with the location-aware content delivery service using Aimulet GH+ based on this location estimation engine because the provided content list was always suitable for the subject's location. Therefore, we confirmed that our location estimation engine had sufficient accuracy to provide appropriate content delivery services for users.

## 4 Proofing Experiments in Exp 2005

### 4.1 Outline of Exp 2005

Expo 2005 Aichi was held in Nagoya, Japan [12] from March 25 to September 25, 2005 (The duration had total of 185 days). The number of pavilions was about 90, and the total number of visitors since the opening was 22 million. Global House was one of the most popular pavilions and was designed to let visitors experience "Nature's Wisdom" that is the theme of Expo 2005. Global House had three parts, Mammoth Laboratory[1], Orange Hall and Blue Hall. In Orange Hall, there was an exhibition zone that explores the history of the human race and the environment from the vantage point of human creativity, i.e., a model of the Yukagir Mammoth, Statue of Dyonisius, Moon Rock, and a model of Voyager.

Japan Association for the 2005 World Exposition required measures for large number of visitors, i.e., information service of exhibits to visitors and reduction of the number of attendants. In Orange hall, maximum 350 visitors enter every 20 minutes from 9:30AM to 8:00 PM. Based on such a requirement, we developed CONSORTS for both content delivery system for visitors with users' mobile devices Aimulet GH and Aimulet GH+ and an exhibition management support service for managers. As a result, we realized information service for visitors and reduced about 50 attendants compared with a preliminary estimate.

### 4.2 Implemented Services in Orange Hall

In Orange Hall of Global House, we provided an integrated exhibition support system based on CONSORTS.

For realization of information service for visitors, we designed and constructed two kinds of users mobile devices, Aimulet GH and Aimulet GH+. Visitors can hear explanations about exhibits in the hall by Aimulet GH and Aimulet GH+. In the entrance of Orange Hall, general visitors received Aimulet GH and wheel chair visitors received Aimulet GH+. In the exit, all visitors must return them.

As for the sensor device that visitors have, an active RFID tag was applied to detect rough location and to count the number of visitors in the exhibit areas in Orange Hall. In Orange Hall, 139 RFID receivers and antennas are set up on the ceiling near exhibits. In development of our system in Orange Hall, we had two purposes. One was content delivery service for individual visitors. The other was exhibition management support for managers of Global House. To achieve both, we developed Aimulet GH and Aimulet GH+ containing an active RFID tag.

### 4.3 Content Delivery Service for Individual Visitors

Aimulet GH is a system using credit-card-sized audio information terminals that provides descriptions of each exhibit in either Japanese or English. Sound data

[1] Mammoth Laboratory housed the frozen remains of the Yukagir Mammoth excavated from the permafrost in Siberia. The Japan Association for the 2005 World Exposition had been making the frozen remains of the Yukagir Mammoth a featured exhibition at EXPO 2005.

**Fig. 7.** Usage of Aimulet GH+ in Orange Hall

and energy are transmitted simultaneously by infrared rays from light resources on the ceiling, eliminating the need for a power supply and enabling the realization of credit-card-sized terminals that measure just 5 millimeters thick and weigh 28 grams.

Aimulet GH+ is a system using PDA and an active RFID tag that provides voice, character, and image of each exhibit either Japanese or English based on the location of a visitor. Content server in the system selects 6 items from all 41 contents for Aimulet GH+ and decides a content list based on the visitor's location. A visitor with Aimulet GH+ can always get explanations of exhibits that are near him/her.

### 4.4 Exhibition Management Support for Managers

In Orange Hall, we provided following exhibition management support services for managers, i) Detection of locations of individual visitors, ii) Research of listening rate of contents, iii) Surveillance of distribution of Aimulet GH and Aimulet GH+, iv) Monitoring of congestion in hall.

Based on active RFID data (RFID receiver IDs detecting a tag ID and their time stamps), the system estimates the location of an individual visitor. In regard to listening rate on Aimulet GH, we can roughly research whether a visitor is listening to a sound content or not, and which content a visitor is listening to. This research is based on the difference of transmission of a tag ID emitted from an active RFID tag. When a visitor dangles the Aimulet GH from his/her neck, the system rarely detects its tag ID because human body absorbs radio wave emitted from an active RFID tag on Aimulet GH. However, when a visitor puts up Aimulet GH to his/her ear to listen a sound content, the system often detects its tag ID. In regard to listening rate on Aimulet GH+, we can research an listening rate of contents easily because all operation of visitors on Aimulet GH+ are sent to the system.

Surveillance of distribution of Aimulet GH and Aimulet GH+ means to count the number of the visitors to Orange Hall and the number of uncollected Aimulet GH and Aimulet GH+. The number of the visitors to Orange Hall of a specific period can be calculated by counting the total number of tag IDs detected in Orange Hall because it is unusual not to detect a tag ID existing in Orange Hall at all. After open hours, the system sometimes detected uncollected Aimulet GHs that staffs left in the hall.

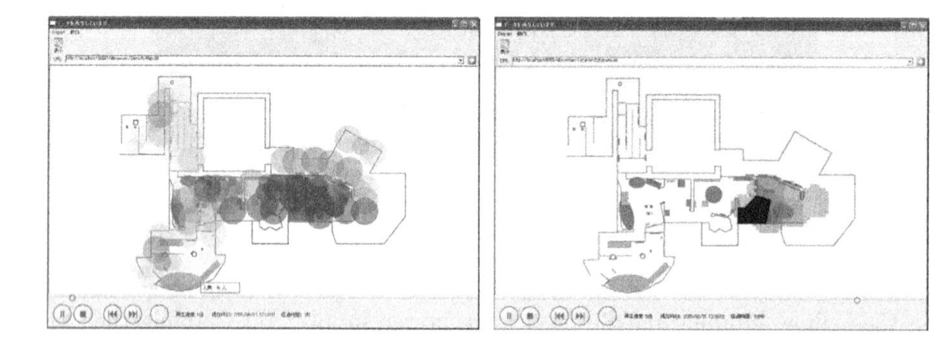

**Fig. 8.** Screenshot of tag density viewer in Orange Hall

**Fig. 9.** Screenshot of user location viewer in Orange Hall

By watching the density of RFID tags in the hall, the managers can monitor congestion of visitors. An example of the tag density viewer was shown in Figure 8. In the screenshot of the viewer, on a ground plan of Orange Hall, a circle means an RFID receiver detecting tag IDs. A darker circle means higher density of RFID tags.

By entering the tag ID to the viewer, the managers can monitor the location of a visitor. An example of our the user location viewer was shown in Figure 9. In the screenshot of the viewer, a darker part means the subarea where the location estimate engine regards as the visitor who uses Aimulet GH or Aimulet GH+ with the tag ID specified existing. Because tag IDs of Aimulet GH and Aimulet GH+ don't connect with personal information of visitors at all, the privacy of the visitors can be protected.

## 5 Discussion

In this section, we compare our context-aware content delivery and related studies, and discuss the advantages of our system.

First, we survey the elemental technology and theoretical studies. From the point of view of context-aware content delivery, our goal is realization of the service that contents are delivered to user's device automatically based on the location of the user. Location estimate with an active RFID system does not require specific actions to users, e.g. to touch an IC card reader with a passive IC card [15]. Wireless LAN systems also can realize such location estimate that does not require additional user's action. Furthermore, the accuracy of recent location estimate with wireless LAN, e.g. EKAHAU [11] and AEROSCOUT [9], is enough to our content delivery service. However, the location estimate with wireless LAN has disadvantage that the system can detect few users simultaneously. On the other hand, one RFID receiver we use can detect 60 tag IDs every second. Actually, our user device is required to be able to work in crowded because our system was planed to be used in Global House, where a maximum of 350 visitors enter every 20 min. In order to develop the system working in crowded, we use an active RFID system.

Although elemental technologies have progressed as described above, context-aware services using them are not performed only in the real world. A ubiquitous communicator was developed for information services in a ubiquitous computing environment. It was used in some verification experiments in shopping areas and underground malls. Shop information was transmitted in cooperation with local shopping areas [17]. However, because this system mainly used a passive RFID system, the services realized by this system differ from location-aware services that are intended to deliver contents to a user's device automatically based on that user's location.

Some researchers have developed a contextual data management platform for context-aware applications in the real world [4,5]. However, these platforms were only confirmed to work well in laboratory experiments; their performance has not been verified in places that ordinary visitors might use them.

In Legoland in Sweden, a missing-child retrieval Wi-Fi location system [9] was introduced. However this system is used only for searching for a missing child. Information services to general visitors are not provided with this system.

Our information platform provided information services to about 10,000 visitors every day during half a year in Expo 2005 Aichi. Moreover, regarding the Expo 2005 Aichi exhibition, the Nature Viewer information terminal was introduced at the Hitachi Group Pavilion [14,13]. Nature Viewer is an information terminal that reads information from a $\mu$-chip built into an exhibition site, and displays it. However, with this service, it was necessary for visitors to touch the $\mu$-chip with a Nature Viewer; it cannot be said that context-awareness, which we sought in our project, was realized.

The Hitachi information system provided information services to visitors only. In contrast, our information platform provided information services not only to visitors but also management support services to managers. Because studies that have realized actual verification experiments and which have simultaneously achieved context-awareness are rare, we can claim that this study has demonstrated both practicality and novelty.

From the standpoint of comparison of ontologies, several spatial ontologies that describe physical space and spatial relations have been proposed, including both geometric (e.g., GPS, GIS) and symbolic representation (e.g., places that are identified by their names) of space: DAML-Space, OpenCyc, SUMO, Region Connection Calculus (RCC). A much-updated ontology was proposed recently by the Digital Enterprise Research Institute (DERI) on WSMO. However, at present, each remains as a proposed description. Moreover, services that use the ontology of location have been the subject of studies with similar aims to ours [6,7,8]. However, the descriptions used in those studies represent simple recognition, e.g., "the floor consists of a room and a passage." These studies did not address performance in the real world.

In another study, a trail provides the contents of context dependence with PDA in a real-time application in the real world [1]. However, the clear separation of parameters and ontology is not established in that study. Our approach not only describes "how we consider location information." The characteristic of our research is mapping the structure of service: the structure of a contents list and the structure of physical area based on the properties of electromagnetic waves including a parameter are shown on a map as a representation of how we recognize location.

## 6 Conclusion

In this paper, the architecture of hierarchical structure including location estimated and service provided with less constraints and parameters is proposed as light-weight ontology. The location-aware content delivery system was developed with Aimulet GH+ which is composed of PDA and an active RFID. The evaluation experiment of our location estimate engine is performed in Akihabara Software Showcase at Information Technology Research Institute. Two services implemented in Orange Hall, Global House were introduced as the result of proofing experiment. One was content delivery service for individual visitors. The other was exhibition management support for managers. As a result of comparison with related researches, we confirmed that our system has both practicality and novelty.

## References

1. Daniel Sonntag: "Towards Interaction Ontologies for Mobile Devices accessing the Semantic Web", 4th International Workshop on "HCI in Mobile Guides", 2005.
2. Sashima, A., Izumi, N., Kurumatani, K.: CONSORTS: A Multiagent Architecture for Service Coordination in Ubiquitous Computing. Multiagent for Mass User Support, LNAI 3012, pp.196-216 (2004).
3. Sashima, A., Izumi, N., Kurumatani, K.: Location-Mediated Service Coordination in Ubiquitous Computing. In Proc. of the Third International Workshop on Ontologies in Agent Systems (OAS-03), AAMAS-03. 39-46 (2003).
4. U. Bandara, M. Minami, M. Hasegawa, M. Inoue, H. Morikawa, T. Aoyama, "Design and Implementation of an Integrated Contextual Data Management Platform for Context-Aware Applications", In Proceedings of Seventh International Symposium on Wireless Personal Multimedia Communications (WPMC'04), Volume 1 pp. 266-270, Abano Terme, Italy, Sept. 2004.

5. Gregory D. Abowd, Agathe Battestini ,Thomas O'Connell "Location Service: A framework for handling multiple location sensing technologies", College of Computing and GVU Center, Georgia Institute for Technology, Atlanta,. Georgia, USA, 2002. (http://www.awarehome.gatech.edu/publications/location_service.pdf)
6. Thibaud Flury, Gilles Privat, Fano Ramparany: "OWL-based location ontology for context-aware services", Artificial Intelligence in Mobile Systems, 2004.
7. Rob Lemmens and Marian de Vries: "Semantic Description of Location Based Web Services Using an Extensible Location Ontology", GI Tage 2004, pp. 261-276, 2004.
8. Harry Chen, Tim Finin, and Anupam Joshi: "An Ontology for Context-Aware Pervasive Computing Environments", Journal of Knowledge Engineering Review, No.3, Vol.18, pp.197-207, 2004.
9. http://www.aeroscout.co.jp/
10. http://www.consorts.org/
11. http://www.ekahau.com/
12. http://www.expo2005.or.jp/
13. http://www.expo2005.or.jp/ml/en/14/index.html
14. http://www.hitachi.co.jp/New/cnews/month/2004/11/1111a.pdf (Japanese)
15. http://www.jreast.co.jp/suica/ (Japanese)
16. http://www.wsmo.org/ontologies/location/
17. http://www.ubin.jp/press/pdf/TEP040915-u01e.pdf

# Adaptive Fuzzy Inventory Control Algorithm for Replenishment Process Optimization in an Uncertain Environment

Davorin Kofjač, Miroljub Kljajić, Andrej Škraba, and Blaž Rodič

University of Maribor, Faculty of Organizational Sciences,
Kidričeva cesta 55a, 4000 Kranj, Slovenia
{davorin.kofjac,miroljub.kljajic,andrej.skraba,blaz.rodic}@fov.uni-mb.si
http://www.uni-mb.si

**Abstract.** This paper presents a real case study of warehouse replenishment process optimization on a selected sample of representative materials. Optimization is performed with simulation model supported by inventory control algorithms. The adaptive fuzzy inventory control algorithm based on fuzzy stock-outs, highest stock level and total cost is introduced. The algorithm is tested and compared to the simulation results of the actual warehouse process and classic inventory control algorithms such as Least-unit cost, Part period balancing and Silver-Meal algorithm. The algorithms are tested on historic data and assessed using the Fuzzy Strategy Assessor (FSA). Simulation results are presented and advantages of fuzzy inventory control algorithm are discussed.

**Keywords:** Replenishment, optimization, fuzzy sets, stochastic model, simulation.

## 1 Introduction

In an organization even with moderate size, there may be thousands of inventory stock keeping units and the main warehouse task is to enable the undisturbed production by assuring the right amount of materials. Several different principles of warehouse optimization are described in [1], [2]. One way of optimizing the warehouse process is to find the right replenishment strategy, while reducing the cost of the warehousing processes to a minimum without stock-outs occurring and warehouse capacity being exceeded. Therefore, the operator is dealing with the decision problem - when to order and how many? This decision problem is vast, especially if we consider the fact, that he has to find the right replenishment strategy for more than 10.000 components stored in the warehouse.

This paper presents a real case study of replenishment process optimization in an automotive company. Optimization is based on an anticipatory simulation model, inventory optimization algorithms and fuzzy sets. In comparison to the other methods, a dynamic analysis of the considered system behavior is the main advantage of testing the strategy with the aid of simulation scenarios [3].

W. Abramowicz (Ed.): BIS 2007, LNCS 4439, pp. 536–548, 2007.
© Springer-Verlag Berlin Heidelberg 2007

Modern business environment is dynamic and uncertain. Inventory control in such an environment is a complex task. The source of uncertainty can be either the demand or the supply process. The demand and supply uncertainty can be divided into time and quantity deviations from the original plan. The time deviation occurs when customers place their orders earlier or later than expected. The quantity deviation occurs when order quantity is different than the one anticipated or delivered. Both deviations can also occur simultaneously. Because of the mentioned uncertainties, a new term must be introduced - a warehouse, which task is to compensate those variations [4]. In other words - it should be a warehouse with adaptive replenishment strategy as deterministic approaches lead to problems in an uncertain environment [5].

Adaptive fuzzy inventory control has been recently researched and is presented in [6], [7], [8], [9]. Our adaptive approach is different from those mentioned, that it uses known heuristic inventory algorithms, e.g. Silver-Meal, and assesses their simulation results and chooses the appropriate replenishment strategy. Assessment is performed using fuzzy number of stock-outs, highest stock level and total cost for inventory control. Demand is stochastic and discontinuous; its quantity is changing and time varying gaps occur between two successive demands. Lead times are long (two to three months) and stochastic. Our approach also optimizes a complex cost function consisting of: ordering cost, take-over cost, transportation cost, physical storage cost, capital cost and manipulation cost, without violation of the following restrictions: warehouse capacity must not be exceeded and stock-outs must not occur.

The paper is structured in the following way: Section 2 describes the warehouse model, Section 3 elaborates replenishment algorithms with examples while simulation results are presented and discussed in Section 4. The conclusion is presented in Section 5.

## 2 Warehouse Problem Formulation

Dealing with problems of warehousing, we encounter several contradictory criteria. An overly large warehouse means a greater amount of stock, greater capital cost and more staff. The space itself is very valuable today. An overly small warehouse can represent possible stock-outs, it demands a reliable supplier etc. The warehouse in our case is used for storing components for further build-in. The components are delivered into the warehouse from different vendors and used in the production line, where they eventually become a part of different finished products. Components belong to different categories according to ABC and XYZ classification, which is described thoroughly in [1].

As the production process and the delivery are stochastic, future stock dynamics cannot be determined with a certainty and have to be anticipated. That is, anticipatory models will act according to the past states of the system and according to the desirable and possible future states [10]. Formally,

$$x_{t+1} = f(x_1, x_2, \ldots, x_t, \hat{x}_s^t) \tag{1}$$

where $\hat{x}_s^t$, $t+1 \leq s \leq t+k(t)$, is a prediction $x_s$ done at time $t$, and $k(t)$ is the size of the forecasting interval at time $t$. If $x$ is the stock variable, then Eqn. 1 is describing its past, present and future dynamics. Anticipation is done by running simulation scenarios, where replenishment algorithms predict future stock dynamics based on production plans. Different algorithms yield different predictions and the algorithm with the best results (regarding Eqn. 5) is used to change the state of a system, which will also affect its future states.

### 2.1 Mathematical Formulation

From control point of view, warehouse optimization problem can be described with the following equation:

$$x(k+1) = x(k) + d(k) - p(k), k = 0, 1, 2, \ldots \quad (2)$$
$$x(0) = x_0$$

where $x(k)$ represents stock variable, $d(k)$ stochastic material delivery, $p(k)$ stochastic production process and $x_0$ the initial stock value at $k = 0$. The delivery function $d(k)$ is delayed for an average time of an order $o(k)$ as defined by Eqn. 3:

$$d(k) = o(k - \varphi(\tau_d)) \quad (3)$$

where $\varphi(\tau_d)$ represents discrete uniform probability density function. Time delays $\tau_d$ are stochastic and tend to be very long for some materials (e.g. two to three months) thus representing a problem, because they are usually much longer than the time period of six weeks in which the production plan can be predicted with a certainty. To compensate the stochastic delivery delay, the replenishment order policy $o(k)$ has to be defined as:

$$o(k) = f(x(k), d(k - \tau_d), p(k + \tau_p)) \quad (4)$$

where $\tau_p$ represents a production plan horizon. It is necessary to find such $o(k)$ to minimize the following cost function:

$$J(o(k)) = \sum_{k=0}^{n} g(c_c x(k), c_p x(k), c_t(w) d(k - \tau_d), c_o o(k), c_v d(k - \tau_d), c_m p(k)) \quad (5)$$
$$x_{min} \leq x(k) \leq x_{max}$$

where $g$ represents cost function consisting of: cost of capital $c_c$, cost of physical storage $c_p$, transportation costs $c_t$, fixed ordering costs $c_o$, costs of taking over the products $c_v$ and cost of manipulation in production process $c_m$. The following restrictions have to be considered:

- maximal warehouse capacity ($x_{max}$) for a specific product must not be exceeded,
- no stock-outs may occur ($x_{min}$).

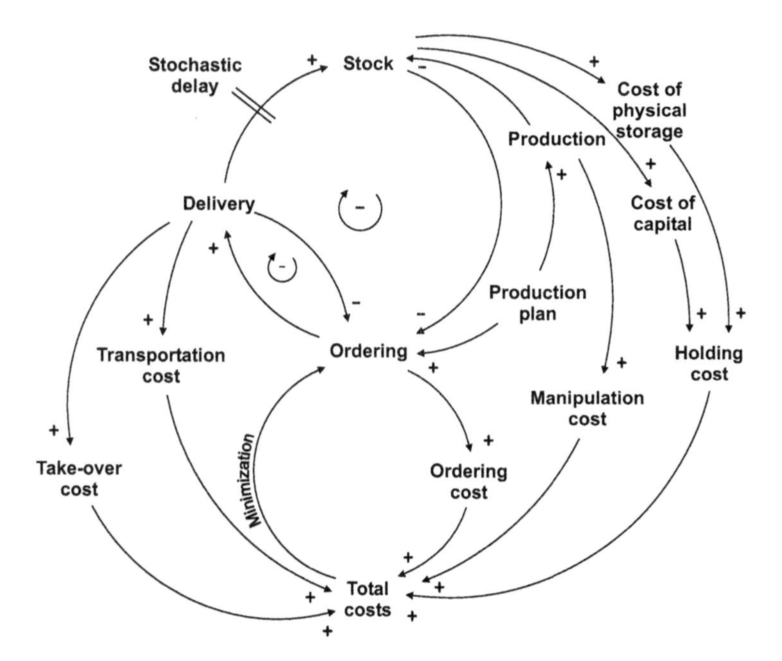


**Fig. 1.** Causal loop diagram (CLD) of the warehouse model for replenishment process optimization

$x_{min}$ can also be represented as a safety stock, where stock level should not fall under the safety stock level.

The transportation cost is explicitly considered in the model. Rather than assuming it to be a part of the fixed ordering cost or to be insignificant [11], we will take the transportation cost as a function of the shipment lot size. In this case transportation cost $c_t$ can be described by the following function:

$$c_t(w) = \begin{cases} 0, w = 0 \\ C_{t1}, w \in (0, W_1] \\ C_{t2}, w \in (W_1, W_2] \\ C_{tn}, w \in (W_{n-1}, W_n] \end{cases} \qquad (6)$$
$$0 < C_{t1} < C_{t2} < \ldots < C_{tn}$$
$$0 < W_1 < W_2 < \ldots < W_n$$

where $C_{t1}$, $C_{t2}$,..., $C_{tn}$ are transportation cost constants, dependent on the transport weight $w$, and $W_1$, $W_2$, ..., $W_n$ are weight constants.

## 2.2 The Warehouse Model

Fig. 1 represents the causal loop diagram (CLD) from which the influences of the warehouse model elements can be observed. The arrow represents the direction of the influence and the "+" or "-" sign its polarity. The "+" sign is used between

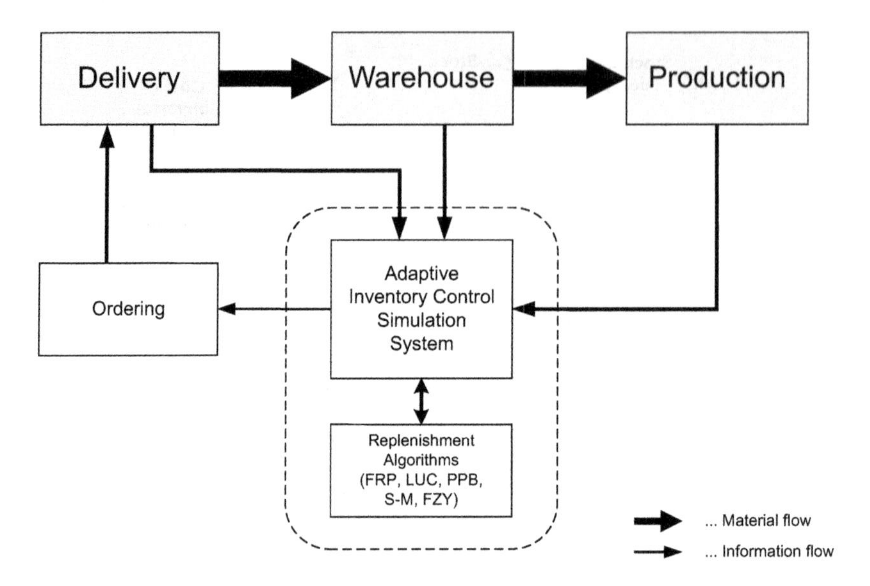


**Fig. 2.** Block model of the warehouse simulation model for replenishment process optimization

two elements, if the causal element has the same influence on the consequential element. The "-" sign is used, if the causal element has opposite influence on the consequential element.

There are two negative feedback loops in the CLD. The first interconnects *Stock*, *Ordering* and *Delivery* and it represents the fact that less is ordered, if the stock level is high. The second interconnects *Delivery* and *Ordering* and represents the concept that less is ordered, if more was ordered before. This loop considers orders, which have not been delivered yet and will have impact on the stock level later.

Fig. 2 presents the simplified warehouse model. It is a warehouse model with added adaptive inventory control simulation system (AICSS) supported by replenishment algorithms. The AICSS gathers information on production plan, stock-on-hand and delivery and runs different simulation scenarios according to the replenishment algorithms to anticipate future stock dynamics. A fuzzy strategy assessor (FSA) is used to assess the results of simulation runs. Replenishment algorithms and FSA are described in the following sections. Based on the FSA assessment, the AICSS offers the ordering process information on how much and when to order. For the detailed description of the warehouse model one should see [12].

## 3 Replenishment Algorithms

Optimization is based on the simulation model described previously and the following inventory control algorithms:

- Fixed review period (FRP),
- Least unit cost (LUC),
- Part period balancing (PPB),
- Silver-Meal algorithm (S-M),
- Fuzzy control algorithm (FZY).

The Fixed review period algorithm (FRP) is similar to the $(R, S)$ system (described in [1]), where, every $R$ units of time, an order is made to adjust the stock level to the order-up-to-level $S$. In contrast to the $(R, S)$ system, $S$ is not a fixed value in the FRP algorithm. The Fixed review period is described in detail in [12], while Least unit cost, Part period balancing and Silver-Meal algorithms are described in detail in [1]. Following is the detailed description of the FZY algorithm.

### 3.1 Fuzzy Control Algorithm

Fuzzy control algorithm uses Fuzzy strategy assessor (FSA) to assess replenishment optimization simulation results. The FSA is based on fuzzy sets introduced by Zadeh [13]. There have been several studies that applied fuzzy sets in inventory control ([14], [15], [16], [17], [18], [9]). These papers presented studies of various fuzzy inference systems where fuzzy demand, inventory level, order quantity, lead times and various fuzzy costs, e.g. shortage cost, were studied and analyzed. This paper is different from those previously mentioned in that it does not deal with shortage cost because it is unknown, but instead investigates the fuzzy number of stock-outs, highest stock level and total cost for inventory control.

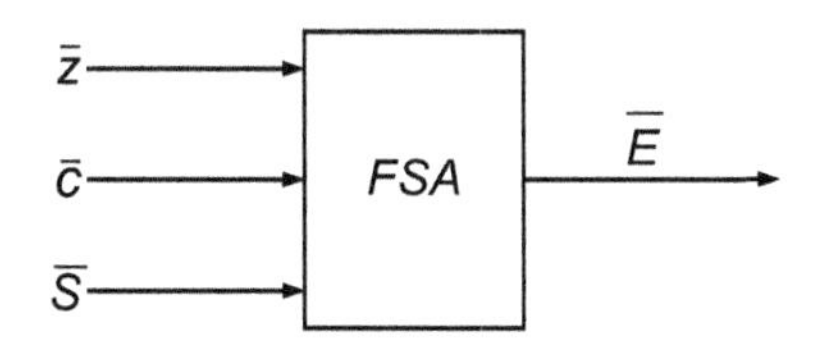


**Fig. 3.** Fuzzy strategy assessor (FSA)

In this case, the value of stock-out or shortage cost is difficult to assess since materials are often a part of several finished products. The shortage cost is not merely the value of material quantity lacking for the production process. The number of stock-outs are the only information available here and difficult to assess regarding costs. Since this problem can be described as soft, fuzzy sets were chosen to assess the replenishment strategies results.

The inputs of the FSA, as shown in Fig. 3, are matrices for number of stock-outs $\bar{z}$, total cost $\bar{c}$ and highest stock level $\bar{S}$, each matrix containing simulation results of various simulation strategies. The output of the FSA is replenishment strategy estimation matrix $\bar{E}$. Matrix elements values for $\bar{z}$, $\bar{c}$ and $\bar{S}$ are normalized in the interval [0,1] using the following equation:

$$b_i = \frac{b_i}{max(\bar{b})}, i = 1, 2, \ldots, n \tag{7}$$

where $\bar{b}$ is either the matrix $\bar{z}$, $\bar{c}$ or $\bar{S}$, $b_i$ is the matrix element, $max(\bar{b})$ is the maximal value among matrix elements and $n$ is the number of elements in the matrix. Matrices are then classified (fuzzified) using equally spaced Gaussian membership functions (GMFs) shown in Fig. 4.

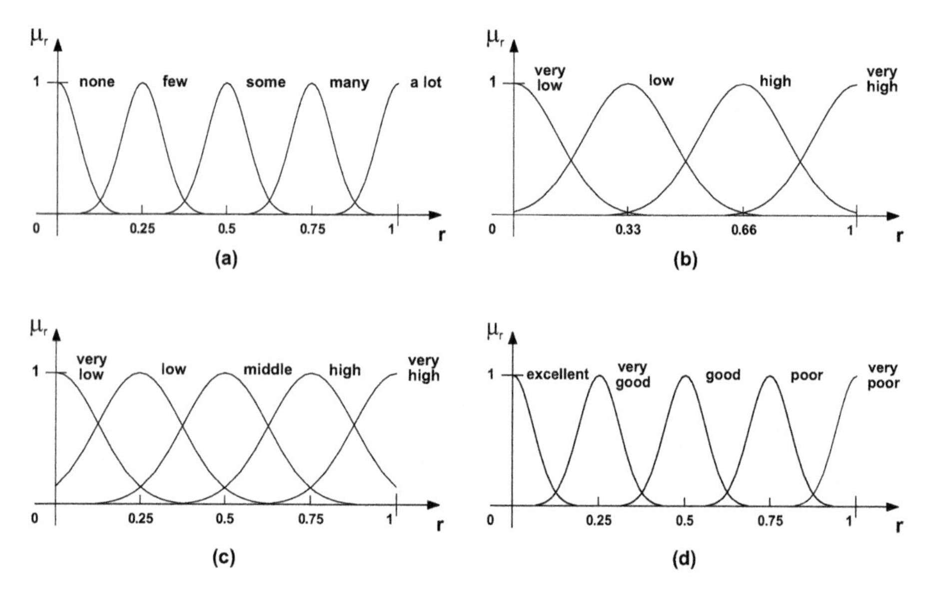


**Fig. 4.** Gaussian membership functions for stock-outs (a), total cost (b), highest stock level (c) and replenishment strategy estimation (d)

The Gaussian curve is given by the following function:

$$f(x) = e^{\frac{-0.5(x-m)^2}{\sigma^2}} \tag{8}$$

where $m$ is mean and $\sigma$ is variance. Means and variances for each GMF are given in Table 1.

**Table 1.** Means and variances of GMFs for each variable used in FSA

| *Variable* | $\sigma$ | $m$ |
|---|---|---|
| z | 0.06 | 0,0.25,0.5,0.75,1 |
| c | 0.15 | 0,0.33,0.66,1 |
| S | 0.12 | 0,0.25,0.5,0.75,1 |
| E | 0.05 | 0,0.25,0.5,0.75,1 |

The FSA inference system contains the expert's rule base consisting of 100 ($5 \cdot 4 \cdot 5 = 100$) rules. Estimation of the $i^{th}$ replenishment strategy is calculated according to the following equation:

$$IF\ z_i\ AND\ c_i\ AND\ S_i\ THEN\ E_i, i = 1, 2, \ldots, n \tag{9}$$

where $n$ is a number of simulation strategies. A few examples of rules are given below:

- If $z_i$ is none and $c_i$ is very low and $S_i$ is very low then $E_i$ is excellent,
- If $z_i$ is many and $c_i$ is very high and $S_i$ is low then $E_i$ is poor.

The rules are used to obtain the final output $\bar{E}$, which is then defuzzified using a *Som* (smallest value of minimum) function in the interval [0,1]. The strategy with the lowest grade is selected as the one yielding the best simulation results. If several strategies achieve the same lowest grade, they are assessed once more. Matrices $\bar{z}$, $\bar{c}$ and $\bar{S}$ now contain only the strategies with the lowest grades and are again normalized using the Eqn. 7 and fuzzified. A new estimation matrix $\bar{E}$ is calculated. This process is repeated until only one strategy has achieved the lowest grade.

The fuzzy control algorithm (FZY) is based on all algorithms mentioned before. All algorithms are run in a simulation and the results are assessed using a fuzzy strategy assessment (FSA), providing the estimation matrix $\bar{E}$, upon which the order time and quantity are selected.

**Table 2.** Numerical example of FZY algorithm estimation

| *Algorithm* | $E_j$ | $q_j$ |
|---|---|---|
| FRP-14 | 0.22 | 137 |
| FRP-21 | 0.24 | 137 |
| FRP-28 | 0.45 | 577 |
| FRP-35 | 0.4 | 777 |
| FRP-42 | 0.17 | 777 |
| FRP-49 | 0.41 | 977 |
| FRP-56 | 0.42 | 977 |
| FRP-63 | 0.46 | 977 |
| FRP-70 | 0.46 | 977 |
| PPB | 0.2 | 777 |
| LUC | 0.46 | 1938 |
| S-M | 0.46 | 577 |

Steps, performed during the simulation of the FZY algorithm, are:

1. Simulate the whole range of algorithms in $m$ iterations
2. Calculate average results of each algorithm
3. Assess average results using FSA to obtain the estimation matrix
4. Select algorithm with the lowest grade

5. Select the first order that occurred in the simulation of the selected algorithm (according to Eqn. 10) and apply it in simulation. The output of the fuzzy control algorithm is an order $o(k)$ as defined by the following equation:

$$o(k) = f(q_j), j = 1, 2, \ldots, n \tag{10}$$

where $j$ represents the chosen strategy, $n$ a number of simulation strategies and $q$ a vector of order quantities of all replenishment algorithms. The numerical example of estimation is provided in Table 2. The algorithm with the lowest assessment is $FRP - 42$ and the order quantity $o(k)$ is 777 pieces.

## 4 Results

The simulation model has been built using Matlab with its Simulink and Fuzzy logic toolboxes. The experiment was performed with the historic data provided by the observed company. Altogether, four materials were examined in this study and their details are described in Table 3.

**Table 3.** Material lead times in weeks, demand mean and demand interval

| *Material* | 1 | 2 | 3 | 4 |
|---|---|---|---|---|
| Lead time | 2-3 | 2-3 | 6-8 | 6-9 |
| Demand mean | 2629.23 | 143.07 | 3500.32 | 1567.27 |
| Demand interval | [0,62800] | [0,744] | [0,22000] | [0,13168] |

The simulation of the actual warehouse process (RP - Real process) was using real data of delivery and demand while the simulation with replenishment algorithms (VP - Virtual process) was using only real demand data. The ordering and delivery process in VP were controlled by replenishment algorithms. The RP simulation was run only once, whereas ten simulation runs were executed for every VP replenishment algorithm. Based on these simulation runs, average costs and average stock-outs were calculated. With several simulation runs and a calculation of average values, we have tried to minimize the influences of the random generator, which represent the stochastic environment. Out of all simulation runs the maximum stock level was considered as the warehouse capacity limit and the strategy with minimal highest stock level was favored.

**Table 4.** Summary of simulation results for all materials

| *Material* | *RP Cost* | *VP Cost* | *Savings (%)* | *VP Strategy* |
|---|---|---|---|---|
| 1 | $1.63 \cdot 10^6$ | $1.52 \cdot 10^6$ | 6.7 | FZY |
| 2 | $3.48 \cdot 10^6$ | $3.11 \cdot 10^6$ | 10.6 | FZY |
| 3 | $1.91 \cdot 10^6$ | $1.88 \cdot 10^6$ | 1.6 | FZY |
| 4 | $2.62 \cdot 10^6$ | $1.89 \cdot 10^6$ | 27.8 | FZY |
| Total | $\Sigma = 9.64 \cdot 10^6$ | $\Sigma = 8.4 \cdot 10^6$ | 12.7 | |

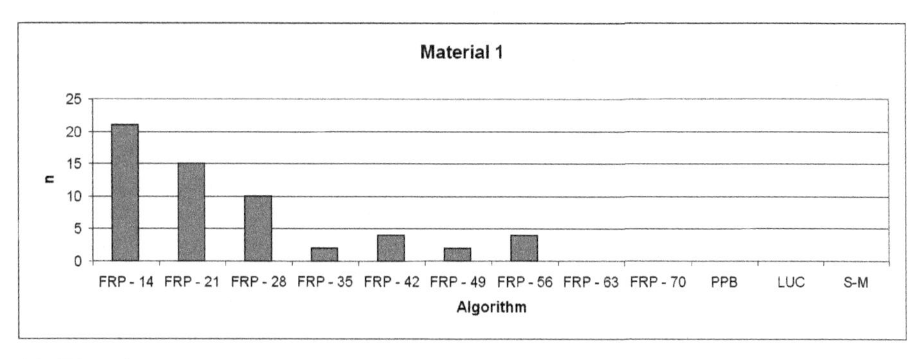


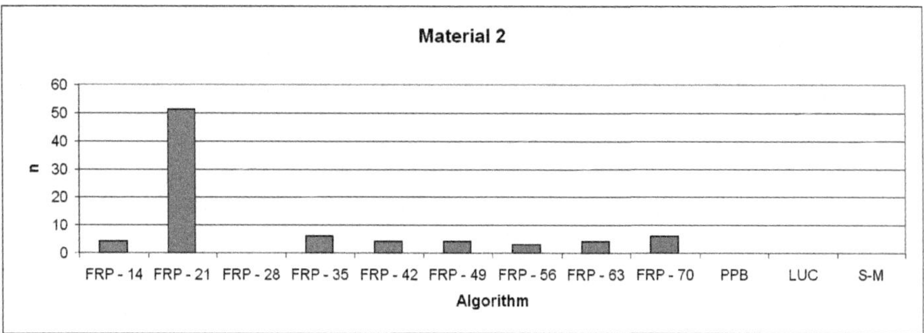


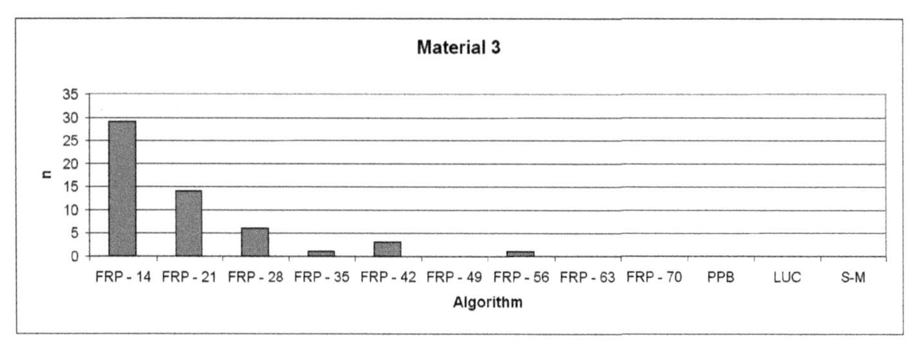


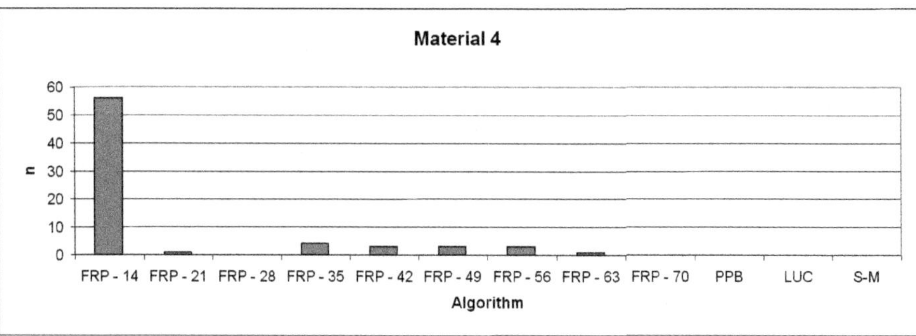


**Fig. 5.** Frequency distributions of algorithms used during FZY algorithm simulation for each material

The simulation was run for a period of seven years in RP and VP, but also with subperiods of 24 weeks in the VP; depending on the algorithm, at each time-step (continuous review) or at each-review period (periodic review), the simulation was run for 24 weeks, since this is the production plan prediction period, thus providing a replenishment schedule until the next review.

A Monte Carlo simulation was used for variation of production plan unreliability. The production plan variability was simulated by perturbations of its quantity every two weeks for a certain amount, e.g. 5%. Variable lead times were simulated by uniform random generator. If stock-outs occurred during the simulation, the missing quantity was transferred into the next period. The safety stock was also considered; it was equal to the average weekly demand.

Results in Table 4 show the comparison of total cost of the RP and the strategy chosen by the FSA among the VP strategies. The results presented in Table 4 indicate that the FZY algorithm was chosen as the one yielding the best results for all cases. FRP, LUC, PPB and S-M algorithms were never chosen. The total cost saving is 12.7%, if all four materials are considered.

Frequency distributions of algorithms used during the simulation of FZY algorithm are presented in Fig. 5. One can notice that predominant algorithm is FRP - 14 days for Materials 1, 3 and 4 while algorithm FRP - 21 days is predominant for Material 2. It is also worth to point out that classic algorithms PPB, LUC and S-M were never chosen during the FZY simulation. Obviously, their heuristics does not work well for the selected materials. The reason for that may be a more complex cost function (Eqn. 5) than the cost function used by classic algorithms in their optimization calculations, which uses only cost of capital and fixed ordering cost.

**Table 5.** Comparison of total cost of FZY ($c_1$) and of predominant algorithm used in FZY simulation ($c_2$)

| *Material* | *Algorithm* 1 | $c_1$ | *Algorithm* 2 | $c_2$ | $\frac{c_2}{c_1}$ |
|---|---|---|---|---|---|
| 1 | FZY | $1.52 \cdot 10^6$ | FRP - 14 | $1.77 \cdot 10^6$ | 1.16 |
| 2 | FZY | $3.11 \cdot 10^6$ | FRP - 21 | $3.93 \cdot 10^6$ | 1.26 |
| 3 | FZY | $1.88 \cdot 10^6$ | FRP - 14 | $2.16 \cdot 10^6$ | 1.15 |
| 4 | FZY | $1.89 \cdot 10^6$ | FRP - 14 | $2.06 \cdot 10^6$ | 1.09 |

Table 5 presents comparison of total cost of FZY ($c_1$) and of predominant algorithm used in FZY simulation ($c_2$). If only predominant algorithm would be used in simulation instead of FZY algorithm, it would produce higher total cost for each material. The total cost would be higher for 16, 26, 15 and 9 percent for Material 1, 2, 3 and 4 respectively. One can deduce that variability of replenishment algorithms produces lower total cost than using strictly one replenishment policy.

## 5 Conclusion

The research presented in this paper is the real case study of replenishment process optimization on historic data of selected materials. Lead times are stochastic and so is demand. The demand has also a high dispersion from the mean.

Simulation results of the FZY algorithm yielded significant cost savings. The introduced adaptive fuzzy inventory control algorithm surpassed results of other algorithms, including classic ones like Least-unit cost, Part period balancing and Silver-Meal. It was also proven, that there is a need for an adaptive algorithms, which is able to follow the changing patterns in demand, because the days of a stable market are long gone.

This paper also indicates that classic inventory control algorithms use inadequate cost function, which is not appropriate for replenishment process optimization for materials selected for this study. The cost function presented in this paper is more complex than the one used by classic algorithms and contains: cost of capital, cost of physical storage, transportation costs, fixed ordering costs, costs of taking over the products and cost of manipulation in production process, while the cost of classic algorithms contains only the cost of capital and fixed ordering cost.

Results of the study are encouraging and the direction of our future research is a thorough validation of the presented warehouse model and its implementation in the company's replenishment process.

**Acknowledgment.** This research was supported by the Ministry of Higher Education, Science and Technology of the Republic of Slovenia (Programme No. UNI-MB-0586-P5-0018). Our sincere thanks goes also to Mr. Valter Rejec and his team.

## References

1. Silver, E.A., Pyke, D.F., Peterson, R.: Inventory management and production planning and scheduling. John Wiley & Sons (1998)
2. Tompkins, A.J., Smith, J.D.: The warehouse management handbook. Thompkins Press (1998)
3. Kljajić, M., Bernik, I., Škraba, A.: Simulation approach to decision assessment in enterprises. Simulation **74** (2000) 199–210
4. Minner, S.: Strategic safety stocks in supply chains. Springer Verlag (2000)
5. Reiner, G., Trcka, M.: Customized supply chain design: Problems and alternatives for a production company in the food industry. a simulation based analysis. International Journal of Production Economics **89** (2004) 217–229
6. Rotshtein, A.P., Rakityanskaya, A.B.: Inventory control as an identification problem based on fuzzy logic. Cybernetics and Systems Analysis **43** (2006) 411–419
7. Samanta, B., Al-Araimi, S.A.: Application of an adaptive neuro-fuzzy inference system in inventory control. International Journal of Smart Engineering System Design **5** (2003) 547–553
8. Shervais, S., Shannon, T.T.: Adaptive critic based adaptation of a fuzzy policy manager for a logistic system. In Smith, M., Gruver, W., Hall, L., eds.: Proceedings of IFSA/NAFIPS. (2001)

9. Xiong, G., Koivisto, H.: Research on fuzzy inventory control under supply chain management environment. In: Proceedings of Computational Science - ICCS 2003. Volume 2658 of Lecture Notes in Computer Science., Springer Verlag (2003) 907–916
10. Rosen, R.: Anticipatory systems. Pergamon Press, New York (1985)
11. Ertogral, K., Darwish, M., Ben-Daya, M.: Production and shipment lot sizing in a vendorbuyer supply chain with transportation cost. European Journal of Operational Research **176** (2007) 1592–1606
12. Kljajić, M., Kofjač, D., Škraba, A., Rejec, V.: Warehouse optimization in uncertain environment. In: Proceedings of the 22nd International Conference of the System Dynamics Society, Oxford, England, UK, System Dynamics Society (2004)
13. Zadeh, L.A.: Fuzzy sets. Information and Control **8** (1965) 338–353
14. Dey, J.K., Kar, S., Maiti, M.: An interactive method for inventory control with fuzzy lead-time and dynamic demand. European Journal of Operational Research **167** (2005) 381–397
15. Katagiri, H., Ishii, H.: Some inventory problems with fuzzy shortage cost. fuzzy sets and systems. Fuzzy Sets and Systems **111** (2000) 87–97
16. Petrovic, D., Roy, R., Petrovic, R.: Modelling and simulation of a supply chain in an uncertain environment. European Journal of Operational Research **109** (1998) 200–309
17. Petrovic, D., Roy, R., Petrovic, R.: Supply chain modelling using fuzzy sets. International Journal of Production Economics **59** (1999) 443–453
18. Petrovic, R., Petrovic, D.: Multicriteria ranking of inventory replenishment policies in the presence of uncertainty in customer demand. International Journal of Production Economics **71** (2001) 439–446

# Post Decision-Making Analysis of the Reengineering Process Supported by Simulation Methods

Miroljub Kljajić[1], Andrej Škraba[1], Mirjana Kljajić Borštnar[1], and Edvard Kolmanič[2]

[1] University of Maribor, Kidričeva cesta 55a, SI-4000 Kranj, Slovenia
[2] Prevent Gradnje IGM, Miklavška cesta 40, SI-2311 Hoče, Slovenia

**Abstract.** This paper analyzes seven years of experience in a concrete production company where reengineering was conducted by employing simulation methodology and AHP methods for decision support. A predictive validation of the simulation model as well as evaluation of the simulation methodology for decision assessment in enterprises was performed by comparing real and expected values. Multiple criteria methods for the simulation scenario selection were applied. The basic advantage of the described approach lies in the interactivity and transparency of the model representation, which is essential for model acceptance by user. The achieved results represent scholars' example of using simulation methodology for solving managerial problems.

**Keywords:** Simulation, reengineering, validation, decision support, enterprise.

## 1 Introduction

The role of simulation methodology in the decision assessment of complex systems is constantly increasing. Human knowledge, the simulation model and decision methodology combined in an integral information system offers a new standard of quality in management problem solving [1]. The simulation model is used as an explanatory tool for a better understanding of the decision process and/or for learning processes in enterprises and in schools. An extensive study on using the simulation method in enterprises can be found in Gopinath and Sawyer [2]. Many successful businesses intensively use simulation as a tool for operational and strategic planning and enterprise resource planning (ERP) [3], [4]. Experiences described in literature [5], [6], [7] emphasize that in a variety of industries actual problems can be solved with computer simulation for different purposes and conditions. At the same time, potential problems can be avoided and operative and strategic business plans may also be tested. Currently the most intensive research efforts are concentrated on a combination of simulation methods and expert systems [8], [9]. Although there is a considerable amount of work devoted to simulation methodology, there is a lack of its application in practice; especially in small- and midium-sized companies. The reason lies not in

W. Abramowicz (Ed.): BIS 2007, LNCS 4439, pp. 549–561, 2007.
© Springer-Verlag Berlin Heidelberg 2007

the methodology itself; the real reason is rather in the problems of methodology transfer to enterprises and the subjective nature of decision-making.

One of the objective problems is model validation, which is very important for any model-based methodology. The validity of the model of a given problem is related to the soundness of the results and its transparency for users. According to Coyle [10], a valid model is one that is well-suited to a purpose and soundly constructed. According to Forrester [5], it is pointless to discuss validation without reference to a particular situation. There is no way to prove the usefulness of the model of complex systems such as enterprises in advance [11]. Coyle and Exelby [12] stressed that there is no such thing as absolute validity - only a degree of confidence, which becomes greater as more and more tests are performed. According to methodology, a valid model is an objective representation of a real system. According to the system approach paradigm, Barlas and Carpenter [13] have suggested that model validation cannot be completely objective, quantitative and formal. Since validity means usefulness with respect to a purpose, model validation must also have subjective, informal and qualitative components.

The second problem, the subjective one, is related to the transparency of the methodology and data presentation [14], preferences of the decision-maker to use a certain decision style and poor communication between methodologist and user. The simulation methodology is a paradigm of problem solving where the personal experiences of users as well as their organizational culture play an important role (e.g., in transition countries: market economy, ownership, etc.). Some of the encountered problems could be overcome with a carefully prepared modelling methodology and selection of a proper simulation package as well as a readiness of the user to explore simulation opportunities. Such is the simulation with animation, which demonstrates the operations of the modelled system and helps participants to recognize the specifics of the presented system [15].

This article describes seven years of production facility operation, which was initially technically determined by the simulation method. Our primary goal is to highlight decision-making analysis of the reengineering process supported by simulation methods. By comparing the predicted and real values, the predictive merit of the simulation model as well as the applicability of the simulation methodology for decision assessment in enterprises was evaluated. In this case the manufacturer of concrete goods had problems with production management as well as with prolonged delivery times due to increasing demand. A simulation model was used for the assessment of investment in a new production line and to test the feasibility of heuristic operational planning [7]. The evaluation criteria and business goals were aquired through the Group Support System (GSS) in connection with the Analytical Hierarchy Process (AHP) method [16], [17]. Visual Interactive Modelling (VIM) and animation of the modelled system [18] are significant advantage compared to the text-oriented simulation interfaces. The model of the process was built with ProModel: a simulation tool designed for Windows environment.

## 2 Simulation Methodology as a Base for Decision Support

Many authors favour the simulation method as a holistic approach for assessment of decision-making [2], [19], [20]; however; user confidence in it is of crucial importance [21]. The main problems of each managerial system are the comprehensiveness of information concerning the state and the environment within appropriate time. This means that a mathematical model of the process and a model of the environment are required. However, it is hard task to get confident model in enterprise processes due to the complex dynamics resulting from the stochastic interaction and delay. Decision-makers, though, cover a broader perspective in problem solving than could be obtained solely through simulation. Simulation in interaction with human experience contributes to the acceptance of simulation methodology by users and emphasizes its merit. The principal representation of the proposed approach that supports man-machine interaction in the evaluation of the decision strategy [22] is shown in Fig. 1.

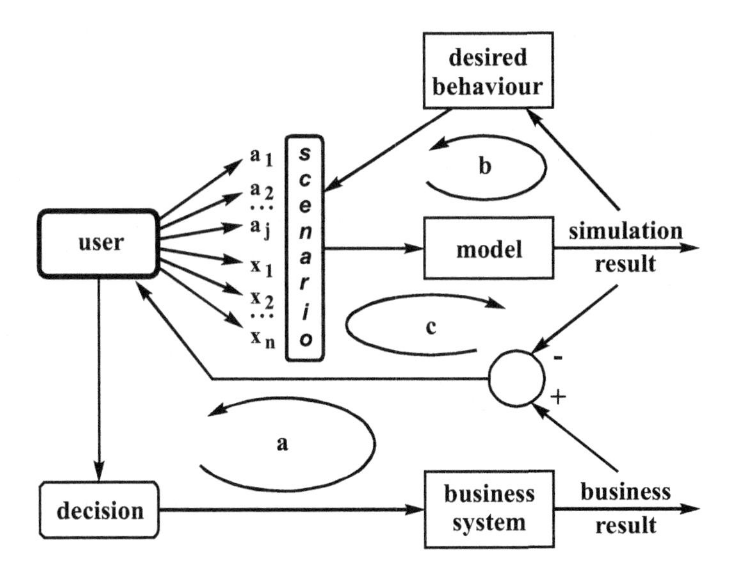


**Fig. 1.** The principle diagram of the simulation methodology for decision support in enterprises

The three basic loops are emphasized on Fig. 1:

a) The causal or feedback loop, representing the result as a consequence of former decision-making, and being a part of management experience and history of the system. From the learning aspect, this loop could be named "learning by experience".

b) The anticipative or intellectual feedback loop, which provides the feedforward information relevant for decision-making. This loop consists of the simulation model of the system, criteria function and scenarios. The simulation scenarios consist of two subsets: a subset of input that anticipates the impact of the environment - the state of nature (or exogenous scenarios) - and a subset of alternatives (or endogenous scenarios). They give answer to the basic question concerning the problem situation for which the answer is being sought. In literature this is known as the what-if analysis. The generation of scenarios of the simulation system that responds to the what-if is based on different scenarios anticipating future impacts of the environment on the system. They usually represent the extrapolation of past behaviour and an expert evaluation of development targets. Variants of business scenarios are evaluated, for example, with the multi-criteria decision function. The most delicate part of this circle is, above all, acceptance of the system simulation methodology by users; this is mainly dependent on transparency of the methodology and the model validity.

c) The a posteriori information loop represents the pragmatic validation of the model concerning model applicability and former decision-making. This loop represents the pragmatic validation of the model. A comparison of prior information concerning the simulated impact of the selected strategy on system behaviour with the actual results achieved allows us to evaluate the value of the model and improve it. In this way learning is enabled on the basis of a priori assumptions on the model and not just on the basis of empirical experiences.

Loops a and b are the basic ones for learning and knowledge acquisition for improved decision-making. Loop c represents the pragmatic validation of the model that supports users' confidence in the simulation methodology. The user is the key element of the three circles because he/she is the one who makes decisions. As most simulation projects necessitate teamwork, considerable attention should also be paid to the presentation of findings in the decision-making process; the problem discussed by Kljajić et al. [23]. This also depicts the trend of current research in the fields of knowledge presentation, expert systems and artificial intelligence. Our analysis is mostly concerned with loop c as explained in the following text.

## 3 Empirical Analysis of Production Line Selection by Simulation

In order to illustrate the problem of model validation we will analyze the results of a solution obtained by employing the simulation methodology [7]. The decision assessment has been organized at two hierarchical levels. The model at the upper level is used for the assessment of an enterprise's strategy (continuous simulation). At the lower level the model is used for discrete event simulation, necessary for operation planning and testing production performance. The concept of simulation is convenient for achieving a balance among different levels through the whole system. In practice, this means that when the discrete-event process is considered, variables are entities (queue, utilization). However when the process is considered as continuous, stock and flow in the system dynamics are entities. The system structure of the simulation model consists of entities connected in a flow diagram. The diagram is sufficiently abstract to allow an understanding of the problem to be solved and precise enough to provide valid experimentation on the model. As soon as one becomes satisfied with the "picture" of the process, one moves to the building of the simulation model. From the decision-making aspect, the state equation of the simulated system is described by Equation (1):

$$y(k+1) = f(y(k), x(k), a);\; k = 0, 1, 2, \ldots, N \tag{1}$$

where $y \in Y$ represents the vector of state variables such as inventory, cash, income, liabilities, backlog, etc., $x_i \in X$ represents the system input: market demand, and $a_j \in A$ represents the control variables (alternatives). The decision strategy was defined as: choose the alternative $a_j$ for the market demands $x_i$ and its probability $p_i \in P$, which satisfies the performance function reflected by the manager's preferences. Performance of alternatives $a_i \in A$ in Equation (1) was obtained through discrete event simulation DES as shown in Fig. 2. Two criteria were considered:

Maximal expected value (of profit) defined by Equation (2):

$$\max EV(a_j) = \sum_i C_{ij} p_i, \tag{2}$$

where $C_{ij}$ represents the values of the $i$-th input at $j$-th alternative, and linear weighted sum of multiple criteria defined by Equation (3):

$$\max J(a_j) = \sum_{r=1}^{m} w_r J_r(a_j), \tag{3}$$

where $w_r$ represents the weight of the $r$-th objective, which reflects the decision-maker's business policy preference. The individual objective $J_r = q(y, x, a)$ in Equation (3) is a function of the state of the system, the state of the market and the chosen alternative in achieving the goal. The multiple criteria and its weighting for the evaluation of scenarios were defined by the decision group using the group support system. Saaty's AHP method [16] was used to determine the relative importance of the objectives $w_r$ and a pair-wise comparison of alternatives $a_j$ for the $r$-th objective.

The described methodology was tested in a medium-sized factory, a manufacturer of concrete goods, for the purpose of reengineering assessment. Due to the increased demand for a specific article and better quality requirements of products, the firm's management considered investing in a new production line. Three suppliers of the new production line equipment besides the existing technology were considered for decision-making. The suppliers are denoted as alternatives: $a_i = a_1; a_2; a_3; a_4$ and their costs in monetary units as: $c_i = 0; 371; 392; 532$ respectively. A brief description of the alternatives would be the following: $a_1$) no change in production facility, $a_2$) new technology, which is a better version of the company's current technology; production is semi-automatic, $a_3$) new technology, unknown to the company; experiences from other companies are positive;

production is semi-automatic, and $a_4$) fully automated production process which can ensure a high quality and quantity of products. For a detailed description of alternatives, refer to Kljajić et al. [7]. An estimation of the state of nature (market demand) $X_i$ and its probability $p(X_i)$ for the next 5 years are shown in Table 1.

**Table 1.** Definition and meaning of the state of nature defined in the beginning of decision-making

| State of nature $X_i$ and its meaning | Probability |
|---|---|
| $X1$ = no change in market demand | 15% |
| $X2 = X1 + 2\%$/year; medium increase in demand | 40% |
| $X3 = X1 + 7\%$/year; high increase in demand | 35% |
| $X4 = X1 - 2\%$/year; medium decrease in demand | 10% |

Probabilities for various states of nature have been estimated by the company's experts by the method of brainstorming conducted in the boardroom and supported by GSS. The decision problem was solved using a computer simulation. A conceptual diagram of the system is shown in Fig. 2. The simulation models of the alternatives were implemented in a ProModel simulation package for the evaluation of technological requirements and management criteria DES (Discrete Event Simulation), which is a powerful and easy-to-use tool for developing a simulation model, especially for production systems. Its visualization of simulation results helps to raise the confidence level of users.

The financial aspect of reengineering was modelled as the continuous simulation model. The block diagram in Fig. 2 shows the main material, financial and information flows of the manufacturing system. Net income is represented as an element dependent on different supplier options simulated on DES. The block Company indicates the ability of the company to support investment in new technology. The behaviour of the enterprise under different market demand for the presumed alternatives with the prescribed criteria was analyzed on the model. This approach provides a unique framework for integrating the functional areas of management - marketing, production, accounting, research and development, and capital investment. The simulation model packed in a user-friendly business simulator enables decision-makers to "experience" different scenarios [24], [25]. Simulation of the continuous model was performed off-line after the DES generated results according to the real data of market demand. Each considered alternative has a different delivery delay from the production process to the customer, which is caused by different production performance and characteristics of the technology used. Delivery delay is not dependent only on main machine performance but also on the technology used for finalization of concrete products. For example, if the technology employs evaporator chambers (i.e. alternative 3), the delay could be less than five days long otherwise delivery delays vary from ten to fifteen days according to the seasonal deviation of outdoor temperature, which was also considered in the model. Depending on the delivery time, cash inflow is also affected, which is of primary concern to managers.

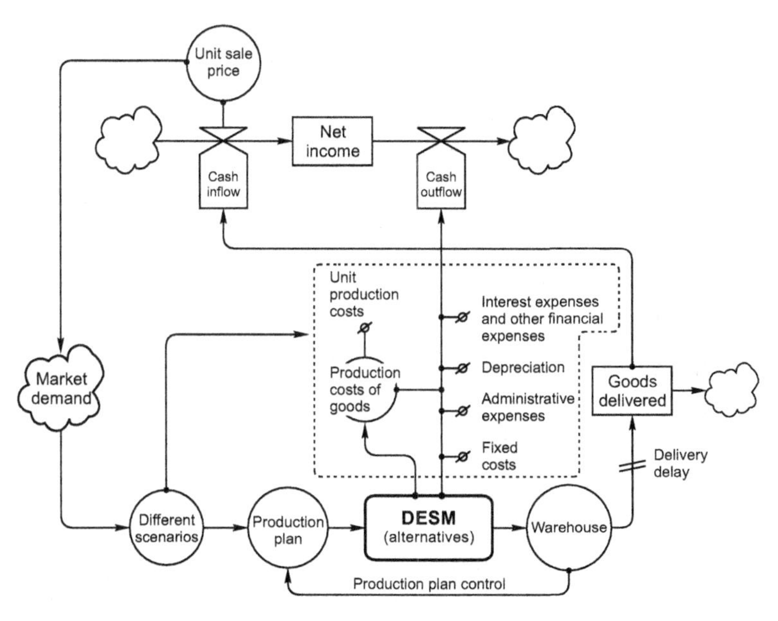


**Fig. 2.** Integral block diagram of the simulation model for decision assessment in production process upgrading

An expert group determined Unit Sale Price and Market Demand Function necessary for different production scenarios. The scenarios are defined as a combination of: Unit Sale Price, Unit Production Costs, Market Demand and Other Operating Expenses. Market demand is defined on the basis of current orders and future estimation, which was determined by the company's expert group. The production plan forms the input for DES with the purpose of evaluating the utilization and capacity constraints of the considered alternative. The importance of choosing an appropriate combination of production capacity in relation to the anticipated demand is of crucial importance for positive system performance. The simulator of the business system allows us to make an analysis of the investment effects, depreciation plan, risk of drop in sales, delivery time and change in sale prices. The model is used for predicting financial and production system efficiency. Four scenarios representing market demand were simulated for each alternative. The expected values of the payoff for alternatives for the 8-year period were computed according to Equation (2).

Several other requirements for the new technology were additionally imposed: Quality of Products, Net Profit, Risk of Company Ruin, Market Demands and Flexibility of Technology. The decision group consisting of enterprise experts carefully determined the relation between the key criterions. Geometric means of each parameter were used for estimation of group decisions. Saaty's AHP method [16] was used to determine the relative importance of objectives $w_r$ and a pairwise comparison of alternatives $a_j$ for the $r$-th objective in Equation (3). A model of the production line of the existing process during a simulation run is shown in Fig. 3. At the top right hand corner of Fig. 3 a simulation clock is shown. Every

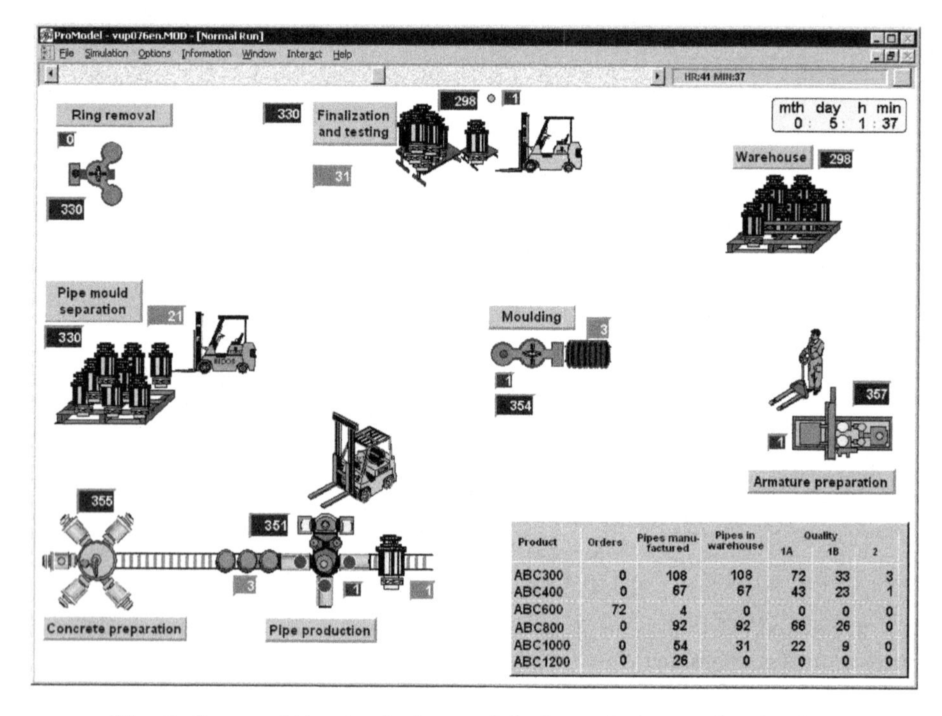


**Fig. 3.** Screen of the simulation model of an existing production line

location has a counter and indicator showing the current status of the location, e.g. blocked, waiting, down, etc. Each location is positioned according to the real layout of the production process for each alternative. At the right-hand bottom corner of the screen is a table with data relevant for production management. The first column lists names of products, whereas the second column lists the number of units on order - backlog. The third column indicates the number of units currently produced while the fourth column represents available units in the warehouse. The last three columns represent the quantity of products of the desired quality.

**Table 2.** The evaluation results of alternatives. Cumulative expected values of profit for 6 and 8 years are shown in the first two rows. In the last row, the multi-criteria decision score is shown.

| | $a_1$ | $a_2$ | $a_3$ | $a_4$ |
|---|---|---|---|---|
| Average $EV(a_j)1999-2004$ | 218 MU | 166 MU | 297 MU | $-2$ MU |
| Average $EV(a_j)1999-2006$ | 279 MU | 390 MU | 483 MU | 291 MU |
| Multicriteria Evaluation $J(a_j)$ | 0.082 | 0.166 | 0.431 | 0.321 |

As a result of the decision-making and final judgment, alternative $a_3$ was chosen according to the Table 2. It was ranked first, evaluated by the expected value and multi-criteria evaluation, considering the period of an 8-year horizon.

For a longer time period, however, alternative $a_4$ was proposed as the best solution, which had been seriously considered for the final judgment. Thus, we will continue to evaluate the results of the first seven years, experience with the selected alternative ($a_3$), bearing in mind that it was chosen with respect to the eight years time horizon. Seven years exploitation of the chosen technology is a reasonable period for post-decision analysis.

## 4 Results and Discussion

Data obtained from the production of concrete goods over the past seven years was used for the model validation of the decision process. Validation was carried out by comparing the business outcomes with the anticipated responses of the business model according to Fig. 1. The following figures will illustrate the model evaluation results according to the seven years experience. The cumulative value of production, predicted and realized, is shown in Fig. 4.

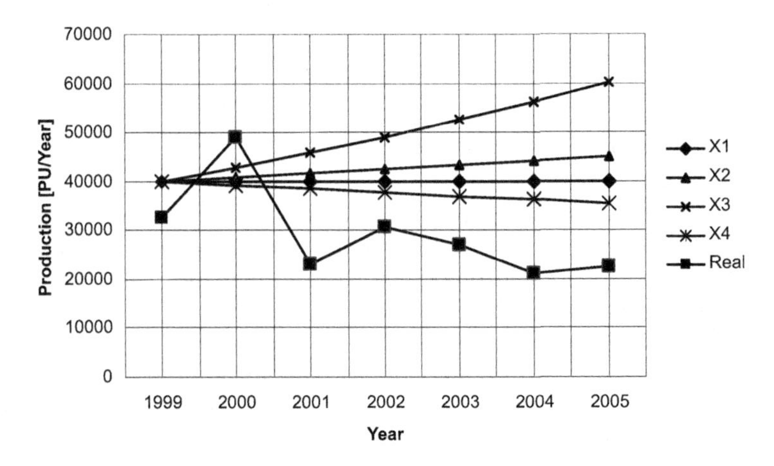


**Fig. 4.** Forecast of cumulative production (X1, X2, X3, X4) and real production in the first seven years

The value of production is stated in production units per year [PU/Year]. We supposed that demand would be equal to production. This is common practice in large concrete goods production, which is accomplished by the adaptive inventory buffering. One can see that in the first two years, the sales figures mainly have grown as predicted in the optimistic scenario $X3$ from Table 1. There is exception only at the beginning of the production period due to the instalment problem of new technology, which took three months to be completed. After that period, another six months were needed to develop smooth running of production process. In the following year, the production rapidly stagnated in 2001, while after this period it started to oscilate up and down with minimum at 2004 and it seems start to rise again. The fall in production in the year 2001 was due to market shrinkage, which is the consequence of a drop in highway construction and privatization process.

Fig. 5 shows the course of actual production on a monthly basis until 2005. The obtained time series were used for the simulation runs and computation of real net incom.

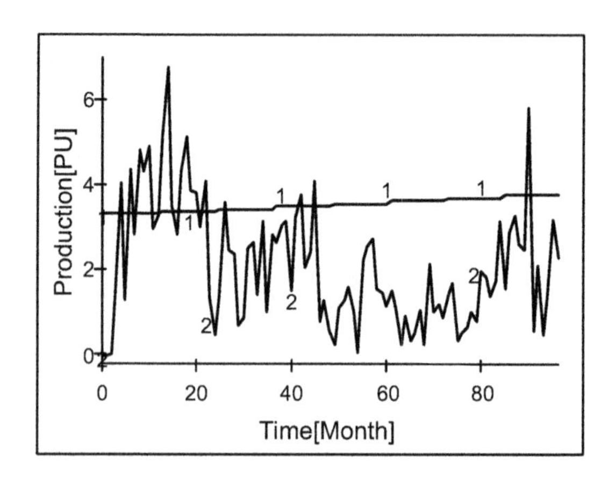


**Fig. 5.** Cumulative production (1) and production by month (2) realized in 84 months

Fig. 6 represents a comparison of EV of predicted Net Income (computed from Table 1; $X1, X2, X3$, and $X4$) and that realized denoted by R. At the beginning of the simulation, one could observe a certain increase of income due to the one-year loan moratorium. The real start-up Net Income marked by R in the first year is slightly lower than that planned due to the problem of production initialization and delayed payments of customers who ordered the goods. This delay in payments was caused by liquidity problems of this particular branch of industry, which had not been considered in the simulation scenarios. Comparing Net Incomes EV and R on Fig. 6, it can be seen that the curves correlate fairly well in the first four years while after that period real Net Income (R) starts to fall down due to drop of production untill 2005. Curve $A4$ represents estimated Net Income in case the decision makers would have selected $a_4$, which was seriously considered for the purchase.

One can learn what would have happened if the user had chosen alternative $a_4$. The highly automated production would ensure a high production volume and automation, but this kind of technology would be too expensive. At the anticipated ratio of demand on the market sales could not cover the financial burden of such volume. This would mean that the company would suffer a financial crisis shortly after the implementation of such technology. Fortunately, with the proposed methodology such a case, although tempting in the process of assessment, was correctly avoided.

The decision assessment supported by a simulation proved to be a very powerful tool in seeking the best alternative, although the problem addressed incorporates

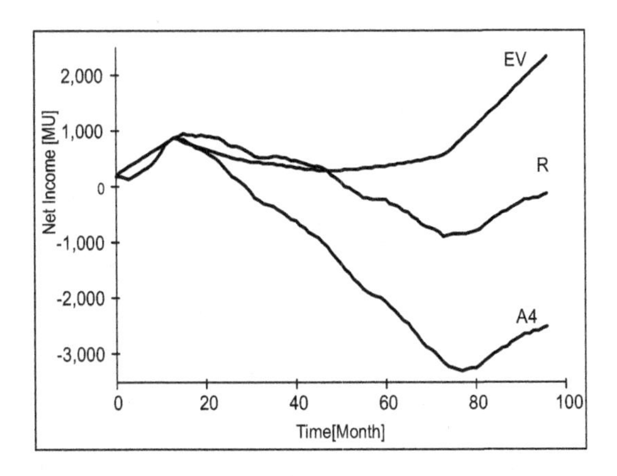


**Fig. 6.** Expected Value of Net Income (EV) for the selected alternative a3, realized Net Income (R) and an estimated Net Income (A4) if alternative a4 was selected

an inaccurate subjective estimation of the market parameters. A pragmatic evaluation of the model shows the importance of a rational approach in the simulation methodology regarding decision assessment. The positive results of the applied methodology are encouraging and the model is used to assess and predict the future behaviour of the system as result of management scenarios. For example, the model can predict the company's potential problems with liquidity in the sector. In this case, the sensitivity of setback of payments is computed and with the proper management action, the negative consequences can be avoided.

## 5 Conclusion

This paper analyses seven years of experience in a concrete production company, where selection of a new production line was conducted by employing simulation methodology. The paper relies on previous work [7], where a simulation procedure for production line selection was described.

The applied simulation system consisting of a continuous and DES model connected with GSS and integrated in the management information system provided useful information for the management. System dynamics was used for modelling the continuous process for the financial aspect, while discrete event simulation was used for production process analyses. Both simulation tools are user friendly and provide visual interactive modelling VIM, which facilitate users in understanding the applied methodology. Multiple criteria methods based on AHP as well as Expected value for the simulation scenario selection for managerial assessment were applied.

The performance of the selected production line in the mentioned work has been followed up. The predictive validation of the simulation model as well as simulation methodology for decision assessment was done by comparing real data with those predicted of the chosen alternative in seven years period. A comparison showed that the gained predictions were a relevant estimation of future company development after the reengineering process was completed. Moreover, post decision analysis shows the quality of a rational decision in favour of alternative $a_3$ compared to the next best competing alternative $a_4$, which was seriously considered.

The basic advantage of the described approach lies in the interactivity and transparency of the model representation, which is essential for model acceptance by the user. By implementing VIM and GSS, decision-makers can better understand the studied process, enhance knowledge about the uncertainty of the scenarios and improve evaluation of alternatives. In this way, the range of bounded rationality in decision-making could be enhanced.

*Acknowledgement.* This research was supported by the Slovenian Research Agency (Grant No. P5-0018).

## References

1. Simon, H.: Models of Man. John Wiley & Sons. New York. (1967)
2. Gopinath, C., Sawyer, J. E.: Exploring the learning from an enterprise simulation. Journal of Management Development. Emerald. Vol. **18**, No. 5 (1999) 477–489
3. Schniederjans, M. J., Kim, G. C.: Implementing enterprise resource planning systems with total quality control and business process reengineering – Survey results. International Journal of Operations & Production Management. Emerald. Vol. **23**, No. 4 (2003) 418–429
4. Muscatello, J. R., Small, M. H., Chen, I. J.: Implementing enterprise resource planning (ERP) systems in small and midsize manufacturing firms. International Journal of Operations & Production Management. Emerald. Vol. **23**, No. 8 (2003) 850–871
5. Forrester, J. W.: Industrial Dynamics – A Response to Ansoff and Slevin. Management Science. **14** 9 (1968) 601–618
6. Homer, B. J.: Why we iterate: Scientific modelling in theory and practice. System Dynamics Review. John Wiley & Sons. Chichester. Vol. **12**, No. 1 (1996) 1–19
7. Kljajić, M., Bernik, I., Škraba, A.: Simulation Approach to Decision Assessment in Enterprises. Simulation. Simulation Councils Inc. (2000) 199–210
8. Dijk, J. N., Jobing, M. J., Warren, J. R., Seely, D., Macri, R.: Visual Interactive Modelling with SimView for Organizational Improvement. Simulation. Simulation Councils Inc. Vol. **67**, No. 2 (1996) 106–120
9. Hall, O. P.: A DSS based simulation model for teaching business strategy and tactics. Hamza M. H., IASTED International Conference on Modelling. Simulation and Optimization. Acta Press. (1996)
10. Coyle, R. G.: System Dynamics Modelling: A Practical Approach. Chapman and Hall. London. (1996)
11. Forrester, J. W.: System Dynamics, Systems Thinking, and Soft OR. System Dynamics Review. John Wiley & Sons. Chichester. Vol. **10**, No. 2–3 (1994)
12. Coyle, R. G., Exelby, D.: The Validation of Commercial System Dynamics Models. System Dynamics Review. Spring. John Wiley & Sons. Chichester. **16** (1) (2000) 27–41

13. Barlas, Y., Carpenter, S.: Philosophical Roots of Model Validation: Two Paradigms. System Dynamics Review. John Wiley & Sons. Chichester. Vol. **6**, No. 2 (1990) 148–166
14. Kahneman, D., Tversky, A.: Prospect theory: An analysis of decision under risk. Econometrica. **47** (1979) 263–291
15. Saltman, R. M.: An Animated Simulation Model for Analyzing On–Street Parking Issues. Simulation. Simulation Councils Inc. San Diego CA. Vol. **69** No. 2 (1997) 79–90
16. Saaty, T. L.: Multicriteria Decision Making: The Analytic Hierarchy Process. RWS Publications. Pittsburg. (1990)
17. Rangone, A.: An analytical hierarchy process framework for comparing the overall performance of manufacturing departments. International Journal of Operations & Production Management. Emerald. No. **8** (1996) 104–119
18. Jain, S., Choong, N. F., Aye, K. M., Luo, M.: Virtual factory: an integrated approach to manufacturing systems modeling. International Journal of Operations & Production Management. Emerald. Vol. **21** No. 5/6 (2001) 594–608
19. Simon, H.: Models of Bounded Rationality. Cambridge MA. The MIT Press. Vol. **3** (1997)
20. Sterman, J. D.: Business Dynamics: Systems Thinking and Modeling for a Complex World. Irwin/McGraw-Hill. Boston MA. (2000)
21. Chen, L.-H., Liaw, S.-Y.: Investigating resource utilization and product competence to improve production management – An empirical study. International Journal of Operations & Production Management. Emerald. Vol. **21**, No. 9 (2001) 1180–1194
22. Kljajić, M.: Theory of Systems. Moderna organizacija. Kranj (1994)
23. Kljajić, M., Škraba, A., Leskovar, R.: Group Exploration of SD Models - Is there a Place for a Feedback Loop in the Decision Process?. System Dynamics Review. John Wiley & Sons. Chichester. (2003) 243–263
24. Anderson, E. G., Morrice, D. J.: A simulation game for teaching services–oriented supply chain management: Does information sharing help managers with service capacity decisions?. Production and Operations Management. Production and Operations Management Society. USA. Vol. **9**, No. 1 (2000) 40–55
25. Noori, H., Chen, C.: Applying scenario–driven strategy to integrate environmental management. Production and Operations Management. Production and Operations Management Society. USA. Vol. **12**, No.: 3 (2003) 353–368

# Contextual Classifier Ensembles

Janina Anna Jakubczyc

University of Economics, Department of Artificial Intelligence Systems, 53-345 Wrocław, ul. Komandorska 118/120
janina.jakubczyc@ae.wroc.pl

**Abstract.** Individual classifiers do not always yield satisfactory results. In the field of data mining, failures are mainly thought to be caused by the limitations inherent in the data itself, which stem from different reasons for creating data files and their various applications. One of the proposed ways of dealing with these kinds of shortcomings is to employ classifier ensembles. Their application involves creating a set of models for the same data file or for different subsets of a specified data file. Although in many cases this approach results in a visible increase of classification accuracy, it considerably complicates, or, in some cases, effectively hinders interpretation of the obtained results. The reasons for this are the methods of defining learning tasks which rely on randomizing. The purpose of this paper is to present an idea for using data contexts to define learning tasks for classifier ensembles. The achieved results are promising.

**Keywords:** classifier ensembles, context, supervised machine learning.

## 1 Introduction

Individual classifiers do not always yield satisfactory results. In the field of data mining, failures are mainly thought to be caused by the limitations inherent in the data itself, which stem from different reasons for creating data files and their various applications [4]. One of the proposed ways of dealing with these kinds of shortcomings is to employ classifier ensembles. Their application involves creating a set of models for the same data file or for different subsets of a specified data file. Although in many cases this approach results in a visible increase of classification accuracy, it considerably complicates, or, in some cases, effectively hinders the results of action prediction taken for creating classifier ensembles and interpretation of the obtained results. The reasons for this are the methods of defining learning tasks which rely on randomizing. The idea for solving the indicated weaknesses is to use contextual classifier ensembles that replace random choices with context choices.

This paper is structured as follows: section 2 introduces classifier ensembles and indicates possible methods of modifying their creation, section 3 contains a proposed solution and the results of the conducted research, and section 4 summarizes the results.

## 2 Rules of Creating Classifier Ensembles

An classifier ensemble is a set of classifiers that organizes new cases by combining the decisions of individual classifiers in a certain manner. The theoretical basis which

W. Abramowicz (Ed.): BIS 2007, LNCS 4439, pp. 562–569, 2007.
© Springer-Verlag Berlin Heidelberg 2007

guarantees a higher accuracy of classification in this type of model is M.J.A. Marquis'a de Condorcet's jury theorem (for an overview and analysis see [5],[16]; for a discussion and an implementation of the theorem in the field of classifier ensembles see [6]).

There are two approaches to creating classifier ensembles. The first one is to create separate models for the same data set (this approach is beyond the scope of this paper). The idea behind the second approach is to define multiple learning tasks within one learning file. This can be accomplished by manipulating either the data set or the learning algorithm's control mechanism ([8], [4], [10]).

Manipulation of the data set consists of methods of sampling the input data, selection of classification features and manipulation of class labels.

The sampling approach involves generation of different learning subsets and creating different classifiers for each of these subsets. Three methods of creating learning files are used. The first one, the so-called bootstrap, involves generating a certain number of learning sets containing the same number of examples as original learning set by applying randomization with replacement. By utilizing this mechanism of selection we can obtain a varied structure of learning sets, therefore enabling us to create diversified classifiers, which, in conjunction, will describe the learning data. The drawback of this method is that we do not know which of the resulting learning files are appropriate to our purposes. Neither do we know how many sets we need to complete before the classifier reaches the presupposed accuracy of classification.

The second method of generating learning files is by employing cross-validation. In this approach the learning files are created on the base of different variants of learning files with a set number of partitions k, which divide the learning file into k parts. Then one of the parts is removed and the algorithm is used to teach the remaining k-1 subsets. The number of resulting classifiers is therefore equal to the number of partitions set. In this case the drawback is the fact that the number k is selected randomly, and that the learning file is thereby divided randomly and not purposefully.

Another method of creating learning files is AdaBoost ([14], [15]). It operates by classifying all examples in subsequent iterations of the classifying algorithm. In order to do so, it utilizes a set of weights assigned to the learning examples. Higher weights are awarded to examples which have not been properly classified by given hypotheses. After each iteration the weights are changed according to classification error. Each iteration produces one classifier. AdaBoost is thriving and is currently available in three variants: aggressive, conservative and inversive [9].

Another technique of creating classifiers, named by its authors Dirtterich and Bakiri error code correction, involves manipulating the labels of examples ([3]). The first step is to randomly divide the class into two subsets, both of which are assigned new labels, i.e. I and II. With the help of a learning algorithm a classifier describing the classes I and II is generated. In the next step an algorithm is run for each class, which learns the original classes contained in the generated learning subset. This two step procedure is run a specified number of times. In this case we are utilizing the hierarchical classifier, which first determines the membership to class I or II, and then assigns them, with the help of the generated classifiers, to the class corresponding to those contained in the primary learning file. This method, according to research conducted by Ricci and Aha [12], Dirtterich and Bakiri [3], Adeva J.J.G.and Calvo R.A [2], Kuncheva [9], Masulli F. and Valentini G., [11] is especially recommended for solving difficult classification problems.

The next stage of implementing classifier ensembles is defining an operating outline that will determine which of the classifiers will take part in generating the results of classification and in what manner these classifiers or their results can be combined together. The basic outline is voting, which can take on the form of: a majority vote, a weighted vote or a weighted majority vote [16].

In a majority voting schema all individual classifiers are valued equally. The result with the highest number of votes is chosen. Whereas in the weighted vote the constituent classifiers are assigned ranks, which can be determined for instance by classification accuracy in the future or some heuristics. Voting schemas are far more numerous, and this field is developing rapidly. They are beginning to appear as an indispensable element for the chosen approach to ensemble classifiers, for instance in the work [16] three possible schemes are proposed as solutions in ensemble classifiers created on the basis of the selection of classification features. These are the following: linear combination of constituent results of the classifier ensemble, winner-takes-all and evidence inference.

## 3 A Proposal for Expanding the Possibilities of Creating Classifier Ensembles

The introduction to classifier ensembles presented above suggests that the dominant mechanism used to create diverse learning files is randomization (with the exception of AdaBoost, the purpose of which lies beyond the scope scientific interests of the author). This approach to creating learning tasks is bereft of intentional and purposeful action. We may therefore conclude that the implementation of random mechanisms deprives us of the possibility to make use of the knowledge of the specific domain as well as information about the data file. This situation induces interpretational limitations in the obtained results.

In the opinion of the author, in spite of the demonstrated effectiveness of classifiers, the case of defining different learning tasks for a specific set of learning information gives rise to the following problems:

- the number of sets which need to be created in order to have a guarantee of a better description
- number of constituent models
- interpretation of differences between various files with learning data (interpretation of individual classifiers)
- interpretation of results
- understanding the impact of the classifier ensembles (which of the classifiers determine the final solution, and why)

Relinquishing the possibility to interpret and understand data and its description is justifiable only when we do not have any other means at our disposal. The author of this paper proposes a new, intentional method of defining different learning tasks, basing on the use of context. This proposal aims to present a contextual classifier ensemble able to solve some of the aforementioned issues. The idea involves creating learning files for classifier ensembles by using identified or recognized contexts of the learning data. The adopted representation of a contextual classifier ensemble and a constituent classifier is a decision tree.

### 3.1 Creating Learning Files Which Utilize Context in Data

The approach proposed in this paper is, in a certain way, a modification of Dirtterich and Bakiri's technique of modifying labels, referred to as error-correcting output codes ([8]). It involves substituting the random partitioning of the learning file with a contextual division according to the identified contextual features. Every contextual feature splits the learning file into subsets, the amount of which is dependent on the structure of values that the contextual feature assumes. A classifier is generated for each subset.

The definitions of elementary (deciding), contextual and context-sensitive features form the basis of discovering context in data. Assuming that: $X_i$ denotes the describing feature, Y – the class feature, $\alpha_i$ – the minimal and $\beta_i$ the maximal context size, p – the probability, $S_{i,j}$ – the set containing all features except $X_i$ and $X_j$, $s_{i,j}$ – the values of all the features in the set $S_{i,j}$. P.Turney in his work [17] presents them as follows:

- a feature is primary iff $\alpha_i=0$; which means that there exist some $x_i$ and y for which $p(X_i=x_i)>0$, such that:

$$p(Y=y|X_i=x_i)=p(Y=y); \quad (1)$$

- a feature is contextual iff $\alpha_i>0$, which means that for all $x_i$ and y the following holds

$$p(Y=y|X_i=x_i)=p(Y=y); \quad (2)$$

- the feature $X_i$ is context-sensitive to the feature $X_j$ iff there exists a subset of features $S'_{i,j}$ of the set $S_{i,j}$, for which there exist some $x_i$, $x_j$, $s'_{i,j}$ and y, for which:

$$p(X_i=x_i;X_j=x_j;S'_{i,j}= s'_{i,j})>0; \quad (3)$$

such that the following conditions hold:

$$p(Y=y|X_i=x_i,X_j=x_j,S'_{i,j}=s'_{i,j}) \neq p(Y=y|X_i=x_i,S'_{i,j}=s'_{i,j}); \quad (4)$$

$$p(Y=y|X_i=x_i,X_j=x_j,S'_{i,j}=s'_{i,j}) \neq p(Y=y|X_j=x_j,S'_{i,j}=s'_{i,j}). \quad (5)$$

The method of creating learning files that are dependent on the identified primary, contextual and context-sensitive features presented as follows:

1. Creation of a decision tree[1] on the basis of the entire learning material; this step allows us to separate the features, which directly influence the classification task from the features, which do not (the non-decision features).
2. Creation of a decision tree for each decision attribute, taking into consideration only the non-decision features; in this step the search for contextual features is conducted among non-decision features, and the search for context-sensitive features – among decision features.

[1] There was applied Quinlan's classifier C4.5 of AITECH's software SPHINX 4.0 (DeTreex), and Decision Tree of SAS's Enterprise Miner 5.2.

3. Identification of pairs of contextual and context-sensitive features that can be used to partition the learning file, according to the assumed level of classification accuracy
4. Partitioning of the learning file according to the structure of values of the contextual feature and its corresponding context-sensitive feature.

## 3.2 Employing the Contextual Classifier Ensemble to Solve Selected Classification Tasks

The classification tasks employed to measure the effectiveness of the contextual classifier ensemble were chosen on the basis of an analysis of research conducted by Dietterich T. ([18]) and Hall et al. ([19].[20]) in the field of classifier ensembles. This research proves that in some cases (7 out of 33) there is no noticeable improvement ('primary', breast-y'), or that it is insufficient with regard to Quinlan's C4.5 individual classifier. (It was assumed that a satisfactory level of classification accuracy exceeds 80%.) These cases are presented in Table 1 (two of the files – 'primary' and breast-y' – were omitted due to their unavailability).

Three of the presented data files only have constant features, which pose serious obstacles for every decision tree building algorithm. An average increase of classification accuracy of approx. 5% is deemed desirable. The occurrence of context in the analyzed data is a necessary condition for creating contextual classifier ensemble.

**Table 1.** Difficult classification tasks

| *Data file* | *Numer of features* | *Numer of numerical features* | *Numer of examples* | *Numer of classes* | *Classification accuracy* | |
|---|---|---|---|---|---|---|
| | | | | | C4.5 | Classifier ensembles |
| glass | 9 | 9 | 214 | 6 | 66% | 73% |
| heart-v | 13 | 5 | 200 | 2 | 72% | 76% |
| sonar | 60 | 60 | 208 | 2 | 71% | 72% |
| vehicle | 18 | 18 | 846 | 4 | 67% | 75% |
| credit-g | 20 | 7 | 1000 | 2 | 71% | 76% |

The research was conducted in the following stages:

1. Identification of contextual, context-sensitive and primary features
2. Partitioning of the learning file according to the identified contexts
3. Creation of the contextual classifier ensemble containing models of decision trees for all contexts

For sake of simplification, the names of features were assigned consecutive numbers. The research was conducted in several stages. The first step was to generate decision trees for each data set. The resulting deciding features for each task are presented in Table 2. The describing features are used to their fullest extent in two tasks: 'vehicle'

(15 out of 18) and 'glass' (6 out of 9). In the remaining instances, the deciding features constitute a minority compared to the non-deciding features.

The next step is to describe each deciding feature using only the non-deciding features (according to the definition of the contextual feature, it does not directly influence class membership; it does, however, have an effect on the decision variable). The task involved the generation of trees for each deciding feature (treated as a class) and the selection of pairs consisting of a deciding feature and a contextual feature. A pair was selected when its classification accuracy exceeded 60%. The selection of contextual features was done by analyzing the generated decision trees for the context-sensitive features. On the basis of an analysis of each of these trees one contextual feature was selected.

The 'vehicle' file is a special case which describes the issue of identifying four vehicles with the help of two-dimensional images taken from nine different angles. Because each of the angles at which the photos were taken is contained in a separate file, it was assumed that they constitute a natural, external context of the description of vehicles. A cumulative comparison of contextual and context-sensitive features is presented in Table 2. Analyzing the results, we conclude that in two of the learning files contextual features were not found, which means that in these cases we cannot apply the contextual classifier ensemble, and therefore we cannot improve the results of the classification accuracy.

**Table 2.** Cumulative comparison of decision, and contextual and context-sensitive features

| *Learning file* | *Numer of features* | *Decison features* | *Contextual features* | *Context-sensitive features* |
|---|---|---|---|---|
| glass | 9 | 3,8.4,6,1,5 | 0 | 0 |
| heart-v | 13 | 13,3,12,10 | 2,9,1,11 | 13,3,12,10 |
| sonar | 60 | 12,20,48,32 | 0 | 0 |
| credit-g | 20 | 1,3,12,4,6 | 7,16,5,15 | 1,3,12,4 |
| vehicle | 18 | 1,2,3,5,6,8,9,10,11,<br>12,13,14,15,17,18 | 7 | 9 |

The contextual features which were identified for the remaining tasks give us the possibility to discern contexts within the learning material and allow us to describe them using separate classification rules. The context is defined by the contextual features and it is these features that determine the ranges for partitioning the learning and testing sets into groups of different contexts. The issue remaining to be solved is thus how to determine the value range of the contextual feature to ensure that it efficiently describes the identified context. It was assumed that the basis will be the information content of the contextual feature in relation to the context-sensitive feature.

The contextualized learning files were created taking into consideration the information content and the principle of preventing the creation of an excessive amount of value ranges (for more details see [7]). The quantity of learning files produced is almost the same, for heart-v and credit-g number is equal to 11 and for vehicle 12.

These separated subsets were used to generate the contextual classifier ensemble for each of the classification tasks. In the preliminary stage this choice seems valid, yet in further iterations it seems important to consider a voting weighted by the strength of the relationship between the appearing contexts. The contextual ensemble classifier, while classifying new cases, not only states the class and accuracy of choice, but also the context of the dominant classifier. It therefore supplies more comprehensive information, which enables correct interpretation.

The comparative results between contextual classifier ensembles (CCE) and classifier ensembles (CE) are contained in Table 3.

**Table 3.** A comparison of the effectiveness of classifier ensembles and contextual classifier ensembles

| *Learning file* | *CE Classification error (%)* | *CCE Classification error (%)* |
|---|---|---|
| heart-v | 27,62% | 18,22% |
| credit-g | 29,21% | 17,44% |
| vehicle | 29,44% | 16,37% |

The demonstrated results point to the conclusion that the efficiency of the contextual classifier ensemble has succeeded in reaching acceptable level, and therefore the extension of the analysis with context seems merited.

## 4 Summary

Contextual classifier ensembles achieve two goals. They increase description precision of a considered problem that results in the increase of classification accuracy. As it was shown, contextual classifier ensembles can be used for difficult classification problem solving, when other methods are unsuccessful.

The second achieved goal is a purposeful way of creating classifier ensembles. Applying data contexts as the basis for generating different learning tasks determines the number of possible learning tasks and the number of classifiers, and gives the opportunity for interpretation of differences among models that are indicated by data contexts.

Further work is being continued on developing scheme voting and more complicated context identification and implementation.

## References

1. Adeva J.J.G., Beresi U.C., Calvo R.A.: Accuracy and Diversity in ECOC Ensembles of Text Categorizers, available: http://citeseer.ist.psu.edu/732806.html, 2000.
2. Adeva J.J.G., Calvo R.A.: A Decomposition Scheme based on Error-Correcting Output Codes for Ensembles of Text Categorizers. In Proc. of the International Conference on Information Technology and Applications (ICITA). IEEE Computer Society, 2005.
3. Dietterich T.G., Bakiri. G.: Solving multiclass learning problems via error-correcting output codes. In Journal of Artificial Intelligence Research, 2:263–286, 1995.

4. Dietterich T.G.: Ensemble methods in Machine Learning. In Proc. of 1[th] International Workshop on Multiple Classifier Systems, 1–15, 2000.
5. Grofman B., Owen G.: Review essay: Condorcet models, avenues for further research. In: Grofman B., Owen G. (eds.): Information Pooling and Group Decision Making, 93–102, Jai Press, Greenwich, CT, 1986.
6. Hansen L., Salamon P.: Neural network ensembles. IEEE Trans. Pattern Analysis and Machine Intelligence, 12:993–1001, 1990.
7. Jakubczyc J.A.: Kontekstowy klasyfikator złożony. In: Niedzielska E., Dudycz H., Dyczkowski M. (eds.): Prace Naukowe Akademii Ekonomicznej: Nowoczesne technologie informacyjne w zarządzaniu, 313-322, Wrocław 2006.
8. Kolen J.K., Pollack J.B.: Back propagation is sensitive to initial conditions. In Advances in Neural Information Processing Systems, Vol. 3, 860-867, CA. Morgan Kaufmann, 1991.
9. Kuncheva L.I.: Using diversity measures for generating error-correcting output codes in classifier ensembles. "Pattern Recognition Letters", (26):83–90, 2005.
10. Kwok S.W., Carter C.: Multiple decision trees. In: Schachter R.D., Levitt T.S., Kannal L.N., Lemmer, J.F.(eds.): Uncertainty in Artificial Intelligence, 327-335, Elsevier Science, Amsterdam, 1990.
11. Masulli F., Valentini G.: An experimental analysis of the dependence among codeword bit errors in ECOC learning machines. Neurocomputing, (57C):189–214, 2004.
12. Ricci F., Aha D.W.: Extending local learners with error-correcting output codes. Technical Report, Naval center for Applied Research in AI, Washington, D.C., 1997.
13. Rousu J.: Efficient Range Partitioning in Classification Learning, Dep.of Computer Science Series of Publications University of Helsinki, A Report A-2001-1, availble: http://citeseer.ist.psu.edu/correct/394680
14. Schapire R.E., Freund Y., Bartlett P., Wee S.L.: Boosting the Margin: A New Explanation for the Effectiveness of Voting Methods. In Proc. of 14[th] International Conference on Machine Learning, 1998.
15. Schapire R.E.: The theoretical Views of Boosting and Applications, Algorithmic Learning Theory. In: Proc. of The 10[th] International Conference ALT '99, Tokyo, Japan, 1999.
16. ShiXin Y.: Feature Selection and Classifier Ensembles: A Study on Hyperspectral Remote Sensing Data, PhD The University of Antwerp, 2003.
17. Turney P.D.: Robust classification with context-sensitive features. In: Industrial and Magineering Applications of Artificial Intelligence and Expert Systems, Edinburgh, Scotland Gordon and Breach, 1993.
18. Dietterich T.G.: An experimental Comparison of Three Methods for Constructing Ensembles of Decision Trees: Bagging, Boosting, and Randomization. Machine Learning, 1-22, 1999.
19. Hall L.O., Bowyer K.W., Banfield R.E., Bhadoria D., Kegelmeyer W.P., Eschrich S.: Comparing Pure Parallel Ensemble creation Techniques against Bagging. In The Third IEEE International Conference on Data Mining, 2003.
20. Banfield R.E., Hall L.O., Bowyer K.W., Bhadoria D., Kegelmeyer W.P., Eschrich S.: A comparison of Ensemble Creation techniques, In Proc. Of the 5th International Conference on Multiple Clasifier Sysstems, Cagliari, Italy, June 2004.

# An Algebraic Algorithm for Structural Validation of Social Protocols

Willy Picard

Department of Information Technology,
The Poznań University of Economics,
ul. Mansfelda 4, 60-854 Poznan, Poland
picard@kti.ae.poznan.pl
http://www.kti.ae.poznan.pl

**Abstract.** Support for human-to-human interactions over a network is still insufficient. In this paper a model for human-to-human collaboration based on the concept of social protocol is presented and formalized. Then, semantical and structural validity of social protocols is defined. Next, an algebraic representation of social protocols is proposed. Based on this algebraic representation of social protocols, an algorithm for structural validation of social protocols is proposed and illustrated by three examples.

**Keywords:** Collaboration modeling, algebraic representation of social protocols, semantical validation, structural validation.

## 1 Introduction

Enterprises are constantly increasing their efforts in order to improve their business processes, which may be explained by the fact that enterprises are exposed to a highly competitive global market. Among the most visible actions associated with this effort towards a better support for better business processes, one may distinguish the current research works concerning Web services and associated standards: high-level languages such as BPEL or WS-Coordination take the service concept one step further by providing a method of defining and supporting workflows and business processes.

However, most of these actions are directed towards interoperable machine-to-machine interactions over a network. Support for *human-to-human interactions* over a network is still insufficient and many research has to be done to provide both theoretical and practical knowledge to this field.

Among various reasons for the weak support for human-to-human interactions, two reasons may be distinguished: first, many *social elements* are involved in the interaction among humans. An example of such a social element may be the roles played by humans during their interactions. Social elements are usually difficult to model, e.g. integrating non-verbal communication to collaboration models. Therefore, their integration to a model of interaction between humans is not easy. A second reason is the *adaptation capabilities* of humans which are not only far more advanced than adaptation capabilities of software entities, but also are not taken into account in existing models for collaboration processes.

W. Abramowicz (Ed.): BIS 2007, LNCS 4439, pp. 570–583, 2007.
© Springer-Verlag Berlin Heidelberg 2007

A model for human-to-human interactions which addresses, at least to some extent, the two characteristics of the interactions between humans is therefore needed. Such a model has already been presented in [1,2]. This model is based on the concept of *social protocol* which may be seen as a model of collaboration processes. A collaboration process may be modelled as a social protocol which describes the potential interactions of collaborators within this process.

In the case of complex collaboration processes, the design of a social protocol modeling these processes may be a complex task. A social protocol may be designed with errors potentially leading to unachievable collaboration processes, i.e. processes in which collaborators are locked and cannot continue their collaboration. Therefore, some techniques to check the validity of social protocols are needed.

In this paper, the concept of *validity* of social protocols is defined. Then, an algorithm for structural validation of social protocols is detailed. This algorithm is based on an algebraic representation of social protocols.

The rest of this paper is organized as follows. In Sect. 2, the concept of social protocol, used to model collaboration processes, is presented. Section 3 then expands on the definition of the concepts of semantical and structural validity of social protocols. Next, an algorithm for structural validation of social protocols is proposed in Sect. 4, and illustrated by three examples in Sect. 5. Then, related works are reviewed in Sect. 6. Finally, Section 7 concludes this paper.

## 2 Modeling Collaboration Processes as Social Protocols

A social protocol aims at modeling a set of collaboration processes, in the same way as as a class models a set of objects in object-oriented programming. In other words, a social protocol may be seen as a model which instances are collaboration processes. Social protocols model collaboration at a group level. The interactions of collaborators are captured by social protocols. Interactions are strongly related with social aspects, such as the role played by collaborators. The proposed model integrates some of these social aspects, which may explain the choice of the term *social protocols.*

### 2.1 Formal Model of Social Protocols

Before social protocols may be formally defined, others concepts must first be defined.

**Definition 1.** *A* role *is a label. Let denote $R$ the set of roles.*

In a given group, a set of roles is played by the collaborators, which means that collaborators are labeled, are associated with given roles. The set of roles $R_p$ for a given protocol $p$ is a subset of $R$, i.e. $R_p \subseteq R$. Collaborators usually play different roles within a given collaboration process. Roles may be associated with collaborators to specify the way they should interact with the rest of the group. Interactions among collaborators are modeled with the concept of *action type.*

**Definition 2.** *An* action type *is an interface of a software entity. Let denote $A$ the set of action types.*

An action may be for instance the execution of a web service, a commit to a CVS repository, the sending of an email. Within a group, collaborators are interacting by executing actions. The execution of actions is a part of the common knowledge of the group, i.e. all collaborators are aware of the execution of an action by one of the members of the group. An action type may be seen as a description of a given action, providing the name and type of parameters required to execute the action as well as the type of the result returned by the action execution.

**Definition 3.** *A* behavioral unit *is a pair $(role, action\_type)$. Let denote $BU$ the set of potential behavioral units. Formally, $BU = R \times A$.*

The concept of behavioral unit comes from the idea that the behavior of a collaborator is to a large extent determined by the role he/she plays. Therefore, roles and action types have to be associated to determine the behavior, i.e. the set of actions that a collaborator playing a given role can perform.

A behavioral $bu = (r, a)$ is said to be executed iff a collaborator labeled with the role $r$ executes the action of the given type $a$. It should be notice that only collaborators labeled with the role $r$ can execute the behavioral unit $bu = (r, a)$.

**Definition 4.** *A* state *is a label associated with a given situation in a collaborative process. Let denote $S$ the set of states.*

In a given protocol $p$, the set of states that may occur $S_p$ is a subset of $S$, i.e $S_p \subseteq S$.

**Definition 5.** *A* transition *is a triplet $(bu, s_{\text{source}}, s_{\text{destination}})$. Let denote $T$ the set of transitions. Formally, $T = BU \times S \times S$.*

In a given protocol $p$, the set of transitions $T_p$ is a subset of $T$, i.e $T_p \subseteq T$.

Now that all concepts underlying social protocols have been formally presented, the concept of social protocol may be defined.

**Definition 6.** *A* social protocol *is a finite state machine. A social protocol consists of $\left\{S_p, {S_p}^{\text{start}}, {S_p}^{\text{end}}, R_p, A_p\right\}$ where ${S_p}^{\text{start}} \subset S_p$ is the set of starting states, ${S_p}^{\text{end}} \subset S_p$ is the set of ending states, ${S_p}^{\text{start}} \cap {S_p}^{\text{end}} = \emptyset$. Let denote $P$ the set of social protocols.*

In a social protocol, collaborators are moving from state to state via the execution of behavioral units. In other words, the execution of behavioral units are transition conditions. As mentioned before, a behavioral unit may be executed only by a collaborator labeled with the appropriate role.

An extended definition of social protocols have been presented in [1]. An application of social protocols to electronic negotiations may be found in [3].

## 2.2 An Example of Social Protocol

The example of social protocol which is presented in this section is oversimplified for readability reasons. It is obvious that social protocols modeling real-world collaboration processes are usually more complex. The chosen collaboration process to be modeled as a social protocol may be described as follows: a set of users are collaborating on the establishment of a "FAQ" document. Some users only asks questions, while others, referred as "experts", may answer the questions. Other users, referred as "managers", may interrupt the work on the FAQ document. The work on the document may be terminated either by a success (the document has been written and the manager estimates that its quality is good enough to be published) or by a failure (the users did not find any way to collaborate and the manager has estimated that the work on the FAQ should be interrupted).

A possible social protocol modeling this collaboration process is presented in Fig. 1.

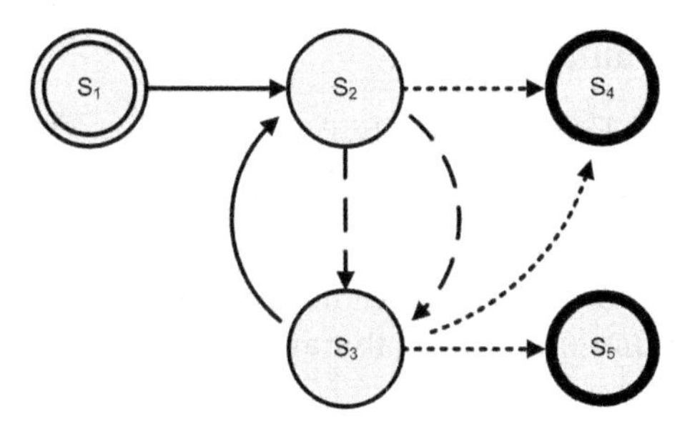


**Fig. 1.** An example of social protocol

In Fig. 1, five states $s_1, \ldots, s_5$ are represented as circles. State $s_1$ is a starting state, states $s_4$ and $s_5$ are ending states. The following states are defined:

- state $s_1$: waiting for a first question;
- state $s_2$: waiting for an answer;
- state $s_3$: waiting for a next question;
- state $s_4$: failed termination;
- state $s_5$: successful termination.

Transitions are represented as arrows, and the line style is associated with the role of the users that may execute a given transition. Continuous line style is used to represent transitions that may be executed by "normal users", fine-dashed style for transitions that may be executed by "experts", and fine-dotted style for transitions that may be executed by "managers". Transitions are summarized in Table 1.

**Table 1.** Transitions for the example of social protocol

| Source state | Destination state | Role | Action |
|---|---|---|---|
| $s_1$ | $s_2$ | Normal | Ask question |
| $s_2$ | $s_3$ | Expert | Answer question |
| $s_2$ | $s_3$ | Expert | Suppress question |
| $s_2$ | $s_4$ | Manager | Failure ending |
| $s_3$ | $s_2$ | Normal | Ask question |
| $s_3$ | $s_4$ | Manager | Failure ending |
| $s_3$ | $s_5$ | Manager | Successful ending |

# 3 Social Protocol Validity

Before the conditions for social protocol validity are presented, the concept of *social protocol validity* should be defined. Two kinds of social protocol validity may be distinguished: a social protocol may be *semantically valid* and/or *structually valid.* Finally, a social protocol is valid iff it is both semantically and structurally valid.

## 3.1 Semantical Validity

A given social protocol is semantically valid iff

1. all transitions leading to an ending state are associated with behavioral units whose actions end the collaboration;
2. no transition leading to a non-ending state is associated with behavioral units whose actions end the collaboration.

The first condition ensures that each transition leading to an ending state actually ends the collaboration. The second condition ensures that the collaboration cannot be "interrupted" by a transition leading to a non-ending state.

The semantical validity of a given social protocol may be relatively easily checked: 1) all behavioral units associated with transitions leading to an ending state should contain only ending actions; 2) all behavioral units containing ending actions should be associated only with transitions leading to ending states.

## 3.2 Structural Validity

A given social protocol is structurally valid iff

1. for each non-starting state $s$, it exists at least one path from one starting state to the state $s$;
2. for each non-ending state $s$, it exists at least one path from the state $s$ to an ending state;
3. for each state $s$ and each behavioral unit $bu$, it exists at most one transition from the state $s$ associated with the behavioral unit $bu$.

The first condition ensures that each state is reachable, i.e. there is no state to which one may not move to from a starting state. The second condition ensures that there is no state from which one may not move to an ending state. The third condition ensures that there is no "ambiguity" in a protocol. An ambiguity may occur in a protocol in the case when it exists many transitions associated to a common behavioral unit, leading from a given state to various states. In such a protocol, the execution of the "shared" behavioral unit may not be performed as it is then impossible to decide which state should be the next one.

While the third condition may be relatively easily checked, checking that the first and second conditions are fulfilled is a more complex task which requires more advanced algorithms.

## 4 Structural Validation of Social Protocols

In this section, an algorithm for structural validation of social protocols is presented. This algorithm is based on an algebraic representation of social protocols.

### 4.1 Algebraic Representation of Social Protocols

Any social protocol may be represented in an algebraic form as a *transition matrix*. The formal definition of the transition matrix requires the definition of the concepts of *sorted states list* and *set of local behavioral units*.

A sorted states list $\Sigma_p$ for a given protocol $p$ is a list containing once all states of the protocol $p$ and such that the first states of the list are starting states of the protocol and the last states of the list are ending states.

**Definition 7.** *A* sorted states list *for a given protocol $p$ is a list $\Sigma_p = \{\sigma_i \in S_p\}$ with $i \in \{1, \ldots, |S_p|\}$ such that:*

- $S_p \cap \Sigma_p = \emptyset$,
- $\forall (i,j) \in \{1, \ldots, |S_p|\}^2, i = j \Leftrightarrow \sigma_i = \sigma_j$,
- $\exists (a,b) \in \{1, \ldots, |S_p|\}^2, 1 \leq a < b \leq |S_p|$ *and*

$$\begin{cases} \forall x \in [1,a], & \sigma_x \in {S_p}^{\text{start}}, \\ \forall x \in ]a,b[, & \sigma_x \in S_p - \left({S_p}^{\text{start}} \cup {S_p}^{\text{end}}\right), \\ \forall x \in [b, |S-p|], & \sigma_x \in {S_p}^{\text{end}}. \end{cases}$$

A *set of local behavioral units* ${\beta_p}^{s,s'}$ for a given protocol $p$ is the set of behavioral units associated with a transition from state $s$ to $s'$. Let denote ${\beta_p}^S$ the set of sets of local behavioral units.

**Definition 8.** *A* set of local behavioral units *from $s$ to $s'$ is a set ${\beta_p}^{s,s'} = bu^{s,s'}$ such that:*

- $\forall bu^{s,s'} \in {\beta_p}^{s,s'}, bu^{s,s'} \in BU_p$,
- $\forall bu^{s,s'} \in {\beta_p}^{s,s'}, \exists t \in T_p$ *such that* $t = (bu^{s,s'}, s, s')$.

Sorted states lists and sets of local behavioral units are required to build a *transition matrix.* A transition matrix $\Theta_p$ is an $|S_p| \times |S_p|$ matrix which elements are sets of local behavioral units laid out according to a sorted states list.

**Definition 9.** *A* transition matrix $\Theta_p$ *is an* $|S_p| \times |S_p|$ *matrix such that* $\Theta_p : \{1, \ldots, |S_p|\} \times \{1, \ldots, |S_p|\} \rightarrow {\beta_p}^S$, $\Theta_p[ij] = {\beta_p}^{\sigma_i, \sigma_j}$.

The elements of a transition matrix are sorted according to a sorted states list, i.e. the first columns and rows are related with starting states, while last columns and rows are related with ending states. Each element of a transition matrix is a set of local behavioral unit for a given source state (in row) and a given destination state (in column).

A *transition cardinality matrix* $\Theta_{p,||}$ may be easily computed from a transition matrix. A transition cardinality matrix is an $|S_p| \times |S_p|$ matrix which elements are the *cardinality* of sets of local behavioral units laid out according to a sorted states list.

**Definition 10.** *A* transition cardinality matrix $\Theta_{p,||}$ *is an* $|S_p| \times |S_p|$ *matrix such that* $\Theta_{p,||} : \{1, \ldots, |S_p|\} \times \{1, \ldots, |S_p|\} \rightarrow \mathbb{N}$, $\Theta_{p,||}[ij] = |{\beta_p}^{\sigma_i, \sigma_j}|$.

Each element of a transition cardinality matrix is the number of transitions from the source state (in row) to the destination state (in column).

### 4.2 State Reachability Computation

The *reachability* of a state $s'$ from state $s$ in a protocol $p$ means that there is a list of transitions in $p$ connecting state $s$ to state $s'$. To formally define the concept of reachability, let's first introduce the concept of *path.*

A path ${\pi_p}^{s,s'}$ from the state $s$ to the state $s'$ is a list of transitions connecting $s$ to $s'$.

**Definition 11.** *A* path *from the state $s$ to the state $s'$, denoted* ${\pi_p}^{s,s'}$, *is such that* ${\pi_p}^{s,s'} = \langle s_1, t_1, s_2, t_2, \ldots, s_{n-1}, t_{n-1}, s_n \rangle$ *with*

- $s_1 = s$,
- $s_n = s'$,
- $\forall i \in [1, n-1]$, $t_n = (bu_n, s_n, s_{n+1})$.

The *length* of a path is defined as the number of its transitions.

A state $s'$ is reachable from state $s$ in a given protocol $p$ iff it exists at least one path from state $s$ to state $s'$.

**Definition 12.** *A state $s'$ is $n$-*reachable *from $s$ iff it exists at least one path of length $n$ from $s$ to $s'$.*

The $n$-reachability of a state $s$ from state $s'$ means that there is a list of exactly $n$ transitions connecting $s$ to $s'$. Let $\pi^{s,s'}_{p,||=n}$ denote the number of paths of length $n$ from $s$ to $s'$ in protocol $p$.

**Definition 13.** *A* path cardinality matrix $\Pi_p$ *is an* $|S_p| \times |S_p|$ *matrix such that*

$$\Pi_p = \sum_{n=1}^{|S_p|-1} \Theta_{p,||}^n$$

The path cardinality matrix contains information about the reachability of states: each element of the path cardinality matrix is the number of paths from the source state (in row) to the destination state (in column).

**Theorem 1.** *A state* $s_j$ *is reachable from* $s_i$ *iff* $\Pi_p[ij] \neq 0$.

*Proof.* The transition cardinality matrix contains information about the 1-reachability of states. As $\Theta_{p,||}[ij] = |\beta_p{}^{\sigma_i,\sigma_j}|$, each element of the transition cardinality matrix is the number of transitions from the source state (in row) to the destination state (in column), i.e. the number of path of length 1.

The number of paths of length 2 may be calculated on the basis of the transition cardinality matrix. Let $s$, $s'$, and $s''$ be three states. The number of paths of length 2 from the state $s$ to the state $s'$ through the state $s''$ equals the number of paths of length 1 from the state $s$ to the state $s''$ multiplied by the number of paths of length 1 from the state $s''$ to the state $s'$. Therefore, the number of paths of length 2 from the state $s$ to the state $s\prime$ equals the sum of the number of paths of length 2 from the state $s$ to the state $s'$ through any state $s'' \in S_p$. Formally,

$$\pi_{p,||=2}^{s,s'} = \sum_{i=1}^{|S_p|} \pi_{p,||=1}^{s,s_i} . \pi_{p,||=1}^{s_i,s'}$$

One may recognize in the former equation the classical multiplication of matrices. Moreover, as $\pi_{p,||=1}^{s_i,s_j} = |\beta_p{}^{\sigma_i,\sigma_j}| = \Theta_{p,||}[ij]$, it may be concluded that $\pi_{p,||=2}^{s,s'} = \Theta_{p,||}^2[ij]$. Therefore, each element of the $\Theta_{p,||}^2$ matrix is the number of paths of length 2 from a source state (in row) to a destination state (in column).

In a similar way, it may be demonstrated that each element of the $\Theta_{p,||}^n$ matrix is the number of paths of length $n$ from a source state (in row) to a destination state (in column).

The reachability of states in a protocol $p$ with $|S_p|$ states may be deduced from the logical sum of $n$-reachability where $n \in [1, |S_p| - 1]$. Indeed, the reachability of a given state $s$ from a state $s'$ means the existence of at least one path from $s'$ to $s$. Moreover, the longest path going through all states only once has a maximal length of $|S_p| - 1$. Therefore, a state $s$ is reachable from state $s'$ iff it exists at least one path from state $s'$ to $s$, of length less or equal to $|S_p| - 1$, i.e. $\Pi_p[ij] \neq 0$. □

### 4.3 Algorithm for Structural Validation

For a given protocol $p$, conditions 1. and 2. presented in Sect. 3.2 may be checked with the following algorithm:

1. Sort the states $s \in S$ from starting states to ending ones as a sorted states list $\Sigma_p = \{\sigma_i \in S_p\}$ with $i \in \{1, \ldots, a, \ldots, b, \ldots, |S_p|\}$ such that $\forall i \in [1, \ldots, a]$, $\sigma_i$ are starting states, and $\forall i \in [b, \ldots, |S_p|]$, $\sigma_i$ are ending states;
2. Compute the transition cardinality matrix $\Theta_{p,||}$ according to the sets of local behavioral units and the sorted states list $\Sigma_p$;
3. Compute the path cardinality matrix $\Pi_{p,||} = \sum_{n=1}^{|S_p|-1} \Theta^n_{p,||}$;
4. Condition 1. is fulfilled $\Leftrightarrow \forall j \in ]a, |S_p|], \exists i \in [1, a]$ such that $\Pi_{p,||}[ij] > 0$.
5. Condition 2. is fulfilled $\Leftrightarrow \forall i \in [1, b[, \exists j \in [b, |S_p|]$ such that $\Pi_{p,||}[ij] > 0$.

# 5 Examples of Structural Validation

In this section, the validity of three protocols is checked to illustrate the algebraic representation of social protocols and the algorithm presented above. In the presented examples, it is assumed that the protocols are semantically valid and that they fulfill the third condition for structural validity. For the three protocols, conditions 1. and 2. for structural validity are checked with the algorithm presented in Sect. 4.3.

In all presented protocols, the states are assumed to be already sorted to improve the readability of the paper. Therefore, the first step of the algorithm for structural validation may be skipped.

## 5.1 Example of Valid Social Protocol

In Fig. 2, a first example of social protocol is presented. Starting states – $\sigma_1$ and $\sigma_2$ – are represented by a double circle, while ending states – $\sigma_5$ and $\sigma_6$ are represented by a bold circle. Transitions are represented as arrows.

For the protocol presented in Fig. 2, the transition cardinality matrix is the following one:

$$\Theta_{p,||} = \begin{bmatrix} 0&0&2&0&0&0 \\ 0&0&0&1&0&0 \\ 0&0&0&0&1&0 \\ 0&0&1&0&0&1 \\ 0&0&0&0&0&0 \\ 0&0&0&0&0&0 \end{bmatrix}$$

As the protocol contains six states, the path cardinality matrix $\Pi_{p,||}$ is the sum of the powers of the transition cardinality matrix from power 1 to power 5, i.e $\Pi_{p,||} = \sum_{n=1}^{|S_p|-1} \Theta^n_{p,||} = \sum_{n=1}^{5} \Theta^n_{p,||} = \Theta_{p,||} + \Theta^2_{p,||} + \ldots + \Theta^5_{p,||}$

By a simple computation, $\Pi_{p,||} = \begin{bmatrix} 0&0&2&0&2&0 \\ 0&0&1&1&1&1 \\ 0&0&0&0&1&0 \\ 0&0&1&0&1&1 \\ 0&0&0&0&0&0 \\ 0&0&0&0&0&0 \end{bmatrix}$

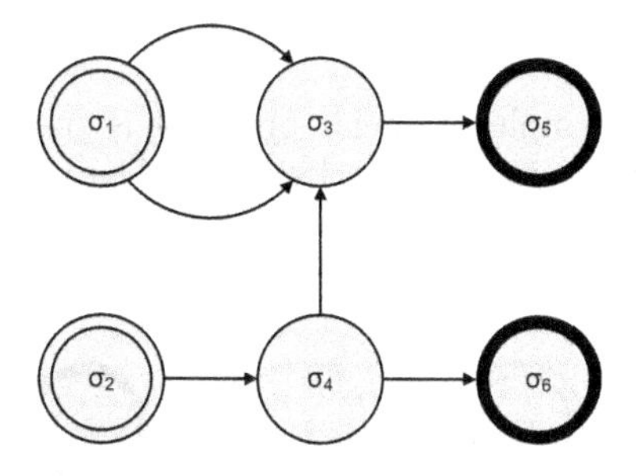


**Fig. 2.** Example of valid social protocol

In Fig. 3, the area of the path cardinality matrix to be checked for the first condition of structural validity is highlighted. This area consists of all elements whose row number is lower or equal than $a = 2$ and whose column number is greater than $a = 2$. If in each column of this area it exists at least one element whose value is greater than 0, the first condition is fulfilled. In this first protocol, the first condition is fulfilled.

$$\Pi_{p,||} = \left[\begin{array}{cc|cccc} 0 & 0 & \mathit{2} & \mathit{0} & \mathit{2} & \mathit{0} \\ 0 & 0 & \mathit{1} & \mathit{1} & \mathit{1} & \mathit{1} \\ \hline 0 & 0 & 0 & 0 & 1 & 0 \\ 0 & 0 & 1 & 0 & 1 & 1 \\ 0 & 0 & 0 & 0 & 0 & 0 \\ 0 & 0 & 0 & 0 & 0 & 0 \end{array}\right]$$

**Fig. 3.** Area to be check for the first condition of structural validity

In Fig. 4, the area of the path cardinality matrix to be checked for the second condition of structural validity is highlighted. This area consists of all elements whose row number is lower than $b = 5$ and whose column number is greater or equal to $b = 5$. If in each row of this area it exists at least one element whose value is greater than 0, the second condition is fulfilled. In this first protocol, the second condition is fulfilled.

$$\Pi_{p,||} = \left[\begin{array}{cccc|cc} 0 & 0 & 2 & 0 & \mathit{2} & \mathit{0} \\ 0 & 0 & 1 & 1 & \mathit{1} & \mathit{1} \\ 0 & 0 & 0 & 0 & \mathit{1} & \mathit{0} \\ 0 & 0 & 1 & 0 & \mathit{1} & \mathit{1} \\ \hline 0 & 0 & 0 & 0 & 0 & 0 \\ 0 & 0 & 0 & 0 & 0 & 0 \end{array}\right]$$

**Fig. 4.** Area to be check for condition 2. of structural validity

### 5.2 Example of a Social Protocol Violating the First Condition

In Fig. 5, a second example of social protocol is presented. This protocol is similar to the protocol presented in Sect. 5.1. The only difference is that the transition from state $\sigma_2$ now leads to $\sigma_1$ instead of $\sigma_4$.

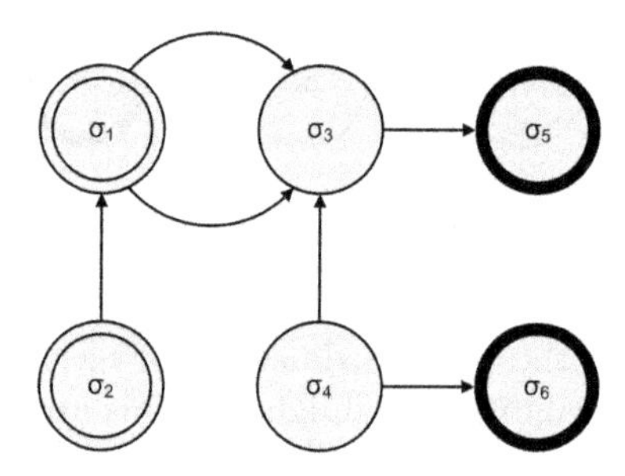


**Fig. 5.** Example of a social protocol violating the first condition

For the protocol presented in Fig. 5, the transition cardinality matrix is the following one:

$$\Theta_{p,||} = \begin{bmatrix} 0 & 0 & 2 & 0 & 0 & 0 \\ 1 & 0 & 0 & 0 & 0 & 0 \\ 0 & 0 & 0 & 0 & 1 & 0 \\ 0 & 0 & 1 & 0 & 0 & 1 \\ 0 & 0 & 0 & 0 & 0 & 0 \\ 0 & 0 & 0 & 0 & 0 & 0 \end{bmatrix}$$

By a simple computation, $\Pi_{p,||} = \left[\begin{array}{cc|cccc} 0 & 0 & 2 & 0 & 2 & 0 \\ 1 & 0 & 2 & 0 & 2 & 0 \\ \hline 0 & 0 & 0 & 0 & 1 & 0 \\ 0 & 0 & 1 & 0 & 1 & 1 \\ 0 & 0 & 0 & 0 & 0 & 0 \\ 0 & 0 & 0 & 0 & 0 & 0 \end{array}\right]$

As it may easily be notice on Fig. 5, this protocol does not fulfill the first condition of structural validity because states $\sigma_4$ and $\sigma_6$ are unreachable from the starting states. The analyzis of the path transition matrix leads to the same conclusion: in the area to be checked for the fist condition, the only value for states $\sigma_4$ and $\sigma_6$ is 0.

### 5.3 Example of a Social Protocol Violating the Second Condition

In Fig. 6, a third example of social protocol is presented. This protocol is similar to the protocol presented in Sect. 5.1. The only difference is that the transition to state $\sigma_5$ now comes from $\sigma_4$ instead of $\sigma_3$.

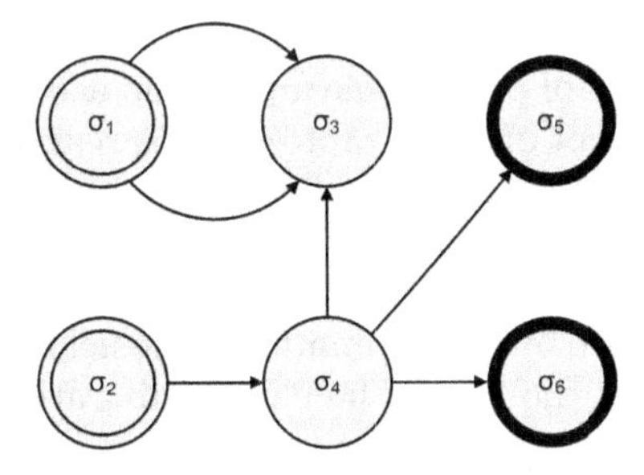


**Fig. 6.** Example of a social protocol violating the second condition

For the protocol presented in Fig. 6, the transition cardinality matrix is the following one:

$$\Theta_{p,||} = \begin{bmatrix} 0&0&2&0&0&0 \\ 0&0&0&1&0&0 \\ 0&0&0&0&\mathbf{0}&0 \\ 0&0&1&0&\mathbf{1}&1 \\ 0&0&0&0&0&0 \\ 0&0&0&0&0&0 \end{bmatrix}$$

By a simple computation, $\Pi_{p,||} = \left[\begin{array}{cccc|cc} 0&0&2&0&\mathbf{0}&\mathbf{0} \\ 0&0&1&1&1&1 \\ 0&0&0&0&\mathbf{0}&\mathbf{0} \\ 0&0&1&0&1&1 \\ \hline 0&0&0&0&0&0 \\ 0&0&0&0&0&0 \end{array}\right]$

As it may easily be notice on Fig. 6, this protocol does not fulfill the second condition of structural validity because no ending state may be reached from states $\sigma_1$ and $\sigma_3$. The analyzis of the path transition matrix leads to the same conclusion: in the area to be checked for the second condition, the only value for states $\sigma_1$ and $\sigma_3$ is 0.

## 6 Related Works

As process modeling is concerned, many works have already been conducted in the research field of workflow modeling and workflow management systems. Many works [4,5,6,7,8,9,10,11] have focused on formal models and conditions under which a modification of an existing – and potentially running – workflow retains workflow validity. However, to our best knowledge, current works concerning workflow adaptation focus on interactions the importance of social aspects, are not or insufficiently taken into account by these works.

Some interesting works have been done in the field of electronic negotiations to model electronic negotiations with the help of negotiation protocols. In [12], it is stated in that, in the field of electronic negotiations, "the protocol is a formal model, often represented by a set of rules, which govern software processing, decision-making and communication tasks, and imposes restrictions on activities through the specification of permissible inputs and actions". One may notice the similarity with the concept of social protocol. The reason for this fact is that the model presented in this paper was originally coming from a work on protocols for electronic negotiations [13]. However, to our knowledge, none of the works concerning negotiation protocols provides mechanisms for protocol validation. Moreover, these works are by nature limited to the field of electronic negotiations which is just a subset of the field of human collaboration.

## 7 Conclusion

While many works are currently done on modeling collaboration processes in which software entities (agents, web services) are involved, modeling collaboration processes in which mainly humans are involved is an area that still requires much attention from the research community. Some of the main issues to be addressed are the social aspects of collaboration and the adaptation capabilities of humans. In this paper the first issue is addressed. The concept of social protocol aims at being a start of answer to the question of computer support for social collaboration. The algorithm for structural validation of social protocols presented in this paper provides protocol designers and/or software supporting social protocol with means of checking the validity of social protocols, which leads to more robust support for social protocols.

The main innovations presented in this paper are 1) the algebraic representation of social protocols, 2) the algorithm for structural validation of social protocols based on their algebraic representation. The proposed concepts have been fully implementated in the *DynG* protocol [14], a social protocol-based platform.

The validation of social protocols is a requirement for 1) the design of robust collaboration models, 2) more advanced support for human-to-human collaboration. Among advanced features, the adaptation of social protocol – i.e. the possibility to modify a collaboration process and its associated social protocol at run-time – is necessary to weaken constraints usually limiting the interaction between collaborators, so that the adaptation capabilities of humans may be integrated in the life of a social protocol. With support of social protocol adaptation, methods for validation of adapted social protocols extending the algorithm presented in this paper are still to be proposed.

## References

1. Picard, W.: Computer support for adaptive human collaboration with negotiable social protocols. In Abramowicz, W., Mayr, H.C., eds.: Proc. of the $9^{th}$ International Conference on Business Information Systems. Volume 85 of LNI., GI (2006) 90–101
2. Picard, W.: Modeling structured non-monolithic collaboration processes. In Camarinha-Matos, L., Afsarmanesh, H., Ortiz, A., eds.: Collaborative Networks and their Breeding Environments, Proc. of the $6^{th}$ IFIP Working Conference on Virtual Enterprises (PRO-VE 2005), Valencia, Spain, Springer (September 2005) 379–386

3. Picard, W.: Towards support systems for non-monolithic electronic negotiations. the contract-group-message model. Journal of Decision Systems: Special Issue on Electronic Negotiations - Models, Systems and Agents **13** (2004) 423–439
4. van der Aalst, W.M.P.: The Application of Petri Nets to Workflow Management. The Journal of Circuits, Systems and Computers **8**(1) (1998) 21–66
5. van der Aalst, W.M.P., Basten, T., Verbeek, H.M.W., Verkoulen, P.A.C., Voorhoeve, M.: Adaptive workflow: On the interplay between flexibility and support. In Filipe, J., ed.: Proc. of the $1^{st}$ International Conference on Enterprise Information Systems. Volume 2., Setúbal, Portugal, Kluwer Academic Publishers (March 1999) 353–360
6. van der Aalst, W.M.P.: Workflow verification: Finding control-flow errors using petri-net-based techniques. In Aalst, W., Desel, J., Oberweis, A., eds.: Business Process Management, Models, Techniques, and Empirical Studies. Volume 1806 of Lecture Notes in Computer Science., Springer (2000) 161–183
7. Sadiq, W., Orlowska, M.E.: Analyzing process models using graph reduction techniques. Information Systems **25**(2) (2000) 117–134
8. Sadiq, S.W., Orlowska, M.E., Sadiq, W., Foulger, C.: Data flow and validation in workflow modelling. In Schewe, K.D., Williams, H.E., eds.: Proceedings of the $15^{th}$ Australasian Database Conference, ADC 2004. Volume 27 of CRPIT., Australian Computer Society (2004) 207–214
9. Sadiq, S.W., Orlowska, M.E., Sadiq, W.: Specification and validation of process constraints for flexible workflows. Information Systems **30**(5) (2005) 349–378
10. ter Hofstede, A.H.M., Orlowska, M.E., Rajapakse, J.: Verification problems in conceptual workflow specifications. Data Knowledge Engineering **24**(3) (1998) 239–256
11. ter Hofstede, A.H.M., Orlowska, M.E., Rajapakse, J.: Verification problems in conceptual workflow specifications. In Thalheim, B., ed.: Conceptual Modeling - ER'96, 15th International Conference on Conceptual Modeling, Cottbus, Germany, October 7-10, 1996, Proceedings. Volume 1157 of Lecture Notes in Computer Science., Springer (1996) 73–88
12. Kersten, G.E., Strecker, S.E., Lawi, K.P.: Protocols for electronic negotiation systems: Theoretical foundations and design issue. In: Proc. of the $5^{th}$ Conference on Electronic Commerce and Web Technologies (ECWeb04), Sarragoza, Spain, IEEE Computer Society (2004)
13. Picard, W., Huriaux, T.: Dyng: A protocol-based prototype for non-monolithic electronic collaboration. Lecture Notes in Computer Science **3865**(CSCW in Design 2005) (2006) 41–50
14. Huriaux, T., Picard, W.: Dyng: a multi-protocol collaborative system. In Funabashi, M., Grzech, A., eds.: Proc. of the $5^{th}$ IFIP International Conference on e-Commerce, e-Business, and e-Government (I3E 2005), Poznań, Poland (2005) 591–605

# Long Tails and Analysis of Knowledge Worker Intranet Browsing Behavior

Peter Géczy, Noriaki Izumi, Shotaro Akaho, and Kôiti Hasida

National Institute of Advanced Industrial Science and Technology (AIST)
Tsukuba and Tokyo, Japan

**Abstract.** We present a formal approach to analysis of human browsing behavior in electronic spaces. An analysis of knowledge workers' interactions on a large corporate intranet have revealed that users form repetitive elemental and complex browsing patterns, use narrow spectrum of resources, and exhibit diminutive exploratory behavior. Knowledge workers had well defined targets and accomplished their browsing tasks via few subgoals. The analyzed aspects of browsing behavior exposed significant long tail characteristics that can be accurately modeled by the introduced novel distribution. The long tail behavioral effects present new challenges and opportunities for business information systems.

## 1 Introduction

*"Nobody has really looked at productivity in white collar work in a scientific way." (Peter Drucker)* [1]. Absence of scientific evidence concerning knowledge worker productivity, efficiency, and their adequate measurement methods has been at the center of the recent managerial discourse [2]. Human dynamics [3] and behavior in electronic spaces [4], [5] have been rapidly gaining importance in a corporate sector. Corporations are eagerly exploiting ways to acquire more behavioral data about customers—primarily for the commercial purposes [6].

Behavioral studies of human interactions in WEB environments are generally time consuming and resource demanding [7]. Only limited attempts have been made toward their automation [8]. Human behavior analysis on WEB is largely performed by analyzing the server-side data (WEB logs) and/or the client-side data from script agents. Data is mined for user click-streams [9] and analyzed using the conventional statistical modeling methods or empirical studies. The empirical investigations [10] provide only rule-based conclusions that are unsuitable for predictive purposes. The statistical approaches have been favoring Markov models with predictive capability [11], however, the higher-order Markov models become exceedingly complex and computationally expensive. Cluster analysis methods [12] and adaptive learning strategies [13] have also been employed, but both have scalability drawbacks. Mining only frequent patterns reduces the computational complexity and improves the speed, however, at the expense of substantial data loss [14].

The presented novel analysis concept encompasses both topological and temporal characteristics of a human behavior in electronic environments, and effectively captures the dimensions of human interactions.

W. Abramowicz (Ed.): BIS 2007, LNCS 4439, pp. 584–597, 2007.
© Springer-Verlag Berlin Heidelberg 2007

## 2 Approach Formulation

We introduce the basic line of inquiry together with the corresponding terminology. Definitions are accompanied by intuitive explanations that help us better understand the concept at a higher formal level.

The click-stream sequences [15] of user page transitions are divided into sessions, and sessions are further divided into subsequences. Division of sequences into subparts is done with respect to the user activity and inactivity. Consider the conventional time-stamp click-stream sequence of the following form: $\{(p_i, t_i)\}_i$, where $p_i$ denotes the visited page $URL_i$ at the time $t_i$. For the purpose of analysis this sequence is converted into the form: $\{(p_i, d_i)\}_i$ where $d_i$ denotes a delay between the consecutive views $p_i \rightarrow p_{i+1}$. User browsing activity $\{(p_i, d_i)\}_i$ is divided into subelements according to the periods of inactivity $d_i$.

**Definition 1.** *(Browsing Session, Subsequence, User Behavior)*
**Browsing session** is a sequence $B = \{(p_i, d_i)\}_i$ where each $d_i \leq T_B$. Browsing session is often in the further text referred to simply as a **session**.
**Subsequence** of a browsing session $B$ is a sequence $S = \{(p_i, dp_i)\}_i$ where each delay $dp_i \leq T_S$, and $\{(p_i, dp_i)\}_i \subset B$.
**User behavior** is a sequence $U = \{(B_i, db_i)\}_i$ of user browsing sessions $BS_i$ separated by delays $db_i$ between consecutive sessions $B_i \rightarrow B_{i+1}$. For simplicity, we consider user behavior to be a set of browsing sessions $U = \{B_i\}$.

The sessions delineate tasks of various complexities users undertake in electronic environments. The subsequences correspond to the session subgoals; e.g. subsequence $S_1$ is a login, $S_2$ – document download, $S_3$ – search for internal resource, etc.

Important issue is determining the appropriate values of $T_B$ and $T_S$ that segment the user activity into the sessions and subsequences. The former research [16] indicated that the student browsing sessions last on average 25.5 minutes. However, we adopt the average maximum attention span of 1 hour as a value for $T_B$. If the user's browsing activity was followed by a period of inactivity greater than 1 hour, it is considered a single session, and the following activity comprises the next session.

Value of $T_S$ is determined dynamically and computed as an average delay in a browsing session: $T_S = \frac{1}{N}\sum_{i=1}^{N} d_i$. If the delays between page views are short, it is useful to bound the value of $T_S$ from below. This is preferable in environments with frame-based and/or script generated pages where numerous logs are recorded in a rapid transition. Since our situation contained both cases, we adjusted the value of $T_S$ by bounding it from below by 30 seconds:

$$T_S = max\left(30, \frac{1}{N}\sum_{i=1}^{N} d_i\right) . \quad (1)$$

Another important aspect is to observe where the user actions are initiated and terminated. That is, to identify the starting and ending points of the subsequences, as well as single user actions.

**Definition 2.** *(Starter, Attractor, Singleton)*
**Starter** is the first navigation point of an element of subsequence or session with length greater that 1.
**Attractor** is the last navigation point of an element of subsequence or session with length greater that 1.
**Singleton** is a navigation point $p$ such that there exist $B$ or $S$ where $|B| = 1$ or $|S| = 1$.

The starters refer to the starting navigation points of user actions, whereas the attractors denote the users' targets. The singletons relate to the single user actions such as use of hotlists (e.g. history or bookmarks) [10].

We can formulate behavioral abstractions simply as the pairs of starters and attractors. Then it is equally important to observe the connecting elements of transitions from one task (or sub-task) to the other.

**Definition 3.** *(SE Elements, Connectors)*
Let $S_i = \{(p_{ik}, dp_{ik})\}_k^N$ and $S_{i+1} = \{(p_{i+1l}, dp_{i+1l})\}_l^M$ be consecutive subsequences, $S_i \rightarrow S_{i+1}$, of a browsing session.
**SE element** (start-end element) of a subsequence $S_i$ is a pair $SE_i = (p_{i1}, p_{iN})$.
**Connector** of subsequences $S_i$ and $S_{i+1}$ is a pair $C_i = (p_{iN}, p_{i+1,1})$.

The SE elements outline the higher order abstractions of user subgoals. Knowing the starting point, users can follow various navigational pathways to reach the target. Focusing on the starting and ending points of user actions eliminates the variance of navigational choices. The connectors indicate the links between elemental browsing patterns. This enables us to observe formation of more complex behavioral patterns as interconnected sequences of elemental patterns.

## 3 Intranet and Data

Data used in this work was a one year period intranet WEB log data of The National Institute of Advanced Industrial Science and Technology (Table 1–left). The majority of users are skilled knowledge workers. The intranet WEB portal had a load balancing architecture comprising of 6 servers providing extensive range of WEB services and documents vital to the organization. Intranet services support managerial, administration and accounting processes, research cooperation with industry and other institutes, databases of research achievements, resource localization and search, attendance verification, and also numerous bulletin boards and document downloads. The institution has a number of branches at various locations throughout the country, thus certain services are decentralized. A size of visible WEB space was approximately 1 GB. Invisible WEB space was considerably larger, but difficult to estimate due to the distributed architecture and constantly changing back-end data.

Daily traffic was substantial and so was the data volume. It is important to note that the data was incomplete. Although some days were completely represented, every month there were missing logs from certain servers. The server side logs also suffered data loss due to caching and proxing. However, because

**Table 1.** Basic information about raw and preprocessed data used in the study

| | | | |
|---|---|---|---|
| Data Volume | ~60 GB | Log Records | 315 005 952 |
| Average Daily Volume | ~54 MB | Clean Log Records | 126 483 295 |
| Number of Servers | 6 | Unique IP Addresses | 22 077 |
| Number of Log Files | 6814 | Unique URLs | 3 015 848 |
| Average File Size | ~9 MB | Scripts | 2 855 549 |
| Time Period | 3/2005 — 4/2006 | HTML Documents | 35 532 |
| | | PDF Documents | 33 305 |
| | | DOC Documents | 4 385 |
| | | Others | 87 077 |

of the large data volume, the missing data only marginally affected the analysis. The WEB servers run the open source Apache server software and the WEB log data was in the combined log format without referrer.

## 4 Data Preprocessing and Cleaning

Starting with a setup description we present the data preprocessing and the initial cleaning. The row data contained large number of task irrelevant logs. Extracted clean data was structured, databased, and linked.

**Setup.** Extraction and analysis of knowledge worker navigation primitives from the intranet WEB logs was performed on a Linux setup with MySQL database as a data storage engine for the preprocessed and processed data. Analytic and processing routines were implemented in various programming languages and optimized for high performance. Processing of large data volumes was computationally and time demanding.

**Preprocessing and Cleaning.** Data fusion of the WEB logs from 6 servers of a load balanced intranet architecture was performed at the preprocessing level. The data was largely contaminated by logs from automatic monitoring software and required filtering. During the initial filtering phase the logs from software monitors, invalid requests, WEB graphics, style sheets, and client-side scripts were eliminated. The access logs from scripts, downloadable and syndicated resources, and documents in various formats were preserved. Information was structured according to the originating IP address, complete URL, base URL, script parameters, date-time stamp, source identification, and basic statistics. Clean raw data was logged into database and appropriately linked.

Approximately 40.15% of the original log records remined after the initial filtering (see Table 1–right). Major access to intranet resources was via scripts (94.68%). Only relatively minor portions of accessible resources were HTML documents (1.18%), PDF documents (1.1%), DOC documents (0.15%), and others (2.89%), such as downloadable software, updates, spreadsheets, syndicated resources, etc. The detected IP address space (22077 unique IPs) consisted of both statically and dynamically assigned IP addresses. Smaller portion of IP addresses were static, and relatively uniquely associable with users.

## 5 Session and Subsequence Extraction

The sessions and subsequences were extracted from clean log records. We observed that data contained machine generated logs. Detection and elimination of the machine generated traffic was carried out during the subsequence extraction.

**Session Extraction.** Preprocessed and databased Apache WEB logs (in combined log format) did not contain referrer information. Click-stream sequences were reconstructed by ordering the logs originating from the unique IP addresses according to the time-stamp information. The ordered log sequences from the specific IP addresses were divided into the browsing sessions as described in Definition 1. The divisor between sessions was the user inactivity period $ds_i$ greater than $T_{BS} = 1\ hour$.

**Table 2.** Observed basic session data statistics

| | |
|---|---|
| Number of Sessions | 3 454 243 |
| Number of Unique Sessions | 2 704 067 |
| Average Number of Sessions per Day | 9 464 |
| Average Session Length | 36 [URL transitions] |
| Average Session Duration | 2 912.23 [s] (48 min 32 sec) |
| Average Page Transition Delay per Session | 81.55 [s] (1 min 22 sec) |
| Average Number of Sessions per IP Address | 156 |
| Maximum | 1 553 |
| Minimum | 1 |

It is noticeable that the knowledge worker sessions on a corporate intranet are on average longer (appx. 48.5 minutes) than those of students (appx. 25.5 minutes) reported in [16]. Average number of 156 sessions per IP address, and large variation in maximum and minimum number of sequences from distinct IP addresses, indicate that association of particular users with distinct IP addresses is relevant only for static IP addresses. Large number (3492) of single sessions only originated from distinct IP addresses due to wide DHCP use. It is possible to employ clustering techniques to identify reasonably diverse groups of users.

**Subsequence Extraction.** Each detected session was analyzed for subsequences as defined in Definition 1. Segmenting element dividing sessions into subsequences was the delay between page transitions $dp_i > T_S$, where $T_S$ was determined according to (1). Lower bound of 30 seconds for the separating inactivity period $dp_i$ was proper.

It has been observed that the sessions contained machine generated subsequences. Periodic machine traffic with inactivity time less than the session separator delay could result in long session sequences. As seen in the histogram of average delays between subsequences (Figure 1-a), there was a disproportionally large number of sessions with average delays between subsequences around 30 minutes and 1 hour. This is indicated by the spikes in the main chart of

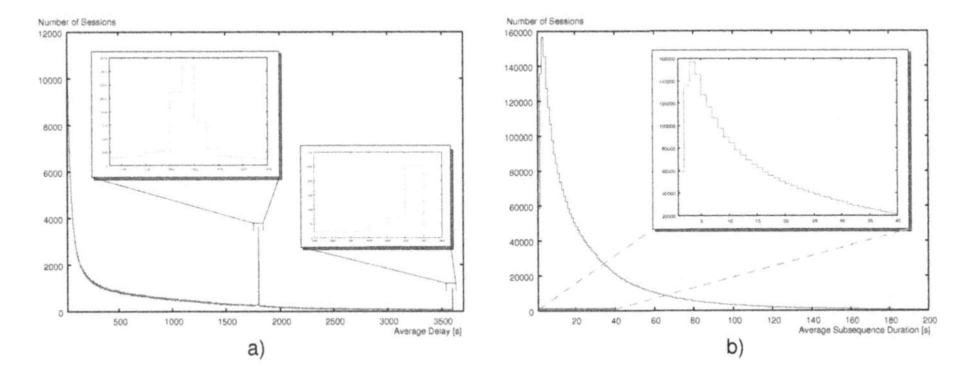


**Fig. 1.** Histograms: **a)** average delay between subsequences in sessions, **b)** average subsequence duration. There are noticeable spikes in chart **a)** around 1800 seconds (30 minutes) and 3600 seconds (1 hour). The detailed view is displayed in subcharts. Temporal variation of spikes corresponds to the peak average subsequence duration in chart **b)**. The spikes with relatively accurate delays between subsequences are due to machine generated traffic.

Figure 1-a. The detailed view (subcharts of Figure 1-a) revealed that the variation in the average delay between subsequences is approximately $\pm$ 3 seconds. It well corresponds to the peak in the histogram of average subsequence duration (Figure 1-b). It is highly unlikely that a human generated traffic would produce this precision (although certain subsequences were legitimate).

The machine generated traffic contaminates the data and should be filtered, since we primarily target a human behavior on the intranet. We filtered two main groups of machine generated subsequences: login subsequences and subsesequences with delay periodicity around 30 minutes and 1 hour.

Every user is required to login into intranet in order to access the services and resources. The login procedure involves a validation and generates several log records with 0 delays. The records vary depending on whether the login was successful or unsuccessful. In both cases the log records and the login related subsequences can be clearly identified and filtered.

The second group of machine generated traffic are the subsequences with periodicity of 30 minutes and 1 hour. Direct way of identifying these subsequences is to search for the sessions with only two subsequences having less than 1 second (or 0 second) duration (machines can generate requests fast and local intranet servers are capable of responding within milliseconds) and delay $ds_i$ between subsequences within the intervals: 1800 and 3600 $\pm$ 3 seconds. It has been discovered that a substantial number of such sessions contained relatively small number (170) of unique subsequences. Furthermore, these subsequences contained only 120 unique URLs. The identified subsequences and URLs were considered to be machine generated and filtered from further analysis. The subsequences with SE elements containing the identified URLs were also filtered.

**Table 3.** Observed basic subsequence data statistics

| | |
|---|---|
| Number of Subsequences | 7 335 577 |
| Number of Valid Subsequences | 3 156 310 |
| Number of Filtered Subsequences | 4 179 267 |
| Number of Unique Subsequences | 3 547 170 |
| Number of Unique Valid Subsequences | 1 644 848 |
| Average Number of Subsequences per Session | 3 |
| Average Subsequence Length | 4.52 [URL transitions] |
| Average Subsequence Duration | 30.68 [s] |
| Average Delay between Subsequences | 388.46 [s] (6 min 28 sec) |

Filtering of detected machine generated subsequences and their URLs significantly reduced the total number of subsequences - by 56.97% (from 7335577 to 3156310), as well as the number of unique subsequences - by 46.37% (from 3547170 to 1644848). Since the login sequences were also filtered, the number of subsequences per session decreased at least by 1. Reduction also occurred in the session lengths due to filtering of identified invalid URLs. Filtering did not significantly affect the duration of subsequences because the logs of machine generated subsequences occurred in rapid transitions with almost 0 durations and delays. It is noticeable that the average subsequence duration (30.68 seconds) is approximately equal to the chosen lower bound for $ds_i$ (30 seconds). This empirically justifies the right choice of lower bound for $T_S$.

## 6 Knowledge Worker Browsing Behavior Analysis

By analyzing the navigation point characteristics (starters, attractors, and singletons) together with the behavioral abstractions (SE elements and connectors) we infer several relevant observations. Exploratory analysis demonstrates usefulness of the approach in elucidating a human browsing behavior in electronic spaces.

### 6.1 Starter, Attractor, and Singleton Analysis

Navigation point characteristics highlight the initial (starters) and the terminal targets (attractors) of knowledge worker activities, and also the single-action behaviors (singletons). The starters, attractors, and singletons were extracted from subsequences.

*Knowledge workers utilized a small spectrum of starting navigation points and targeted relatively small number of resources during their browsing.* The set of starters, i.e. the initial navigation points of knowledge workers' (sub-)goals, was approximately 3.84% of total navigation points. Although the set of unique attractors, i.e. (sub-)goal targets, was approximately three times higher than the set of initial navigation points, it is still relatively minor portion (appx. 9.55% of unique URLs). Knowledge workers aimed at relatively few resources.

**Table 4.** Statistics for starters, attractors, and singletons

| | Starters | Attractors | Singletons |
|---|---|---|---|
| Total | 7 335 577 | 7 335 577 | 1 326 954 |
| Valid | 2 392 541 | 2 392 541 | 763 769 |
| Filtered | 4 943 936 | 4 943 936 | 563 185 |
| Unique | 187 452 | 1 540 093 | 58 036 |
| Unique Valid | 115 770 | 288 075 | 57 894 |

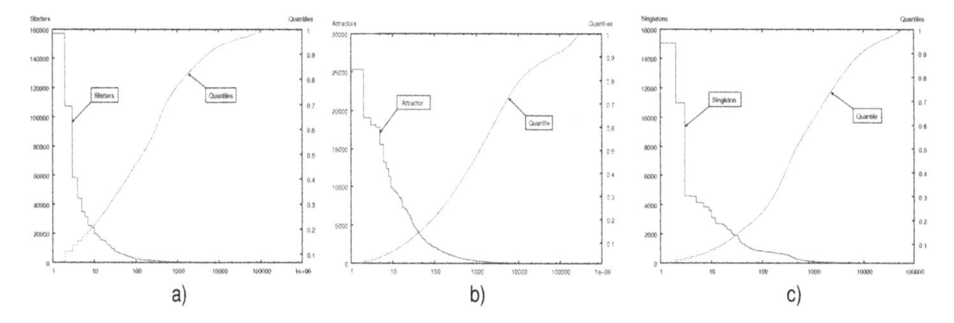


**Fig. 2.** Histograms and quantiles: **a)** starters, **b)** attractors, and **c)** singletons. Right y-axis contains a quantile scale. X-axis is in a logarithmic scale.

*Few resources were perceived of value to be bookmarked.* Number of unique single user actions was minuscule. Single actions, such as use of hotlists [10], followed by delays greater than 1 hour are represented by the singletons. Unique singletons accounted for only 1.92% of navigation points. If only small number of starters and/or attractors was perceived useful, there is a possibility that they were bookmarked and accessed directly in the following browsing experiences.

*Knowledge workers had focused interests and exhibited diminutive exploratory behavior.* A narrow spectrum of starters, attractors, and singletons was frequently used. The histograms and quantile characteristics of starters, attractors, and singletons (see Figure 2) indicate that higher frequency of occurrences is concentrated to relatively small number of elements. Approximately ten starters and singletons, and fifty attractors were very frequent. About one hundred starters and singletons, and one thousand attractors were relatively frequent. The quantile analysis (Figure 2) reveals that ten starters (appx. 0.0086% of unique valid starters) and singletons (appx. 0.017% of unique valid singletons), and fifty frequent attractors (appx. 0.017% of unique valid attractors) accounted for about 20% of total occurrences. One hundred starters (appx. 0.086% of unique valid starters) and one thousand attractors (appx. 0.35% of unique valid attractors) constituted about 45% and 48% of total occurrences, respectively. Analogously, one hundred twenty singletons (appx. 0.21% of unique valid singletons) compounded to about 37% of total occurences.

### 6.2 SE Element and Connector Analysis

These components serve as higher order abstractions of knowledge worker behavior. The SE elements represent the starting and ending points of subsequences, or corresponding elemental patterns. The connectors delineate transitions between pattern primitives, and thus formation of more complex patterns.

Extraction of the SE elements of subsequences and the connectors between subsequences is relatively straightforward. The SE elements and connectors also undergone filtering. If the invalid URLs were present in at least one element of a pair, the respective SE element and/or connector was marked as invalid.

**Table 5.** Statistics for SE Elements and connectors

| | SE Elements | Connectors |
|---|---|---|
| Total | 7 335 577 | 3 952 429 |
| Valid | 2 392 541 | 2 346 438 |
| Filtered | 4 943 936 | 1 605 991 |
| Unique | 1 540 093 | 1 142 700 |
| Unique Valid | 1 072 340 | 898 896 |

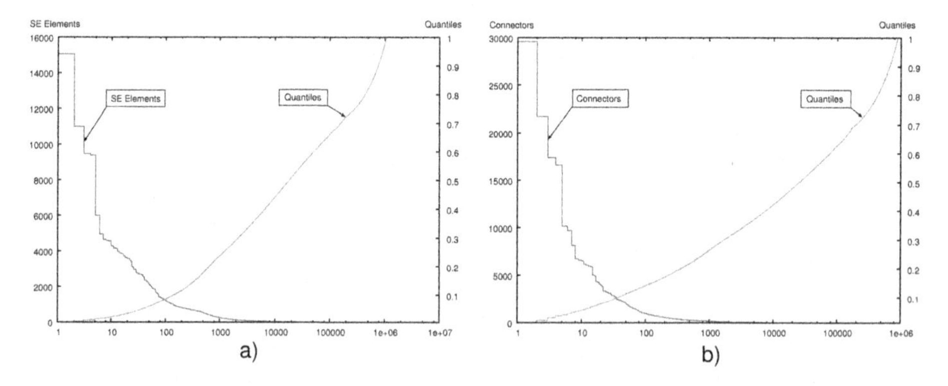


**Fig. 3.** Histograms and quantiles: **a)** SE elements, and **b)** connectors. Right y-axis contains a quantile scale. X-axis is in a logarithmic scale.

There is a noticeable reduction of the SE elements and connectors due to the filtering. The number of SE elements decreased by 56.97% (from 7335577 to 3156310) and connectors by 40.63% (from 3952429 to 2346438). Similarly, reduction is evident in the number of unique SE elements (30.37%: from 1540093 to 1072340) and connectors (21.34%: from 1142700 to 898896).

*Frequent users knew their targets and navigational paths to reach them.* Duration of subsequences in sessions was short - with the peak in the interval of two to five seconds (see histogram in Figure 1-b). During such short period users were able to navigate through four to five pages on average (see Table 3) in order to reach the target. Since there was approximately one second per page transition, there was virtually no time to thoroughly scan the page. Therefore it is reasonable to assume the knowledge workers knew where the next navigational point was located on the given page and proceed directly there. There was little exploratory behavior.

*Session objective was accomplished via few subgoals.* Average session (after filtering) contained three subsequences (see Table 3) where each subsequence can be considered a separate action and/or subgoal. Average knowledge worker spent about 30 seconds to reach the subgoal/resource, and additional 6.5 minutes before taking another action. Considering the number of unique valid subsequences (about 1.6 million) the complete population of users had relatively wide spectrum of browsing patterns. However, the narrow explored intranet space of a single user suggests large diversification.

*Small number of SE elements and connectors was frequently repetitive.* The histogram and quantile charts in Figure 3 depict re-occurrence of SE elements and connectors. Approximately thirty SE elements and twenty connectors were very frequent (refer to left histogram curves of Figure 3). These thirty SE elements (appx. 0.0028% of unique valid SE elements) and twenty connectors (appx. 0.0022% of unique valid connectors) accounted for about 20% of total observations (see right quantile curves of Figure 3).

*Knowledge workers formed frequent elemental and complex browsing patterns.* Strong repetition of the SE elements indicates that knowledge workers often initiated their browsing actions from the same navigation point and targeted the same resource. This underlines the elemental pattern formation. Relatively small number of elemental browsing patterns was frequently repeated. Re-occurrence of connectors suggests that after completing a browsing sub-task, by reaching the desired target, they proceeded to the frequent starting point of following sub-task(s). Frequently repeating elemental patterns interlinked with frequent transitions to other elemental sub-task highlights formation of more complex browsing patterns. Although the number of highly repetitive SE elements and connectors was small, knowledge workers exposed a spectrum of behavioral diversity in the elemental as well as more complex behavioral patterns.

Formation of behavioral browsing patterns positively correlates with the short peak average duration of subsequences (3 seconds). Knowledge workers with formed browsing patterns exhibited relatively fast page transitions. They also displayed shorter delays between subsequences.

Management and support of knowledge workers intranet activities can be provided at various levels. *Browsing Management:* assistance tools that optimize knowledge worker browsing behavior and access in alignment with the organizational goals and structure can be implemented. *Attention Management:* adequate information selection and filtering that is brought to relevant knowledge worker attention in a timely and appropriate manner can be performed. *Personalization Management:* personalized support for frequent routine tasks according to the organizational policies can be provided, so the knowledge workers can concentrate on creative tasks that have the potential to generate new value.

## 7 Long Tails of Knowledge Worker Browsing Behavior

The term *long tail* colloquially refers to a feature of statistical distributions where the *head* contains a small number of high frequency elements that gradually progresses to the *long tail* of low frequency elements (Figure 4). The mass of a long tail can substantially outweigh the mass of a head.

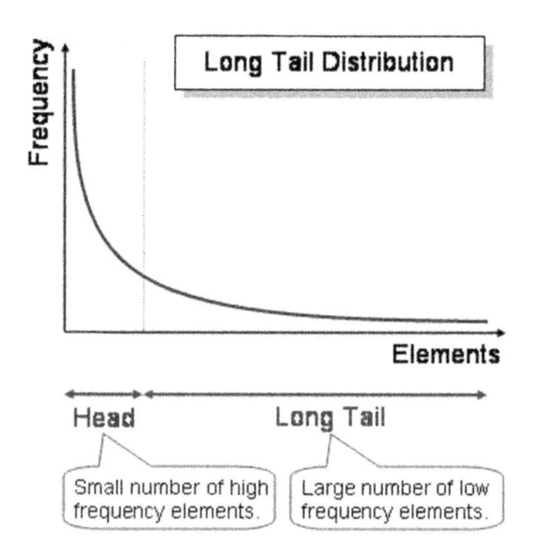


**Fig. 4.** Depiction of a long tail distribution

Long tails are the *modus operandi* in modern business models. Particularly internet related operations ranging from advertising through marketing to sales all attempt to utilize the power of long tails. To explain the principle, consider for example a conventional bookstore v.s. an internet based bookstore operation. A conventional bookstore, due to its storage limitations, targets the high demand items. That is, it operates in the head range of the market offering a relatively small volume of items in a high demand. An internet based bookstore, on the other hand, is not limited by physically storing the items (it keeps only electronic records and mediates transactions between buyer and seller). Thus it can expand its offering to a large volume of items in a low demand, and operate in the long tail range. The overall sales volume and revenue accumulated from a wide range of low demand items can substantially exceed sales volume and revenues generated by a narrow range of high demand items.

The former analysis indicates that the long tail characteristics are evident in knowledge worker browsing behavior. All histograms of starters, attractors, and singletons show long tails. The elemental behavioral abstractions, that is the SE elements, and their connectors, throughout which users form more complex behavioral patterns, equally display long tails. Furthermore, even the complete sessions have this attribute. (Note that the histogram charts have x-axis in a logarithmic scale. It allows us to observe the details of heads of distributions.)

*If the long tails are the common denominator of a human browsing behavior in electronic spaces, what is the underlying functional law that accurately captures it?* Conventionally, the heavy tails in human dynamics are modeled by Pareto distribution [17] (in economics Pareto distribution is also used to model a wealth allocation among individuals). However, the results of our analysis suggest better and more accurate novel distribution.

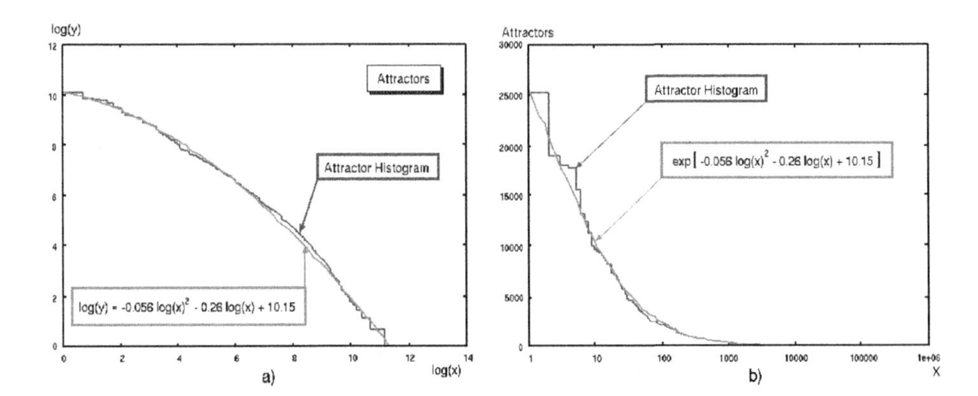


**Fig. 5.** Long tail analysis in attractor histogram: a) log-log plot, b) normal plot with x-axis in a logarithmic scale. Log-log plot clearly shows inverted quadratic characteristics. The distribution is well approximated by the LPE p.d.f. function $f(x) = \exp\left[-0.056\log(x)^2 - 0.26\log(x) + 10.15\right]$.

The novel distribution that efficiently captures the long tail features of a human browsing behavior in WEB environments is derived from the analysis of log-log plots. Figure 5-a shows a log-log plot of attractor histogram. It is evident that the curve has a quadratic shape. Plots of the other histograms have the same quadratic appearance. Nonlinearity is the reason why Pareto distribution (and other well known long tail distributions) is unsuitable since it only captures a linear dependency. Models employing the conventional distributions may display systematic deviations.

Expressing the quadratic characteristics of a log-log plot in an analytic form leads to the formula:

$$\log(y) = \sum_{i=0}^{2} \theta_i \log(x)^i .$$

Eliminating the logarithm on the left-hand-side of the equation, and presenting the generalized polynomial form results in the following expression:

$$f(x;\theta) = \exp\left[\sum_{i=0}^{n} \theta_i \log(x)^i\right] . \tag{2}$$

Naturally, even more generalized form can be obtained by not limiting $i$ to the non-negative integers, but considering it to be a real, $i \in R$.

The derived log-polynomial-exponential (LPE) function (2) appropriately represents the observed long tail dynamics of user browsing behavior. Although the general n-th order polynomial can be considered, the second order form was sufficient for modeling our observations (see Figure 5-b). When using the second order polynomial form, the common concave shape depicted in Figure 5-a suggests that the quadratic term will always be negative, $\theta_2 \in R^-$, and the offset at the origin always positive, $\theta_0 \in R^+$. One can also notice that LPE p.d.f. (2) is base independent. The estimation of parameters $\theta$ can be done by applying various statistical inference techniques.

Long tail characteristics of a human behavior in electronic spaces present new challenges and opportunities for the next generation business information systems. Particularly active areas of inquiry are linked to personalization, collaborative engineering and filtering, and recommender systems. The common aim of the approaches is to categorize or cluster a user population to groups with respect to similarities (e.g. buying habits, searched resources, interests, etc.) and infer/take business actions targeted for the specific user or customer groups. The approaches employ clustering and/or classification algorithms that are based on segmentation with respect to the high frequency components of heads. However, if the mass of a long tail outweighs the mass of a head, the effective user coverage by the features extractable from a head may be substantially smaller than the user coverage by a spectrum of features extractable from a long tail. Effective behaviorally centered user clustering and classification algorithms should exploit the power of long tails in combination with the resources acquired from heads.

## 8 Conclusions and Future Work

A novel formal approach to analysis of human browsing behavior in electronic spaces has been introduced. Extraction of navigational primitives, behavioral abstractions and their connecting elements, and analysis of knowledge worker navigational activities on corporate intranet were performed. Exploratory analysis revealed several important behavioral aspects. Knowledge workers had focused interests and explored diminutive range of intranet resources. They had well defined targets and knew how to achieve them. The browsing objectives were accomplished via few subgoals. In using the intranet platform knowledge workers formed elemental and complex browsing patterns that were repeatedly applied. General browsing strategy of knowledge workers was remembering the starting point and recalling the navigational path to the target.

Significant long tail characteristics have been exposed in all analyzed aspects of knowledge worker browsing behavior. A novel distribution that accurately models it has been derived. The long tail attributes of human browsing behavior have important implications in personalization trends of business information systems, and call for novel approaches.

Future work targets exploration of novel and efficient approaches to behavioral clustering in connection with the content based analysis. Results should be utilized for design and development of the next generation personalization systems and tools.

## Acknowledgment

The authors would like to thank Tsukuba Advanced Computing Center (TACC) for providing raw WEB log data.


## References


1. B. Schlender. Peter Drucker sets us straight. *Fortune*, (December 29, 2003), http://www.fortune.com.
2. T.H. Davenport. *Thinking for a Living - How to Get Better Performance and Results from Knowledge Workers.* Harvard Business School Press, Boston, 2005.
3. A.-L. Barabasi. The origin of bursts and heavy tails in human dynamics. *Nature*, 435:207–211, 2005.
4. Y-H. Park and P.S. Fader. Modeling browsing behavior at multiple websites. *Marketing Science*, 23:280–303, 2004.
5. P. Géczy, S. Akaho, N. Izumi, and K. Hasida. Navigation space formalism and exploration of knowledge worker behavior. In G. Kotsis, D. Taniar, E. Pardede, and I.K. Ibrahim, Eds., *Information Integration and Web-based Applications and Services*, pp. 163–172, OCG, Vienna, 2006.
6. W.W. Moe. Buying, searching, or browsing: Differentiating between online shoppers using in-store navigational clickstream. *Journal of Consumer Psychology*, 13:29–39, 2003.
7. R. Benbunan-Fich. Using protocol analysis to evaluate the usability of a commercial web site. *Information and Management*, 39:151–163, 2001.
8. K.L. Norman and E. Panizzi. Levels of automation and user participation in usability testing. *Interacting with Computers*, 18:246–264, 2006.
9. R.E. Bucklin and C. Sismeiro. A model of web site browsing behavior estimated on clickstream data. *Journal of Marketing Research*, 40:249–267, 2003.
10. M.V. Thakor, W. Borsuk, and M. Kalamas. Hotlists and web browsing behavior–an empirical investigation. *Journal of Business Research*, 57:776–786, 2004.
11. M. Deshpande and G. Karypis. Selective markov models for predicting web page accesses. *ACM Transactions on Internet Technology*, 4:163–184, 2004.
12. H. Wu, M. Gordon, K. DeMaagd, and W. Fan. Mining web navigaitons for intelligence. *Decision Support Systems*, 41:574–591, 2006.
13. I. Zukerman and D.W. Albrecht. Predictive statistical models for user modeling. *User Modeling and User-Adapted Interaction*, 11:5–18, 2001.
14. J. Jozefowska, A. Lawrynowicz, and T. Lukaszewski. Faster frequent pattern mining from the semantic web. *Intelligent Information Processing and Web Mining, Advances in Soft Computing*, pp. 121–130, 2006.
15. P. Géczy, S. Akaho, N. Izumi, and K. Hasida. Extraction and analysis of knowledge worker activities on intranet. In U. Reimer and D. Karagiannis, Eds., *Practical Aspects of Knowledge Management*, pp. 73–85, Springer-Verlag, Heidelberg, 2006.
16. L. Catledge and J. Pitkow. Characterizing browsing strategies in the world wide web. *Computer Networks and ISDN Systems*, 27:1065–1073, 1995.
17. A. Vazquez, J.G. Oliveira, Z. Dezso, K.-I. Goh, I. Kondor, and A.-L. Barabasi. Modeling bursts and heavy tails in human dynamics. *Physical Review*, E73:036127(19), 2006.

# A Trustworthy Email System Based on Instant Messaging

Wenmao Zhang[1], Jun Bi[2], Miao Zhang[2], and Zheng Qin[1]

[1] Software School, Tsinghua University, Beijing 100084, China
[2] Network Research Center, Tsinghua University, Beijing 100084, China

**Abstract.** Email service is a killer application in the Internet. Due to the problem of SPAM, building a trustworthy and secure email service is an important research topic. This paper presents a new email system called SureMsg, which is based on Extensible Messaging and Presence Protocol (XMPP). In addition to the security mechanism provided by XMPP, a reputation mechanism is also designed for SureMsg. Therefore, SureMsg enhances the user authentication and is feasible for current Email service transition.

**Keywords:** Email, SPAM, Instant Messaging, XMPP.

## 1 Introduction

Email service is a killer application in the Internet. At present, SMTP based email system encounters many problems, especially SPAM. [1] indicates that about 50%~90% of daily emails are SPAM. Many solutions [2, 3, 4] have been proposed, but they aren't effective in every case and may make the ham emails as spam, which makes serious influence to users. The SMTP-based email service has been widely used for a long time. As a result, it is very difficult to improve or deploy new technologies on such a huge 'legacy system'.

Though prevailing work concentrates on updating existing anti-spam methods, some diverts their focus to new method based on trust and reputation mechanism [5, 6, 7]. With the support of reputation mechanism, we could get a trustworthy email service environment and make anti-spam more effective. But in such a spoofable SMTP environment, reputation mechanism can't work effectively. Even with authentication technologies, the cost of deploying them on such a legacy system are too high to be ignored, which makes it hardly apply to daily use. Thus, making a trustworthy email environment in SMTP network will be very hard.

Building a mail service based on new standard is another solution, and people have made some progress. In this paper, we introduce *SureMsg*: a trustworthy email system based on XMPP protocol and enhanced with trust and reputation mechanism. By leveraging the security mechanisms provided by XMPP and the characteristics of instant messaging (IM) service, a trustworthy email system is established with the help of reputation mechanism.

W. Abramowicz (Ed.): BIS 2007, LNCS 4439, pp. 598–609, 2007.
© Springer-Verlag Berlin Heidelberg 2007

The paper is organized as follows: Section 2 discusses the relevant work on new email system related to XMPP. In Section 3, the system architecture of SureMsg is described. In Section 4, we describe the basic idea on how to build trustworthy and reputation mechanism. In Section 5, we introduce the basic principle on transition from current email system. Section 6 concludes the paper.

## 2 Related Work

XMPP is a protocol for streaming XML elements in order to exchange structured information close to real time between any two network endpoints [8]. It is a XML based protocol and provides fully asynchronous data transferring, which is widely used in IM services. The XMPP has built in security mechanisms in order to prevent spoofing. Since 2000, there are already 10,000 public XMPP servers running around the world, let alone many internal XMPP servers in companies [9]. As it is such a huge trustworthy network since all servers support anti-spoofing technology, the XMPP network is suitable to build a new mail service. Up to now, two new email systems related to XMPP were proposed: Internet Mail 2000 (IM2000) [9] and Instant Mail (IMail) [10].

In IM2000, receiver-pull model is used for mail delivery instead of sender-push one in SMTP. Mail is stored on sender-side and receivers initiate to pull the mail. This kind of mail delivery model makes senders know whether the receiver has taken over the mail, on the other hand, since spammers need to store and manage emails on their own mail servers (only receivers accepted ones will be transferring); it becomes relatively easier to prevent spammers from sending countless spam.

IMail is another gateway-based solution in building new mail service based on XMPP. It makes use of WEBDAV network to store the emails' attachments and decrease the burden of XMPP server.

## 3 Architecture of SureMsg

### 3.1 SureMsg Overview

SureMsg service has such features:

- Providing mail service based on XMPP protocol.
- In order to increase the efficiency of mail transferring, SureMsg uses two main strategies in mail transferring from senders to receivers:
    - Exploiting end-to-end data transfer just like file transferring in IM service, together with server relay method.
    - The message will be segmented when it is transferred through server, and assembled on receiver side.
- Providing an anti-spam method by leveraging contact-list mechanism based on IM service and the reputation network building on XMPP network.

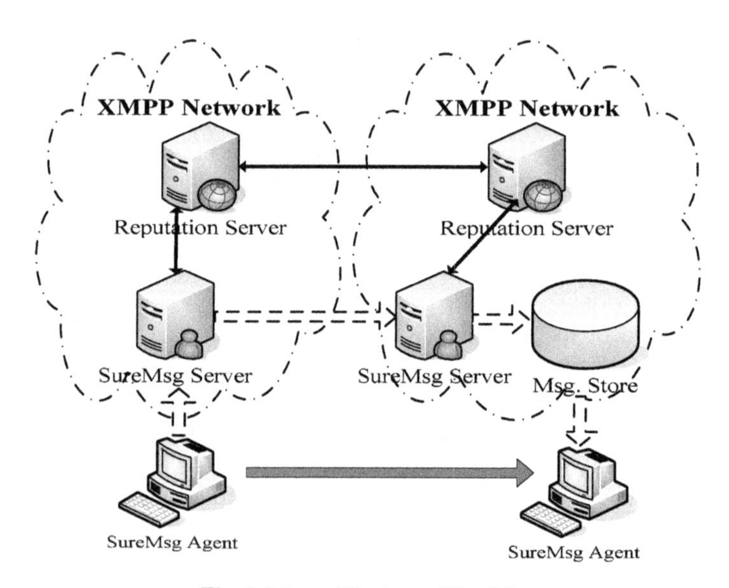


**Fig. 1.** The architecture of SureMsg

Figure 1 shows the architecture of SureMsg. Entities included in SureMsg are:

- **SureMsg Server.** It takes the main responsibility of mail services, and communicates with Reputation Server, providing reputation information and requiring reputation score from Reputation Server.
- **SureMsg Client.** It is used to edit, send and receive emails (or blocks). It also provides interfaces for users to configure blacklist and reputation threshold information which is useful in anti-spamming. Here, blacklisting method is supported by extending the contact-list mechanism provided by IM service.
- **Mail Store.** It stores emails for receivers who are offline and can't get email at that moment.
- **Reputation Server.** It collects reputation information from SureMsg Server and other Reputation Servers in remote domain, computing reputation score for senders in local domain and distributing necessary information to the request.

Receivers configure the blacklist and reputation threshold information through SureMsg Client and such information will be stored on SureMsg Server. Before sending out mail, the sender client will request the server first (using Info/Query mechanism provided by XMPP) to make sure that mail is allowed by receivers. The response to senders depends on the receiver's blacklist and the reputation threshold information that stored on server. If the sender is not in blacklist, or if he is a stranger to receiver but his reputation score getting from local Reputation Server is higher than threshold, server will response to sender that receiver wish to pull the mail. After that, server requests sender to pull the mail, and mail will be transferred either by end-to-end or server-relay mode, depending on whether the receiver is online or the end-to-end transferring is allowed (clients may behind NAT or firewall). When segmenting/assembling the message, the responsibility of maintaining the state of message blocks and splitting/recomposing them will be taken by sender and receiver clients.

### 3.2 SureMsg Server

The server describing here includes SureMsg Server and Reputation Server entities, as shown in Figure 2. The Reputation Server can be implemented as a server-side component plugging into current extended XMPP server. The server's main responsibilities are: (1) authenticating the user to avoid spoofing; (2) relaying negotiation information between clients; (3) recording the action about sender (how many of his mails are rejected by receivers); (4) storing the offline mail (or packets) for receivers; (5) maintaining blacklist and reputation threshold information that configured by receivers; (6) collecting reputation information, computing reputation score and distributing them to the one who required the score.

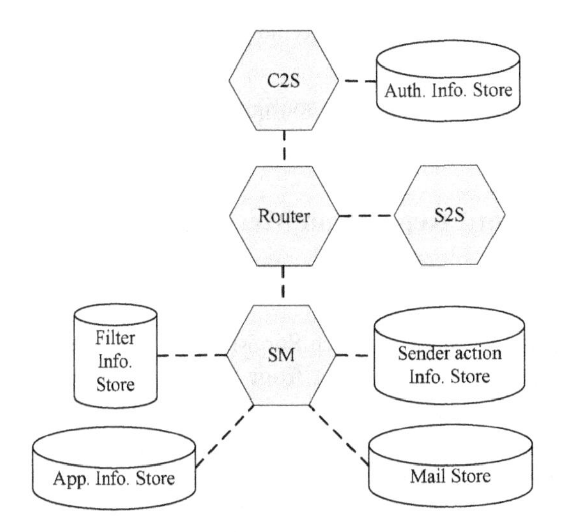


**Fig. 2.** The architecture of SureMsg server

Our server is designed on current XMPP-based IM server. Figure 2 shows the architecture of SureMsg server that extended by normal XMPP IM server: a normal XMPP server has such functional modules: C2S (Client-to-Server) and S2S (Server-to-Server) take the responsibility of communicating and authenticating with clients and remote server respectively. SM (Session-Management) module focuses on main IM services and Router takes charge of inter-module communication.

We extend functions and add new modules to support our SureMsg services: we make use of C2S and C2S to support authenticating users and remote servers, and the SM module are extended to maintain senders' action information and support filtering mail-sending requests depending on blacklist and reputation threshold. Also, SM maintains senders' action information and offline mails (or packets) for receivers, which stored in Sender Action Information Store and Mail Store respectively.

### 3.3 SureMsg Client

Clients take charge of: (1) editing, sending and receiving email (or email packets); (2) segmenting mails or assembling the segements; (3) maintaining state of email packets;

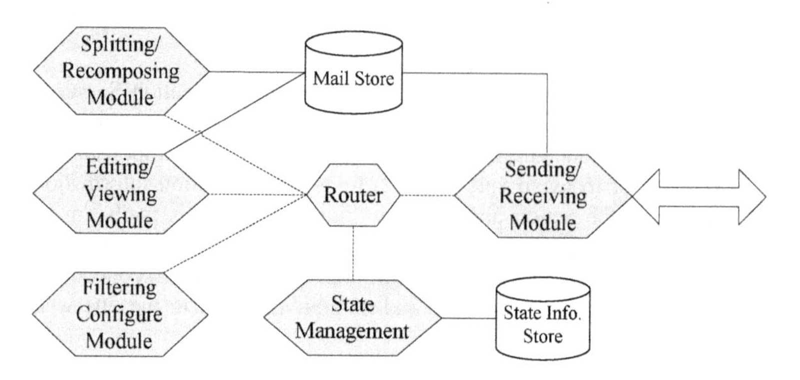


**Fig. 3.** The architecture of SureMsg Client

(4) supporting interface to users for configuring blacklist and reputation threshold information. Figure 3 shows the architecture of client.

# 4 Trustworthy and Reputation Mechanism in SureMsg

## 4.1 Basic Algorithm to Reputation Service

To compute a reputation value for each SureMsg user, we need to create a reputation network. SureMsg firstly collects data from contact-list information and aggregates them into a single whitelist based reputation network. And secondly, SureMsg forms a single blacklist based reputation network in the same way.

We calculate two kinds of values in SureMsg system: local trust value and global reputation value. When stranger *C* want to communicate with A, C's local trust value (if it exists) will be calculated first to *A*. Otherwise, a global one will be calculated.

When calculating the global reputation value to a SureMsg user, we follow such principle: a user who is added by many of the users, whose reputation values are high, will surely be a guy with high reputation value. Such principle is similar with the one that Google PageRank followed. Here, we transit the whitelist network to a 0-1 matrix, where $a_{ij}$=1 means user *j* has added user *i* as his friend. From this matrix, we will use power-iteration algorithm to calculate basic reputation value for every user, just like Google PageRank did.

At the same time, we will get a vote-list from blacklist network. Every entry in this list represents how many votes this user has got from others. The more users vote to him, the higher position he will be in this list.

Based on basic reputation value from whitelist network and vote-list from blacklist network, we can calculate the global reputation value for every user. Commonly, normal user will get high basic reputation value and low position in vote-list. Sometimes, we meet with the attack of malicious collectives, which may disturb the accuracy of the reputation mechanism. We will discuss how to avoid such attack later.

The calculation of the local trust value in SureMsg is based on such principles:

- Trust transitivity [12]. If *A* trusts *B* (*B* exists in *A*'s contact-list) and *B* trusts *C*, then we can confirm that *A* will also trust *C* on the recommendation from *B*.
- Compared with the user who has lower reputation value, Users with higher reputation value will influence more heavily when rating to other user.

Based on such principles, we can represent the algorithms like this:

- Find all achievable paths from *A* to *C* (such as *A->B->C*), named as $S_{path}$.
- For each path in $S_{path}$ , calculating the weighted average $E_w = \sum w_i \times T_i$ , where *i* is the node from *C* to the node next to *A* in this path, and $T_i$ is the global reputation value to node i, and weighted value $w_i = 2^{-(hop(i, C))}$, where *hop(i, C)* is the number of hop from node *i* to *C*.
- For each $E_w$ to path, calculating the weighted average value $LT_{ac} = \sum w'_I \times E_w(i)$, where $w'_I$ is the weighted value for $E_w(i)$, and the higher of $E_w(i)$, the larger $w'_I$ is. Finally, the $LT_{ac}$ is the local trust value to *C* (relative to *A*).

### 4.2 The Architecture to Reputation Service

The Reputation service in SureMsg may distribute in a multi-domain environment. When considering about such scenario, the architecture will be showed in Figure 4: A trustable **Register Server** is used to record the agents that provide reputation service. It also distributes computing task to necessary registered agents to compute reputation value. The agents register themselves to and accept computing task from Register Server. They will communicate with compute reputation value coordinately.

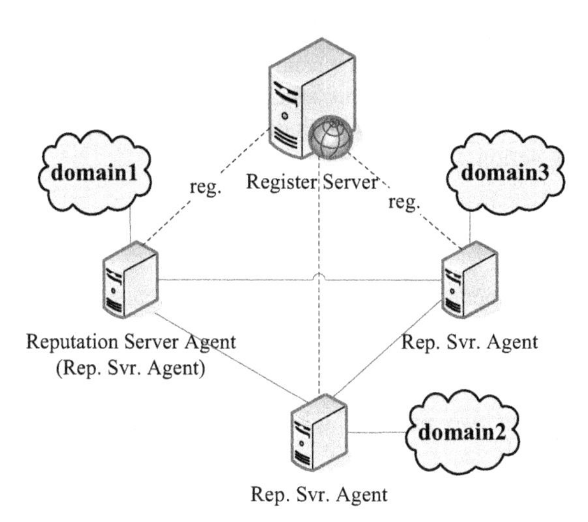


**Fig. 4.** Architecture for reputation system in one-domain environment

### 4.3 Related Problems

In this section, we will discuss some attacks that SureMsg may encounter in reputation service, and show our solutions to these attacks:

- **Malicious collectives attack.** Spammers may unite to form some malicious collectives to boost some users to increase their reputation values or impute some victims to decrease their reputation values.
- **One time used identity attack.** [7] shows that 95% of the spammer addresses were used only once.
- **Zombie users attack.** The users who are infected by virus or worms may send spam (always with virus and worms) to his contact-list users.

As we described in Section 3.2, SureMsg server records the sender's action (how many letters were blocked by receivers, $N_{reject}$). We will use this information to identify whether a user is boosted or imputed. Table 1 shows the method how to counteract malicious collectives attack.

**Table 1.** Method to counteract with the malicious collectives attack

| | $N_{reject} > T$ | $0 \leqslant N_{reject} < T$ |
|---|---|---|
| $S_{white\text{-}hign}$ && $S_{black\text{-}low}$ | Abnormal (Boosted) | Normal (high) |
| $S_{white\text{-}hign}$ && $S_{black\text{-}high}$ | Abnormal (Boosted) | Abnormal (Imputed) |
| $S_{white\text{-}low}$ && $S_{black\text{-}high}$ | Normal (low) | Abnormal (Imputed) |
| $S_{white\text{-}low}$ && $S_{black\text{-}low}$ | Normal (low) | Normal (an initial user) |

For one time used identity attack, we can increase the cost of registering new account, on the other hand, limiting the sending speed will resist to such attack.

Current solutions to resist to zombie attack are not good enough. Commonly, anti-virus applications are always used and challenge-response method is useful to check robot auto-sending action.

# 5 Interoperation and Transition Mechanism

As SMTP network has been existed for more than 20 years, it is impossible for a new email service to replace SMTP service in a short term. As a result, interoperation with current email service is necessary. And what's more, an incremental deployment mechanism to transit the traditional service to our new one is also needed. This section is divided into two parts, which describe the interoperation and transition mechanism respectively.

### 5.1 Interoperation Mechanism

#### 5.1.1 Solution

There are two main solutions for heterogeneous services to interoperate with each other: multi-protocols client and server-side gateway. By using multi-protocols client, end users are convenient to communicate with others in different service domains. But it increases the burden of client maintenance and doesn't have enough stimulation to service provider to take further transition. Compared to the multi-protocols client, the latter solution server-side gateway achieves the server-to-server co-operation and made for further service transition. But some security issues should be considered in such solution.

Our solution to achieve interoperating between SureMsg and current email service is based on server-side gateway. Figure 5 shows the basic idea of interoperation between SureMsg and Email service.

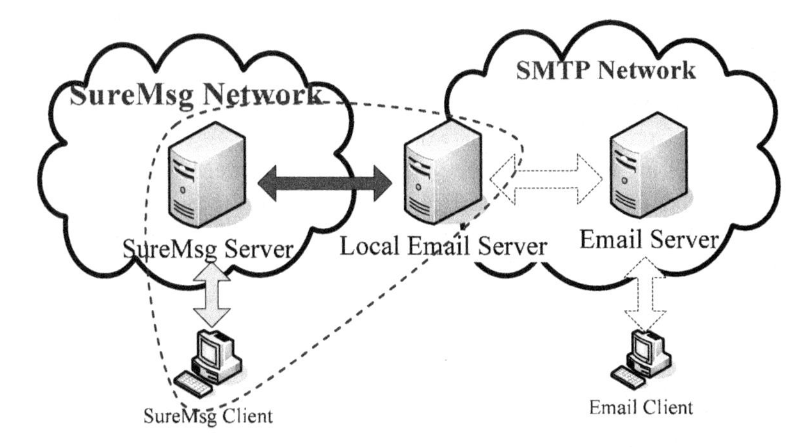


**Fig. 5.** Solution to interoperation between SureMsg and email service

An email server (called **Local Email Server, LES**) is built for SureMsg users to communicate with SMTP-based email network. It is controlled by SureMsg server and exchange information with SureMsg server and other email servers, just like a gateway between two service networks. With the help of LES server, SureMsg users wouldn't be aware of what kind of user they are communicating with, because the interoperation is transparent to them.

### 5.1.2 Related Problems

As concerned above, security issues should be considered in server-side gateway solution. In our interoperation mechanism, the deficiencies in SMTP network may be injected into our service. The main security problem is that spammers may forge legal SureMsg address or email address those controlled by LES service domain, and send sender-spoofed SPAM to SureMsg users. Such problem is serious since the receiver may realize legal senders as spammer and put them into blacklist.

How to filter such sender-spoofed emails in our LES server is the key issue. The first step is, the emails received by LES server, should be divided into two catalogs: those sent by legal users in LES service domain, and the spoofed ones. Our solution is to keep the SureMsg users from emailing with each other through their own LES service domain. The only way to communicate with each other is the SureMsg service. After such disposal, when LES server receives emails whose sender addresses contain LES server domain name, they will be flagged as SPAM by LES server.

Following the forbidden disposal, there are two 'how to' issues should be addressed: (1) how to recognize sender-spoofed emails in LES server and, (2) how to keep SureMsg users from emailing with each other by using LES addresses. Since SureMsg service is pull-model based, sender should send query first before sending mails. When server receives such query, they could retrieve the receiver address to check if it is an LES address. Similarly, LES servers should retrieve sender address to check whether it is a LES address in order to distinguish sender-spoofed email.

Our solution to these 'how to' issues is, every SureMsg server maintain an information list (called SureMsgInfoList) that contains our trusted servers who support SureMsg service. Every entry in SureMsgInfoList is a two-parameter tuple: the domain name of SureMsg server (sureMsgDomainName) and the domain name of LES server (emailDomainName) controlled by SureMsg server, whose domain name is sureMsgDomainName. We compare the sender (receiver) address with every tuple in SuerMsgInfoList to see whether they are matching.

When considering how to collect information in SureMsgInfoList, manual or automatic methods can both be used. We can leverage the Service Discovery [11] method supported by XMPP protocol to let server automatically discover whether other servers support SureMsg service.

## 5.2 Transition Mechanism

### 5.2.1 Design Considerations

Most of the current solutions to transition from email system are based on extending SMTP protocol. IM2000, together with Imail, proposes to extend email header and add new functions to SMTP entities. Such solution may increase the transition complexity. But traditional email service providers are very conservative and cautious, they won't agree to change existing protocol unless they find the advantage of new service and get benefits from transition.

Our solution to transition is based on the following principles:

- Bootstrapping the transition. We should divide the process of transition into several phases. Different goals should be accomplished according to the situation of the time in different phases.
- Harmonizing various profit factors. As far as transition is concerned, end users are not the only focus. The profit of those service providers (SP) who provide traditional email service shouldn't be ignored. We should provide an incentive mechanism in order to encourage them to transit from current systems

### 5.2.2 Solution

We divide transition into three phases: initial-phase, metaphase and anaphase:

- Initial-phase

In initial phase, SureMsg network is still small, and then users are mainly using email service. Email SPs won't provide interoperation with our service proactively. In this phase, we should focus on our own service and the end users.

Our aim in this phase is to make more users shifted into SureMsg service from traditional email service, so that a much larger SureMsg network could be formed. At the beginning of this phase, we will adopt 'multi-protocols client' solution to current XMPP based IM users to extend our service, and then, we will build our gateway-based interoperation solution to communicate proactively with email network so that our service could be accepted by email users.

- Metaphase

In this phase, SureMsg network grows larger; email users are shifting into our service. To avoid the loss of end users, email SPs are considering to interoperate with SureMsg service. Email SPs are the most important object we should focus on.

Our aim is to encourage email SPs to deploy SureMsg service. Three advantages will be gained after deploying our services: (1) Avoid the loss of current end users (2) Avoid the threat of SPAM (3) Attract more users into his service domain. Figure 6 shows the solution to deployment of SureMsg service in email network.

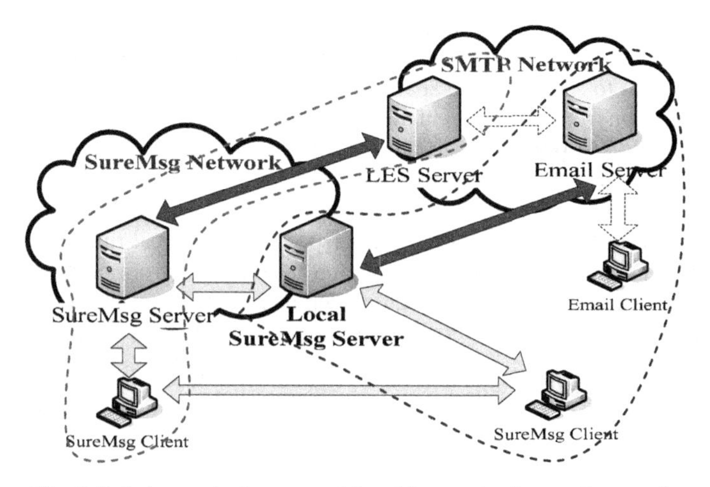


**Fig. 6.** Solution to deployment of SureMsg service in email network

As Figure 6 described, we add a local SureMsg server (LSS) controlled by email server to communicate with other SureMsg servers. Such solution is similar to our interoperation mechanism described above, so our solution is much more feasible than extending SMTP protocol. And our solution is incremental deployment in that traditional email service, and could be used normally with no affection.

- Anaphase

In this phase, SureMsg service has formed a large-scale, allied network. In traditional email network, many email SPs have deployed our solution to support interoperation with SureMsg service. They will follow the tendency to transit to pure SureMsg service. We hope more email SPs could transit into pure SureMsg network. The solution to achieve such aim not only involves the technical aspect, but also the business and policy aspects, which are beyond the scope of this paper.

## 6 Conclusion

In this paper, we introduce SureMsg, a trustworthy mail service based on XMPP and support reputation-based anti-spam solution. Compared with traditional SMTP based email service, SureMsg has following advantages:

- Guarantees the authenticity of sender's identity.
- The reputation network we built will be more reliable and effective.

Comparing with other new mail systems (IM2000 and IMail), our solution have below advantages:

- **On design.** SureMsg using both end-to-end transferring and segment/assemble mechanism, which increases the efficiency of mail delivery and decreases the burden on server.
- **On anti-spam.** IMail solution has little discussion on this aspect, and IM2000 uses receiver-pull model to limit the spammer of sending spam, but the solution may be complicated when dealing with users in gray-list (neither in white nor black-list). As a result, receivers may deal with it manually. In SureMsg, we first use black-list to filter blocked spammer and if disabled we leverage the reputation-based solution to filter low reputation score strangers.
- **On transition support.** Firstly, SureMsg service has huge user base, since it is built on current XMPP based IM networks, those who use IM will be our basic users. Secondly, the mail delivery strategy will reduce the burden on server and won't undermine much on IM original performance (real-time, etc). These two aspects strengthen transition. Our solution to transition from current email system avoids extending the SMTP protocol which increases the feasibility of deployment. And it supports incremental deployment as described in above section.

The prototype of SureMsg is implemented and being tested in China Education and Research Network (CERNET). The incentive mechanism for deployment from current email service to SureMsg service will be further explored.

## References

1. Claburn, T.: Big guns aim at spam. Information Week. (2004)
2. Carrreras, X., M´arquez, L.: Boosting trees for anti-spam email filtering. In Proceedings of RANLP-01, 4th International Conference on Recent Advances in Natural Language Processing, Tzigov Chark, BG (2001)
3. Cranor, L., Lamacchia, B.: Spam!. In Communications of the ACM, (1998) 41:74–83
4. Graham, P.: Better Bayesian filtering. http://www.paulgraham.com/better.html. (2003)
5. Golbeck, J., Hendler, J.: Reputation Network Analysis for Email Filtering. In Proc. of the Conference on Email and Anti-Spam (CEAS), Mountain View, CA, USA (2004)
6. Kamvar, S., Schlosser, M., Garcia-Molina, H.: The EigenTrust Algorithm for Reputation Management in P2P Networks. In Proc. of the 12th Intl. WWW Conference (2003)
7. Chirita, P.A., Diederich, J., Nejdl, W.: MailRank: Using Ranking for Spam Detection. In CIKM'05. Bremen Germany (2005)
8. Saint-Andre, P.: Extensible Messaging and Presence Protocol (XMPP): Core, RFC 3920 (2004)

9. Devrieze, S.: http://users.telenet.be/s.devrieze/imail/imail.pdf
10. Pollard, J.B.: http://homepages.tesco.net/J.deBoynePollard/Proposals/IM2000/
11. Saint-Andre, P.: http://www.jabber.org/jeps/jep-0030.html
12. Jøsang, A., Ismail, R., Boyd, C.: A Survey of Trust and Reputation Systems for Online Service Provision. Decision Support Systems (2005). http://dx.doi.org/10.1016/j.dss.2005.05.019

# Towards Operationalizing Strategic Alignment of IT by Usage of Software Engineering Methods

## An Enterprise–Modelling Oriented Approach to IT Governance

Bernd Tilg, Joanna Chimiak-Opoka, Chris Lenz, and Ruth Breu

Quality Engineering Research Group
Institute of Computer Science, University of Innsbruck
Technikerstrasse 21a, A–6020 Innsbruck
csaa086@uibk.ac.at

**Abstract.** This paper presents an enterprise–modelling oriented approach to IT Governance. Main requirements for a IT Governance Framework are presented and applicability is shown by means of operationalizing Strategic Alignment — a main focus area of IT Governance. Two frameworks are briefly described, namely PRO$^2$SA (domain-specific for Strategic Business Alignment) and the general–purpose framework SQUAM for metamodelling, enterprise modelling and model-analysis. Key aspects of this approach are the information enrichment of enterprise models (mainly with key figures) and the integration into a Business Intelligence Suite for comprehensive business analysis.

**Keywords:** IT Governance, Strategic Business Alignment, IT Management Framework, Business Intelligence, Model Driven Software Development.

## 1 Background and Motivation

This prefacing chapter motivates a new approach of a model– and business–driven analysis framework to integrate strategic topics of IT Management into corporate management based on software engineering methods. Our primary business goals focus on communication improvements between IT's and corporate's management and processing fundamental strategical and tactical questions of IT Governance[1] for decision–making in top management. Technical objectives focus on the appliance of existing information systems methods and –techniques (especially software engineering) for corporate management issues as well as the implementation of a domain–specific Analysis Framework PRO$^2$SA (***Pro****cess– and* ***Pro****ject–oriented* ***S****trategic* ***A****lignment*, chapter 2) and a general purpose Prototype SQUAM (***S****ystem for* ***Qu****ality* ***A****ssessment of* ***M****odels*, chapter 3).

Foremost we discuss recent requirements and the changing world of IT–Management. With actual shortfalls and problems in mind we identify essential requirements for this intended integration and respectively the operationalizing of Strategic Business Alignment as a main part of IT Governance. In the following chapters we summarize the frameworks PRO$^2$SA and SQUAM.

[1] At the moment we concentrate on topics of Strategic Business Alignment.

W. Abramowicz (Ed.): BIS 2007, LNCS 4439, pp. 610–625, 2007.
© Springer-Verlag Berlin Heidelberg 2007

### 1.1 New Challenges of IT–Management

In the last years requirements for enterprise–wide strategic information technology have changed massively — IT is facing its greatest challenges of the last decades. The increasing interconnectedness of business and information systems poses new ways to master the strategically use of IT in corporations. In the late Nineties — caused by the booming Internet — investments in IT seldom were questioned. However, in the last years management's perception of information technology has changed. Not only proposed investments, but also existing IT–oriented solutions are put into question. IT–managers have to legitimate their investments and argue IT's contribution to corporation's business success.

IT's strategical role in companies is actually discussed very controversial — the overall significance of IT in companies is critically questioned. On the one hand IT is seen as a separate strategical dimension of business which must be adequate managed (e.g. [1,2,3]). On the other hand N.G. Carr provoked a still ongoing discussion by his statement "*IT doesn't matter*" [4]. Because of standardization and overall availability IT's strategical position becomes less important, so that information technology is getting more and more a commodity factor in business rather than a potential for strategic differentiation. Not astonishing that this statement triggered a intensive discussion on this topic (e.g. [5]) . For many years the topic *productivity paradoxon* has been widely discussed in business and science communities — which impact has IT on business productivity and how can it be measured. Some empirical studies have been made, but outcomes and corresponding conclusions have been very controversial and so the discussion remains open until now (e.g. [6,7,8]). With the issues *value of IT* or *IT value delivery* nowadays a similar topic is discussed in the context of IT Governance ([9,10,11]). In the last years some new laws and directives had a great impact on IT requirements in companies. To establish compliance (Sarbanes–Oxley in the United States or Basel II in Europe) enormous efforts have to be done [12].

Gardner [13] states that "*for 30 years IT changed the world by creating new industries, altering business and dramatically improving performance. In 2006, CIO attention will grow IT´s contribution in new ways by applying IT to meet the challenge of external forces. Until now, IT has been changing the world. Now CIOs face a future where the world is changing IT*" and the IT–Governance Global Status Report [14] constitutes as the #1 key finding that "*IT is more critical to business than ever: For 87% of the participants, IT is quite to very important to the delivery of the corporate strategy and vision. For 63% of the respondents, IT is regularly or always on the board's agenda*".

### 1.2 Strategic Alignment in the Context of IT Governance

In the last years the term *IT Governance* has been internationally established for the strategical embedding and the organizational orientation of information technology in companies. This topic is broadly discussed in research institutes and in business context. However, a exact definition of IT Governance is still missing — although meaning and importance for companies is beyond question. In accordance to [15] the concept of IT Governance has evolved from *Corporate Governance* and *Strategic Information Systems Planning* and that different researchers focus on different aspects of the concept and may fail to encapsulate the true nature of IT governance. Twelve definitions of IT Governance have been analysed and a new *definitive* definition extracted: "*IT–Governance is the strategic alignment of IT with the business such that maximum business value is achieved through the development and maintenance of effective IT control and accountability, performance management and risk management.*" According to [16] one of the primary aspects of IT Governance is about decision–making in Top Management: "*IT governance is about IT decision–making: The preparation for, making of and implementation of decisions regarding goals, processes, people and technology on a tactical and strategic level*".

According to these definitions of IT Governance the next question is: What is the scope of duties? The main issues of IT Governance are defined in [17] and [15]:

**Strategic alignment** focuses on ensuring the linkage of business and IT plans; on defining, maintaining and validating the IT value proposition; and on aligning IT operations with enterprise operations.

**Value delivery** is about executing the value proposition throughout the delivery cycle, ensuring that IT delivers the promised benefits against the strategy, concentrating on optimising costs and proving the intrinsic value of IT.

**Resource management** is about the optimal investment in, and the proper management of, critical IT resources: applications, information, infrastructure and people. Key issues relate to the optimisation of knowledge and infrastructure.

**Risk management** requires risk awareness by senior corporate officers, a clear understanding of the enterprise's appetite for risk, understanding of compliance requirements, transparency about the significant risks to the enterprise, and embedding of risk management responsibilities into the organisation.

**Performance measurement** tracks and monitors strategy implementation, project completion, resource usage, process performance and service delivery, using, for example, balanced scorecards that translate strategy into action to achieve goals measurable beyond conventional accounting.

**Control and Accountability** is about the existence and practical application of a proper control and accountability system.

A major aspect in the discussion of IT Governance is the responsibility of the Management Board, which is stressed in appropriate definitions: "*IT governance is the organizational capacity exercised by the board, executive management and IT management to control the formulation and implementation of IT strategy and in this way ensure the fusion of business and IT*" [18]. Similar definitions can be found in [19,3,20].

Summarized, these mentioned topics have in common, that a efficiently communication between IT Management (CIO) and board should be aimed. Due to the fact that these two management domains have totally different knowledge bases, terminologies and methods, it doesn´t astonish that a efficiently and accurate communication seems complicated. It remains not only the question, how this can be accomplished in a efficient and economic way, but also which technologies and methods are adequate to support IT- and corporate management in decision–making. Until now there is no continuous approach established, with which it is possible to support IT–related management decisions by the usage of established software engineering methods and techniques.

## 2 The PRO$^2$SA Framework

The discussion in chapter 1 depicts the necessity of an integrated and flexible IT Governance Framework. The overall objective is the implementation of a model– and business–driven analysis environment based on established software engineering methods to achieve the desired results — a software–engineering–oriented way to IT Governance. On the basis of the prefacing discussion some essential business–oriented requirements for operationalizing Strategic Alignment in an enterprise–wide context (see also [21]) can be defined:

**R1:** Strategic Alignment and Value Delivery are major focus areas[2] of IT Governance and are massively affected by each other.
$\Rightarrow$ Strategic Alignment has to be examined in the broader context of IT Governance to integrate it into corporate governance.

**R2:** IT Governance is in the responsibility of the Board of Management [17] and a great part of it is about decision making [22].
$\Rightarrow$ To *narrow the gap* between IT– and Enterprise–Management not only communication has to be improved but also analysis and reports have to speak the *language of management* much more than technical issues.

**R3:** IT delivers a indirect value proposition to business success. IT's main part is to enable, optimize or innovate core business processes of a company [20]. In the majority of cases corporation's considerations effect business processes.
$\Rightarrow$ To provide an added value, Strategic Alignment has to consider alongside to overall business strategy also current and future interaction of IT on business processes.

**R4:** Company specific strategy– and decision–making–processes play a prominent role to integrate IT Governance and Corporate Governance.
$\Rightarrow$ Strategic Alignment should take specific company requirements into account and therefore exists a great need for flexibility and adaptability to company–specific standards. Requirements and business environments will steadily and rapidly change.

### 2.1 Fundamental Concepts

It appears to be not enough to prepare management reports and recommendations on Strategic Business Alignment issues sporadically at defined milestones. Rather, continuous analysis and steadily support for decision–making is needed in communication between CIO and Enterprise Management. In short, a consistent and adequate information basis, a nearly seamless methodical integration and a analysis– and reporting–system at a highly aggregated, compact and concise form is necessary. A model based framework for Strategic Business Alignment that aims to bridge the business and technical perspectives on this topic and facilitates the described requirements seems appropriate. Stated below are the main concepts and assumptions of PRO$^2$SA:

[2] See mapping of IT processes to IT Governance focus areas [17, Appendix II].

**Software Engineering:** In the last years intensive standardization in software engineering took place. OMG's[3] efforts in enhancing UML (Unified Modeling Language) involved intensive interest in industrial software development. It is conceivable that in near future UML is not only used for object–oriented software development, but also used for modelling of business processes, IT landscapes and further more business–oriented domains. Methodically we use UML and extension–mechanisms (UML–Profiles) for modelling aspects and technically we focus on the Eclipse Software–Development–Environment with its enormous set of plug–ins (chapter 3).

**Enterprise Modelling:** One main idea is a formal approach to an extensible IT Governance Framework by usage of enterprise models as a fundamental information base and the provision of a model repository. One advantage of this approach is, that in companies already existing model information can be reused. It's not seldom that in different departments of a company with varying tool–support and varying abstraction valuable model–oriented information is on hand, e.g. business process models, IT architecture models, application portfolio. The usage of these available *information pool* as a fundamental base for model driven analysis is of great advantage and allows a economical proceeding.

**Metamodel:** This approach is in some aspects similar to *conventional* Business Engineering and Enterprise Modelling methods. Needed flexibility, adaptability and extensibility is established by the underlying metamodels. The general purpose framework SQUAM is suitable for different business domains (e.g. Strategic Business Alignment in framework PRO$^2$SA) and for each domain a separate metamodel can be used.

**Modelling and Information Enrichment:** Based on domain–specific metamodels available models can be reused, new models can be created and the information is stored in a *Model Repository*. An essential issue is the enrichment of the stored model information, especially with key figures. These *information enriched* models are the fundamental base for reporting and analysis in aggregated and compact form.

**Strategic Alignment Model:** The Strategic Alignment Model (SAM), developed by Henderson and Venkatraman [23] is a widely discussed and accepted methodical framework. On the other hand a major difficulty of SAM is, that it is abstract and mostly implemented at an informal level [24,25,26]. The framework PRO$^2$SA provides different Alignment Perspectives (strategy execution, technology transformation, competitive potential, service level) according to SAM.

**Integration in Business Intelligence:** Integration with other Management Information Systems of the corporation is established through an integration in a company's Business Intelligence Suite and respectively by usage of a Balanced Scorecard oriented controlling tool [27,28].

**Generation of Management–Charts:** Management seldom speaks in technical or in IT–related terms. In fact, the *Language of Management* often refers to strategies, business goals and controlling-related information (e.g. key figures, balanced scorecard, ...). Nevertheless, to provide Management with list–oriented reports or an

[3] Object Management Group, http://www.omg.org

incoherent accumulation of key figures isn't promising because of complex and comprehensive information context. To achieve a better and precise understanding it seems necessary to provide Management–Charts and –Reports on a highly aggregated, compact but concise form, which include strategy– and business-related information with key figures.

Summarized in a nutshell, we are using software engineering techniques in development and metamodelling, *information–enriched* enterprise models as a basis of enterprise–wide information and last but not least the integration into Business Intelligence for adequate reporting and analysis in the form of generated Management–Charts and –Reports as foundation for management planning and decision–making. Straight forward to this key concepts and methodical environment the appliance of PRO$^2$SA consists of the main steps illustrated in Fig. 1 — from the development of a metamodel till decision–making and feedback.

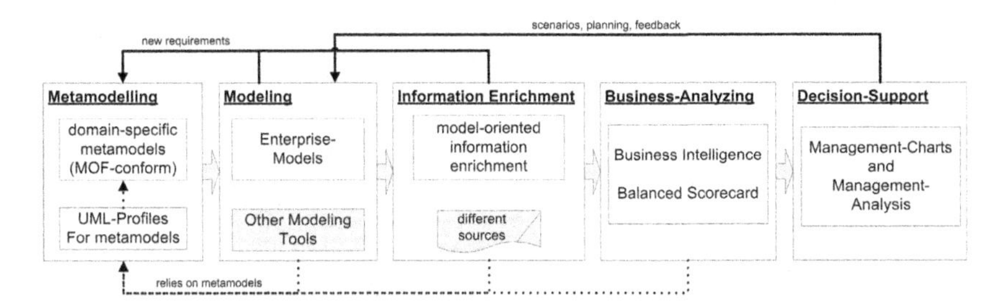


**Fig. 1.** Main steps of PRO$^2$SA

## 2.2 PRO$^2$SA Metamodel

The first task in integrating these two worlds of methodologies is to identify the main domain areas of Enterprise Management. At a first glance at applied and discussed state–of–the–art methods in management this topic seems confusing and complex, too many approaches are discussed. However, on a closer examination we identified three core management domains: Strategy–Management and Corporate Culture (S&C), Organizational Development (OD) and Organizational Transformation (OT). The first domain includes typical topics of overall goal setting, long–term planning and strategic controlling. The second domain includes the established approaches of business process–management, –optimization and organizational structures. The latter domain organizes the ongoing business transformation, the realization of business strategies — mainly by accomplishing projects with great focus on change management.

These three primary management domains are supplemented by a information technology[4] (IT) domain, so that we organize this framework on these fundamental four core domains of enterprise management. A additional metamodel is defined to integrate the prior four domains. Having defined these domains we have to identify and focus on

[4] Which itself consists of a couple of subdomains: e.g. strategic information management (SIM), software development (SD), requirements engineering (RE), Infrastructure (ISt), Enterprise Architecture Integration (EAI), . . .

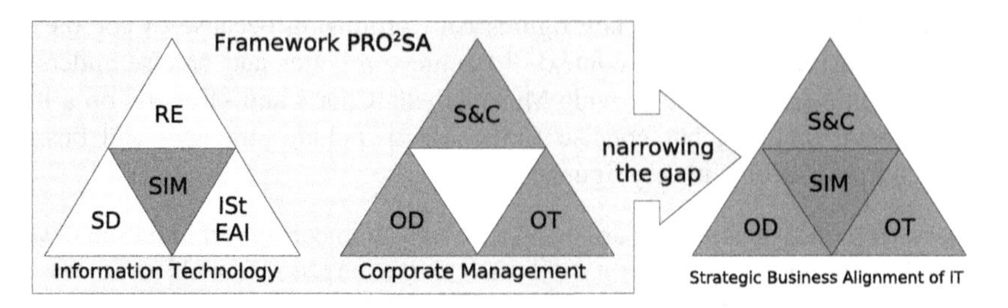


**Fig. 2.** 'narrowing the gap' between IT and Management

appropriate applied methods in these fields. In Table 1 the corresponding methods for these domains are briefly summarized.

**Table 1.** PRO$^2$SAmanagement domains and corresponding methods

| Domains | Applied Methods |
|---|---|
| Strategy Management | goal setting, vision– and mission–statements, Balanced Scorecard, critical success factors, ... |
| Organizational Development | Business Process Management, Process Re–Engineering and Optimization, Organizational Structures ... |
| Organizational Transformation | Project–Management, Change–Management, ... |
| Information Technology | Enterprise Architecture Management, IT–Architecture in the large, Information System Landscapes, ... |
| Integration of Domains | Business Intelligence, key figures, critical success factors, modeling inter–domain relationships |

With each of these metamodels and corresponding UML–Profiles a domain–specific (modelling) language (DSL) is defined [29]. On the one hand enterprise models can be created with UML–compliant tools and on the other hand existing models can be imported by specific tool–adapters [30].

A main objective is the reference to the metamodels in all subsequent steps. As well as modelling and information enrichment and also the formulation of queries in business analysing relies semantically on the defined metamodels.

### 2.3 Information Enrichment and Business Intelligence

The next step is to fill the model repository with models. Technically this is accomplished through usage of tool–adapters and mapping the model element–types to the metamodel. Enterprise models alone don´t provide the necessary information for corporate management. Therefore we provide ways to enrich models with additional information. Key figures play an important role, they can be defined for single model–elements and associations. Advantages and integration aspects in a Business Intelligence Suite are described in [27].

We have to find a proper language for executive management, which has to be abstracted from technology details. Rather, in the foreground have to be a nexus to business strategy, business processes, customer–oriented aspects and finally of course also financial aspects (see Balanced Scorecard [31]). These can be achieved by highly aggregated and customized analysis, reports and visualized by proper charts. These Management–Charts and –Reports are suitable for members of the management board and provide meaningful information on a abstracted and aggregated level in a compact and clear form. 'Software Cartography' [32] and 'Solution Maps'[5] are examples for visualizing complex landscapes of information systems in a compact and concise form, but nevertheless containing enough aggregated information for a precise overview, discussion and decision–support.

# 3 The SQUAM Prototype

A prototypic implementation was created as a general purpose System for Quality Assessment of Models (SQUAM) [33]. In subsequent sections the architecture design, some technical details of the implementation and a usage scenario are given.

## 3.1 The Architecture

The architecture of the prototype is modular, at the topmost level three components can be distinguished: a modelling environment, a model data repository, and an analysis tool (c.f. Fig. 3). As stated in section 2.2 the meta modelling is a crucial part of our framework and thus properties of all components reflect a certain meta model. In the modelling environment validity of models is determined by the meta model, the structure of the model data repository reflects the meta model, and the expression definitions used in the analysis module are based on the meta model.

*The Modelling Environment.* The heterogeneity of the modelling environment is a major issue in the enterprise model design. In our framework models from diverse modelling tools and conformant to different notations can be integrated. A constraint at the conceptual level is notation conformance to the Meta–Object Facility (MOF) [34] and model conformance to a given meta model. At the technical level the problem is solved by a set of adaptors implementing the repository interface for a selected set of modelling tools (c.f. Fig. 3). The syntactical correctness of models, i.e. their conformance to the meta model, is checked before their storage in the repository.

*The Model Data Repository.* The enterprise model design is a distributed process with users from different departments involved in it. The goal of the model data repository is to provide a central storage for all models. The repository is used to store model elements and expressions (c.f. Fig. 3). The structure of the repository is automatically determined by the meta model of a domain under consideration and the meta model of the expression management system [35].

*The Analysis Tool.* The core of the SQUAM framework is the analysis module, where the semantical correctness can be checked. The analysis tool provides mechanism to store definitions of constraints, queries and metrics. The regression test component is

[5] http://www.sap.com/solutions/businessmaps/index.epx

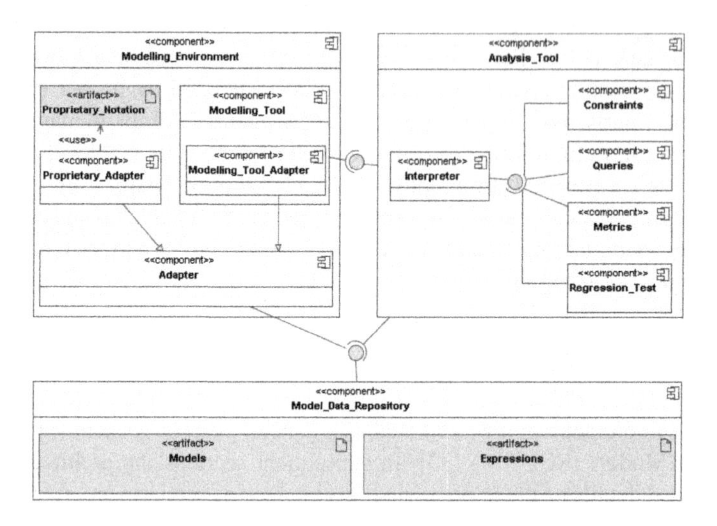


**Fig. 3.** The architecture design of the SQUAM framework

designed for recurrent quality checks, and the interpreter is used for evaluation of all types of expressions on the base of the content of the model data repository. As all model notations are MOF conformant we decided to use Object Constraint Language (OCL) [36] as a query language. Moreover in [35] we showed that the OCL 2.0 is expressive enough for model analysis. Recently we integrated charts into the analysis tool for easier interpretation of metrics.

*Business Intelligence Integration — The Business Analysing Tool.* The core of many Business Intelligence platforms (BI) is a *datawarehouse* [37] with additional features for interactive– and paper–oriented reporting and analysis as well as possibilities to management–oriented visualization (e.g. dashboards, cockpits). A main advantage of datawarehouses is the underlying star– or snowflake-schema (OLAP – online analytical processing) which allows efficient interactive analysis (aggregation, drill-down, sclice and dice, filter, usage of hierarchies, ...) of the stored data — in PRO$^2$SA the data base are enterprise models with enriched information.

In terms of Business Intelligence the stored model information (OLTP – online transactional processing) has to be mapped to a OLAP-cube (see Fig. 4) so that full functionality of BI analysis tools are available. Fig. 7 shows a portion of an OLAP schema, which includes parts of the model repository.

Our business-analysis tool is currently being implemented with Pentaho[6] [28] — this is a open and comprehensive Business Intelligence Suite which provides all features to implement and generate Management–Reports and –Charts.

### 3.2 The Implementation

The implementation of SQUAM follows the client–server paradigm, therefore it consists of the Model Data Repository Server and the Base Repository Client (c.f. Fig. 5).

[6] http://www.pentaho.org/

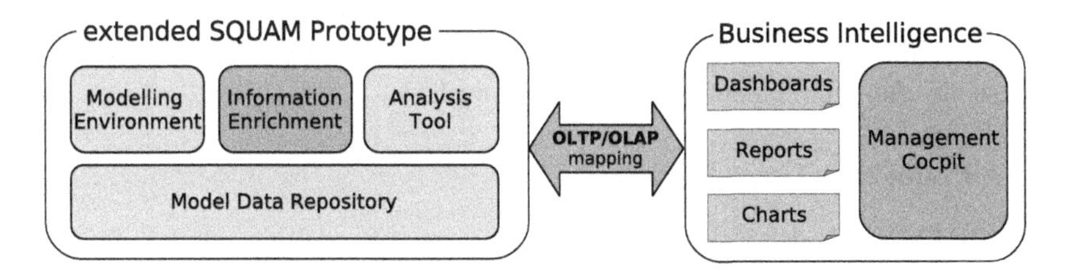


**Fig. 4.** Integration of SQUAM and BI Suite

In general the persistence layer of the application is based on the Eclipse Modeling Framework (EMF[7]). All used model element instances are encapsulated in resources. The repository server is able to save the resources on a database or in a version control system. To save the models on a **database** the client can connect to the server via the Connected Data Objects Framework (CDO[8]). This gives the advantage of multi–user support with change notifications to keep all clients up to date. The client can also connect to a **version control system** like subversion (SVN[9]), with the advantage of versioning, change histories, merging and locking mechanisms. The multi–user support is given only implicit, because of the merging and locking functionalities of the SVN.

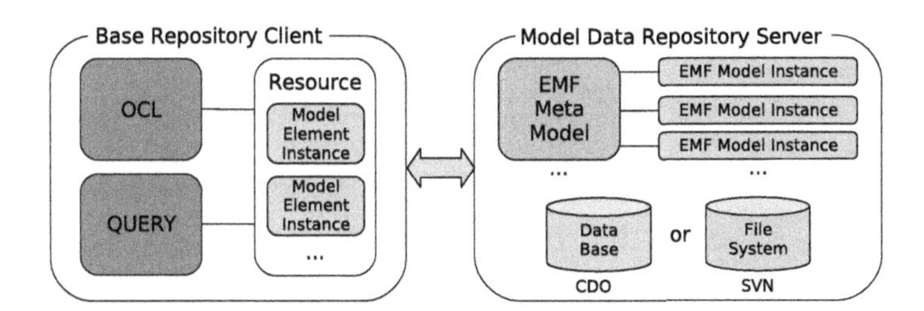


**Fig. 5.** The implementation components of the SQUAM tool

All repository clients uses as core the Base Repository Client with the functionalities of managing models in the repository, and querying the models with OCL[10] and QUERY[11] interpreters. These interpreters are able to evaluate the OCL and QUERY expressions against the EMF models. The interpreters provide also an interface to share the evaluated results to multiple listeners (the observer pattern [38]), which gives the possibility to represent the results in various ways (e.g. text–based and chart–based, c.f. (E) and (V) in Fig. 6, respectively).

The described architecture allows building separate applications for modelling or analysis of models as well as applications which integrate both modelling and analysis.

[7] http://www.eclipse.org/emf/

[8] http://www.eclipse.org/emft/projects/cdo/

[9] http://subversion.tigris.org/

[10] http://www.eclipse.org/emft/projects/OCL/

[11] http://www.eclipse.org/emft/projects/QUERY/

The Base Repository Client can also be used within the modelling environment, e.g. it can be integrated into a modelling tool like MagicDraw[12]. The integration is possible because the client implementation relies on plain Java.

### 3.3 A Framework Usage Scenario

The SQUAM method consists of four phases: meta modelling, modelling, model–analysis and business-analysis. The process is iterative and some activities from all phases can be proceeded in different order. In subsequent paragraphs we describe which activities belong to each phase.

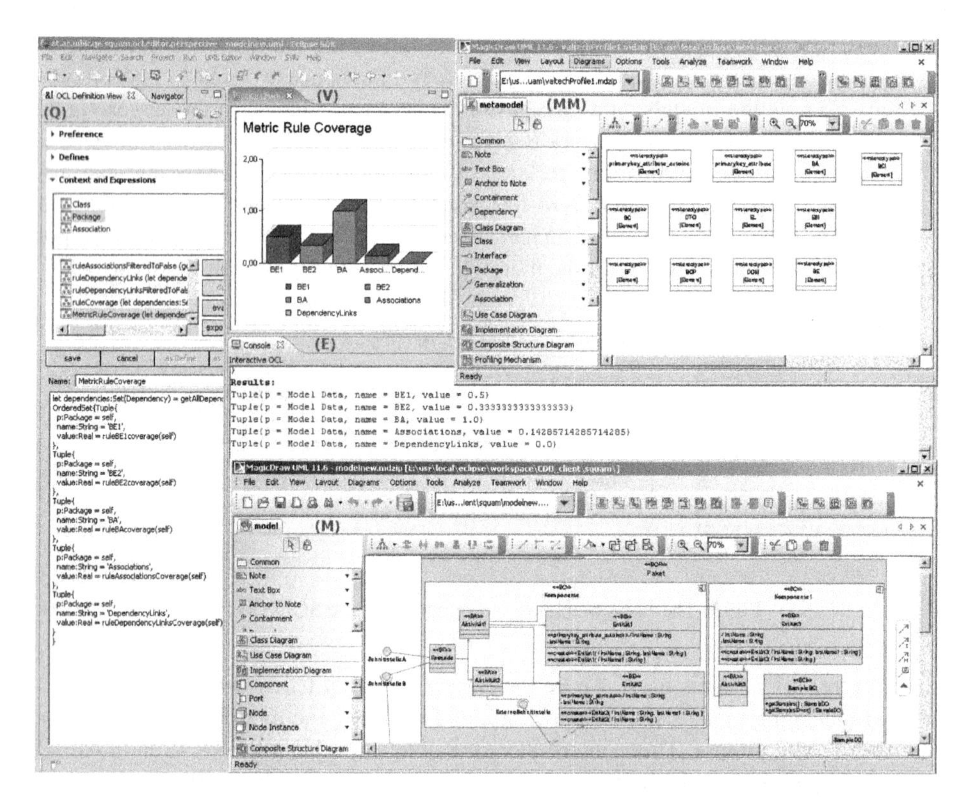


**Fig. 6.** The SQUAM framework tools

**Meta Modelling.** The meta modelling phase is the initial step of the whole modelling and analysis process. In this step **a core meta model** for a domain under consideration is designed. The meta model is considered to be a Domain Specific Language (DSL), where the domain can be a business or a technical one. In PRO$^2$SA project we deal with enterprise modelling (section 2.1) and enterprise architecture is a domain under consideration.

[12] http://www.magicdraw.com/

Based on the core meta model the initial structure of the model data repository is determined and activities from two other phases can be started. Nevertheless the core meta model is designed at the beginning future **extensions** can be made later. The extension mechanism can be used to attach key figures to model elements (section 2.3).

From the technical view point, the meta model must be EMF conform, and therefore a MOF conform model. The meta model can be created from scratch or based on existing notation, for which EMF implementations are provided. One of such notations is UML, which can be restricted or extended by use of the UML Profile mechanism. The development of a new EMF meta model or an UML Profile can be done within the modelling environment. e.g. in modelling tool like MagicDraw (c.f. (MM) in Fig. 6), and later converted to an EMF model.

To extend the model with informations like key figures, it is also necessary to extend the meta model to allow aggregation of this additional information. Therefore cross–references between different meta models are allowed, the core meta model can be extended by a cross–reference to a key figures meta model.

**Modelling.** In the modelling phase models are created in diverse modelling tools in a distributed, multi–user environment and saved into the central model data repository. The meta model determines syntactical **constraints** on models, which can be checked already in a modelling tool.

The modelling phase gives feedback to the meta modelling phase by meta model extension requests.

The syntactical correctness of models can be checked within the adapters provided for the diverse modelling tools (c.f. Fig. 3). Each adapter is based on the Base Repository Client (c.f. Fig. 5), which hooks the features of the repository access and the interpreter functionalities up to the modelling tool. Additionally the plug–in mechanism of modelling tools, like MagicDraw (c.f. (M) in Fig. 6), allows restrictions on a set of used model elements to elements defined in a certain UML Profile [39,33].

**Model–Analysis.** In the model–analysis phase **model-oriented queries** are defined and evaluated. Queries are defined in OCL based on a certain DSL. They are evaluated over the content of the model data repository and provide *aggregated* and *filtered* information on model elements to allow analysis from different perspectives. Moreover **metrics** of diverse types can be evaluated and visualised with charts. In PRO$^2$SA project the key figures attached to model elements can be used for metric evaluation.

If the data provided by the model data repository is not sufficient for analysis, changes at the meta model and model level can be requested.

Our model–analysis tool (c.f. Fig. 6) is implemented as an Eclipse plug–in, and is based on the Base Repository Client (c.f. Fig. 5). This tool consists of various independent views which are connected by defined interfaces:

**The Define View** (c.f. (Q) in Fig. 6) can manage OCL expressions and store them in the model repository.

**The Evaluation View** (c.f. (E) in Fig. 6) executes OCL expressions, which can be typed into a console or selected in the Define View, thus new and predefined expressions can be evaluated. The OCL interpreter provides the evaluation results via an interface to all result listeners. The Evaluation View acts as a text–based result listener.

**The Charting View** (c.f. (V) in Fig. 6) is a chart based result listener, and represents the evaluation results of metrics in a graphical way.

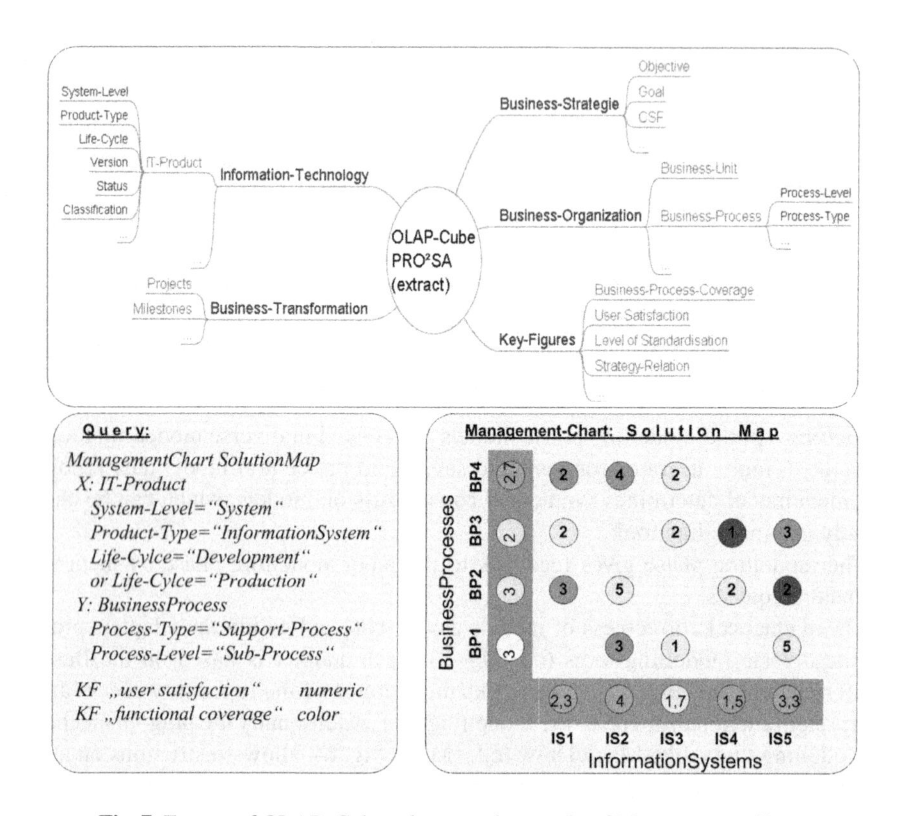


**Fig. 7.** Extract of OLAP–Cube schema and example of Management–Chart

**Business–Analysis.** In the business-analysis phase **business-oriented and domain-specific queries** are defined and evaluated. This phase allows interaction and communication with corporate management by usage of Business Intelligence or Balanced Scorecard.

Fig. 7 depicts a simple example for business–analysis: In the upper part a portion of the OLAP–Cube design is displayed. The chart-example illustrates the formulation of a query and the generated Management-Chart of type *Solution Map*. This chart permits an overview of the relationships between information systems and supported business processes.

## 4 Conclusion

This paper briefly presents an enterprise–modelling oriented approach to IT Governance. By means of operationalizing Strategic Business Alignment (a main focus area of IT Governance) the applicability is shown. Management-related analysis and charts are generated on the basis of underlying enterprise model, which are stored in a model repository. All phases and tools rely on a previously defined metamodel — one of the first tasks.

One key aspects of this approach is the information enrichment of enterprise models. Yet another major role plays the integration into a Business Intelligence platform, which provides technically a great potential in reporting and analysis. Furthermore another advantage is the usage of already stored enterprise information — e.g. datamarts for controlling, cost accounting, asset management, ...

Both frameworks PRO$^2$SA and SQUAM are work in progress. Currently we are developing full support for the OCL management system and the regression test component for automation of quality checks for the SQUAM framework. We plan to carry out more case studies to determine more requirements for model assessments queries and define patterns for query definitions.

Also the integration of Eclipse-based SQUAM framework with the open Business Intelligence Suite PENTAHO is under development. First outcomes are very promising.

## References


1. Buchta, D., Eul, M., Schulte-Croonenberg, H.: Strategisches IT-Management — Wert steigern, Leistung steuern, Kosten senken. Gabler Verlag (2004)
2. Zarnekow, R., Brenner, W., Grohmann, H.H.H.: Informationsmanagement — Konzepte und Strategien für die Praxis. dpunkt.verlag, Heidelberg (2004)
3. Weill, P., Ross, J.W.: IT-Governance. 1 edn. Harvard Business School Press, Boston (2004)
4. Carr, N.G.: It doesn't matter. Harvard Business Review **81**(5) (May 2003) 41–49
5. Stewart, T.A.e.: Does it matter? an hbr debate. Harvard Business Review **Web Exclusive** (June 2003) 1–17
6. Brynjolfsson, E., Hitt, L.M.: Beyond productivity paradox; computers are the catalyst for bigger changes. Communications of the ACM **41**(8) (August 1998) 49–55
7. Brynjolfsson, E.: The it productivity gap; roi valuation. Optimize **Issue 21**(Issue 21) (July 2003)
8. Piller, F.T.: Das produktivitätsparadoxon der informationstechnologie aus betriebswirtschaftlicher sicht. Wirtschaftspolitische Blätter **1998** (Winter 1998 1998) Kommentar und Ergänzungen zum Diskussionsschwerpunkt der WirtschaftspolitischenBlätter, Heft 1 / 1998.
9. Rau, S.E., Bye, B.S.: Are you getting value from your it? Journal of Business Strategy (May/June 2003) 16–20
10. ISACA: Optimising value creation of it-investments. Technical report, ISACA-Organisation, IT-Governance-Institute (2005)
11. ISACA: Measuring and demonstrating the value of it. Technical report, ISACA-Organisation, IT-Governance-Institute (2005)
12. ITGI: IT Control Objectives for Sarbanes-Oxley - The role of IT in design and implementation of internal control over financial reporting. 2nd edn. IT-Governance Institute (2006)
13. Gartner: Growing it's contribution: The 2006 cio agenda. pdf (2006)
14. PricewaterhouseCoopers: It governance global status report - 2006. Technical report, IT Governance Institute (2006)
15. Webb, P., Pollard, C.E., Ridley, G.: Attempting to define it governance: Wisdom or folly? In: HICSS, IEEE Computer Society (2006)

16. Simonsson, M., Hultgren, E.: Administrative systems and operation support systems - a comparison of IT governance maturity. In: Proceedings of the CIGRÉ International Colloquium on Telecommunications and Informatics for the Power Industry. (June 2005)
17. ISACA: Cobit 4. PDF (11 2005)
18. Van Grembergen, W.: Introduction to the minitrack "it governance and its mechanisms" hicss 2004. In: System Sciences, 2004. Proceedings of the 37th Annual Hawaii International Conference on. (5-8 Jan. 2004) 231–231
19. ISACA: Board briefing on it governance. book (2003)
20. GSE: It-governance. Technical report, Guide Share Europe (2004)
21. Tilg, B.: Anforderungen an ein modellbasiertes strategic business alignment der informationstechnologie. In Cremers, A.B., Manthey, R., Martini, P., Steinhage, V., eds.: GI Jahrestagung (2). Volume 68 of LNI., GI (2005) 506–510
22. Simonsson, M., Johnson, P.: Defining IT governance - a consolidation of literature. Technical report, Royal Institute of Technology (KTH) (November 2005)
23. Henderson, J.C., Venkatraman, N.: Strategic alignment: Leveraging information technology for transforming organizations. IBM Systems Journal **32**(1) (1993) 4–16
24. Henderson, J.C., Venkatraman, N., Oldach, S.: 2. [25] 21– SOWI 505-COMP.
25. Luftman, J.N.: Competing in the Information Age; Strategic Alignment in Practice. Oxford University Press (1996) SOWI 505-COMP.
26. Avison, D., Jones, J., Powell, P., Wilson, D.: Using and validating the strategic alignment model. Volume 13. (2004) 223–246
27. Tilg, B., Breu, R.: Pro2sa — ein modellgetriebener ansatz zur integration des strategic business alignments in die business intelligence. In Lehner, F., Nösekabel, H., Kleinschmidt, P., eds.: GI Jahrestagung (2). Volume Band 2., GITO (2006) 97–109
28. Tilg, B., Hechenblaikner, C., Breu, R.: Integration der open business intelligence-suite pentaho in ein modellgetriebenes analyse-framework am beispiel des strategic alignments. In Schelp, J., Winter, R., Frank, U., Rieger, B., Turowski, K., eds.: Lecture Notes in Informatics, Proceedings, Volume 90, Integration, Informationslogistik und Architektur, DW2006. Volume P-90 of LNI., Gesellschaft für Informatik (September 2006) 125–140
29. Stahl, T., Völter, M.: Modellgetriebene Softwareentwicklung; Techniken, Engineering, Management. 1. auflage edn. dpunkt.verlag (2005)
30. Chimiak-Opoka, J., Giesinger, G., Innerhofer-Oberperfler, F., Tilg, B.: Tool–supported systematic model assessment. Volume P–82 of Lecture Notes in Informatics (LNI)—Proceedings., Gesellschaft fuer Informatik (2006) 183–192
31. Kaplan, R.S., Norton, D.P.: The balanced scorecard — measures that drive performance. Harvard Business Review **January-February** (1992) 71–79
32. Matthes, F., Wittenburg, A.: Softwarekartographie: Visualisierung von anwendungslandschaften und ihrer schnittstellen. In: GI Jahrestagung (2). (2004) 71–75
33. Chimiak-Opoka, J., Giesinger, G., Innerhofer-Oberperfler, F., Tilg, B.: Tool–Supported Systematic Model Assessment. In Mayr, H., Breu, R., eds.: Modellierung 2006, 22.-24. März 2006, Innsbruck, Tirol, Austria, Proceedings. Volume 82 of Lecture Notes in Informatics (LNI)—Proceedings., Gesellschaft fuer Informatik (2006) 183–192
34. OMG: Meta-Object Facility (MOF)—Specification of version 2.0 (Jan 2006) `http://www.omg.org/technology/documents/formal/MOF_Core.htm`.
35. Chimiak-Opoka, J., Lenz, C.: Use of OCL in a Model Assessment Framework. Number TUD-FI06-04-Sept. 2006, Genova (Oct 2006) 53–67 ISSN 1430-211X, OCLApps workshop co-located with MoDELS 2006: 9th International Conference on Model-Driven Engineering Languages and Systems (formerly the UML series of conferences).

36. Warmer, J., Kleppe, A.G.: The Object Constraint Language—Precise Modeling with UML. first edn. (1999)
37. Inmon, W.H.: Building the data warehouse (4th ed.). John Wiley & Sons, Inc., New York, NY, USA (2005)
38. Gamma, E., Helm, R., Johnson, R., Vlissides, J.: Entwurfsmuster. Addison Weseley (1995)
39. Breu, R., Chimiak-Opoka, J.: Towards Systematic Model Assessment. In Akoka, J., et al., eds.: Perspectives in Conceptual Modeling: ER 2005 Workshops CAOIS, BP-UML, CoMoGIS, eCOMO, and QoIS, Klagenfurt, Austria, October 24-28. Volume 3770 of Lecture Notes in Computer Science., Springer-Verlag (October 2005) 398–409

# A Modelling Approach for Dynamic and Complex Capacities in Production Control Systems

Wilhelm Dangelmaier, Benjamin Klöpper, Thorsten Timm, and Daniel Brüggemann

Heinz Nixdorf Institute, University of Paderborn,
Fürstenallee 11, 33102 Paderborn, Germany
{whd,kloepper,timm,dbruegg}@hni.upb.de
http://www.hni.upb.de/cim

**Abstract.** In this paper we introduce some aspects of the development process of a production planning tool for a leading European car manufacturer. In this project we had to face a gap between theoretical problem definition in manufacturing planning and control and the actual requirements of the dispatcher. Especially the determination of production capacities and product processing times was a severe problem. The software system had to support the derivation of these important variables from shift plans, factory calendars and exceptional events. In order to implement this we defined a formal model that is a combination of the theoretical and practical view on manufacturing problem. The *Model for Serial Manufacturing* allows the dispatcher to provide up-to-date information in an easier way then provided by standard ERP systems about the production systems and transfers the information automatically to planning algorithms. Thus the production planning and control is always performed on the most recent information.

**Keywords:** Production Control, Modeling, Implementation of PPC-Systems, Mathematical Models.

## 1 Introduction

Flexible allocation of human resources holds the potential to be a significant competitive advantage in many industries due to nonuniform market situations. This flexibility induces highly dynamic capacity situations. Most mathematical PPC (Production Planing and Control) problem definitions abstract from the derivation of capacity from factory calendars. ERP systems take this information into account, but the task of modelling a production system—including a appropriate time model—is a very complex task only performed by highly qualified and thus expensive experts. To provide up-to-date information about the current capacity situation is unattainable when using standard ERP software.

W. Abramowicz (Ed.): BIS 2007, LNCS 4439, pp. 626–637, 2007.
© Springer-Verlag Berlin Heidelberg 2007

In this paper our solution to this problem, used in a newly developed PPC-System called OOPUS (germ. Objektorientierte Planung und Steuerung - object-oriented planning and control), will be described in detail. This information system was developed at the Fraunhofer Application Centre for logistic-oriented business administration (Fraunhofer-ALB) in cooperation with the chair for information systems, especially CIM of the University of Paderborn. Nowadays OOPUS is the leading production control system within two plants.

The paper is structured as follows: In the next section we describe the problems which occur when up-to-date information has to be provided to PPC systems. In the third section we describe the general solution we used in OOPUS to tackle these problems. This solution is based on the *Model for Serial Manufacturing* introduced in the fourth section. Section 5 describes the user-friendly graphical interface. In the sixth section it is shown that the *Model of Manufacturing* can be used to map all relevant information provided by the dispatcher on classical PPC problem definitions. Section 7 gives a conclusion.

## 2 Problem Statement

PPC encompasses several subproblems [1]. Figure 1 shows a commonly accepted classification for PPC tasks. For each of the problems defined in this classification,

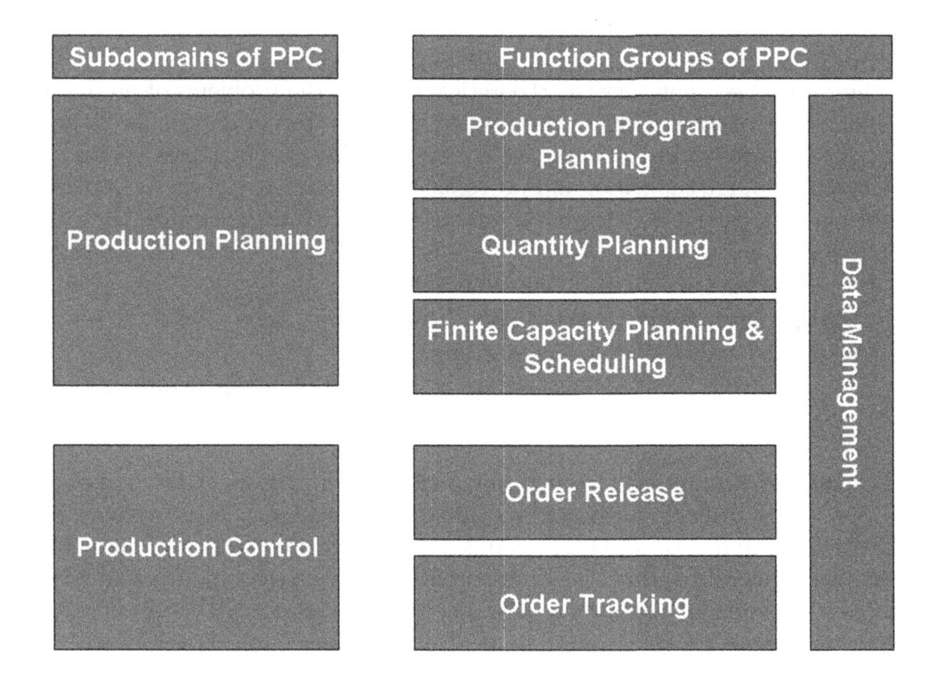


**Fig. 1.** Production Planning and Control Tasks

various mathematical problem definitions are available. In this paper we will focus on the problem of quantity planning.

For capacitated single level lot sizing problems (for examples see [3], [4], [5], [6]) the following variables are relevant, if stock costs and set-up costs are not considered:

- $d_{kt}$ demand for product $k$ in period $t$
- $p_{kt}$ production cost for product $k$ in period $t$
- $tb_k$ processing time per unit for product $k$
- $b_{jt}$ capacity of ressource $j$ in period $t$
- $y_{kt}$ inventory of product $k$

For multilevel problems (for examples see [7], [8], [5], [9], [10]) the set of variables has to be extended by the following variables:

- $a_{ki}$ direct demand coefficient of product $k$ according to product $i$ ($k$ is used on the next level to build product $i$)
- $z_k$ minimal lead time for product $k$
- $N_k$ index set of all successors of product $k$ (superior product or next processing step)
- $K_j$ index set for all processing steps performed by ressource $j$

These variables represent the theoretical view on the problems of production planning and control. Dispatchers solving these problems in real life have an entirely different view. To meet their requirements, the variables $b_{jt}$ and $tb_k$ describing resource capacities and processing times are not sufficient. In real life problems these variables depend on a large number of influences like:

- the calendar, valid for a particular resource
- working times, defined by shifts and breaks
- exceptional events (e.g. works meetings)
- reduced output rates due to absence of staff (e.g. different teams, sickness)
- and many more...

Dispatchers are not able to determine the values of $b_{jt}$ and $tb_k$. A possible solution to this problem is offered by standard ERP systems. These systems (e.g. SAP ECC 5.0 and SAP APO [11], SAGE [12] or Navision [13]) cover many PPC tasks like demand planning, lot sizing or scheduling using standard algorithms. The task of modelling a manufacturing system in such an ERP system is very complex and requires expensive experts. Not all influences on capacities and processing times can be included in the ERP master data without customization, which leads to even higher costs. Summarizing, providing up-to-date information about the production system is unattainable when standard ERP implementations are used.

In order to implement the OOPUS PPC system including an automated quantity planning and detailed scheduling (incl. to-the-minute starting and finishing times of lots) the first step was to close the gap between the theoretical and practical approach. Thus we developed the *Model for Serial Manufacturing* introduced in the next chapter.

## 3 The Modeling Process

The main objective of OOPUS is to provide powerful planning capabilities based on up-to-date information about the production system. Thus we have to meet partially conflicting requirements: the requirement of a well-defined problem definition in order to apply powerful planning algorithms and the requirement of every day life in a company. Calculating capacities and processing times considering all exceptions which may occur in a company is very complex and has to be sufficiently supported to use any PPC effectively. Thus we separated the two requirements and defined a formal model, which includes all relevant exceptions and implements an automated mapping process to a well defined mathematical problem definition. Figure 2 shows the procedure of model generation in OOPUS. In the first step the formal *Model for Serial Manufacturing* was defined on basis of the practical requirements of the clients and formal problem definitions. Thus, the *Model for Serial Manufacturing* provides a mapping between all relevant information and influences in the plants and the mathematical formulation of production control problems. The *Model for Serial Manufacturing* was first transformed into a relational model and then into a database definition. Supported by a user friendly master data editor, the dispatchers are enabled to model their manufacturing process and keep planning parameters up-to-date.

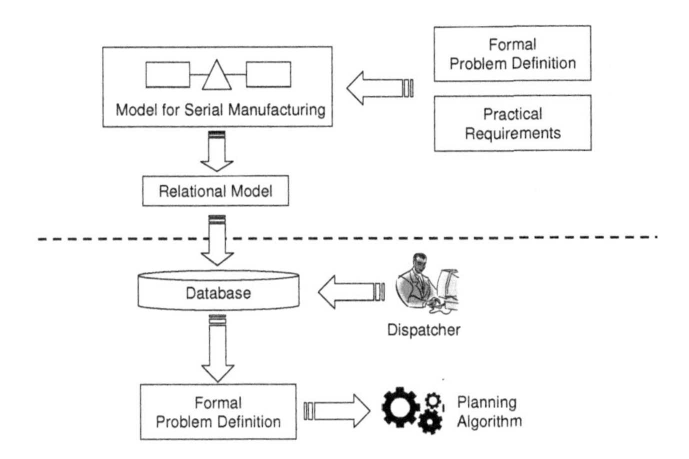


**Fig. 2.** Process of Modeling the Planning Problem

## 4 The Model for Serial Manufacturing

In this section we introduce the fundamental elements of our modelling approach. The *Model for Serial Manufacturing* allows to model a multilevel manufacturing process in a flexible and extensible way. The *Model for Serial Manufacturing* encompasses five element types to model different manufacturing and organisational structures:

- production stages
- buffers
- lines
- planning groups

Another important element of the model are products and intermediate products. Usually it is possible to split serial production over the entire product range into several stages. In these stages, similar production steps are performed. Examples are parts department, pre-assembly and final assembly. We define:

- $PS$ as the set of production stages
- $\prec$ as an ordering relation over $PS$. The minimal and maximal elements of $PS$ according to $\prec$ are $ps_{min} = 0$ and $ps_{max}$

Within a production stage, lines represent specific resources, which are used for the processing of products and intermediate products. Within the *Model for Serial Manufacturing*, a production stage is treated as a single step process. This means a line represents a combination of machines and each product or intermediate product passes through exactly one group. This concept of production stage and line is tailored for the requirements of serial production.

Line classes pool production lines, which have similar properties and can process similar products or product groups:

- $L$ is the set of production lines
- $LC$ is the set of line classes
  - $\forall lc \in LC : lc \subseteq L$; a line class is a set containing lines
  - $\forall lc, lc' \in LC, lc \neq lc' : lc \cap lc' = \emptyset$; each line is assinged to only one line class
  - $L = \bigcup_{lc \in LC} lc$; each line is at least assigned to one line class

The transfer of products and intermediate products from one production stage to another is modeled by buffers. A buffer can be understood as a stock with limited capacity or just the transportation process between production stages. We define:

- $P$ as the set of products and intermediate products
- $B$ as the set of buffer

Planning groups overlay the previously introduced model elements. They do not describe the structure or the process of manufacturing, but organise the planning process:

- PG is the set of planning groups
- $\forall pg \in PG : pg \subseteq LC$; a planning group is a set containing line classes
- $\forall pg, pg' \in PG, pg \neq pg' : pg \cap pg' = \emptyset$; each line class is only assigned to one planning group
- $LC = \bigcup_{pg \in PG} pg$; each line is at least assigned to one planning group

Each planning group is assigned to one dispatcher who is responsible for the planning of the corresponding line classes. Figure 3 shows an example.

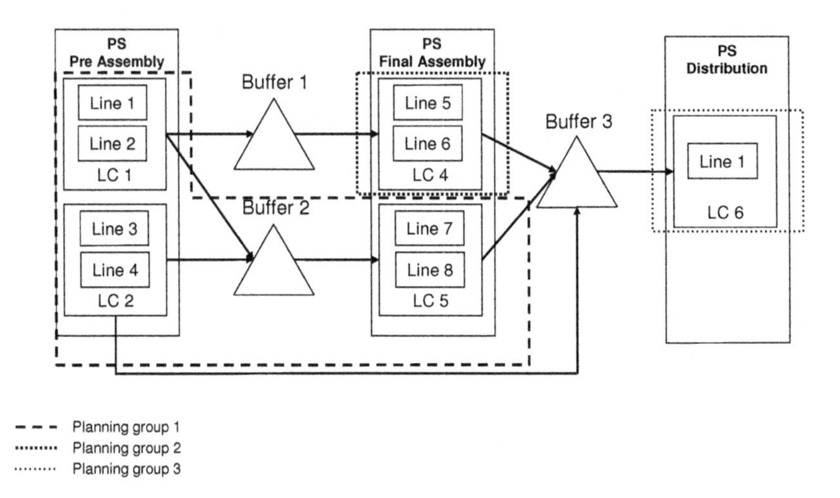


**Fig. 3.** Example for the Structure of a Manufacturing System

### 4.1 The Time Model

A generic shift model forms the main part of the time model. A shift model encompasses:

- A sequence $SC = (1...sc_{max})$ of shift classes
- A time set $T = 1...1440$; 1440 denotes the number of minutes per day
- A set of weekdays $W = \{1..7\}$
- A set of shifts $S$, where
  - $Start : S \rightarrow T$ returns the start time of a shift
  - $End : S \rightarrow T$ return the end time of a shift
  - $\forall s \in S : Start(s) < End(s)$
  - $\forall s \in S : Duration(s) = End(s) - Start(s)$
- A set BR of breaks, where
  - $Breaks : S \rightarrow \mathcal{P}(BR)$ returns the breaks of a shift
  - $Start : BR \rightarrow T$ returns the start time of a break
  - $End : BR \rightarrow T$ returns the end time of a break
  - $\forall br \in BR : Start(br) < End(br)$

  - $\forall br \in BR : Duration(br) = End(br) - Start(br)$
  - $\forall s \in S, \forall br \in Breaks(s) : Start(br) \geq Start(s) \wedge End(br) \leq End(s)$
- $type : S \rightarrow SC$ returns the class of a shift

The shift plan for a week is defined by two functions:

- $sp : SC \times W \rightarrow S$ returns for each tuple of shift class and weekday the assigned shift
- $productive : sp \rightarrow \{True, False\}$ returns if production is possible during a shift
- $SP$ denotes the set of all shift plans

Equation 1 defines a predecessor-successor relationship between the elements of a shift plan $sp$.

$$successor(sp(sc, w)) = \begin{cases} sp(sc + 1, w),\ if\ sc < sc_{\max} \\ sp(1, w + 1),\ if\ sc = sc_{\max}\ and\ w < 7 \\ sp(1, 1),\ else \end{cases} \quad (1)$$

There is a number of restrictions according to shift plans, which assure a consistent data base for planning algorithms:

- $\forall (sc, w, s) \in sp : type(s) = sc$ only shifts of the corresponding type can be assigned to a tuple of shift class and weekday
- $sp(sc, w) = successor(z(sc', w')) \rightarrow start(sp(sc, w) = end(sp(sc', w'))$, start and end time of shifts are consistent

The shift plans are supplemented by a calendar. A calendar specifies if a day is a workday or not:

- $calendar : calendardays \rightarrow \{true, false\}$
- $C$ denotes the set of all calendars

In order to define a fine granular time model, it is possible to assign calendars and shift plans to lines:

- $calendarAssignment : L \rightarrow C$
- $shiftplanAssignment : L \times CW \rightarrow SP$, where CW denotes all calendar weeks

Which shift is valid for a certain point of time can be deduced from the shift plan of a calendar week:

$$Shift : L \times CW \times W \times T \rightarrow S \quad (2)$$

With these functions it is possible to deduce if it is possible to produce something on a certain line on a certain calendar day at a certain time. The productivity information from the calendar overrides the information from the shift plan. An additional adjustment is possible by so called capacity deviations:

$$labor\_utilization : SC \times L \times CW \times W \rightarrow \mathbb{R} \quad (3)$$

The function *labor_utilization* returns which percentage of the standard utilization is achieved during a specific shift. A *labor_utilization* of 1 results in the standard processing times (see section 4.2).

### 4.2 Material Flow

The flow of material as modelled by a construct called *material flow element* $mf$:

- $MF \subset PS \times PS \times P$; $vf \in PS \times PS \times F$

A *material flow element* is a thus a combination of production stage and product. A product or intermediate product is produced in the first production stage and delivered to the second one. In the OOPUS planning concept no loops in the production are allowed. Thus the ordering relation $\prec$ is used to assure this condition. For each material flow element the following condition must be true:

$$Projection(mf, 2) \prec Projection(mf, 1) \quad (4)$$

In addition to the flow through the production stages it must be defined on which lines a product can be processed. This is defined by the following relation:

$$Processed \subset MF \times L, \; fb \in Processed \quad (5)$$

In order to map the *Model for Serial Manufacturing* on mathematical optimization models for each line, the set of processable products or material flow elements must be known:

$$Processes(l) = \{mf | l \in Processed(mf)\}; \; Processes : L \rightarrow \mathcal{P}(MF) \quad (6)$$

Another important variable in multilevel, multi period and multi product problems is lead time. In the *Model for Serial Manufacturing*, the lead time of an intermediate product is the transfer time between two productions stages. It is possible to model lead time in shifts or in minutes:

$$Leadtime_{\min} : LC \times LC \times MF \rightarrow \mathbb{N} \quad (7)$$

$$Leadtime_{shift} : LC \times LC \times MF \rightarrow \mathbb{N} \quad (8)$$

$$Leadtime_{shift}(lc, lc',, mf) \geq 0 \rightarrow Leadtime_{\min}(lc, lc', mf) = \infty \quad (9)$$

Equation 9 expresses that a leadtime in minutes is only considered when no lead time in shifts is indicated.

### 4.3 Capacity

Two requirements have to be meet by the modelling of capacities in OOPUS:

- In order to consider parallel machines it is required to model processing time depending on the production line
- The capacity provided by a production line is individual for each shift

In order to fullfil these requirements it is possible to define a standard processing time in minutes for each shift class, production line and material flow element:

$$ProcessingTime_{stand} : Processed \times SC \rightarrow \mathbb{N} \tag{10}$$

With the *labor_utilization* it is possible to calculate the effective processing time in certain shift:

$$\begin{aligned} &ProcessingTime_{eff}(fb, s, v, kw, w) = \\ &ProcessingTime(fb, type(s)) \cdot labor\_utilization(type(s), l, cw, w) \end{aligned} \tag{11}$$

The available processing time of a production line in a certain period is derived by:

$$\begin{aligned} &Shiftduration_{net}(l, cw, w, sc) = \\ &Duration(Shift(l, cw, w, sc)) - \sum_{br \in Break(Shift(l,cw,w,sc))} duration(br) \end{aligned} \tag{12}$$

With knowledge of the processing time and the net shift duration for a given production on a given day in a given shift class, all required variables for PPC-Planning problems consired by OOPUS are defined.

# 5 Graphical User Interface

All elements of the *Model for Serial Manufacturing* were formulated as sets, functions and relations. Thus, it is possible to transfer the *Model for Serial Manufacturing* into a relational model and thus into a redundancy free database scheme [14]. The corresponding database is used for data storage. A graphical

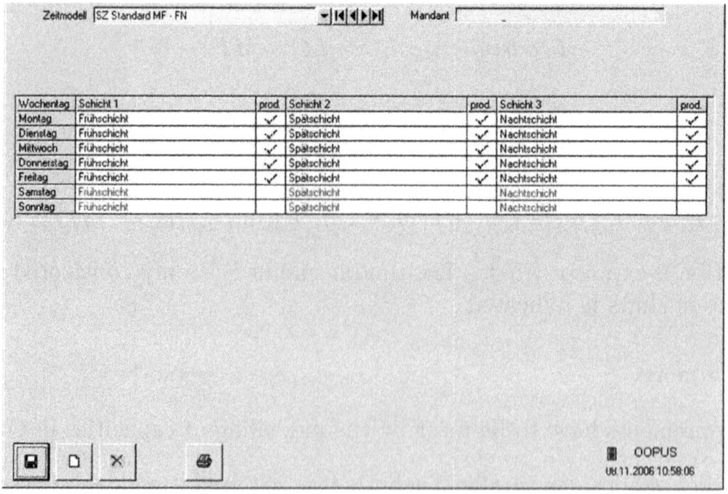

Zeitmodell SZ Standard MF - FN   Mandant

| Wochentag | Schicht 1 | prod. | Schicht 2 | prod. | Schicht 3 | prod. |
|---|---|---|---|---|---|---|
| Montag | Frühschicht | ✓ | Spätschicht | ✓ | Nachtschicht | ✓ |
| Dienstag | Frühschicht | ✓ | Spätschicht | ✓ | Nachtschicht | ✓ |
| Mittwoch | Frühschicht | ✓ | Spätschicht | ✓ | Nachtschicht | ✓ |
| Donnerstag | Frühschicht | ✓ | Spätschicht | ✓ | Nachtschicht | ✓ |
| Freitag | Frühschicht | ✓ | Spätschicht | ✓ | Nachtschicht | ✓ |
| Samstag | Frühschicht | | Spätschicht | | Nachtschicht | |
| Sonntag | Frühschicht | | Spätschicht | | Nachtschicht | |

OOPUS

**Fig. 4.** Screenshot of the time model

user interface allows the dispatcher to enter up-to-date information about the current working times, which are adapted for the current demand situation. Figure 4 shows an exemplary form in which shift plans can be created. The definition of the database scheme and mechanisms in the graphical user interface assure a consistent modeling of the production system and the capacity situation.

The entire *Model for Serial Manufacturing* including several additional planning parameters such as minimal/maximal lot size for each product $k$, maximal quantity of product $k$ in $x$ days, palette size, standard resources, preferred shift classes and many more can be built up using seven forms. Most of the forms are product or resource centric and thus correspond to the dispatcher's way of thinking.

# 6 Mapping to Formal Problem Definition

With the *Model for Serial Manufacturing* it is possible to map the data from the database to mathematical problem definitions, which can be used for solving PPC-Problems with standard algorithm. In this section we demonstrate how the information from the *Model for Serial Manufacturing* is mapped to the MLCLSP [2]. The OOPUS version of the MLCLSP is given by:

$$\text{Minimize } Z = \sum_{vf \in VF} \sum_{t=1} \sum_{v \in V} (y_{vf,t}) \tag{13}$$

s.t.

$$y_{mf,t} = y_{mf,t-1} + \sum_{l \in L} q_{mf,t,l} - \sum_{mf' \in N_{mf}} (a_{mf,mf'} \cdot \sum_{l \in L} q_{mf,t-lead(mf',mf),l}) - d_{mf,t} \quad \forall l \in L; \forall t \in T \tag{14}$$

$$\sum_{mf \in processes(l)} (tb_{mf,t,l} \cdot q_{mf,t,l}) \leq b_{l,t} \quad \forall l \in L; \forall t \in T \tag{15}$$

$$q_{mf,t,l} \geq 0 \quad \forall mf \in MF; \forall t \in T \tag{16}$$

$$q_{mf,0,l} = 0; \; q_{mf,T,l} = 0 \quad \forall mf \in MF; \forall t \in T \tag{17}$$

$$y_{mf,t} \geq 0 \quad \forall mf \in MF; \forall t \in T; \forall l \in L \tag{18}$$

The parameters in this mathematical optimization model are:

- $a_{mf,mf}$ *the direct demand coefficient*, derived from the bill of material in legacy systems
- $b_{l,t}$ the capacity of resource $j$ in period $t$, derived from $capacity(l, cw, w, sc)$
- $d_{mf,t}$ the primary demand for material flow element $k$ in period t, derived via interface from clients
- $N_{mf}$ the index set of sucessors of material flow element $vf$, derived from the bill of material in legacy systems
- $q_{mf,t,v}$ lotsize for material flow element $k$ in period $t$ on resource $j$, variable to determine
- $T$, the planning horizon, a planning parameter
- $tb_{mf,t,v}$ the processing time of material flow element vf in period $t$ on machine v, derived from $processingTime_{eff}(fb, s, v, kw, w)$
- $y_{mf,t}$ the inventory of material flow element at the end of period $t$, variable to determine

This modified mathematical problem is the basis for the quantity planning in the current version OOPUS. From the list above it can be seen, that all information can either be derived from the *Model for Serial Manufacturing* or is a variable to be determined (the only exception are the bill of materials, legacy systems are used to determine the direct demand coefficients and the index sets of the successors). Since the *Model for Serial Manufacturing* is mapped into the database, it is possible to create a new problem definition with up-to-date data for each planning run.

## 7 Conclusion

In our opinion several conclusion can be drawn from the OOPUS development process. The most important insight is that up-to-date information is at least as important as well performing optimization algorithms. Especially the process of providing this up-to-date information is not sufficiently supported in most PPC and ERP tools. The modelling process we developed in the OOPUS-Project can be easily adopted to other PPC implementations. The modelling process is characterized by the following steps:

1. identify the actual requirements of the dispatchers
2. find the most suitable formal problem definitions for the production system
3. define a flexible formal model like the *Model for Serial Manufacturing*
4. transfer this model into a database scheme
5. provide easy access to the database

In this way, the problem of up-to-date information is solved in methodical and verifiable way. At the same time, standard algorithms can be used to solve the PPC problems tackled by the developed tool. OOPUS is today the leading tool in production planning of two plants.

# References


1. Higgins, P. Le Roy, P., Tierney, L.: Manufacturing Planning and Control: Beyond MRP II. Springer Verlag, Berlin, Heidelberg, New York (1996)
2. Tempelmeier, H., Kuhn, H.: Flexible Manufacturing Systems: Decision Support for Design and Operation. Wiley-Interscience, New York (1993)
3. Salomon, M.: Deterministic Lotsizing Models for Production Planning. Springer Verlag, Berlin, Heidelberg, New York (1991)
4. Suerie, C.: Time Continuity in Discrete Time Models - New Approaches for Production Planning in Process Industries. Springer, Verlag, Berlin, Heidelberg, New York (2005)
5. Haase, K.: Lotsizing and scheduling for production planning. In: Lecture Notes in Economics and Mathematical Systems, Vol. 408. Springer-Verlag, Berlin, Heidelberg, New York (1994)
6. Kimms, A.: Multi-Level Lot Sizing and Scheduling. Physica Verlag, Heidelberg (1997)
7. Afentakis, P.,Gavish, B.: Optimal Lot-Sizing Algorithms for complex product structures. Operations Research, Vol. 34 (1986) 237-249,
8. Billington, P., McClain, J. Thomas, L.: Heuristics for multilevel lotsizing with a bottleneck. Management Science, Vol. 32, (1986) 989-1006
9. Ingold, T.: Multi-level Lot Sizing: Feasible Sequential Decisions and Flexible Lagrangean-Based Heuristics. Doktorarbeit, Institute for Computer Science, University of Freiburg (Swiss) (1998)
10. Belvaux, G., Laurence A.W.: Modelling Practical Lot-Sizing Problems as Mixed-Integer Programs. Management Science, Vol. 47 (2001) 993-1007
11. Balla, J.: Production Planning with SAP APO-PP/DS, SAP PRESS (2006)
12. Wallace, T.F., Kremzar, M.H.: ERP:Making It Happen: The Implementers' Guide to Success with Enterprise Resource Planning, Wiley & Son, Chichester (2001)
13. Diffenderfer, P.M., El-Assai, S.: Microsoft Navision 4.0: Jump Start to Optimisation, GWV-Vieweg, Wiesbaden (2005)
14. Connolly, T.M. C. Begg, C.: Database Systems: A Practical Approach to Design, Implementation, and Management. Addison-Wesley, Boston (2001)

# Product Line Architecture for RFID-Enabled Applications*,**

Mikyeong Moon and Keunhyuk Yeom

Department of Computer Engineering, Pusan National University
San-30 Changjeon Dong, Geumjeong Gu, Busan, 609-735, Korea
{mkmoon,yeom}@pusan.ac.kr

**Abstract.** Radio Frequency Identification (RFID) is an established technology and has the potential, in a variety of applications, to significantly reduce cost and improve performance. RFID may dramatically change an organization's capacity to obtain real-time information concerning the location and properties of tagged people or objects. However, simply adding RFID to an existing process is a losing proposition. The entire process should be reconsidered in order to take advantage of real-time inventory data and the near real-time tracking and management of inventory. As RFID-enabled applications will fulfill similar tasks across a range of processes adapted to use the data gained from RFID tags, they can be considered as software products derived from a common infrastructure and assets that capture specific abstractions in the domain. That is, it may be appropriate to design RFID-enabled applications as elements of a product line. This paper discusses product line architecture for RFID-enabled applications. In developing this architecture, common activities are identified among the RFID-enabled applications and the variability in the common activities is analyzed in detail using variation point concepts. A product line architecture explicitly representing commonality and variability is described using UML activity diagrams. Sharing a common architecture and reusing assets to deploy recurrent services may be considered an advantage in terms of economic significance and overall quality.

**Keywords:** RFID, RFID-enabled application, software product line, product line architecture.

## 1 Introduction

Radio Frequency Identification (RFID) technology is considered the next step in the revolution affecting supply-chain management, retail, and beyond. At present, the reliability of RFID hardware is improving, the EPCglobal body is developing widely accepted standards [1, 2], and middleware firms are providing software to link RFID

---

* This work was supported by the Brain Korea 21 Project in 2007.

** This work was supported by the Korea Research Foundation Grant funded by the Korean Government (MOEHRD) (The Regional Research Universities Program/Research Center for Logistics Information Technology).

W. Abramowicz (Ed.): BIS 2007, LNCS 4439, pp. 638–651, 2007.
© Springer-Verlag Berlin Heidelberg 2007

data with business applications. Therefore, firms must modify relevant business processes to gain the full benefits from RFID [3, 4, 5, 6]. Although RFID middleware deletes duplicate readings from the same tag and helps manage the flow of data, developers are required to implement systems that derive meaningful high-level data, containing information that is more useful for the application than the simple RFID data itself. In implementing the RFID-enabled application, the developer must collect RFID data, access the data server to retrieve reference information for the RFID data, and process business logic. Most RFID-enabled applications will include these activities in processes adapted to use the raw data gained from RFID tags. Therefore, these activities may be considered as common components for the whole family of applications sharing the same architecture. As a result, RFID-enabled applications can easily be constructed by reusing them. However, actual management research and academic management literature on the use of RFID-enabled applications is still scarce. Work related to RFID-enabled applications focuses mainly on case studies and discussions of business opportunities. In order to develop RFID-enabled applications effectively, it is necessary to systematically identify common activities and to define a variability mechanism.

Software product-line engineering is a method that prepares for future reuse and supports seamless reuse in the application development process. In particular, a software product line defines a Product Line Architecture (PLA), shared by the products, and a set of reusable components that implements a considerable part of the products functionalities. Analyzing the commonality and variability between products in a product line is one of the essential concerns that must be considered when building a product line. In this paper, we suggest a product-line engineering approach that systematically supports RFID-enabled application development. More concretely, we describe a method for developing PLAs for RFID-enabled applications that reflects the characteristics of process-oriented software. The architecture is described using UML activity diagrams, which are typically used for business process modeling, for modeling the logic captured by a single use case or usage scenario, or for modeling the detailed logic of a business rule. The architecture can explicitly represent common and variable activities in RFID-enabled applications. It also identifies variation points in the activities. Using this method, the development of RFID-enabled applications can be supported effectively. Further, the method enhances flexibility and reduces the expenses related to building new RFID-enabled applications.

## 2 Background

Logistics Information Technology (LIT) is the Korean national project for developing the next generation of logistics information technology. The research center for LIT has developed a prototype, Version 1.0, of the LIT RFID system [7]. In particular, several RFID-enabled applications, such as the u-PNU (Pusan National University) library system, an RFID blood management system, and an RFID warehouse management system [8], were developed and demonstrated using components developed by the LIT project, namely the RFID middleware, the Electronic Product Code Information Service (EPCIS), and the Object Naming Service (ONS). This clearly showed that RFID-enabled applications include common components for a whole family of applications sharing the same architecture, and specialized components that are specific for an individual application.

### 2.1 Software Product Line

Software produce line engineering supports seamless reuse in the application development process by understanding and controlling their common and distinguishing characteristics [9]. The general product lines process is based on the reusability of core assets: requirements [10], architecture [11], and components. Commonality and variability analysis is the key to producing core assets in all product-line development processes. Weiss and Lai [12] define *variability* in product-line development as "an assumption about how members of a family may differ from each other". Fig. 1 shows a metamodel for representing the fundamental concepts of variability.

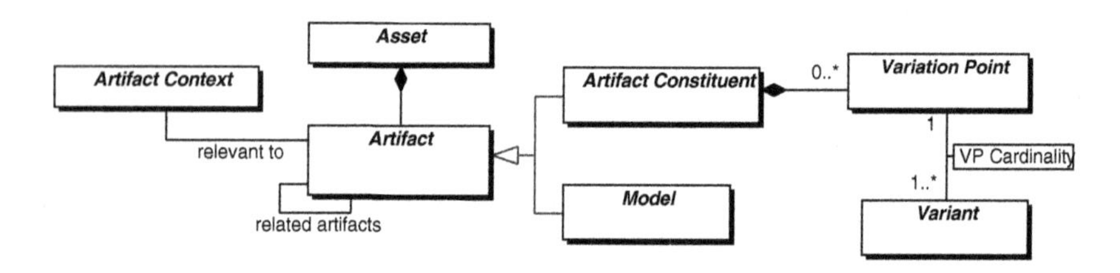


**Fig. 1.** Variability model for a product line

An asset provides a collection of artifacts. An artifact is a work product that can be created, stored, and manipulated by asset producers, consumers, and tools. An artifact may have a relationship to another artifact. It is specialized as the form of an artifact constituent and a model. The artifact constituents are the elements for specifying models in a specific domain. An artifact may be relevant to a particular artifact context such as a requirement, design, implementation, or test context. An artifact context helps explain the meaning of the elements in the artifact. An artifact constituent may have a variability point that needs to be altered by the asset consumer. It describes where, and in what respect, the artifact can be modified. Each variant is one way of realizing a particular variability and binding it in a concrete way. A variation point can have one or more variants. The variation point cardinality, denoted as association class *VP cardinality* in Fig. 1, tells an application developer how many variants can be applied to the variation point.

In this paper, we will describe a product-line architecture that is the artifacts of a domain design context.

### 2.2 RFID-Enabled (Business) Applications

The RFID system is generally partitioned into three tiers: the reader layer, the RFID middleware layer, and the application layer. RFID middleware receives reader event streams (tags) from one or more RFID readers. It collects, filters, and cleanses these reader events to make them available to the RFID applications. Included in its information are the logical reader name, tag value, direction, and time [13]. The application developer must collect RFID data, access the data server to retrieve reference information of RFID data, and process business logic to implement the RFID applications. That is, application developers must be conversant with RFID knowledge and communication techniques; substantial applications should involve additional codes, rather than just business logic, to process RFID data.

For example, the scenario for the *warehousing of products* in an RFID-enabled warehouse management system (WMS)[1] is as follows (Fig. 2):

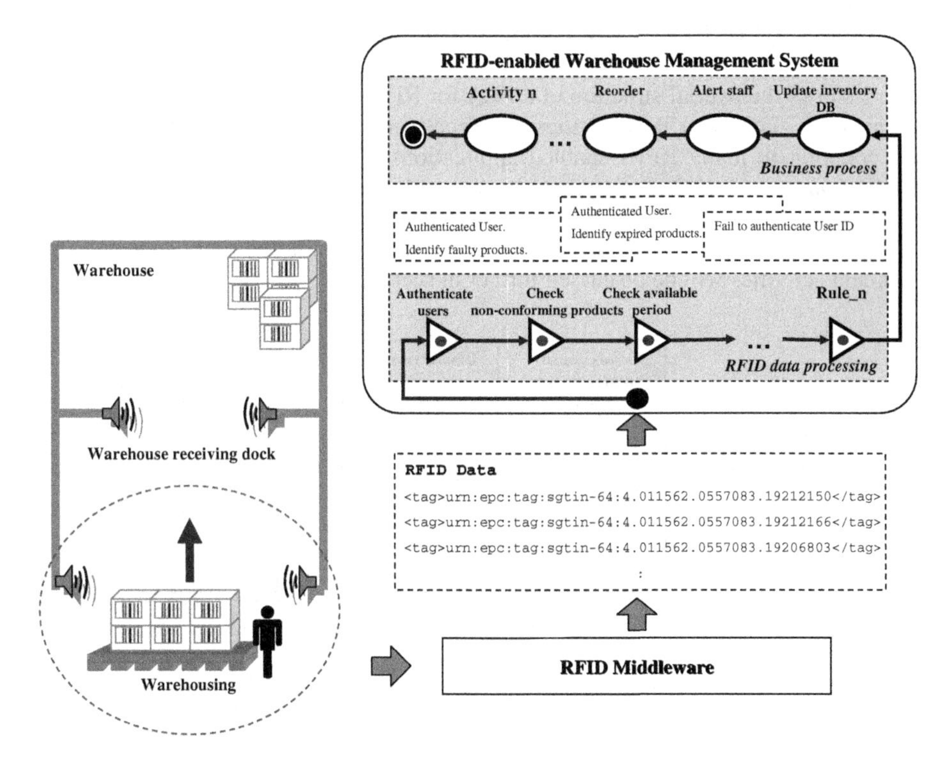


**Fig. 2.** An example scenario of the RFID-enabled warehouse management system

1. A trailer reaches the warehouse receiving dock, a forklift goes inside the trailer and pulls out a pallet, the forklift then carries the pallet to the staging area where an RFID reader reads the RFID data from the product tag. The reader passes the data to RFID middleware, which collects the data, filters them, and removes duplicates. The middleware passes the RFID data to the RFID-enabled WMS.
2. The RFID-enabled WMS receives the RFID data from the RFID middleware and communicates with the reference data servers to obtain the

[1] This application was used to demonstrate an RFID middleware development, which was exhibited in Busan exhibition and convention center BEXCO at 2006/02/16~17.

product details. It checks the product information with the business rules, which are user defined, and signals that correct product has been received. The products are then accepted and stored in the warehouse. It also signals if the wrong product is received and the product is rejected and removed from the warehouse. The forklift then carries the pallet from the staging area. RFID data are analyzed using the business rules. These processes can detect non conforming products in real time and quickly correct mistakes and problems in the RFID system.

## 3 Two-Layered PLA for RFID-Enabled Applications

Fig. 3 shows the overall structure of a PLA for RFID-enabled applications. The lower layer in Fig. 3 contains the activities that process the raw RFID data. These activities are common to many RFID-enabled applications. Even though they are defined as ***RFID Common Activities***, they may have variabilities that need to be altered by the application developer. In addition, the common flow of RFID common activities is defined. In this paper, variability is analyzed using the variation point concept. These common activities will be discussed further in section 4.

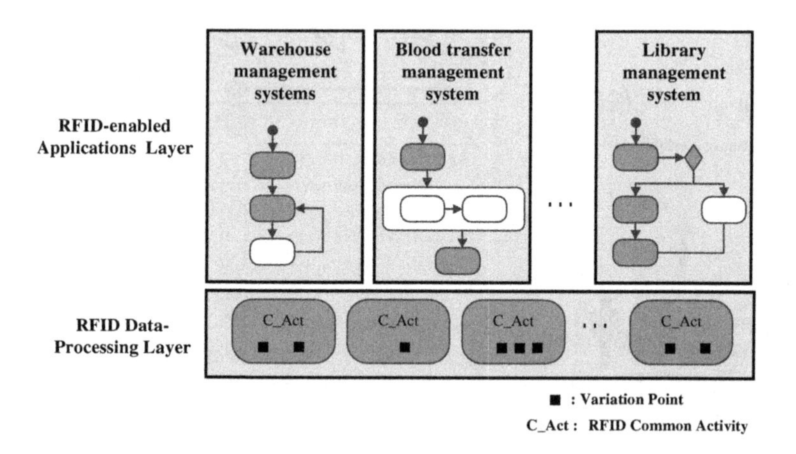


**Fig. 3.** Two-layered structure for RFID PLAs

The upper layer of Fig. 3 shows PLA models for RFID-enabled applications, with common activities of gray boxes and variable activities of white boxes. It contains the RFID common activities identified in the lower layer, common activities for an individual domain, and variant (application-specific) activities. This layer will be discussed further in section 5.

## 4 RFID Data-Processing Layer

The RFID common activities are identified in the domain requirements analysis step. They include properties such as commonality and variability that were analyzed in this step. Fig. 4 shows a partial Primitive Requirement (PR) [10] matrix of RFID-enabled applications, where the PRs identified are listed in the first column and the context names of the RFID-enabled applications are shown in other column headers. An "O" at the intersection of $PR_i$ and $context_j$ indicates that the $PR_i$ is found in $context_j$. An "X" indicates that $context_j$ does not have the $PR_i$.

| PR No. | PR | ratio/CV_property | context1 | context2 | context3 | context4 |
|---|---|---|---|---|---|---|
| PR1 | Define the ECspec | 100% | O | O | O | O |
| PR2 | Undefine the ECspec | 100% | O | O | O | O |
| PR3 | Subscribe the ECspec | 100% | O | O | O | O |
| PR3.1 | Request an event cycle using subscribe method | 100% | O | O | O | O |
| PR3.2 | Request an event cycle using poll method | 100% | O | O | O | O |
| PR3.3 | Request an event cycle using immediate method | 100% | O | O | O | O |
| PR4 | Unsubscribe the ECspec | 100% | O | O | O | O |
| PR5 | Identify the count of EPCs | 100% | O | O | O | O |
| PR6 | Retrieve reference data of EPC | 100% | O | O | O | O |
| PR6.1 | Retrieve expiry date of EPC | 75% | O | O | X | O |
| PR6.2 | Retrieve price information of EPC | 50% | O | O | X | X |
| PR6.3 | Retrieve location information of EPC | 100% | O | O | O | O |
| PR6.4 | Retrieve manufacturer information of EPC | 75% | O | X | O | O |

**Fig. 4.** PR-context matrix of the RFID-enabled applications

The determination of whether a PR can be reused is based on its frequency of appearance in the PR matrix (i.e., the commonality ratio of the PR). Using this matrix, two or more different PRs with similar functionalities can be generalized into one PR by abstracting the differences into variability. For example, as Fig. 4 indicates, PR3.1, PR3.2, and PR3.3 can be generalized into a single PR, *Subscribe the ECspec with variability*.

## 4.1 RFID Common Activities

The RFID common activities can be derived from the PR-context matrix during domain requirements analysis. Each identified RFID common activity is categorized as one of three types:

- ***Trigger activity***
  A trigger activity specifies processes that request RFID data. It consists of elements that describe the RFID data of interest and the RFID reader-control information related to the events, such as a start or stop trigger for an event cycle, the repeat period, and the duration. In addition, it specifies an element that receives RFID data from the RFID middleware. TimeTrigger events make it easy for developers to build two classes of RFID-enabled application: real-time applications and batch-oriented applications.

- ***Reference activities***
  In general, rules are statements of knowledge derived from other knowledge by using an inference or a mathematical calculation. In an RFID-enabled application, to apply the business rule to RFID data, the reference data related to the RFID data

must be accessed. These reference data are retrieved from an information service. An EPCIS is the networked database that stores the additional data associated with the tagged object [14]; it provides a standard interface for access and permanent storage of EPC-related data, for read-and-write access by authorized parties. EPCglobal defines an ONS [15] and EPCISs to exchange product level information in the networks for RFID data and product data. A reference activity contains EPCIS activity, ONS activity, and EPCIS Discovery Service (DS) activity, which retrieve data from EPCIS, ONS, and EPCIS DS [16], respectively. In addition, it specifies a computation activity for mathematical calculation. In this paper, only the EPCIS activity will be discussed.

- ***Rule manager activities***

A business rule constrains some aspects of the business that are related to the RFID data and the reference data. To prevent inaccurate data from being transmitted to applications, the business rules deal beforehand with erroneous or missing information. Rule manager activities process the business rules that are required in the applications. These are composed of several condition and action activities. A condition activity specifies the RFID business conditions that enable filtering of irregular RFID data. An action activity specifies the processes that notify the application of the subscribed events, or that specify the invocation of actions in response to an event. If the conditions of the rule are not satisfied, the rule execution notifies an exception. For example, if a shipment of 24 cases is expected but only 20 tags are read when it arrives, the system can send an alert so that the operator can check the pallet. It automatically triggers any alerts that were incorporated into the business rules.

### 4.2 Variation Points in RFID Common Activities

A variation point that needs to be altered by the asset consumer should describe where, and in what respect, the activity can be modified. The variation points included in each RFID common activity are listed in Table 1. In addition, the RFID

**Table 1.** The variation points in RFID common activities

| RFID common activity | | Variation Point |
|---|---|---|
| Trigger activity | TimeTrigger event | Setting of trigger time |
| | Trigger activity | Event request method<br>Variables for receiving event |
| Reference activity | EPCIS activity | Query statements<br>getEPCSchema / getEPCAttribute / getEPCClassSchema / getEPCClassAttribute<br>Parameters required in a query statement<br>EPC / Schema / Xpath / EPCClass |
| Rule manager activity | Condition activity | The business rule |
| | Action activity | The RFID business event name<br>The result of the corresponding business rule<br>A related data component |

common activities are analyzed and modeled using a UML2 activity diagram. For example, the variation-point analysis model for the ECPIS activity is shown in Fig. 5. The variation points are explicitly denoted by the stereotype <<v.p>> in the activity diagram. If the variation points are bound at runtime, the binding values (i.e., variants) are parameterized in the activity diagram.

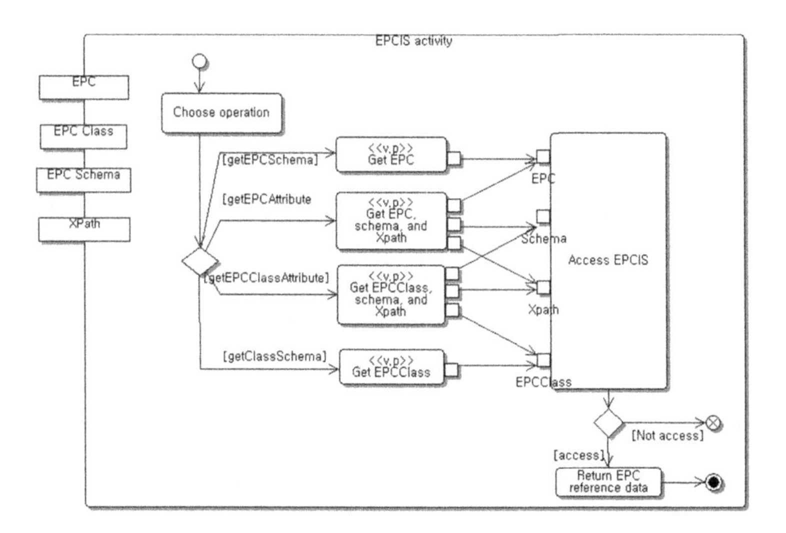


**Fig. 5.** The variation-point analysis model for EPCIS activity

## 4.3 The Common Flow with RFID Common Activities

The common flow with RFID common activities becomes a base architecture for the PLA of RFID-enabled applications. This is shown in Fig. 6 as a UML activity diagram. It is composed of the RFID common activities previously described. The diagram is divided into three partitions (called swim lanes) by using vertical lines, which separate the activity categories. Each activity is embodied in the PLA of a specific RFID domain. The stereotype in each activity specifies how many instances can be realized from the RFID common activity in the PLA. That is, stereotype <<0..*>> implies that the activity may be realized at least zero times in a PLA. For example, the activity "*Retrieve reference data of EPC*" may be realized more than once or may not be realized at all in the domain of RFID-enabled *Smart Shelf* applications. In addition, an RFID-enabled application may be a real-time triggered application or a batch-oriented application. In the former case, the applications are automatically executed whenever items with an RFID tag are read by the RFID reader. In the latter case, the applications may have the function of setting a time interval for periodic scanning. The time interval can range from once at a particular time every day to once every second. Thus, the TimeTrigger event is represented using time-event notation. "*Check business rule*", "*Trigger alert*", and "*Trigger business process*" within rule manager activity are realized as a package.

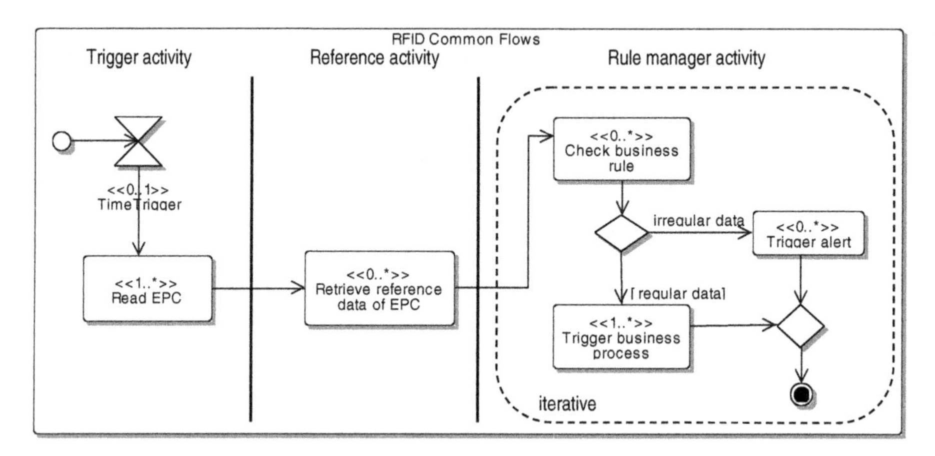


**Fig. 6.** The common flow with RFID common activities

# 5 RFID Business Application Layer

This layer allows developers to quickly create specialized vertical solutions (PLAs) across a wide range of RFID-enabled applications. We describe a PLA in detail using a case study of RFID-enabled Smart Shelf applications, in other words smart-shelf domain. The smart shelves are equipped with readers that inform the store staff when the shelves have to be replenished. The system provides complete data acquisition control over the products on the shelves with real-time event triggering and automatic inventory capture. Furthermore, smart shelves automatically recognize when the expiration date has been exceeded, and inform the staff accordingly. The PLA is developed in two phases as follows:

## 5.1 Identification of Domain Activities

Firstly, most of the activities that should be processed in RFID-enabled applications are identified. The activities can be driven by the embodiment of RFID common activities or by analysis of the specific RFID domain. In the smart shelf domain example, the "*Check business rule*" RFID common activity is embodied into three activities: "*Check product quantity*", "*Check expiration date of product*", and "*Check the misplaced product*" as shown in Fig. 7. In addition, a smart shelf can provide an alert when it is time to restock or when a potential shoplifting situation arises because an unusual number of products are removed simultaneously. Therefore, there are "*Alert staff with picking list*", "*Alert staff with removing list*," and "*Alert staff with rearrangement message*" activities driven by the "*Trigger alert*" RFID common activity. The variation points in RFID common activities are bound to variants according to the requirements of RFID-enabled applications. In the smart shelf domain, the variation points in EPCIS activity are bound to *getEPCAttribute* as a query statement, and to *EPC*, *schema*, and *xpath* as parameters. Those activities that are embodied from RFID common activities are explicitly represented using the stereotype <<RFID>>. The activities that are extracted by analyzing the RFID domain are "*Update inventory bookstock*" activity, "*Find product type in backroom*" activity, and etc.

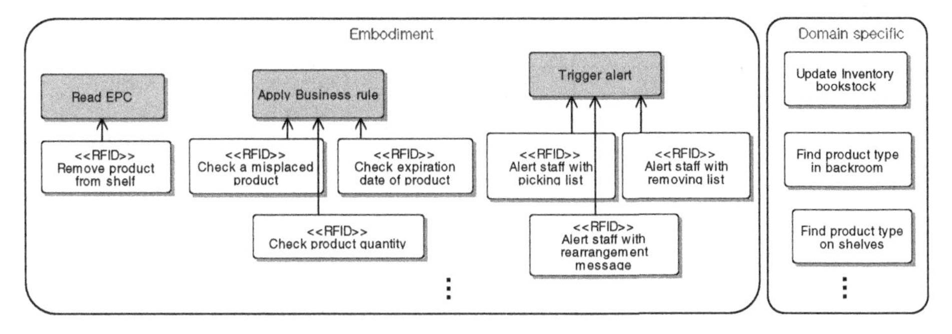


**Fig. 7.** The two types of activity identified in the smart shelf domain

## 5.2 PLA for RFID-Enabled Applications

The PLA should explicitly represent commonality and variability. The base constituent for representing the PLAs in this paper is a domain activity. We define the domain activity as activity involving property. The domain activity may or may not appear in application architecture depending on the extent to which the domain activity has the property of being common or optional. As shown in Table 2, this domain activity property has four possible types.

**Table 2.** Property types of domain activity

| Property type | Description | Stereotype | Notation |
|---|---|---|---|
| Common | The activity should be realized in most applications. | <<common>> Can be omitted. | <<common>> |
| Optional | The activity may only be realized in a specific application. | <<optional>> | <<optional>> |
| Generalized common | Two or more different domain activities with similar functionalities are generalized into one domain activity by abstracting the differences into variability. In addition, the property of the generalized domain activity is common. | • <<g-common>> for generalized domain activity (can be omitted)<br>• <<variant>> for domain activities generalized<br>• <<default>> for preset variant domain activity which will always be followed | <<g-common, 1..n>> <<variant>> <<variant>> |
| Generalized optional | The property of the generalized domain activity is optional. | • <<g-optional>> for generalized domain activity<br>• <<variant>> for domain activities generalized | <<g-optional, 0..n>> <<variant>> <<variant>> |

**Table 3.** The stereotypes used in PLAs

| | Property | Activity | Generalized activity | Variant activity |
|---|---|---|---|---|
| RFID common activity | common | <<RFID>> | <<RFID>><br><<1..n>> | <<RFID>><br><<variant>> |
| | optional | <<RFID>><br><<optional>> | <<RFID>><br><<optional>><br><<1..n>> | <<RFID>><br><<variant>> |
| Domain specific | common | - | <<0..n>> | <<variant>> |
| | optional | << optional >> | << optional>><br><<0..n>> | <<variant>> |

The variability of domain activities is explicitly represented in the PLA by the stereotypes described in Table 2. To model the relationships between the generalized activity and variant activities, we use a dependency relationship. The <<abstraction>>

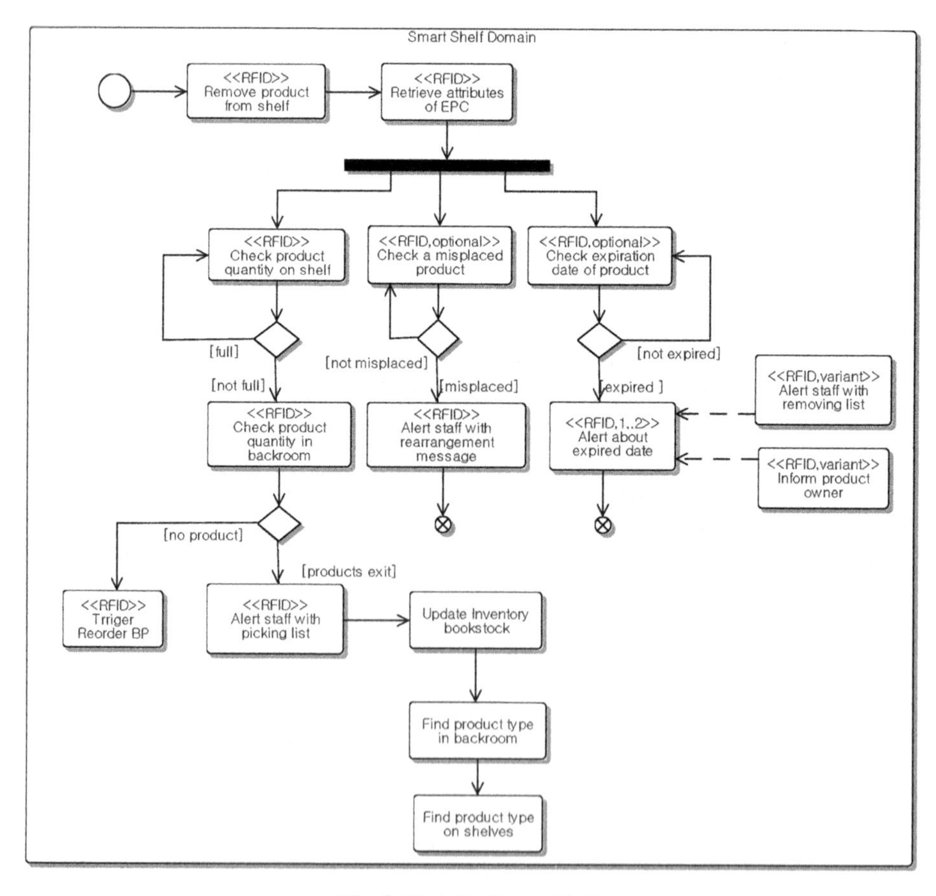


**Fig. 8.** PLA for Smart Shelf

relationship, which is one type of dependency, can be used because the relationships involve abstraction. In addition, because a generalized domain activity can have one or many variant domain activities, it is necessary to indicate how many variant domain activities can be realized in a specific application. This information is also described by using a stereotype in a generalized domain activity. All stereotypes used in the PLA are summarized in Table 3. Stereotypes that may be omitted, such as <<common>> and <<g-common>>, are not described.

Fig. 8 shows a partial PLA for the smart shelf domain. Because this application is triggered in real time, a TimeTrigger element is not realized in this domain. That is, if a product is removed from or added to the shelf, the RFID reader detects the movement and triggers the application. The domain activities driven from the embodiment of RFID common activities are denoted by stereotype <<RFID>>. The common properties of the domain activities are represented without stereotypes. Only several applications in the smart shelf domain may have the function of managing expiration dates, so the property "*Check expiration date of product*" is determined as optional. The PLA is designed to automatically trigger any alerts that were incorporated into the business rules. Further, the application may inform its owner or alert store staff directly when an RFID-tagged product is approaching its expiration date. Thus, the functions are realized as variant activities, and the generalized activity "*Alert about expired date*" is related to its variants. At this time, at least one variant *must* be realized in a specific application. The stereotype <<1..2>> implies the count of realizable the variants in a specific application.

# 6 Related Work

***Representing variability in the PLA for process oriented applications.*** Referring especially to product line architecture, an important task is to analyze the domain and to identify the commonalities and variabilities of the domain. The FORM (Feature-Oriented Reuse Method) [17] was developed as an extension of the FODA (Feature-Oriented Domain Analysis) method [18]. The main characteristic of FORM is its four-layer decomposition, which describes different points of view on product development. However, this does not address explicitly the variations in the reference architecture, and entails complexity when many variants must be represented. In [19], variation is represented using patterns associated with discriminants. A discriminant has three types: single, multiple, and optional, and is closely related to the division of feature properties into mandatory, optional, and alternative. It does, however, not emphasize characteristics of variation at the design level. Gomma explained the product line design phase in connection with features [20]. He strived to describe of design models with explicit variations in structural and dynamic views. However, all alternative variants appear at the same level in those models, so they are complex even in the simple case study. Because model elements that have common or optional properties may be modified when applied, variants should be modeled separately at the detailed level.

***Methodology for supporting the development of RFID-enabled applications.*** RFID has recently gained enormous attention in various industry sectors, in the media, and in academic research. However, management research and academic management literature on the use of RFID is still scarce. The publications that cover RFID applications focus mainly on case studies and discussions of business opportunities. Vendors such as Sun Microsystems [3], IBM [4], Oracle [5], and Microsoft [6], have been extending their application development and middleware technology stacks to handle RFID. In order to realize the time and cost benefits associated with RFID, certain business processes should be changed. Existing business processes should adapt to using data gained from RFID tags. To the best of our knowledge, few existing approaches focus on an RFID technology integration of development methodologies i.e., focusing on how RFID-enabled applications are organized and developed.

## 7 Conclusions and Future Work

PLA can play an important role in managing the complexity of software and in reducing its cost of development and maintenance. In addition, RFID technology can be used to significantly improve the efficiency of business processes by providing the capabilities of automatic identification and data capture. Obviously, RFID technology and applications based on product line are still developing. This paper has proposed an approach to developing PLAs in the RFID domain. We first identified the common activities in the RFID domain and analyzed their variability. Then we suggested a modeling approach that explicitly represented these variabilities in the PLA model. Using this PLA, we have developed several RFID-enabled applications. These results show that RFID-enabled applications do not have to involve additional analysis of RFID data processing, thereby substantially reducing the cost of developing and managing RFID applications.

Currently, we are working with a company on implementing RFID-enabled logistics, based on the technique of the proposed PLA. Our future research activities will include an extension to the RFID PLA, which will enable processing of RFID readers and other types of sensors, such as temperature, humidity, shock, and location.

## References


[1] EPCglobal, “The EPCglobal Architecture Framework,” EPCglobal Final Version, July 2005.

[2] EPCglobal Inc., http://www.epcglobalinc.org

[3] Sun Microsystems, http://www.sun.com/software/sol utions/rfid/

[4] IBM, http://www306.ibm.com/software/pervasive/w_ rfid_premises_server/, December 2004.

[5] Oracle, http://www.oracle.com/technology/products/ iaswe/edge_server

[6] Microsoft, http://www.microsoft.com/business/insigh ts/about/aboutus.aspx

[7] Research Center Logistics Information Technology, http://lit.pusan.ac.kr

[8] M. Moon, Y. Kim, and K. Yeom, “Contextual Events Framework in RFID System”, In proceedings of third International Conference on Information Technology (IEEE Computer Society) pp. 586-587, 2006.

[9] D. Muthig and C. Atkinson, "Model-Driven Product Line Architecture", In proceedings of second Software Product Line Conference, Springer Lecture Notes in Computer Science Vol.2379, pp.110-129, Aug. 2002.
[10] M. Moon, K. Yeom, and H.S. Chae, "An Approach to Developing Domain Requirements as a Core Asset Based on Commonality and Variability in a Product Line," IEEE Transactions on Software Engineering, vol. 31, no. 7, pp.551-569, Jul. 2005.
[11] M. Moon, H.S. Chae, and K. Yeom, "A Metamodel Approach to Architecture Variability in a Product Line", In Proceeding of 9th International Conference on Software Reuse, ICSR2006, Italy, pp.115-126, 2006.
[12] Weiss, D.M., Lai, C.T.R., *Software Product-Line Engineering: A Family Based Software Development Process*, Addison-Wesley, ISBN 0-201-694387, 1999.
[13] K. Traub, S. Bent, T. Osinski, S. N. Peretz, S. Rehling, S. Rosenthal, and B. Tracey, "The Application Level Events (ALE) Specification, Version 1.0," EPCglobal Proposed Standard, Feb. 2005.
[14] EPCglobal Inc., "EPC Information Services (EPCIS) Version 1.0 Specification," EPCglobal Working Draft, June 2005.
[15] EPCglobal Inc., Object Naming Service (ONS), Version 1.0, Oct. 2005.
[16] EPCglobal Inc., The EPCglobal Network and The Global Data Synchronization Network (GDSN): Understanding the Information & the Information Networks, Oct. 2004.
[17] K. Kang, S. Kim, J. Lee, and K. Kim, "FORM: A Feature-Oriented Reuse Method with Domain Specific Reference Architectures," Annals of Software Engineering, vol. 5, pp.143-168, 1998.
[18] K. Kang, S. Cohen, J. Hess, W. Novak, and S. Peterson, "Feature-Oriented Domain Analysis (FODA) Feasibility Study," Technical Report CMU/SEI-90-TR-21, Software Engineering Institute, Carnegie Mellon University, Nov. 1990.
[19] B. Keepence and M. Mannion, "Using patterns to model variability in product families," IEEE Software, vol. 16, no. 4, pp.102-108, 1999.
[20] H. Gomma, *Designing Software Product Lines with UML, From Use Cases to Pattern-Based Software Architectures*, Addison-Wesley, 2004.

# Author Index

# Lecture Notes in Computer Science

For information about Vols. 1–4353

please contact your bookseller or Springer

Vol. 4453: T. Speed, H. Huang (Eds.), Research in Computational Molecular Biology. XVI, 550 pages. 2007. (Sublibrary LNBI).

Vol. 4450: T. Okamoto, X. Wang (Eds.), Public Key Cryptography – PKC 2007. XIII, 491 pages. 2007.

Vol. 4448: M. Giacobini (Ed.), Applications of Evolutionary Computing. XXIII, 755 pages. 2007.

Vol. 4447: E. Marchiori, J.H. Moore, J.C. Rajapakse (Eds.), Evolutionary Computation,Machine Learning and Data Mining in Bioinformatics. XI, 302 pages. 2007.

Vol. 4446: C. Cotta, J. van Hemert (Eds.), Evolutionary Computation in Combinatorial Optimization. XII, 241 pages. 2007.

Vol. 4445: M. Ebner, M. O'Neill, A. Ekárt, L. Vanneschi, A.I. Esparcia-Alcázar (Eds.), Genetic Programming. XI, 382 pages. 2007.

Vol. 4444: T. Reps, M. Sagiv, J. Bauer (Eds.), Program Analysis and Compilation, Theory and Practice. X, 361 pages. 2007.

Vol. 4443: R. Kotagiri, P.R. Krishna, M. Mohania, E. Nantajeewarawat (Eds.), Advances in Databases: Concepts, Systems and Applications. XXI, 1126 pages. 2007.

Vol. 4440: B. Liblit, Cooperative Bug Isolation. XV, 101 pages. 2007.

Vol. 4439: W. Abramowicz (Ed.), Business Information Systems. XV, 654 pages. 2007.

Vol. 4438: L. Maicher, A. Sigel, L.M. Garshol (Eds.), Leveraging the Semantics of Topic Maps. X, 257 pages. 2007. (Sublibrary LNAI).

Vol. 4433: E. Şahin, W.M. Spears, A.F.T. Winfield (Eds.), Swarm Robotics. XII, 221 pages. 2007.

Vol. 4432: B. Beliczynski, A. Dzielinski, M. Iwanowski, B. Ribeiro (Eds.), Adaptive and Natural Computing Algorithms, Part II. XXVI, 761 pages. 2007.

Vol. 4431: B. Beliczynski, A. Dzielinski, M. Iwanowski, B. Ribeiro (Eds.), Adaptive and Natural Computing Algorithms, Part I. XXV, 851 pages. 2007.

Vol. 4430: C.C. Yang, D. Zeng, M. Chau, K. Chang, Q. Yang, X. Cheng, J. Wang, F.-Y. Wang, H. Chen (Eds.), Intelligence and Security Informatics. XII, 330 pages. 2007.

Vol. 4429: R. Lu, J.H. Siekmann, C. Ullrich (Eds.), Cognitive Systems. X, 161 pages. 2007. (Sublibrary LNAI).

Vol. 4427: S. Uhlig, K. Papagiannaki, O. Bonaventure (Eds.), Passive and Active Network Measurement. XI, 274 pages. 2007.

Vol. 4426: Z.-H. Zhou, H. Li, Q. Yang (Eds.), Advances in Knowledge Discovery and Data Mining. XXV, 1161 pages. 2007. (Sublibrary LNAI).

Vol. 4425: G. Amati, C. Carpineto, G. Romano (Eds.), Advances in Information Retrieval. XIX, 759 pages. 2007.

Vol. 4424: O. Grumberg, M. Huth (Eds.), Tools and Algorithms for the Construction and Analysis of Systems. XX, 738 pages. 2007.

Vol. 4423: H. Seidl (Ed.), Foundations of Software Science and Computational Structures. XVI, 379 pages. 2007.

Vol. 4422: M.B. Dwyer, A. Lopes (Eds.), Fundamental Approaches to Software Engineering. XV, 440 pages. 2007.

Vol. 4421: R. De Nicola (Ed.), Programming Languages and Systems. XVII, 538 pages. 2007.

Vol. 4420: S. Krishnamurthi, M. Odersky (Eds.), Compiler Construction. XIV, 233 pages. 2007.

Vol. 4419: P.C. Diniz, E. Marques, K. Bertels, M.M. Fernandes, J.M.P. Cardoso (Eds.), Reconfigurable Computing: Architectures, Tools and Applications. XIV, 391 pages. 2007.

Vol. 4418: A. Gagalowicz, W. Philips (Eds.), Computer Vision/Computer Graphics Collaboration Techniques. XV, 620 pages. 2007.

Vol. 4416: A. Bemporad, A. Bicchi, G. Buttazzo (Eds.), Hybrid Systems: Computation and Control. XVII, 797 pages. 2007.

Vol. 4415: P. Lukowicz, L. Thiele, G. Tröster (Eds.), Architecture of Computing Systems - ARCS 2007. X, 297 pages. 2007.

Vol. 4414: S. Hochreiter, R. Wagner (Eds.), Bioinformatics Research and Development. XVI, 482 pages. 2007. (Sublibrary LNBI).

Vol. 4412: F. Stajano, H.J. Kim, J.-S. Chae, S.-D. Kim (Eds.), Ubiquitous Convergence Technology. XI, 302 pages. 2007.

Vol. 4411: R.H. Bordini, M. Dastani, J. Dix, A.E.F. Seghrouchni (Eds.), Programming Multi-Agent Systems. XIV, 249 pages. 2007. (Sublibrary LNAI).

Vol. 4410: A. Branco (Ed.), Anaphora: Analysis, Algorithms and Applications. X, 191 pages. 2007. (Sublibrary LNAI).

Vol. 4409: J.L. Fiadeiro, P.-Y. Schobbens (Eds.), Recent Trends in Algebraic Development Techniques. VII, 171 pages. 2007.

Vol. 4407: G. Puebla (Ed.), Logic-Based Program Synthesis and Transformation. VIII, 237 pages. 2007.

Vol. 4406: W. De Meuter (Ed.), Advance in Smaltalk. VII, 157 pages. 2007.

Vol. 4405: L. Padgham, F. Zambonelli (Eds.), Agent-Oriented Software Engineering VII. XII, 225 pages. 2007.

Vol. 4403: S. Obayashi, K. Deb, C. Poloni, T. Hiroyasu, T. Murata (Eds.), Evolutionary Multi-Criterion Optimization. XIX, 954 pages. 2007.

Vol. 4401: N. Guelfi, D. Buchs (Eds.), Rapid Integration of Software Engineering Techniques. IX, 177 pages. 2007.

Vol. 4400: J.F. Peters, A. Skowron, V.W. Marek, E. Orłowska, R. Słowiński, W. Ziarko (Eds.), Transactions on Rough Sets VII, Part II. X, 381 pages. 2007.

Vol. 4399: T. Kovacs, X. Llorà, K. Takadama, P.L. Lanzi, W. Stolzmann, S.W. Wilson (Eds.), Learning Classifier Systems. XII, 345 pages. 2007. (Sublibrary LNAI).

Vol. 4398: S. Marchand-Maillet, E. Bruno, A. Nürnberger, M. Detyniecki (Eds.), Adaptive Multimedia Retrieval: User, Context, and Feedback. XI, 269 pages. 2007.

Vol. 4397: C. Stephanidis, M. Pieper (Eds.), Universal Access in Ambient Intelligence Environments. XV, 467 pages. 2007.

Vol. 4396: J. García-Vidal, L. Cerdà-Alabern (Eds.), Wireless Systems and Mobility in Next Generation Internet. IX, 271 pages. 2007.

Vol. 4395: M. Daydé, J.M.L.M. Palma, Á.L.G.A. Coutinho, E. Pacitti, J.C. Lopes (Eds.), High Performance Computing for Computational Science - VECPAR 2006. XXIV, 721 pages. 2007.

Vol. 4394: A. Gelbukh (Ed.), Computational Linguistics and Intelligent Text Processing. XVI, 648 pages. 2007.

Vol. 4393: W. Thomas, P. Weil (Eds.), STACS 2007. XVIII, 708 pages. 2007.

Vol. 4392: S.P. Vadhan (Ed.), Theory of Cryptography. XI, 595 pages. 2007.

Vol. 4391: Y. Stylianou, M. Faundez-Zanuy, A. Esposito (Eds.), Progress in Nonlinear Speech Processing. XII, 269 pages. 2007.

Vol. 4390: S.O. Kuznetsov, S. Schmidt (Eds.), Formal Concept Analysis. X, 329 pages. 2007. (Sublibrary LNAI).

Vol. 4389: D. Weyns, H.V.D. Parunak, F. Michel (Eds.), Environments for Multi-Agent Systems III. X, 273 pages. 2007. (Sublibrary LNAI).

Vol. 4385: K. Coninx, K. Luyten, K.A. Schneider (Eds.), Task Models and Diagrams for Users Interface Design. XI, 355 pages. 2007.

Vol. 4384: T. Washio, K. Satoh, H. Takeda, A. Inokuchi (Eds.), New Frontiers in Artificial Intelligence. IX, 401 pages. 2007. (Sublibrary LNAI).

Vol. 4383: E. Bin, A. Ziv, S. Ur (Eds.), Hardware and Software, Verification and Testing. XII, 235 pages. 2007.

Vol. 4381: J. Akiyama, W.Y.C. Chen, M. Kano, X. Li, Q. Yu (Eds.), Discrete Geometry, Combinatorics and Graph Theory. XI, 289 pages. 2007.

Vol. 4380: S. Spaccapietra, P. Atzeni, F. Fages, M.-S. Hacid, M. Kifer, J. Mylopoulos, B. Pernici, P. Shvaiko, J. Trujillo, I. Zaihrayeu (Eds.), Journal on Data Semantics VIII. XV, 219 pages. 2007.

Vol. 4379: M. Südholt, C. Consel (Eds.), Object-Oriented Technology. VIII, 157 pages. 2007.

Vol. 4378: I. Virbitskaite, A. Voronkov (Eds.), Perspectives of Systems Informatics. XIV, 496 pages. 2007.

Vol. 4377: M. Abe (Ed.), Topics in Cryptology – CT-RSA 2007. XI, 403 pages. 2006.

Vol. 4376: E. Frachtenberg, U. Schwiegelshohn (Eds.), Job Scheduling Strategies for Parallel Processing. VII, 257 pages. 2007.

Vol. 4374: J.F. Peters, A. Skowron, I. Düntsch, J. Grzymała-Busse, E. Orłowska, L. Polkowski (Eds.), Transactions on Rough Sets VI, Part I. XII, 499 pages. 2007.

Vol. 4373: K. Langendoen, T. Voigt (Eds.), Wireless Sensor Networks. XIII, 358 pages. 2007.

Vol. 4372: M. Kaufmann, D. Wagner (Eds.), Graph Drawing. XIV, 454 pages. 2007.

Vol. 4371: K. Inoue, K. Satoh, F. Toni (Eds.), Computational Logic in Multi-Agent Systems. X, 315 pages. 2007. (Sublibrary LNAI).

Vol. 4370: P.P Lévy, B. Le Grand, F. Poulet, M. Soto, L. Darago, L. Toubiana, J.-F. Vibert (Eds.), Pixelization Paradigm. XV, 279 pages. 2007.

Vol. 4369: M. Umeda, A. Wolf, O. Bartenstein, U. Geske, D. Seipel, O. Takata (Eds.), Declarative Programming for Knowledge Management. X, 229 pages. 2006. (Sublibrary LNAI).

Vol. 4368: T. Erlebach, C. Kaklamanis (Eds.), Approximation and Online Algorithms. X, 345 pages. 2007.

Vol. 4367: K. De Bosschere, D. Kaeli, P. Stenström, D. Whalley, T. Ungerer (Eds.), High Performance Embedded Architectures and Compilers. XI, 307 pages. 2007.

Vol. 4366: K. Tuyls, R. Westra, Y. Saeys, A. Nowé (Eds.), Knowledge Discovery and Emergent Complexity in Bioinformatics. IX, 183 pages. 2007. (Sublibrary LNBI).

Vol. 4364: T. Kühne (Ed.), Models in Software Engineering. XI, 332 pages. 2007.

Vol. 4362: J. van Leeuwen, G.F. Italiano, W. van der Hoek, C. Meinel, H. Sack, F. Plášil (Eds.), SOFSEM 2007: Theory and Practice of Computer Science. XXI, 937 pages. 2007.

Vol. 4361: H.J. Hoogeboom, G. Păun, G. Rozenberg, A. Salomaa (Eds.), Membrane Computing. IX, 555 pages. 2006.

Vol. 4360: W. Dubitzky, A. Schuster, P.M.A. Sloot, M. Schroeder, M. Romberg (Eds.), Distributed, High-Performance and Grid Computing in Computational Biology. X, 192 pages. 2007. (Sublibrary LNBI).

Vol. 4358: R. Vidal, A. Heyden, Y. Ma (Eds.), Dynamical Vision. IX, 329 pages. 2007.

Vol. 4357: L. Buttyán, V. Gligor, D. Westhoff (Eds.), Security and Privacy in Ad-Hoc and Sensor Networks. X, 193 pages. 2006.

Vol. 4355: J. Julliand, O. Kouchnarenko (Eds.), B 2007: Formal Specification and Development in B. XIII, 293 pages. 2006.

Vol. 4354: M. Hanus (Ed.), Practical Aspects of Declarative Languages. X, 335 pages. 2006.